*Edited by Gustaaf Van Tendeloo,
Dirk Van Dyck, and Stephen J. Pennycook*

Handbook of Nanoscopy

Further Reading

Ohser, J. Schladitz, K.

3D Images of Materials Structures

Processing and Analysis

2009
Hardcover
ISBN: 978-3-527-31203-0

Codd, S. L., Seymour, J. D. (eds.)

Magnetic Resonance Microscopy

Spatially Resolved NMR Techniques and Applications

2009
Hardcover
ISBN: 978-3-527-32008-0

Maev, R. G.

Acoustic Microscopy

Fundamentals and Applications

2008
Hardcover
ISBN: 978-3-527-40744-6

Fukumura, H., Irie, M., Iwasawa, Y., Masuhara, H., Uosaki, K. (eds.)

Molecular Nano Dynamics

Vol. I: Spectroscopic Methods and Nanostructures/Vol. II: Active Surfaces, Single Crystals and Single Biocells

2009
Hardcover
ISBN: 978-3-527-32017-2

Roters, F., Eisenlohr, P. Bieler, T. R., Raabe, D.

Crystal Plasticity Finite Element Methods

in Materials Science and Engineering

2010
Hardcover
ISBN: 978-3-527-32447-7

Guo, J. (ed.)

X-Rays in Nanoscience

Spectroscopy, Spectromicroscopy, and Scattering Techniques

2010
Hardcover
ISBN: 978-3-527-32288-6

Tsukruk, V., Singamaneni, S.

Scanning Probe Interrogation of Soft Matter

2012
Hardcover
ISBN: 978-3-527-32743-0

Edited by Gustaaf Van Tendeloo, Dirk Van Dyck, and Stephen J. Pennycook

Handbook of Nanoscopy

Volume 1

WILEY-VCH Verlag GmbH & Co. KGaA

The Editors

Prof. Gustaaf Van Tendeloo
Univ. of Antwerp (RUCA)
EMAT
Groenenborgerlaan 171
2020 Antwerp
Belgium

Prof. Dirk Van Dyck
Univ. of Antwerp (RUCA)
EMAT
Groenenborgerlaan 171
2020 Antwerp
Belgium

Prof. Dr. Stephen J. Pennycook
Oak Ridge National Lab.
Condensed Matter Science Div.
Oak Ridge, TN 37831-6030
USA

All books published by **Wiley-VCH** are carefully produced. Nevertheless, authors, editors, and publisher do not warrant the information contained in these books, including this book, to be free of errors. Readers are advised to keep in mind that statements, data, illustrations, procedural details or other items may inadvertently be inaccurate.

Library of Congress Card No.: applied for

British Library Cataloguing-in-Publication Data
A catalogue record for this book is available from the British Library.

Bibliographic information published by the Deutsche Nationalbibliothek
The Deutsche Nationalbibliothek lists this publication in the Deutsche Nationalbibliografie; detailed bibliographic data are available on the Internet at <http://dnb.d-nb.de>.

© 2012 Wiley-VCH Verlag & Co. KGaA, Boschstr. 12, 69469 Weinheim, Germany

All rights reserved (including those of translation into other languages). No part of this book may be reproduced in any form – by photoprinting, microfilm, or any other means – nor transmitted or translated into a machine language without written permission from the publishers. Registered names, trademarks, etc. used in this book, even when not specifically marked as such, are not to be considered unprotected by law.

Cover Design Adam-Design, Weinheim
Typesetting Laserwords Private Limited, Chennai, India
Printing and Binding betz-druck GmbH, Darmstadt

Printed in the Federal Republic of Germany
Printed on acid-free paper

Print ISBN: 978-3-527-31706-6
ePDF ISBN: 978-3-527-64188-8
oBook ISBN: 978-3-527-64186-4
ePub ISBN: 978-3-527-64187-1
Mobi ISBN: 978-3-527-64189-5

Contents to Volume 1

Preface *XVII*
List of Contributors *XIX*

The Past, the Present, and the Future of Nanoscopy *1*
Gustaav Van Tendeloo and Dirk Van Dyck

Part I   **Methods** *9*

1	**Transmission Electron Microscopy** *11*	
	Marc De Graef	
1.1	Introduction *11*	
1.2	The Instrument *12*	
1.2.1	General Layout *12*	
1.2.2	Lenses and Lens Aberrations *14*	
1.3	Imaging and Diffraction Modes *16*	
1.3.1	Important Diffraction Geometries *17*	
1.3.2	Important Imaging Modes *19*	
1.4	Dynamical Diffraction Theory *19*	
1.4.1	Perfect Crystal Theory *21*	
1.4.1.1	Fourier Space Approach *21*	
1.4.1.2	Real-Space Approach *23*	
1.4.1.3	Bloch Wave Approach *24*	
1.4.2	Example Dynamical Computations *26*	
1.4.2.1	Analytical Two-Beam Solutions *26*	
1.4.2.2	Numerical Multibeam Approaches *30*	
1.4.2.3	Other Dynamical Scattering Phenomena *35*	
1.4.3	Defect Images *37*	
1.4.3.1	Theory *38*	
1.4.3.2	Defect Image Simulations *41*	
	References *42*	

2	**Atomic Resolution Electron Microscopy** 45	
	Dirk Van Dyck	
2.1	Introduction 45	
2.1.1	Atoms: the Alphabet of Matter 45	
2.1.2	The Ideal Experiment 46	
2.1.3	Why Imaging? 49	
2.1.4	Why Electron Microscopy? 50	
2.2	Principles of Linear Image Formation 51	
2.2.1	Real Imaging 51	
2.2.2	Coherent Imaging 53	
2.3	Imaging in the Electron Microscope 54	
2.3.1	Theory of Abbe 54	
2.3.2	Incoherent Effects 57	
2.3.3	Imaging at Optimum Defocus: Phase Contrast Microscopy 58	
2.3.4	Resolution 59	
2.4	Experimental HREM 60	
2.4.1	Aligning the Microscope 60	
2.4.2	The Specimen 61	
2.4.3	Interpretation of the High-Resolution Images 63	
2.5	Quantitative HREM 65	
2.5.1	Model-Based Fitting 65	
2.5.2	Phase Retrieval 65	
2.5.3	Exit Wave Reconstruction 67	
2.5.4	Structure Retrieval: Channeling Theory 67	
2.5.5	Resolving versus Refining 70	
	Appendix 2.A: Interaction of the Electron with a Thin Object 72	
	Appendix 2.B: Multislice Method 73	
	Appendix 2.C: Quantum Mechanical Approach 75	
	References 78	
3	**Ultrahigh-Resolution Transmission Electron Microscopy at Negative Spherical Aberration** 81	
	Knut W. Urban, Juri Barthel, Lothar Houben, Chun-Lin Jia, Markus Lentzen, Andreas Thust, and Karsten Tillmann	
3.1	Introduction 81	
3.2	The Principles of Atomic-Resolution Imaging 83	
3.2.1	Resolution and Point Spread 83	
3.2.2	Contrast 84	
3.2.3	Enhanced Contrast under Negative Spherical Aberration Conditions 87	
3.2.4	NCSI Imaging for Higher Sample Thicknesses 91	
3.3	Inversion of the Imaging Process 94	
3.4	Case Study: $SrTiO_3$ 98	
3.5	Practical Examples of Application of NCSI Imaging 102	
	References 105	

4	**Z-Contrast Imaging** 109
	Stephen J. Pennycook, Anrew R. Lupini, Albina Y. Borisevich, and Mark P. Oxley
4.1	Recent Progress 109
4.2	Introduction to the Instrument 113
4.3	Imaging in the STEM 116
4.3.1	Probe Formation 116
4.3.2	The Ronchigram 123
4.3.3	Reciprocity between TEM and STEM 125
4.3.4	Coherent and Incoherent Imaging 129
4.3.5	Dynamical Diffraction 134
4.3.6	Depth Sectioning 135
4.3.7	Image Simulation and Quantification 140
4.4	Future Outlook 144
	Acknowledgments 145
	References 146
5	**Electron Holography** 153
	Hannes Lichte
	General Idea 153
5.1	Image-Plane Off-Axis Holography Using the Electron Biprism 154
5.1.1	Recording a Hologram 154
5.1.2	Reconstruction of the Electron Wave 156
5.2	Properties of the Reconstructed Wave 157
5.2.1	Time Averaging 157
5.2.2	Inelastic Filtering 159
5.2.3	Basis for Recovering the Object Exit Wave 160
5.2.4	Amplitude Image 160
5.2.5	Phase Image 160
5.2.6	Field of View 161
5.2.7	Lateral Resolution 161
5.2.7.1	Fringe Spacing 161
5.2.7.2	Optimizing the Paths of Rays for Holography 162
5.2.8	Digitization of the Image Wave 162
5.2.9	Signal Resolution–Signal/Noise Properties 163
5.2.10	Amount of Information in the Reconstructed Wave 165
5.3	Holographic Investigations 166
5.3.1	Electric Fields 166
5.3.1.1	Structure Potentials 166
5.3.1.2	Intrinsic Electric Fields 169
5.3.2	Magnetic Fields 173
5.3.2.1	Distinction between Electric and Magnetic Phase Shift 176
5.3.3	Holography at Atomic Dimensions 180
5.3.3.1	Special Aspects for Acquisition of Atomic Resolution Holograms 183
5.3.3.2	Lateral Resolution: Fringe Spacing 183

5.3.3.3	Width of Hologram, Number of Fringes, and Pixel Number of CCD Camera	183
5.3.3.4	Adaptation of the Hologram Geometry	184
5.3.3.5	Optimum Focus of Objective Lens	184
5.3.3.6	Demands on Signal Resolution	185
5.4	Special Techniques	197
5.4.1	Holographic Tomography	199
5.4.2	Dark-Field Holography	202
5.4.3	Inelastic Holography	204
5.5	Summary	209
	Acknowledgments	212
	References	213

6 Lorentz Microscopy and Electron Holography of Magnetic Materials 221
Rafal E. Dunin-Borkowski, Takeshi Kasama, Marco Beleggia, and Giulio Pozzi

6.1	Introduction	221
6.2	Lorentz Microscopy	221
6.2.1	Historical Background	221
6.2.2	Imaging Modes	222
6.2.3	Applications	224
6.3	Off-Axis Electron Holography	227
6.3.1	Historical Background	227
6.3.2	Basis and Governing Equations	228
6.3.3	Experimental Requirements	233
6.3.4	Magnetic and Mean Inner Potential Contributions to the Phase	235
6.3.5	Applications	239
6.3.6	Quantitative Measurement of Magnetic Moments Using Electron Holography	244
6.4	Discussion and Conclusions	246
	Acknowledgments	247
	References	247

7 Electron Tomography 253
Paul Anthony Midgley and Sara Bals

7.1	History and Background	253
7.1.1	Introduction to Nanoscale Systems	253
7.1.2	Tomography	253
7.2	Theory of Tomography	255
7.2.1	Real Space Reconstruction Using Backprojection	256
7.3	Electron Tomography, Missing Wedge, and Imaging Modes	259
7.4	STEM Tomography and Applications	261
7.5	Hollow-Cone DF Tomography	264
7.6	Diffraction Contrast Tomography	266

7.7	Electron Holographic Tomography	268
7.8	Inelastic Electron Tomography	270
7.9	Advanced Reconstruction Techniques	271
7.10	Quantification and Atomic Resolution Tomography	273
	Acknowledgments	274
	References	274

8 Statistical Parameter Estimation Theory – A Tool for Quantitative Electron Microscopy 281

Sandra Van Aert

8.1	Introduction	281
8.2	Methodology	282
8.2.1	Aim of Statistical Parameter Estimation Theory	282
8.2.2	Parametric Statistical Model of the Observations	283
8.2.3	Properties of Estimators	285
8.2.4	Attainable Precision	286
8.2.5	Maximum Likelihood Estimation	288
8.2.5.1	The Need for a Good Starting Model	290
8.2.6	Model Assessment	291
8.2.7	Confidence Regions and Intervals	292
8.3	Electron Microscopy Applications	292
8.3.1	Resolution versus Precision	293
8.3.2	Atom Column Position Measurement	294
8.3.3	Model-Based Quantification of Electron Energy Loss Spectra	296
8.3.4	Quantitative Atomic Resolution Mapping using High-Angle Annular Dark Field Scanning Transmission Electron Microscopy	298
8.3.5	Statistical Experimental Design	300
8.4	Conclusions	303
	Acknowledgments	304
	References	305

9 Dynamic Transmission Electron Microscopy 309

Nigel D. Browning, Geoffrey H. Campbell, James E. Evans, Thomas B. LaGrange, Katherine L. Jungjohann, Judy S. Kim, Daniel J. Masiel, and Bryan W. Reed

9.1	Introduction	309
9.2	Time-Resolved Studies Using Electrons	311
9.2.1	Brightness, Emittance, and Coherence	314
9.2.2	Single-Shot Space–Time Resolution Trade-Offs	316
9.3	Building a DTEM	320
9.3.1	The Base Microscope and Experimental Method	320
9.3.2	Current Performance of Single-Shot DTEM	323
9.4	Applications of DTEM	324
9.4.1	Reactive Nanolaminate Films	324
9.4.2	Experimental Methods	326

9.4.3	Diffraction Results	327
9.4.4	Imaging Results	329
9.4.5	Discussion	331
9.5	Future Developments for DTEM	333
9.5.1	Arbitrary Waveform Generation Laser	333
9.5.2	Acquiring High Time Resolution Movies	334
9.5.3	Aberration Correction	335
9.5.4	In Situ Liquid Stages	336
9.5.5	Novel Electron Sources	338
9.5.6	Pulse Compression	339
9.6	Conclusions	339
	Acknowledgments	340
	References	340
10	**Transmission Electron Microscopy as Nanolab**	**345**
	Frans D. Tichelaar, Marijn A. van Huis, and Henny W. Zandbergen	
10.1	TEM and Measuring the Electrical Properties	346
10.1.1	TEM and Measuring Electrical Properties. Example 1: Electromigration	347
10.1.2	TEM and Measuring Electrical Properties. Example 2: Carbon Nanotubes	348
10.2	TEM with MEMS-Based Heaters	349
10.2.1	TEM with MEMS-Based Heaters. Example 1: Graphene at Various Temperatures	351
10.2.2	TEM with MEMS-Based Heaters. Example 2: Morphological Changes on Au Nanoparticles	352
10.3	TEM with Gas Nanoreactors	353
10.3.1	TEM with Gas Nanoreactors. Example 1: Hydrogen Storage Materials	354
10.3.2	TEM with Gas Nanoreactors. Example 2: STEM Imaging of a Layer of Gold Nanoparticles at 1 bar	355
10.4	TEM with Liquid Nanoreactors	356
10.4.1	TEM with Liquid Nanoreactors. Example 1: Cu Electrodeposition	357
10.4.2	TEM with Liquid Nanoreactors. Example 2: Nucleation, Growth, and Motion of Small Particles	359
10.5	TEM and Measuring Optical Properties	359
10.6	Sample Preparation for Nanolab Experiments	363
10.6.1	Sculpting with the Electron Beam	365
10.6.2	Sculpting with the Ga Ion Beam	368
10.6.3	Sculpting with the Helium Ion Beam	368
	References	370

11	**Atomic-Resolution Environmental Transmission Electron Microscopy** *375*	
	Pratibha L. Gai and Edward D. Boyes	
11.1	Introduction *375*	
11.2	Atomic-Resolution ETEM *376*	
11.3	Development of Atomic-Resolution ETEM *377*	
11.4	Experimental Procedures *380*	
11.5	Applications with Examples *383*	
11.6	Nanoparticles and Catalytic Materials *383*	
11.6.1	Dynamic Nanoparticle Shape Modifications, Electronic Structures of Promoted Systems, and Dynamic Oxidation States *383*	
11.7	Oxides *388*	
11.8	*In situ* Atomic Scale Twinning Transformations in Metal Carbides *389*	
11.9	Dynamic Electron Energy Loss Spectroscopy *389*	
11.10	Technological Benefits of Atomic-Resolution ETEM *391*	
11.11	Other Advances *392*	
11.12	Reactions in the Liquid Phase *392*	
11.13	*In situ* Studies with Aberration Correction *393*	
11.14	Examples and Discussion *396*	
11.15	Applications to Biofuels *398*	
11.16	Conclusions *401*	
	Acknowledgments *402*	
	References *402*	
12	**Speckles in Images and Diffraction Patterns** *405*	
	Michael M. J. Treacy	
12.1	Introduction *405*	
12.2	What Is Speckle? *406*	
12.3	What Causes Speckle? *408*	
12.4	Diffuse Scattering *410*	
12.5	From Bragg Reflections to Speckle *411*	
12.6	Coherence *417*	
12.6.1	Temporal Coherence *418*	
12.6.2	Coherence Length *418*	
12.6.3	Spatial Coherence *419*	
12.6.4	Coherence Volume *421*	
12.7	Fluctuation Electron Microscopy *421*	
12.7.1	Measuring Speckle *424*	
12.8	Variance versus Mean *428*	
12.9	Speckle Statistics *430*	
12.10	Possible Future Directions for Electron Speckle Analysis *433*	
	References *435*	

13	**Coherent Electron Diffractive Imaging** 437	
	J.M. Zuo and Weijie Huang	
13.1	Introduction 437	
13.2	Coherent Nanoarea Electron Diffraction 439	
13.3	The Noncrystallographic Phase Problem 443	
13.4	Coherent Diffractive Imaging of Finite Objects 445	
13.4.1	Brief History of Coherent Diffractive Imaging 446	
13.4.2	Oversampling 447	
13.4.3	Sampling Experimental Diffraction Patterns and the Field of View 449	
13.4.4	Requirements on Beam Coherence 450	
13.4.5	Phase Retrieval Algorithms 451	
13.4.6	Use of Image Information 454	
13.4.7	Simulation Study of Phase Retrieval Algorithms 455	
13.5	Phasing Experimental Diffraction Pattern 461	
13.5.1	Processing of Experimental Diffraction Patterns and Images 462	
13.5.2	Phasing CdS Quantum Dots 464	
13.6	Conclusions 470	
	Acknowledgments 470	
	References 470	
14	**Sample Preparation Techniques for Transmission Electron Microscopy** 473	
	Vasfi Burak Özdöl, Vesna Srot, and Peter A. van Aken	
14.1	Introduction 473	
14.2	Indirect Preparation Methods 474	
14.3	Direct Preparation Methods 475	
14.3.1	Preliminary Preparation Techniques 475	
14.3.2	Cleavage Techniques 477	
14.3.3	Chemical and Electrolytic Methods 478	
14.3.4	Ion Beam Milling 479	
14.3.5	Tripod Polishing 483	
14.3.6	Ultramicrotomy 485	
14.3.7	Focused Ion Beam Milling 488	
14.3.8	Combination of Different Preparation Methods 491	
14.4	Summary 493	
	Acknowledgments 494	
	References 495	
15	**Scanning Probe Microscopy – History, Background, and State of the Art** 499	
	Ralf Heiderhoff and Ludwig Josef Balk	
15.1	Introduction 499	
15.2	Detecting Evanescent Waves by Near-Field Microscopy: Scanning Tunneling Microscopy 503	

15.3	Interaction of Tip–Sample Electrons Detected by Scanning Near-Field Optical Microscopy and Atomic Force Microscopy 505	
15.4	Methods for the Detection of Electric/Electronic Sample Properties 510	
15.5	Methods for the Detection of Electromechanical and Thermoelastic Quantities 516	
15.6	Advanced SFM/SEM Microscopy 522	
	Acknowledgments 533	
	References 533	
16	**Scanning Probe Microscopy—Forces and Currents in the Nanoscale World** 539	
	Brian J. Rodriguez, Roger Proksch, Peter Maksymovych, and Sergei V. Kalinin	
16.1	Introduction 539	
16.2	Scanning Probe Microscopy—the Science of Localized Probes 541	
16.2.1	Local Probe Methods before SPM 541	
16.2.2	Positioning, Probes, and Detectors 543	
16.2.2.1	Positioning 543	
16.2.2.2	Detectors 545	
16.2.2.3	Probes 547	
16.2.3	Data Processing and Acquisition 550	
16.2.3.1	Classical Detection Strategies in SPM 550	
16.2.3.2	Multifrequency Methods 552	
16.2.3.3	Real-Space Methods 554	
16.3	Scanning Tunneling Microscopy and Related Techniques 554	
16.3.1	The Role of Tip Effects in Tunneling Conductance and Inelastic Phenomena 559	
16.3.2	Electron Tunneling in Correlated Electron Materials 560	
16.3.3	Tunneling into Low Conducting Surfaces 562	
16.3.4	Time-Resolved Studies 562	
16.4	Force-Based SPM Measurements 563	
16.4.1	Contact, Intermittent, and Noncontact AFM 564	
16.4.2	Force–Distance Spectroscopy and Spectroscopic Imaging 567	
16.4.3	Decoupling Interactions 572	
16.4.3.1	Dual-Pass Methods 572	
16.4.3.2	Modulation Approaches 573	
16.4.3.3	Decoupling through Different Mechanical Degrees of Freedom 574	
16.5	Voltage Modulation SPMs 574	
16.5.1	Kelvin Probe Force Microscopy and Scanning Impedance Microscopy 574	
16.5.2	Transport Imaging of Active Device Structures 577	
16.5.3	Piezoresponse Force Microscopy and Electrochemical Strain Microscopy 579	
16.6	Current Measurements in SPM 583	

16.6.1	DC Current 583
16.6.2	AC Current 584
16.6.3	Scanning Capacitance Microscopy 586
16.7	Emergent SPM Methods 587
16.7.1	Thermal SPM Imaging 587
16.7.2	Microwave Microscopy 588
16.7.3	Mass Spectrometric Imaging in SPM 592
16.8	Manipulation of Matter by SPM 593
16.9	Perspectives 594
16.9.1	Platforms, Probes, and Detectors 595
16.9.2	Dynamic Detection 596
16.9.3	Multidimensional Data Analysis and Interpretation 597
16.9.4	Combined Methods 598
	Acknowledgments 599
	References 599

17 Scanning Beam Methods 615
David Joy

17.1	Scanning Microscopy 615
17.1.1	Introduction 615
17.1.2	Instrumentation 617
17.1.3	Performance 619
17.1.4	Modes of Operation 623
17.1.4.1	Secondary Electron Imaging 623
17.1.4.2	Backscattered Electrons and Ions 628
17.1.4.3	Electron Backscatter Diffraction Patterns 634
17.1.4.4	Electron Beam–Induced Conductivity (EBIC) 635
17.1.4.5	Image Artifacts 636
17.2	Conclusions 642
	References 642

18 Fundamentals of the Focused Ion Beam System 645
Nan Yao

18.1	Focused Ion Beam Principles 645
18.1.1	Introduction 645
18.1.2	The Ion Beam 646
18.1.2.1	Focusing the Ion Beam 646
18.1.2.2	Beam Overlap 647
18.1.3	Interactions of Ions with Matter 648
18.1.3.1	Interatomic Potentials 649
18.1.3.2	Binary Scattering and Recoil 650
18.1.3.3	Cross Section 650
18.1.3.4	Energy Loss 651
18.1.4	Detection of Electron and Ion Signals 652
18.2	FIB Techniques 655

18.2.1	Milling 655	
18.2.1.1	TEM Sample Preparation 657	
18.2.2	Deposition 661	
18.2.3	Implantation 663	
18.2.4	Imaging 664	
	Acknowledgments 667	
	References 667	
	Further Reading 669	

Contents to Volume 2

Preface *XIX*
List of Contributors *XXI*

19 **Low-Energy Electron Microscopy** 673
Ernst Bauer

20 **Spin-Polarized Low-Energy Electron Microscopy** 697
Ernst Bauer

21 **Imaging Secondary Ion Mass Spectroscopy** 709
Katie L. Moore, Markus Schröder, and Chris R. M. Grovenor

22 **Soft X-Ray Imaging and Spectromicroscopy** 745
Adam P. Hitchcock

23 **Atom Probe Tomography: Principle and Applications** 793
Frederic Danoix and François Vurpillot

24 **Signal and Noise Maximum Likelihood Estimation in MRI** 833
Jan Sijbers

25 **3-D Surface Reconstruction from Stereo Scanning Electron Microscopy Images** 855
Shafik Huq, Andreas Koschan, and Mongi Abidi

Part II Applications 877

26 **Nanoparticles** *879*
Miguel López-Haro, Juan José Delgado, Juan Carlos Hernández-Garrido, Juan de Dios López-Castro, César Mira, Susana Trasobares, Ana Belén Hungría, José Antonio Pérez-Omil, and José Juan Calvino

27 **Nanowires and Nanotubes** *961*
Yong Ding and Zhong Lin Wang

28 **Carbon Nanoforms** *995*
Carla Bittencourt and Gustaaf Van Tendeloo

29 **Metals and Alloys** *1071*
Dominique Schryvers

30 ***In situ* Transmission Electron Microscopy on Metals** *1099*
J.Th.M. De Hosson

31 **Semiconductors and Semiconducting Devices** *1153*
Hugo Bender

32 **Complex Oxide Materials** *1179*
Maria Varela, Timothy J. Pennycook, Jaume Gazquez, Albina Y. Borisevich, Sokrates T. Pantelides, and Stephen J. Pennycook

33 **Application of Transmission Electron Microscopy in the Research of Inorganic Photovoltaic Materials** *1213*
Yanfa Yan

34 **Polymers** *1247*
Joachim Loos

35 **Ferroic and Multiferroic Materials** *1273*
Ekhard Salje

36 **Three-Dimensional Imaging of Biomaterials with Electron Tomography** *1303*
Montserrat Bárcena, Roman I. Koning, and Abraham J. Koster

37 **Small Organic Molecules and Higher Homologs** *1335*
Ute Kolb and Tatiana E. Gorelik

Index *1381*

Preface

Since the edition of the previous "Handbook of Microscopy" in 1997 the world of microscopy has gone through a significant transition.

In electron microscopy the introduction of aberration correctors has pushed the resolution down to the sub-Angstrom regime, detectors are able to detect single electrons, spectrometers are able to record spectra from single atoms. Moreover the object space is increased which allows to integrate these techniques in the same instrument under full computer support without compromising on the performance. Thus apart from the increased resolution from microscopy to nanoscopy, even towards picoscopy, the EM is gradually transforming from an imaging device into a true nanoscale laboratory that delivers reliable quantitative data on the nanoscale close to the physical and technical limits. In parallel, scanning probe methods have undergone a similar evolution towards increased functionality, flexibility and integration.

As a consequence the whole field of microscopy is gradually shifting from the instrument to the application, from describing to measuring and to understanding the structure/property relations, from nanoscopy to nanology.

But these instruments will need a different generation of nanoscopists who need not only to master the increased flexibility and multifunctionality of the instruments, but to choose and combine the experimental possibilities to fit the material problem to be investigated.

It is the purpose of this new edition of the "Handbook of Nanoscopy" to provide an ideal reference base of knowledge for the future user.

Volume 1 elaborates on the basic principles underlying the different nanoscopical methods with a critical analysis of the merits, drawbacks and future prospects. Volume 2 focuses on a broad category of materials from the viewpoint of how the different nanoscopical measurements can contribute to solving materials structures and problems.

The handbook is written in a very readable style at a level of a general audience. Whenever relevant for deepening the knowledge, proper references are given.

Gustaaf Van Tendeloo, Dirk Van Dyck, and Stephen J. Pennycook

List of Contributors

Mongi Abidi
The University of Tennessee
Min H. Kao
Department of Electrical
Engineering and Computer
Science
Imaging Robotics and Intelligent
Systems (IRIS) Lab
209 Ferris Hall
Knoxville
TN 37996-2100
USA

Ludwig Josef Balk
Bergische Universität Wuppertal
Fachbereich Elektronik
Informationstechnik
Medientechnik
Lehrstuhl für Elektronische
Bauelemente
Rainer-Gruenter-Str. 21
42119 Wuppertal
Germany

Sara Bals
University of Antwerp
Department of Physics
EMAT
Groenenborgerlaan 171
2020 Antwerp
Belgium

Montserrat Bárcena
Leiden University Medical Center
Department of Molecular Cell
Biology
Section Electron Microscopy
Einthovenweg 20
2333 ZC
The Netherlands

Juri Barthel
Forschungszentrum Jülich
GmbH
Peter Grünberg Institute and
Ernst Ruska Centre for
Microscopy and Spectroscopy
with Electrons
D-52425 Jülich
Germany

Ernst Bauer
Arizona State University
Department of Physics
Tempe
AZ 85287-1504
USA

Marco Beleggia
Technical University of Denmark
Center for Electron Nanoscopy
DK-2800 Kongens Lyngby
Denmark

Hugo Bender
Imec
Kapeldreef 75
Leuven 3001
Belgium

Carla Bittencourt
University of Antwerp
EMAT
Groenenborgerlaan 171
B-2020 Antwerp
Belgium

Albina Y. Borisevich
Oak Ridge National Laboratory
Materials Science and Technology
Division
Oak Ridge
TN 37831-6071
USA

Edward D. Boyes
The University of York
The York JEOL Nanocentre
Departments of Physics
Helix House
Heslington
York, YO10 5BR
UK

and

The University of York
The York JEOL Nanocentre
Department of Electronics
Helix House
Heslington
York, YO10 5BR
UK

Nigel D. Browning
Lawrence Livermore National
Laboratory
Condensed Matter and Materials
Division
Physical and Life Sciences
Directorate
7000 East Avenue
Livermore
CA 94550
USA

and

University of California-Davis
Department of Chemical
Engineering and Materials
Science
One Shields Ave
Davis, CA 95616
USA

and

University of California-Davis
Department of Molecular and
Cellular Biology
One Shields Ave
Davis, CA 95616
USA

and

Pacific Northwest National
Laboratory
902 Battelle Boulevard
Richland
WA 99352
USA

José Juan Calvino
Facultad de Ciencias de la
Universidad de Cádiz
Departamento de Ciencia de los
Materiales e Ingeniería
Metalúrgica y Química
Inorgánica
Campus Rio San Pedro
Puerto Real
11510-Cádiz
Spain

Geoffrey H. Campbell
Lawrence Livermore National
Laboratory
Condensed Matter and Materials
Division
Physical and Life Sciences
Directorate
7000 East Avenue
Livermore
CA 94550
USA

Frederic Danoix
Université de Rouen
Groupe de Physique des
Matériaux, UMR CNRS 6634
Site universitaire du Madrillet
Saint Etienne du Rouvray
76801
France

Juan José Delgado
Facultad de Ciencias de la
Universidad de Cádiz
Departamento de Ciencia de los
Materiales e Ingeniería
Metalúrgica y Química
Inorgánica
Campus Rio San Pedro
Puerto Real
11510-Cádiz
Spain

Marc De Graef
Carnegie Mellon University
Materials Science and
Engineering Department
5000 Forbes Avenue
Pittsburgh
PA 15213-3890
USA

J.Th.M. De Hosson
University of Groningen
Department of Applied Physics
Zernike Institute for Advanced
Materials and Materials
Innovation Institute
Nijenborgh 4
9747 AG Groningen
The Netherlands

Yong Ding
School of Materials Science and
Engineering
Georgia Institute of Technology
Atlanta
GA 30332-0245
USA

Rafal E. Dunin-Borkowski
Forschungszentrum Jülich
GmbH
Peter Grünberg Institute and
Ernst Ruska Centre for
Microscopy and Spectroscopy
with Electrons
D-52425 Jülich
Germany

and

Technical University of Denmark
Center for Electron Nanoscopy
DK-2800 Kongens
Lyngby
Denmark

James E. Evans
Lawrence Livermore National Laboratory
Condensed Matter and Materials Division
Physical and Life Sciences Directorate
7000 East Avenue
Livermore
CA 94550
USA

and

University of California-Davis
Department of Molecular and Cellular Biology
One Shields Ave
Davis
CA 95616
USA

and

Pacific Northwest National Laboratory
902 Battelle Boulevard
Richland
WA 99352
USA

Pratibha L. Gai
The University of York
The York JEOL Nanocentre
Department of Chemistry
Helix House
Heslington
York, YO10 5BR
UK

and

The University of York
The York JEOL Nanocentre
Department of Physics
Helix House
Heslington
York, YO10 5BR
UK

Jaume Gazquez
Oak Ridge National Laboratory
Materials Science and Technology Division
Oak Ridge
TN 37831-6071
USA

and

Universidad Complutense de Madrid
Departamento de Fisica Aplicada III Avda.
Complutense s/n
28040 Madrid
Spain

and

Instituto de Ciencia de Materiales de Barcelona-CSIC
Campus de la UAB
08193 Bellaterra
Spain

Tatiana E. Gorelik
Johannes Gutenberg-Universität
Mainz
Institut für Physikalische Chemie
Welderweg 11
55099 Mainz
Germany

Chris R. M. Grovenor
University of Oxford
Department of Materials
Parks Road
Oxford OX1 3PH
UK

Ralf Heiderhoff
Bergische Universität Wuppertal
Fachbereich Elektronik
Informationstechnik
Medientechnik
Lehrstuhl für Elektronische
Bauelemente
Rainer-Gruenter-Str. 21
42119 Wuppertal
Germany

Juan Carlos Hernández-Garrido
Facultad de Ciencias de la
Universidad de Cádiz
Departamento de Ciencia de los
Materiales e Ingeniería
Metalúrgica y Química
Inorgánica
Campus Rio San Pedro
Puerto Real
11510-Cádiz
Spain

Adam P. Hitchcock
McMaster University
Department of Chemistry and
Chemical Biology
Brockhouse Institute for
Materials Research
1280 Main Street West
Hamilton
ON L8S 4M1
Canada

Lothar Houben
Forschungszentrum Jülich
GmbH
Peter Grünberg Institute and
Ernst Ruska Centre for
Microscopy and Spectroscopy
with Electrons
D-52425 Jülich
Germany

Weijie Huang
University of Illinois at
Urbana-Champaign
Department of Materials Science
and Engineering and Materials
Research Laboratory
Urbana
IL 61801
USA

and

Carl Zeiss SMT Inc.
One Corporation Way
Peabody
MA 01960
USA

Ana Belén Hungría
Facultad de Ciencias de la
Universidad de Cádiz
Departamento de Ciencia de los
Materiales e Ingeniería
Metalúrgica y Química
Inorgánica
Campus Rio San Pedro
Puerto Real
11510-Cádiz
Spain

Shafik Huq
The University of Tennessee
Min H. Kao
Department of Electrical
Engineering and Computer
Science
Imaging, Robotics, and
Intelligent Systems (IRIS) Lab
209 Ferris Hall
Knoxville
TN 37996-2100
USA

Chun-Lin Jia
Forschungszentrum Jülich
GmbH
Peter Grünberg Institute and
Ernst Ruska Centre for
Microscopy and Spectroscopy
with Electrons
D-52425 Jülich
Germany

and

Xi'an Jiaotong University
International Centre for
Dielectrics Research (ICDR)
School of Electronic and
Information Engineering
28 Xianning West Road
Xi'an 710049
China

David Joy
University of Tennessee
Science and Engineering
Research Facility
Knoxville
TN 37996-2200
USA

and

Center for Nano Material Science
Oak Ridge National Laboratory
Oak Ridge
TN 37831
USA

Katherine L. Jungjohann
University of California-Davis
Department of Chemical
Engineering and Materials
Science
One Shields Ave
Davis, CA 95616
USA

Sergei V. Kalinin
Oak Ridge National Laboratory
Oak Ridge
TN 37922
USA

Takeshi Kasama
Technical University of Denmark
Center for Electron Nanoscopy
DK-2800 Kongens Lyngby
Denmark

Judy S. Kim
University of California-Davis
Department of Chemical
Engineering and Materials
Science
One Shields Ave
Davis
CA 95616
USA

Ute Kolb
Johannes Gutenberg-Universität
Mainz
Institut für Physikalische Chemie
Welderweg 11
55099 Mainz
Germany

Roman I. Koning
Leiden University Medical Center
Department of Molecular Cell
Biology
Section Electron Microscopy
Einthovenweg 20
2333 ZC
Leiden
The Netherlands

Andreas Koschan
The University of Tennessee
Min H. Kao
Department of Electrical
Engineering and Computer
Science
Imaging Robotics and Intelligent
Systems (IRIS) Lab
330 Ferris Hall
Knoxville
TN 37996-2100
USA

Abraham J. Koster
Leiden University Medical Center
Department of Molecular Cell
Biology
Section Electron Microscopy
Einthovenweg 20
2333 ZC
Leiden
The Netherlands

Thomas B. LaGrange
Lawrence Livermore National
Laboratory
Condensed Matter and Materials
Division
Physical and Life Sciences
Directorate
7000 East Avenue
Livermore
CA 94550
USA

Markus Lentzen
Forschungszentrum Jülich
GmbH
Peter Grünberg Institute and
Ernst Ruska Centre for
Microscopy and Spectroscopy
with Electrons
D-52425 Jülich
Germany

Hannes Lichte
Technische Universität Dresden
Triebenberg Laboratory
Institute for Structure Physics
01062 Dresden
Germany

Joachim Loos
University of Glasgow
School of Physics and Astronomy
Kelvin Building
Glasgow G12 8QQ
Scotland
UK

Juan de Dios López-Castro
Facultad de Ciencias de la
Universidad de Cádiz
Departamento de Ciencia de los
Materiales e Ingeniería
Metalúrgica y Química
Inorgánica
Campus Rio San Pedro
Puerto Real
11510-Cádiz
Spain

Andrew R. Lupini
Oak Ridge National Laboratory
Materials Science and Technology
Division
Oak Ridge
TN 37831-6071
USA

Peter Maksymovych
Oak Ridge National Laboratory
Oak Ridge
TN 37922
USA

Daniel J. Masiel
University of California-Davis
Department of Chemical
Engineering and Materials
Science
One Shields Ave
Davis
CA 95616
USA

Paul Anthony Midgley
University of Cambridge
Department of Material Science
and Metallurgy
Pembroke Street
Cambridge CB2 3QZ
UK

César Mira
Facultad de Ciencias de la
Universidad de Cádiz
Departamento de Ciencia de los
Materiales e Ingeniería
Metalúrgica y Química
Inorgánica
Campus Rio San Pedro
Puerto Real
11510-Cádiz
Spain

Katie L. Moore
University of Oxford
Department of Materials
Parks Road
Oxford OX1 3PH
UK

Mark P. Oxley
Oak Ridge National Laboratory
Materials Science and Technology
Division
Oak Ridge
TN 37831-6071
USA

and

Vanderbilt University
Department of Physics and
Astronomy
Nashville
TN 37235
USA

Vasfi Burak Özdöl
Stuttgart Center for Electron
Microscopy
Max Planck Institute for
Intelligent Systems
Heisenbergstr. 3
70569 Stuttgart
Germany

Sokrates T. Pantelides
Oak Ridge National Laboratory
Materials Science & Technology
Divsion
1 Bethel
Valley Road
Oak Ridge
TN 37831-6071
USA

and

Vanderbilt University
Department of Physics and
Astronomy
Nashville
TN 37235
USA

Stephen J. Pennycook
Oak Ridge National Laboratory
Materials Science and Technology
Division
Oak Ridge
TN 37831-6071
USA

and

Vanderbilt University
Department of Physics and
Astronomy
Nashville
TN 37235
USA

Timothy J. Pennycook
Oak Ridge National Laboratory
Materials Science & Technology
Divsion
1 Bethel Valley Road
Oak Ridge
TN 37831-6071
USA

and

Vanderbilt University
Department of Physics and
Astronomy
Nashville
TN 37235
USA

José Antonio Pérez-Omil
Facultad de Ciencias de la
Universidad de Cádiz
Departamento de Ciencia de los
Materiales e Ingeniería
Metalúrgica y Química
Inorgánica
Campus Rio San Pedro
Puerto Real
11510-Cádiz
Spain

Giulio Pozzi
Universita' di Bologna
Dipartimento di Fisica
V. le B. Pichat 6/2
40127 Bologna
Italy

Roger Proksch
Asylum Research
Santa Barbara
CA 93117
USA

Bryan W. Reed
Lawrence Livermore National Laboratory
Condensed Matter and Materials Division
Physical and Life Sciences Directorate
7000 East Avenue
Livermore
CA 94550
USA

Brian J. Rodriguez
University College Dublin
Conway Institute of Biomolecular and Biomedical Research and School of Physics
Belfield
Dublin 4
Ireland

Ekhard Salje
Cambrige University
Department Earth Science
Downing Street
Cambridge CB2 3EQ
UK

Markus Schröder
University of Oxford
Department of Materials
Parks Road
Oxford
OX1 3PH
UK

Dominique Schryvers
University of Antwerp
EMAT
Groenenborgerlaan 171
B-2020 Antwerp
Belgium

Jan Sijbers
University of Antwerp (CDE)
Vision Lab
Department of Physics
Universiteitsplein 1 (N.1.13)
B-2610 Wilrijk
Belgium

Vesna Srot
Stuttgart Center for Electron Microscopy
Max Planck Institute for Metals Research
Heisenbergstr. 3
70569 Stuttgart
Germany

Andreas Thust
Forschungszentrum Jülich GmbH
Peter Grünberg Institute and Ernst Ruska Centre for Microscopy and Spectroscopy with Electrons
D-52425 Jülich
Germany

Frans D. Tichelaar
Delft University of Technology
Applied Sciences
Kavli Institute of Nanoscience
Lorentzweg 1
NL-2628CJ Delft
The Netherlands

Karsten Tillmann
Forschungszentrum Jülich GmbH
Peter Grünberg Institute and Ernst Ruska Centre for Microscopy and Spectroscopy with Electrons
D-52425 Jülich
Germany

Susana Trasobares
Facultad de Ciencias de la
Universidad de Cádiz
Departamento de Ciencia de los
Materiales e Ingeniería
Metalúrgica y Química
Inorgánica
Campus Rio San Pedro
Puerto Real
11510-Cádiz
Spain

Michael M. J. Treacy
Arizona State University
Department of Physics
Bateman Building
B-147
Tyler Mall
Tempe
AZ 85287-1504
USA

Knut W. Urban
Forschungszentrum Jülich
GmbH
Peter Grünberg Institute and
Ernst Ruska Centre for
Microscopy and Spectroscopy
with Electrons
D-52425 Jülich
Germany

Sandra Van Aert
University of Antwerp
Electron Microscopy for Materials
Research (EMAT)
Groenenborgerlaan 171
2020 Antwerp
Belgium

Peter A. Van Aken
Stuttgart Center for Electron
Microscopy
Max Planck Institute for
Intelligent Systems
Heisenbergstr. 3
70569 Stuttgart
Germany

Dirk Van Dyck
University of Antwerp (UA)
EMAT
Groenenborgerlaan 171
2020 Antwerpen
Belgium

Marijn A. van Huis
Delft University of Technology
Applied Sciences
Kavli Institute of Nanoscience
Lorentzweg 1
NL-2628CJ Delft
The Netherlands

Gustaaf Van Tendeloo
University of Antwerp
EMAT
Groenenborgerlaan 171
B-2020 Antwerp
Belgium

Maria Varela
Materials Science & Technology
Division
Oak Ridge National Laboratory
1 Bethel Valley Road
Oak Ridge
TN 37831-6071
USA

François Vurpillot
Université de Rouen
Groupe de Physique des
Matériaux
UMR CNRS 6634
Site universitaire du Madrillet
Saint Etienne du Rouvray
76801
France

Zhong Lin Wang
School of Materials Science and
Engineering
Georgia Institute of Technology
Atlanta
GA 30332-0245
USA

Yanfa Yan
Department of Physics and
Astronomy
The University of Toledo
2801 Bancroft street
Toledo
Ohio 43606
USA

Nan Yao
Princeton University
Princeton Institute for the Science
and Technology of Materials
120 Bowen Hall
70 Prospect Avenue
Princeton
NJ 08540
USA

Henny W. Zandbergen
Delft University of Technology
Applied Sciences
Kavli Institute of Nanoscience
Lorentzweg 1
NL-2628CJ Delft
The Netherlands

J.M. Zuo
University of Illinois at
Urbana-Champaign
Department of Materials Science
and Engineering and Materials
Research Laboratory
Urbana
IL 61801
USA

The Past, the Present, and the Future of Nanoscopy

Gustaaf Van Tendeloo and Dirk Van Dyck

The Past

Science stems from the curiosity of humankind. We all want to look behind the curtain and discover new things that have not been exposed yet. We explore the world, like a child, with our five senses. For the present book, we mainly, although not exclusively, focus on the eyesight. With our eyes, we explore the world around us at different scales: from the stars and the cosmos down to the sub-millimeter scale. However, details below 0.1 mm are hardly visible to the naked eye. In order to improve our eyesight, we use a magnifying glass, but for further magnifications, we have to use combined lenses, which form an instrument that we know as a microscope.

The first multilens microscopes were built in the seventeenth century by Jan Swammerdam and Robert Hooke, but it was the Dutchman Antoni van Leeuwenhoek who captured the interest of scientists for the new technique [1]. He was able to produce such strong lenses that the magnification could go up to several hundred times. This evidently created a whole new inside in the "invisible" world of micron-scale details and organisms such as bacteria and cells. van Leeuwenhoek was not a scientist by education, but he got large recognition in the scientific world because of his amazing results. Nevertheless, he kept the secret of producing such strong lenses for himself until he died. Once the interest of the scientists was captured, the optical microscope evolved in the eighteenth and nineteenth centuries through the introduction of more lenses to minimize aberrations.

The resolution of "classical optical microscopy" kept improving, but at the end of the nineteenth century, Ernst Karl Abbe stated that the resolution of an optical microscope is actually limited by the wavelength of light; that is, of the order of about half a micrometer.

At the end of the nineteenth century and in the beginning of twentieth century, a number of things changed. Thomson discovered the electron, Einstein introduced relativity theory, and quantum mechanics got solid foundations, thanks to Planck, Bohr, de Broglie, Heisenberg, Schrodinger, and Dirac. Particularly, the description of the particle – wave duality for accelerated electrons was important for the

Handbook of Nanoscopy, First Edition. Edited by Gustaaf Van Tendeloo, Dirk Van Dyck, and Stephen J. Pennycook.
© 2012 Wiley-VCH Verlag GmbH & Co. KGaA. Published 2012 by Wiley-VCH Verlag GmbH & Co. KGaA.

discovery of the "electron microscope" in 1931 by Ernst Ruska. Based on the fact that electrons could be deflected by an electrostatic or an electromagnetic field, Max Knoll and Ernst Ruska demonstrated a two-stage magnification of a simple object with a magnification of only 17 times. The first experiments with a real electron microscope were really disappointing from a scientific point of view. Although already in 1933 Ruska had no problem proving a resolution of 50 nm, considerably better than the resolution of an optical microscope, virtually all materials were burnt to a cinder under the electron beam and the interest of scientists faded away.

In that same period, Marton and others found a way to avoid severe burning of the samples, and the interest of industry to build a commercial instrument revived. In 1939, Siemens and Halske promoted the first commercial electron microscope, but it was only after the second World War that the real success of electron microscopy started and that biologists, chemists, physicists, materials scientists, and engineers got interested in the developments and applications [2, 3]. Technology of electron microscopy progressed very fast, and resolution was one of the key issues. Different European, American, and Japanese companies were involved in the race for the better resolution. In the middle of the 1950s, Menter [4] impressed the scientific community by showing lattice resolution and lattice imperfections with a resolution of about 1 nm. The first atomic resolution images in transmission electron microscopy TEM of heavy atoms such as thorium or gold appeared in the beginning of the 1970s [5–7]. This triggered a whole new research on defect studies at an atomic or nearly atomic scale, with S. Iijima being one of the pioneers in the field of solid state chemistry with his high-resolution studies of Nb_2O_5-based materials [8]. Independently, atomic resolution images of heavy atoms on a carbon support were obtained by Crewe and coworkers in Chicago using a high-resolution scanning transmission electron microscope (STEM) [9].

During the 1980s and the 1990s, the resolution steadily improved, and by the turn of the century, the instrumental resolution of most commercial instruments approached 0.1 nm. At this value, the limit seemed to be reached since spherical aberration and chromatic aberration (and often also the sample) limited further progress. However, the introduction of spherical-aberration-corrected lenses [10] opened a new world of subangstrom resolution and improved signal-to-noise ratio. A state-of-the-art description is given by Urban [11].

A remarkable fact is that in the 1960s and the 1970s, researchers as well as commercial companies pushed the accelerating voltage of microscopes, in order to benefit from the decreased wavelength of the electrons for voltages of 1 MeV or higher. This increased voltage had the extra effect that the penetration depth increased and that much thicker samples could be analyzed. Famous high-voltage centers were Toulouse, Cambridge, Osaka, Argonne, Berkeley, etc. However, nowadays, with the interest in nanostructured materials, nanotubes, and single-sheet materials such as graphene where the radiation damage at higher voltages becomes important, the tendency is more toward lower voltages: 40 keV and even lower, maintaining a subangstrom resolution.

Although TEM and STEM are now mainly used to study atomic details, the first technique that allowed atomic resolution was field emission microscopy or field ion microscopy (FIM). Cooper and Müller [12] showed already in 1958 clean atomic patterns of an iron needle. For a long time, the delicate sample preparation and the limited sample region that could be imaged hampered the applications, but recently, three-dimensional imaging using the atom probe technique and improved sample preparation by focused ion beam (FIB) have revived the technique.

Surface imaging for a long time has been the field of optical microscopy and later scanning electron microscopy (SEM), but atomic resolution has never been obtained. For the first technique, the wavelength is the limiting factor, and for SEM, the probe size and the signal-to-noise ratio were the limiting factors. However, with the introduction of scanning near-field optical microscopy (SNOM), the resolution limit has been pushed forward. By placing a detector very close to the sample, the resolution is limited by the size of the detector aperture rather than by the wavelength of the illuminating light. In this way, a lateral resolution of 20 nm has been demonstrated [13].

New developments in SEM and scanning ion microscopy (SIM) have revived the "atomic resolution dream," and recently, Zhu showed atomic resolution using secondary electrons in an electron microscope [14].

However, in the 1980s, scanning tunneling microscopy (STM) and later atomic force microscopy (AFM) was invented and fine tuned by Binnig et al. [15] in order to produce atomically sharp images of (metallic) surfaces. It is worth noting that in 1986, more than 50 years after his invention of the electron microscope, Ernst Ruska obtained the Nobel Prize together with the inventors of the STM, Binnig and Roher.

The Present and the Future

From nanoscopy to nanology

The reason why technological advancements increase almost exponentially with time, as demonstrated by the well-known curve of Moore, is because they can profit directly from current-state technology. This is also the case with the improvement in the performance of microscopes. But every process of technological improvement will ultimately bounce against the physical limits. And in nanoscopy, we are now in a stage where we see these limits at the horizon. However, this does not mean the end of the road. In the future, nanoscopy will continuously evolve to nanology, from observing and describing to understanding. And understanding structure–property relations means interaction between experimentalists and theorists, which is based on the language of numbers such as atom positions or other structural parameters. In the future, images will thus become merely intermediate dataplanes from which these numbers can be extracted quantitatively. Thus, more important than the performance specifications of the instrument will be the precision ("error bar") on the fitted parameters. For instance, as shown in Ref. [16], the precision to which the position of an atom can be determined is not only proportional to the "resolution" of the instrument but also inversely related to the square root of the number of imaging particles that interact with the atom. Thus, if one wants to design a better microscope or a better experimental setting, one has to take account of both factors.

Electron microscopy

With the development of aberration correctors [17], the resolution of the electron microscope is not limited anymore by the quality of the lenses but by the "width" of the atom itself, which is determined by the electrostatic potential and the thermal motion of the atom. Furthermore, when the electron collides with an atom at very close distance, it will inevitably transfer energy to the motion of the atom, which influences its position. This puts an ultimate limit to the resolution as the closest distance that nature allows to approach an atom without altering its position. This limit is of the order of 0.2 Å [18]. If that limit is reached, the images contain all the information that can be obtained with electrons, which makes HREM superior to diffraction since the experimental data are in the same real space where one wants to use this information.

In HAADF–STEM imaging [19], where the main part of the signal is caused by phonon scattering, which is very localized at the atom core, the ultimate resolution is mainly limited by the size of the probe, which is also in the subangstrom regime. An advantage of ADF STEM is that images can be interpreted more easily in terms of the atomic number of the atoms, but at the disadvantage of a weaker signal.

Imaging can also be combined with spectroscopy both in TEM (EFTEM) and in STEM (EELS), and spectral image maps can be obtained with atomic resolution [19, 20].

Most atomic resolution work is still done in two dimensions in which one can only observe the projected structure of the object. But the ultimate goal is to achieve 3D electron tomography with atomic resolution [21]. This dream is not yet achieved, mainly because of demanding flexibility and stability requirements of the object holders. Miniaturized MEMS holders can hopefully overcome these difficulties. There also remain theoretical difficulties in combining multiple scattering with tomographic schemes.

Another important technological achievement is that advanced detectors are able to detect single electrons so that the signal-to-noise ratio is ultimately determined by the unavoidable quantum statistics, which again is a function of the number of interacting particles. The main limitations on the number of imaging particles are the electron source, the recording time, and the stability of the object. The present electron sources are still far from the physical limits imposed by phase space so that one can still expect improvement [22]. The recording time is mainly limited by the mechanical stability of the holders. Also, there is a big improvement that can be expected from miniaturized piezo holders as has been used in SPM.

But a very important limitation can be the radiation damage in the specimen caused by the incident electrons. Knock-on damage in which atoms are displaced can be avoided by using lower electron energy. And both TEM and STEM instruments are moving toward lower accelerating voltages below 50 keV, and with the use of advanced chromatic aberration correctors [17], it is possible to keep the resolution at the atomic level. But ionization damage that occurs mainly in organic and biological objects cannot be avoided and only reduced by cooling (cryoprotection). In that case, the resolution is not anymore limited by the instrument but by the object and can only be improved by averaging over many identical objects such as in single-particle cryo-EM.

An alternative to electron microscopy that gets a lot of attention is the use of femtosecond X-ray pulses generated by a free electron laser (XFEL) [30]. If the pulse length is shorter than the time needed to destroy the structure, one can obtain a diffraction pattern from a single undamaged particle. Methods are being developed with success to bring the particles successively in the beam at a very high rate, to align the diffraction patterns on the fly, and to average them [21]. Although the original hope is to use the technique on single particles, the best demonstration up till now has been on small microcrystals of a virus.

And finally, it is worth mentioning that for many years, specimen preparation methods for electron microscopy kept on operating on the macroscale, which was anachronistic in comparison with the resolution performance and requirements of the EM. But during the past decade, we have seen an immensely growing impact of FIB preparation methods [24].

4D microsocpy

An important new development is ultrafast microscopy in which stroboscopic images can be made at nanosecond time intervals, which opens new ways to study the dynamics of various processes [23].

But the most exciting new development, stimulated by the Nobel Prize winner Ahmad Zewail is the so-called 4D microscopy in which femtosecond electron pulses are generated. At this level of time resolution, it becomes possible to use the same femtosecond laser pulse to radiate the object and to emit the electron pulse so that the imaging is quantum mechanically linked to the local process [25].

Nanolab

Another challenge that is far from being reached is to perform atomic resolution electron microscopy in realistic conditions (*in situ* EM) and to combine it with other physical measurements (nanolaboratory). Recently, there has been a major improvement in the development in the atomic resolution environmental transmission electron microscope (ETEM) under controlled environments to investigate gas–solid reactions *in situ* in which the dynamic nanostructure of the solid in working environments of gas and temperatures can be monitored in real time, under environmental pressures up to 1 bar and at high temperatures [26].

Aberration correctors make it possible not only to increase the working place inside the EM but also to visualize the images at atomic resolution without processing so as to observe dynamical processes in real-time atomic resolution and exactly at the position of the structural discontinuities (surfaces, defects, . . .) where they occur. The disposal of such instrument will give a boost to the field; SPM holders have already been installed in EM's since the 1980s. But nowadays, they are used to investigate surfaces and as extremely fine probes for local mechanical and electrical measurements, which transforms the specimen chamber in the electron microscope into a nanolaboratory [27, 28]. Recently, in the "atomscope" project, one even inserts an atom probe in a TEM so as to combine structural and chemical investigations on the same specimen and at the nanoscale [29].

LEEM

It is unlikely that LEEM will be able to reach the level of the atomic dimensions because the fundamental limits are determined by wavelength and numerical aperture. With decreasing wavelength, (increasing energy) the backscattering decreases rapidly, up to 2 orders of magnitude, which requires long acquisition times associated with practical difficulties (specimen and electronics stability). Increasing the numerical aperture would require correction of higher order aberrations, both of the objective lens and of the beam separator, an unlikely endeavor. The second limit is the pressure limit. *In situ* microscopy at relatively high pressure as in TEM is not possible because of the high field between objective lens and specimen. Nevertheless, surface processes at high pressures will be done in high-pressure cells attached to the specimen chamber, allowing before/after-process studies. Time-resolved LEEM and SPLEEM with pulsed beams is certainly a possibility, but probably not at time resolutions of practical interest. Magnetic imaging with SPLEEM is certainly still far from its limits. The new high-brightness, high-polarization spin-polarized electron gun should extend this field considerably.

There is a trend to combine LEEM with synchrotron radiation photoemission electron microscopy in order to complement the structural information with chemical, electronic, and magnetic information. LEEM does not have one of the driving forces of TEM biology because most of these specimens are not crystalline and suitable for UHV. LEEM cannot study powder samples, which are important, for example, in catalysis and in environmental studies. Nevertheless, further growth of LEEM and techniques associated with it (SPLEEM, PEEM) can be expected due to the improvements of resolution, brightness, and ease of operation, which is expected to broaden the field of their applications considerably. When comparing LEEM with TEM, it should not be forgotten that LEEM is not used to study externally prepared small specimens but is basically a technique in which the emphasis is not on maximum lateral resolution but analytical versatility. This will also determine its future.

Scanning probe microscopy (SPM)

The future of SPM strongly depends on the development and how far manufacturers implement user-friendly, easy-to-handle systems. Replacement of optical microscopes, which are still mostly used in medicine, biology, material science, and engineering, or even of electron microscopes will not happen in the next years. Beam techniques quite often have a lot of advantages, such as scan velocities or the ability of performing noncontact measurements in comparison to the use of mechanically driven probes. In contrast, using a mechanical tip, a lot of interaction mechanisms can be measured under near-field conditions with high predefined resolution independent of detected wavelength simultaneously with the topography. The future process strongly depends on the market and will be different for STM, SNOM, and AFM (SFM). While the application areas of STM and SNOM stagnate, the capabilities of SFM-based techniques increase exponentially. Highest resolution analyses will be carried out using AFM, which offers new insights into the atomic structure and symmetry of surface atoms and tips. Of course, there is a limitation in scan velocities because of small interaction volumes, and therefore a lot of efforts will be made using multiprobe systems and arrays of different tips detecting different sample properties.

The fact that the used probes can either be used as sensors or as actuators is promising for future applications. Functionalized actuators, which can deliberately modify samples at a nanometer scale such as nanoscalpels or pipettes modifying DNA structures, will revolutionize medicine and biology. Functionalized detector arrays will give access to a vast variety of material properties simultaneously.

Furthermore, the progress in combining classic microscopes and SPM is already detectable. There are many application fields applicable in microbiology/medicine, material science/physics, and chemical/electrical engineering integrating, for example, FIB, SEM, and SPM in one system. Furthermore, quite often, no or a minimum sample preparation is necessary using ESEM/SPM hybrid systems because both SPM as well as ESEM allow immediate characterization of low-conducting materials as well as living biomaterials.

A survey of future perspectives on SPM is also given in Ref. [31].

Acknowledgments

We acknowledge input from many authors of the chapters of the handbook: Pennycook, Danoix, Gai, Bauer, Balk, and John Spence.

References

1. Egerton, F.N. (1968) Leeuwenhoek as a founder of animal demography. *J. Hist. Biol.*, **1**, 1–22.
2. Mulvey, T. (1974) *Science and Technology under the Microscope*, Inaugural Lecture, published by the University of Aston, Birmingham.
3. Mulvey, T. (1973) Forty years of electron microscopy. *Phys. Bull.*, **24**, 147–154.
4. Menter, J.W. (1956) The direct study by electron microscopy of crystal lattices and their imperfections. *Proc. R. Soc. A*, **236**, 119–135.
5. Formanek, H., Muller, M., Hahn, M.H., and Koller, T. (1971) Visualisation of single heavy atoms with electron microscope; *Naturwissenschaftliche*, **58**, 339–344.
6. Ottensmeyer, F.P., Schmidt, E.E., Jack, T., and Powel, J. (1972) Molecular architecture – optical treatment of dark field electron micrographs of atoms. *J. Ultrastruct. Res.*, **46**, 546–555.
7. Hashimoto, H., Kumao, A., Hino, K., Yotsumoto, H., and Ono, A. (1971) Images of Thorium atoms in transmission electron microscopy. *J. Appl. Phys.*, **10**, 1115–1116.
8. (a) Iijima, S. (1973) Direct observation of lattice defects in H-NB_2O_5 by high resolution electron microscopy. *Acta Cryst.*, **A29**, 18–14; (b) Iijima, S. and Allpress, J.G. (1973) High resolution electron microscopy of $TiO_2 \cdot 7Nb_2O_5$. *J. Solid State Chem.*, **7**, 94–105.

9. Crewe, A.V., Wall, J., and Langmore, J. (1970) Visibility of single atoms. *Science*, **168**, 1338–1340.
10. Haider, M., Rose, H., Uhlemann, S., Kabius, B., and Urban, K. (1998) Electron microscopy image enhanced. *Nature*, **392**, 768–771.
11. Urban, K.W. (2008) Studying atomic structures by aberration-corrected transmission electron microscopy. *Science*, **321**, 506–510.
12. Cooper, E.C. and Müller, E.W. (1958) Field desorption by alternating fields – an improved technique for field emission microscopy. *Rev. Sci. Instrum.*, **29**, 309–312.
13. Oshikane, Y., Kataoka, T., Okuda, M., Hara, S., Inoue, H., and Nakano, M. (2007) Observation of nanostructure by scanning near-field optical microscope with small sphere probe. *Sci. Technol. Adv. Mater.*, **8**, 181–185.
14. Zhu, Y., Inada, H., Nakamura, K., and Wall, J. (2009) Imaging single atoms using secondary electrons with an aberration-corrected electron microscope. *Nat. Mater.*, **8**, 808–812.
15. Binnig, G., Rohrer, H., Gerber, C., and Weibel, E. (1982) Surface studies by scanning tunneling microscopy. *Phys. Rev. Lett.*, **49**, 57–61.
16. Van Aert, S. (2012) Statistical parameter estimation theory – a tool for quantitative electron microscopy, *Handbook of Nanoscopy*, Wiley-VCH Verlag GmbH, Weinheim.
17. Rose, H. (1990) Outline of a spherically corrected semiplanatic medium-voltage transmission electron microscope. *Optik*, **85**, 19–24.
18. Van Dyck, D. (2012) Atomic resolution electron microscopy, *Handbook of Nanoscopy*, Wiley-VCH Verlag GmbH, Weinheim.
19. Pennycook, S. (2012) Z contrast imaging, *Handbook of Nanoscopy*, Wiley-VCH Verlag GmbH, Weinheim.
20. Botton, G. (2012) EELS and energy filtered microscopy, *Handbook of Nanoscopy*, Wiley-VCH Verlag GmbH, Weinheim.
21. Midgley, P. and Bals, S. (2012) Electron tomography, *Handbook of Nanoscopy*, Wiley-VCH Verlag GmbH, Weinheim.
22. Spence, J.C.H., Qian, W., and Silverman, M.P. (1994) Electron source brightness and degeneracy from Fresnel fringes in field emission point protection microscopy. *J. Vac. Sci. Technol. A*, **12**, 543–547.
23. Browning, N.D. et al. (2012) Dynamic transmission electron microscopy, *Handbook of Nanoscopy*, Wiley-VCH Verlag GmbH, Weinheim.
24. Yao, N. (2012) Fundamentals of the focused ion beam system, *Handbook of Nanoscopy*, Wiley-VCH Verlag GmbH, Weinheim.
25. Zewail, A.H. and Thomas, J.M. (2010) *4D Electron Microscopy Imaging in Space and Time*, Imperial College Press, London.
26. Gai, P.L. and Boyes, E.D. (2012) Atomic resolution-environmental transmission electron microscopy and applications, *Handbook of Nanoscopy*, Wiley-VCH Verlag GmbH, Weinheim.
27. Ding, Y. and Wang, Z.L. (2012) Nanowires and nanotubes, *Handbook of Nanoscopy*, Wiley-VCH Verlag GmbH, Weinheim.
28. Tichelaar, F.D., van Huis, M.A., and Zandbergen, H.W. (2012) TEM as nanolab, *Handbook of Nanoscopy*, Wiley-VCH Verlag GmbH, Weinheim.
29. Danoxi, F., and Vurpillot, F., (2012) Atom probe tomography: principle and applications, *Handbook of Nanoscopy*, Wiley-VCH Verlag GmbH, Weinheim.
30. Chapman, H.N. et al. (2011) Femtosecond X-ray protein nanocrystallography. *Nature*, **470**, 73–81.
31. Rodriguez, B.J., Proksch, R., Maksymovych, P., and Kalinin, S.V. (2012) Scanning probe microscopy – forces and currents in the nanoscale world, *Handbook of Nanoscopy*, Wiley-VCH Verlag GmbH, Weinheim.

Part I
Methods

1
Transmission Electron Microscopy
Marc De Graef

1.1
Introduction

The first transmission electron microscope (TEM) was constructed in Berlin in the early 1930s by Max Knoll and Ernst Ruska [1, 2] and became available commercially only a few years later [3]. Ruska's 1933 microscope had only three lenses, with a maximum magnification of $12,000\times$ and a resolution of about 50 nm, significantly better than the best optical microscope of that time. While the basic principles of the TEM have not changed much since, modern instruments have benefitted from rapid improvements in electronics, leading to more stable lens currents and high voltage supplies, which, in turn, have made possible the practical implementation of theoretical electron optics ideas, such as energy filters and aberration correctors. An extensive description of the history of TEM can be found in [3] and [4].

In this Chapter,[1] we will first review the main components of a basic transmission electron microscope (Section 1.2), including magnetic lenses and lens aberrations. Then we introduce the most important imaging and diffraction modes in Section 1.3), including bright field/dark field imaging and convergent beam electron diffraction (CBED). The standard dynamical diffraction theory is developed in Section 1.4, including the Fourier space approach, the real-space approach, and the Bloch wave formalism. These are then applied to two-beam (1.4.2.1) and multibeam (1.4.2.2) diffraction conditions in perfect crystals. Other diffraction phenomena, such as HOLZ lines and Kikuchi lines, are briefly discussed in Section 1.4.2.3. The chapter then concludes with Section 1.4.3, which describes the image formation theory for lattice defects.

---

1) The following figures in this Chapter are based on figures in the author's book *Introduction to Conventional Transmission Electron Microscopy* and are reproduced with permission from Cambridge University Press, who holds the copyright to the original figures: 1–19.

Handbook of Nanoscopy, First Edition. Edited by Gustaaf Van Tendeloo, Dirk Van Dyck, and Stephen J. Pennycook.
© 2012 Wiley-VCH Verlag GmbH & Co. KGaA. Published 2012 by Wiley-VCH Verlag GmbH & Co. KGaA.

1.2 The Instrument

1.2.1 General Layout

Typically, a TEM can be divided into five main regions, as illustrated in Figure 1.1: electron gun, illumination, interaction, magnification, and observation. The electron gun and illumination sections consist of all the components needed to generate an electron beam with a specified geometry and project it onto or scan it across the sample. In a modern TEM, the following components typically are present:

- The *electron gun* generates a partially coherent beam of electrons with wave length λ (see Table 1.1) for a given accelerating voltage E:

$$\lambda = \frac{h}{\sqrt{2m_0 e \hat{\Psi}}} \quad \text{with} \quad \hat{\Psi} \equiv E(1 + \frac{e}{2m_0 c^2} E) \tag{1.1}$$

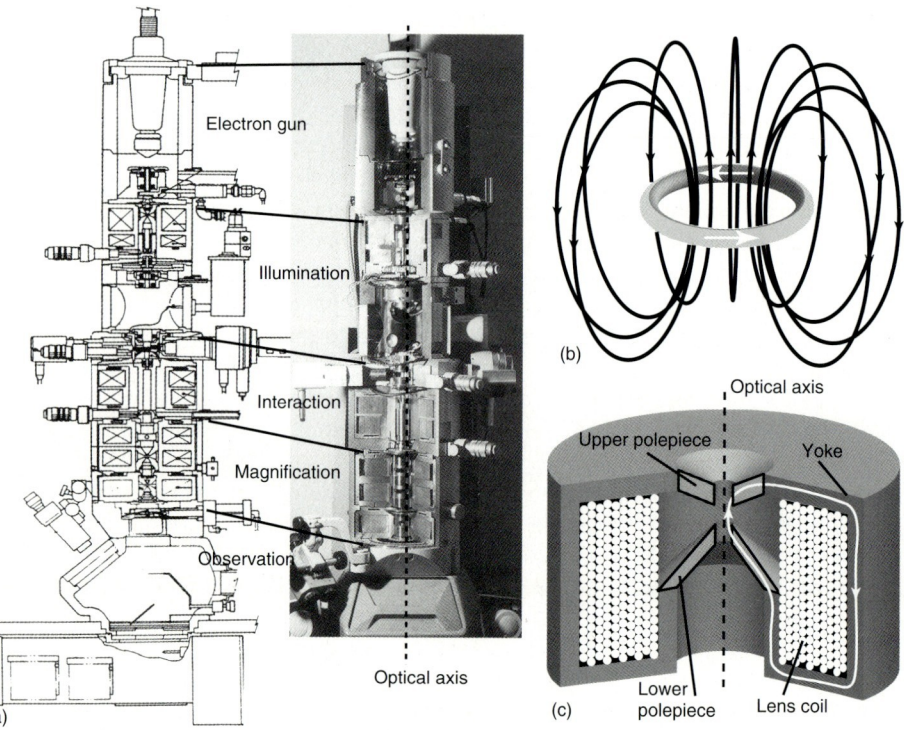

Figure 1.1 (a) Schematic of a JEOL 120CX microscope along with a photograph of a cross-section of an actual microscope, indicating the five major regions of the column. (b,c) Schematics of a typical round magnetic lens (Figure adapted from Figs. 3.2 and 3.3 in [5]).

Table 1.1 Relativistic acceleration potential $\hat{\Psi}$, electron wave length λ, wave number $K_0 = \frac{1}{\lambda}$, mass ratio $\gamma = m/m_0$, relative velocity $\beta = \frac{v}{c}$, and interaction constant σ for various acceleration voltages E. (Table adapted from Ref. [5].)

E (kV)	$\hat{\Psi}$ (V)	λ (pm)	K_0 (nm^{-1})	m/m_0	$\beta = v/c$	σ (V^{-1} nm^{-1})
100	109 784	3.701	270.165	1.196	0.548	0.009244
200	239 139	2.508	398.734	1.391	0.695	0.007288
300	388 062	1.969	507.937	1.587	0.777	0.006526
400	556 556	1.644	608.293	1.783	0.828	0.006121
1000	1978 475	0.872	1146.895	2.957	0.941	0.005385

There are three main gun types available commercially: thermionic emission (heated W or LaB$_6$ filament, no applied field), Schottky emission (heated ZrO-coated W filament, applied electric field), and cold field emission (room temperature filament, strong applied field); details can be found in [6, 7].

- The energy spread ΔE of an electron gun can be reduced by means of a *monochromator*; depending on the gun type, the energy spread typically ranges between 0.5 and 1 eV. A monochromator can reduce the energy spread to around 0.1 eV, resulting in significant improvements in both imaging and analytical work [8, 9].
- The beam leaving the gun passes through two or more *condensor lenses*, which demagnify the source and project it onto the sample. Typically, a fixed beam limiting aperture and a variable condensor aperture are present to eliminate high angle electrons and define the beam convergence angle, respectively. Beam tilt and shift coils as well as scanning and stigmator coils complete the illumination section.

The electron beam interacts with the sample inside the *objective lens*, which surrounds the sample, as shown in Figure 1.2a. The sample is thus located inside the lens magnetic field, which is generated by the surrounding coil (as illustrated in Figure 1.1b,c). The objective lens has a back focal plane (BFP), located very close to the sample plane, and an image plane. In each of these planes, an aperture can be inserted to select either one or more diffracted beams (diffraction aperture) or a region of the image (selected area aperture); both apertures are indicated in Figure 1.2b.

At the bottom of the column, a diverse array of electron detectors can be mounted. The traditional observation mode employs a fluorescent screen and photographic plates. More quantitative observations are made possible by the use of a charge coupled device camera or an imaging plate [10, 11]. In addition to these standard detector systems, the microscope column may have an imaging energy filter [12] or an electron energy loss spectrometer [13].

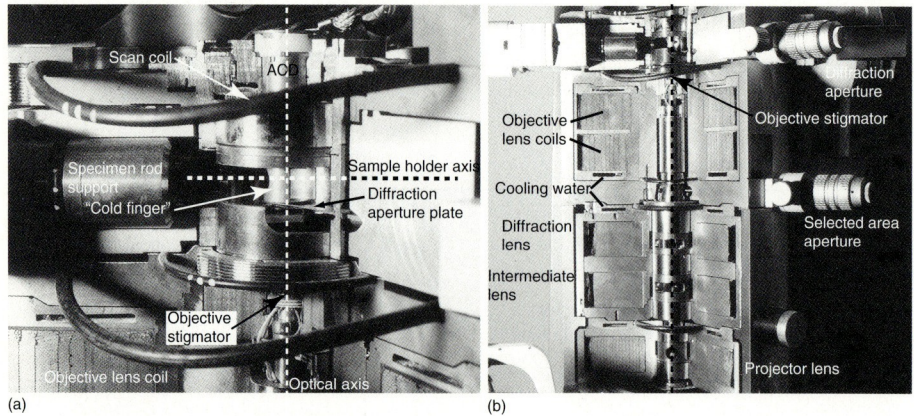

Figure 1.2 (a) Cross-section of the objective lens of a JEOL 120 CX microscope, with major components labeled, (b) Photograph of the objective lens and magnifying lenses (Adapted from Fig. 3.32 in [5]).

1.2.2
Lenses and Lens Aberrations

When an electron beam interacts with a thin foil, both elastic and inelastic scattering processes take place, which, in general, cause the scattered electrons to exit the sample in directions that are different from the incident beam direction. It is the purpose of the objective lens to bend those diverging electron trajectories back toward the optical axis and bring them to a cross-over in the BFP, where the diffraction pattern is formed; the image is then formed in a plane well below the BFP, and the remaining lenses magnify and project the image onto the detector system. For an ideal lens, all electrons that leave a certain point inside the sample will be focused onto a single point in the image plane. For light optical applications, it is possible to fabricate imaging elements with minimal or even vanishing aberrations by judiciously combining convex and concave lens elements. In electron optics, on the other hand, Scherzer [14] showed in 1936 that it is not possible to eliminate lens aberrations by combining round magnetic lenses. As a consequence, TEM instrument design was, for many decades, hindered by the presence of significant spherical and chromatic aberrations. The recent availability of fast computers and highly stable electronic circuitry has made it possible to build and employ aberration correctors; by necessity, these correctors make use of nonround optical elements, such as quadrupoles and sextupoles [15–17].

From an electron-optical point of view, an ideal round magnetic lens can be shown to be an "analog Fourier transformer," that is, the lens produces the Fourier transform of the object wave front in its BFP; in the image plane, the inverse Fourier transform of the BFP wave can be found. In mathematical notation, with ψ_o the object wave, and $\mathcal{F}$ the Fourier transform operator, we have (for Fraunhofer diffraction conditions)

$$\psi_{bfp}(\mathbf{q}) = \mathcal{F}_\mathbf{q}[\psi_o(\mathbf{r})] \tag{1.2}$$

In the image plane, we find (ignoring image magnification, rotation and inversion):

$$\psi_i(\mathbf{r}) = \mathcal{F}_\mathbf{r}^{-1}[\psi_{bfp}(\mathbf{q})] \tag{1.3}$$

For an ideal lens, we must have $\psi_i = \psi_o$; in practice, a magnetic lens has aberrations, which can be described by means of a point spread function $T(\mathbf{r})$ or its Fourier transform, the transfer function $T(\mathbf{q})$. The above relations are then written as (with $\otimes$ the convolution operator):

$$\psi_{bfp}(\mathbf{q}) = \mathcal{F}_\mathbf{q}[\psi_o(\mathbf{r}) \otimes T(\mathbf{r})] = \mathcal{F}_\mathbf{q}[\psi_o(\mathbf{r})]T(\mathbf{q}) \tag{1.4}$$

and

$$\psi_i(\mathbf{r}) = \mathcal{F}_\mathbf{r}^{-1}[\psi_{bfp}(\mathbf{q})] = \psi_o(\mathbf{r}) \otimes T(\mathbf{r}) \tag{1.5}$$

The modulus-squared of the BFP wave, $|\psi_{bfp}(\mathbf{q})|^2$, is known as the *diffraction pattern*. The image intensity is given by the modulus-squared of the image wave function:

$$I(\mathbf{r}) = |\psi_i(\mathbf{r})|^2 = |\psi_o(\mathbf{r}) \otimes T(\mathbf{r})|^2 \tag{1.6}$$

Since the Dirac delta-function is the identity function for the convolution operator, it follows that, for an ideal lens, the point spread function is equal to the delta-function, $T(\mathbf{r}) = \delta(\mathbf{r})$. It is the task of the microscope designer to create an electron-optical system for which the point spread function is as close to a delta-function as possible.

Since the objective lens provides the highest lateral magnification (and hence the highest angular demagnification) of all the lenses in the column, it is sufficient to consider only the objective lens aberrations. The primary lens aberrations are known as the Seidel aberrations or third-order aberrations; the designation "third-order" stems from the fact that a perfect spherical wave is of second order in the coordinates and that the lowest order deviation from a spherical wave front must hence be of third order. Denoting the angle between an electron trajectory and the optical axis by α, the five Seidel aberrations are *spherical aberration* (proportional to α^3), *coma* (proportional to α^2), *astigmatism* and *field curvature* (linear in α), and *distortion* (independent of α). Figure 1.3 schematically illustrates the two most relevant aberrations for conventional microscopy observations: spherical aberration, in which electrons with larger α have a shorter focal length, and astigmatism, in which the focal length varies depending on the plane in which the electrons travel. Distortions (barrel or pincushion) are only important at low magnifications. Coma and astigmatism can be fully corrected by the microscope operator using beam tilt coils and stigmators, respectively. For an extensive discussion of lens aberrations, we refer the reader to Chapters 21 through 31 in [18].

In addition to the Seidel aberrations, one must also consider *chromatic aberration*, which depends on the stabilities of the electron gun and the microscope lenses.

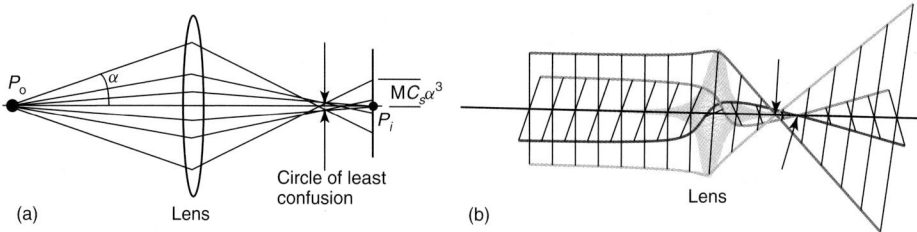

Figure 1.3 (a) Schematic of the electron trajectories in the presence of spherical aberration; (b) electron trajectories in the presence of astigmatism (Adapted from Figs. 3.13 and 3.16 in [5]).

Chromatic aberration gives rise to an uncertainty in the knowledge of the precise location of the focal plane; this uncertainty is known as the *defocus spread* and depends on the intrinsic energy spread of the electrons leaving the filament and on the variances of the accelerating voltage and the objective lens current. Furthermore, the defocus spread may be increased by electron energy losses inside the specimen. The effect of the chromatic aberration may be reduced by the use of a postgun monochromator [9] and/or an imaging energy filter [12].

In recent years, it has become possible to correct both spherical and chromatic aberrations, in addition to a series of higher order aberrations (such as threefold and fourfold astigmatisms). Modern microscopes can be equipped with image and/or probe correctors that allow the user to dial in a desired value of the spherical and/or chromatic aberration parameters. More information about aberration-corrected microscopy can be found in [17, 19, 20].

1.3
Imaging and Diffraction Modes

Diffraction is, to a large extent, governed by geometry; the relation between the positions of the source, sample, and detector(s) fixes the outcome of a diffraction experiment, so it is important to understand clearly how various imaging and diffraction geometries in TEM are related to each other. In a TEM, the source and detector are usually placed on opposite sides of the sample; this is known as the *Laue geometry*; if the source and detector are on the same side, the geometry is known as the *Bragg geometry*. In the following sections, we restrict the discussion to the Laue geometry only.

The Bragg equation

$$2d_{hkl} \sin \theta = \lambda \qquad (1.7)$$

describes how radiation with wave length λ undergoes constructive interference when it is reflected by a set of lattice planes with interplanar spacing d_{hkl}; the diffraction angle between the incident and outgoing waves equals twice the Bragg angle θ. In TEM, owing to the small electron wave length λ, the typical Bragg angle is only a few milliradians (mrad); for instance, at 200 kV ($\lambda = 2.508$ pm), the diffraction angle for the (200) planes in aluminum (with lattice parameter $a = 0.4049$ nm) is only $2\theta = 12.38$ mrad (0.71°). Therefore, electron scattering in a TEM is essentially a forward scattering process, and the lattice planes that give rise to diffracted beams are planes whose normals are nearly perpendicular to the incident beam direction. Furthermore, the small wave length translates into an Ewald sphere with a radius that is significantly larger than that for a typical x-ray diffraction (XRD) experiment; for Cu-Kα radiation with $\lambda = 0.15428$ nm, the Ewald sphere radius is 6.482 nm^{-1} versus 398.7 nm^{-1} for 200 kV electrons. Since the crystal lattice and its reciprocal lattice do not change in going from XRD to TEM, in TEM diffraction experiments, the number of reciprocal lattice points that can potentially give rise to a diffracted beam is very large; in fact, it is not unusual to have many tens or hundreds of diffracted beams simultaneously excited. This fact, combined with the existence of both an objective lens BFP and an image plane, gives rise to a large number of observation modes; some of the basic modes are discussed in the following sections.

1.3.1
Important Diffraction Geometries

When the imaging lenses of a TEM have the objective lens BFP as their object plane, a magnified version of the intensity distribution in the BFP, known as a diffraction pattern, is projected onto the detector plane. There are two main diffraction pattern types: those acquired with a parallel beam and those acquired with a converged beam, as illustrated in Figure 1.4. For parallel illumination, obtained with an underfocused condenser lens, a symmetric spot pattern is obtained when the incident beam lies along a zone axis. For a given beam direction [uvw], the reciprocal lattice points (hkl) that lie in the plane tangent to the Ewald sphere satisfy the zone equation $hu + kv + lw = 0$. All the spots in a zone axis diffraction pattern (ZADP) can be indexed as linear combinations of two short noncollinear reciprocal lattice vectors, as shown in the titanium [11.0] zone axis pattern of Figure 1.4c. Usually, a measurement of the length of a few short reciprocal vectors and the angles between them is sufficient for an unambiguous ZADP indexing. The reciprocal lattice points for a thin foil are not mathematical points; one can show that the shape of a reciprocal lattice point equals the Fourier transform of the 3D shape of the illuminated region. In practice, this means that for a round illuminated region in a thin foil, the reciprocal lattice point is a cylindrical rod with rounded end caps, oriented along the foil normal. The larger the illuminated region, the narrower the "relrod" (*reciprocal lattice rod*); this can be achieved by means of second condenser lens defocusing (underfocus).

When a focused electron beam is used (focused on the foil surface), each diffraction spot becomes a circular disk, as shown in Figure 1.4b. The beam convergence angle θ_c can be changed in discrete steps using the condenser

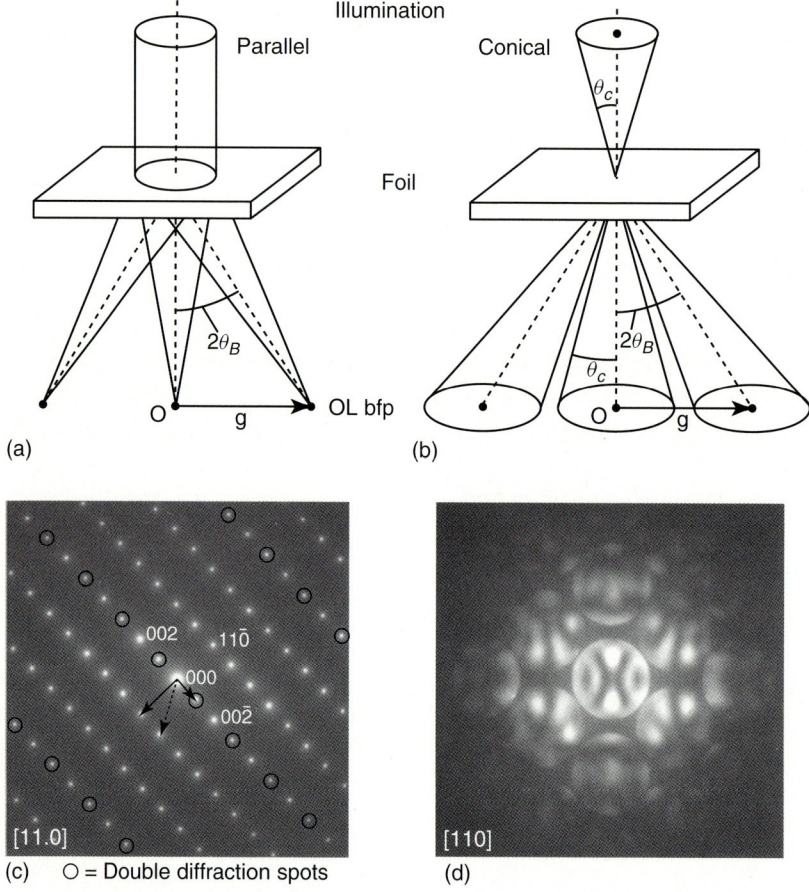

Figure 1.4 (a) Schematic illustration of parallel illumination, leading to a diffraction spot pattern in the objective lens back focal plane; (b) schematic of converged illumination, leading to diffraction disks in the bfp. (c,d) Experimental diffraction patterns: (c) is the [11.0] zone axis pattern of Ti, recorded at 200 kV, and (d) is a [110] convergent beam pattern of GaAs, recorded at 120 kV. (Figure adapted from Figs. 4.18 and 4.25 in [5]).

aperture, or continuously in microscopes with more than two condenser lenses. An example CBED pattern for [110] GaAs is shown in Figure 1.4d.

By tilting the sample, one can obtain a diffraction condition in which a single row of reflections becomes tangential to the Ewald sphere. This case is known as a systematic row (SR) orientation, and each reciprocal lattice point can be written as an integer multiple of a basic vector, that is, $\mathbf{g} = n\mathbf{G}$. The SR orientation is the preferred diffraction condition for the study of defects, as will be discussed further in Section 1.4.3. In particular, when the foil is tilted so that only a single reflection of the SR is strongly excited (in addition to the transmitted beam), one obtains the so-called two-beam orientation, for which analytical solutions to the electron scattering equations can be formulated (Section 1.4.2.1).

1.3.2
Important Imaging Modes

The diffraction aperture, located in the objective lens BFP, can be used to select one or more reciprocal lattice points (diffracted beams) with which to create an image. Owing to the increasing effect of lens aberrations with increasing angle of the electron trajectories with respect to the optical axis, it is important to ensure that the beam(s) used for imaging are positioned symmetrically with respect to this axis. If a single beam is used, one must tilt the incident beam so that the diffracted beam falls along the axis; for multiple beams, the optical axis should coincide with the "center-of-mass" of the beams. In conventional TEM observations, one usually selects a single beam, either the transmitted beam or one of the diffracted beams, to create a bright field or dark field image, respectively. For high-resolution observations, one employs a larger diameter aperture to allow several beams to contribute to the image.

Figure 1.5a shows a bright field image of a bent Ti thin foil; the dark bands intersecting near the center of the image are known as bend contours. When the selected area aperture is placed on the bend center (the intersection of multiple bend contours), one obtains a ZADP similar to the one shown in Figure 1.4c. When the aperture is moved to cover only one bend contour, the diffraction pattern consists of a single row of reflections, an SR, as shown in Figure 1.5b. Near the edge of a thin foil, the images corresponding to an SR diffraction condition are usually similar to those shown in Figure 1.5c,d. The bend contour is nearly perpendicular to the foil edge and shows a number of intensity oscillations; the origin of these oscillations are explained in Section 1.4.2.1.

1.4
Dynamical Diffraction Theory

In 1928, H.A. Bethe published the first theoretical description of the scattering of electrons traversing a crystal lattice [21]. In the years since, several different formulations of electron scattering have emerged [22–27], mostly based on the concept of a *plane wave*. In this section, we will review briefly the principal descriptions of elastic electron scattering, namely the Darwin–Howie–Whelan equations, the real-space formalism, and the Bloch wave approach.

It is convenient to describe the crystalline sample by means of an *electrostatic lattice potential*, $V(\mathbf{r})$, which is the solution to Poison's equation:

$$\Delta V(\mathbf{r}) = -\frac{|e|}{\epsilon_0} \left(\rho_n(\mathbf{r}) - \rho_e(\mathbf{r}) \right) \tag{1.8}$$

where e is the electron charge,[2] ϵ_0 the vacuum permittivity, and ρ_n and ρ_e are the nuclear and electron charge densities, respectively; Δ is the Laplacian operator.

---

2) In SI units, $|e| = 1.602\,177 \times 10^{-19}$ C, $\epsilon_0 = 8.854\,187\,817 \times 10^{-12}$ Fm^{-1}.

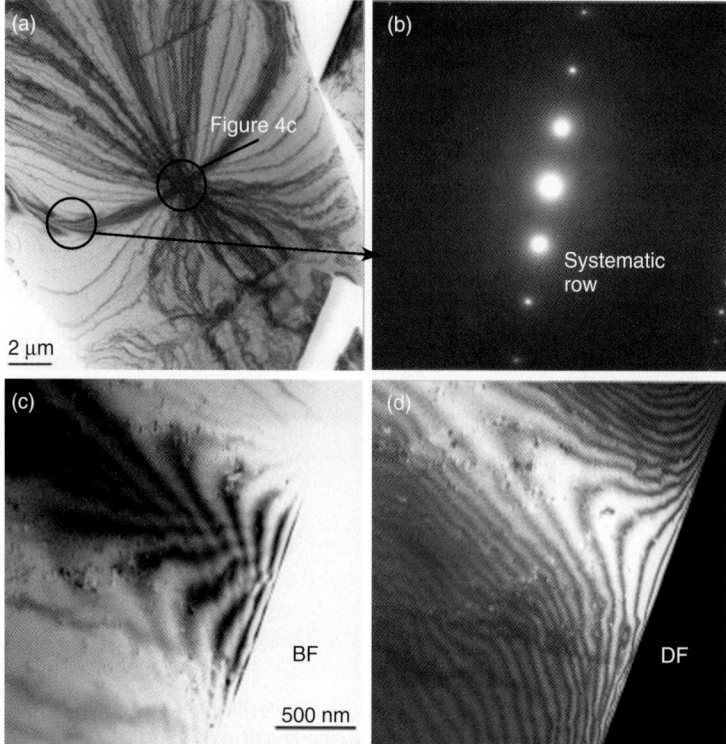

Figure 1.5 (a) Bright field image of a bend center in a bent Ti foil, with two selected area aperture positions indicated by circles; (b) systematic row diffraction pattern obtained using one of the bend contours in (a); and (c,d) bright field (BF) and dark field (DF) images of a bend contour near the edge of the foil. (Adapted from Figs. 4.18 and 4.19 in [5]).

Since $V(\mathbf{r})$ is a periodic function, we introduce its Fourier expansion:

$$V(\mathbf{r}) = \sum_{\mathbf{g}} V_{\mathbf{g}} e^{2\pi i \mathbf{g} \cdot \mathbf{r}} \tag{1.9}$$

where **g** is a reciprocal lattice vector. The Fourier coefficients $V_{\mathbf{g}}$ can be expressed in terms of the atomic scattering factors, $f^e(s)$ (with $s = |\mathbf{g}|/2$) as

$$V_{\mathbf{g}} = \frac{1}{\Omega} \sum_{j=1}^{N_a} f_j^e(s) \sum_{(D|\mathbf{t})} e^{-2\pi i \mathbf{g} \cdot (D|\mathbf{t})[\mathbf{r}_j]} \tag{1.10}$$

where Ω is the unit cell volume, N_a is the number of sites in the asymmetric unit, $(D|\mathbf{t})$ is the Seitz symbol for a space group symmetry operator, and $\mathbf{r}_j$ is an atom position vector. The atomic scattering factors, which typically include a vibrational Debye-Waller factor, can be computed from tabulated expansions [28, 29].

Once the lattice potential is known, it is straightforward to derive the geometry of the elastic scattering process. For an incident electron with wave vector $\mathbf{k}$ ($|\mathbf{k}| = 1/\lambda$, with λ the relativistic electron wave length), the incident wave function is given by

$$|\psi_\mathbf{k}\rangle = e^{2\pi i \mathbf{k} \cdot \mathbf{r}} \tag{1.11}$$

where we use Dirac's bra-ket notation [30]. The probability amplitude, P, that this electron will be scattered elastically by the potential $V(\mathbf{r})$ into the direction $\mathbf{k}'$ is

$$P = \langle \psi_{\mathbf{k}'}|V(\mathbf{r})|\psi_\mathbf{k}\rangle = \sum_\mathbf{g} V_\mathbf{g} \langle \psi_{\mathbf{k}'}|e^{2\pi i \mathbf{g}\cdot\mathbf{r}}|\psi_\mathbf{k}\rangle = \sum_\mathbf{g} V_\mathbf{g} \iiint d\mathbf{r}\, e^{2\pi i(\mathbf{k}+\mathbf{g}-\mathbf{k}')\cdot\mathbf{r}};$$

$$= \sum_\mathbf{g} V_\mathbf{g} \delta(\mathbf{k}+\mathbf{g}-\mathbf{k}') \tag{1.12}$$

This probability amplitude is only nonzero when $\mathbf{k}' = \mathbf{k} + \mathbf{g}$, which is the reciprocal space version of Bragg's equation $2 d_{hkl} \sin\theta = \lambda$.

The probability amplitude of an elastic electron scattering process is described by the wave function $\Psi(\mathbf{r})$, which is the solution to the following Schrödinger equation:

$$\Delta \Psi + 4\pi^2 k_0^2 \Psi = -4\pi^2 \left[U + iU'\right] \Psi \tag{1.13}$$

where k_0 is the wave number corrected for refraction, $U(\mathbf{r}) = V(\mathbf{r}) \times 2me/h^2$ (with $m = \gamma m_0$ the relativistic electron mass), and $U'(\mathbf{r}) = V'(\mathbf{r}) \times 2me/h^2$ is a (positive-valued) phenomenological absorption potential [31] with the same periodicity as the regular lattice potential. Equation (1.13) is the starting equation for all elastic dynamical diffraction theories. In the following section, we will rewrite this equation in a number of useful but different formulations for a perfect crystalline thin foil. In Section 1.4.3, we introduce defects with either continuous or discontinuous displacement fields.

1.4.1
Perfect Crystal Theory

Consider a perfect thin foil with parallel top and bottom faces and an incident electron beam with wave vector $\mathbf{k}_0 = |k_0|\mathbf{e}_z$ (corrected for refraction). Equation (1.13) can be solved by making an assumption (ansatz) about the mathematical form of the wave function $\Psi(\mathbf{r})$. In the following subsections, three commonly used approaches are discussed in some detail.

1.4.1.1 Fourier Space Approach
We make use of the fact that, far from the crystal, the elastically scattered electrons travel as plane waves in the directions predicted by the Bragg equation $\mathbf{k}' = \mathbf{k}_0 + \mathbf{g}$, and we assign a function $\psi_\mathbf{g}(\mathbf{r})$ to each diffracted beam, resulting in the following starting wave function:

$$\Psi(\mathbf{r}) = \sum_\mathbf{g} \psi_\mathbf{g}(\mathbf{r}) e^{2\pi i(\mathbf{k}_0+\mathbf{g})\cdot\mathbf{r}} \tag{1.14}$$

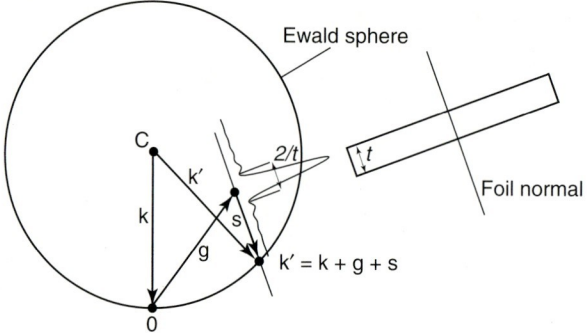

Figure 1.6 Graphical definition of the excitation error vector **s** in terms of the incident and diffracted wave vectors **k** and **k'** and the reciprocal lattice vector **g**. The excitation error is measured along the relrod, which is parallel to the foil normal. (Figure adapted from Fig. 2.11 in [5]).

The Fourier space ansatz Eq. (1.14) leads to a system of coupled first-order differential equations known as the *Darwin–Howie–Whelan equations*:

$$\frac{d\psi_g}{dz} - 2\pi i s_g \psi_g = i\pi \sum_{g'} \frac{e^{i\theta_{g-g'}}}{q_{g-g'}} \psi_{g'} \tag{1.15}$$

In these equations, s_g is the excitation error or deviation parameter, which measures the distance of the reciprocal lattice point **g** to the Ewald sphere along a direction parallel to the foil normal (see Figure 1.6); θ_g is the phase angle corresponding to the Fourier coefficient of the lattice potential $U_g = |U_g|e^{i\theta_g}$; and $q_{g-g'}$ is a measure for the probability that an electron will be scattered elastically from the beam **g'** into the beam **g**. It is formally defined as

$$\frac{1}{q_g} \equiv \frac{1}{\xi_g} + i\frac{e^{i(\theta'_g - \theta_g)}}{\xi'_g} \tag{1.16}$$

the *extinction distances* ξ_g and *anomalous absorption lengths* ξ'_g are defined as

$$\frac{1}{\xi_g} \equiv \frac{|U_g|}{|k_0 + g|\cos\alpha} \qquad \frac{1}{\xi'_g} \equiv \frac{|U'_g|}{|k_0 + g|\cos\alpha} \tag{1.17}$$

α is the angle between the beam direction and $k_0 + g$. Note that the quantity ξ'_0 is known as the *normal absorption length*; it represents the attenuation of a beam as it travels through the foil and is common to all beams.

In principle, one can solve this system of differential equations for a given set of initial conditions (typically $\psi_0 = 1$ and $\psi_g = 0$ for all other beams) using any of a variety of numerical solver approaches (see Section 1.4.2). It is more practical, and it leads to deeper insight, to rewrite Eq. 1.15 in matrix form. Employing the following substitution

$$\psi_g(z) = S_g(z)e^{-\pi\frac{z}{\xi'_0}} \tag{1.18}$$

1.4 Dynamical Diffraction Theory

renumbering the reciprocal lattice points from 0 to $N-1$, and introducing a column vector $\mathbf{S} = (S_0, \ldots, S_n, \ldots, S_{N-1})^T$, leads to

$$\frac{d\mathbf{S}(z)}{dz} = i\mathcal{A}(\mathbf{r})\mathbf{S}(z) \tag{1.19}$$

the diagonal of the *structure matrix* $\mathcal{A}$ contains the beam excitation errors, whereas the off-diagonal elements contain the scattering factors:

$$\begin{aligned}\mathcal{A}_{nn} &= 2\pi s_n \\ \mathcal{A}_{nn'} &= \frac{\pi}{q_{n-n'}} \quad n \neq n'\end{aligned} \tag{1.20}$$

This formulation has the advantage that the general solution can be written down (at least formally) as

$$\mathbf{S}(z_0) = e^{i\mathcal{A}z_0}\mathbf{S}(0) \equiv \mathcal{S}(z_0)\mathbf{S}(0) \tag{1.21}$$

where z_0 is the foil thickness and the matrix $\mathcal{S}$ is the so-called *scattering matrix*. Note that the geometry of the scattering problem (i.e., the excitation errors $s_\mathbf{g}$) is neatly separated from the scattering strengths, which is found in the off-diagonal entries of the structure matrix $\mathcal{A}$. This separation is common to all elastic scattering approaches and becomes useful in numerical solution approaches (see Section 1.4.2). In fact, we can split the structural matrix $\mathcal{A}$ into a diagonal matrix $\mathcal{T}$, with $\mathcal{T}_{mn} = 2\pi s_m \delta_{mn}$, and an off-diagonal matrix $\mathcal{V}$, with $\mathcal{V}_{mn} = (1 - \delta_{mn})\pi/q_{m-n}$, where δ_{mn} is the identity matrix. When we combine this split with the Zassenhaus theorem [32], we obtain, for small foil thickness ϵ

$$\mathbf{S}(\epsilon) \approx e^{i\mathcal{T}\epsilon}e^{i\mathcal{V}\epsilon}\mathbf{S}(0) \tag{1.22}$$

If the two exponential factors can be computed, and repeating this operation n times, such that $z_0 = n\epsilon$, an efficient slice-based method is obtained for the numerical computation of the diffracted amplitudes.

1.4.1.2 Real-Space Approach

In the real-space approach, the starting wave function is described exclusively in real space as a modulated plane wave

$$\Psi(\mathbf{r}) = \psi(\mathbf{r})\, e^{2\pi i \mathbf{K}_0 \cdot \mathbf{r}} \tag{1.23}$$

where $\mathbf{K}_0$ is the electron wave vector in vacuum. Substitution into Eq. (1.13) leads to

$$\frac{\partial \psi}{\partial z} = (\overline{\Delta} + \overline{\mathcal{V}})\psi \quad \left[\text{compare to } \frac{d\mathbf{S}}{dz} = i(\mathcal{T} + \mathcal{V})\mathbf{S} \text{ in Fourier space}\right] \tag{1.24}$$

where

$$\overline{\Delta} \equiv \frac{i}{4\pi k_{0z}}\left[\Delta_{xy} + i4\pi \mathbf{k}_{xy} \cdot \nabla_{xy}\right]; \tag{1.25a}$$

$$\overline{\mathcal{V}} \equiv \frac{i\pi}{k_{0z}}\left[U + iU'\right]. \tag{1.25b}$$

Note that the differential operator $\overline{\Delta}$ operates in a plane normal to the electron beam. The solution to the real-space equation is obtained by considering separately the effect of the two operators on the right hand side of the equation. For a thin crystal slice of thickness ϵ, the *phase grating* equation

$$\frac{\partial \psi}{\partial z} = \overline{V}\psi \tag{1.26}$$

has a solution given by

$$\psi(x,y,\epsilon) = e^{\int_0^\epsilon \overline{V}(x,y,z)\,dz} = e^{i\sigma \int_0^\epsilon V_c(x,y,z)\,dz} \equiv e^{i\sigma V_p(x,y)} \tag{1.27}$$

where the subscript c on the potential indicates that the complex potential is used, and $\sigma = 2\pi\gamma m_0 e\lambda/h^2$ is the *interaction constant*. The potential $V_p(x,y)$ is the *projected potential* of the slice. For a thin slice, the phase grating equation shows that the wave function is equal to a pure phase factor; that is, the thin slice acts as a pure phase object.

The second equation

$$\frac{\partial \psi}{\partial z} = \overline{\Delta}\psi \tag{1.28}$$

is known as the *propagator equation*; its solution is more complicated and is given formally by:

$$\psi(x,y,\epsilon) = e^{\epsilon\overline{\Delta}}\psi(x,y,0) \tag{1.29}$$

The propagator equation is a complex diffusion equation and describes how electrons propagate laterally from the slice at $z = 0$ to the slice at $z = \epsilon$. It can be shown that

$$e^{\overline{\Delta}\epsilon}\psi = \mathcal{F}^{-1}\left[e^{-\pi i\lambda\epsilon q^2}\right] \otimes \psi \tag{1.30}$$

The factor before the convolution sign $\otimes$ is known as the *Fresnel propagator*.

The solution to the complete real-space equation is then given to a good approximation by alternating the phase grating and propagator solutions for a thin slice:

$$\psi(x,y,z_0) = e^{\overline{\Delta}\epsilon}e^{i\sigma V_p^n}\ldots e^{\overline{\Delta}\epsilon}e^{i\sigma V_p^2}e^{\overline{\Delta}\epsilon}e^{i\sigma V_p^1}\psi(x,y,0) \tag{1.31}$$

where the number of pairs of exponentials increases with decreasing ϵ, such that the product $n\epsilon = z_0$ is constant. Note that this equation is formally similar to the Fourier space Eq. (1.22).

1.4.1.3 Bloch Wave Approach

In the Bloch wave approach, the wave function is assumed to be a superposition of waves with the periodicity of the lattice (Bloch waves):

$$\Psi(\mathbf{r}) = \sum_{\mathbf{g}} C_{\mathbf{g}}\, e^{2\pi i(\mathbf{k}+\mathbf{g})\cdot\mathbf{r}} \tag{1.32}$$

where the wave vector $\mathbf{k}$ and the Bloch wave coefficient $C_{\mathbf{g}}$ are to be determined. Note that this approach does not make any a priori assumptions about the directions in which the diffracted electrons will travel, which is fundamentally different from the Fourier space ansatz discussed before. Substitution of this ansatz in Eq. (1.13) leads to the following set of N equations (with N the number of contributing beams):

$$[k_0^2 - (\mathbf{k}+\mathbf{g})^2] C_\mathbf{g} + \sum_{\mathbf{h}\neq\mathbf{g}} U_{\mathbf{g}-\mathbf{h}} C_\mathbf{h} = 0 \qquad (1.33)$$

This set of equations, one for each **g**, is *exact*, that is, no approximations have been used so far. The equations relate the wave vector **k** of the Bloch wave to the energy of the incident electron (through k_0), hence they are *dispersion relations*. The equations can be written in matrix form:

$$\begin{pmatrix} k_0^2 - k^2 & U_{0-\mathbf{g}} & \cdots & U_{0-\mathbf{m}} \\ U_{\mathbf{g}-0} & k_0^2 - (\mathbf{k}+\mathbf{g})^2 & \cdots & U_{\mathbf{g}-\mathbf{m}} \\ \vdots & \vdots & \vdots & \vdots \\ U_{\mathbf{m}-0} & U_{\mathbf{m}-\mathbf{g}} & \cdots & k_0^2 - (\mathbf{k}+\mathbf{m})^2 \end{pmatrix} \begin{pmatrix} C_0 \\ C_\mathbf{g} \\ \vdots \\ C_\mathbf{m} \end{pmatrix} = 0 \qquad (1.34)$$

This equation can only have nontrivial solutions if the determinant of the matrix vanishes; this gives rise to a *characteristic polynomial equation* of order 2N in k, which has 2N roots. It is customary to label these roots with a superscript (j). For each root there is a wave vector $\mathbf{k}^{(j)}$, and a set of Bloch wave coefficients $C_\mathbf{g}^{(j)}$, so that the total wave function becomes a superposition of Bloch waves:

$$\Psi(\mathbf{r}) = \sum_j \alpha^{(j)} \sum_\mathbf{g} C_\mathbf{g}^{(j)} e^{2\pi i (\mathbf{k}^{(j)}+\mathbf{g})\cdot\mathbf{r}} \qquad (1.35)$$

where the coefficients $\alpha^{(j)}$ are the Bloch wave excitation amplitudes, which are determined from the boundary conditions at the sample entrance surface. One can show that half of the roots corresponds to waves traveling in the direction antiparallel to the incident beam (backscattered waves); ignoring these waves, the equations reduce to

$$2k_0 s_\mathbf{g} C_\mathbf{g}^{(j)} + \sum_{\mathbf{h}\neq\mathbf{g}} U_{\mathbf{g}-\mathbf{h}} C_\mathbf{h}^{(j)} = 2k_n \gamma^{(j)} C_\mathbf{g}^{(j)} \qquad (1.36)$$

where $s_\mathbf{g}$ is the excitation error, k_n is the normal wave vector component, and continuity conditions have been imposed in the form $\mathbf{k}^{(j)} = \mathbf{k}_0 + \lambda^{(j)} \mathbf{n}$, with **n** the unit foil normal. The left hand side of this equation can once again be written as a matrix, $\mathcal{M}$, so that the final equation becomes

$$\mathcal{M} \mathbf{C}^{(j)} = 2k_n \gamma^{(j)} \mathbf{C}^{(j)} \qquad (1.37)$$

This is a general eigenvalue problem, with a nonsymmetric complex matrix $\mathcal{M}$. In the presence of absorption, the potential coefficients $U_{\mathbf{g}-\mathbf{h}}$ must be replaced by $U_{\mathbf{g}-\mathbf{h}} + i U'_{\mathbf{g}-\mathbf{h}}$, and the eigenvalues are usually written as $\lambda^{(j)} = \gamma^{(j)} + i q^{(j)}$; the numbers $q^{(j)}$ are the Bloch wave absorption coefficients.

Note that all three approaches discussed so far solve the same starting equation, hence it is possible to transform each formalism into the other two. The equivalence of these formalisms, and a few additional ones, is described in detail in [27, 33]

1.4.2
Example Dynamical Computations

1.4.2.1 Analytical Two-Beam Solutions

The two-beam case, defined in Section 1.3.1, is of particular interest because it is one of the only cases for which explicit analytical solutions can be worked out. If we assume that the crystal foil is brought into a two-beam orientation, such that reflections **0** and **g** are the only ones of significant intensity, then, using the substitutions

$$\psi_0 = T(z)\, e^{\alpha z} \tag{1.38a}$$
$$\psi_g = S(z)\, e^{i\theta_g} e^{\alpha z} \tag{1.38b}$$

with

$$\alpha = \pi i \left(\frac{i}{\xi_0'} + s_g \right) = -\frac{\pi}{\xi_0'} + \pi i s_g \tag{1.39}$$

the Darwin–Howie–Whelan Eq. 1.15 can be written as

$$\frac{dT}{dz} + \pi i s_g T = \frac{i\pi}{q_{-g}} S \tag{1.40a}$$

$$\frac{dS}{dz} - \pi i s_g S = \frac{i\pi}{q_g} T \tag{1.40a}$$

where T is the transmitted and S the scattered amplitude.

These equations are formally identical to the equations for two coupled pendulums, and the general solution for the initial condition $T(0) = 1$, $S(0) = 0$ is given by

$$T(s, z) = \cos(\pi \sigma z) - \frac{i s_g}{\sigma} \sin(\pi \sigma z) \tag{1.41a}$$

$$S(s, z) = \frac{i}{q_g \sigma} \sin(\pi \sigma z) \tag{1.41b}$$

where

$$\sigma = \sqrt{s_g^2 + \frac{1}{q_g q_{-g}}} \tag{1.42}$$

It is useful to introduce two dimensionless quantities: the normalized thickness $z_\xi \equiv z/\xi_g$ and the normalized excitation error $w \equiv s_g \xi_g$. If we define a standard parameter space $(w, z_\xi$ with $-4 \leq w \leq +4$ and $0 \leq z_\xi \leq 6$, as shown in Figure 1.7b (corresponding to the foil shape shown in Figure 1.7a), then, by taking the modulus-squared of Eqns. (1.41a) and (1.41b), we obtain the bright field and dark field image intensities shown in Figure 1.7c. Note that the bright field image is asymmetric with respect to the Bragg condition $w = 0$, whereas the dark field image is symmetric. For negative excitation errors, anomalous absorption occurs, described by the absorption length ξ_g', whereas for positive excitation errors, the transmitted intensity is higher than what it would be in the presence of normal absorption (ξ_0) only.

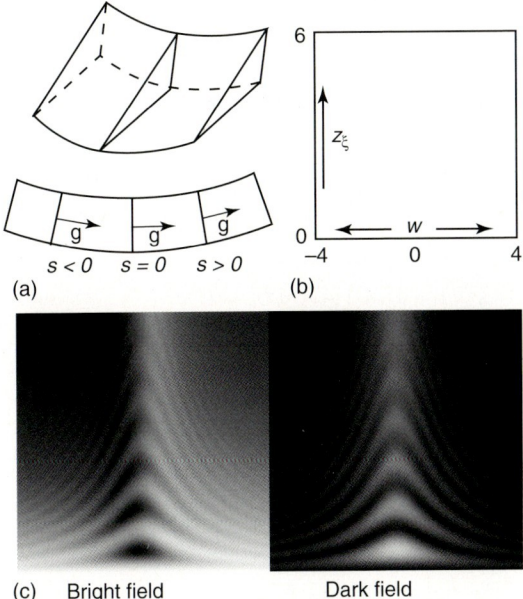

Figure 1.7 (a) Schematic of a bent foil, illustrating the change in the orientation of the plane normal **g**. (b) Definition of the standard parameter space, with normalized excitation error w on the horizontal axis and normalized thickness z_ξ along the vertical axis. (c,d) The two-beam bright field and dark field intensity profiles for the standard parameter space with normal and anomalous absorption lengths equal to $10\xi_g$. (Figure adapted from Figs. 6.2 and 6.3 in [5]).

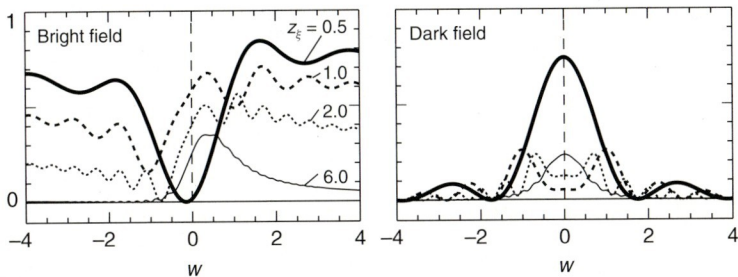

Figure 1.8 Bright field (a) and dark field (b) intensity rocking curves for four different normalized thickness values. (Figure adapted from Fig. 6.4 in [5]).

Figure 1.8 shows the bright field and dark field intensity profiles (rocking curves) for four different normalized thickness values z_ξ. Note that the number of oscillations in the bright field profile increases with thickness, but the amplitude of the oscillations decreases. The bright field asymmetry is clearly visible in the $z_\xi = 6$ rocking curve.

The two-beam case can also be analyzed using the Bloch wave formalism. The eigenvalue equation to be solved for this case is given by

$$\begin{pmatrix} iU_0' & U_{-g} + iU_{-g}' \\ U_g + iU_g' & 2k_0 s_g + iU_0' \end{pmatrix} \begin{pmatrix} C_0^{(j)} \\ C_g^{(j)} \end{pmatrix} = 2k_n \Gamma^{(j)} \begin{pmatrix} C_0^{(j)} \\ C_g^{(j)} \end{pmatrix} \quad (1.43)$$

For a centrosymmetric crystal we have $U_g = U_{-g}$ and $U_g' = U_{-g}'$, and the determinantal equation can be solved (ignoring quadratic terms in U'/U) and results in

$$2k_n \Gamma^{(j)} \approx |U_g| \left[w + i\frac{U_0'}{|U_g|} \pm \sqrt{1 + w^2 + i\frac{2}{|U_g|}(wU_0' + U_g')} \right]$$

Converting the potential coefficients to absorption lengths and extinction distances, we find for the real and imaginary parts of the eigenvalues

$$\gamma^{(j)} = \frac{1}{2\xi_g}\left[w \pm \sqrt{1 + w^2} \right] \quad (1.44a)$$

$$q^{(j)} = \frac{1}{2}\left[\frac{1}{\xi_0'} \pm \frac{1}{\xi_g'\sqrt{1 + w^2}} \right] \quad (1.44b)$$

The real and imaginary parts of the eigenvalues are shown in Figure 1.9a,c, along with the Bloch wave excitation amplitudes $\alpha^{(j)}$ in Figure 1.9b. When Bloch wave (1) is strongly excited (for negative excitation error), it is also strongly absorbed since $q^{(1)} > q^{(2)}$; for Bloch wave (2), the reverse is true, and this wave is only weakly absorbed for positive excitation errors. As a consequence, the bright field rocking curve becomes asymmetric, as noted before. The Bloch wave approach leads to the consideration of two different types of waves: Type I waves have their probability maxima at the atom locations, and are hence strongly absorbed, whereas Type II waves peak in between the atoms and are hence only weakly absorbed (Figure 1.9d). In Bragg orientation, the absorption parameters $q^{(j)}$ can be written in terms of the conventional absorption lengths as

$$q^{(1)} = \frac{1}{2}\left[\frac{1}{\xi_0'} + \frac{1}{\xi_g'} \right] \quad (1.45a)$$

$$q^{(2)} = \frac{1}{2}\left[\frac{1}{\xi_0'} - \frac{1}{\xi_g'} \right] \quad (1.45b)$$

Since $q^{(2)}$ must be positive,[3] we must have

$$\frac{1}{\xi_0'} \geq \frac{1}{\xi_g'}$$

which means that ξ_g' must be greater than ξ_0' [34]. Experiments show that typically $\xi_g' \approx 2\xi_0'$ and also that $\xi_g \approx 0.1\xi_0'$ (e.g. [35]).

3) A negative value for $q^{(2)}$ would give rise to an *increase* with crystal thickness z_0 of the number of electrons in Bloch wave (2).

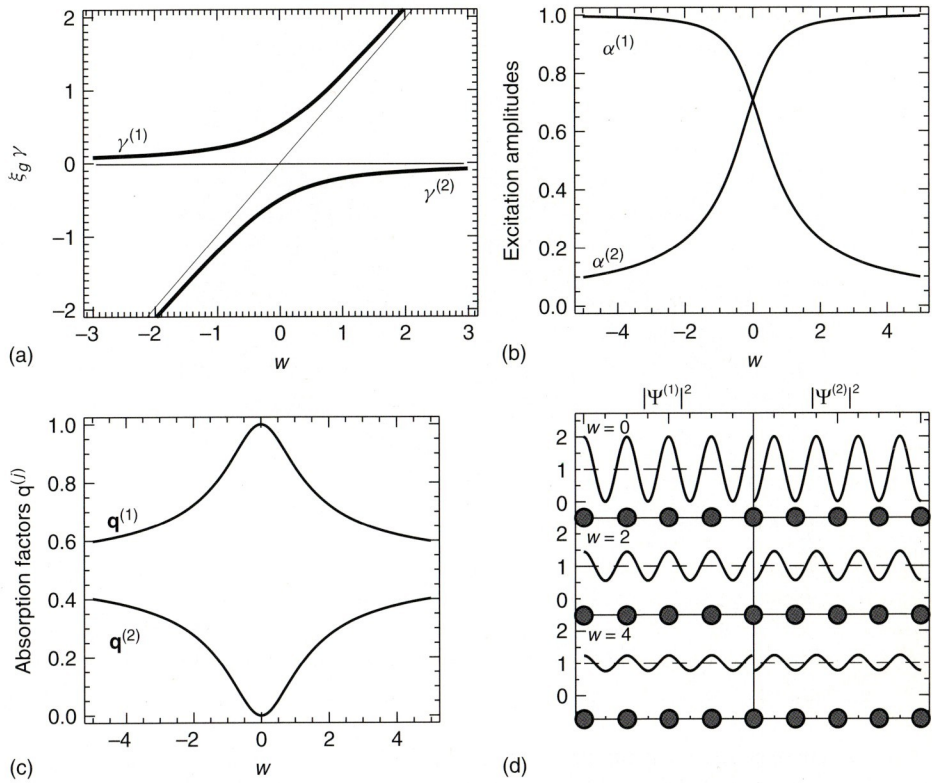

Figure 1.9 (a) Graphical representation of the two-beam eigenvalues $\gamma^{(1)}$ and $\gamma^{(2)}$ (multiplied by the extinction distance ξ_g) as a function of the dimensionless parameter $w = s_g \xi_g$. (b) Two-beam excitation amplitudes $\alpha^{(j)}$. (c) Bloch wave absorption parameters versus dimensionless parameter w. (d) Type I and II Bloch wave intensities for $w = 0, 2, 4$, relative to the atom locations (spheres). (Figure adapted from Figs. 6.6 and 6.7 in [5]).

We conclude this section with a brief analysis of CBED in the two-beam case; Figure 1.4b shows the schematic setup. At each point inside the transmitted disk, the incident wave vector has a different tangential component $\mathbf{k}_t$. Since a beam tilt is equivalent to a specimen tilt in the opposite direction, a two-beam CBED pattern contains the same information as the two-beam rocking curves. The microscope condensor settings determine the beam divergence angle θ_c, and the sample orientation determines where inside the disk the Bragg condition is satisfied; the disks can hence be regarded as "windows" through which a portion of the respective rocking curves can be observed. In the two-beam case, the diffraction condition is constant when the beam is tilted perpendicular to the reciprocal lattice vector $\mathbf{g}$, so that the CBED pattern shows straight fringes across the disk.

Figure 1.10 shows a series of two-beam CBED patterns for the $(11.\bar{2})$ reflection of Ti as a function of increasing foil thickness (a) through (d). Panel (f) shows the result

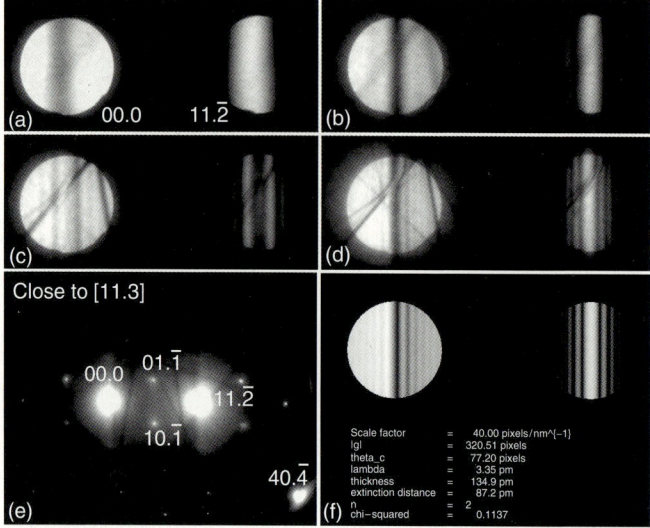

Figure 1.10 Experimental two-beam CBED patterns for the (11.$\bar{2}$) reflection of Ti (Philips EM420 operated at 120 kV). The thickness increases from (a) to (d). The diffraction pattern in (e) shows that, in addition to the two main reflections, the (40.$\bar{4}$) reflection is also strongly excited; this gives rise to the diagonal dark lines running through both bright field and dark field disks in (a) through (d). (f) Result of a two-beam simulation for the fitted parameters listed. (Adapted from Fig. 6.19 in [5]).

of a dynamical two-beam simulation with parameters fitted to the experimental pattern in (d). From the analytical expression for the dark field intensity in Eq. (1.41b), one can determine the location of intensity minima in the diffracted disk from the following relation:

$$s_i^2 = \frac{(n+i)^2}{z_0^2} - \frac{1}{\xi_g^2} \qquad (1.46)$$

where n is an integer labeling the first minimum, and i labels consecutive minima. From the measured values of s_i and the locations of the minima, one can estimate both the sample thickness z_0 and the extinction distance ξ_g. For the pattern in Figure 1.10d, linear regression analysis shows that $n = 2$, $z_0 = 134.9$ nm, and $\xi_g = 87.2$ nm, which compares favorably with the theoretical extinction distance of 86.47 nm for the {2$\bar{1}$.2} family of planes (computed from the Weickenmeier–Kohl atomic scattering factors for neutral atoms [29] at 120 kV accelerating voltage). More details on the use of CBED patterns to determine the foil thickness can be found in [36].

1.4.2.2 Numerical Multibeam Approaches

In the multibeam case, we distinguish between the SR condition, in which a single row of reflections is excited, and the zone axis case, in which a plane of reciprocal lattice points is tangent to the Ewald sphere. The SR case is illustrated in Figure 1.11a

1.4 Dynamical Diffraction Theory

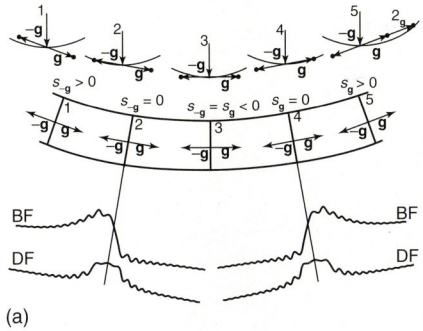
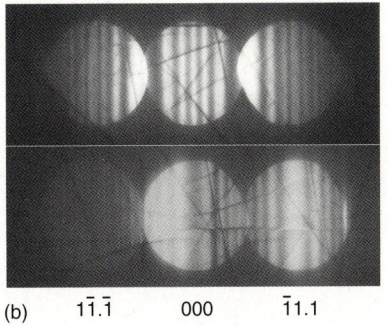

Figure 1.11 (a) Illustration of the systematic row case for a bent foil, giving rise to a bend contour; (b) systematic row CBED patterns for the $(\bar{1}1.1)$ row in Ti (200 kV, 5.5 mrad beam convergence). The top pattern was recorded in the symmetric orientation, with $s_{1\bar{1}.\bar{1}} = s_{\bar{1}1.1}$, whereas the second row is taken near Bragg orientation for $(\bar{1}1.1)$. (Figure adapted from Figs. 7.1 and 7.2 in [5]).

for a bent foil. In position 4, the Bragg condition is satisfied for the planes **g**; if this were a two-beam situation, then the BF and DF intensity distributions as a function of s_g would be similar to the ones shown at the bottom right. At location 2, the Bragg condition is satisfied for the planes $-\mathbf{g}$, and the corresponding two-beam intensity distributions are shown at the bottom left. Combining these two-beam cases into an SR leads to the presence of anomalous absorption (a dark band) for all foil orientations for which both s_{-g} and s_g are negative, that is, both reciprocal lattice points are *outside* the Ewald sphere. The Bragg condition is satisfied at both edges of the bend contour, for each of the reflections $-\mathbf{g}$ and $+\mathbf{g}$. Dark field images with either reflection will show a symmetric intensity distribution offset from the center of the bend contour. This simple sketch explains the most important feature of bend contours.

Figure 1.11b shows two CBED patterns recorded at foil locations corresponding to 3 (top row) and 4 (bottom row) of Figure 1.11a. The systematic row is the $(\bar{1}1.1)$ row of Ti, recorded at 200 kV with a beam convergence angle of 5.5 mrad. The CBED pattern is symmetric (apart from lines running across the diffraction disks) with respect to the center of the 000 disk, corresponding to the symmetric condition $s_{1\bar{1}.\bar{1}} = s_{\bar{1}1.1}$. In the bottom row, the Bragg condition is nearly satisfied for the $(\bar{1}1.1)$ reflection, and the $(1\bar{1}.\bar{1})$ reflection has a rather low intensity overall, justifying the use of the two-beam model for the intensity distributions in the other two disks.

Image simulations for the SR case are carried out by solving the dynamical diffraction equations numerically; any of the approaches introduced in Section 1.4.1 can be used. Figure 1.12 shows simulated Ti (00.2) SR bright field and dark field images for a seven-beam simulation. These images are to be compared to those shown in Figure 1.5c,d. The dark lines parallel to the central bend contour in (a) are known as Bragg lines, and correspond to foil locations where the higher order SR reflections, such as $2\mathbf{g}$ and $-3\mathbf{g}$, are in Bragg condition.

The Bloch wave simulation approach is of particular interest, since it allows for the direct computation of the scattered wave amplitudes at any depth inside the foil,

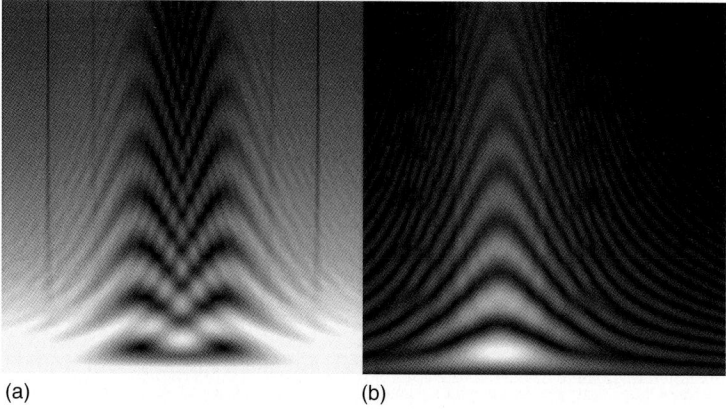

Figure 1.12 (a) Bright field and (b) dark field systematic row simulations of a bend contour (Ti (00.2) at 200 kV) intersecting the foil edge. (Adapted from Fig. 7.5 in [5]).

whereas the Fourier space approach essentially computes the far-field scattered amplitudes. Figure 1.13 shows the Bloch wave eigenvalues (a), excitation amplitudes (b), and absorption factors (c) for a nine-beam (101) SR in $BaTiO_3$. The horizontal axis denotes the tangential wave vector component in units of the reciprocal lattice vector length; for the Bragg orientation, we have $k_t/|\mathbf{g}| = -1/2$. Note the slight asymmetry (arrow) in the excitation amplitudes for Bloch wave 2, caused by the small displacement of the Ti atom away from the center of the unit cell.

For the zone axis case, the same arguments can be made as for Figure 1.11a, except that now there are reciprocal lattice points in a plane tangential to the Ewald sphere instead of just along a single line. When the excitation errors of all reflections are negative, the transmitted intensity becomes very low, due to anomalous absorption; for each pair of reflections $\mathbf{g}$ and $-\mathbf{g}$ of the zone axis pattern, there is a corresponding bend contour in the bright field image, as illustrated in Figure 1.14a,b. This figure shows, as a function of the position of the selected area aperture with respect to the bend center and bend contours, the corresponding diffraction pattern. As the aperture moves away from the bend center, the diffraction pattern becomes more similar to an SR pattern, in particular when the aperture is centered on one of the bend contours (as in (e), (g), and (i)). In (k), the aperture is moved off to the side of one of the bend contours, and one obtains a near two-beam situation.

The zone axis CBED pattern in Figure 1.15a was simulated using a 35-beam Bloch wave approach (120 kV, 5.1 mrad beam convergence angle, 120 nm foil thickness). This pattern should be compared to the experimental pattern in Figure 1.4d. While there is a clear horizontal mirror plane across the center disk, the vertical mirror plane is absent because of the noncentrosymmetric nature of the GaAs crystal structure. CBED patterns are predominantly used for the determination of crystal symmetry (both point group and space group). For additional information, the reader is referred to Refs [37–39].

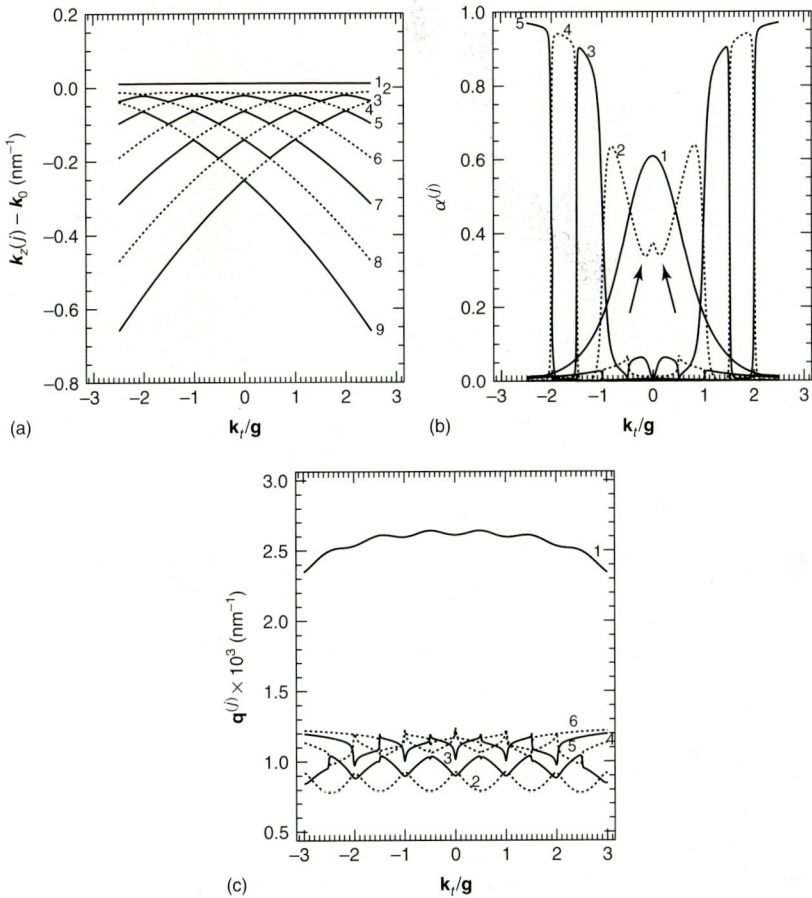

Figure 1.13 (a) Bloch wave eigenvalues, (b) excitation amplitudes, and (c) absorption coefficients for a 9-beam computation for the BaTiO$_3$ (101) systematic row (200 kV). (Adapted from Fig. 7.8 in [5]).

The multibeam dynamical simulation approach can be applied both to diffraction patterns, as in the GaAs CBED example above, and to images, such as the [211] bend center bright field image for copper in Figure 1.15b. The simulated image in (c) was obtained via a 47-beam Bloch wave calculation (200 kV) for a foil with a thickness of 40 nm at the bottom of the figure and 110 nm at the top. While the experimental bend center is distorted because of local bending of the foil, the overall characteristics of the pattern as well as most of the small intensity details are reproduced correctly in the simulated image.

We conclude this section with a few references to the literature on multibeam dynamical scattering. For the Bloch wave approach, the review articles by Metherell [24] and Humphreys [40] provide good starting points. CBED pattern formation and

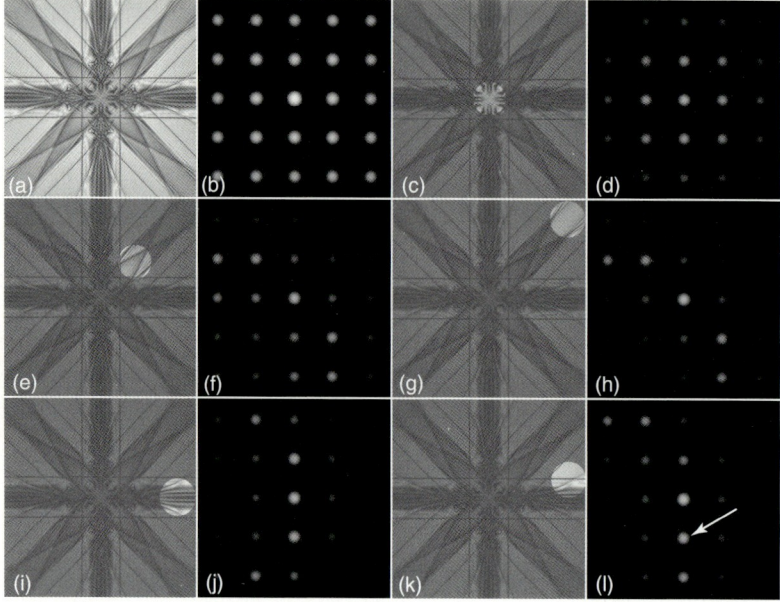

Figure 1.14 Forty nine beam bright field zone axis images (Bloch wave simulation) for a bent copper foil of thickness 100 nm (200 kV). Each image-diffraction pattern pair corresponds to a different position of the selected area aperture across the bend center. (Figure adapted from Fig. 7.17 in [5]).

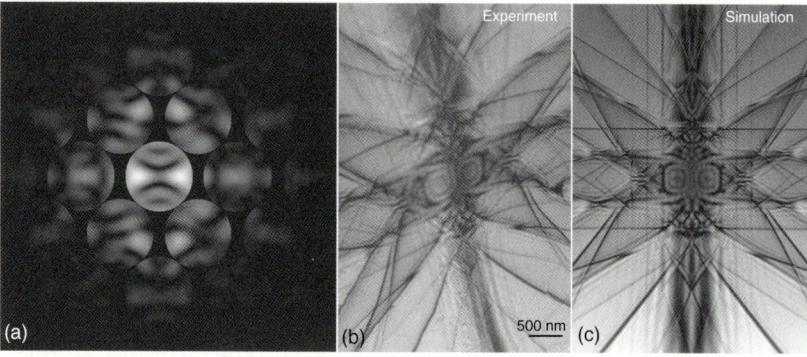

Figure 1.15 (a) Simulated zone axis CBED pattern for the [110] zone of GaAs, to be compared with the experimental pattern in Figure 1.4d. (b) Experimental and (c) simulated (47 beams) bend center bright field images for the [211] zone axis of copper (200 kV). (Adapted from Figs. 7.19 and 7.23 in [5]).

computations were reviewed by Bird [41], while Goodman and Moodie [33] reviewed a number of equivalent elastic scattering formalisms. Additional information on the simulation of multibeam patterns and images can be found in [42–46].

1.4.2.3 Other Dynamical Scattering Phenomena

In the preceding sections, we described standard dynamical scattering events involving SR or zone axis orientations, and all reflections corresponded to reciprocal planes/lines that contain the origin. It is important to note, however, that because of the large size of the Ewald sphere, it is likely that reciprocal lattice points from planes parallel to the zone axis plane will lie close to the Ewald sphere, as shown in Figure 1.16a. Such planes are known as higher order Laue zones (HOLZ). In the CBED mode, the curved Ewald sphere intersects the diffraction disks in the HOLZ planes along a narrow (bright or excess) line, and the corresponding deficit line shows up as a nearly straight dark line in the central diffraction disk (Figure 1.16b).

An example CBED pattern for the [10.4] zone axis of Ti is shown in Figure 1.16c, with a schematic explanation in (d). Only reciprocal lattice points close to the HOLZ rings are shown, and the six points near the top of the first ring correspond to the plane normals and bright lines in (c). The central disk is shown in (e), along with the indexing of the dark HOLZ lines in (f). The positions of HOLZ lines are very sensitive to small changes in the lattice parameters [47], so these patterns can be used to determine the presence and magnitude of local lattice strains. From a dynamical scattering theory point of view, it is straightforward to take HOLZ reflections into account, either directly (which can lead to a large dynamical matrix) or via the use of Bloch waves and Bethe potentials, which incorporate only the strongest beams in the dynamical matrix and treat weaker beams by means of perturbation theory.

Inelastic scattering events typically are taken into account by the imaginary part of the lattice potential. This is a phenomenological approach, in that the details of the inelastic events, as well as their time dependence, are ignored and replaced by the absorptive form factors [29, 48] and absorption lengths. To fully account for inelastic events requires the use of the time-dependent Schrödinger equation (e.g., [49]). Inelastic and elastic scattering events can both occur for a given electron; in particular, it is possible for an electron to undergo an inelastic event, thereby losing a portion of its energy, and subsequently undergo an elastic scattering event by a given set of planes. Since the inelastic event sends the electron off in an arbitrary direction, one obtains a continuous background intensity inbetween the Bragg reflections. Superimposed on this background are linear intensity features, known as Kikuchi lines [50]. These lines occur when inelastically scattered electrons are subsequently scattered elastically by a set of planes **g** onto a conical surface at an angle θ (the Bragg angle) with respect to the planes, as shown in Figure 1.17. A pair of excess (E) and deficit (D) cones arise for each reciprocal lattice point, and the intersection of these cones with the (nearly flat) Ewald sphere gives rise to lines in the diffraction plane. The lines are only visible for sufficiently thick foils, and their location can be used to accurately determine the value of the excitation error

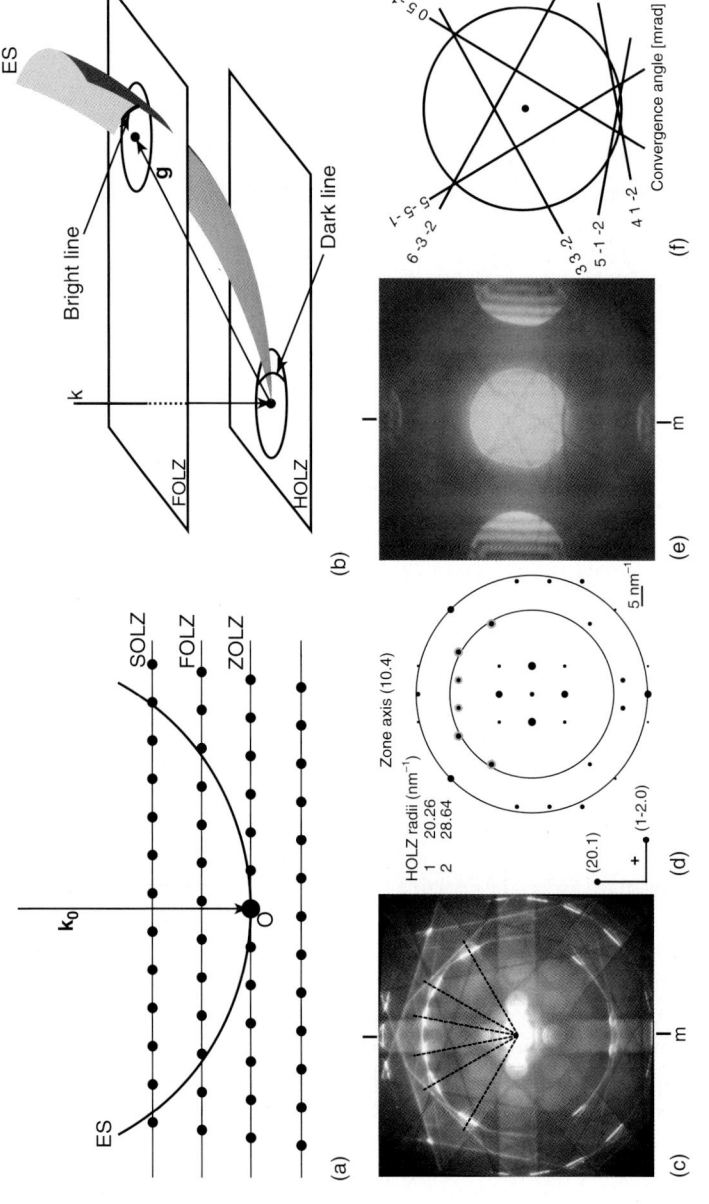

Figure 1.16 (a) Schematic illustration of the geometry of higher order Laue zones; (b) illustration of the origin of dark HOLZ lines in the central diffraction disk (exaggerated Ewald sphere curvature); (c) experimental [10.4] CBED pattern (Ti, 200 kV, beam convergence 5.1 mrad) with the central area shown enlarged in (d); (e) locations of the strongest reflections and the HOLZ rings; and (f) indices of the main HOLZ lines in the central disk. (Figure adapted from Figs. 3.37, 9.15, and 9.17 in [5]).

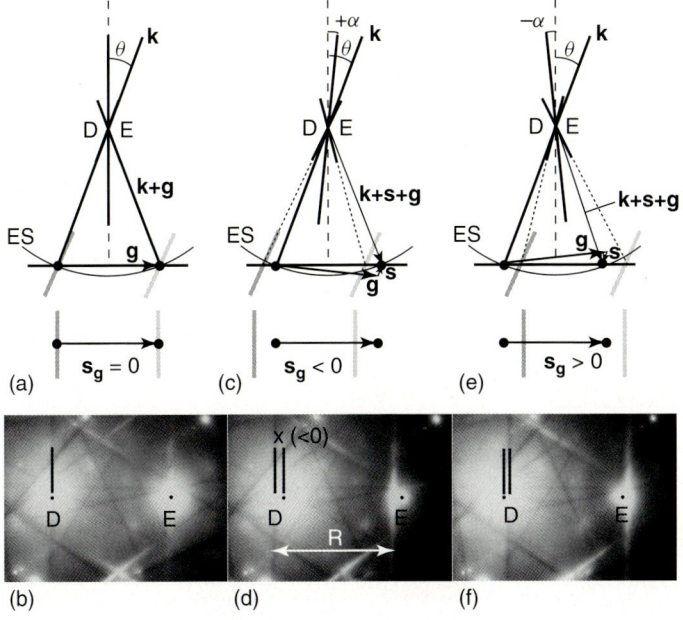

Figure 1.17 Illustration of the location of Kikuchi lines with respect to reciprocal lattice points for (a) $s_g = 0$, (b) $s_g < 0$, and (c) $s_g > 0$. (b), (d) and (f) Experimental patterns for the (10.3) reflection of Ti at 200 kV. (Figure adapted from Fig. 9.21 in [5]).

s_g. An example of the dynamical simulation of Kikuchi lines in MgO can be found in [51].

1.4.3
Defect Images

The preceding sections have dealt exclusively with defect-free, perfect crystal foils. In reality, many materials exhibit a variety of defects, and the TEM is one of a small number of tools perfectly suited for the study of defects in crystals. A defect can be defined in general terms as a (0-D, 1-D, 2-D, or 3-D) region inside the material where one or more space group symmetry elements are broken. For instance, at the free surface of a crystal foil, the translational symmetry normal to the surface is broken; at a dislocation core, the translational symmetry is also broken in the direction of the Burgers vector; at a stacking fault, one portion of the crystal is translated with respect to the other by a subunit cell translation vector; at an antiphase boundary (APB), two crystal regions with ordering on different sublattices are in contact; and so on. In the following sections, we will first analyze the standard defect image formation theory and then describe numerical procedures for defect image simulations.

1.4.3.1 Theory

In general, crystalline defects are described by a displacement field, $\mathbf{R}(\mathbf{r})$; this field may be a continuous long-range field, or it can be constant with one or more jump discontinuities. If linear elasticity theory holds, then the displacement fields of nearby defects can be added linearly. If we assume that the displacements are small compared to the unit cell dimensions, then it is reasonable to argue that the electrostatic lattice potential at a displaced location in the deformed crystal is very nearly identical to that at the original location in the underformed crystal; in mathematical form this means that $V'(\mathbf{r}) \approx V(\mathbf{r} - \mathbf{R}(\mathbf{r}))$. In Fourier space, the Fourier coefficients of the distorted lattice potential, $V'_\mathbf{g}$, are equal to the original Fourier coefficients multiplied by phase-shift factors that depend on the displacement field:

$$V'_\mathbf{g} = V_\mathbf{g} e^{-i\alpha_\mathbf{g}(\mathbf{r})} \tag{1.47}$$

where $\alpha_\mathbf{g}(\mathbf{r}) = 2\pi \mathbf{g} \cdot \mathbf{R}(\mathbf{r})$. Note that, wherever α is an integer multiple of 2π, the original undistorted Fourier coefficients are recovered.

It is straightforward to add this phase shift to the potential in the derivation of the DHW equations, so that the Eqs. (1.15) is modified in the presence of defects; using the substitution $\psi_\mathbf{g} = S_\mathbf{g} \exp[-\pi z/\xi'_0]$, we find

$$\frac{dS_\mathbf{g}}{dz} = 2\pi i s_\mathbf{g} S_\mathbf{g} + i\pi \sum_{\mathbf{g'}} \frac{e^{-i\alpha_{\mathbf{g}-\mathbf{g'}}(\mathbf{r})}}{q_{\mathbf{g}-\mathbf{g'}}} S_{\mathbf{g'}} \tag{1.48}$$

The term α now includes *all* displacement fields in the foil.

If the displacement field is continuous and differentiable, then this equation can be rewritten as

$$\frac{dS_\mathbf{g}}{dz} = 2\pi i \left(s_\mathbf{g} + \mathbf{g} \cdot \frac{d\mathbf{R}}{dz} \right) S_\mathbf{g} + i\pi \sum_{\mathbf{g'}} \frac{S_{\mathbf{g'}}}{q_{\mathbf{g}-\mathbf{g'}}} \tag{1.49}$$

Comparing this equation to Eq. (1.15), we notice that the excitation error, $s_\mathbf{g}$, of the undistorted crystal has been replaced by an effective excitation error:

$$s_\mathbf{g}^{\text{eff}} = s_\mathbf{g} + \frac{d(\mathbf{g} \cdot \mathbf{R}(\mathbf{r}))}{dz} \tag{1.50}$$

The derivative can vanish in two cases: (1) if the displacement field is constant along the beam (z) direction and (2) if the dot product $\mathbf{g} \cdot \mathbf{R}$ vanishes. In the former case, we find that a defect with a displacement field that does not vary with depth in the crystal will be invisible. In the latter case, we find that if the displacements are confined to lattice planes $\mathbf{g}$, then the defect will not affect elastic scattering from those planes. Since the excitation error $s_\mathbf{g}$ describes the distance of the reciprocal lattice point to the Ewald sphere, the local bending of lattice planes in a displacement field corresponds to a position-dependent change in this distance.

Figure 1.18a shows a schematic illustration of a bend contour with a nearby defect; while the overall excitation error $s_\mathbf{g}$ is positive in the region of the defect, locally some parts of the defect may have a displacement field such that the

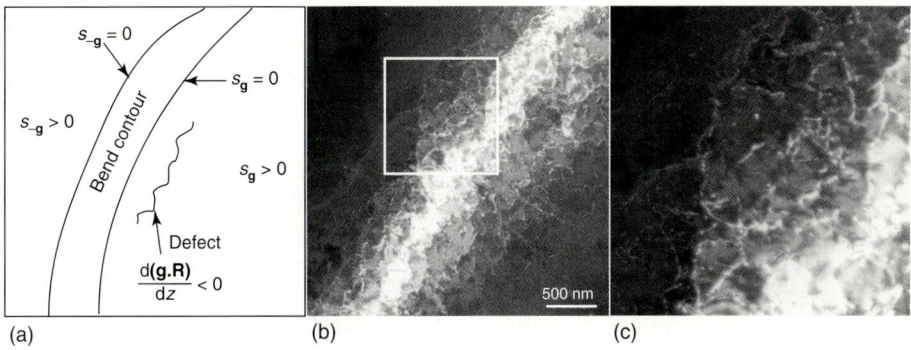

Figure 1.18 (a) Schematic of a defect near a bend contour. (b) Dark field ($\bar{2}1.0$) image for Ti (200 kV) showing clear defect diffraction contrast well outside of the bend contour. (c) Enlarged view of the area outlined in (b). (Adapted from Fig. 8.1 in [5]).

second term on the right hand side in Eq. (1.50) is negative and cancels the first term, thereby generating strong diffraction contrast in the image. An example of such contrast is shown in Figure 1.18b and the enlarged image in (c); dislocation diffraction contrast is clearly visible both inside the bend contour and away from the contour. It should be noted that the dislocation image width varies from broad near the contour to narrow far away from the contour; the larger the s_g the larger the derivative needed to bring the effective excitation error close to zero. Such large derivatives are found only close to the dislocation core, which accounts for the narrowing of the image with distance from the bend contour.

It is customary to study defects by looking at one set of planes at a time, that is, by using the two-beam approximation/orientation. For a particular two-beam orientation, a lattice defect described by a displacement field $\mathbf{R}(\mathbf{r})$ will be visible only if $\mathbf{g} \cdot \mathbf{R}$ varies along the beam direction. This is known as the *visibility criterion*. For dislocations with Burgers vector $\mathbf{b}$, in elastically isotropic materials, the displacement field can be written (in a polar coordinate system (r, θ) normal to the line direction $\mathbf{u}$) as

$$\mathbf{R}(r,\theta) = \frac{1}{2\pi}\left[\mathbf{b}\theta + \mathbf{b}_e \frac{\sin 2\theta}{4(1-\nu)} + \frac{\mathbf{b} \times \mathbf{u}}{4(1-\nu)}\left\{(2-4\nu)\ln r + \cos\theta\right\}\right] \quad (1.51)$$

where ν is the Poisson's ratio and $\mathbf{b}_e$ is the edge component of the Burgers vector. For a pure screw dislocation, the expression reduces to $\mathbf{R}(r,\theta) = \mathbf{b}\theta/2\pi$, so that $\alpha \sim \mathbf{g} \cdot \mathbf{b}$. In other words, a screw dislocation is invisible for all planes $\mathbf{g}$ that contain the screw's line direction. This leads to a simple experimental procedure to determine the Burgers vector of a screw dislocation: find two (or more) two-beam conditions for which the image contrast vanishes, and take the cross product of the corresponding plane normals to find the line direction. Knowledge of the crystal structure then allows for the determination of the length of the Burgers vector. For a pure edge dislocation, both $\mathbf{g} \cdot \mathbf{b} = 0$ and $\mathbf{g} \cdot (\mathbf{b} \times \mathbf{u}) = 0$ must be satisfied for the dislocation contrast to vanish. For materials with anisotropic elastic properties,

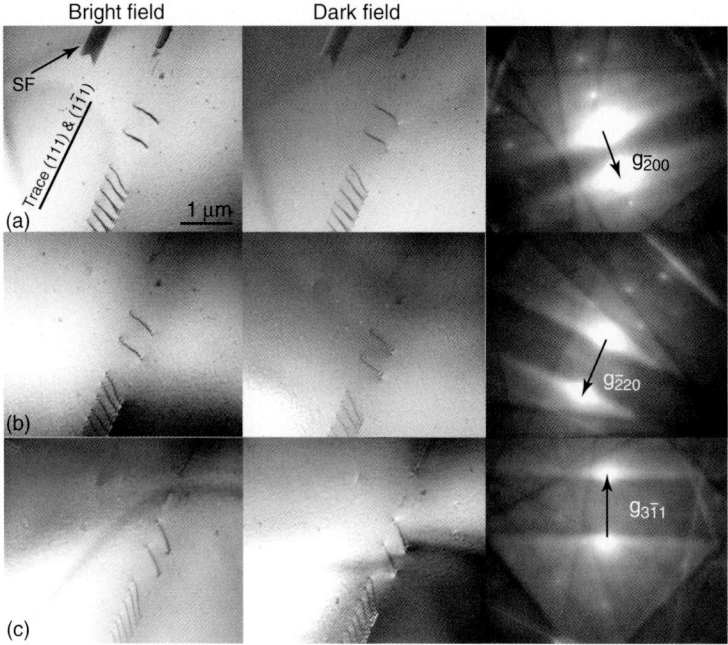

Figure 1.19 Bright field/dark field image pairs along with two-beam diffraction patterns of a series of dislocation pairs and a stacking fault (SF) in Cu-15 at% Al (200 kV). In (b) and (c), the stacking fault is invisible; in (c) the pairs of narrowly spaced partial dislocations are visible. (Figure adapted from Fig. 8.5 in [5]).

there is no simple closed-form expression for the displacement field, and all computations must be carried out numerically, as described in the next section.

Figure 1.19 shows a series of bright field/dark field image pairs of a row of partial dislocation pairs in Cu-15 at % Al, along with a stacking fault (SF) near the top. The dislocation contrast depends strongly on the diffraction vector. In (c), the SF contrast between the partial dislocations vanishes, and one can observe the partials themselves. For displacement fields with jump discontinuities, such as SFs and APBs, one can write down an analytical expression for the two-beam case. Starting from Eqs. (1.41a) and (1.41b), one writes down the scattering matrix for a perfect crystal of thickness z_0 and excitation error s:

$$\mathcal{S}(s, z_0) \equiv e^{-\frac{\pi}{\xi_0'} z_0} \begin{pmatrix} T_0 & S_0 e^{-i\theta_{\mathbf{g}}} \\ S_0 e^{i\theta_{\mathbf{g}}} & T_0^{(-)} \end{pmatrix} \quad (1.52)$$

where the $(-)$ sign on T_0 indicates that the negative, $-s$, of the excitation error should be used. It can be shown [52] that the defect can be described by a phase-shift matrix:

$$\mathcal{P}(\alpha_{\mathbf{g}})) = \begin{pmatrix} 1 & 0 \\ 0 & e^{i\alpha_{\mathbf{g}}} \end{pmatrix} \quad (1.53)$$

so that the defect image amplitudes in bright field and dark field as given by

$$\begin{pmatrix} \psi_0 \\ \psi_g \end{pmatrix} = \mathcal{P}(-\alpha_g)\mathcal{S}_2\mathcal{P}(\alpha_g)\mathcal{S}_1 \begin{pmatrix} 1 \\ 0 \end{pmatrix} \quad (1.54)$$

Repeating this computation for each column as a function of the fault depth then results in the bright field and dark field images for the planar fault. Overlapping faults can easily be dealt with by subdividing the column into more than two segments and employing the appropriate defect phase-shift matrices $\mathcal{P}(\alpha_g)$. For a description of characteristic contrast for planar faults we refer the reader to [5, 52, 53].

1.4.3.2 Defect Image Simulations

Any of the scattering approaches described in Section 1.4 can be expanded to include defects. In this final section, we employ the scattering matrix formalism along with the column approximation; each beam electron is assumed to remain inside a rectangular column parallel to the incident beam direction. Each column corresponds to an image pixel, and is independent of its neighbors, so that image simulations consist of repeatedly solving the dynamical equations for an array of columns.

For each column, the solution to the dynamical equations is written as a product of scattering matrices: $\mathbf{S}(t) = \mathcal{S}_n(\epsilon)\ldots\mathcal{S}_2(\epsilon)\mathcal{S}_1(\epsilon)\mathbf{S}(0)$, where $t = n\epsilon$ is the column height. Each scattering matrix can be precomputed, because the defect phase factor is a periodic function; for instance, for an SR orientation $n\mathbf{G}$, we have

$$e^{-i\alpha_{\mathbf{g}-\mathbf{g}'}(\mathbf{r})} \rightarrow e^{-2\pi i(n-n')\mathbf{G}\cdot\mathbf{R}_t(\mathbf{r})} \quad (1.55)$$

with $\mathbf{R}_t$ the total defect displacement vector. One can precompute scattering matrices for all possible values of $\mathbf{G}\cdot\mathbf{R}_t$ mod 1, and then, for each column slice, select the appropriate scattering matrix corresponding to the local value of $\mathbf{G}\cdot\mathbf{R}_t$. This leads to a fast computational algorithm (e.g., [54]), since matrix multiplication can be performed much faster than typical differential equation solver algorithms.

The example in Figure 1.20 illustrates that it is possible to include multiple defect types in a single simulation. Bright field and dark field images are shown for a seven-beam SR ($\mathbf{G} = \mathbf{g}_{220}$) in fcc-copper. The accelerating voltage is $V = 200$ kV, the foil thickness is constant at 125 nm, and a slight bending of the foil is included by means of an effective displacement field $\mathbf{R}(\mathbf{r}) = zs_\mathbf{G}(x,y)\mathbf{G}^*$, where $\mathbf{G}\cdot\mathbf{G}^* = 1$. The simulation includes 3 SFs on close-packed planes, 6 perfect dislocations, 50 spherical inclusions, and 50 voids of varying sizes and depths inside the foil. The simulation can easily be carried out for other diffraction conditions, such as weak-beam (shown in Figure 1.20c,d) and zone axis orientations, for parallel illumination as well as for converged probe (STEM) illumination, and takes into account the full crystallography of the material and the defects.

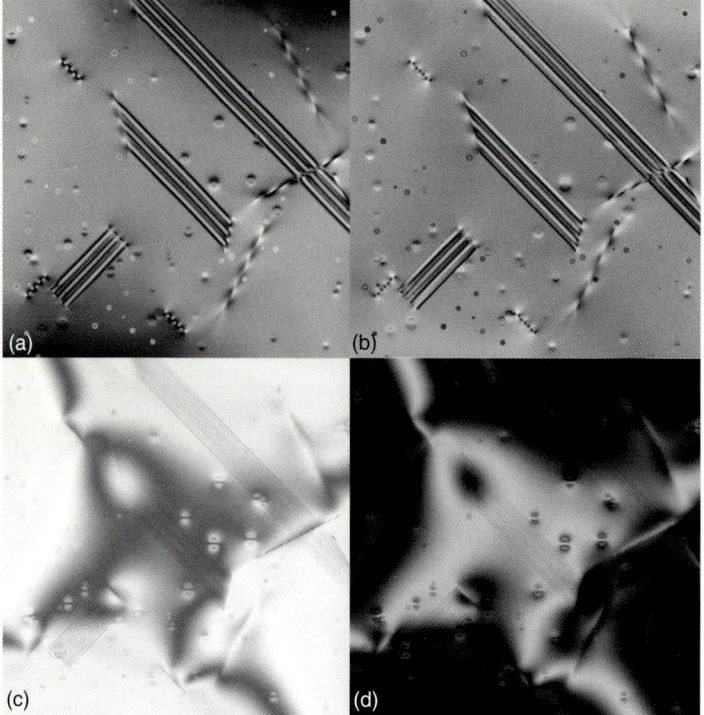

Figure 1.20 (a,b): Bright field/dark field image pair for the $\mathbf{g}_{220}$ systematic row of Cu (200 kV), using a 7-beam dynamical scattering matrix simulation for a random configuration of 6 perfect dislocations, 3 stacking faults bounded by partials, 50 voids, and 50 random spherical inclusions. (c,d) Bright field and (660) dark field images for the $s_{660} = 0$ weak beam condition.

References

1. Knoll, M. and Ruska, E. (1932) The electron microscope. *Z. Phys.*, **78**, 318–339.
2. Ruska, E. (1934) Advances in building and performance of the magnetic electron microscope. *Z. Phys.*, **87**, 580–602.
3. Hawkes, P. (1985) The beginnings of electron microscopy, in *Advances in Electronics and Electron Physics*, vol. 16, (supplement), Academic Press, Orlando.
4. Fujita, H. (1986) *History of Electron Microscopes*, Komiyama Printing Co., Ltd.
5. De Graef, M. (2003) *Introduction to Conventional Transmission Electron Microscopy*, Cambridge University Press.
6. Fransen, M., Can Rooy, T., Tiemeijer, P., Overwijk, M., Faber, J., and Kruit, P. (1999) On the electron-optical properties of the ZrO/W Schottky electron emitter, in *Advances in Imaging and Electron Physics*, vol. 111, (ed. P. Hawkes), Academic Press, New York, pp. 92–167.
7. Kasper, E. (1982) Field electron emission systems, in *Advances in Imaging and Electron Physics*, vol. 8, (ed. P. Hawkes), Academic Press, New York, pp. 207–260.
8. Browning, N., Arslan, I., Erni, R., Idrobo, J., Ziegler, A., Bradley, J., Dai, Z., Stach, E., and Bleloch, A. (2006)

Monochromators and aberration correctors: taking EELS to new levels of energy and spatial resolution. *J. Phys: Conf. Ser.*, **26**, 59–64.

9. Freitag, B., Kujawa, S., Mul, P., Rignalda, J., and Tiemeijer, P. (2005) Breaking the spherical and chromatic aberration barrier in transmission electron microscopy. *Ultramicroscopy*, **102**, 209–214.

10. de Ruijter, W. (1995) Imaging properties and applications of slow-scan charge-coupled device cameras suitable for electron microscopy. *Micron*, **26**, 247–275.

11. Zuo, J. (1996) Electron detection characteristics of slow-scan CCD camera. *Ultramicroscopy*, **66**, 21–33.

12. Krivanek, O., Gubbens, A., Dellby, N., and Meyer, C. (1992) Design and applications of a post-column imaging filter, in *50th Annual Proceedings Electron Microscopy Society of America* (eds G.W. Bailey, J. Bentley, and S.A. Small), San Francisco Press, Inc., pp. 1192–1193.

13. Egerton, R. (1996) *Electron Energy-loss Spectroscopy in the Electron Microscope*, 2nd edn, Plenum Press, New York.

14. Scherzer, O. (1936) Über einige fehler von elektronenlinsen. *Z. Phys.*, **101**, 593–603.

15. Rose, H. (1971) Properties of spherically corrected achromatic electron-lenses. *Optik*, **33**, 1–24.

16. Crewe, A. and Kopf, D. (1980) A sextupole system for the correction of spherical aberration. *Optik*, **55**, 1–10.

17. Haider, M. and Uhlemann, S. (1997) Seeing is not always believing: reduction of artifacts by an improved point resolution with a spherical aberration corrected 200 kv transmission electron microscope. *Microsc. Microanal.*, **3**, 1179–1180.

18. Hawkes, P. and Kasper, E. (1989) *Principles of Electron Optics: Applied Geometrical Optics*, vol. 1, Academic Press.

19. Tillmann, K., Houben, L., Thust, A., and Urban, K. (2006) Spherical-aberration correction in tandem with the restoration of the exit-plane wavefunction: synergetic tools for the imaging of lattice imperfections in crystalline solids

at atomic resolution. *J. Mater. Sci.*, **41**, 4420–4433.

20. Erni, R. (2010) *Aberration-corrected imaging in transmission electron microscopy: an introduction*, Imperial College Press.

21. Bethe, H. (1928) Theorie der beugung von elektronen an Kristallen. *Ann. Phys.*, **87**, 55–129.

22. Cowley, J. and Moodie, A. (1957) The scattering of electrons by atoms and crystals. I. A new theoretical approach. *Acta Crystallogr.*, **10**, 609–619.

23. Howie, A. and Whelan, M. (1961) Diffraction contrast of electron microscope images of crystal lattice defects. II the development of a dynamical theory. *Proc. R. Soc. London*, **A263**, 217–237.

24. Metherell, A. (1975) Diffraction of electrons by perfect crystals, in *Electron Microscopy in Materials Science* (eds U. Valdre and E. Ruedl), Commission of European Communities, Luxembourg, pp. 401–552.

25. Sturkey, L. (1962) The calculation of electron diffraction intensities. *Proc. Phys. Soc.*, **80**, 321–354.

26. Tournarie, M. (1961) Théorie dynamique rigoureuse de la propagation cohérente des electrons à travers une lame cristalline absorbante. *C. R. Hebd. Seances Acad. Sci.*, **252**, 2862–2864.

27. van Dyck, D. (1975) The path integral formalism as a new description for the diffraction of high-energy electrons in crystals. *Phys. Status Solidi B*, **72**, 312–336.

28. Doyle, P. and Turner, P. (1968) Relativistic hartree-fock x-ray and electron scattering factors. *Acta Crystallogr., Sect. A*, **24**, 390–397.

29. Weickenmeier, A. and Kohl, H. (1991) Computation of absorptive form factors for high-energy electron diffraction. *Acta Crystallogr., Sect. A*, **47**, 590–597.

30. Dirac, P. (1947) *The Principles of Quantum Mechanics*, 3rd edn, Clarendon Press, Oxford.

31. Gevers, R., Blank, H., and Amelinckx, S. (1966) Extension of the howie-whelan equations for electron diffraction to non-centro symmetrical crystals. *Phys. Status Solidi B*, **13**, 449–465.

32. Suzuki, M. (1977) On the convergence of exponential operators – the

zassenhaus formula, BCH formula and systematic approximants. *Commun. Math. Phys.*, **57**, 193–200.

33. Goodman, P. and Moodie, A. (1974) Numerical evaluation of N-Beam wave functions in electron scattering by the multi-slice method. *Acta Crystallogr., Sect. A*, **30**, 280–290.

34. Hashimoto, H., Howie, A., and Whelan, M. (1962) Anomalous electron absorption effects in metal foils: theory and comparison with experiment. *Proc. R. Soc. London, Ser. A*, **269**, 80–103.

35. Hashimoto, H. (1964) Energy dependence of extinction distance and transmissive power for electron waves. *J. Appl. Phys.*, **35**, 277–290.

36. Delille, D., Pantel, R., and Van Cappellen, E. (2001) Crystal thickness and extinction distance determination using energy filtered CBED pattern intensity measurement and dynamical diffraction theory fitting. *Ultramicroscopy*, **87**, 5–18.

37. Buxton, B., Eades, J., Steeds, J., and Rackham, G. (1976) The symmetry of electron diffraction zone axis patterns. *Philos. Trans. R. Soc.*, **281**, 171–194.

38. Mansfield, J. (1984) *Convergent Beam Electron Diffraction of Alloy Phases*, Adam Hilger, Bristol.

39. Tanaka, M., Sekii, H., and Nagasawa, T. (1983) Space-group determination by dynamic extinction in convergent-beam electron diffraction. *Acta Crystallogr., Sect. A*, **39**, 825–837.

40. Humphreys, C. (1979) The scattering of fast electrons by crystals. *Rep. Prog. Phys.*, **42**, 1825–1887.

41. Bird, D. (1989) Theory of zone axis electron diffraction. *J. Electron Microsc. Tech.*, **13**, 77–97.

42. Coene, W., van Dyck, D., van Tendeloo, G., and Van Landuyt, J. (1985) Computer Simulation of High-Energy Electron Scattering by Non-Periodic Objects. The Real Space Patching Method as an Alternative to the Periodic Continuation Technique. *Philos. Mag. A*, **52**, 127–143.

43. Janssens, K., Vanhellemont, J., De Graef, M., and Van der Biest, O. (1992) SIMCON : a versatile software package for the simulation of electron diffraction contrast images of arbitrary displacement fields. *Ultramicroscopy*, **45**, 323–335.

44. Kilaas, R. (1987) Interactive software for simulation of high resolution tem images. 22nd Annual Conference of the Microbeam Analysis Society, Kona Hawaii, 13–17 July, pp. 293–300.

45. Krakow, W. and O'Keefe, M. (1989) *Computer Simulation of Electron Microscope Diffraction and Images*, TMS, Warrendale, PA.

46. Stadelmann, P. (1987) EMS - A software package for electron diffraction analysis and HREM image simulation in materials science. *Ultramicroscopy*, **21**, 131–146.

47. Jones, P., Rackham, G., and Steeds, J. (1977) Higher order laue zone effects in electron diffraction and their use in lattice parameter determination. *Proc. R. Soc. London*, **A354**, 197–222.

48. Bird, D. (1990) Absorption in high energy electron diffraction from non-centrosymmetric crystals. *Acta Crystallogr., Sect. A*, **46**, 208–214.

49. Yoshioka, H. (1957) Effect of inelastic waves on electron diffraction. *J. Phys. Soc. Jpn.*, **12**, 618–628.

50. Kikuchi, S. (1928) Diffraction of cathode rays by Mica. *Jpn. J. Phys.*, **5**, 83–96.

51. Omoto, K., Tsuda, K., and Tanaka, M. (2002) Simulations of kikuchi patterns due to thermal diffuse scattering on MgO crystals. *J. Electron Microsc.*, **51**, 67–78.

52. Amelinckx, S. and van Landuyt, J. (1976) Contrast effects at planar interfaces, in *Electron Microscopy in Mineralogy* (ed. H. Wenk), Springer-Verlag, pp. 68–112.

53. Williams, D. and Carter, C. (1996) *Transmission Electron Microscopy, A Textbook for Materials Science*, Plenum Press, New York.

54. Thölén, A. (1970) A rapid method for obtaining electron microscope contrast maps of various lattice defects. *Philos. Mag.*, **22**, 175–182.

2
Atomic Resolution Electron Microscopy

Dirk Van Dyck

2.1
Introduction

2.1.1
Atoms: the Alphabet of Matter

> It would be very easy to make an analysis of any complicated chemical substance; all one would have to do would be to look at it and see where the atoms are. The only trouble is that the electron microscope is one hundred times too poor. I put this out as a challenge: Is there no way to make the electron microscope more powerful? (Richard Feynman: *There's plenty of room at the bottom: an invitation to enter a new field of physics* (1959))

> In 1947, the electron microscope had produced a hundredfold improvement on the resolving power of the best light microscopes, and yet it was disappointing, because it had stopped short of resolving atomic lattices. (Dennis Gabor: *Nobel Lecture* (1971))

> 'All things are made of atoms' is a sentence with an enormous amount of information about the world. (Richard Feynman: *Six easy pieces* (1995))

We are witness to an exciting era in which nanoscience gradually evolves from describing to understanding and designing.

Science has made it possible not only to fabricate and characterize materials and devices on the nanoscale but also to understand and predict their properties. In the future, this interplay between theory and experiment will further lead to fabrication of nanostructures with designed properties (Figure 2.1). But this interplay needs a quantitative communication language. Fortunately, nature itself provides the ideal language since matter consists of discrete atoms and all the structure–property relationships are unambigeously coded in the positions of these atoms. In terms of communication theory, the atoms are the "alphabet" of nature and the atomic positions are the "messages" between theorists and experimentalists.

What is the precision on the atom positions that one will need to understand and predict structure–property relationships? Figure 2.2 shows a graph in which the

Handbook of Nanoscopy, First Edition. Edited by Gustaaf Van Tendeloo, Dirk Van Dyck, and Stephen J. Pennycook.
© 2012 Wiley-VCH Verlag GmbH & Co. KGaA. Published 2012 by Wiley-VCH Verlag GmbH & Co. KGaA.

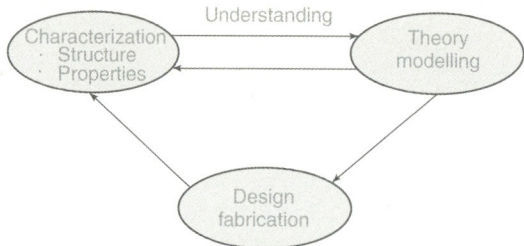

Figure 2.1 Scheme of the future interplay between experiment – theory – design.

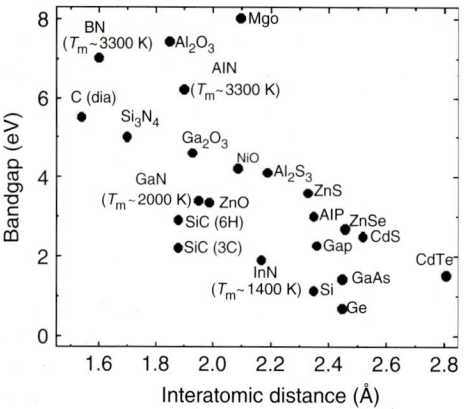

Figure 2.2 Bandgap versus interatomic distance for a number of simple semiconductors and isolators. (Courtesy: C. Kisielowski.)

bandgap of a range of semiconductors and isolators is plotted against the distance between neighboring atoms. From this roughly linear relationship, we see that even a change in interatomic distance in the order of 1 pm (0.01 Å) will alter the energy gap by about 50 meV, which is of the same order of magnitude as the melting heat of ice. Thus for designing the optical properties of semiconductors (bandgap engineering) and for the properties of nanostructures in general, one must be able to measure and control the atom positions to a precision of the order of picometers.

2.1.2
The Ideal Experiment

The only way to obtain information from an object is by interaction with particles. For this purpose, the state of the particles has to be determined before and after the interaction and from this relationship one can then deduce information about the object. Ideally, a source should emit particles with a well-calibrated direction and speed, but in practice one has to compromise this with brightness. After interaction with the object and transfer through the instrument, the particles hit

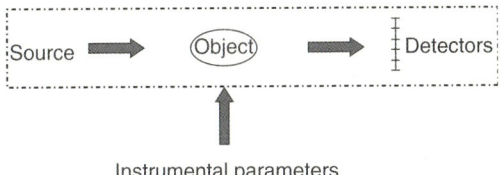

Figure 2.3 Conceptual scheme of an experiment.

a pixel in a detector (Figure 2.3). The newest generation of detectors are able to detect single X-ray photons or electrons (at energies used for atomic structure determination). Every pixel in the detector can be read out independently and will give the number of detected particles. Thus the outcome of the experiment is a set of integer numbers. This has an important consequence. If the model for the interaction with the object is accurately known quantum mechanics enables to predict the outcome (intensity) of an experiment in terms of the statistics of the imaging particles, the deviation between theory and experiment is only caused by the counting noise. Since the statistics of the counting noise is well known, we are approaching the "ideal experiment" in which all the available information is exploited and the error bar on the final results is only determined by the counting noise due to the finite number of particles.

The coordinates of the atoms in an object can be estimated quantitatively by fitting the experimental data with computer simulations. These simulations are based on a theoretical model for the experiment. The atom coordinates are introduced in the model as unknown parameters. The theoretical signal in every pixel can be expressed as a function of the parameters of the model. These parameters can then be estimated by comparing (fitting) the theoretical and the experimental values. A necessary mathematical condition is that the number of independent experimental data exceeds the number of unknown parameters. If this condition is not met, one has more variables than equations, which give an infinite number of solutions (uniqueness problem). This puts an important limit on the required resolution of the experiment.

Let us now analyze the information content of an experiment in terms of independent data per atom. We consider a perfect crystal in 3D but the conclusions can be generalized for nonperiodic objects and for 2D projections.

The outcome of an experiment can be either in Fourier space (diffraction) or in real space (imaging), but since both spaces are mutually linked by a Fourier transform, the information content in both spaces is the same.

In case of a perfect crystal, the Fourier space is discrete and thus easier to analyze. In a diffraction experiment, every reflection can be recorded as an independent measurement. In order to determine the atom positions in the crystal unambiguously, the number of reflections must be larger than the number of atom coordinates. Let us investigate the consequences. Around an atom, we can roughly define three spherical symmetrical zones a, r, and s with respective radii ρ_a, ρ_r, and ρ_s (Figure 2.4a). Since at this stage we only intend to derive some simple rules of

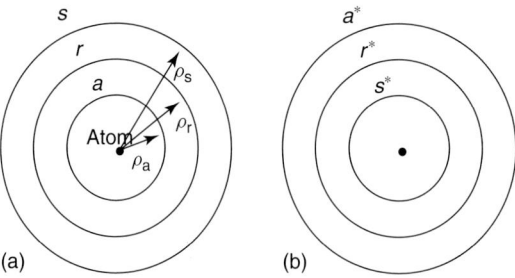

Figure 2.4 (a) Three characteristic zones centered around the atom and (b) the corresponding zones in Fourier space.

thumb, we do not consider the detailed shape of these zones (an exact analysis is possible).

ρ_a = the interaction radius of the atom. For X-ray scattering, this is the radius of the electron charge distribution of the atom; for electron scattering, it is the radius of the electrostatic potential, which includes also the nucleus of the atom. It is typically of the order of 0.5 Å.

ρ_r = the "resolution" of the experiment. In a diffraction experiment, the resolution is diffraction limited and of the order of the wavelength. In an imaging experiment, ρ_r is the radius of the (3D) point-spread function (PSF) (see further) of the microscope.

ρ_s = the radius of the average space occupied by the atom in the crystal (total volume divided by the number of atoms). It is typically of the order of 2–3 Å.

The Fourier transform of these three zones then also yield three spherical symmetrical zones in the diffraction space, respectively, a^*, r^*, and s^* with respective radii $(1/\rho_a)$, $(1/\rho_r)$, and $(1/\rho_s)$ (Figure 2.4b). Let us now assume that the atom in Figure 2.4 is the "average" atom of the unit cell so that also ρ_a, ρ_r, and ρ_s are average numbers.

In a *diffraction experiment* with sufficiently small wavelength, one has $\rho_r < \rho_a$. In that case, the diffraction pattern is limited by the envelope of the average scattering factor of the atoms. All the reflections of the diffraction pattern are then confined in the zone a^* so that the number of reflections is proportional to $(1/\rho_a)^3$. Now the number of independent reflections (data) per atom is given by the ratio $(\rho_s/\rho_a)^3$. The necessary condition for quantitative determination of the atom coordinates is then $(\rho_s/\rho_a)^3 > 3$.

Since in real crystals, $(\rho_s/\rho_a)^3$ is of the order of 10–20, the number of independent data exceeds largely the number of parameters and the atom coordinates can be determined unambiguously. This is thus a direct consequence of the atomicity principle that "atoms are small discrete points relative to the space between them" [1]. If this condition is met, we say that the atoms can be "resolved" and that their positions can be determined quantitatively by fitting of the model with the observed intensities (Figure 2.5). This is common practice in X-ray crystallography. The goodness of fit is the well-known R factor. An advantage of X-ray diffraction is that a huge number of identical unit cells contribute to the Bragg reflections

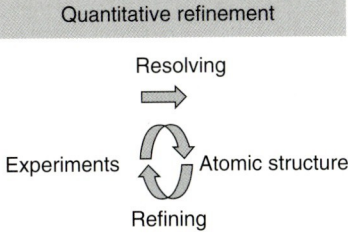

Figure 2.5 Scheme for quantitative refinement of an atomic structure.

which result in a very large signal-to-noise ratio and very precise values for the atom coordinates, typically of the order of 0.1 pm. The disadvantage of a diffraction experiment is that it can only be applied in case of a (nearly) perfect crystal and does not provide phase information, and, in case of disordered objects, it can only yield statistical information.

In an *imaging experiment* the PSF is usually larger than the interaction volume of the atom so that $\rho_r > \rho_a$. Now the Fourier components are confined in the zone r and the necessary condition for quantitative determination of the atom coordinates is then $(\rho_s/\rho_r)^3 > 3$. This puts a limit to the required resolution of the microscope. If this requirement is met, we can also state that the atoms can be "resolved" and that their positions can be determined quantitatively by fitting of the model with the observed intensities. The advantage of an imaging experiment is that it can also be used for nonperiodical objects and that it provides phase information.

With the newest generation of electron microscopes, we see that the gap between ρ_r and ρ_a is gradually closing, which means that the atom itself becomes the final resolution limit and the resolution in the image is the same as that in the diffraction pattern. In that case there is no more "superresolution" to be gained from electron diffraction or diffractive (lensless) imaging [4].

The results derived above also apply to the 2D plane. The diffraction pattern then corresponds to a particular zone plane and the crystal structure is then projected along the zone axis. Every atom in projection now has two coordinates: ρ_a is the radius of the individual atom in projection and ρ_s that of the average space of the atom. But since all the atoms of the 3D structure are now projected in the 2D plane, the average space per atom can be small so that the condition $(\rho_s/\rho_a)^2 > 2$ only holds for low-order zone axes. For aperiodic structures such as amorphous objects, the structure cannot be resolved from a single projection, and one has to combine many projections in a tomographic scheme [2].

2.1.3
Why Imaging?

If one disposes of a perfect crystal, a diffraction experiment has the advantage that a very large number of identical unit cells contribute to the diffraction pattern so that one can determine the intensities with a large signal-to-noise ratio. The problem however is that the phases of the diffracted beams (Fourier components) are not determined. Fortunately, here the atomic principle provides a solution since the atomic peaks pin the waves of the different Fourier components in real space, which enables to derive relations between the phases of the various reflections. In this manner, approximate phases can be estimated for a number of reflections ("direct methods") [3] which, by Fourier transformation, yield an approximate starting structure. In the subsequent refinement step, the atom coordinates can then be determined with a precision of the order of 1 pm.

In case of a crystalline object with finite and known shape ("support") such as a crystalline nanoparticle, one can use diffractive imaging [4] in which one uses a computational scheme that iterates between real space and Fourier space to pin the phases of the diffracted beams within the known support. However, in case the information of the shape is obtained from images, the accuracy of the phases is again limited by the resolution of the microscope.

The advantage of imaging is that the images contain the phase information of the Fourier components up to the resolution of the microscope. In case of a perfect crystal in which the diffraction pattern extends beyond the resolution of the microscope, one can use a hybrid method by which one first "resolves" an approximate structure by imaging, which is then subsequently refined using the diffraction pattern [5]. However, since most nanostructures are nonperiodical, diffraction methods cannot be applied and one has to rely on imaging alone.

2.1.4
Why Electron Microscopy?

Most nanostructures are aperiodic. Therefore one cannot exploit the redundancy of a large number of identical units as in crystals (except for single-particle cryo electron microscopy (EM)) and one disposes only of the particles that have interacted with the single nanostructure. For this reason, electrons are by far the most appropriate imaging particles because they interact with the electrostatic potential of the atom (electrons and nucleus) and this interaction is orders of magnitude stronger than that of X-rays and neutrons even if one takes the radiation damage into account [1, 6]. A disadvantage of this strong interaction is that multiple scattering becomes dominant and the interpretation of the experimental data requires more elaborate theoretical and computational efforts. However, today the progress in the theory and simulations has also reached the stage of a full quantitative agreement so that there is no need to avoid strong scattering conditions.

Since electrons are charged particles, they can be deflected in an electrostatic or magnetic field which makes it possible to construct electron lenses so as to complement the information in Fourier space (diffraction pattern) with information in real space (image) that is more closely related to the atomic structure that one wants to determine. And since the resolution of the newest electron microscopes is sufficient to visualize single atoms, it has become possible to use high resolution electron microscopy (HREM) images for quantitative refinement of the atom positions. Moreover, because of their large kinetic energy, individual electrons can be detected with high efficiency in novel detectors such as charge-coupled device (CCD) cameras so that all information be captured and atom positions can be determined with the highest attainable precision, only limited by the counting noise. In that case, the ultimate limiting factor is the number of imaging particles available in a given observation time, which can be limited by the brightness of the source, by the stability of the instrument, or by the radiation damage in the object. A further advantage is that the electron beam can be focused by lenses and combined with a bright field emission source (FEG) to yield a higher brightness than the X-ray beams in a synchrotron.

As argued in 2.12, with the newest generation of electron microscopes, one enters a situation in which the ultimate resolution is determined by the atom itself and imaging reaches the same "superresolution" as electron diffraction [7].

2.2
Principles of Linear Image Formation

2.2.1
Real Imaging

The quality of an imaging device such as a microscope or a telescope can be judged by the image it makes of a sharp point. This is called the *point-spread function*(PSF). In the field of signal processing this is called the *"impulse response function."*

Let us now consider a very simple case of real imaging of a real object such as the picture in Figure 2.6. Every pixel in an object can be considered as an independent point. If we now assume that the imaging is linear, the image of an assembly of pixels is the same as the assembly of the images of the pixels. Thus every pixel in the image $f(\mathbf{r})$ is blurred into a PSF $P(\mathbf{r})$. Furthermore, one can assume that the imaging characteristics are translation invariant, so that the shape of the PSF is independent of the position of the pixel.

$$I(\mathbf{r}) = \sum_n F(\mathbf{r}_n) P(\mathbf{r} - \mathbf{r}_n) \qquad (2.1)$$

This result expresses mathematically that the final image is the weighted sum of the PSF. If we now take the limit at which the points are taken infinitesimally close together, the sum in Eq. (2.1) becomes an integral.

$$I(\mathbf{r}) = \int F(\mathbf{r}') P(\mathbf{r} - \mathbf{r}') \, d\mathbf{r}' \qquad (2.2)$$

which is mathematically called a *convolution product*.

$$I(\mathbf{r}) = F(\mathbf{r}) * P(\mathbf{r}) \qquad (2.3)$$

This result is valid in 1D, 2D, or even in 3D (tomography) imaging.

The blurring limits the resolution of the imaging device. Indeed, when two points are imaged with a distance smaller than the "width" of the PSF, their images will overlap so that they become visually indistinguishable. The *resolution*, defined as the smallest distance that can be resolved, is then related to the width of the PSF.

Figure 2.6 Principle of linear image formation from top to bottom: original image (Centre for Electron Microscopy, University of Antwerp); Gaussian PSF with different sizes; blurred image, deblurred image.

It is very informative to describe the imaging process in Fourier space. Let us call the Fourier transforms of $F(\mathbf{r})$, $P(\mathbf{r})$, and $I(\mathbf{r})$ respectively $F(\mathbf{g})$, $P(\mathbf{g})$, and $I(\mathbf{g})$. The convolution theorem states that the Fourier transform of a convolution product is a normal product. If we thus Fourier transform, we obtain

$$I(\mathbf{g}) = F(\mathbf{g}) \cdot P(\mathbf{g}) \qquad (2.4)$$

Note that henceforth the vector $\mathbf{r}$ will describe the position in real space and $\mathbf{q}$ the vectors in Fourier space. In case of 2D imaging we might further decompose $\mathbf{r}$ into $(\mathbf{R}, z)$, where $\mathbf{R}$ is the 2D vector in the plane of projection and z is the perpendicular direction. The interpretation of Eq. (2.4) is rather simple. $F(\mathbf{g})$ represents the content of the object in the spatial frequency domain (or the Fourier domain) and $P(\mathbf{g})$ is the transfer function. Every imaging device can thus also be characterized by its transfer function (band filter) $P(\mathbf{g})$, which describes the magnitude and phase with which a spatial frequency component $F(\mathbf{g})$ is transferred through the device. This is shown in Figure 2.7. The noise level, N, is also schematically indicated.

The *resolution* ρ of the instrument is defined from the cutoff $1/\rho$ between signal and noise beyond which no spatial information is transferred. This is the type of resolution in the sense as defined by Rayleigh. It is inversely related to the "width" of the PSF. If the transfer function were constant (i.e., perfectly flat) in the whole

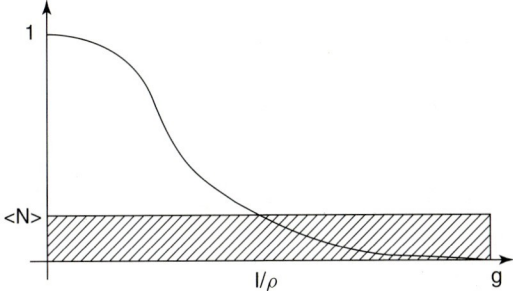

Figure 2.7 Transfer function.

spatial frequency range, the impulse response function would be a delta function so that $I(\mathbf{r}) = F(\mathbf{r})$.

If the PSF is known, the original image can be restored up to the resolution ρ by the following steps: Fourier transform $I(\mathbf{r})$ to $I(\mathbf{g})$, multiply $I(\mathbf{g})$ by $1/P(\mathbf{g})$ and Fourier transform back to $I(\mathbf{r})$. This is called *image restoration or deblurring*. $1/P(\mathbf{g})$ is called a *deconvolution filter*. However, a problem occurs for these values of $\mathbf{g}$ for which the transfer function is zero, since dividing the noise by zero will yield unreliable results. A modified type of a deconvolution operator that takes care of this problem is the so-called Wiener filter. The attainable resolution after deblurring depends on the resolution of the imaging device. Figure 2.6 shows an example of image restoration.

In many cases, an image is formed through many imaging steps or devices. Each of these steps (if linear) has its own PSF. For instance, the image of a star through a telescope can be blurred by the atmosphere, by the telescope, and by the photoplate or camera. Let us denote the respective PSFs of the successive steps by $P_1(\mathbf{r})$, $P_2(\mathbf{r})$, $P_3(\mathbf{r})$,...; then the final PSF is given by

$$P(\mathbf{r}) = P_1(\mathbf{r}) * P_2(\mathbf{r}) * P_3(\mathbf{r}) \tag{2.5}$$

and its Fourier transform

$$P(\mathbf{g}) = P_1(\mathbf{g}) * P_2(\mathbf{g}) * P_3(\mathbf{g}) \tag{2.6}$$

The total transfer function is thus the product of the respective transfer functions. The resolution is then mainly limited by the weakest step in the imaging chain.

2.2.2
Coherent Imaging

In the case of coherent imaging as in electron microscopy, the object and the PSF are complex wave functions having an amplitude component and a phase component.

Also in this case, the complex image wave $\psi_{im}(\mathbf{r})$ can be described in real space as a convolution product of the complex object function $\psi(\mathbf{r})$ and the complex

PSF $P(\mathbf{r})$

$$\psi_{\text{im}}(\mathbf{r}) = \psi(\mathbf{r}) * P(\mathbf{r}) \tag{2.7}$$

or in Fourier Space

$$\psi_{\text{im}}(\mathbf{g}) = \psi(\mathbf{g}) * P(\mathbf{g}) \tag{2.8}$$

When the image is recorded, only the image intensity is measured

$$I(\mathbf{r}) = |\psi_{\text{im}}(\mathbf{r})|^2 = |\psi(\mathbf{r}) * P(\mathbf{r})|^2 \tag{2.9}$$

Thus, at that stage, the phase information is lost. If one could retrieve the image phase by "holographic" methods, it would be possible to deconvolute the transfer of the microscope and to reconstruct the object wave and enhance the resolution.

2.3
Imaging in the Electron Microscope

2.3.1
Theory of Abbe

We now briefly recall the imaging theory of Abbe [8]. The main property of a lens is to focus a parallel beam into a point of the back focal plane of the lens. If a lens is placed behind a diffracting object, each parallel diffracted beam is focused into another point of the back focal plane, whose position is given by the reciprocal vector $\mathbf{g}$ characterizing the diffracted beam. The wavefunction $\psi(\mathbf{R})$ at the exit face of the object can be considered as a planar source of spherical waves (Huyghen's principle) ($\mathbf{R}$ is taken in the plane of the exit face). The amplitude of the diffracted wave in the direction given by the reciprocal vector $\mathbf{g}$ (or spatial frequency) is given by the Fourier transform of the object function, that is,

$$\psi(\mathbf{g}) = F_{\mathbf{g}} \psi(\mathbf{R}) \tag{2.10}$$

The intensity distribution in the diffraction pattern is given by $|\psi(\mathbf{g})|^2$. The back focal plane visualizes the square of the Fourier transform (i.e., the diffraction pattern) of the object. If the object is periodic, the diffraction pattern will consist of sharp spots. A continuous object will give rise to a continuous diffraction pattern. In the second stage of the imaging process, the back focal plane acts, in its turn, as a set of Huyghens sources of spherical waves that interfere, through a system of lenses, in the image plane (Figure 2.8). This stage in the imaging process is described by an inverse Fourier transform that reconstructs the object function $\psi(\mathbf{R})$ (usually enlarged) in the image plane. The intensity in the image plane is then given by $|\psi(\mathbf{R})|^2$. During the second step in the image formation, which is described by the inverse Fourier transform, the electron beam $\mathbf{g}$ undergoes a phase shift $\chi(\mathbf{g})$ with respect to the central beam caused by spherical aberration and defocus.

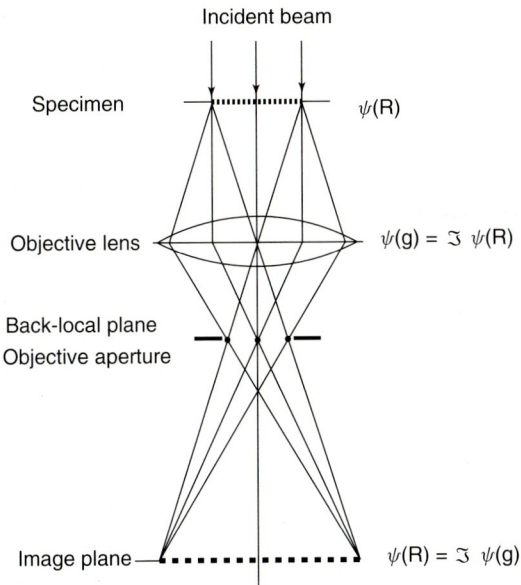

Figure 2.8 Schematic representation of the image formation by the objective lens in a transmission electron microscope. The corresponding mathematical operations are indicated (see text).

Then the transfer function in $P(\mathbf{g})$ is given by

$$P(\mathbf{g}) = A(\mathbf{g}) \exp\left[-i\chi(\mathbf{g})\right] \tag{2.11}$$

$A(\mathbf{g})$ represents the physical aperture with radius g_A selecting the imaging beams:

thus $A(\mathbf{g}) = \begin{matrix} 1 & \text{for } |g| \leq g_A \\ 0 & \text{for } |g| > g_A \end{matrix}$.

The total phase shift due to spherical aberration and defocus is

$$\chi(\mathbf{g}) = \frac{1}{2}\pi C_s \lambda^3 g^4 + \pi \varepsilon \lambda g^2 \tag{2.12}$$

with C_s: the spherical aberration coefficient, ε: the defocus, λ: the wavelength.

The phase shift $\chi(\mathbf{g})$ increases with $\mathbf{g}$. At large spatial frequencies higher-order aberrations can also become important.

The imaginary parts of the PSF and the transfer function of an HREM are shown in Figure 2.9.

For a hypothetical ideal pointlike object, the object wave would be a delta function so that the image wave is $\psi(\mathbf{r}) = P(\mathbf{r})$, that is, the microscope would reveal the point spread function $P(\mathbf{r})$. For a perfect electron microscope, the transfer function would be constant (i.e., perfectly flat) over the whole spatial frequency range so that the PSF would be a delta function and hence the wavefunction in the image plane represents exactly the wavefunction of the object. However, a perfect electron microscope will not be useful for the following reason: as shown in Appendix 2.A a thin object acts

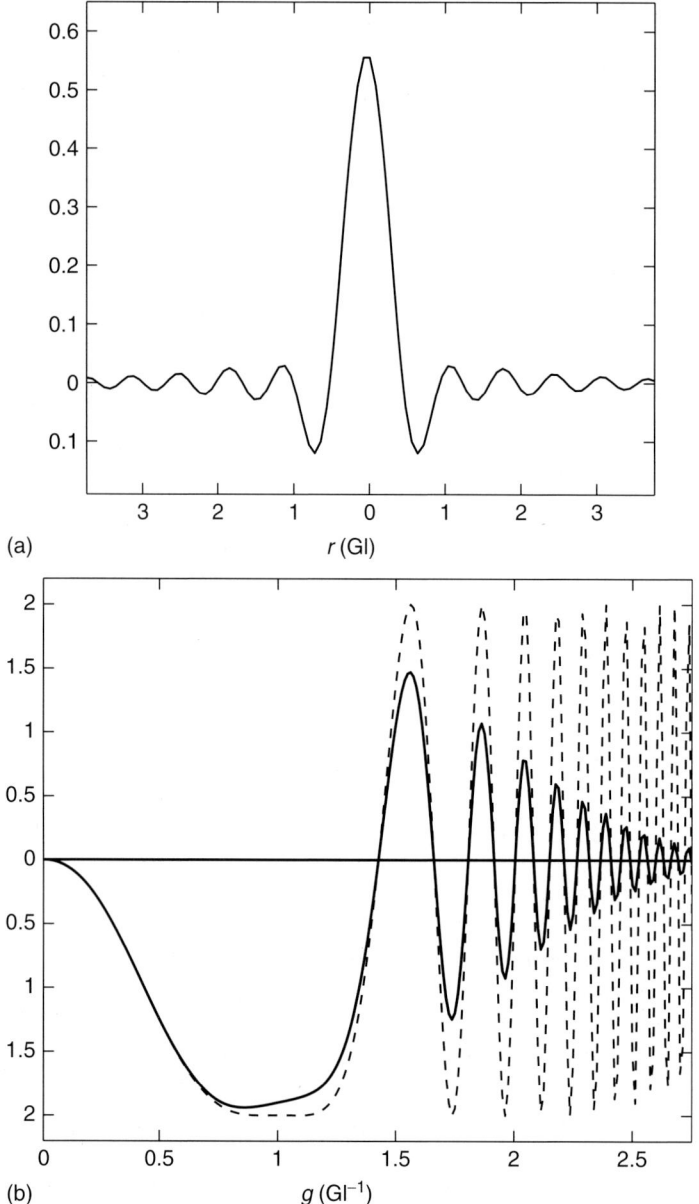

Figure 2.9 (a) Imaginary parts of the point-spread function and (b) the transfer function of an electron microscope. Dotted line: without incoherent damping. Solid line: with incoherent damping. The axes are denoted in Glaser units (see Section 2.3.4).

as a phase object. The intensity behind a phase object is constant so that the image would not show any detail. Thus for very thin objects, aberrations are a necessity.

However, in practice, the transfer function cannot be made arbitrarily flat as is shown in Figure 2.9b and the PSF has a finite width as shown in Figure 2.9a. Hence, the object wavefunction $\psi(\mathbf{R})$ is then smeared out (blurred) over the width of the peak. This width can then be considered as a measure for the resolution in the sense as originally defined by Rayleigh. The width of this peak is the inverse of the width of the constant plateau of the transfer function in Figure 2.9. In fact, the constant phase of the spatial frequencies **g** assures that this information is transferred forward, that is, retains a local relation to the structure. All information beyond this plateau is still contributing to the image but with a wrong phase. It is present in the long "tails" of the PSF and it is thus redistributed over a larger area in the image plane.

One has to note, however, that, on recording, only the intensity of the image is detected and the phase of the image wave is lost. The intensity in the image plane is then given by $|\psi(\mathbf{R})^2|$. In order to deconvolute the image wave so as to restore the object wave, one first has to retrieve the phase of the image phase. This can be done by using a holographic technique. Once the image phase and thus the whole image wave is known, one can deconvolute in the same way as above. This is discussed in Sections 2.5.2 and 2.5.3.

2.3.2
Incoherent Effects

In HREM, the intensity in the image plane is given by Eq. (2.9)

$$I_{HREM}(\mathbf{r}) = |\psi(\mathbf{r}) * P(\mathbf{r})|^2 \tag{2.13}$$

where $*$ represents convolution, $\psi(\mathbf{r})$ is the exit wave of the object, and $P(\mathbf{r})$ is the complex PSF of the electron microscope. If the electron microscope is subject to fluctuations during the recording of the image, we have to average the image intensity over the various states of the microscope. We can write this average as $\langle\rangle_M$. In practice, the fluctuation in the strength (focus) of the electron microscope is called *temporal incoherence* and the averaging over the illumination conditions is called *spatial incoherence*. For details, we refer to [9]. However, also the object can be subject to variations during the recording of the image so that the intensity has to be averaged over the different states of the object. We write this average as $\langle\rangle_O$. We can also assume that the states of the microscope and of the object are uncorrelated. Hence the total averaged intensity is given by

$$I(\mathbf{r}) = \langle\langle|\psi(\mathbf{r}) * P(\mathbf{r})|^2\rangle_O\rangle_M \tag{2.14}$$

However, the numerical calculation of these averages requires a repetitive calculation for the coherent image intensity over all the states of microscope and object which is very time consuming. Therefore, the usual way to speed up this process is by performing the averages of the microscope and the object separately

$$I(\mathbf{r}) = |\langle\psi(\mathbf{r})\rangle_O * \langle P(\mathbf{r})\rangle_M|^2 \tag{2.15}$$

In this way, the PSF of the microscope is replaced by an effective PSF $\langle P(\mathbf{r})\rangle$ and the object wave is replaced by an effective object wave $\langle \psi(\mathbf{r})\rangle$. Then Eq. (2.14) can be considered again as a coherent imaging process where an "averaged" microscope images an "averaged" object. Since both microscope and object become much "smoother," the image simulations can be done with much fewer sampling points. However, it is clear that the operations for averaging and imaging do not commute

$$\langle\langle|\psi(\mathbf{r}) * P(\mathbf{r})|^2\rangle_O\rangle_M \neq |\langle\psi(\mathbf{r})\rangle_O * \langle P(\mathbf{r})\rangle_M|^2 \tag{2.16}$$

so that the approximation (Eq. (2.14)) is only correct for very thin objects. A more correct treatment is discussed in [9].

In case of a weak object, the averaging over the incoherent fluctuations of the microscope results in a damping envelope of the phase transfer function [9].

$$P(\mathbf{g}) = A(\mathbf{g}) \exp[-i\chi\mathbf{g}] D(\alpha, \Delta, \mathbf{g}) \tag{2.17}$$

This is called the *coherent approximation*.

The effect of the damping envelope is shown in Figure 2.9.

2.3.3
Imaging at Optimum Defocus: Phase Contrast Microscopy

In an ideal microscope, the image would exactly represent the object function and the image intensity for a pure phase object function would be

$$|\psi_{im}(\mathbf{R})^2| = |\psi(\mathbf{R})|^2 = |\exp[i\phi(\mathbf{R})]|^2 = 1 \tag{2.18}$$

that is, the image would show no contrast. This can be compared with imaging a glass plate with variable thickness in an ideal optical microscope. Assuming a weak phase object (WPO) one has

$$\phi(\mathbf{R}) \ll 1$$

so that

$$\psi(\mathbf{R}) \approx 1 + i\phi(\mathbf{R}) \tag{2.19}$$

The constant term 1 contributes to the central beam (zeroth Fourier component), whereas the term $i\phi$ mainly contributes to the diffracted beams. If the phases of the diffracted beams can be shifted over $\frac{\pi}{2}$ with respect to the central beam, the amplitudes of the diffracted beams are multiplied with $\exp(i\frac{\pi}{2})$. Hence the image wave $i\phi(\mathbf{R})$ becomes

$$\psi_{im}(\mathbf{R}) \approx 1 - \phi(\mathbf{R}) \approx \exp[-\phi(\mathbf{R})] \tag{2.20}$$

that is, the phase object now acts as an amplitude object. The image intensity is then

$$|\psi_{im}(\mathbf{R})|^2 \approx 1 - 2\phi(\mathbf{R}) \tag{2.21}$$

which is a direct representation of the phase of the object. In optical microscopy, this has been achieved by F. Zernike by shifting the central beam through a quarter wavelength plate. In electron microscopy, the optimal imaging can be achieved by making the transfer function as constant as possible. From Eq. (2.11), it is clear that oscillations occur due to spherical aberration and defocus. However, the effect of spherical aberration which, in a sense, makes the objective lens too strong for the most inclined beams, can be compensated to some extent by slightly underfocusing the lens. The optimum defocus value (also called *Scherzer defocus*) for which the plateau width is maximal, is given by

$$\varepsilon = -1.2\,(\lambda C_s)^{1/2} = -1.2\,\text{Sch} \tag{2.22}$$

with $1\,\text{Sch} = (\lambda C_s)^{1/2}$ the Scherzer unit.

The transfer function for this situation is depicted in Figure 2.9b. The phase shift $\chi(g)$ is nearly equal to $-\pi/2$ for a large range of spatial coordinates g which is exactly the condition for phase contrast microscopy. Furthermore, as is shown in Appendix 2.A, a thin object acts as a phase object in which the phase is proportional to the projected potential of the object so that the image contrast for a very thin object can be interpreted directly in terms of the projected structure of the object.

2.3.4
Resolution

As shown above in the phase contrast mode at optimum focus, the high-resolution image directly reveals the projected potential, that is, the structure, of the object provided the object is very thin. All spatial frequencies g with a nearly constant phase shift are transferred forward from object to image. Hence the resolution can be obtained from the first zero of the transfer function (Eq. (2.12)) as

$$\rho_s = \frac{1}{g} \approx 0.65 C_s^{1/4} \lambda^{3/4} = 0.65\,\text{Gl} \tag{2.23}$$

with $\text{Gl} = C_s^{1/4} \lambda^{3/4}$ the Glaser unit. This value is generally accepted as the standard definition of the point resolution of an electron microscope. It is also equal to the width of the PSF (Figure 2.9a). The information beyond the intersection ρ_s is transferred with a nonconstant phase and, as a consequence, is redistributed over a larger image area.

Note that even for C_s-corrected microscopes, Eq. (2.23) still applies and one can still work at optimum focus. However, at very low C_s values, the fifth order aberration becomes important so that the passband (resolution) can be extended by compensating this aberration with a negative defocus and a negative spherical aberration. Details are given in [10].

As discussed in Section 2.5.2, it is possible to retrieve the phase of the image wave. Furthermore, since the complex PSF of the electron microscope is known, it can be deconvoluted from Eq. (2.7) so as to reconstruct the electron wave at the exit face of the object. This is called *exit wave reconstruction*. Historically, this was the reason that Gabor developed holography [11] in the hope to be able to push the resolution of the electron microscope so as to reveal the individual atoms.

By exit wave reconstruction, the phase oscillations in the transfer function can be corrected and the resolution is not limited by the point resolution but by the "information limit," which can be defined as the finest detail that can be resolved by the instrument. It corresponds to the maximal spatial frequency that is still transmitted with appreciable intensity. For a thin specimen, this limit is mainly determined by the temporal incoherence and spatial incoherence. In principle, spatial incoherence can be reduced using a coherent FEG. If temporal incoherence is predominant, the resolution can be estimated from the damping envelope (Eq. (2.7)) as

$$\rho_I = \frac{1}{g} = \left(\frac{\pi \lambda \Delta}{2}\right)^{1/2} \tag{2.24}$$

with the defocus spread

$$\Delta = C_c \sqrt{\left(\frac{\Delta V}{V}\right)^2 + \left(\frac{\Delta E}{V}\right)^2 + 4\left(\frac{\Delta I}{I}\right)^2} \tag{2.25}$$

C_c is the chromatic aberration function (typically 10^{-3} m), ΔV is the fluctuation in the incident voltage, ΔE is the thermal energy spread of the electrons, and $(\Delta I/I)$ is the relative fluctuation of the lens current. In advanced electron microscopes, the information limit can be pushed below 0.5 Å.

However, as discussed in Section 2.1.5, the ultimate resolution is limited by the finite "width" of the atom. In case of electron channeling (Section 2.5.4), in which the atoms along a column act as focusers, the exit wave can be more sharply peaked, which also improves the ultimate resolution.

2.4
Experimental HREM

2.4.1
Aligning the Microscope

Before starting high-resolution work, it is necessary to determine the most important optical parameters of the instrument for later use in image simulation and reconstruction. For very high resolution, the standard correction procedure for the aberrations is not sufficient and methods have been developed for automatic alignment. One of the commonly used methods is to calculate the Fourier transform of an image of an amorphous object (diffractogram), which acts as a kind of white noise object. The diffractogram represents the contrast transfer function (Figure 2.9) corresponding to the particular focus. By tilting the specimen in different directions, one then obtains a series of diffractograms, called a *Zemlin tableau*, which allows to calculate the main aberration constants (Figure 2.10). In advanced electron microscopes, this is done semiautomatically.

Figure 2.10 A Zemlin tableau showing the diffractograms for an amorphous object for various beam tilt conditions. This tableau can be used to determine the most important instrumental aberrations.

2.4.2
The Specimen

The main requirement for atomic resolution electron microscopy is that the specimen should be sufficiently thin, that is, less than about 10 nm; and clean crystalline specimens with a unit cell with two large and one small lattice parameter are most ideal for HREM. In that case the reciprocal lattice consists of dense planes (Laue zones) which are largely separated. Such crystals can be oriented with their short axis parallel to the incident beam so that the nearly flat Ewald sphere touches the Laue zone through the origin (Figure 2.11) and a large number of diffracted beams are excited simultaneously and maximal information is present in the image. In this situation, the electrons propagate parallel to a zone axis, that is, parallel to the atom rows. A possible interpretation of the images in terms of the projected structure can be meaningful only in this way. The same argument holds also for crystals with defects (Figure 2.12).

After finding a suitably thin part with the proper orientation, one has to adjust the focus. When the specimen is very thin, the zero focus corresponds to minimal contrast. Maximal contrast appears close to the optimum defocus (Eq. (2.22)). Even in aberration-corrected microscopes, one can, on going through focus, reverse the contrast. In practice, since the focus is not exactly known, especially in the case of thicker specimens, one has to take a series of images at gradually different focus settings, recorded approximately around the optimum defocus. This is called a *through focus series*.

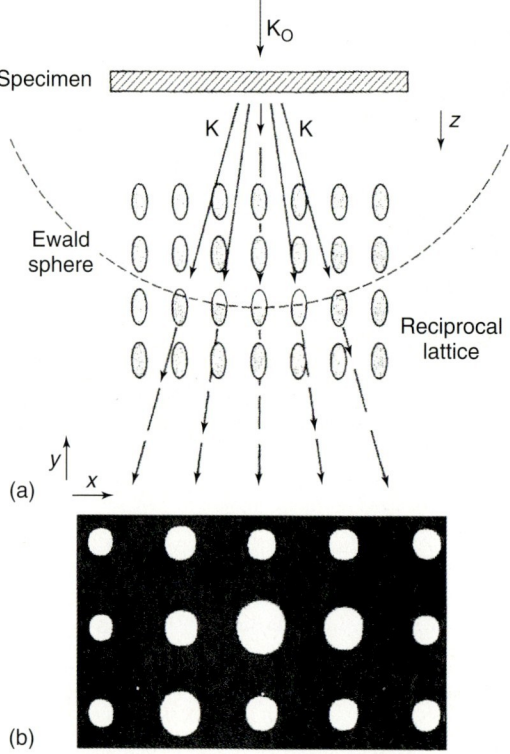

Figure 2.11 Formation of the diffraction pattern. The simultaneously excited electron beams can be used for the image formation.

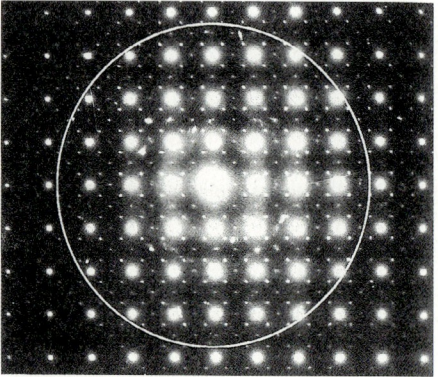

Figure 2.12 Typical diffraction pattern used for structure imaging. The aperture, selecting the contributing beams is also indicated.

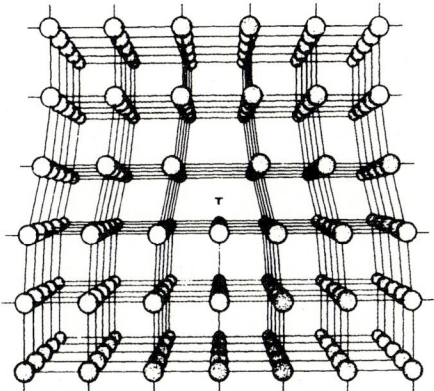

Figure 2.13 Structure model of a crystal containing a dislocation viewed along the atomic columns.

2.4.3
Interpretation of the High-Resolution Images

In Section 2.3.3, we have shown that, at optimum defocus up to the point resolution of the electron microscope, the high-resolution image of a thin object can be integrated directly in terms of the projected structure. This is clear in Figure 2.14, which shows a series of images at different focus values for a $Ti_2Nb_{10}O_{29}$ perovskite

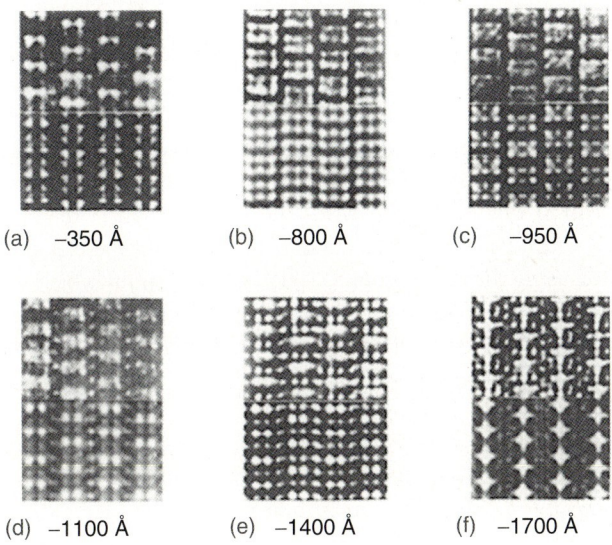

(a) −350 Å   (b) −800 Å   (c) −950 Å

(d) −1100 Å   (e) −1400 Å   (f) −1700 Å

Figure 2.14 (a–c) Comparison of experimental images (Courtesy: S. Iijima) and (d–f) computer-simulated images for $Ti_2Nb_{10}O_{29}$ as a function of defocus.

Figure 2.15 Schematic representation of the unit cell of $Ti_2Nb_{10}O_{29}$ consisting of corner-sharing NbO_6 octahedra with the Ti-atoms in tetrahedral sites.

experimental images (a–c) (Courtesy: S. Iijima) [12] and computer-simulated images (d–f). Furthermore, close to the optimum focus (which is the case at about −800 Å), the image clearly corresponds with the projected structure of the perovskite in Figure 2.15 in which the octahedrons and the open tunnels are clearly revealed. Although these results are very old (in fact they were the first HREM images ever), they are still very informative.

In general, the interpretation of high-resolution images never appears to be trivial. The only way out remains in the comparison of the experimental images with those calculated for various trial structures. During the imaging process, the electrons undergo three distinct interactions. Each of these interactions is known and can be calculated by the computer. First, the electron scatters dynamically in the crystal. This interaction can be simulated using the multislice methods [13]. The multislice theory can be derived from an optical approach (Appendix 2.B) or a quantummechanical approach (Appendix 2.C). However, as an input to the program, one has to specify all the object parameters such as unit cell, position, and type of cell atoms, thermal atom factors (Debye–Waller factors), object orientation, and thickness. The result of this calculation yields the wavefunction at the exit face of the crystal. In a second step, the formation of the image in the electron microscope is simulated using the expressions (11.11) and (11.12), for which all the instrumental parameters have to be specified. Finally, the electron intensity in the image plane is calculated by squaring the wavefunction. Different commercial software packages exist for high-resolution image simulations. References are given in [14].[1]

If image simulation is used for visual comparison, it can only be used if the number of plausible models is very limited. Direct methods, which extract the information from the images in a direct way so as to be used as input for further quantitative refinement, are a better way to go.

---

1) Multislice Computer Program, High Resolution Electron Microscope Facility, Arizona State University, Tempe, Arizona 85281.

2.5
Quantitative HREM

2.5.1
Model-Based Fitting

In principle, one is usually not so interested in high-resolution images as such, but rather in the object under study. High-resolution images are then to be considered as data planes from which the structural information has to be extracted in a quantitative way. This can be done as follows: one has a model for the object and for the imaging process, including electron–object interaction, microscope transfer, and image detection. The model contains parameters that have to be determined by the experiment. This can be done by optimizing the fit between the theoretical images and the experimental images. The goodness of the fit is evaluated using a matching criterion such as maximum likelihood or R-factor (cf. X-ray crystallography). For each set of parameters, one can calculate this fitness function and search for the optimal fit by varying all parameters. The optimal fit then yields the best estimates for the parameters of the model that can be derived from the experiment. In a sense, one is searching for an optimum of the fitness function in the parameter space, the dimension of which is equal to the number of parameters. The object model that describes the interaction with the electrons should describe the electrostatic potential, which is the assembly of the electrostatic potentials of the constituting atoms. Since for each atom type the electrostatic potential is known, the model parameters then reduce to atom types, positions, and thermal atom factors.

A major problem now is that the object information can be strongly delocalized by the image transfer in the electron microscope (Figure 2.9) so that the influence of the model parameters of the object is completely scrambled in the high-resolution images and the dimension of the parameter space is much too high to be feasible for model-based filtering. The only way out is to find a method that "unscrambles" the many parameters so as to provide a pathway to the global optimum. In a sense, such a direct method must thus "resolve" the atoms so as to yield an approximate atomic structure that can then be used as a seed for further quantitative refinement by fitting with the original experimental data. In X-ray crystallography, where the information of all the atom positions is also scrambled in the intensities of the reflections of the diffraction patterns, direct methods have been developed that enable to get sufficient information on the phases of the reflections so as to yield an approximate starting structure [3]. In electron microscopy, the information about the atom positions is scrambled by the blurring due to the electron–object interaction and due to the imaging in the electron microscope.

2.5.2
Phase Retrieval

Undoing the scrambling from object to image consists of three stages. First, one has to reconstruct the wavefunction in the image plane (phase retrieval), then one has to reconstruct the exit wave of the object, and one has to "invert" the scattering in the object so as to retrieve the object structure (Figure 2.16).

The phase problem can be solved by holographic methods. Two methods exist for this purpose: off-axis holography and focus variation. In off-axis holography, the beam is split by an electrostatic biprism into a reference beam and a beam that traverses the object. Interference of both beams in the image plane then yields fringes, the positions of which yield the phase information. In the focus variation method, which is a kind of in-line holography, the focus is used as a controllable parameter so as to yield focus values from which both amplitude and phase information can be extracted [15, 16]. Images are captured at very close focus values so as to collect all information in the three-dimensional image space. Each image contains linear as well as nonlinear information. By filtering out the linear information, the phase can be retrieved.

A simple way to describe this reconstruction is the following. Both for weak objects Eq. (2.19) as for thick objects Eq. (2.31) and using Eq. (2.7), the wave in the image plane is

$$\psi_{im}(\mathbf{R}) = 1 + \theta(\mathbf{R}) \tag{2.26}$$

with

$$\theta(\mathbf{R}) = \psi(\mathbf{R}) * P(\mathbf{R}) \tag{2.27}$$

where $\psi(\mathbf{R})$ is the interaction wave and $P(\mathbf{R})$ is the PSF of the microscope. If we now defocus the image wave over a defocus distance ε, then

$$\psi(\mathbf{R}, \varepsilon) = 1 + \theta(\mathbf{R}) * P(\mathbf{R}, \varepsilon) \tag{2.28}$$

with $P(\mathbf{R}, \varepsilon)$ the defocus propagator. For the image intensity, we now have

$$I(\mathbf{R}, \varepsilon) = 1 + \theta(\mathbf{R}) * P(\mathbf{R}, \varepsilon) + \theta^* (\mathbf{R}) * P^* (\mathbf{R}, \varepsilon) + [\theta(\mathbf{R}) * P(\mathbf{R}, \varepsilon)]^2 \tag{2.29}$$

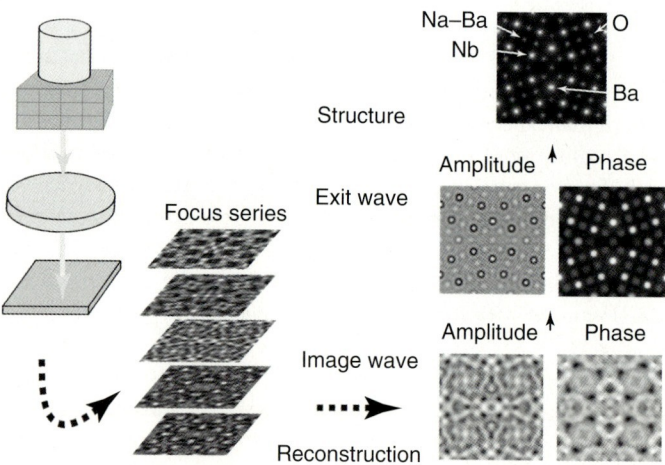

Figure 2.16 Schematic representation of the phase retrieval procedure.

If we now take a series of images at focus values ε_n and backpropagate them to $\varepsilon = 0$ and take the average, we get

$$\frac{1}{N} \sum_n I(\mathbf{R}, \varepsilon_n) * P(\mathbf{R}, -\varepsilon_n) = 1 + \theta(\mathbf{R}) + \theta\left(\frac{1}{N}\right) \tag{2.30}$$

Thus, in a sense, we have linearized the imaging by a factor N, which already gives a very good estimate of $\theta(\mathbf{R})$ and from Eq. (2.27). In this way we have solved the phase problem and reconstructed the wave function in the image plane. From this wave we can reconstruct the exit wave of the object by deconvolution of the complex point spread function of the electron microscope. The precision can be improved further by including also the nonlinear contribution of Eq. (2.29) in the fitting.

Focus variation is more accurate for high spatial frequencies, whereas off-axis holography is more accurate for lower spatial frequencies but puts higher demands on the number of pixels and the coherence.

2.5.3
Exit Wave Reconstruction

As is clear from Eq. (2.7), the exit wave of the object $\psi(\mathbf{r})$ can be calculated from the wavefunction in the image plane by deconvoluting the PSF of the microscope. This procedure is straightforward, provided the proper parameters describing the transfer function (such as the spherical aberration constant C_s).

Figure 2.17 shows the exit wave of an object of $YBa_2Cu_4O_8$ (high TC superconductor), which was historically the first experimental result obtained with the focus variation method [16b].

It should be noted that, once the exit wave is reconstructed, it is in principle possible to recalculate all the images of the Fourier series that perfectly fit in the experimental images within the noise level so that the reconstructed exit wave contains all experimentally attainable object information.

2.5.4
Structure Retrieval: Channeling Theory

The final step consists in retrieving the projected structure of the object from the wavefunction at the exit face. If the object is thin enough to act as a phase object, the phase is proportional to the electrostatic potential of the structure, projected along the beam direction so that the retrieval is straightforward. If the object is thicker, the problem is more complicated.

It is possible however to obtain an approximate structure if the object is a crystal viewed along a zone axis, in which the incident beam is parallel to the atom columns. It can be shown that in such a case, the electrons are trapped in the positive electrostatic potential of the atom columns, which then act as channels (Figure 2.18). If the distance between the columns is not too small, a one-to-one correspondence between the wavefunction at the exit face and the column structure of the crystal is maintained. Within the columns, the electrons

2 Atomic Resolution Electron Microscopy

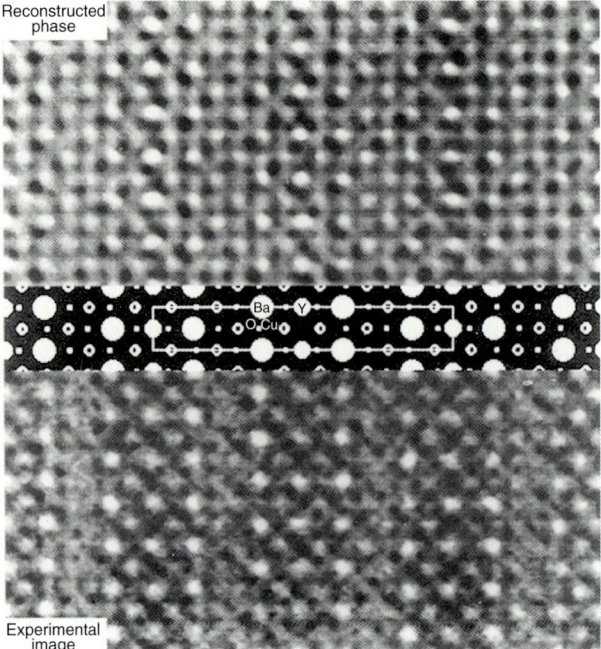

Figure 2.17 Experimentally reconstructed exit wave for YBa_2CuO_8. Top: reconstructed phase; center: structure model; bottom: experimental image.

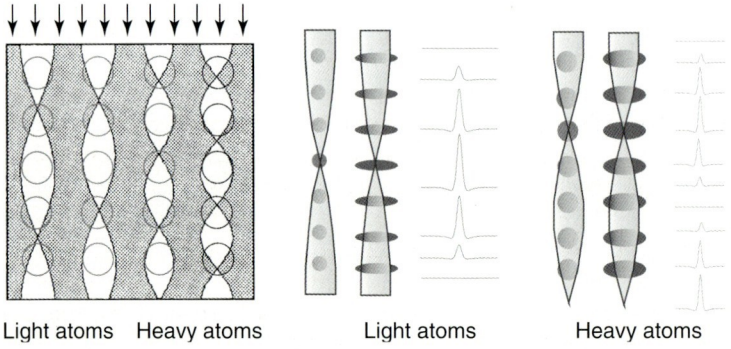

Figure 2.18 Schematic representation of electron channeling through atom columns with different mass densities.

oscillate as a function of depth without however leaving the column. Hence the classical picture of electrons traversing the crystal as planelike waves in the direction of the Bragg beams, which historically stems from X-ray diffraction, is in fact misleading. This is called *electron channeling* [17]. Every atom in the column acts as a small lens so that the electron wave is focused and defocused at periodical depths. When the thickness of the object is exactly equal to one period, the exit wave of the column is exactly equal to the entrance wave as if the column has "disappeared." For this reason, one period is called the *dynamical extinction distance*.

Electron channeling has also a number of other advantages. Because of the focusing effect, the exit wave can be more sharply peaked than the individual atom. Hence the Fourier transform is broader, which means that the electrons are scattered to larger angles and consequently the resolution is improved. Another advantage is that, for particular thicknesses, the scattering of a light column can be enhanced as compared to that of a heavy column, whereas in X-ray diffraction, the signal of a light column will remain relatively weak. It is important to note that channeling is not a property of a crystal, but occurs even in an isolated column and is not much affected by the neighboring columns, provided the columns do not overlap. Hence the one-to-one relationship is still present in case of defects such as translation interfaces or dislocations provided they are oriented with the atom columns parallel to the incident beam.

The basic result is that the wavefunction at the exit face of a column is expressed as

$$\psi(\mathbf{R}, z) = 1 + \left[\exp\left(-i\pi \frac{E}{E_0} kz\right) - 1\right] \phi(\mathbf{R}) \qquad (2.31)$$

For a full treatment we refer to [19]. This result holds for each isolated column. In a sense, the whole wavefunction is uniquely determined by the eigenstate $\phi(\mathbf{R})$ of the Hamiltonian of the projected column and its energy E, which are both functions of the "density" of the column and the crystal thickness z. It is clear from Eq. (2.31) that the exit wave is peaked at the center of the column and varies periodically with depth. The periodicity is inversely related to the "density" of the column. In this way, the exit wave still retains a one-to-one correspondence with the projected structure. Furthermore, it is possible to parameterize the exit wave in terms of the atomic number Z and the interatomic distance d of the atoms constituting the column [19]. This enables to retrieve the projected structure of the object from matching with the exit wave. In practice, it is possible to retrieve the positions of the columns with high accuracy (1 pm) and to obtain a rough estimate of the density of the columns.

For most cases, this expression is sufficiently accurate except for thick objects containing heavy atoms, where other higher order states will become more important when the distance between adjacent atom columns decreases. It turns out to be more convenient to subtract the entrance wave from the exit wave. We call this the *object wave*.

From Eq. (2.31), it follows that the amplitude is peaked at the atom column position and that it varies periodically with depth and that the phase, which is a constant over the column, is proportional to the average mass density of the column. The phase linearly increases with depth. The amplitude can be used to determine the positions of the atom columns and the phase can be used to determine the composition of the atom column.

Figure 2.19 shows an experimentally reconstructed exit wave from which the entrance wave is subtracted. As expected from Eq. (2.31), the amplitude is then

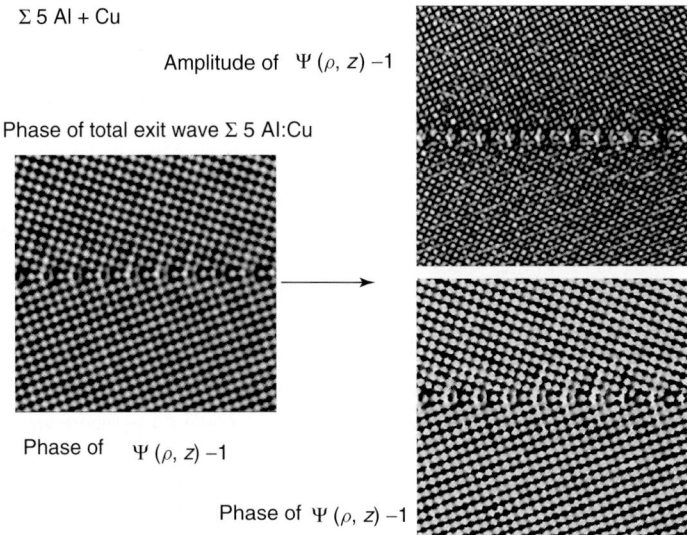

Figure 2.19 Experimentally reconstructed exit wave (Courtesy C. Kisielowski, J.R. Jinschek (NCEM, Berkeley)).

clearly peaked at the positions of the columns and the phase is constant over the column and is a measure of the "weight" of the column.

A convenient way to visualize the effect of electrons passing through a column is by plotting each pixel of the complex exit wave, which is located at a projected atom column position, in an Argand plot [20]. This is a representation in which each pixel is plotted as a point in a complex plane with its x-coordinate corresponding to the real pixel value and the y-coordinate corresponding to the imaginary pixel value. As can be derived from Eq. (2.31), the pixels at the exit face of a column should all be located on a circle that passes through the point (1,0) representing the reference wave. As shown in Figure 2.20, an increase in the mass of the column shifts the point along that "mass" circle and defocusing the exit wave shifts the point along the defocus circle. Thus by accurately analyzing the Argand plot, one can determine both mass and vertical position of the columns. Figure 2.21 shows experimental results for Au[100].

2.5.5
Resolving versus Refining

Once the individual atoms can be resolved, their position can be resolved accurately. From statistics, it can be shown that the ultimate precision (standard deviation) on the atom position is given by the simple rule $\sigma = \rho/\sqrt{N}$, where σ is the standard deviation, ρ the resolution and N the number of imaging particles [21]. Hence, for a resolution of 1 Å and 10 000 interacting electrons, the precision can

2.5 Quantitative HREM

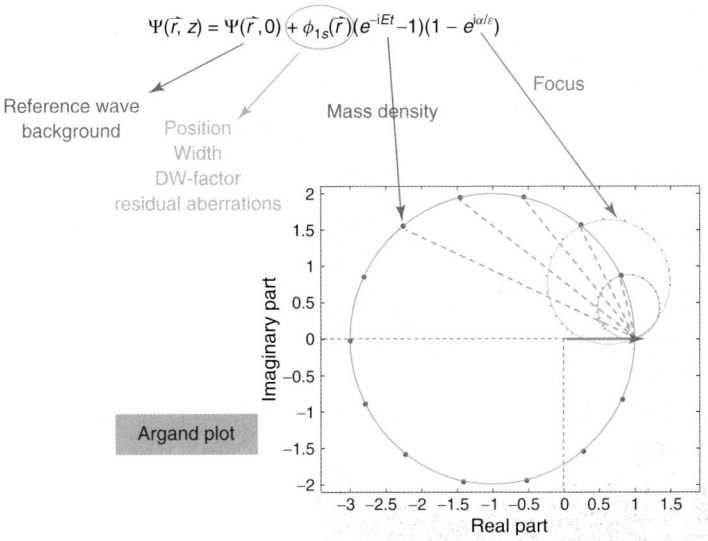

Figure 2.20 Representation of the exit wave in an Argand plot. The "mass circle" and the "defocus circle" are indicated. (Courtesy A. Wang)

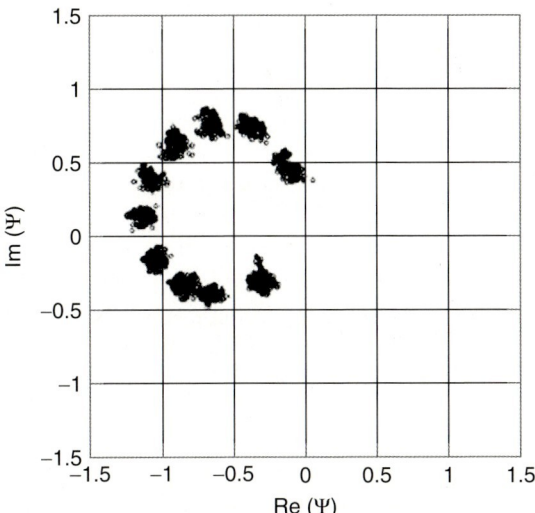

Figure 2.21 Argand plot of Au[110]. From this plot, the number of atoms in the individual Au columns can be counted. (Courtesy J. Jinczek.)

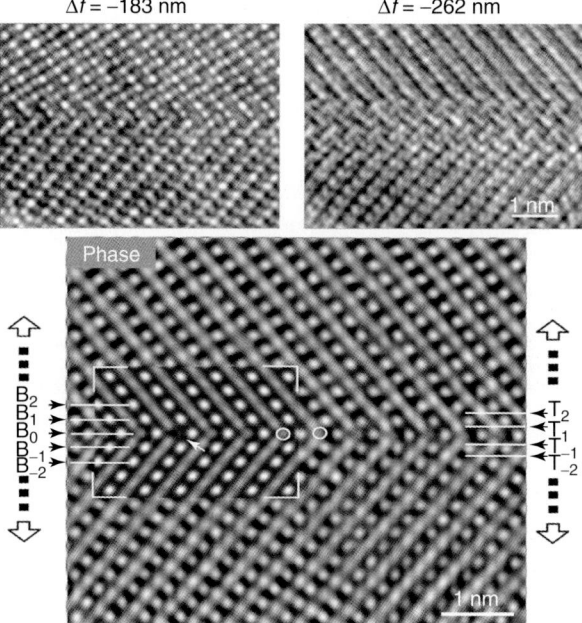

Figure 2.22 Exit wave of BaTiO$_3$ with Σ3 (111) twin boundary. Bottom: exit wave. Inset: fitted structure. (Courtesy of Jia and Thust [22].)

be 0.01 Å, which is sufficient to explain the structure–property relationships (see Section 2.1.1). Figure 2.22 shows the exit wave of BaTiO$_3$ reconstructed from a focal series (two focal images are shown in the inset). By careful fitting of the positions of the columns the following results were obtained [22].

	Ti-Ti (pm)	Ba-Ba (pm)
Geometric	232	232
Experiment	270	216
Theory	266	214

The experimental results are compared with theoretical results, obtained from ab initio calculations, and the differences are of the order of 0.02 Å (2 pm).

Appendix 2.A: Interaction of the Electron with a Thin Object

We now follow a classical approach.

The nonrelativistic expression for the wavelength of an electron accelerated by an electrostatic potential E is given by

$$\lambda = \frac{h}{\sqrt{2meE}} \tag{2.A.1}$$

with h the Planck constant, m the electron mass, and e the electron charge.

During the motion through an object with local potential $V(x,y,z)$ the wavelength will vary with the position of the electron as

$$\lambda'(\mathbf{r}) = \frac{h}{\sqrt{2me[E + V(\mathbf{r})]}} \tag{2.A.2}$$

For thin phase objects and large accelerating potentials, the assumption can be made that the electron keeps traveling along the z-direction so that by propagation through a slice dz the electron suffers a phase shift.

$$d\chi(\mathbf{R}, z) = 2\pi\, dz \left(\frac{1}{\lambda'} \frac{1}{\lambda}\right) = V(\mathbf{R}, z)\, dz \tag{2.A.3}$$

With $\sigma = \pi/\lambda E$, so that the total phase shift is given by:

$$\chi(\mathbf{R}) = \sigma \int V(\mathbf{R}, z)\, dz = \sigma V_p(\mathbf{R}) \tag{2.A.4}$$

where $V_p(\mathbf{R})$ represents the potential of the specimen projected along the z-direction. Under this assumption, the specimen acts as a pure phase object with transmission function

$$\psi(\mathbf{R}) = \exp\left[i\sigma V_p(\mathbf{R})\right] \tag{2.A.5}$$

In case the object is very thin, one has

$$\psi(\mathbf{R}) \approx 1 + i\sigma V_p(\mathbf{R}) \tag{2.A.6}$$

This is the weak phase object approximation (WPO).

Appendix 2.B: Multislice Method

Although the multislice formula can be derived from quantum-mechanical principles, we follow a simplified version of the more intuitive original optical approach. Consider a plane wave, incident on a thin specimen foil and nearly perpendicular to the incident beam direction z. If the specimen is sufficiently thin, we can assume the electron to move approximately parallel to z so that the specimen acts a pure phase object with transmission function (Eq. (2.A.5))

$$\psi(\mathbf{R}) = \exp\left[i\sigma V_p(\mathbf{R})\right] \tag{2.B.7}$$

A thick specimen can now be subdivided into thin slices, perpendicular to the incident beam direction. The potential of each slice is projected into a plane which acts as a two-dimensional phase object. Each point (**R**) of the exit plane of the first

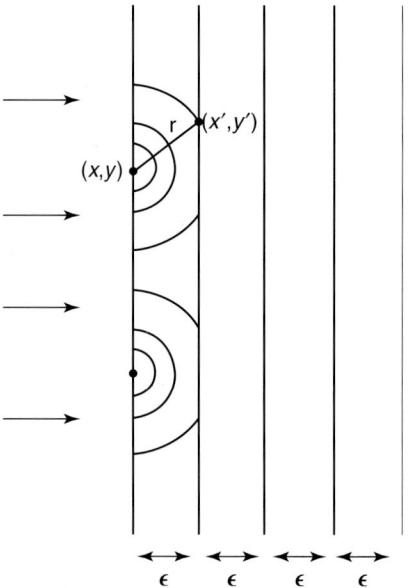

Figure 2.23 Schematic representation of the propagation effect of electrons between successive slices of thickness ε.

slice can be considered as a Huyghens source for a secondary spherical wave with amplitude $\psi(x, y)$ (Figure 2.23).

From each point of this slice, the electron can freely propagate to the next slice as a spherical wave. In the forward-scattering approximation (Fresnel approximation), the spherical wavefront is approximated by a paraboloidal wavefront (a complex gaussian function) so that this step apart from constant factors, can be written as a convolution product:

$$\psi(\mathbf{R}) = \exp\left[i\sigma V_p(\mathbf{R})\right] \cdot \exp\left[\frac{i\pi k(\mathbf{R}^2)}{\varepsilon}\right] \quad (2.\text{B}.8)$$

The propagation through the vacuum gap from one slice to the next is thus described by a convolution product in which each point source of the previous slice contributes to the wavefunction in each point of the next slice. The motion of an electron through the whole specimen can now be described by an alternating of phase object transmissions (multiplications) and vacuum propagations (convolutions). In the limit of the slice thickness ε tending to zero, this multislice expression converges to the exact solution of the nonrelativistic Schrödinger equation in the forward-scattering approximation. The multislice expression can also be deduced from a correct quantum-mechanical approach [23].

Appendix 2.C: Quantum Mechanical Approach

As is clear from Eq. (2.7), the calculation of the image wavefunction $\psi(\mathbf{R})$ requires the knowledge of $\psi(\mathbf{R})$, that is, the wavefunction at the exit face of the object [18]. This can be obtained by numerically solving the Schrödinger equation in the object. For convenience, we now follow a simplified more intuitive approach, which leads to the correct results.

If we assume that the fast electron, in the direction of propagation (z-axis) behaves as a classical particle with velocity $v = hk/m$ we can consider the z-axis as a time axis with

$$t = \frac{mz}{hk} \tag{2.C.9}$$

Hence we can start from the time-dependent Schrödinger equation

$$\psi(\mathbf{g}, \varepsilon) = \{\psi(\mathbf{g}, 0) \exp[i\sigma V_g]\} \exp\left[\frac{i\pi g^2 \varepsilon}{k}\right] \tag{2.C.10}$$

with

$$H = -\frac{\hbar^2}{2m}\Delta_\mathbf{R} - eU(\mathbf{R}, t) \tag{2.C.11}$$

with $U(\mathbf{R}, t)$ the electrostatic crystal potential, m and k the relativistic electron mass and wavelength, and $\Delta_\mathbf{R}$ the Laplacian operator acting in the plane $(\mathbf{R})$ perpendicular to z.

Using Eq. (2.C.9), we then have

$$\frac{\partial \psi(\mathbf{R}, z)}{\partial z} = \frac{i}{4\pi k}(\Delta_\mathbf{R} + V(\mathbf{R}, z))\psi(\mathbf{R}, z) \tag{2.C.12}$$

with

$$V(\mathbf{R}, z) = \frac{2me}{\hbar^2}U(\mathbf{R}, z) \tag{2.C.13}$$

This is the well-known high-energy equation in real space that can also be derived from the stationary Schrödinger equation in the forward-scattering approximation [23].

In high-resolution electron microscopy of crystalline objects, the object is usually oriented along a zone axis, so that the electrons are traveling parallel to the atom columns. If the periodicity along the column direction is not too large (less than 1–2 nm), the fast electron does not feel this variation. In fact it sees the potential as constant along z. In other words, the effect of higher order Laue zones or upper layer lines is negligible. This is the projection approximation, which is usually valid for most high-resolution conditions. Now Eq. (2.C.12) becomes

$$\frac{\partial \psi(\mathbf{R}, z)}{\partial z} = \frac{i}{4\pi k}(\Delta_\mathbf{R} + V(\mathbf{R}))\psi(\mathbf{R}, z) \tag{2.C.14}$$

with

$$V(\mathbf{R}) = \frac{2me}{\hbar}\frac{1}{z}\int_0^z U(\mathbf{R}, z)\,dz \tag{2.C.15}$$

the potential, averaged (projected) along z. In the time-dependent Schrödinger picture (Eq. (2.C.10)), the electron walks as a function of time in a two-dimensional potential of projected atom columns.

Equation (2.C.14) can also be transformed to reciprocal space.

Assuming V(**R**) to be periodic in two dimensions, we can expand it in Fourier series

$$V(\mathbf{R}) = \sum_{\mathbf{g}} V_{\mathbf{g}} \exp 2\pi i \mathbf{g} \cdot \mathbf{R} \tag{2.C.16}$$

with **g** in the zone plane. $V_{\mathbf{g}}$ are known as *structure factors*. Similarly, we have

$$\psi(\mathbf{R}) = \sum_{\mathbf{g}} \psi_{\mathbf{g}}(z) \exp 2\pi i \mathbf{g} \cdot \mathbf{R} \tag{2.C.17}$$

$\psi_{\mathbf{g}}(z)$ represents the amplitude of the beam **g** at a depth z. Substitution in Eq. (2.C.14) then yields

$$\frac{\delta \psi(\mathbf{R}, z)}{\delta z} = i\pi \left[2s_{\mathbf{g}} \psi_{\mathbf{g}}(z) + \sum_{\mathbf{g}'} V_{\mathbf{g}-\mathbf{g}'} \psi_{\mathbf{g}'}(z) \right] \tag{2.C.18}$$

with

$$s_{\mathbf{g}} = \frac{g^2}{2k} \tag{2.C.19}$$

the excitation error, which is approximately equal to the distance between the reciprocal node **g** and the Ewald sphere, measured along z. This system of coupled first-order differential equations has been derived in the early 1960s (for references, see [18]). Most of the image simulation programs are based on a numerical solution of the dynamical equation in real space 1, or reciprocal space 2, or a combination of both.

The dynamical Eq. (2.C.12) or (2.C.14) is a mixture of two equations, each representing a different physical process

$$\frac{\partial \psi(\mathbf{R}, z)}{\partial z} = \frac{i}{4\pi k} \Delta_{\mathbf{R}} \psi(\mathbf{R}, z) \tag{2.C.20}$$

is a complex diffusion-type of equation, which represent the free electron propagation and whose solution can be represented formally as

$$\psi(\mathbf{R}, z) = \exp\left(\frac{i\Delta_{\mathbf{R}} z}{4\pi k}\right) \psi(\mathbf{R}, 0) \tag{2.C.21}$$

It can also be written as a convolution product

$$\psi(\mathbf{R}, z) = \exp\frac{i\pi k R^2}{z} * \psi(\mathbf{R}, 0) \tag{2.C.22}$$

The other part of Eq. 2.C.14 is a differential equation

$$\frac{\partial \psi(\mathbf{R}, z)}{\partial z} = \frac{i}{4\pi k} V(\mathbf{R}, z) \psi(\mathbf{R}, z) \tag{2.C.23}$$

which represents the scattering of the electron by the crystal potential. It can be readily integrated in real space, yielding

$$\psi(\mathbf{R}, z) = \exp\left(\frac{i}{4\pi k} V(\mathbf{R}) z\right) \psi(\mathbf{R}, 0) \tag{2.C.24}$$

with $V(\mathbf{R})$ the projected potential as defined in Eq. (2.C.15).

The wavefunction in real space is multiplied by a phase factor that is proportional to the electrostatic potential of the object projected along z, called the *phase object function*. The solution of the complete dynamical Eq. (2.C.14) can be written formally as

$$\psi(\mathbf{R}, z) = \exp\left(\frac{i}{2\pi k}[\Delta_\mathbf{R} + V(\mathbf{R})] z\right) \psi(R, 0) \tag{2.C.25}$$

For the explicit calculation, slice methods are the most appropriate. Here the crystal is cut into thin slices with thickness ε perpendicular to the incident beam.

If the slice thickness is sufficiently small, the solution within one slice is approximated by

$$\psi(\mathbf{R}, z+\varepsilon) = \exp\left(\frac{i}{2\pi k}\Delta_\mathbf{R}\varepsilon\right) \exp\left(\frac{i}{4\pi k} V(\mathbf{R}) \varepsilon\right) \psi(R, z) \tag{2.C.26}$$

or explicitly

$$\psi(\mathbf{R}, z+\varepsilon) = \exp\frac{i\pi k R^2}{\varepsilon} \left\{ \exp\left(\frac{i}{4\pi k} V(\mathbf{R}) \varepsilon\right) \psi(\mathbf{R}, 0) \right\} \tag{2.C.27}$$

This expression is essentially the same as Eq. (2.C.10), which was derived from an optical approach. In practice, the wavefunction is sampled in a network of closely spaced points. In each point, the wavefunction is multiplied with the phase object function. Then the wavefunction is propagated to the next slice, and so on. Calling N the number of sampling points, the phase object requires a calculation time proportional to N^2. In reciprocal space, direct and convolution products are interchanged yielding

$$\psi_g(z+\varepsilon) = \exp\frac{i\pi g^2 \varepsilon}{k} \left[\Im_g \left(\exp\frac{i}{4\pi k} V(\mathbf{R}) \varepsilon \right) \psi_g(z) \right] \tag{2.C.28}$$

This expression is the same as Eq. (2.C.10). Now the calculation time of the propagation is proportional to N the number of beams, whereas the scattering in the phase object gives a calculation time proportional to N^2.

In order to speed up the calculation, the phase object is calculated in real space, and the propagator in reciprocal space [18]. Between each a fast Fourier transform is performed, the calculation time of which is only proportional to $N \log_2 N$. In the standard slice programs, the object is assumed to be a perfect crystal. Defects are treated by the periodic continuation method in which the defect is artificially repeated so as to create an artificial supercrystal.

In the real space method, proposed in [14c], the whole calculation is performed in real space but due to the forward scattering of the electrons, the propagation effect is limited to a local area so that the calculation time remains proportional to N. This is particularly interesting for treating extended or aperiodic structures.

References

1. Henderson, R. (2004) Realising the potential of cryo-electron microscopy. *Q. Rev. Biophys.*, **37**, 3–13.
2. Midgley, P. and Bals, S. (2012) Electron tomography, *Handbook of Nanoscopy*, Wiley-VCH Verlag GmbH, Weinheim.
3. Hauptman, H. (1986) The direct methods of X-ray crystallography. *Science*, **233** (4760), 178–183.
4. Zuo, J.M. and Huang, W. (2012) Coherent electron diffractive imaging, *Handbook of Nanoscopy*, Wiley-VCH Verlag GmbH, Weinheim.
5. Fu-Son, M., Hai-Fu, F., and Fang-Hua, L. (1986) Image processing in HREM using the direct method. II. Image deconvolution. *Acta Crystallogr.*, **A42**, 353–356.
6. Howells, M.R. *et al.* (2009) An assessment of the resolution limitation due to radiation-damage in X-ray diffraction microscopy. *J. Electron. Spectrosc. Relat. Phenom.*, **170**, 4–11.
7. Van Tendeloo, G., Pennycook, S., and Van Dyck, D. (2012) Introduction, *Handbook of Nanoscopy*, Wiley-VCH Verlag GmbH, Weinheim.
8. (a) Scherzer, O. (1949) The theoretical resolution limit of the electron microscope *J. Appl. Phys.*, **20**, 20; (b) Spence, J.C.H. (1988) *Experimental High Resolution Electron Microscopy*, Clarendon Press.
9. Van Dyck, D. (2011) Persistent misconceptions about incoherence in electron microscopy. *Ultramicroscopy*, **111**, 894–200.
10. Urban, K. *et al.* Ultra-high resolution transmission electron microscopy at negative spherical aberration, *Handbook of Nanoscopy*, Wiley-VCH Verlag GmbH, Weinheim.
11. (a) Gabor, D. (1948) A new microscopic principle. *Nature*, **161**, 777–778; (b) Gabor, D. (1949) Microscopy by reconstructed wavefronts. *Proc. R. Soc. A*, **197**, 454–487.
12. Cowley, J. and Iijima, S. (1972) Electron-microscope image contrast for thin crystals. *Z. Naturforsch., Part A: Astrophys. Phys. Phys. Chem.*, **A27**, 445.
13. Cowley, J.M. and Moodie, A.F. (1957) The scattering of electrons by atoms and crystals: a new theoretical approach. *Acta Crystallogr.*, **10**, 609–619.
14. (a) Stadelman, P.A. (1987) EMS - a software package for electron diffraction analysis and HREM image simulation in materials science. *Ultramicroscopy*, **21**, 131–146; (b) Kilaas, R. and Gronsky, R. (1982) Real space image simulation in high resolution electron microscopy. *Ultramicroscopy*, **11**, 289–298; (c) Van Dyck, D. and Coene, W. (1984) The real space method for dynamical electron diffraction calculations in high resolution electron microscopy. I. Principles of the method. *Ultramicroscopy*, **15**, 29–40; (d) Van Dyck, D. and Coene, W. (1984) The real space method for dynamical electron diffraction calculations in high resolution electron microscopy. II. Critical analysis of the dependency on the input paramters. *Ultramicroscopy*, **15**, 41–50; (e) Van Dyck, D. and Coene, W. (1984) The real space method for dynamical electron diffraction calculations in high resolution electron microscopy. III. A computational algorithm for the electron propagation with its practical applications. *Ultramicroscopy*, **15**, 287–300.
15. (a) Schiske, P. (1973) Image processing using additional statistical information about the object, in *Image Processing of Computer-aided Design in Electron Optics* (ed. P.W. Hawkes); Academic Press, p. London, 82–90; (b) Saxton W.O. (1986) Focal series restoration in HREM, in Proceedings of the XIth International Congress on Electron Microscopy, Kyoto.
16. (a) Van Dyck, D. and Op de Beeck, M. (1990) New direct methods for phase and structure retrieval in HREM. *Proceedings XIIth International Congress for Electron Microscopy (Seattle)*, San Francisco Press Inc., pp. 26–27; (b) Coene, W., Janssen, G., Op de Beeck, M., and Van Dyck, D. (1992) Phase retrieval through focus variation for ultra resolution in field

emission trans-mission electron microscopy. *Phys. Rev. Lett.*, **69**, 3743–3746; (c) Op de Beeck, M., Van Dyck, D., and Coene, W. (1995) Focal series wave function reconstruction in HRTEM, in *Electron Holography* (eds. A. Tonomura et al.), North Holland-Elsevier, pp. 307–316, ISBN: 0-444-82051-5.

17. (a) Fujimoto, F. (1978) Periodicity of crystal structure images in electron microscopy with crystal thickness. *Phys. Stat. Sol.*, **A45**, 99–106; (b) Kambe, K., Lehmpfuhl, G., and Fujimoto, F. (1974) Interpretation of electron channeling by the dynamical theory of electron diffraction. *Z. Naturforsch.*, **29**, 1034–1044.

18. Van Dyck, D. (1985) in *Advances in Electronics and Electron Physics*, vol. 65 (ed. P. Hawkes), Academic Press, p. 295.

19. (a) Van Dyck, D. and Op de Beeck, M. (1996) A simple intuitive theory for electron diffraction. *Ultramicroscopy*, **64**, 99–107; (b) Geuens, P. and Van Dyck, D. (2002) The s-state model: a weak base for HRTEM. *Ultramicroscopy*, **93**, 179–198.

20. (a) Sinkler, W. and Marks, L.D. (1999) Dynamical direct methods for everyone. *Ultramicroscopy*, 75,251–268; (b) Wang, A., Chen, F.R., Van Dyck, D., and Van Aert, S. (2010) Direct structure inversion from exit wave. *Ultramicroscopy*, **110**, 527–534.

21. Van Aert, S. Statistical parameter estimation theory – a tool for quantitative electron microscopy, *Handbook of Nanoscopy*, (2012) Wiley-VCH Verlag GmbH, Weinheim.

22. Jia, C.L. and Thust, A. (1999) Investigation of atomic displacements at a $\Sigma 3\{111\}$ twin boundary in $BaTiO_3$ by means of phase-retrieval electron microscopy. *Phys. Rev. Lett.*, **82**, 5052–5055.

23. Van Dyck, D. (1979) Improved methods for high-speed calculations of electron microscopic structure images. *Phys. Status Solidi A*, **52**, 283–292.

24. Ishizuka, K. and Uyeda, N. (1977) A new theoretical and practical approach to the multislice method. *Acta Crystallogr.*, **A33**, 740–749.

3
Ultrahigh-Resolution Transmission Electron Microscopy at Negative Spherical Aberration

Knut W. Urban, Juri Barthel, Lothar Houben, Chun-Lin Jia, Markus Lentzen, Andreas Thust, and Karsten Tillmann

3.1
Introduction

With the realization of the Rose-corrector, the old dream of electron optics to be able to construct spherical-aberration-corrected lens systems has come true [1–3]. Aberration-corrected transmission electron microscopy offers a great variety of new or expanded research opportunities for physics, chemistry, and materials science. Of these, a prominent field is that of investigations in atomic dimensions. With certain exceptions, most of the earlier high-resolution work was carried out in conventional noncorrected instruments, in spite of the fact that the images looked like atomic, was in reality not truly atomically resolving. Owing to aberration-induced contrast delocalization, a given atomic column does also contribute intensity to the positions of neighboring columns in the image. As a consequence, atomic concentrations and lateral shifts on the scale of an individual atomic column could not be measured. In most cases the images merely showed the crystallographic *structure* of a sample; this means a collective *nonlocal* property. Today things have changed; genuine atomic resolution has become available, allowing measurement of individual atomic properties. However, in order to exploit the new possibilities, investigations have to become more quantitative, much beyond the level that appeared sufficient in pre-aberration correction times. But it is worth making the effort: aberration-corrected transmission electron microscopy allows us to measure atomic positions and lateral shifts with close to picometer precision. This changes the electron microscope from a predominantly structure-oriented tool into a physical measurement instrument.

A prominent part in this progress from conventional high-resolution to ultrahigh-resolution work is played by the improvement of both image contrast

---

Handbook of Nanoscopy, First Edition. Edited by Gustaaf Van Tendeloo, Dirk Van Dyck, and Stephen J. Pennycook.
© 2012 Wiley-VCH Verlag GmbH & Co. KGaA. Published 2012 by Wiley-VCH Verlag GmbH & Co. KGaA.

as well as signal to noise ratio achievable by employing aberration-corrected optics. The common reference for high-resolution microscopy is Scherzer's phase-contrast theory [4]. Formulated at a time when, at a fixed value of the objective-lens spherical aberration parameter, the only variable to optimize contrast was the objective-lens focal length, this theory provides a value for the lens defocus, yielding optimal contrast up to a certain optical resolution (Scherzer resolution). Unfortunately, the Scherzer focus setting leads to a rather high value of contrast delocalization, higher than the value of the Scherzer resolution itself [5, 6]. This dramatically limits the practical value of Scherzer's conditions. Take a crystal, the lattice parameter of which is just large enough to image, for a given Scherzer resolution, the atom columns separately. In this case, the actual information supplied by the image for a given atom column is rather limited and unspecific. The reason is that the intensity at the position of this column is the sum of its own intensity plus the intensity of the delocalized images of the neighboring columns at this position. Although this may be tolerable for work in which one is interested only in the crystallographic structure, it is totally inadequate for genuine atomic resolution. In microscopes equipped with aberration-correctors, in addition to defocus aberration, the other symmetric aberrations, the spherical aberration of third and fifth order, can be treated as variables suited to optimize contrast not only for optimal resolution but also for minimal contrast delocalization [6, 7]. This is one of the great advantages of aberration-corrected transmission electron microscopy.

Scherzer's theory [4] and also Lentzen's contrast theory for aberration-corrected instruments [7] are linear theories assuming ideally weak objects. However, in materials science most objects are strong scatterers. The possibility to adjust the spherical aberration including the sign allows tackling the problem that the nonlinear terms in treatments for stronger objects make a contribution of opposite sign thus in effect weakening the contrast. This can be considered a second substantial advantage of transmission electron microscopes with aberration-correcting optics and provides the background for the negative spherical-aberration imaging (NCSI) technique discovered by Jia in 2001. NCSI offers substantially sharper contrast maxima, and it proved to be particularly advantageous for imaging low-nuclear-charge light atom species close to heavy atoms [8].

This section is organized as follows. In Section 3.2, the principles of atomic imaging, including NCSI, is treated first. An extra section, Section 3.3, is dedicated to the numerical inversion of the imaging process as a prerequisite for ultrahigh-resolution work. This is followed, in Section 3.4, by a theoretical case study of $SrTiO_3$, in which the key features of NCSI are compared to imaging with positive spherical aberration. In Section 3.5, following a guide to the literature, measurement of the atom positions in ferroelectric 180° inversion domain walls in lead zirconate titanate (PZT) will be treated as an example of the ultrahigh-resolution application of NCSI. The research opportunities that opened up for materials science by atomic-resolution studies employing the NCSI technique have been reviewed by Urban [9, 10] and by Urban et al. [11, 12].

3.2
The Principles of Atomic-Resolution Imaging

3.2.1
Resolution and Point Spread

The goal of atomic-resolution work is to measure the set of individual atom positions in a sample $X = \{r_1, r_2, r_3, \ldots\}$. We are treating here the case of a crystalline sample. We employ an electron wave field represented for simplicity by a single plane wave

$$\psi_0(\mathbf{r}) = \exp(2\pi i \mathbf{k}_0 \cdot \mathbf{r}) \tag{3.1}$$

incident on the upper sample surface. Here, $\mathbf{k}_0$ denotes the wave vector whose modulus is the inverse of the electron wave length λ; $\mathbf{r}$ denotes the general position vector. On its way through the specimen, this wave field interacts with the atomic potential $V(\mathbf{r})$. At electron energies of typically 200–300 keV, the electron wave function $\psi(\mathbf{r})$ in the crystal is given as a solution of the Dirac equation subject to the boundary conditions at the surface. In a small-angle scattering approximation, spin polarization can be neglected and the equation adopts a Schrödiger-type form with relativistically corrected mass and wavelength [13–15].

The wave function at the lower plane of the specimen, the exit-plane wave function ψ_e, contains all the information on the specimen that the electrons can supply us with. In the general case, this information is not directly accessible but encoded in a complicated way determined by the quantum mechanical interaction. In the potential-free space, the exit-plane wave field can be described as a superposition of plane waves. Written as a Fourier integral we thus obtain

$$\psi(\mathbf{r}) = \int_g \psi(\mathbf{g}) \exp(2\pi i \mathbf{g} \cdot \mathbf{r}) \, d\mathbf{g} \tag{3.2}$$

with the modulus of the reciprocal vector $\mathbf{g}$ representing the spatial frequency. This wave field is then the object of the objective lens of the microscope. The intensity distribution in the image plane is given by an equivalent to the Poynting vector, which, again in a small-angle approximation, is proportional to the electron probability density, that is, the absolute square of the wave function,

$$I(\mathbf{r}) \propto \frac{ih}{4\pi m}(\psi \nabla \psi^* - \psi^* \nabla \psi) \propto \psi^* \psi \tag{3.3}$$

where h is Planck's constant and m is the relativistically corrected electron mass.

If the lens has optical aberrations, these have the effect that the individual components $\psi(\mathbf{g})$ of the exit-plane wave function are multiplied by the phase factor

$$\exp(-2\pi i \chi(\mathbf{g})) \tag{3.4}$$

where

$$\chi(\mathbf{g}) = \tfrac{1}{2} Z \lambda g^2 + \tfrac{1}{4} C_S \lambda^3 g^4 + \ldots \tag{3.5}$$

is the wave-aberration function (expressions comprising up the fifth-order aberration terms can be found in Refs. [3, 16, 17]). The first term is the lens defocus aberration, with Z denoting the defocus parameter [13].[1] The second term is due to third-order spherical aberration, where C_S is the spherical aberration parameter.

The presence of aberrations has the consequence that a point in the object is not imaged into a sharp corresponding point in the image plane but rather into an error or point-spread disk whose radius is given by the point spread function, which also characterizes contrast delocalization

$$R = \max \left| \frac{\partial \chi}{\partial g} \right| = \max \left| Z\lambda g + C_S \lambda^3 g^3 + \ldots \right|, g \text{ in } [0;\, g_{\max}] \tag{3.6}$$

where the maximum has to be taken over the whole range of spatial frequencies, up to $g_{\max}$, contributing to the image [13, 6, 18]. In the fully aberration-corrected case, point spread becomes zero. Any finite values of Z and C_S (e.g., applied when taking focal series or for optimizing contrast) will lead to finite values of R. However, as shown below, the values of R can be kept sufficiently small, not to affect most practical high-resolution work.

3.2.2
Contrast

Because numerical solutions of the quantum mechanical scattering problem can now be readily obtained on a computer and comfortable software packages are available (e.g., [19, 20]), contrast should be discussed on the basis of a state-of-the art numerical treatment. That in the following we nevertheless start with the classical approach to treat the simplified cases of weak phase and weak amplitude objects serves a schematic understanding of contrast formation. On the other hand, such a discussion also allows us to point out the severe limitations of such treatments.

We start the discussion with phase contrast. This means that access to the specimen structure is obtained by exploiting the information contained in the locally varying phase of the exit-plane wave field. As in light microscopy, under Zernike phase-contrast conditions [21, 22], the problem arises that the atomic phase contrast has to be converted into amplitude contrast. While in light microscopy the phase shifts are small, the phase shifts in electron microscopy, depending on atomic number and specimen thickness, can be quite large. Nevertheless, for the sake of illustration we apply for the moment the weak phase object approximation (WPO).

---

1) We follow here the usual sign convention based on a consideration of the phase at a fixed position of the image plane. If the excitation of the objective lens is weakened compared to the ideal value appertaining to the Gaussian focus f_0, the lens is called *underfocused* and the focal length increases to become $f = f_0 + \Delta f$. If the specimen to lens distance is left unchanged, the phase in the image plane is less advanced compared to the Gaussian case, and the defocus parameter Z in Eq. (3.5) adopts a negative value. In this case, a (hypothetical) focused image would be formed behind the fixed image plane. If the excitation of the lens is increased, the lens is said to be *overfocused*, its focal length becomes shorter (Δf is negative), and the focused image would occur in front of the fixed image plane. As a consequence, the phase in the image plane is advanced and Z adopts a positive value.

The conditions can be illustrated schematically in the Gaussian complex number plane. The incident wave is characterized by a vector along the real axis. The diffracted wave is represented by a short vector along the imaginary axis, taking account the fact that the basic physical phase shift of a diffracted wave is $\pi/2$ with respect to the incident wave. The sum of both is a vector rotated in mathematically positive direction by a small angle but of essentially the same amplitude as that of the incident wave. In light microscopy, an additional $\pi/2$ phase shift is imposed on the scattered wave by the so-called $\lambda/4$ plate. Now, summing up yields a shorter resulting vector and a corresponding intensity reduction. As a result, the scattering regions give dark contrast on a bright background. This is called *positive phase contrast* [14].

Phase contrast is governed by the phase-contrast transfer function (PCTF) $\sin 2\pi \chi(\mathbf{g})$, and the spatial-frequency-dependent phase-shifting properties of the lens aberration function can be exploited to produce an equivalent to the Zernike phase plate. Scherzer's optimized PCTF is displayed in Figure 3.1a for an uncorrected instrument. The corresponding PCTF for an aberration-corrected instrument (Figure 3.1b), still maintaining the WPO approximation, was derived by Lentzen et al. [6], taking advantage of the fact that now the defocus and the third-order spherical aberration are available for optimization of χ. How this optimization is done depends on the preferences. First of all, we want to eliminate the contrast oscillations between g_S and g_I. Here, g_S is the spatial frequency corresponding to Scherzer's point resolution, and g_I denotes the spatial frequency marking the information limit determined by partial temporal coherence of the illumination system. This in turn is limited by the energy spread of the electron source and by fluctuations of the electron energy and the objective lens current [23, 24]. The second goal is to reduce point spread to a fraction of g_I^{-1}. Finally, we have to take care that the low-contrast (poor-transfer) region at small spatial frequencies is

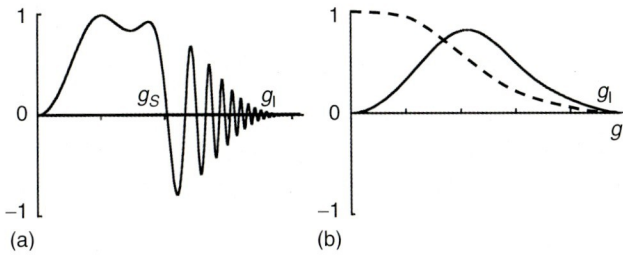

Figure 3.1 The value of the phase contrast transfer function (PCTF) as function of spatial frequency g. (a) Plot of $-\sin 2\pi \chi(\mathbf{g})$ under Scherzer conditions for an uncorrected microscope. $\chi(\mathbf{g})$ is the wave aberration function (Eq. (3.5)). g_S marks the Scherzer point resolution, g_I is the information limit defined by the spatial frequency where contrast damping due to partial temporal coherence drops to a value of $1/e^2$. (b) Plot of $\sin 2\pi \chi(\mathbf{g})$ (solid line) for NCSI conditions in an aberration-corrected instrument (see text). The region of optimized contrast expands up to the information limit. Note the bell shape indicating that transfer is far from ideal. The amplitude contrast transfer function $\cos 2\pi \chi(\mathbf{g})$ is also given (broken line).

Table 3.1 Typical optimum values for the spherical aberration parameter $C_{S,opt}$, the defocus parameter Z_{opt}, and the radius of the point spread function R_{opt}, calculated by employing Eqs (3.7) to (3.9) for the electron energy E and the information limit g_I.

E (keV)	g_I^{-1} (nm)	$C_{S,opt}$ (µm)	Z_{opt} (nm)	R_{opt} (nm)
200	0.12	31.0	−10	0.07
300	0.07	7.5	−4.5	0.04
300	0.05	1.9	−2.3	0.03

kept as narrow as possible. There is a trade-off in the sense that increasing g_I is widening the low-spatial-frequency gap. The optimal settings for phase contrast in an aberration-corrected instrument are then given by

$$C_{S,opt} = +\frac{64}{27}\left(\lambda^3 g_I^4\right)^{-1} \tag{3.7}$$

$$Z_{opt} = -\frac{16}{9}\left(\lambda g_I^2\right)^{-1} \tag{3.8}$$

$$R_{opt} = \frac{16}{27} g_I^{-1} \tag{3.9}$$

Values for $C_{S,opt}$, Z_{opt}, and R_{opt} are given in Table 3.1. With typical values of 0.5–1.2 mm for C_S in an uncorrected instrument, it is evident that the residual values required for optimal contrast are only a small percentage of the original C_S values. A corresponding treatment including variable fifth-order spherical aberration is given by Lentzen [25].

We point out that the PCTF for the aberration-corrected case has a shape that is far from ideal. Admittedly, the transfer function does not show any zeros or oscillations and thus fulfills one of the central goals of constructing aberration-corrected optics. However, the transfer is very sensitively dependent on spatial frequency. Recalling that microscopy is essentially a two-Fourier transform process, we have to keep in mind that the microscope acts as a highly nonlinear filter, with the consequence that the images are formed on the basis of a wave field severely modified by inadequate contrast transfer. A number of contrast artifacts have been reported in the literature, which can be traced back to the particular shape of the PCTF [7, 26]. In another case, it was found that the width of the low-spatial-frequency gap that has to be tolerated in order to maximize g_I is too high to allow imaging of effects of chemical-bonding-induced charge redistributions in nitrogen-doped graphene [27]. All these cases indicate that high-resolution imaging is incomplete without taking focal series (whereby the region of low-contrast transfer at low spatial frequencies can be reduced) and performing a numerical backward calculation from images to structure.

3.2.3
Enhanced Contrast under Negative Spherical Aberration Conditions

So far the treatment of Lentzen *et al.* [6] is still conservative with respect to the direction of the Zernike phase shift adjusted to induce contrast. This means that under the conditions of Eqs. (3.7) and (3.8) positive phase contrast is obtained by combining a positive value of C_S with an objective lens underfocus (negative value of Z). In the following, we shall demonstrate that much improved contrast can be obtained for a wide range of imaging conditions inverting the sign of the terms in Eqs. (3.7) and (3.8.) This yields atom positions appearing bright on a dark background. According to the above definition, this is "negative phase contrast." It is brought about by an overcompensation of the original value of the spherical aberration of the objective lens, that is, the contrast is due to imaging under negative spherical-aberration conditions.

This will be illustrated via image simulations for the case of strontium titanate ($SrTiO_3$). The material has a cubic perovskite structure with a lattice parameter of 0.3905 nm. Figure 3.2a shows a perspective view of the unit cell. Figure 3.2b displays the projection along the [110] crystal direction. In this viewing direction, three types of atomic columns are distinct, which are occupied alternatively with strontium and oxygen, with titanium, and with oxygen atoms. Figure 3.3 shows an experimental image taken along the [110] direction. All atomic positions are visible, including oxygen. Before, oxygen could never be seen directly in electron microscopic images. It was only accessible via the exit-plane wave function reconstruction technique [28–30]. Figure 3.4 shows simulated images for C_S values of 0 (a), +40 μm (b), and −40 μm (c) for different defocus values (horizontal axis) and sample thicknesses (vertical axis) at an electron energy of 200 keV [20].[2)] Figure 3.4d shows a comparison

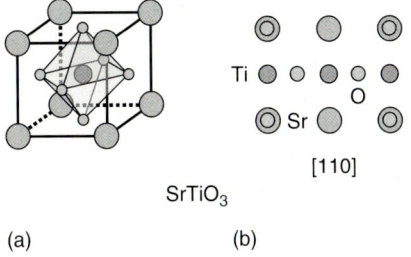

(a)　　　　　　　　(b)

Figure 3.2 $SrTiO_3$. (a) Perspective view of the unit cell. (b) Projection along the [110] crystal direction. In this viewing direction, three types of atomic columns are distinct, which are occupied alternatively with strontium and oxygen, with titanium, and with oxygen atoms.

2) All images were calculated with the electron beam parallel to the [110] zone axis using the MacTempas software package.

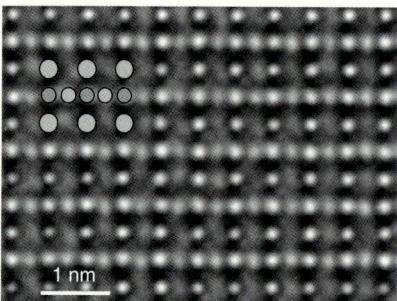

Figure 3.3 Experimental image of SrTiO$_3$ taken along the [110] zone axis employing the NCSI technique (FEI Titan 80–300 with imaging corrector, operated at 300 keV). All three atomic species are visible (compare inset) at bright contrast on a dark background.

of positive C_S combined with underfocus and negative C_S combined with overfocus. Clearly, oxygen cannot be imaged in positive phase contrast, meaning that this setup does not supply us with full atomic details in spite of the fact that the optical resolution of about 0.12 nm is sufficient to resolve the oxygen–titanium atom separation of 0.138 nm. The enhanced contrast under NCSI conditions is obvious.

Within the framework of the WPO, approximation inverting the sign of $C_{S,\text{opt}}$ and Z_{opt} changes the contrast from positive to negative phase contrast but does not yield any contrast enhancement. This can be concluded when comparing Figure 3.4b,c for very small specimen thicknesses for which the WPO approximation can be considered adequate. In order to arrive at an understanding of the contrast enhancement, two assumptions of the linear contrast theory have to be abandoned. The first is the WPO approximation. In fact, calculating the phase angles and amplitudes for different reflections for SrTiO$_3$, we find that for realistic sample thicknesses this approximation is inadequate and only a fully dynamical treatment of the electron scattering and imaging problem can provide an adequate description. Furthermore, we have to discuss *amplitude contrast*, which so far was entirely neglected.

In order to estimate the change in phase contrast intensity predicted by an expansion of the WPO treatment to the case of a "not so weak" object, we write for the wave function in the object plane

$$\psi_{\text{obj}}(\mathbf{r}) = \psi_0 + \pi i \lambda U(\mathbf{r}) t \qquad (3.10)$$

where $U(\mathbf{r})$ denotes the projected crystal potential and t the specimen thickness. This is altered by the application of the phase plate to the exit wave. Thus we obtain in the image plane,

$$\psi_{\text{im}}(\mathbf{r}) = \psi_0 \mp \pi \lambda U(\mathbf{r}) t \qquad (3.11)$$

employing a coefficient of $+i$ for positive phase contrast and $-i$ for negative phase contrast. For the "not so weak" object we maintain the intensity calculation terms

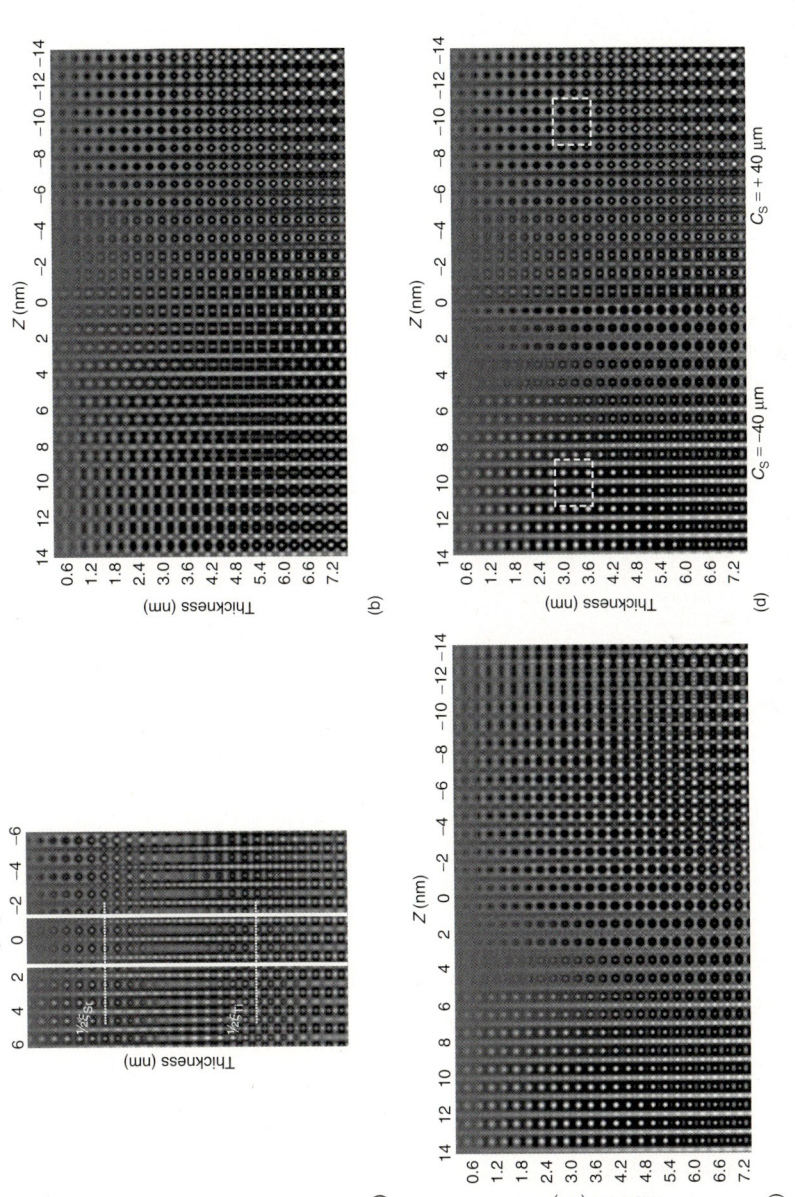

Figure 3.4 Simulated images for SrTiO$_3$ [110] at 200 keV. The composite shows a single unit cell in [110] projection for different defocus values Z and sample thicknesses at an electron energy of 200 keV for the spherical aberration parameter C_S = 0 nm, +40 μm, and −40 μm (a–c). (d) Direct comparison of the positive and negative C_S situation, where in the −C_S case overfocus (positive values of Z) and in the positive C_S case the usual underfocus (negative values of Z) is applied. The frame is used as a guide for the eye for same sample thickness.

up to second order, that is,

$$I(\mathbf{r}) = \psi_0^2 \mp 2\pi \psi_0 \lambda U(\mathbf{r}) t + (\pi \lambda U(\mathbf{r}) t)^2 \qquad (3.12)$$

A common phase of ψ_0 and $\psi_{sc}(\mathbf{r})$ has been chosen to set ψ_0 to a real value. The comparison of these two cases shows that the linear contribution and the quadratic contribution have different signs for positive phase contrast. The local intensity modulation at an atom column site becomes weaker on increasing the strength of the object. On the other hand, the linear contribution and the quadratic contribution have the same sign for negative phase contrast, and the local intensity modulation at an atom column site becomes stronger. In other words, setting up a negative value of spherical aberration combined with an overfocus enhances the atomic phase contrast compared to a setting with positive spherical aberration and underfocus.

Now to amplitude contrast: if the exit-plane wave function has a locally varying amplitude structure, consequentially amplitude contrast occurs. In the classic contrast treatments (e.g., [13]), only amplitude changes due to electron absorption are considered. This is not what we are dealing with. As we shall see, amplitude structure occurs for crystalline specimens as a result of electron diffraction channeling, that is, because of the elastic quantum mechanics inside the specimen. In the linear theory, amplitude contrast is proportional to $\cos 2\pi \chi(\mathbf{g})$, the amplitude contrast transfer function (ACTF). Therefore, in the case of $\chi(\mathbf{g}) = 0$, that is, full compensation of spherical aberration and zero defocus, no phase contrast but optimal amplitude contrast occurs. The ACTF for NCSI conditions is displayed schematically in Figure 3.1b.

For more insight into amplitude contrast in crystalline specimens, we neglect phase contrast for the moment, that is, we carry out a contrast calculation for $C_S = 0$ and $Z = 0$. For defect-free ideal crystals, we can perform a Bloch wave calculation. This approach is taken in the classic treatment of electron diffraction channeling [31–33]. Electron diffraction channeling describes phenomenologically an oscillatory motion of the electrons as schematically depicted in Figure 3.5 for three different sample thicknesses. This can be understood as follows. At the specimen entrance surface, the electrons are spatially uniformly distributed. While the wave field penetrates into the specimen, the positively charged atom strings interact with the electrons, attractively concentrating the electrons after some distance from the entrance surface on the atom positions. Subsequently the electrons fan out again yielding an electron current density distribution after a certain distance, which is similar to that at the surface.

In the Bloch wave formalism, it is straightforward to calculate the extinction distance

$$\xi = \left(k^{(i)} - k^{(j)}\right)^{-1} \qquad (3.13)$$

where $k^{(i)}$ and $k^{(j)}$ denote the eigenvalues of the two most excited Bloch states [13]. Solving the corresponding eigenvalue problem for 200 keV electrons, we find $\xi_{SrO} \approx 14$ nm for a SrO atom column and $\xi_{Ti} \approx 38$ nm for a Ti column. Comparing this result with the atom column intensities displayed in Figure 3.4a, we find that

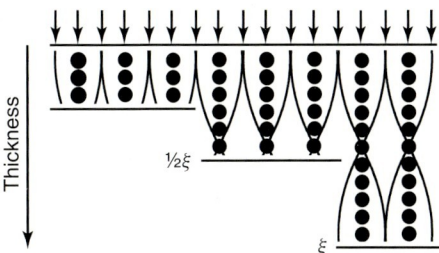

Figure 3.5 Schematic illustrating the effect of electron diffraction channeling on the local electron current density distribution for three different sample thicknesses. At the sample entrance surface, the current density distribution is assumed to be laterally uniform. The charged atoms (the atomic potential) "focus" the electrons toward the atom strings. This leads to enhanced current density at the atom positions with a maximum at a depth of odd multiples of $\xi/2$, where ξ is the extinction distance, at the expense of the current density in between the atoms. This induces an "amplitude structure" of the sample. At large sample thicknesses the current density at the atom positions decreases again, and ideally the current density becomes uniform again at depth ξ. Note that the extinction distance depends on the atomic species forming the atom string.

the atom positions appear brightest at depths of odd multiples of half the extinction distance, that is, at about 7 nm for Sr and at about 19 nm for Ti. For oxygen, the extinction distance is so large that for the moderate sample thicknesses used in high-resolution work, the intensity of oxygen atom spots in the images is always in the increasing-intensity range. This means that at $C_S = 0$ and $Z = 0$, the contrast is essentially determined by amplitude contrast induced by electron diffraction channeling.

Now it becomes evident that the strong contrast under NCSI conditions is due to additive contributions of both *amplitude* and *phase* contrast. On the one hand, because of amplitude contrast the intensity increases, at moderate specimen thicknesses, at the atom positions. The negative phase contrast, leading to bright-atom contrast at these positions too, further enhances the intensity there. On the other hand, the diffraction channeling effect reduces the electron density in between the atom positions. In this way the contrast is further enhanced.

3.2.4
NCSI Imaging for Higher Sample Thicknesses

In the previous section, we have demonstrated that the enhancement of contrast gained by adjusting the spherical-aberration parameter to a negative value and combining it with an overfocus arises from an additive contribution of bright phase contrast and bright amplitude contrast. However, the sign of phase contrast depends on sample thickness. Assuming for simplicity a case of a primary incident beam and a single scattered beam, the phase of the scattered beam at small sample thickness starts with $\pi/2$. But as the wave field advances, the specimen's scattering potential induces an additional depth-dependent phase shift. For a phase of π of the scattered beam the maximum positive phase contrast is achieved (without the use of a Zernike phase plate). For a phase $>3\pi/2$, the contrast of the scattering centers changes from dark to bright since at this angle the vectors of the primary beam and the scattered beam are additive. The increase of the phase change per unit depth is proportional to the scattering potential. As a consequence, in order to always obtain optimal positive or negative phase contrast, a sample-thickness- and nuclear-charge-dependent pair of values for the spherical-aberration parameter and the defocusing has to be adjusted. We point out that this is not a particular property of NCSI. As soon as the advancement in the phase of the scattered beam is taken into account, optimal contrast always requires optimization of the aberration function, which means that constructing the equivalent of a Zernike phase-shift plate, advancing the phase in such a way that a total phase shift of $(2m-1)\pi/2$ is achieved with respect to the primary beam, where m is an odd integer for positive phase contrast and an even integer for negative phase contrast. The optimal conditions for bright-atom contrast have been studied by Lentzen [7] in the framework of a simple independent-atom string model for describing the diffraction channeling effect [34]. In the following, we present a brief sketch of this treatment.

The electron wave $\psi_e(\mathbf{r})$ at the exit plane of the specimen is written as the sum of a direct, unscattered wave ψ_0 and a scattered wave $\psi_s = \psi_e(\mathbf{r}) - \psi_0$. Properly adjusting the aberration function of the objective lens, a constant phase $-2\pi\chi_0$ is added to the phase of the scattered wave, which is equivalent to the action of the Zernike phase plate. Then the wave function in the image plane $\psi_i(\mathbf{r})$ can be written as

$$\psi_i(\mathbf{r}) = \psi_0 + (\psi_e(\mathbf{r}) - \psi_0)\exp(-2\pi i\chi_0) \tag{3.14}$$

Optimal contrast is obtained for the following settings for the defocus

$$Z = \frac{8\chi_0}{\lambda g_I^2} \tag{3.15}$$

and the spherical aberration parameter

$$C_S = -\frac{40\chi_0}{3\lambda^3 g_I^4} \tag{3.16}$$

In the framework of the independent-atom string model, the expressions for the direct and the scattered waves, respectively, are

$$\psi_0 = \cos\tau + i\left(\frac{U_0}{k}\xi - 1\right)\sin\tau \tag{3.17}$$

$$\psi_s = i\frac{(U(\mathbf{r}) - U_0)\xi}{k}\sin\tau \tag{3.18}$$

$\tau = \pi t/\xi$, and the modulus of the wave vector $k = 1/\lambda$. $U(\mathbf{r})$ is related by the scattering potential $V(\mathbf{r})$ via $U(\mathbf{r}) = 2me/h^2 V(\mathbf{r})$, with U_0 and e denoting the mean inner potential and the elementary charge, respectively. Defining the characteristic parameter

$$c = \frac{U_0\xi}{k} - 1 \tag{3.19}$$

the image intensity distribution for coherent illumination becomes

$$I(\mathbf{r}) = |\psi_i|^2 = \cos^2 \tau + c^2 \sin^2 \tau - 2\frac{(U(\mathbf{r}) - U_0)\xi}{k} \sin \tau$$
$$\times \operatorname{Im}\left((\cos \tau - ic \sin \tau) \exp(-2\pi i \chi_0)\right)$$
$$+ \left(\frac{U(\mathbf{r}) - U_0}{k}\right)^2 \sin^2 \tau \qquad (3.20)$$

The first term is the intensity of the direct wave providing uniform background intensity. The second term describes the linear interference between the direct and the scattered wave. The last term describes the intensity of the scattered wave. The maximum contrast at the atom positions occurs if the modulus of the expression in the second line adopts a maximum value and if its sign is opposite to that of $\sin \tau$. For the aberration interval of $-\frac{1}{2} < \chi_0 < \frac{1}{2}$ both conditions are met for

$$\tan 2\pi \chi_0 = (c \tan \tau)^{-1} \qquad (3.21)$$

and

$$\tan \tau \sin 2\pi \chi_0 > 0 \qquad (3.22)$$

For a given material, the optimal contrast condition is then obtained employing the proper values for U_0 and ξ by calculating the value of the characteristic parameter c and solving Eq. (3.21) for χ_0 subject to fulfilling the condition Eq. (3.22). Both electron diffraction channeling and the image intensity have a thickness period equal to ξ. The phase of the scattered wave with respect to the direct wave starts with $\pi/2$ (corresponding to $\lambda/4$). It remains positive in the first half of the extinction period and turns negative in the second half, approaching $-\pi/2$ at ξ. For $c < 0$, the phase increases in the first half of the extinction period from $\pi/2$ to π and in the second half from $-\pi$ to $-\pi/2$. For $c > 0$, the phase decreases in the first half of the extinction period from $\pi/2$ to 0 and in the second half from 0 to $-\pi/2$. The conditions Eqs. (3.21) and (3.22) determine how the equivalent of the Zernike phase plate compensates the phase of the scattered wave favorably (Figure 3.6). For the first half of the extinction period χ_0 is positive; for the second half it is negative. Hence the approximation of the Zernike phase plate by defocus and spherical aberration, Eqs. (3.14) and (3.16), yields an overfocus combined with a negative value for C_S for the first half of the extinction distance and an underfocus combined with a positive C_S for the second half. Approaching a specimen thickness just smaller than half the extinction distance, the favorable compensation is achieved for $c < 0$ with a defocus

$$Z = \left(2\lambda g_I^2\right)^{-1} \qquad (3.23)$$

and a value of the spherical aberration of

$$C_S = -20 \left(3\lambda^3 g_I^4\right)^{-1} \qquad (3.24)$$

For a specimen thickness just larger than half the extinction distance, the favorable compensation is achieved with the same settings but with the respective signs

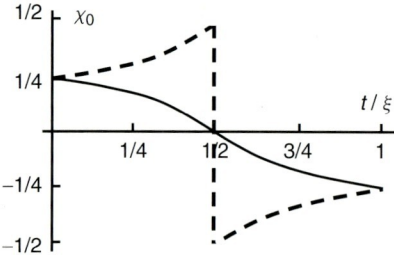

Figure 3.6 Optimal value of the aberration function χ_0 (see text) compensating the phase of the scattered wave in order to obtain maximum bright phase contrast for different sample thicknesses t in units of the extinction distance ξ. Adapted from Ref. [7].

inversed. For $c > 0$, the favorable compensation at half the extinction distance is achieved with $Z = 0$ and $C_S = 0$.

We note that the theoretical treatments as well as the experiments show that the contrast conditions are rather robust with respect to a variation of defocus and specimen thickness. We further note that the adjusted value of C_S also can vary over a wide range. Even a deviation of 30% of the calculated optimal value does not destroy the high-resolution contrast, although it certainly changes the absolute contrast behavior. The robustness of the bright-atom contrast conditions with respect to variations of C_S, Z, and t can be explained in the two-level channeling model described above by the relatively weak phase variation of the scattered wave and by the tolerance limits for the respective aberration settings [3, 35]. Although the absolute values of the imaging parameters are important for quantitative contrast evaluation, the NCSI mode *per se* does not make high demands with respect to the adjustment of particular values of these parameters.

3.3
Inversion of the Imaging Process

The images obtained in the transmission electron microscope with or without aberration-correcting optics do not generally supply us with the wanted information on the atom positions. In order to exploit the full potential of aberration-corrected atomic-resolution electron microscopy, the numerical inversion of this imaging process is mandatory. This has been done before and has also been the basis of high-resolution work in the pre-aberration correcting era. However, it has to be pointed out that the exactness of the involved numerical calculations has to be much higher than that considered satisfactory at times when the main interest was in crystallographic structure rather than in ultrahigh-precision individual atomic position measurements.

The inversion of the imaging process is carried out in two steps [28, 36–38]. The first is the reconstruction of the electron exit-plane wave function. In its generally applied form, this is done by employing the focus-variation technique: employing a charge-coupled device (CCD) camera, a series of (typically about 20) images is taken by varying the objective lens focus about a central value in steps, which, for example, in the work described in the following are based on an equidistant focus change of about 2 nm. These images form the set of primary experimental data that have to be corrected for the nonlinear transfer characteristics of the CCD camera employed for recording the images [39]. Such a correction is essential since the detector modulation-transfer function (MTF) has a strong effect on the experimental intensity versus spatial frequency characteristics in the images and contributes the major part to the so-called Stobbs factor [40]. The corrected set of data then forms the input for the numerical exit-plane wave function reconstruction employing one of the state-of-the art software codes employing maximum-likelihood methods.

For wave-function retrieval, the *actual* values of the optical parameters that appear in the aberration function are used in the calculation. Therefore, it is recommended to measure these aberrations employing the Zemlin tableau technique [3] not only before but also after the acquisition of the focal series of images. Since this often requires shifting the specimen location to an amorphous area or even exchanging the specimen for an amorphous test sample, it is likely that the measured aberration values deviate from those applying during image acquisition. The only way out of this dilemma is to consider the imaging itself as a quantum mechanical experiment by which the aberrations can be measured as long as we have a structure the precise atomic coordinates of which are known. In reality this means that we have to carry out a full and self-consistent run of the whole image-process inversion in which the exit-plane wave function is calculated with the residual aberrations taken as free parameters.

As opposed to a widespread assumption in the literature, the reconstructed ψ_e does not in general supply us with a direct representation of $\mathbf{X}$, the set of atomic coordinates. Mapping of neither the real nor the imaginary part of ψ_e gives us unambiguously the correct atomic sites. As described in the previous sections, depending on the sample thickness, the contrast varies dramatically and in a complex fashion, which cannot be understood intuitively. Furthermore, the ψ_e obtained is the wave function of the "real" case. This means that a tilted specimen yields as a result a "tilted" wave function, and the amplitude and phase distributions are those of the "real" case, that is, they depend on the actual specimen thickness. Of course, the set of wanted coordinates $\mathbf{X}$ must be universal and therefore independent of these rather arbitrary imaging circumstances. Therefore, as long as the imaging parameters are not precisely known, the reconstructed wave function is not yet sufficient for picometer precision microscopy. This means that the backward calculation has to be continued by performing the second step. As usual, this is done by choosing a first-guess model for which the relativistically corrected Schrödinger equation is solved to obtain the exit-plane wave function. Since neither the sample thickness nor the beam tilt is in general known with sufficient accuracy, these have to be introduced as free parameters in an iterative fit of the model in order to reach an optimal match between calculated and experimental exit-plane wave functions. Only after a self-consistent set of atomic coordinates and imaging parameters is obtained may we consider the problem solved.

The inversion of the imaging process as it was just described looks rather canonical. Nevertheless, concerning picometer precision electron microscopy, there are issues that have to receive much greater attention than was standard in the past. First of all, it is quite likely that the original set of images is affected by small residual aberration values; in particular, this holds true for *twofold astigmatism* and *coma*. These have to be corrected in the backward calculation by expanding the iteration to comprise both steps. Another problem to be tackled is the limited precision of the alignment of the direction of the incident electron wave field with a principal low-index crystal direction. Neither the angular sensitivity of the intensity distribution in the diffraction pattern nor that in the image is sufficiently high for a precise adjustment of sample orientation. In addition, there are technical limits of even the best specimen goniometers available today, and there is always a risk of local specimen bending. Therefore, specimen tilting angles in the order of about 10 mrad can hardly be avoided. On the other hand, for realistic sample thicknesses, specimen tilts in the order of 1 mrad result in projection-related geometrical distortions of the atomic images affecting the precision of the atom position measurements. Moreover, the observable shifts of the contrast maxima inside unit cells containing different atom species cannot simply be related to each other by means of a purely geometrical factor representing the effect of the tilting angle. This can be explained by nuclear-charge dependence of electron diffraction channeling and of the corresponding shift of atomic maxima on specimen tilt. An example is given in Figure 3.7. If not properly taken into account in the numerical calculations, these different shifts of the contrast maxima for the same specimen tilting angle will lead to erroneous results. This is just another case that shows that the images, even when they resemble so closely the projected crystal structure as is the case in Figure 3.7, are not really images in the conventional sense. Only after a proper deconvolution by inversion of the two-step image formation process are our results representative of the actual atomic structure.

In the past, quantum mechanical and optical image calculations concentrated mainly on the strong atomic maxima in the images. This may be sufficient for structure investigations that concern, as already mentioned, a collective nonlocal property. But for investigations concerning individual atom column properties, this is in general not sufficient, and beyond that this would mean that we give away valuable information. Figure 3.8 shows experimental and matching simulation results concerning the behavior of the diffuse background intensity in the micrographs on changing the sample thickness and the specimen tilt. We find that the effect of tilting angles as low as 0.5 mrad can be readily detected and thus be taken into account in order to be able to realize picometer precision.

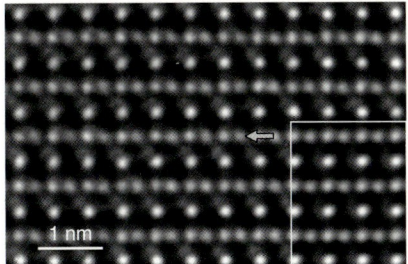

Figure 3.7 NCSI image of a SrTiO$_3$ sample taken under conditions (300 keV) where the direction of the incident electron wave field is tilted by about 7 mrad with respect to the crystallographic [110] direction. The resulting shifts of the SrO, Ti, and O atom positions (arrow) in the image are substantially different. Note that the vertical Ti atom row is no longer colinear with the SrO atom rows and that the horizontal position of a given O atom (arrow) is no longer in the exact center between the two neighboring Ti atoms. This means that the arrangement of contrast maxima in the image cannot be related in a simple way to the geometrically projected atom structure. The inset (right) was calculated for $Z = 3.0$ nm, $t = 8.3$ nm, $C_S = -15$ μm, tilting angle 7 mrad, coefficient of twofold astigmatism $A_1 = 3$ nm, of threefold astigmatism $A_2 = 40$ nm, and of coma $B_2 = 30$ nm; see Ref. [3].

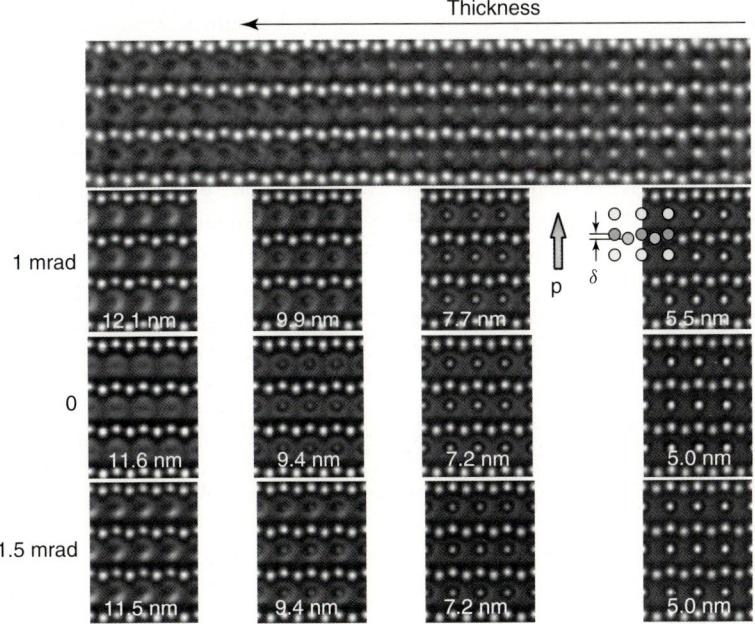

Figure 3.8 Experimental and matching simulation results concerning the behavior of the diffuse background intensity in the micrographs on changing the sample thickness and the specimen tilt (300 keV). The effect of tilting angles as low as 0.5 mrad can be readily detected and thus be taken into account in a high-quality image calculation.

3.4
Case Study: SrTiO$_3$

For quantitative atomic-resolution work it is mandatory to have an adequate optical resolution available in order to be able to separate closely spaced atom columns. Once these columns are separated, it is no longer the resolution that counts but the *precision* at which the position of an atom column or its lateral shift can be measured. Depending on the achieved signal to noise ratio of the images, this precision can be better by more than an order of magnitude than the optical resolution [41–43].

In the following, we compare by image simulations the conventional imaging mode employing a positive value of the spherical aberration coefficient (PCSI) with that achieved under NCSI conditions with respect to the obtained contrast and measurement precision of atomic column positions [44]. The images were calculated for an aberration-corrected 300 kV instrument employing optical imaging parameters for an instrumental information limit close to 0.08 nm [6, 35, 24]. Optimal NCSI contrast is obtained under these conditions (compare Section 3.4) with a spherical aberration value of $C_S = -15$ µm in combination with a defocus value of $Z = +6$ nm. The calculated intensity values were corrected on the basis of the measured actual CCD camera MTF. In addition, the effect of small electronic and mechanical instabilities of the instrument was taken into account by an image convolution using a Gaussian vibration function with a 1/e width of 0.03 nm.

Figure 3.9 shows a thickness series of images calculated for the two alternative imaging modes. The NCSI mode leads to bright-atom contrast on a dark background. This contrast is preserved up to a sample thickness of 7.7 nm and beyond. On the other hand, the PCSI mode results in dark atom contrast for relatively thin objects ($t < 4.4$ nm), while bright peaks appear at the Sr atom positions for higher thicknesses. For a quantitative comparison, the images are normalized to a mean intensity of 1, such that the standard deviation of the intensity reflects the image contrast. Figure 3.10 displays the image contrast as a function of thickness. The contrast resulting from the two modes exhibits the same value at an object thickness of 1.1 nm. In the case of the NCSI mode, the contrast grows with increasing thickness up to about 5 nm, reaching a saturation value of about 0.27. The contrast increase of the PCSI mode is much lower, and a saturation plateau already occurs in the thickness region between 3 and 7 nm. For the thickness range of 3–7 nm typical for high-resolution work in SrTiO$_3$, the NCSI contrast is on average stronger by about a factor of 2 than the PCSI contrast. We note that this comparison neglects the fact that under PCSI conditions and larger sample thicknesses the different atom species appear at different contrasts (bright on a dark background for Sr and complicated gray-scale contrast for Ti and O), which is not very convenient for quantitative studies, in particular, those where measurements on Ti and O are intended.

The obtainable signal to noise ratio depends directly on the image intensity recorded at an atomic column position and determines the precision of position and occupancy measurements. The intensity values for the three types of columns,

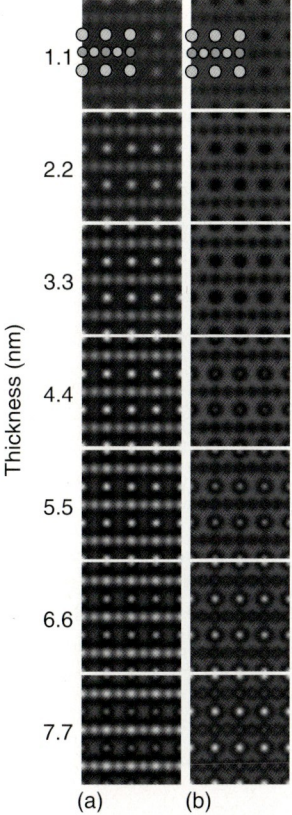

Figure 3.9 Simulated thickness series of images calculated for the negative spherical-aberration imaging (NCSI) mode (a) and the conventional PCSI imaging mode (b). The NCSI mode leads to bright-atom contrast on a dark background. This contrast is preserved up to a sample thickness of 7.7 nm and beyond. PCSI conditions result in dark atom contrast for relatively thin objects ($t < 4.4$ nm), while bright peaks appear at the Sr atom positions for higher thicknesses. In the upper part of the figure, the atom positions of SrTiO$_3$ are given schematically [44].

SrO, Ti, and O are plotted in Figure 3.11 for a sample thickness of 3.3 nm. At this thickness, the image taken under PCSI conditions has essentially reached its plateau contrast level, while the already much superior contrast level in the NCSI case increases further at larger thickness. Figure 3.11a shows the signal intensity associated with the SrO columns, and Figure 3.11b, the intensity of the Ti and O columns. The mean intensity one is denoted by a dotted line. For a quantitative comparison, the ratio of the column-based signal strength between the two imaging modes is defined by the ratio of the respective intensity extrema at the column positions, given by $(I_{max} - 1)_{NCSI}/(1 - I_{min})_{PCSI}$. The calculated results are 3.6 for SrO, 2.4 for Ti, and 2.1 for O. These ratios of the extremal values

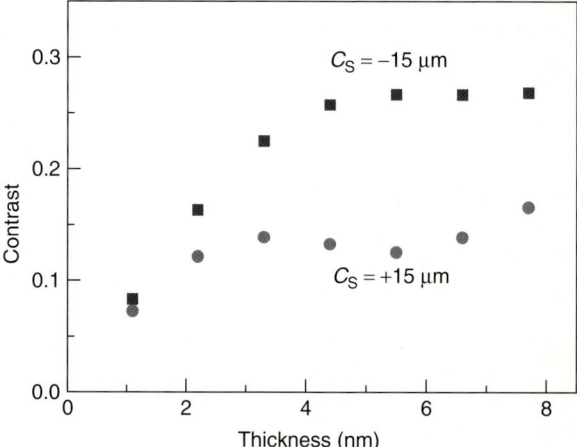

Figure 3.10 The image contrast (defined by the standard deviation from the mean intensity) as a function of specimen thickness. In the case of NCSI (squares), the contrast is growing with increasing thickness up to about 5 nm, where it reaches a saturation value of about 0.27. The contrast increase of the PCSI mode (filled circles) is much lower, and saturation occurs already in the thickness region between 3 and 7 nm [44].

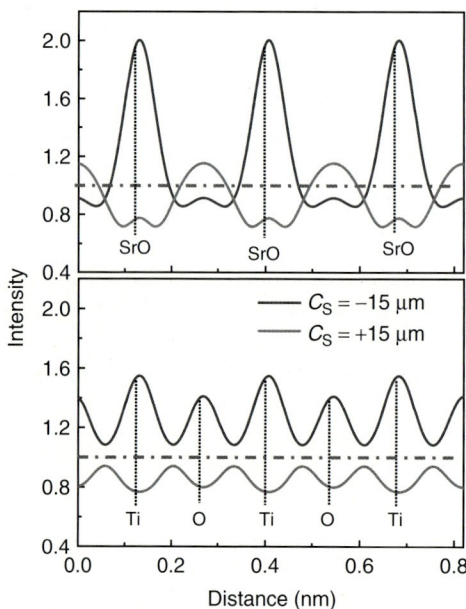

Figure 3.11 Plot of the intensity values for SrO, Ti, and O atom positions for a sample thickness of 3.3 nm. High-amplitude black line: NCSI. Low-amplitude grey line: PCSI. The mean intensity 1 is denoted by a dotted line [44].

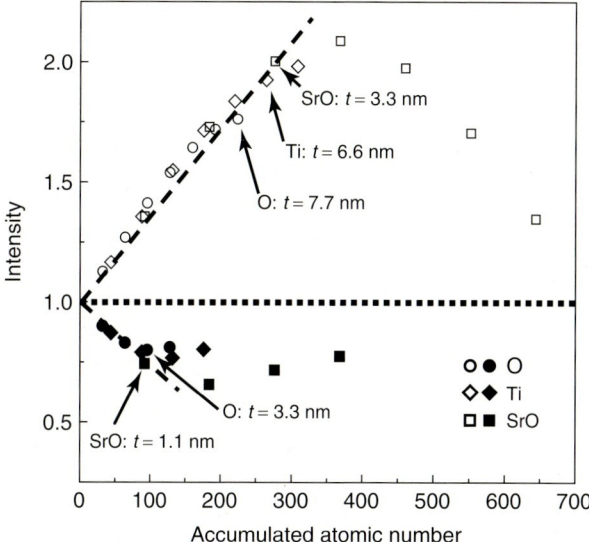

Figure 3.12 Dependence of the image intensity on the atomic number accumulated per individual atomic column along the viewing direction. Along the [110] direction, a unit cell period of SrTiO$_3$ includes one strontium plus one oxygen atom in the SrO column, one titanium atom in the Ti column, and two oxygen atoms in the oxygen column. The atomic numbers accumulated over a single unit cell period are thus 46 (SrO), 22 (Ti), and 16 (O), respectively [44].

between the NCSI and the PCSI modes are for all individual column types even larger than the corresponding ratio of the overall image contrast of 1.6. This reflects the much stronger "focused" shape of the atomic column images in the NCSI mode.

Figure 3.12 shows the dependence of the image intensity on the atomic number accumulated per individual atomic column along the viewing direction. It is noted that along the [110] direction, a unit cell period of SrTiO$_3$ includes one strontium plus one oxygen atom in the SrO column, one Ti atom in the Ti column, and two oxygen atoms in the oxygen column. The atomic numbers accumulated over a single unit cell period are thus 46, 22, and 16, respectively. For the NCSI mode, the image intensity of all columns follows essentially a linear dependence on the total atomic number accumulated up to a thickness of at least 3.3 nm. Beyond this thickness, the linear relation is still valid for the lighter Ti and O columns, whereas the linearity is lost for the SrO columns because of their shorter extinction length. In the case of the PCSI mode, the linearity between column intensity and the accumulated atomic number is already lost for all column types at a thickness of 3.3 nm. Most importantly, the linear dependence of the column intensity on the accumulated atomic number exhibits also a higher slope in the NCSI mode than in PCSI mode.

3.5
Practical Examples of Application of NCSI Imaging

The first atomic-resolution studies employing the NCSI technique were carried out in $SrTiO_3$ and in $YBa_2Cu_3O_7$ [8]. Subsequently, the technique was employed to measure the oxygen occupancy in $\Sigma 3\{111\}$ twin boundaries in $BaTiO_3$ thin films [45]. Further studies employing the NCSI technique concerned the reconstructed 90° tilt grain boundary in $YBa_2Cu_3O_7$ [43], the atomic structure of the core of dislocations in $SrTiO_3$ [46, 47], of defects in GaAs [48] and of the $Si/SrTiO_3$ interface [49]. Using NCSI, the stability and dynamics of graphene were studied [50]. The effect of a single dislocation in the $SrTiO_3$ substrate on the structure of epitaxial PZT was investigated [51]. Another study employing the NCSI mode yielded the high-precision atomic structure of the interface between $SrTiO_3$ and $LaAlO_3$ [52]. A study of the core structure of dislocations in plastically deformed sapphire [53] was carried out, and the atomic surface termination, structural relaxation, and electronic structure of the polar (111) surface were investigated in Co_3O_4 [54]. Mapping the polarization dipole distribution measured by atom position measurements of Ti and O atoms close to a $PZT/SrTiO_3$ interface allowed demonstration that closure-domain structures closing the electric flux by continuous dipole rotation occur in ferroelectrics [55].

For illustration, we reproduce here the results of the study on the atomic structure of 180° inversion domain walls in PZT, that is, $Pb(Zr_{0.2}Ti_{0.8})O_3$ [56]. PZT is a widely used technical ferroelectric, used, for instance, in the form of thin films in ferroelectric memories. Data storage is performed by switching the ferroelectric polarization. The polarization state migrates through the specimen, and areas of different polarization are separated by domain walls in which the state of polarization changes from one direction to the other. Figure 3.13 displays an NCSI electron micrograph of PZT sandwiched between two $SrTiO_3$ layers. The insets are magnifications that allow identification of the individual atom species and a closer look at the particular position of the atomic sites. Evidence for the polarized state can be obtained by inspection of the individual atom positions in the unit cell. In the inset on the left, the Zr/Ti (mixed) sites are shifted toward the upper vertical Pb atom row. The oxygen atoms are shifted even further. As a result, they are no longer collinear with the Zr/Ti atom row. The situation is depicted schematically in the inset. According to definition, the polarization vector **p** points downward. In the inset at the lower right, the shifts are in the opposite direction, and as a consequence, the polarization vector is inverted. In between we have a 180° inversion domain wall (broken line).

The oblique parts of the wall are made up of transversal segments and very short longitudinal segments. They are essentially uncharged since the electric fields characterized by the polarization vectors of the adjoining boundary segments just cancel each other pair wise. In the horizontal domain-wall segments, the polarization vectors are, across the wall, meeting head to head. As a result, the electric fields do not cancel and the boundary is charged. Since we have genuine atomic resolution, we have access to each individual atomic position in this [110]

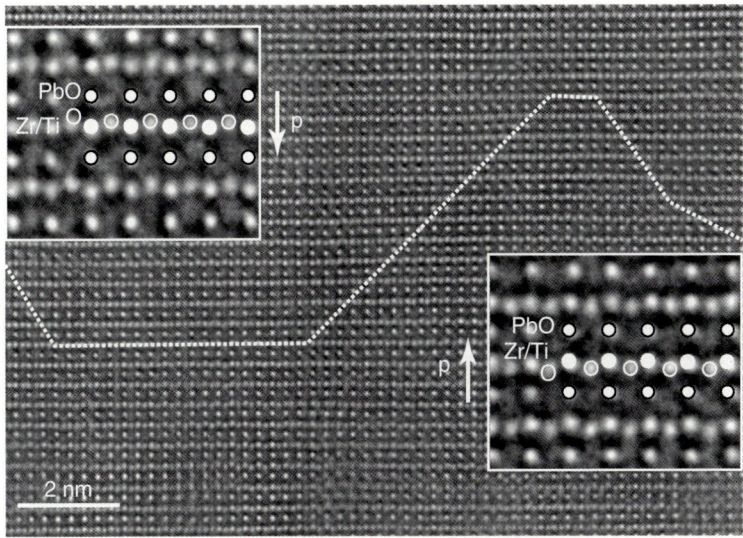

Figure 3.13 Pb(Zr$_{0.2}$Ti$_{0.8}$)O$_3$ imaged along the [110] direction. The inset on the left shows that the horizontal Zr/Ti atom rows are shifted toward the respective Pb atom row above the Zr/Ti row. Oxygen is shifted even more, thus becoming no longer co-linear with the Zr/Ti rows. This indicates that the material is ferroelectrically polarized. The polarization vector **p** points downward. The inset on the right shows opposite atomic shifts. The direction of the polarization vector there is upward. The dotted line shows the appertaining ferroelectric inversion domain wall. With respect to the atomic structure, the inclined domain-wall sections consist of vertical transversal and horizontal longitudinal domain-wall segments. As a result they are uncharged. The horizontal sections are longitudinal domain walls, which are charged [56].

projection. This allows making individual measurements of the atomic shifts. It is found that the transversal walls are only a single projected unit cell thick (Figure 3.14). Figure 3.15a shows for a longitudinal domain wall the results of the measurements of the atomic shifts of the O and Zr/Ti atoms out of their symmetric positions with single-atom column resolution. The Gaussian regression analysis indicates a precision of better than 5 p.m. (for a 95% confidence level). This wall is rather extended, about 10 lattice constants. From these measurements, we can infer that this kind of domain wall reduces the field energy by increasing its width substantially over that of the charge-neutral transversal wall. Figure 3.15b shows the value of the macroscopic spontaneous polarization P_S calculated from the measured atomic shifts employing calculated values [57] for the effective charges of the ions. This demonstrates that by exploiting the potential of the atomically resolving ultrahigh-resolution techniques we can determine local physical properties such as the spontaneous polarization directly from measurements of shifts of the individual atom positions. This fulfills an old dream in materials science to be able to obtain a direct link between atomic level information and macroscopic properties.

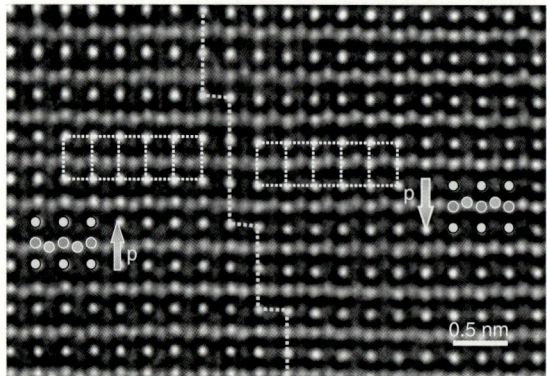

Figure 3.14 A 180° domain-wall segment of mixed type seen edge-on. The arrows "**p**" indicate the opposite polarization directions across the domain wall. The parallelograms denote the segments of transversal domain wall. The vertical dotted line marks the central plane of the domain wall. The horizontal dotted lines trace projected unit cells on either side of the domain wall. Indicated by the shift of the oxygen atoms, "up" on the left and "down" on the right of the central plane indicates directly that the width of the wall is a single <110> projected unit cell wide [56].

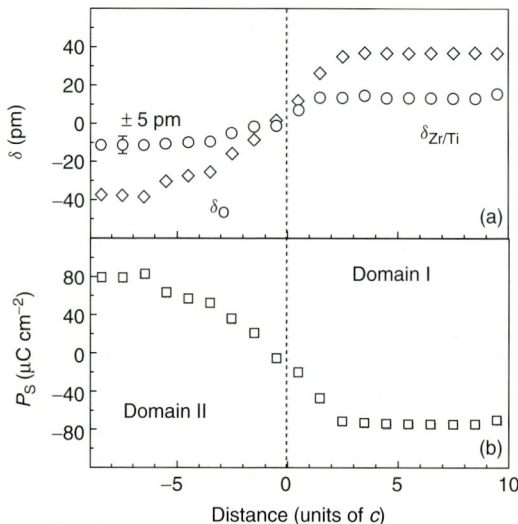

Figure 3.15 (a) Longitudinal inversion domain-wall atom shifts measured on the atomic sites and averaged for better statistics. $\delta_{Zr/Ti}$ denotes the upward shift of the Zr/Ti atom positions toward Pb (cf. inset on the left in Figure 3.13) as a function of distance (in units of the crystallographic c lattice parameter) from the domain wall center. δ_O denotes the corresponding oxygen atom shift. The domain wall width amounting to 10 unit cell distances is much wider than the transversal wall sections. This reduces the electric field energy. (b) The macroscopic spontaneous polarization P_S calculated on the basis of measuring the individual atomic shifts [56].

References

1. Rose, H. (1990) Outline of a spherically corrected semiaplanatic medium-voltage transmission electron microscope. *Optik*, **85**, 19–24.
2. Haider, M., Uhlemann, S., Schwan, E., Rose, H., Kabius, B., and Urban, K. (1998) Electron microscopy image enhanced. *Nature*, **392**, 768–769.
3. Uhlemann, S. and Haider, M. (1998) Residual wave aberrations in the first spherical aberration corrected transmission electron microscope. *Ultramicroscopy*, **72**, 109–119.
4. Scherzer, O. (1949) The theoretical resolution limit of the electron microscope. *J. Appl. Phys.*, **20**, 20–29.
5. Lichte, H. (1991) Optimum focus for taking electron holograms. *Ultramicroscopy*, **38**, 13–22.
6. Lentzen, M., Jahnen, B., Jia, C.-L., Thust, A., Tillmann, K., and Urban, K. (2002) High-resolution imaging with an aberration corrected transmission electron microscope. *Ultramicroscopy*, **92**, 233–242.
7. Lentzen, M. (2004) The tuning of a Zernike phase plate with defocus and variable spherical aberration and its use in HRTEM imaging. *Ultramicroscopy*, **99**, 211–220.
8. Jia, C.-L., Lentzen, M., and Urban, K. (2003) Atomic-resolution imaging of oxygen in perovskite ceramics. *Science*, **299**, 870–873.
9. Urban, K.W. (2008) Studying atomic structures by aberration-corrected transmission electron microscopy. *Science*, **321**, 506–510.
10. Urban, K.W. (2009) Is science prepared for atomic resolution electron microscopy. *Nat. Mater.*, **8**, 260–262.
11. Urban, K.W., Houben, L., Jia, C.-L., Lentzen, M., Mi, S.-B., Tillmann, K., and Thust, A. (2008) in *Advances in Imaging and Electron Physics*, vol. 153 (ed. P. Hawkes), Academic Press, Oxford, pp. 320–344.
12. Urban, K.W., Jia, C.-L., Houben, L., Lentzen, M., Mi, S.-B., and Tillmann, K. (2009) Negative spherical aberration ultrahigh-resolution imaging in corrected transmission electron microscopy. *Philos. Trans. R. Soc. A*, **367**, 3735–3753.
13. Reimer, L. (1984) *Transmission Electron Microscopy*, Springer, Berlin.
14. Williams, D.B. and Carter, C.B. (2009) *Transmission Electron Microscopy*, 2nd edn, Springer, New York.
15. Spence, J.C.H. (2007) *High Resolution Electron Microscopy*, 3rd edn, Oxford University Press, New York.
16. Saxton, W.O. (2000) A new way of measuring microscope aberrations. *Ultramicroscopy*, **81**, 41–45.
17. Lentzen, M. (2006) Progress in aberration-corrected high-resolution transmission electron microscopy using hardware aberration correction. *Microsc. Microanal.*, **12**, 191–205.
18. Born, M. and Wolf, E. (1999) *Principles of Optics*, 7th edn, Cambridge University Press, Cambridge, UK.
19. Stadelmann, P.A. (1987) EMS – a software package for electron diffraction analysis and HREM image simulation in materials science. *Ultramicroscopy*, **21**, 131–145.
20. Kilaas, R. (1987) in *Proceedings of the 45th Annual EMSA Meeting* (ed. G.W. Bailey), San Francisco Press, San Francisco, CA, pp. 66–67.
21. Zernike, F. (1942) Phase contrast, a new method for the microscopic observation of transparent objects, Part I. *Physica*, **9**, 686–698.
22. Zernike, F. (1942) Phase contrast, a new method for the microscopic observation of transparent objects, Part II. *Physica*, **9**, 974–986.
23. Hanszen, K.-J. and Trepte, L. (1971) The influence of voltage and current fluctuations and of a finite energy width of the electrons on contrast and resolution in electron microscopy. *Optik*, **32**, 519–538.
24. Barthel, J. and Thust, A. (2008) Quantification of the information limit of transmission electron microscopes. *Phys. Rev. Lett.*, **101**, 200801-1–200801-4.
25. Lentzen, M. (2008) Contrast transfer and resolution limits for sub-angström high-resolution transmission electron

microscopy. *Microsc. Microanal.*, **14**, 16–26.

26. Zhang, Z. and Kaiser, U. (2009) Structural imaging of β-Si$_3$N$_4$ by spherical aberration-corrected high-resolution transmission electron microscopy. *Ultramicroscopy*, **109**, 1114–1120.

27. Meyer, J.C., Kurasch, S., Park, H.J., Skakalova, V., Künzel, D., Groß, A., Chuvilin, A., Algara-Siller, G., Roth, S., Iwasaki, T., Starke, U., Smet, J.H., and Kaiser, U. (2011) Experimental analysis of charge redistribution due to chemical bonding by high-resolution transmission electron microscopy. *Nat. Mater.*, **10**, 209–215.

28. Coene, W., Janssen, G., Op de Beeck, M., and van Dyck, D. (1992) Phase retrieval through focus variation for ultra-resolution in field emission transmission electron microscopy. *Phys. Rev. Lett.*, **69**, 3743–3746.

29. Jia, C.-L. and Thust, A. (1999) Investigation of atom displacements at a $\sum 3$ {111} twin boundary in BaTiO$_3$ by means of phase-retrieval electron microscopy. *Phys. Rev. Lett.*, **82**, 5052–5055.

30. Kisielowski, C., Hetherington, C.J.D., Wang, Y.C., Kilaas, R., O'Keefe, M.A., and Thust, A. (2001) Imaging columns of the light elements carbon, nitrogen and oxygen with sub- Angström resolution. *Ultramicroscopy*, **89**, 243–263.

31. Howie, A. (1966) Diffraction channelling of fast electrons and positrons in crystals. *Philos. Mag.*, **14**, 223–237.

32. Urban, K. and Yoshida, N. (1979) The effect of electron diffraction channelling on the displacement of atoms in electron-irradiated crystals. *Radiat. Eff. Defects Solids*, **42**, 1–15.

33. van Dyck, D. and Op de Beeck, M. (1996) A simple intuitive theory for electron diffraction. *Ultramicroscopy*, **64**, 199–2007.

34. Lentzen, M. and Urban, K. (2000) Reconstruction of the projected crystal potential in transmission electron microscopy by means of a maximum-likelihood refinement algorithm. *Acta Crystallogr. A*, **56**, 235–247.

35. Jia, C.-L., Lentzen, M., and Urban, K. (2004) High resolution transmission electron microscopy using negative spherical aberration. *Microsc. Microanal.*, **10**, 174–184.

36. Coene, W.M.J., Thust, A., Op de Beeck, M., and Van Dyck, D. (1996) Maximum-likelihood method for focus-variation image reconstruction in high resolution transmission electron microscopy. *Ultramicroscopy*, **64**, 109–135.

37. Thust, A., Coene, W.M.J., Op de Beeck, M., and Van Dyck, D. (1996) Focal-series reconstruction in HRTEM: simulation studies on non-periodic objects. *Ultramicroscopy*, **64**, 211–230.

38. Thust, A., Overwijk, M.H.F., Coene, W.M.J., and Lentzen, M. (1996) Numerical correction of lens aberrations in phase-retrieval HRTEM. *Ultramicroscopy*, **64**, 249–264.

39. Thust, A. (2009) High-resolution transmission electron microscopy on an absolute contrast scale. *Phys. Rev. Lett.*, **102**, 220801-1–220801-4.

40. Hÿtch, M. and Stobbs, W. (1994) Quantitative comparison of high resolution TEM images with image simulations. *Ultramicroscopy*, **53**, 191–203.

41. Den Dekker, A.J., Van Aert, S., van den Bos, A., and Van Dyck, D. (2005) Maximum likelihood estimation of structure parameters from high resolution electron microscopy images. Part I: a theoretical framework. *Ultramicroscopy*, **104**, 83–106.

42. Van Aert, S., den Dekker, A.J., van den Bos, A., Van Dyck, D., and Chen, J.H. (2005) Maximum likelihood estimation of structure parameters from high resolution electron microscopy images. Part II: a practical example. *Ultramicroscopy*, **104**, 107–125.

43. Houben, L., Thust, A., and Urban, K. (2006) Atomic-precision determination of the reconstruction of a 90° tilt boundary in YBa$_2$Cu$_3$O$_7$ by aberration corrected HRTEM. *Ultramicroscopy*, **106**, 200–214.

44. Jia, C.-L., Houben, L., Thust, A., and Barthel, J. (2010) On the benefit of the negative spherical-aberration imaging technique for quantitative HRTEM. *Ultramicroscopy*, **110**, 500–505.

45. Jia, C.-L. and Urban, K. (2004) Atomic-resolution measurement of oxygen concentration in oxide materials. *Science*, **303**, 2001–2004.
46. Jia, C.-L., Thust, A., and Urban, K. (2005) Atomic-scale analysis of the oxygen configuration at a SrTiO3 dislocation core. *Phys. Rev. Lett.*, **95**, 225506-1–225506-4.
47. Jia, C.-L., Houben, L., and Urban, K. (2006) Atom vacancies at a screw dislocation core in SrTiO$_3$. *Philos. Mag. Lett.*, **86**, 683–690.
48. Tillmann, K., Thust, A., and Urban, K. (2004) Spherical aberration correction in tandem with exit-plane wave function reconstruction: interlocking tools for the atomic scale imaging of lattice defects in GaAs. *Microsc. Microanal.*, **10**, 185–198.
49. Mi, S.-B., Jia, C.-L., Vaithyanathan, V., Houben, L., Schubert, J., Schlom, D.G., and Urban, K. (2008) Atomic structure of the interface between SrTiO3 thin films and Si (001) substrates. *Appl. Phys. Lett.*, **93**, 101913-1–101913-3.
50. Girit, Ç.Ö., Meyer, J.C., Erni, R., Rossel, M.D., Kisielowski, C., Yang, L., Park, C.H., Crommie, M.F., Cohen, M.L., Louie, S.G., and Zettl, A. (2009) Graphene at the edge: stability and dynamics. *Science*, **323**, 1705–1708.
51. Jia, C.-L., Mi, S.-B., Urban, K., Vrejoiu, I., Alexe, M., and Hesse, D. (2009) Effect of a single dislocation in a heterostructure layer on the local polarization of a ferroelectric layer. *Phys. Rev. Lett.*, **102**, 117601-1–117601-4.
52. Jia, C.-L., Mi, S.-B., Faley, M., Poppe, U., Schubert, J., and Urban, K. (2009) Oxygen octahedron reconstruction in the SrTiO$_3$ / LaAlO$_3$ heterointerfaces investigated using aberration-corrected ultrahigh-resolution transmission electron microscopy. *Phys. Rev. B*, **79**, 81405(R)-1–81405(R)-4.
53. Heuer, A.H., Jia, C.-L., and Lagerlöf, K.P.D. (2010) The core structure of basal dislocations in deformed sapphire (a-Al$_2$O$_3$). *Science*, **330**, 1227–1231.
54. Yu, R., Hu, L.H., Cheng, Z.Y., Li, Y.D., Ye, Q., and Zhu, J. (2010) Direct sub-angström measurement of surfaces of oxide particles. *Phys. Rev. Lett.*, **105**, 226101-1–226101-4.
55. Jia, C.-L., Urban, K., Vrejoiu, I., Alexe, M., and Hesse, D. (2011) *Science*, **331**, 1420–1423.
56. Jia, C.-L., Mi, S.B., Urban, K., Vrejoiu, I., Alexe, M., and Hesse, D. (2008) Atomic-scale study of electric dipoles near charged and uncharged domain walls in ferroelectric films. *Nat. Mater.*, **7**, 57–61.
57. Zhong, W., King-Smith, R.D., and Vanderbilt, D. (1994) Giant LO-TO splittings in perovskite ferroelectrics. *Phys. Rev. Lett.*, **72**, 3618–3621.

4
Z-Contrast Imaging

Stephen J. Pennycook, Anrew R. Lupini, Albina Y. Borisevich, and Mark P. Oxley

4.1
Recent Progress

The decade since the first edition of the "Handbook of Microscopy" has seen a revolution in the capabilities of the transmission electron microscope (TEM), through the successful correction of lens aberrations. While the physical origins of the aberrations arise intrinsically through the need to use electromagnetic fields to focus electrons [1], and possible means for aberration correction were proposed over 70 years ago [2], it is only recently that aberration correctors have led to improved microscope resolution. The first success was with the scanning electron microscope [3] followed by successful correction of aberrations in the TEM [4–6] and the scanning transmission electron microscope (STEM) [7, 8]. The reason for the almost 50 year delay is technological: correcting aberrations is achieved through multipole lenses and all elements must be controlled simultaneously to a precision of 1 part in 10^6 or 10^7. It is similar to having 40–60 focus controls that all need to be precisely adjusted. Human beings cannot do such tasks, but the modern computer can.

Aberration correction has enormously benefited both TEM and STEM. In TEM it has removed the major source of image delocalization arising from the imaging lenses. Now the only delocalization is that due to scattering within the specimen, as indeed is the case with STEM annular dark-field (ADF) imaging. The TEM image has become much more local so that light atom columns can be quantified in position and occupancy to levels that were unimaginable before aberration correction [9, 10], see chapter 3 by Urban. However, in the case of the STEM, aberration correction has not only improved the resolution of the image, it has largely overcome that other historic limitation of STEM, which is noise. By making the probe smaller, the peak intensity is correspondingly increased, and the signal-to-noise ratio of the image, or of an electron energy loss spectrum, is much improved. High-angle annular dark-field (HAADF) imaging, also referred to as *Z-contrast imaging* owing to its sensitivity to atomic number Z, showed not only improved resolution, but better contrast and less noise. With the introduction of third-order correctors, STEM Z-contrast imaging achieved sub-Angstrom resolution in a crystal

Handbook of Nanoscopy, First Edition. Edited by Gustaaf Van Tendeloo, Dirk Van Dyck, and Stephen J. Pennycook.
© 2012 Wiley-VCH Verlag GmbH & Co. KGaA. Published 2012 by Wiley-VCH Verlag GmbH & Co. KGaA.

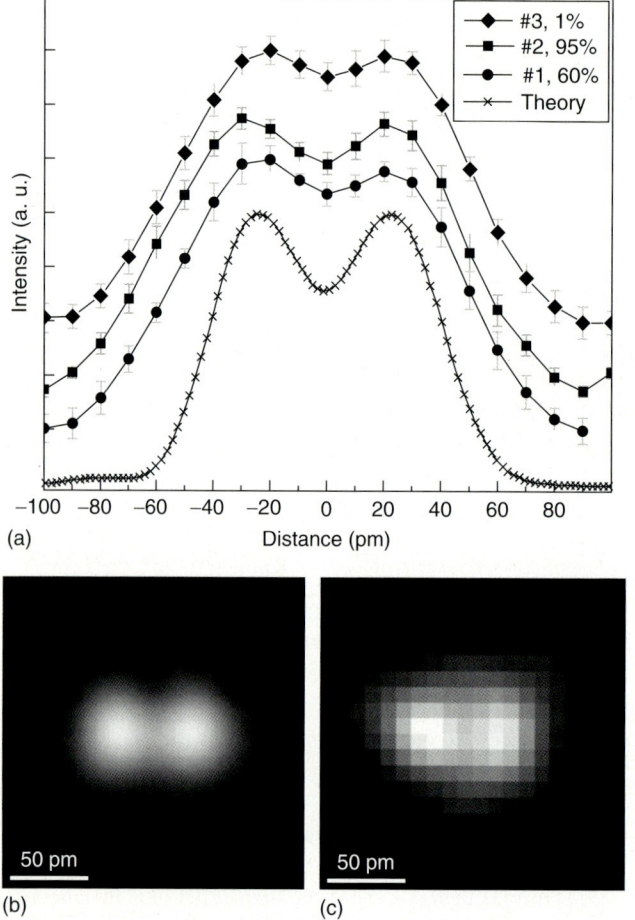

Figure 4.1 (a) Representative intensity profiles averaged over 12–13 dumbbells of a Z-contrast image from Ge ⟨114⟩ showing resolution of the 0.47 Å spacing, taken with the TEAM 0.5 microscope, an FEI Titan 80–300 equipped with a high-brightness Schottky gun and CEOS fifth-order corrector [16]. Solid line is the theoretical curve. (b) Calculated dumbbell image. (c) Averaged experimental image. (Reproduced from Ref. [13].)

by resolving the dumbbells in Si ⟨112⟩ [11]. The same year, the spectroscopic identification of a single atom in a crystal was achieved with atomic resolution [12]. With the introduction of fifth-order correctors, STEM Z-contrast imaging has surpassed the goal of 0.5 Å by imaging the dumbbells in Ge ⟨114⟩ spaced just 0.47 Å apart [13] as shown in Figure 4.1. STEM now holds the record for image resolution over the TEM, which is in accordance with optical physics, since an incoherent Z-contrast image should have higher resolution than a coherent phase contrast image [14, 15].

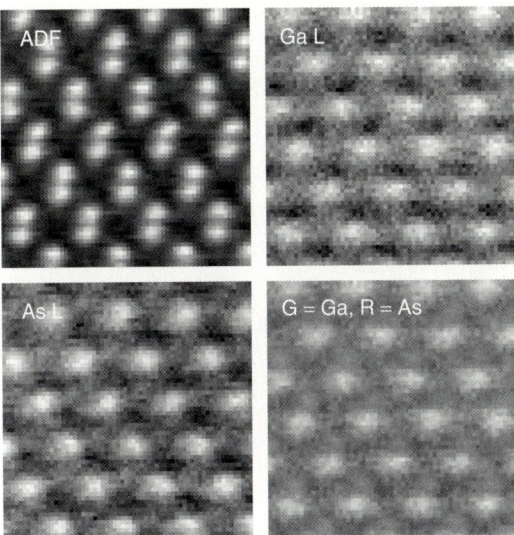

Figure 4.2 Spectroscopic imaging of GaAs in the ⟨110⟩ projection comparing the ADF image to the Ga- and As-L spectroscopic images, obtained on the Nion UltraSTEM with a fifth-order aberration corrector operating at 100 kV. Images are 64 × 64 pixels, with collection time 0.02 s per pixel and a beam current of approximately 100 pA, after noise reduction by principal component analysis [20]. (Data from M. Varela.)

Electron energy loss spectroscopy (EELS) has also benefited from the latest generation of aberration correctors. Rather than pushing to the limit of resolution, aberration correction also allows much higher probe currents to be focused into atomic-scale probes. Very approximately, each generation of aberration correction (e.g., uncorrected to third-order) can either halve the probe size or increase the probe current by a factor of 10. Somewhat less improvement is seen from third-order to fifth-order correction as chromatic effects start to become important, especially so with lower accelerating voltages. However, together with more efficient coupling optics from the specimen into the spectrometer, the fifth-order correctors allow 100% collection efficiency for many edges, with the result that two-dimensional EELS mapping is achievable at atomic resolution [17–19]. Figure 4.2 gives an example of the spectroscopic imaging of GaAs.

The benefit of aberration correction is very obvious in the improved imaging of single atoms. The field emission gun STEM was invented to image single atoms with the goal of DNA sequencing by Crewe and his collaborators [21–23]. Their spectacular images of single heavy atoms were in fact the first to be obtained by an electron microscope. However, these microscopes were designed for biology with very small pole piece gaps giving low aberrations (0.4 mm) enabling probe sizes of around 2.5 Å to be achieved at an accelerating voltage of 42.5 kV [24]. Commercial instruments designed for materials science typically had larger gaps to allow high specimen tilts, and probe sizes in this range could only be achieved at higher accelerating voltages [25, 26]. Figure 4.3 compares images of Pt atoms

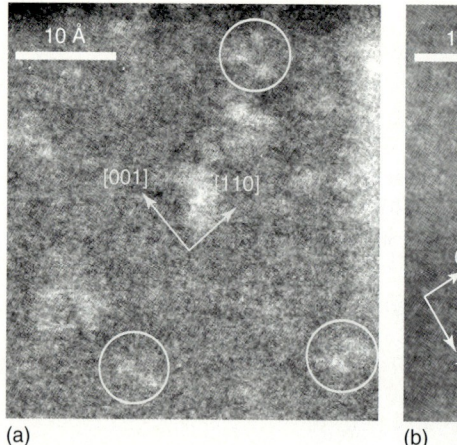

Figure 4.3 Imaging of Pt atoms on γ-alumina with a 300 kV STEM (a) before and (b) after aberration correction. Some Pt trimers and dimers are just visible in (a) but individual atoms are much more visible in (b). Some faint contrast from the γ-alumina lattice is also present in (b) as it was not far from a zone axis. (Images reproduced from (a) [27], (b) [28].)

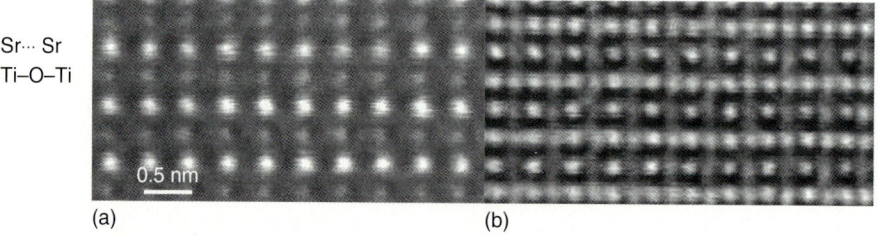

Figure 4.4 (a) Z-contrast and (b) phase contrast images of ⟨110⟩ SrTiO$_3$ taken with a VG Microscopes HB603U with Nion aberration corrector operating at 300 kV, using defocus of +2 and +6 nm, respectively (raw data). O columns are just detectable in between the Ti columns in (a). (Reproduced from Ref. [29], see also [30].)

supported on γ-alumina obtained with a 300 kV STEM before and after installation of a third-order aberration corrector. The improved visibility is quite dramatic.

Light atoms scatter much less than heavy atoms and have usually been invisible in a Z-contrast image until aberration correction, especially in the presence of adjacent heavy atom columns. With a third-order corrector, columns of oxygen atoms became just visible in between the heavier columns in SrTiO$_3$ (Figure 4.4a). However, they are still easier to see in the simultaneous phase contrast image (Figure 4.4b), although there are spurious features between the Sr columns that do not correspond to atomic columns at all.

The most remarkable example of the Z-contrast imaging of light atoms has recently been shown with the imaging of monolayer BN [31], using a fifth-order aberration corrector. Such samples can be challenging to image not only because they consist of sheets of single very light (and therefore weakly scattering atoms), but because the same electrons that we use to form the image may also damage the sample. The route chosen to reduce the damage in this example was to lower the accelerating voltage to 60 kV. Even at this lower accelerating voltage, the enhanced resolution provided by aberration correction not only allows individual B and N atoms to be directly resolved but further allows their identification directly from their different scattering cross sections (Figure 4.5). The higher atomic number of the N atoms causes them to appear brighter than the B atoms, even in the raw data. This result is remarkable, because in phase contrast imaging the contrast difference is weak and the lattice polarity has only been distinguished by exit wave reconstruction methods [32–34]. Furthermore, Figure 4.5 shows several sites where the intensity does not correspond to either B or N. Since the intensity in the image depends on the atomic number, the hypothesis is that atoms with intermediate intensity should correspond to C, whereas those with higher intensity must correspond to an element with a higher atomic number, such as O. To confirm this hypothesis, a histogram of intensities was taken that confirms that the brighter atoms are indeed O, as shown in Figure 4.6. Note that it was important to remove the probe tails by a Fourier filter to obtain such good statistical separation between the elements. Furthermore, the error bars indicate that, in principle, Z-contrast images could detect H atoms; however, they are likely to move around very fast under the beam so imaging individual H atoms is likely to remain a major challenge for the future. These images were obtained with a fifth-order corrected Nion UltraSTEM using a cold field emission gun. The small energy spread of the gun allowed this level of resolution to be achieved at only 60 kV accelerating voltage, which is below the threshold for knock-on damage. Nevertheless, some defects were still observed to be created by the beam; in fact, the carbon ring was previously a hole in the BN film which was filled by migrating C atoms.

Another recent development important for the imaging of light atoms is that of annular bright-field (ABF) imaging [35–37]. First proposed by Rose [38], the image is obtained by an annular detector that collects typically the outer half of the bright-field disk. The resulting image is a phase contrast image, but because of the range of interfering angles, shows fairly incoherent characteristics, that is, atoms show dark contrast independent of specimen thickness over a substantial range. The advantage of low-angle scattering is that Z-contrast is suppressed allowing both light and heavy atoms to be visualized in the same image.

4.2
Introduction to the Instrument

The basic components of the STEM are illustrated in Figure 4.7. The microscope is designed to form a small focused probe on the specimen and images are obtained

114 | *4 Z-Contrast Imaging*

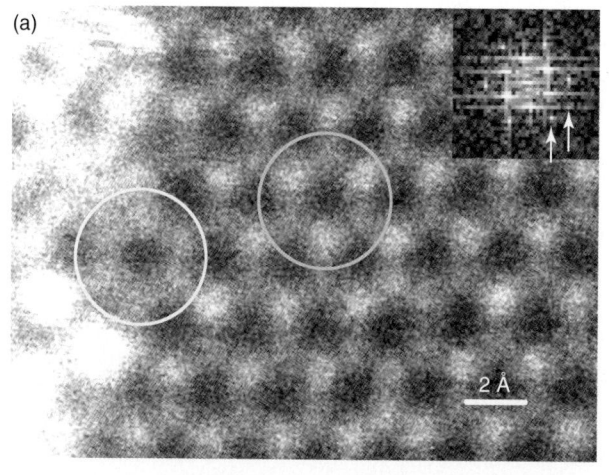

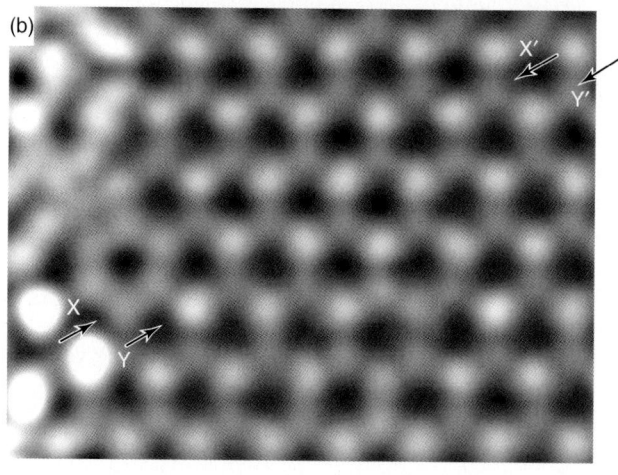

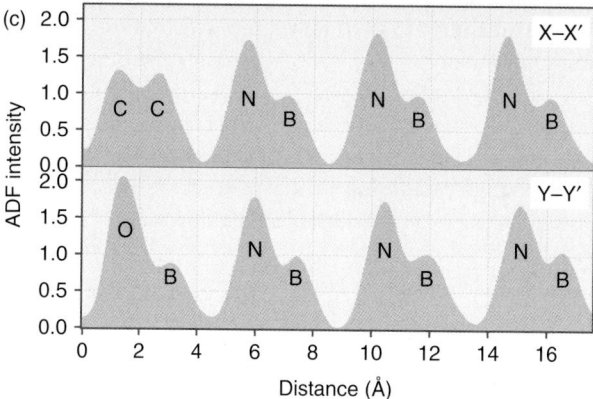

serially as the probe is scanned pixel-by-pixel through the use of appropriate detectors. Shown are a HAADF detector normally used for Z-contrast imaging, a bright-field detector, and an EELS detector. These latter detectors can be used simultaneously with the HAADF detector allowing pixel-by-pixel comparison of data, a major advantage of the STEM compared to TEM. If the probe is stopped, then point analysis is possible either by EELS or with an X-ray detector (not shown). If the probe is small enough, then the analyzed volume can be a single column [39, 40]. Alternatively, the probe can be scanned in a line along a plane of atoms parallel to an interface to reduce exposure to the beam [41].

The most important lens in the microscope is the objective lens, which is the part that actually focuses the electrons onto the sample. The resolution of the microscope is dominated by this one lens since, as seen in Figure 4.7, this is the lens where the electrons travel at the highest angles to the axis, and the aberrations increase dramatically with angle as we will discuss later. The aberration corrector is used to compensate for those aberrations. Design and optimization of the objective lens was one of the most important aspects of building an electron microscope for many years. The small size of the pole piece necessary to produce small aberrations has historically limited the size of samples that can be inserted, stage control particularly specimen tilt, and possibilities for *in situ* measurement. It is likely that aberration correction, as well as enhancing the resolution, will allow some developments in those areas also.

The flexibility of the STEM arises from the fact that the resolution-controlling optics are before the specimen, while the detector apertures that determine contrast and the nature of the image (coherent or incoherent) are after the specimen. The probe is a demagnified image of the source, formed by a set of condenser lenses, the aberration corrector, and the objective lens, often called the *probe-forming lens* because its aberrations dominate at the high demagnifications needed for atomic resolution. If three condenser lenses are used, it is possible to vary probe angle while maintaining the same overall demagnification (and therefore probe current) or alternatively to vary probe current at the same probe angle. Once aligned, it is normal to maintain the same specimen height from sample to sample and region to region, so as not to affect the illumination conditions. The post-specimen optics compress the emerging electron beam into the spectrometer and control the angles of the various detectors.

Figure 4.5 Z-contrast STEM images of single-layer boron nitride. (a) As recorded, (b) rotated and corrected for distortion, and Fourier filtered to reduce noise and probe tails. (c) Line profiles through the locations marked in (b), normalized to one for a single B atom. Inset at top right in (a) shows the Fourier transform of an image area away from the thicker regions. The two arrows point to ($01\bar{2}0$) and ($20\bar{2}0$) reflections of the hexagonal BN that correspond to recorded spacings of 1.26 and 1.09 Å. Taken with a Nion UltraSTEM operating at 60 kV. (Reproduced from Ref. [31].)

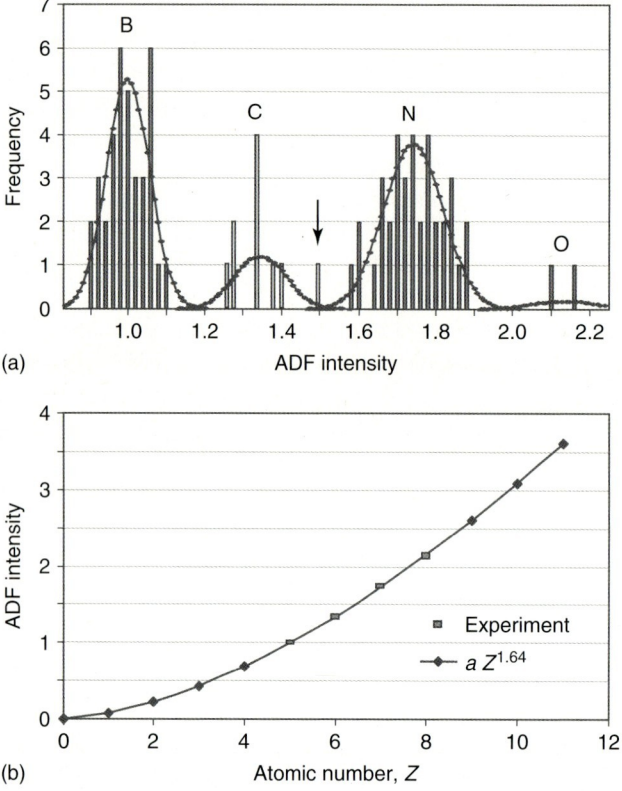

Figure 4.6 (a) Histogram of intensities of the atomic images in the monolayer area of Figure 4.5b. (b) Plot of the average intensities of the different types of atoms versus their atomic number, Z. The heights of the rectangles shown for B, C, N, and O correspond to the experimental error in determining the mean of each atomic type's intensity distribution. Only the atom arrowed in (a) cannot be unambiguously distinguished. (Reproduced from Ref. [31].)

4.3
Imaging in the STEM

4.3.1
Probe Formation

Let us assume for the moment that we have a monochromatic source of infinite brightness and arrange the condenser lenses to produce a very high demagnification onto the specimen. In this case, the image of the source will be very small and the size of the beam will depend only on the probe-forming aperture (through diffraction) and on the aberrations of the system. For a monochromatic source, we can ignore chromatic aberrations and concentrate on geometric aberrations.

4.3 Imaging in the STEM

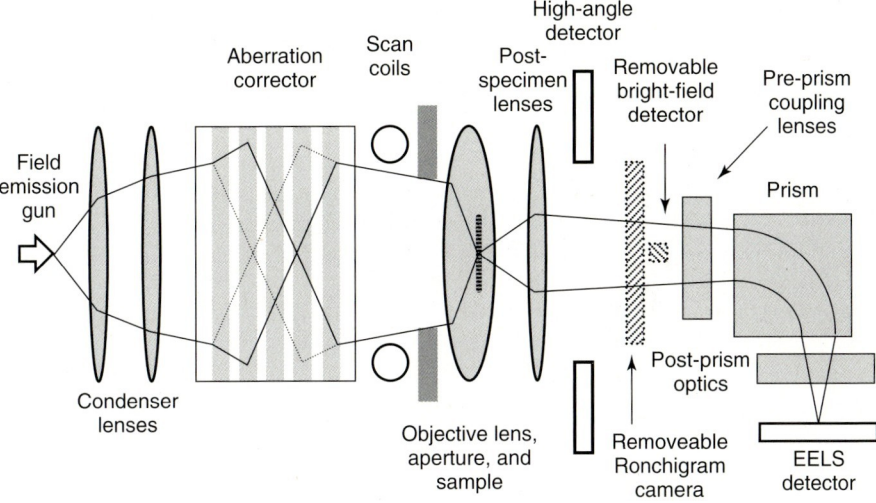

Figure 4.7 Schematic showing the main components of a high-resolution dedicated STEM. (Adapted from Ref. [42].)

To minimize diffraction the aperture should be as large as possible, but then the geometric aberrations of the optical elements increase, and the smallest probe always involves a balancing of these two effects. Ideally, what we need the illumination system to do is to form a perfectly spherical wave front that will then collapse into a point on the specimen, limited only by diffraction. The aberration is the error in optical path length between the actual wave front and the ideal spherical wave. The aberration coefficients are expressed as a power series in angle θ, the first few rotationally symmetric terms being

$$\gamma(\theta) = \frac{1}{2}\Delta f \theta^2 + \frac{1}{4}C_S\theta^4 + \frac{1}{6}C_5\theta^6 + \frac{1}{8}C_7\theta^8 + \ldots \qquad (4.1)$$

where Δf is defocus, and C_S, C_5, and C_7 are the coefficients of third-, fifth-, and seventh-order spherical aberration respectively. The aberration γ has dimensions of length; in units of radians, it is just $\chi = 2\pi\gamma/\lambda$. For round magnetic lenses, the spherical aberration coefficients are all positive, a point first noted long ago by Scherzer [1]. This means that high-angle beams are focused more strongly than beams near the axis, as shown schematically in Figure 4.8. The rays close to the axis, where the aberrations are negligible, cross the optic axis at a point called the *Gaussian focus*, which defines the reference point from which the lens defocus Δf is defined. It is clear that the smallest beam size is not in the Gaussian plane as it is in the absence of aberrations, but at a small distance behind. This position can be placed at the specimen through a small defocus (weakening) of the lens. This is the principle of aberration balancing. The negative defocus contribution in Eq. (4.1) partially compensates the positive spherical aberration term over a limited range of angles.

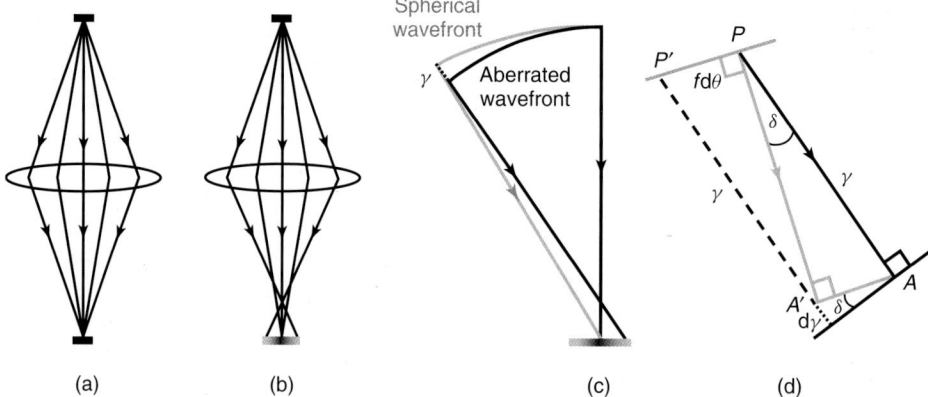

Figure 4.8 Image formation by a perfect lens (a) and an aberrated lens (b). (c) Definition of the aberration γ as the difference in path length between the perfect spherical wave front (gray) and the actual wave front (black) along the direction of the ray (dotted line). (d) Enlarged view showing the angular deviation δ is proportional to the gradient of the aberration function $d\gamma/d\theta$.

Equation (4.1) gives the aberration as a length, which is defined as the distance from a point P on the perfect wave front (Figure 4.8d) to the point A on the aberrated wave front along the trajectory of the ray. Now we can see there is a simple approximate relationship between the angular deviation of the aberrated ray, δ, and its lateral deviation. Considering an infinitesimal θ, we can consider the perfect and aberration path lengths γ and $\gamma + d\gamma$ to be parallel, and draw A A' parallel to P P' which is equal to, $f d\theta$. We therefore have $\delta = (1/f) d\gamma/d\theta$. Hence, the angular deviation is given by

$$\delta(\theta) = \frac{1}{f}\left(\Delta f \theta + C_S \theta^3 + C_5 \theta^5 + C_7 \theta^7 + \ldots\right) \tag{4.2}$$

which is the reason for naming the terms as third-order, fifth-order, and so on.

The probe amplitude distribution can now be calculated by superposition of all the contributing rays, an integration over the objective (probe-forming) aperture. Defining **R** as a two-dimensional transverse coordinate in the focal plane, at a position Δf from Gaussian focus, the probe amplitude distribution is given by

$$P(\mathbf{R}) = \int A(\mathbf{K}) e^{2\pi i \mathbf{K} \cdot \mathbf{R}} e^{-i\chi(\mathbf{K})} d\mathbf{K} \tag{4.3}$$

where the two-dimensional vector **K** is the transverse wave vector with magnitude $|\mathbf{K}| = \theta/\lambda$ and A is a circular aperture function with $A(\mathbf{K}) = 1$ inside the aperture and zero outside. The factor $e^{-i\chi(\mathbf{K})}$ is the phase change due to the aberrations, where the aberration function is now expressed as a function of **K**. Considering only the round aberrations, we have

$$\chi(\mathbf{K}) = \pi \left(\Delta f K^2 \lambda + \frac{1}{2} C_S K^4 \lambda^3 + \frac{1}{3} C_5 K^6 \lambda^5 \ldots\right) \tag{4.4}$$

4.3 Imaging in the STEM

Equation (4.3) is therefore the Fourier transform of the aberrated wave front, converting the amplitude distribution from reciprocal space to real space. The probe intensity distribution is obtained by squaring

$$p^2(\mathbf{R}) = \left| \int A(\mathbf{K}) e^{2\pi i \mathbf{K} \cdot \mathbf{R}} e^{-i\chi(\mathbf{K})} d\mathbf{K} \right|^2 \quad (4.5)$$

The choice of aperture is important, too small and the probe broadens by diffraction, too large and it broadens owing to spherical aberration. In the uncorrected case, the optimum aperture introduces one wavelength of spherical aberration at its perimeter [15, 43]. If α_{opt} denotes the semiangle of the optimum aperture, this gives

$$\alpha_{opt} = \left(\frac{4\lambda}{C_S} \right)^{1/2} \quad (4.6)$$

and the defocus is set to compensate this contribution at the aperture rim, that is,

$$\Delta f = -(C_S \lambda)^{1/2} \quad (4.7)$$

As can be seen in Figure 4.9, the probe intensity profile under these conditions is not too different from the corresponding Airy disk for the same size aperture without aberrations. There is a reduction of peak intensity which is shifted into the probe tails, but the first minimum remains close to the first zero of the Airy disk. For this reason, the optimum resolution is defined in an analogous way to that of

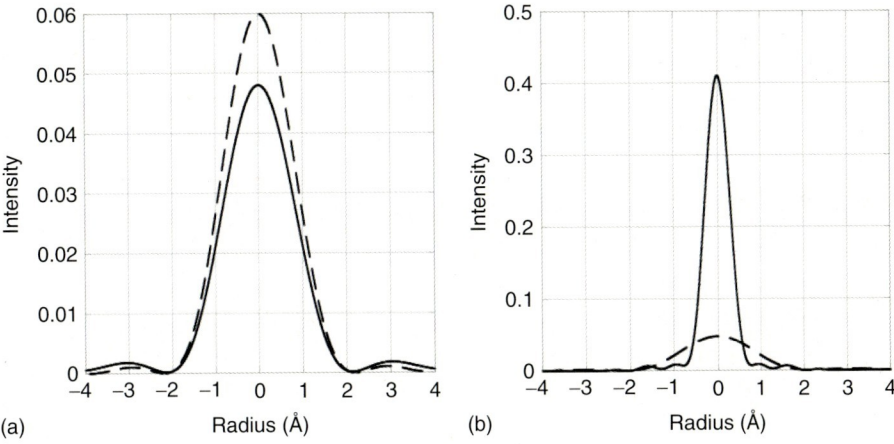

Figure 4.9 (a) Plot of the optimum probe intensity for a 100 kV microscope with C_S = 1.3 mm (solid line) compared to the ideal Airy disk distribution (dashed line). For these conditions, the optimum aperture is 10.3 mrad and optimum defocus is −69 nm.
(b) Plot of the optimum probe intensity after aberration correction (note change in vertical scale), with the uncorrected probe on the same scale for comparison (dashed line). Note the sixfold increase in peak intensity and corresponding decrease in FWHM. Probes are scaled to the same total current through the aperture. Optimum conditions in (b) are α_{opt} = 30 mrad, $C_{S opt}$ = −59 μm, and Δf_{opt} = 9.5 nm.

the Airy disk, as

$$d_{opt} = \frac{0.61\lambda}{\alpha_{opt}} = 0.43\lambda^{3/4}C_S^{1/4} \qquad (4.8)$$

For a 100 kV microscope with $C_S = 1.3$ mm, $d_{opt} = 2.2$ Å. The full width half maximum (FWHM) of the probe intensity profile is 1.8 Å.

If the third-order aberrations are corrected, then the limiting aberrations are of fifth order and we find [44] an optimum aperture of

$$\alpha_{opt} = \left(\frac{12\lambda}{C_5}\right)^{1/6} \qquad (4.9)$$

an optimum C_S of

$$C_{Sopt} = -2.4\left(C_5^2\lambda\right)^{1/3} \qquad (4.10)$$

and an optimum defocus of

$$\Delta f_{opt} = \left(C_5\lambda^2\right)^{1/3} \qquad (4.11)$$

Note how the positive C_5 is most effectively balanced by making the next lower-order aberration, C_S, negative, and the defocus positive. This maintains a phase closest to optimum over the range of the aperture, and again the aberrations add to zero at the aperture edge. If d_n denotes the limiting resolution due to nth-order aberrations, we obtain

$$d_5 = 0.4\lambda^{5/6}C_5^{1/6} \qquad (4.12)$$

which, for $C_5 = 63$ mm, gives $d_5 = 0.75$ Å. The FWHM of the probe is now 0.64 Å. The principle can be extended to the case where seventh-order aberrations are limiting, in which case C_5 would be made negative, C_S, positive, and the defocus negative [44]. Note that aberration balancing becomes much more complicated if nonround aberrations are considered [45, 46].

If these probes sound smaller than achieved typically in practice it is because a number of other factors have not been included. First is the need to have sufficient current in the probe for the task at hand, that is, we cannot have infinite demagnification from source to probe or we would have zero current. Imaging tolerates a smaller probe current than spectroscopy, but nevertheless, it is typical to have a source size comparable to the geometric size of the probe. A useful expression for the increase in probe size with probe current is given by Krivanek et al. [47]

$$d = \left(1 + \frac{1.1I}{\beta\lambda^2}\right)^{1/2} d_{opt} \qquad (4.13)$$

where d_{opt} is the geometric optimum probe and β is the gun brightness. A typical value of the normalized brightness for a cold field emission gun is 1×10^8 A (m^2 sr V)$^{-1}$, whereas for a Schottky source, it is usually $\beta = 2 \times 10^7$ A (m^2 sr V)$^{-1}$ [16], although recently, new Schottky designs have been reported to give values above 10^8 A (m^2 sr V)$^{-1}$ [48]. These values give the curves plotted in

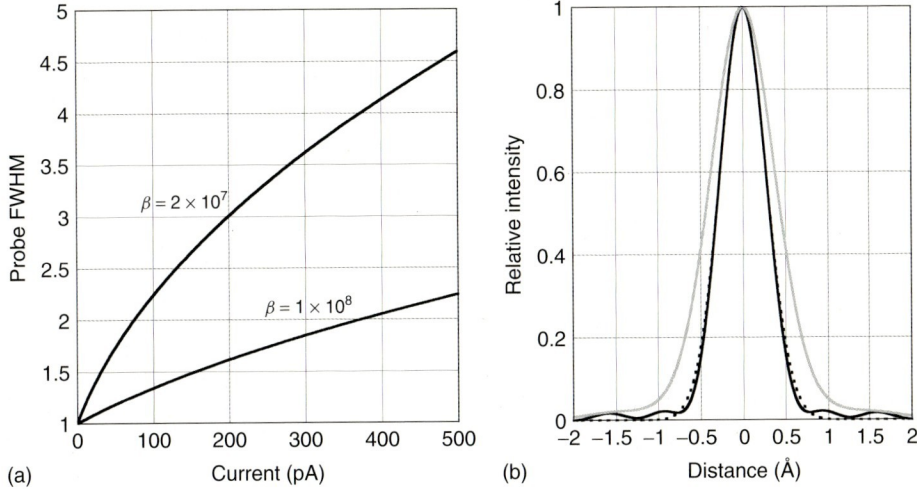

Figure 4.10 (a) Relative increase in probe size with probe current at 100 kV in units of d_{opt} for two values of gun brightness, 1×10^8 A (m² sr V)$^{-1}$ typical of a cold field emission source and 2×10^7 A (m² sr V)$^{-1}$ typical for a Schottky source. (b) Comparison of the ideal aberration-balanced probe of Figure 4.9b, black line, a Gaussian source size with the same FWHM of 0.64 Å, dotted line, and the convolution of the two, grey line, giving a probe size of around 1 Å for a probe current of about 180 pA.

Figure 4.10, showing the increase in probe size with current relative to the optimum probe, d_{opt}. With the higher brightness, 180 pA can be focused into a probe size of 1 Å, whereas with the lower brightness, only 36 pA can be achieved, five times smaller. Clearly, gun brightness is a key issue in STEM imaging or spectroscopy. Figure 4.10b shows the effect of a source size equal to the 0.64 Å FWHM of the ideal aberration-balanced probe, when the FWHM increases to around 1 Å.

The second contributor to probe broadening is chromatic aberration, which introduces a spread in focus values Δf given by

$$\Delta f = \frac{C_C \Delta E}{E} \qquad (4.14)$$

where C_C is the coefficient of chromatic aberration, ΔE is the energy spread, and E is the accelerating voltage. Since the energy spread from the gun is mostly due to the intrinsic energy spread of the cold field emission source, the spread in focus values becomes worse for lower accelerating voltages. The main effect is to transfer intensity from the peak of the probe to the tails, as shown in Figure 4.11. Note that the FWHM of the probe is unchanged, which means that the resolution will not be affected, but the contrast will be reduced owing to the extended probe tails. This is a reflection of the fact that the energy spread does not limit the resolution in STEM as it does in TEM [49]. The reason is that the highest spatial frequencies

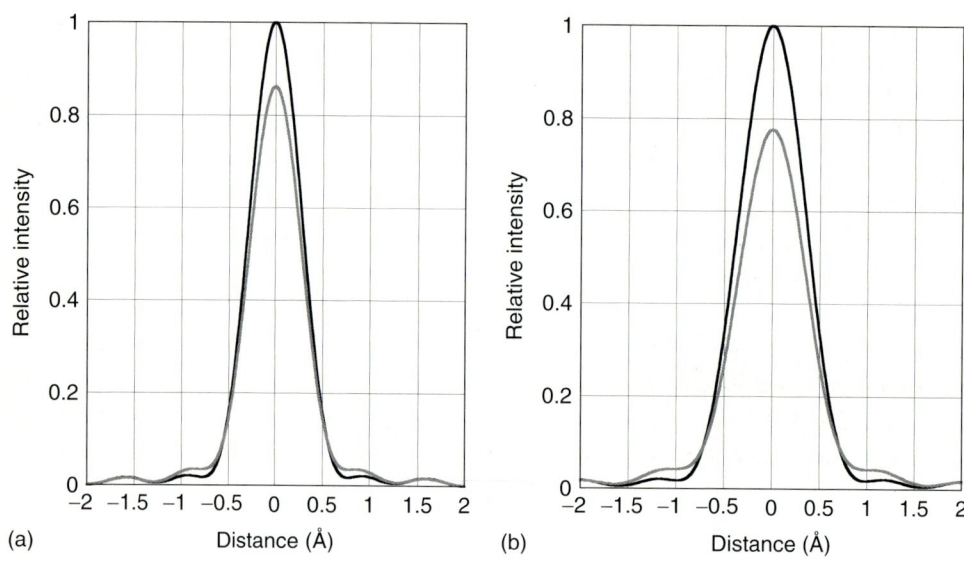

Figure 4.11 Effect of an energy spread of 0.3 eV on the optimum probes at (a) 100 kV and (b) 60 kV accelerating voltage for a third-order aberration-corrected system with $C_C = 1.6$ mm.

in the ADF image come from interference of beams in the outer regions of the probe-forming aperture, which are symmetrically positioned either side of the optic axis and equally affected by chromatic aberration. It is the lower spatial frequencies that are suppressed, resulting in the loss of contrast.

Another important contribution to probe broadening is from nonround aberrations. Ideally these would be zero up to nth order for an nth-order corrector, but in practice cannot be reduced below certain values [16, 44, 45, 47, 50, 51]. In practice, as we saw earlier for round terms, low-order nonround aberrations can be used to balance higher-order aberrations of the same symmetry, so that the limiting higher-order nonround terms are those that have no lower-order aberrations of the same symmetry, and so cannot be compensated. Figure 4.12 compares an aberration-free probe calculated for the UltraSTEM operating at 60 kV with a probe calculated using typical aberrations measured in practice. Although the FWHM is not much altered, the peak intensity is reduced by over a factor of 2, the lost intensity being distributed into tails. The probe-forming aperture size was restricted to 31 mrad as for the imaging of BN in Figure 4.5, which is seen to give little additional broadening due to chromatic aberration. The effect of chromatic aberration alone is shown in Figure 4.13 for the 31 mrad aperture and also for a 50 mrad aperture which is expected to be achievable on the basis of geometric aberrations only [45]. With the larger aperture, chromatic aberration effects are more severe.

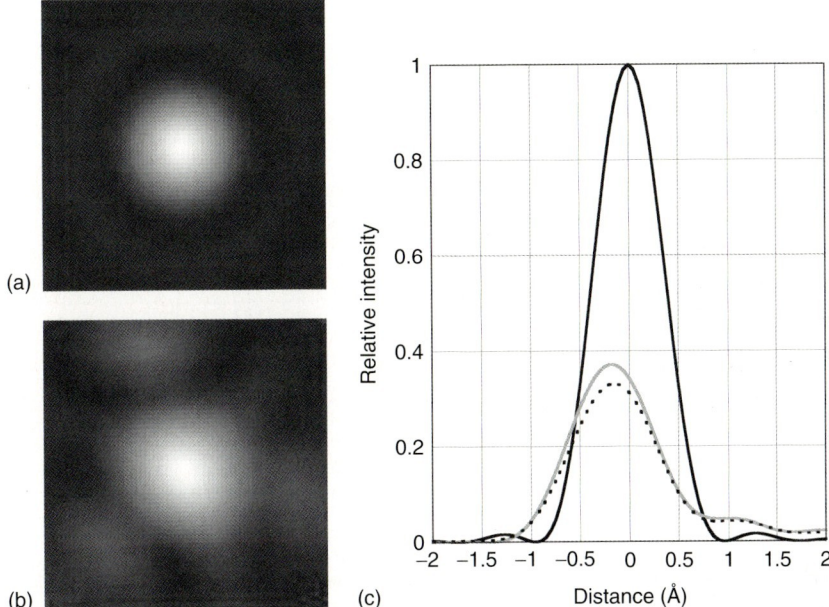

Figure 4.12 Probes calculated for the Nion UltraSTEM operating at 60 kV and 31 mrad probe-forming aperture. (a) Aberration free, (b) with a typical set of measured aberrations and uncorrected seventh-order aberrations [45]. (c) Horizontal intensity profiles through the probes shown in (a), black line, (b), grey line, and the additional effect of a 0.3 eV energy spread on (b) shown dotted, using $C_C = 1.6$ mm.

4.3.2
The Ronchigram

The electron Ronchigram is a kind of shadow image obtained in a (S)TEM that closely resembles a form of optical lens test developed by Ronchi [52], which is useful because it can provide fast and accurate diagnosis of lens aberrations. A Ronchigram is obtained by focusing a converging beam at or near the plane of a sample and observing the diffraction plane. It is an extremely interesting form of image because it is somewhere between a real space image, where position in the image corresponds to position on the sample, and a diffraction pattern, where position corresponds to angle at the sample, and Ronchigrams share characteristics of both. Since the Ronchigram is viewed in the diffraction plane, clearly the position at which a ray is detected will depend on the angle at which it went through the aperture and the diffractive properties of the sample. In the limit of a very small probe-forming aperture, the Ronchigram resembles a diffraction pattern. Using a large aperture is equivalent to superposing such patterns at different angles, which results in a convolution, broadening the spots into disks. These disks overlap and interfere and this simple picture works very well for a crystalline sample [53, 54].

A slightly different way of thinking about the Ronchigram is as the transmitted shadow of the sample, constructed from rays that pass through the sample at

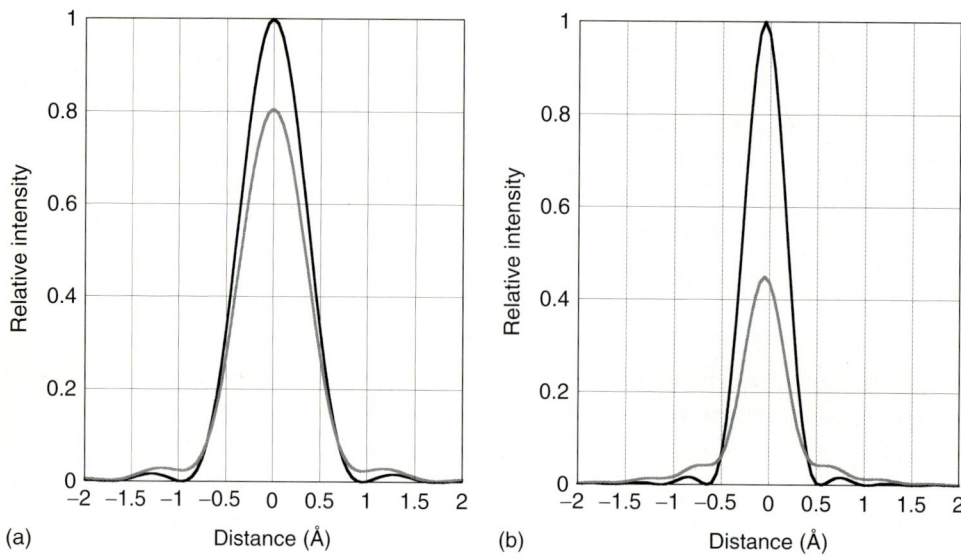

Figure 4.13 Increasing influence of chromatic aberration with increasing aperture size, (a) 31 mrad and (b) 50 mrad. Black lines are the aberration-free probes, grey lines include the effect of a 0.3 eV energy spread for $C_C = 1.6$ mm, 60 kV accelerating voltage.

different positions. This picture works well for amorphous materials and for situations with a large aperture and significant defocus. Therefore we have two rather different limits for considering the Ronchigram; however, we can see that there is a coupling between these limits. Recall, the position at which a ray goes through the sample depends on the gradient of the aberration function (Figure 4.8d). Hence for nonzero aberrations (including focus as an aberration) and a finite aperture size, there is a coupling between position and angle for rays at the sample. Thus the Ronchigram will resemble a distorted (and possibly diffracted) shadow of the sample, where the distortions depend on the lens aberrations. If the probe is focused on the specimen, the Ronchigram is mostly a coherent diffraction pattern and with a thin amorphous sample can give a clear indication of the region of flat phase contributing to the formation of a coherent probe [53, 55] as shown schematically in Figure 4.14a. The unaberrated (or aberration-corrected) rays near the center of the pattern all pass through the same region of the specimen and show the same contrast.

At larger angles, the aberrated rays pass through different regions of the sample and contrast appears in the Ronchigram. This is why the Ronchigram is so useful in aligning an electron microscope. Rays at different angles see different regions of sample. In this mode, the Ronchigram is very sensitive to angular deviations, which show up as distortions in the image; hence it forms the basis of many schemes for aberration correction [7, 8, 56–58].

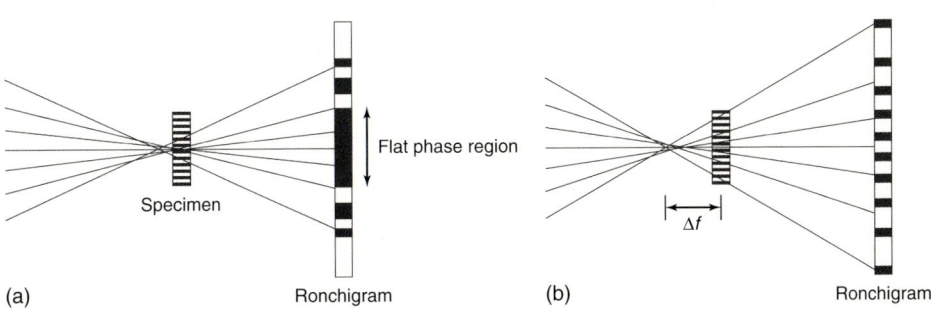

Figure 4.14 (a) Schematic showing the formation of a Ronchigram with the probe focused on the specimen. A region of constant contrast delineates the angular extent of constant phase, and represents the area to be selected by the objective aperture for probe formation. (b) Formation of a shadow image by defocusing the probe.

Far from the focus, a Ronchigram looks much like a shadow of the sample, as shown in Figure 4.14b. Figure 4.15 compares Ronchigrams as a function of focus for amorphous and crystalline specimens. Distortions in the image are due to the aberrations (in particular the uncorrected fifth-order terms) and are most apparent when underfocused. The fringes in the crystalline Ronchigram correspond to atomic planes and are also distorted by the aberrations. The approximate angular scale and the position for an aperture, about the same size as the patch of uniform phase, are indicated.

4.3.3
Reciprocity between TEM and STEM

There is a useful and interesting fundamental relationship between the TEM and the STEM that arises because image contrast in electron microscopy is predominantly due to elastic scattering. A general property of elastic scattering amplitudes is that they exhibit time reversal symmetry [59], which means that the amplitude for scattering from $\mathbf{k}$ to $\mathbf{k}'$ is equal to the amplitude for scattering in the reverse direction from $-\mathbf{k}'$ to $-\mathbf{k}$,

$$f(\mathbf{k}', \mathbf{k}) = f(-\mathbf{k}, -\mathbf{k}') \qquad (4.15)$$

Figure 4.16 shows a schematic ray diagram for a single image point in STEM and TEM. In TEM, we use a nearly parallel beam to illuminate an area of specimen and the objective lens focuses emerging waves onto a screen. In STEM, the field emission source is formed into a probe focused onto the specimen, and for bright-field imaging, the detector is a small axial aperture. The directions of the rays are reversed from the TEM situation. However, elastic scattering mechanisms will give the same image in the two cases. We see that the STEM is a time-reversed form of the TEM. The difference is all image points are obtained simultaneously in TEM but sequentially in STEM through scanning the probe. An alternative expression

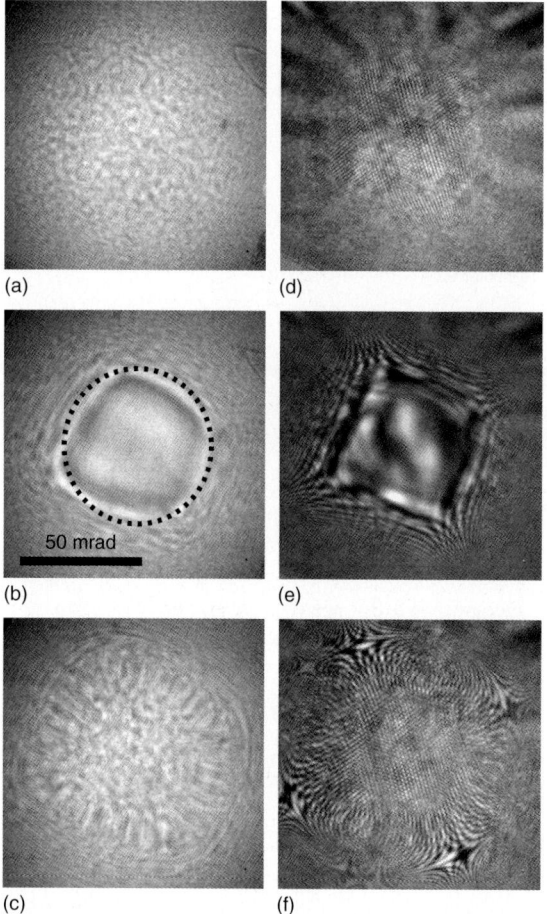

Figure 4.15 Electron Ronchigrams recorded at 300 kV on the aberration-corrected VG Microscopes HB603U. (a–c) Ronchigrams from an amorphous material, a C-film. (d–f) Ronchigrams from a crystal of GaAs. (a,d) Overfocused by approximately 250 nm, (b,e) in focus, (c,f) approximately 250 nm underfocused. In the case of the amorphous sample, the region of flat phase after aberration correction is readily visible.

of this reciprocity principle is that the amplitude scattered from a source at point A to a detector at point B is the same as for a source at B and a detector at A [60, 61]. This remains true independent of the number of optical elements involved, for example, whether or not there is an aberration corrector in the two microscopes as shown in the figure, and also applies in the presence of multiple scattering and absorptive processes [61].

In the case of inelastic scattering, the specimen changes from its ground state to an excited state. Reciprocity would only apply in this case if in the time-reversed situation the sample was initially in the same excited state. However, this obviously is unlikely to be the case.

4.3 Imaging in the STEM

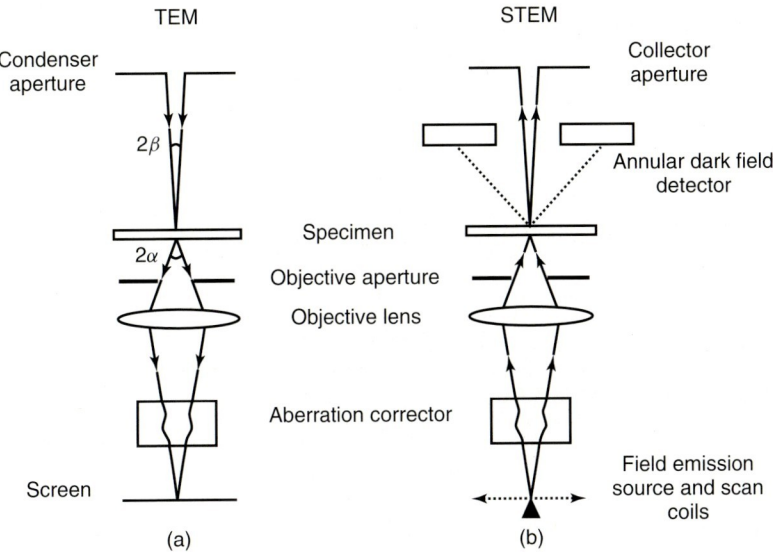

Figure 4.16 Ray diagrams for (a) the TEM and (b) the STEM, showing the reciprocal nature of the optical pathways. The TEM image is obtained in parallel, the STEM image pixel-by-pixel by scanning the probe. The STEM also provides simultaneous annular dark-field (ADF) imaging. Actual microscopes have several additional lenses and the beam-limiting aperture positions may differ. (Reproduced from Ref. [62].)

Reciprocity allows us to understand a major benefit of aberration correction in the STEM, the availability of a high-quality phase contrast image. Contrast transfer functions and damping envelopes due to beam divergence in the TEM (finite collector aperture in STEM) are shown in Figure 4.17, before and after aberration correction. Before aberration correction, the intrinsically positive C_S is optimally balanced by a negative defocus, creating a passband followed by a rapidly oscillating transfer function. The passband is negative, as the defocus term dominates at small angles; hence the phase change is negative and weak phase changes appear as dark regions. The effect of beam divergence is to average images over a range of incident beam angles, so if the range of angles becomes comparable to the period of the oscillations, they will be damped out. A damping factor can be defined on the basis of the gradient of the aberration function [63, 64], which including round terms up to fifth order gives

$$D_\alpha = \exp\left[-\pi^2\alpha^2 K^2 \left(\Delta f + \lambda^2 K^2 C_S + \lambda^4 K^4 C_5\right)^2\right] \quad (4.16)$$

where the angular distribution is modeled as a Gaussian distribution with standard deviation α. In the absence of aberration correction, this leads to a rapid attenuation of the transfer as shown in Figure 4.17a. It is also possible to describe the effect of energy spread through a damping factor

$$D_E = \exp\left[-0.5\pi^2\lambda^2 K^4 \Delta^2\right] \quad (4.17)$$

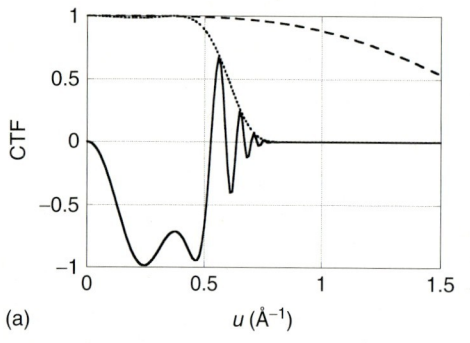 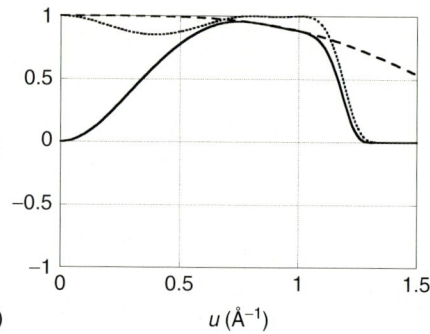

Figure 4.17 (a) Contrast transfer functions (CTFs) for the uncorrected 300 kV HB603U microscope, solid line, with $C_S = 1.0$ mm, $C_C = 1.6$ mm, and $\Delta f = -74$ nm, showing the damping envelopes introduced by a beam divergence $\alpha = 0.5$ mrad (dotted line) and an energy spread of 0.35 eV (dashed line). (b) CTF after aberration correction, solid line, ($C_S = -65$ μm, $C_5 = 100$ mm, $C_C = 1.6$ mm, $\Delta f = 10$ nm), with the damping envelopes introduced by a beam divergence $\alpha = 5$ mrad (dotted line) and an energy spread of 0.35 eV (dashed line).

where Δ is the standard deviation of the distribution of focus values, which is related to energy spread by

$$\Delta = \frac{C_C \Delta E}{2.35 E} \tag{4.18}$$

Here ΔE is the FWHM of the energy distribution of the gun, E is the accelerating voltage, and the factor of 2.35 is to convert from a FWHM to a standard deviation. For a cold field emission gun, the energy spread is not a serious limitation, even after aberration correction, except for low accelerating voltages.

Another important consideration is the current available for the phase contrast image. Before correction, the collector aperture in STEM had to be very restricted in size to be equivalent to the nearly parallel illumination of the TEM. With a probe size of around 10 mrad and a collector aperture around 1 mrad or less, it is clear that very little of the probe current could contribute to the phase contrast image, and consequently such images were usually too noisy to be of any use. However, after aberration correction, the situation is quite different. The passband is opened up, the fringes are much more widely spaced, and as a result, not only is the potential resolution greatly enhanced but the collector aperture can be opened up an order of magnitude while maintaining a good contrast transfer function (CTF), as shown in Figure 4.17b.

The important result is that phase contrast images are now of comparable quality to those obtained in the aberration-corrected TEM, and there is a significant advantage to having both images simultaneously, with pixel-to-pixel correlation. One can even perform through-focal series of simultaneous images. An example of the imaging of individual Pd atoms in an activated carbon material of interest for hydrogen storage is shown in Figure 4.18, with the corresponding CTFs. The Pd atoms are only visible in the ADF image, but the graphitic fringes are much more

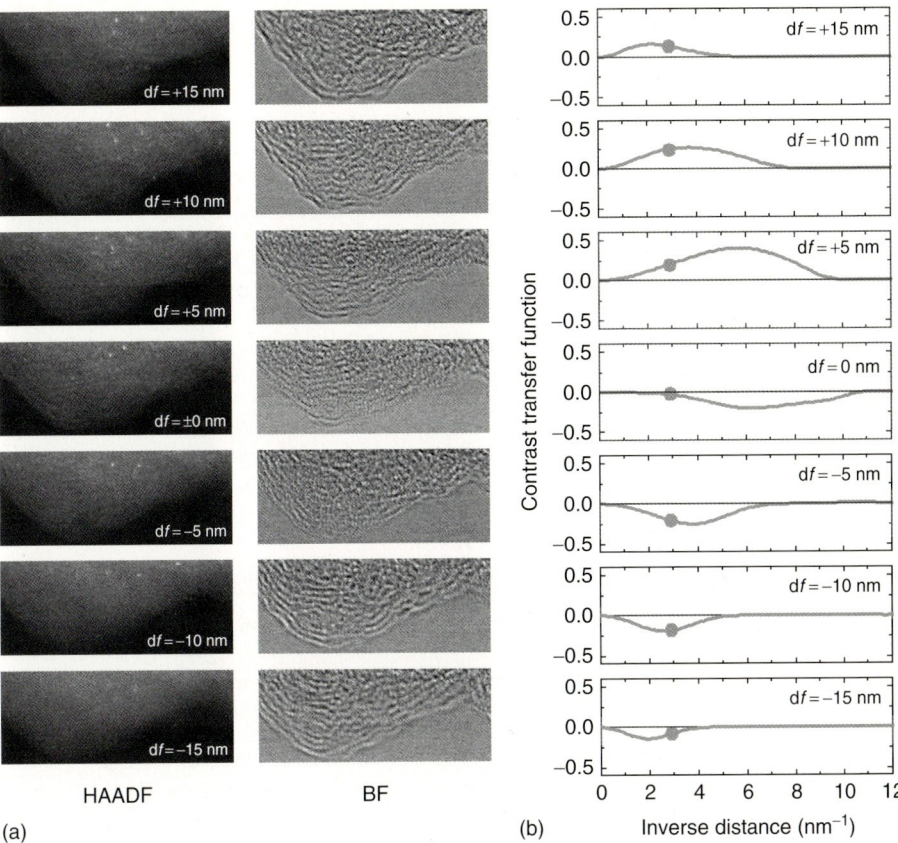

Figure 4.18 (a) Sections of seven HAADF and BF images that were simultaneously acquired as a through-focal series. The defocus is calibrated with respect to Gaussian focus. (b) Calculated contrast transfer functions for BF STEM imaging ($C_S = -37$ μm, $C_5 = 100$ mm, $C_C = 1.6$ mm, $\alpha = 10$ mrad). The characteristic inverse lattice distance for graphene layers is marked to guide the eye. At the defocus, where the Pd atoms are visible in the ADF image, the graphitic layers show dark in the BF image. (Reproduced from van Benthem et al., 2011.)

easily visible in the simultaneous phase contrast images. Note that the conditions used here are different from those in Figure 4.17. With a C_S of −37 μm, the best probe is now at a negative defocus, when carbon atoms show dark in the simultaneous phase contrast image.

4.3.4
Coherent and Incoherent Imaging

We now place our probe over a specimen, which is assumed to be very thin, so that it can be represented by the convenient transmission function. The specimen is represented as a projected potential $V(\mathbf{R})$, which, for an incident plane wave, produces a phase shift of

$$\varphi(\mathbf{R}) = e^{i\sigma V(\mathbf{R})} \tag{4.19}$$

where $\sigma = \pi/\lambda E$ is the interaction constant. For an incident probe, the exit face wave function is therefore

$$\psi(\mathbf{R}) = \varphi(\mathbf{R}) P(\mathbf{R}) \tag{4.20}$$

To allow for the probe to scan, we need to introduce a scan coordinate $\mathbf{R}_0$ that locates the center of the probe; then the wave function emerging from the specimen is written as

$$\psi(\mathbf{R}, \mathbf{R}_0) = \varphi(\mathbf{R}) P(\mathbf{R} - \mathbf{R}_0) \tag{4.21}$$

To describe the intensity in the plane of the annular detector, defined by the vector $\mathbf{K}_f$ we take the Fourier transform of the exit wave with respect to $\mathbf{K}_f$

$$\psi(\mathbf{K}_f) = \int e^{-2\pi i \mathbf{K}_f \cdot \mathbf{R}} \varphi(\mathbf{R}) P(\mathbf{R} - \mathbf{R}_0) \, d\mathbf{R} \tag{4.22}$$

Now we can take the square to find the intensity in the detector plane and integrate over a detector to find the form of the image,

$$I(\mathbf{R}_0) = \int \left| \int e^{-2\pi i \mathbf{K}_f \cdot \mathbf{R}} \varphi(\mathbf{R}) P(\mathbf{R} - \mathbf{R}_0) \, d\mathbf{R} \right|^2 d\mathbf{K}_f \tag{4.23}$$

If we first consider a point detector at $\mathbf{K}_f = 0$ we immediately recover the form of a bright-field phase contrast image,

$$I(\mathbf{R}_0) = \left| \int \varphi(\mathbf{R}) P(\mathbf{R} - \mathbf{R}_0) \, d\mathbf{R} \right|^2 \tag{4.24}$$

The integral is in the form of a convolution, and the intensity is the square of the convolution,

$$I(\mathbf{R}_0) = |\varphi(\mathbf{R}_0) \otimes P(\mathbf{R}_0)|^2 \tag{4.25}$$

Atoms can therefore look bright or dark depending on the phase of P, a key characteristic of phase contrast imaging. This is exactly the expression we would obtain for parallel illumination in the TEM, so we have an explicit demonstration of the reciprocity principle at work.

At the other extreme, the case of an infinite detector, we also find a simple form of image. Expanding the square in Eq. (4.20) we obtain

$$I(\mathbf{R}_0) = \int\int\int e^{-2\pi i \mathbf{K}_f \cdot (\mathbf{R}-\mathbf{R}')} \varphi(\mathbf{R}) \varphi*(\mathbf{R}') P(\mathbf{R}-\mathbf{R}_0) P*(\mathbf{R}'-\mathbf{R}) \, d\mathbf{R} \, d\mathbf{R}' \, d\mathbf{K}_f \tag{4.26}$$

Now, since

$$\int e^{-2\pi i \mathbf{K}_f \cdot (\mathbf{R}-\mathbf{R}')} \, d\mathbf{K}_f = \delta(\mathbf{R} - \mathbf{R}') \tag{4.27}$$

we can integrate over $\mathbf{R}'$ in Eq. (4.23) to obtain

$$I(\mathbf{R}_0) = \int |\varphi(\mathbf{R})|^2 |P(\mathbf{R} - \mathbf{R}_0)|^2 \, d\mathbf{R} \tag{4.28}$$

which again is in the form of a convolution but now of two positive quantities. Hence contrast reversals are no longer possible. This is the form of an ideal incoherent image,

$$I(\mathbf{R}_0) = |\varphi(\mathbf{R}_0)|^2 \otimes |P(\mathbf{R}_0)|^2 \tag{4.29}$$

a convolution of intensities. An example of such an imaging mode is EELS imaging, provided a large collection angle is used to collect all of the scattering [65, 66]. In cases of partial signal collection, the situation is more complex, and nonintuitive effects can arise from the nonlocal nature of the inelastic scattering probabilities [67–70].

An ADF detector can be considered an approximation to the case of infinite collection, provided that it samples a range of reciprocal space that is large compared to the (reciprocal) scale of atomic spacings. That is, in the case of a crystal, it must average over a large number of diffracted beams. This is achieved in practice by using a high-angle annular detector, with a large central hole [71, 72]. The large hole is needed because the atomic scattering factors fall off with increasing angle, and if the hole is too small, the total intensity becomes dominated by a few low-angle reflections and there is insufficient averaging to achieve incoherence. With increasing size of the hole, there are more beams around the perimeter of the hole, and the detected intensity becomes closer to that expected for incoherent scattering. If the inner angle of the ADF detector is

$$\theta_i = 1.22 \frac{\lambda}{a} \tag{4.30}$$

where a is the interatomic spacing in the transverse plane, then the image intensity will be within 5% of the incoherent result [71].

In this case, the detector can be included in the expression for incoherent imaging. Defining a detector function $D(\mathbf{K}_f)$, which is unity over the ADF detector and zero elsewhere, we can see that the detector becomes a spatial frequency filter, in the sense that only the high spatial frequencies of the object, those which are intercepted by the detector, can contribute to the image intensity. This has important consequences for image resolution, because increasing the detector inner angle means that the atomic potentials that contribute to the image become sharper, as illustrated in Figure 4.19.

Note how the full atomic potential, as used in bright-field imaging, overlaps significantly. With decreasing spacing there is therefore a fundamental limit to resolution set by the intrinsic width of the atomic potential [74]. This is not the case with the ADF detector: atoms are smaller when viewed in ADF STEM. However, there is a penalty to be paid. Increasing the inner detector angle means that there is less scattering falling on the detector, and signal-to-noise ratio can become the resolution-limiting factor [73]. In real space, the detector function is given by the

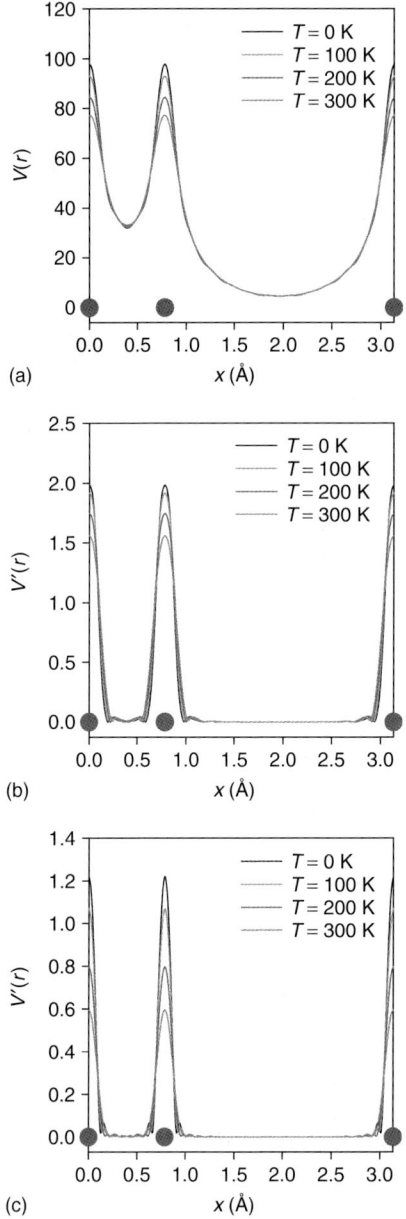

Figure 4.19 Illustration of the filtering effect of the ADF detector: Profiles of the projected scattering potential in Si ⟨112⟩ for different detectors and temperatures. (a) Full elastic scattering potential as applicable for bright-field imaging. The potential for scattering to high angles is narrower. (b) 45–200 mrad. (c) 90–200 mrad. (Adapted from Ref. [73].)

Fourier transform

$$D(\mathbf{R}) = \int e^{2\pi i \mathbf{K}_f \cdot (\mathbf{R}' - \mathbf{R})} D(\mathbf{K}_f) \, d\mathbf{K}_f = \delta(\mathbf{R} - \mathbf{R}') - \frac{2J_1(K_i, \mathbf{R} - \mathbf{R}')}{K_i(\mathbf{R} - \mathbf{R}')} \quad (4.31)$$

and Eq. (4.22) becomes

$$I(\mathbf{R}_0) = |\varphi(\mathbf{R}_0) \otimes D(\mathbf{R}_0)|^2 \otimes |P(\mathbf{R}_0)|^2 \quad (4.32)$$

Another fundamental difference in the resolution of the ADF image compared to the bright-field image arises from the different nature of the two images. As first pointed out by Lord Rayleigh in the context of light microscopy, incoherent imaging has a significant advantage in resolution because it is based on intensities, not amplitudes. The square of an amplitude distribution is always sharper than the amplitude distribution itself, by a factor of about $\sqrt{2}$ for an Airy disk. This is reflected in the well-known resolution expressions for coherent and incoherent imaging [15], that for coherent imaging being given by $0.66\lambda^{3/4}C_S^{1/4}$, approximately 50% worse than the expression for incoherent imaging which has the prefactor 0.43 (see Eq. (4.8)).

Incoherent imaging also results in a simple form of transfer function, which is referred to as an *optical or modulation transfer function*. If we Fourier transform Eq. (4.22) we obtain

$$I(\mathbf{Q}) = |\varphi(\mathbf{Q})|^2 \times |P(\mathbf{Q})|^2 \quad (4.33)$$

where $\mathbf{Q}$ represents spatial frequency. We see that the transfer function for incoherent imaging is just the Fourier transform of the probe intensity profile. Figure 4.20 shows the optical transfer functions corresponding to the coherent phase CTFs of Figure 4.18. Note the lack of contrast reversals and the extended transfer at high spatial frequencies, by approximately a factor of 2.

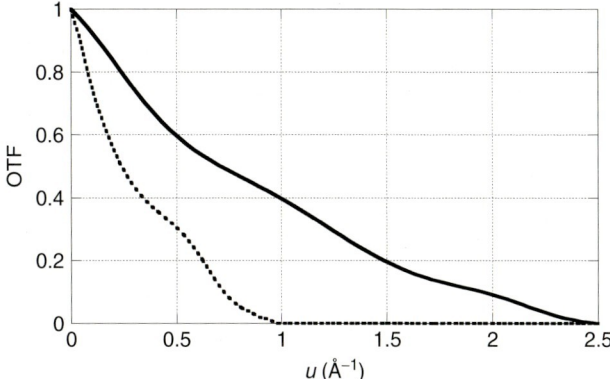

Figure 4.20 Optical transfer functions corresponding to the coherent phase contrast transfer functions of Figure 4.17, except that optimum focus for incoherent imaging was used, −44 nm in the uncorrected case (dotted line) and 7.3 nm in the corrected case (solid line).

4.3.5
Dynamical Diffraction

So far we have only considered weakly scattering thin-phase objects to illustrate the different nature of the Z-contrast image compared to a phase contrast image. One of the most remarkable aspects of Z-contrast imaging is its relative insensitivity to dynamical diffraction. The form of the image is almost independent of thickness, largely just suffering a loss of contrast in thicker samples. Atoms remain white, with a brightness that reflects their Z, although variations in thermal vibration amplitude, or static displacements can also produce image contrast. This simple dependence on specimen thickness allows a relatively intuitive interpretation of a Z-contrast image, and is quite different from the thickness dependence of a coherent image, which can show thickness fringes or contrast reversals. The reason for this useful behavior of the Z-contrast image is that the high-angle detector does not see all the probe as it scatters through a crystal. It only sees that part of the probe that is in the vicinity of the nucleus, where it can undergo high-angle (Rutherford) scattering.

The situation is best understood through a Bloch wave picture of dynamical scattering. If the probe is decomposed into a set of two-dimensional Bloch states with cylindrical symmetry centered on the atomic columns, then the contribution of each to the ADF image can be calculated [75–79]. The 1s state is most highly bound, and is preferentially effective at causing high-angle scattering, as it has high amplitude close to the nucleus. The other more weakly bound states, such as 2s or 2p states, contribute to low-angle scattering, but are less effective at high-angle scattering. In Si⟨110⟩ for example it is the interference between 1s and 2p Bloch states that gives the well-known thickness oscillations of bright-field imaging [80]. However, with the 2p states ineffective at high-angle scattering the Z-contrast image is dominated by the 1s states and the interference is greatly suppressed.

Another important difference with a phase contrast image is that the high-angle scattering is incoherent, owing to the thermal motion of the atoms. In crystals, sharp diffraction spots are found at angles given by $\theta = \lambda/d$. Clearly, at a sufficiently high angle, the planar spacing d becomes comparable to the thermal vibration amplitude. There will be no well-defined planes and the crystal will appear the same as a random arrangement of atoms. Therefore high-angle scattering is called *thermal diffuse scattering*. Once generated, it is improbable that the high-angle scattering will be rescattered back into the zone axis. For this reason, the Z-contrast image can be considered as a thickness-integrated image of the high-angle scattering, quite different from the situation with a coherent phase contrast image, which is an image of the exit wave emerging from the crystal at low angles, dominated by coherent scattering.

However, this is not to imply that the Z-contrast image samples all thicknesses equally. The 1s state preferentially generates high-angle scattering because it has a high amplitude near the peak of the atomic potential. Once the electrons are scattered to high angles, they do not return to the 1s state, and it is therefore effectively absorbed. The result is less current flowing along the column of interest, and in a sufficiently thick crystal, the high-angle scattered beams will intercept adjacent columns, an effect referred to as *beam broadening*. A column of atoms is therefore not equally sampled along its length. It also takes some distance for a probe to focus on a column before it is scattered away, the distance depending on the strength of the columnar potential. These effects result in regions near the top of a column being preferentially sampled by the beam, an effect that is stronger with columns of higher Z. Figure 4.21 shows the depth dependence of the probe intensity for the La, MnO, and O columns in $LaMnO_3$, for two different probe convergence angles, 10 mrad, typical of a noncorrected probe, and 30 mrad, typical of an aberration-corrected probe. With increasing channeling strength (higher Z), the intensity focuses faster onto the column and depletes faster, and the effect is enhanced at higher convergence angles.

Examining the total probe intensity in this manner can give a misleading view of the behavior of the HAADF image, however, since not the whole probe contributes to the image, as discussed before. In Figure 4.22, we compare the total probe intensity to the HAADF scattering as a function of thickness along the MnO column of $LaMnO_3$. For each thickness, we show a small area of 3×3 unit cells so that the beam broadening can be appreciated. Perhaps contrary to intuition, the beam broadening is reduced for the larger convergence angle, seen especially clearly for $z = 150$ Å. The figure shows that 20% of the peak HAADF signal is still being generated at that depth for the 30 mrad case, whereas with a 10 mrad probe only 15% survives; in addition, significant HAADF intensity is generated by the neighboring La columns. It appears that with the larger convergence angle high-angle scattered electrons are less likely to be captured by an adjacent column and the resulting image shows reduced beam broadening.

The analogous situation for the EELS potential is given in Figure 4.23, where it can be seen that the more delocalized potential compared to the HAADF potential (Figure 4.23b,c) leads to increased beam broadening at $z = 150$ Å. In addition, now the highest fractional signal generation at 150 Å depth is found with the smaller convergence angle.

4.3.6
Depth Sectioning

The larger convergence angles of the aberration-corrected probe not only increase the lateral resolution but actually improve the depth resolution also, bringing it into the nanometer range, less than the typical thickness of a specimen for the first time. It then becomes possible to perform optical sectioning by a focal series, obtaining sections of the sample at different depths gaining information on the three-dimensional properties of the sample [82–84]. An example of the optical sectioning of catalyst clusters is shown in Figure 4.24.

The depth resolution can be defined by a Rayleigh criterion, as the distance along the axis of the probe from the peak to the first zero [85]

$$d_z = 2\frac{\lambda}{\alpha^2} \qquad (4.34)$$

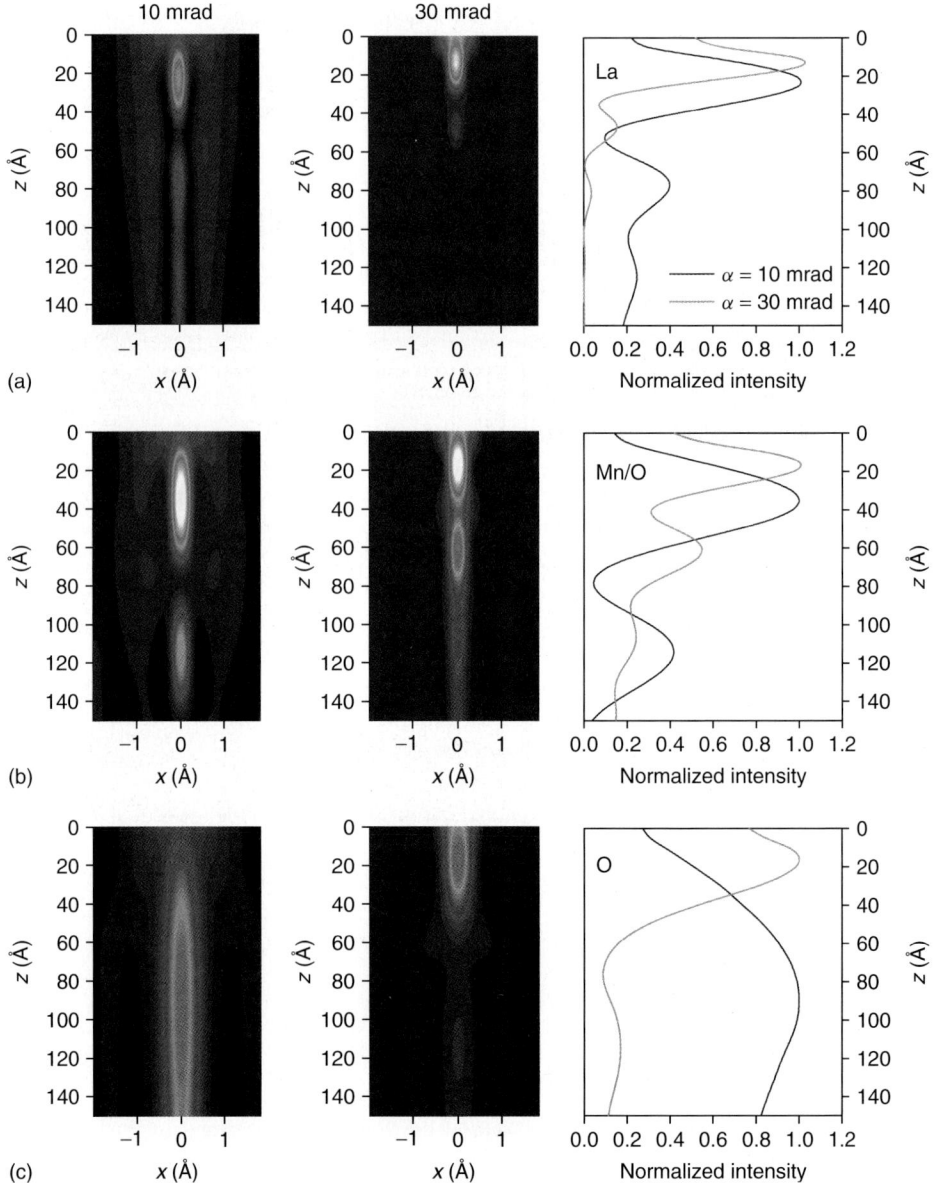

Figure 4.21 Depth dependence of the probe intensity for the La (a), MnO (b), and O (c) columns in LaMnO$_3$ (using a cubic structure) for probe convergence angles of 10 mrad (left) and 30 mrad (right). Calculations assume aberration-free probes, detector angles of 70–300 mrad, and 100 kV accelerating voltage. (Adapted from Ref. [70].)

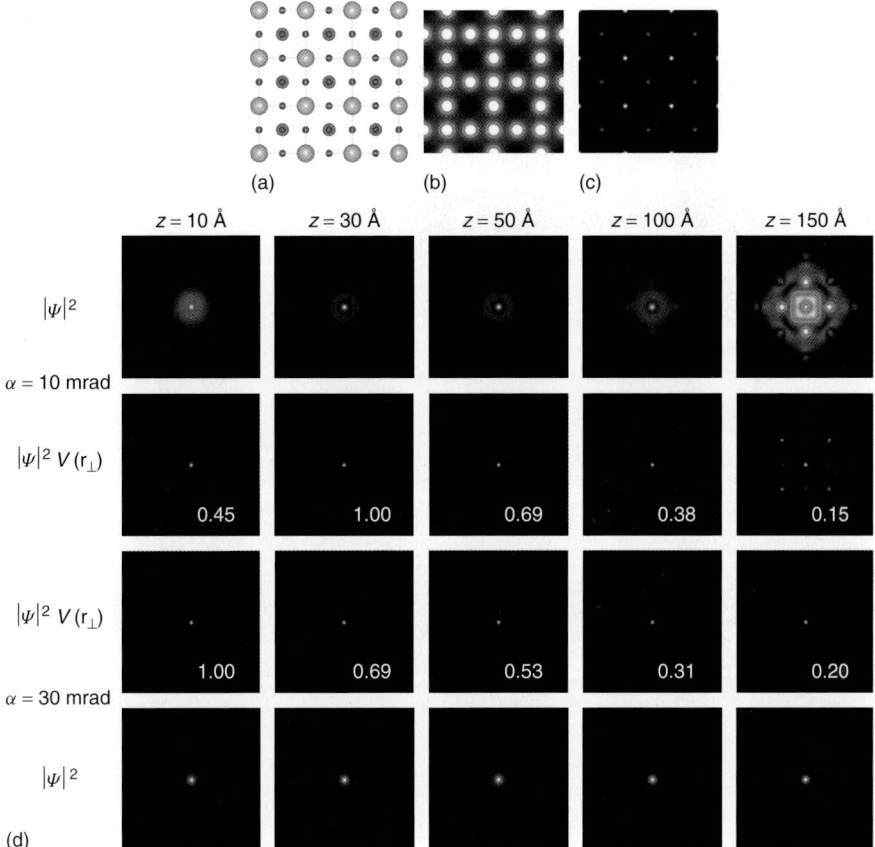

Figure 4.22 (a) Cubic structure model used for dynamical simulations of LaMnO$_3$, La as light grey spheres, Mn as dark grey spheres, and O as small black spheres. (b) O K-shell EELS projected potential for 100 keV accelerating voltage and a 40 mrad detector semiangle. (c) HAADF-projected potential for 100 keV and a detector spanning 70–300 mrad. (d) The distribution of total probe intensity $|\psi|^2$ and HAADF scattering $|\psi|^2 V(\mathbf{r}_\perp)$ at different depths along the MnO column for probe convergence angles of 10 mrad (upper two rows) and 30 mrad (lower two rows). Numbers represent the fractional signal generated at that depth normalized to the maximum. Each image represents an area of 3 × 3 unit cells. For the 10 mrad convergence angle at 150 Å depth, the probe intensity has spread significantly to the adjacent La columns and they contribute intensity to the HAADF signal. For a 30 mrad convergence angle, this is not seen.

By this definition, the depth resolution is a factor of 2 larger than the depth of field, and spacing two objects vertically by this distance results in a clear dip in the intensity between the two objects as the focus is changed. One disadvantage of the simple optical sectioning technique is the large disparity between the lateral and depth resolution. Provided individual high-Z atoms are spaced far apart in z, their depth coordinate can extracted with a precision of 0.2 nm, over an order

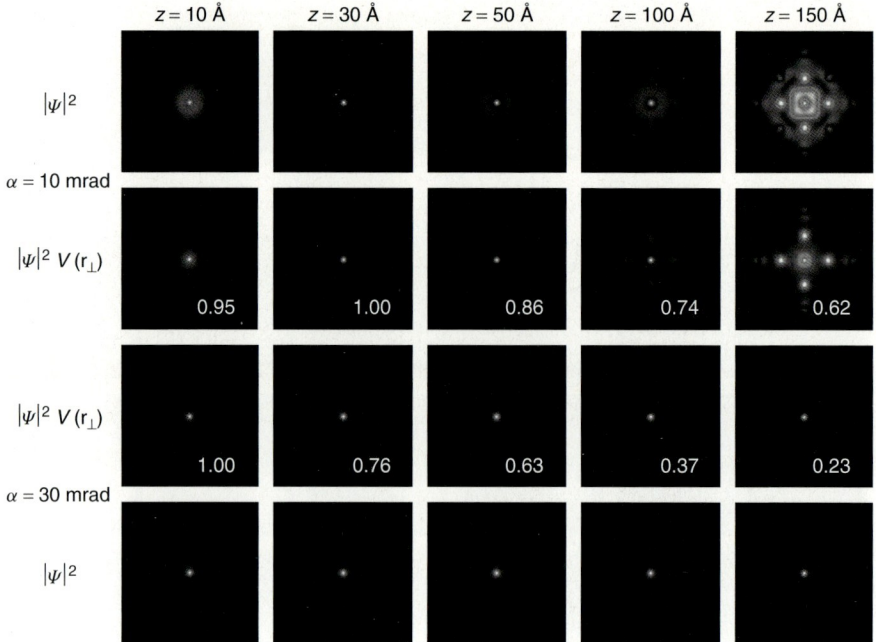

Figure 4.23 The distribution of total probe intensity $|\psi|^2$ and EELS scattering $|\psi|^2 V(r_\perp)$ at different depths along the MnO column for probe convergence angles of 10 mrad (upper two rows) and 30 mrad (lower two rows). The more delocalized EELS potential leads to higher beam broadening than the HAADF signal, and now greater signal generation at large thicknesses occur with the smaller convergence angle. (Reproduced from Ref. [81].)

of magnitude smaller than the depth resolution itself (Figure 4.24). Another disadvantage, however, is that the depth resolution degrades with finite particle size. If an object is of diameter d, then it is clear that its image intensity cannot change significantly until the probe size is comparable to d, and, in this case, the depth resolution becomes [86–88]

$$d_z = \frac{d}{\alpha} \qquad (4.35)$$

For the conditions of Figure 4.24 the theoretical depth resolution is $d_z = 7.4$ nm, whereas for a 10 nm diameter particle it is over 400 nm. This limitation is attributable to the missing wedge when viewing in only a single direction, and can be overcome with tilt series tomography [89, 90]. Although atomic lateral resolution has not yet been achieved, three-dimensional resolution below 1 nm has been demonstrated [91]. An alternative means to avoid the problem is through a true confocal mode of operation [86, 92–96].

Depth sectioning has also been demonstrated in aligned crystals provided they are not too strongly channeling, when the dominance of the 1s state in generating high-angle scattering can be overcome by all the plane waves around the perimeter

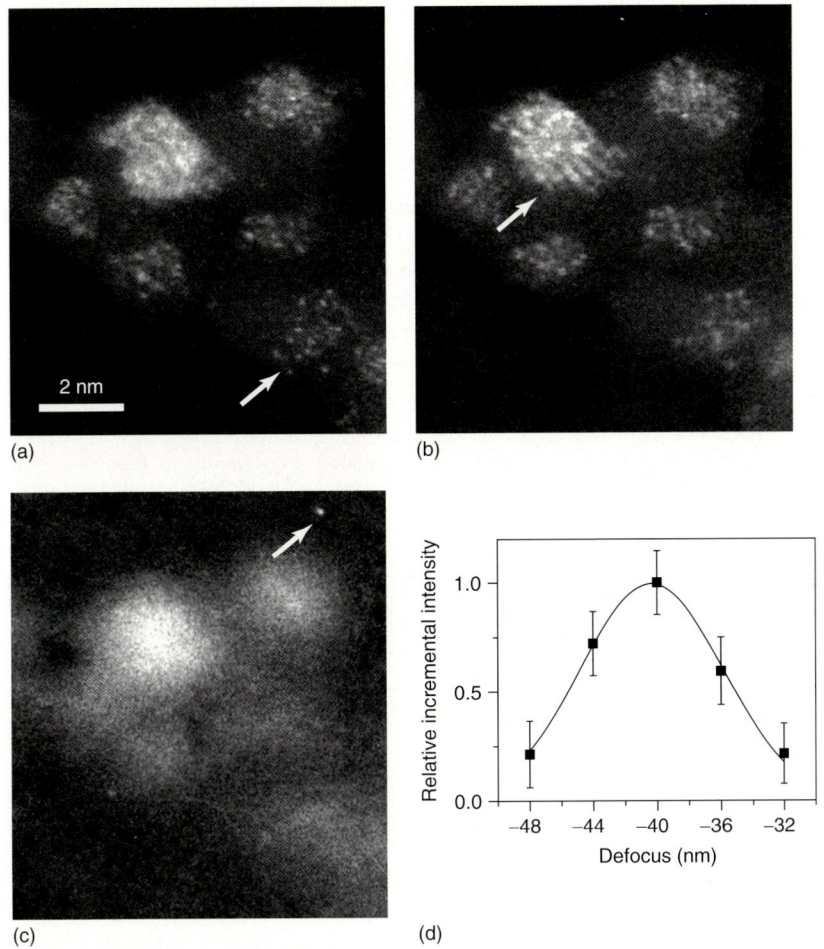

Figure 4.24 Three frames from a through-focal series of Z-contrast images from a Pt/Ru catalyst on a γ-alumina support, at defocus values of (a) −12 nm, (b) −16 nm, (c) −40 nm from the initial defocus setting. Arrows point to regions in focus, and in (c) a single atom is seen in focus on the carbon support film. (d) Integrated intensity of the Pt atom seen in (c) above the level of the carbon film as a function of defocus, compared to a Gaussian fit. The FWHM of the fit is 12 nm but the precision of the location of the peak intensity is 0.2 nm with 95% confidence. Results obtained with a 300 kV STEM with $\alpha = 23$ mrad. (Reproduced from Ref. [82].)

of the probe-forming aperture. These peripheral beams are so far from the zone axis that they propagate through the crystal essentially as plane waves and all come to a focus at a specific depth, much as they would in free space. The probe can be thought of as comprising a channeling part (the 1s state part near the zone axis) and a nonchanneling part (the outer region of the objective aperture). Which wins depends on the strength of the columnar potential [97]. Depth sectioning has worked in Si ⟨110⟩, where interstitial Au atoms have been located in various configurations inside a Si nanowire [98] and also in Si ⟨100⟩, where Bi dopants were located in depth [99].

4.3.7
Image Simulation and Quantification

Aberration correction has given improved resolution and signal-to-noise ratio, sensitivity to light atoms, and more localized images in TEM and STEM. One might therefore think that with more intuitive images, simulations are less needed, but in fact, the opposite is true. The aberration-corrected images can be interpreted to such great precision that the need for accurate simulations has if anything increased. For example, it is now feasible to track ferroelectric displacements and octahedral rotations across interfaces and domain boundaries to an accuracy of just a few picometers, both in TEM [9, 10] and in STEM [100]. Image simulations are critical to distinguish real displacements from any spurious effects due to crystal tilt or residual aberrations, for example.

Methods for image simulation fall into two major methodologies, either a multislice algorithm in which the wave function is propagated slice by slice through the crystal, or the Bloch wave method, which solves the Schrodinger equation for propagation of the electrons by an eigenvalue method. The Bloch wave method decomposes the full wave function into a set of stationary (Bloch) states each of which propagates independently through the crystal. It has the advantage that once the Bloch states are known, the full wave function can be simply reconstructed at any depth, and then the inelastic scattering generated at any depth is given by $|\psi|^2 V(\mathbf{r}_\perp)$, as in Figures 4.22 and 4.23. Using the appropriate absorptive potential, any inelastic signal, thermal diffuse, X-ray, or EELS can be simulated [101, 102].

In the Bloch wave method, it is usual to include an imaginary component to the crystal potential to account for thermal diffuse scattering; then the elastic wave function becomes depleted in intensity with increasing thickness. This is the reason for the decay of the intensity oscillations as can be seen in Figure 4.21. In actual fact, the high-angle scattered electrons are not lost, but continue to propagate through the crystal. They are diffuse since they have undergone phonon scattering, which makes them incoherent with the elastic wave function, but they can still scatter to high angles when they will still contribute to the HAADF image. For this reason, the Bloch wave method can become inaccurate at large thicknesses.

The frozen phonon method of image simulation overcomes this problem by treating the crystal by the static lattice approximation [103, 104]. Since the time the electron is inside the crystal is much shorter than the phonon vibration period, each electron sees essentially a static lattice, with the phonon vibrations frozen, but each sees a different static lattice. The frozen phonon approach calculates the diffuse scattering by averaging many coherent scattering calculations over different configurations of atomic displacements, usually calculated in an Einstein model [105]. A phonon dispersion model can also be used with similar results [106]. Although this treatment appears to ignore the loss or gain of thermal energy on phonon absorption or emission, in fact, it can be shown to be formally

4.3 Imaging in the STEM | 141

equivalent to a full quantum mechanical model of phonon scattering [107–109]. The method should therefore give the most accurate results, and indeed, fully quantitative agreement between simulated and experimental images has recently been achieved, first for the STEM HAADF image [110, 111] then also for STEM bright-field images [112, 113]. Hence the historic Stobbs factor [114] is no longer present, and we can be confident that the physics of electron scattering is sufficiently described by present methods.

As an example of the importance of image simulation in quantitative image analysis, we present the case of column shape analysis in $BiFeO_3$ [115]. Precise analysis of atomic column shapes in Z-contrast STEM images is able to reveal polarization and octahedral tilt behavior across domain walls at thicknesses far higher than those needed for phase contrast methods. In rhombohedral $BiFeO_3$, the ⟨110⟩ pseudocubic orientation is best for the observation of octahedral tilt patterns, giving two distinct projections as shown in Figure 4.25.

The O positions in these two projections are slightly different. While O contributes little to the image, its presence affects the precise shape of the Bi columns. In the $[001]_{rh}$ projection the O displacements show a checkerboard symmetry, which is reflected in a slight asymmetry of the Bi column shape, most notable in the thicker specimen. In the $[01\bar{1}]_{rh}$ projection, all the Bi columns appear identical. These shape changes were quantified using principal component analysis (PCA), as shown in Figure 4.26. The first eigenshape captured the basic round shape of the column, while higher eigenshapes captured deviations from roundness. A spatial map of component 2 produced the expected checkerboard pattern of the oxygen positions.

In Figure 4.27, the method is applied to a domain wall. There is a slight change in the appearance of both HAADF and bright-field images across the domain wall. PCA analysis of the HAADF image produces three strong components, the first again being the average column shape. The second component picks up the change in unit cell shape across the domain wall, and shows a step-like change in contrast

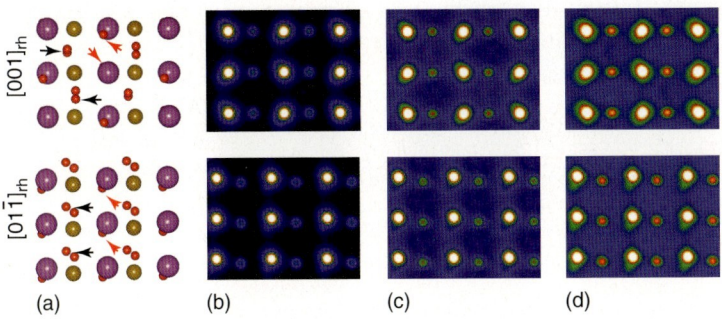

Figure 4.25 Schematic models (a) and HAADF image simulations performed with the Bloch wave method for different sample thicknesses ((b) 10 nm, (c) 20 nm, and (d) 30 nm) of the two distinct $BiFeO_3$ projections corresponding to the ⟨110⟩ pseudocubic direction: $[001]_{rh}$ (top panel) and $[01\bar{1}]_{rh}$ (bottom panel). Red and black arrows in (a) point to O positions, Bi is purple and Fe is brown. (Reproduced from Ref. [115].)

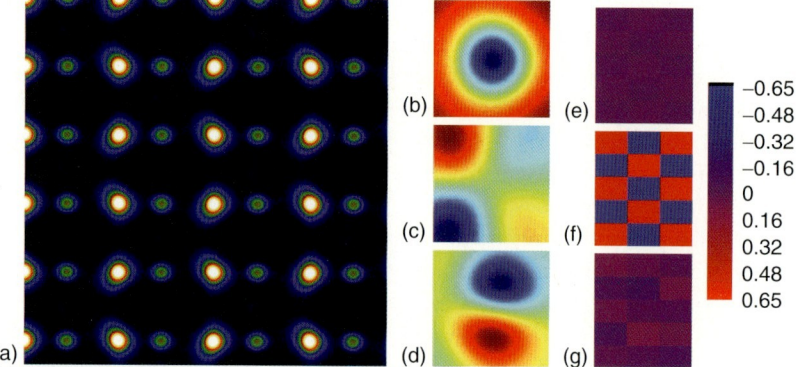

Figure 4.26 Shape PCA analysis of a simulated ADF image in the [001]$_{rh}$ orientation (a); first three eigenshapes (b–d); and the corresponding weight factor maps (e–g). Color scale is given to the right of the weight factor maps. (Reproduced from Ref. [115].)

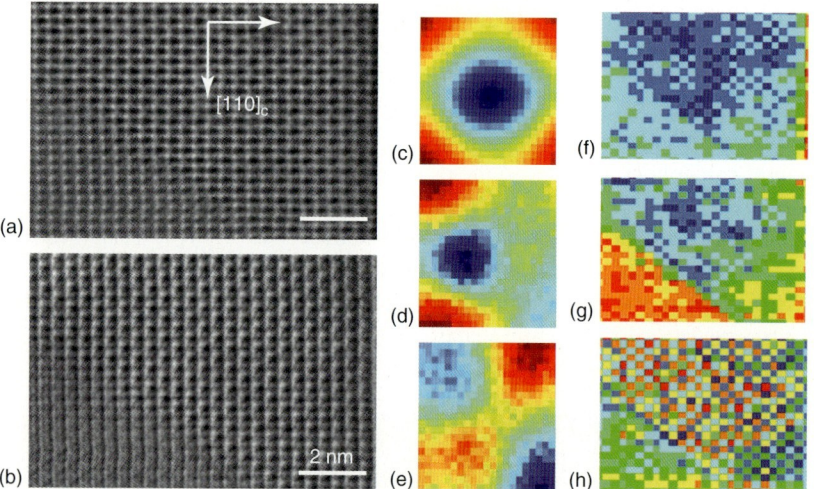

Figure 4.27 Shape PCA analysis of experimental data: (a) HAADF image used for analysis, (b) simultaneously acquired BF image showing domain contrast, (c–e) first three eigenshapes, and (f–h) the corresponding weight factor maps. (Reproduced from Ref. [115].)

at the wall. The third component picks up the checkerboard pattern of O columns in the top right-hand domain, which is not seen in the other domain. This identifies the upper-right domain as [001]$_{rh}$ type projection with a checkerboard pattern of tilts, while the lower left domain appears to be a $\left[01\bar{1}\right]_{rh}$ type projection with a uniform spatial distribution of tilts.

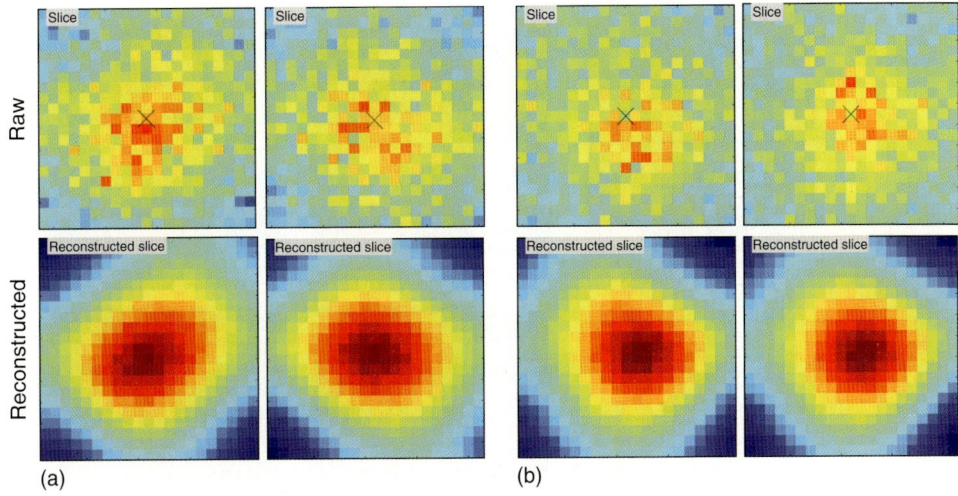

Figure 4.28 Raw (top) and PCA reconstructed (bottom) image slices taken from the image in Figure 4.27a. Panel (a) shows two atoms from the upper-right domain located on different checkerboard sublattices; their elliptical character and different "tilt" angles are apparent. Panel (b) shows two atoms from the bottom-left domain; atomic columns here are all similar in shape to each other, but different from those in (a). (Reproduced from Ref. [115].)

This remarkable sensitivity to small changes in column shape arises from the excellent noise reduction properties of PCA, which is shown in Figure 4.28 by comparing the raw image with the PCA reconstructed images for two columns in the top and bottom domains of Figure 4.27. In Figure 4.28a, the asymmetry is observable in the raw image data from the upper domain, while in Figure 4.28b, all columns show a similar shape from the lower grain. After noise reduction by PCA, the difference is much easier to see, and allows quantitative extraction of octahedral tilts.

The ability to simulate the image not only validates the analysis procedure, but can also give insight into the physics of the imaging, specifically, why is the Bi column shape distorted by the presence of the O column, is it due to scattering by the column itself (Z-contrast) or is it due to a channeling effect? Figure 4.29 shows that it is not due to scattering, as might be expected from the large Z difference between Bi ($Z = 83$) and O ($Z = 8$). Simulating the image without the O atoms produced only round columns, but including the scattering potential of the O columns also produced only round columns. Including the O atoms in calculation of the elastic wave, but not including their HAADF scattering, reproduced the observed elongated columns just as in the full simulation, proving that it is the effect of the O columns on the channeling along the Bi columns that is the source of the shape changes.

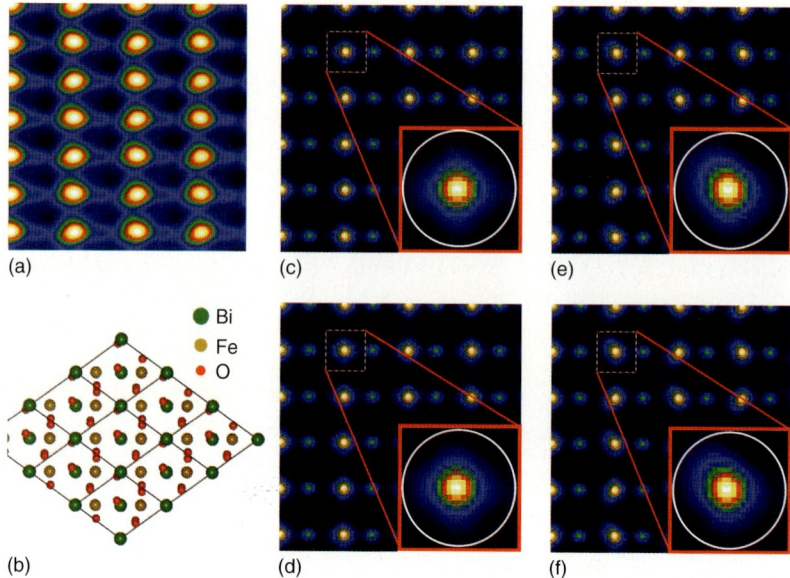

Figure 4.29 HAADF image of BiFeO$_3$ in the [001]$_{rh}$ zone axis orientation. (a) Experimental image acquired on the VG Microscopes HB603U fitted with a Nion aberration corrector. The image has nonorthogonal scan distortion removed and is Fourier filtered. (b) Projected structure. (c) Simulated image with the O atoms next to the Bi columns removed. (d) Same as (c) but with the O atoms included in the calculation of the elastic wave function. (e) Same as (c) but with the ADF scattering potential of the O atoms included. (f) Full calculation. (Reproduced from Ref. [116].)

4.4
Future Outlook

Over the last 10 years, Z-contrast STEM has become a mainstream technique available on most commercial TEM/STEM columns, and consequently has become much more widely applied across numerous areas of materials science. Several reviews of applications have appeared [30, 117–120] and also a dedicated book [121]; see also chapter 32 by Varela *et al.* in this handbook. Aberration correction has stimulated the construction of more stable microscope columns [16, 45, 122] and rooms [123], and the development of new drift correction procedures [124, 125] to mitigate the long acquisition times still necessary for spectroscopic imaging.

As far as aberration correction is concerned, it is likely that chromatic aberration correction will be of higher priority than the development of seventh-order correctors owing to their increasing complexity and the fact that low-voltage operation is becoming increasingly important to avoiding knock-on damage in light materials. Graphene, polymers, and Li-containing battery materials are a major focus at present and can only be usefully examined at low voltages. However, chromatic aberration becomes more limiting at low voltages [44, 126] and the possibility of a combined spherical and chromatic aberration corrector for the STEM is very appealing [127, 128]. We might also anticipate better correction algorithms, perhaps the development of real-time autotuning based on the Ronchigram, a signal that can be acquired very rapidly, maybe even during a long image or spectrum acquisition [57, 58].

Another area that is likely to grow is *in situ* microscopy. New hot or cold stages are under development [129] as well as the ability to image in gas or liquid environments [130, 131]. We can also expect to see more demand for the correlation of STEM Z-contrast imaging and EELS with other signals that directly reflect local functionality. For example, electron-beam-induced-current (EBIC) is a signal that can be used to map charge collection efficiency in solar cells. With the smaller, brighter probes of an aberration-corrected STEM, it may be feasible to obtain EBIC data from areas thin enough to compare with atomic-resolution imaging or EELS data, providing new insights into structure/property correlations. Cathodoluminescence is another such signal that could give new information on the role of defects on light generation in solid-state lighting materials. We are also seeing the first reports combining scanning probe microscopy with STEM to probe electrical transport and ferroelectric switching with simultaneous atomic resolution, maybe in both signals simultaneously. One area that still remains to be solved is extracting atomic coordinates and species with three-dimensional atomic resolution. It appears that this will not be possible without mechanically tilting the specimen, so will likely remain an engineering challenge for some time to come.

The last 10 years has surely been one of the most exciting times in the history of electron microscopy. Now that we are able to explore materials at the atomic scale with unprecedented capabilities, the next 10 years seem set to be equally or even more exciting.

Acknowledgments

The authors are grateful to M. Varela, P. D. Nellist, K. Sohlberg, M. F. Chisholm, O. L. Krivanek, V. Nicolosi, T. J. Pennycook, G. J. Corbin, N. Dellby, M. F. Murfitt, C. S. Own, Z. S. Szilagyi, S. T. Pantelides, K. van Benthem, Y. Peng, H. J. Chang, M. Huijben, S. Okamoto, M. K. Niranjan, J. D. Burton, E. Y. Tsymbal, Y. H. Chu, P. Yu, R. Ramesh, S. V. Kalinin, W. M. Sides, and J. T. Luck for collaboration in this research, which was supported by the US Department of Energy, Office of Basic Energy Sciences, Materials Sciences and Engineering Division. This manuscript has been authored by UT-Battelle, LLC, under Contract No. DE-AC05-00OR22725 with the U.S. Department of Energy. The United States Government retains and the publisher, by accepting the article for publication, acknowledges that the United States Government retains a non-exclusive, paid-up, irrevocable, world-wide license to publish or reproduce the published form of this manuscript, or allow others to do so, for United States Government purposes.

References

1. Scherzer, O. (1936) Über einige Fehler von Elektronenlinsen. *Z. Phys.*, **101**, 114–132.
2. Scherzer, O. (1947) Sparische und chromatische Korrektur von Electronen-linsen. *Optik*, **2**, 114–132.
3. Zach, J. and Haider, M. (1995) Aberration correction in a low voltage SEM by a multipole corrector. *Nucl. Inst. Methods A*, **363**, 316–325.
4. Haider, M., Rose, H., Uhlemann, S., Kabius, B., and Urban, K. (1998) Towards 0.1 nm resolution with the first spherically corrected transmission electron microscope. *J. Electron Microsc.*, **47**, 395–405.
5. Haider, M., Rose, H., Uhlemann, S., Schwan, E., Kabius, B., and Urban, K. (1998) A spherical-aberration-corrected 200 kV transmission electron microscope. *Ultramicroscopy*, **75**, 53–60.
6. Haider, M., Uhlemann, S., Schwan, E., Rose, H., Kabius, B., and Urban, K. (1998) Electron microscopy image enhanced. *Nature*, **392**, 768–769.
7. Krivanek, O.L., Dellby, N., and Lupini, A.R. (1999) Towards sub-angstrom electron beams. *Ultramicroscopy*, **78**, 1–11.
8. Dellby, N., Krivanek, O.L., Nellist, P.D., Batson, P.E., and Lupini, A.R. (2001) Progress in aberration-corrected scanning transmission electron microscopy. *J. Electron Microsc.*, **50**, 177–185.
9. Jia, C.L., Lentzen, M., and Urban, K. (2003) Atomic-resolution imaging of oxygen in perovskite ceramics. *Science*, **299**, 870–873.
10. Jia, C.L. and Urban, K. (2004) Atomic-resolution measurement of oxygen concentration in oxide materials. *Science*, **303**, 2001–2004.
11. Nellist, P.D., Chisholm, M.F., Dellby, N., Krivanek, O.L., Murfitt, M.F., Szilagyi, Z.S., Lupini, A.R., Borisevich, A., Sides, W.H., and Pennycook, S.J. (2004) Direct sub-angstrom imaging of a crystal lattice. *Science*, **305**, 1741–1741.
12. Varela, M., Findlay, S.D., Lupini, A.R., Christen, H.M., Borisevich, A.Y., Dellby, N., Krivanek, O.L., Nellist, P.D., Oxley, M.P., Allen, L.J., and Pennycook, S.J. (2004) Spectroscopic imaging of single atoms within a bulk solid. *Phys. Rev. Lett.*, **92**, 095502.
13. Erni, R., Rossell, M.D., Kisielowski, C., and Dahmen, U. (2009) Atomic-resolution imaging with a sub-50-pm electron probe. *Phys. Rev. Lett.*, **102**, 096101.
14. Rayleigh, L. (1896) On the theory of optical images with special reference to the microscope. *Philos. Mag.*, **42** (5), 167–195.
15. Scherzer, O. (1949) The theoretical resolution limit of the electron microscope. *J. Appl. Phys.*, **20**, 20–29.
16. Müller, H., Uhlemann, S., Hartel, P., and Haider, M. (2006) Advancing the hexapole Cs-corrector for the scanning transmission electron microscope. *Microsc. Microanal.*, **12**, 442–455.
17. Bosman, M., Keast, V.J., Garcia-Munoz, J.L., D'Alfonso, A.J., Findlay, S.D., and Allen, L.J. (2007) Two-dimensional mapping of chemical information at atomic resolution. *Phys. Rev. Lett.*, **99**, 086102.
18. Kimoto, K., Asaka, T., Nagai, T., Saito, M., Matsui, Y., and Ishizuka, K. (2007) Element-selective imaging of atomic columns in a crystal using STEM and EELS. *Nature*, **450**, 702–704.
19. Muller, D.A., Kourkoutis, L.F., Murfitt, M., Song, J.H., Hwang, H.Y., Silcox, J., Dellby, N., and Krivanek, O.L. (2008) Atomic-scale chemical imaging of composition and bonding by aberration-corrected microscopy. *Science*, **319**, 1073–1076.
20. Varela, M., Oxley, M.P., Luo, W., Tao, J., Watanabe, M., Lupini, A.R., Pantelides, S.T., and Pennycook, S.J. (2009) Atomic-resolution imaging of oxidation states in manganites. *Phys. Rev. B*, **79**, 085117.
21. Crewe, A.V. and Wall, J. (1970) A scanning microscope with 5 Å resolution. *J. Mol. Biol.*, **48**, 375–393.
22. Crewe, A.V., Wall, J., and Langmore, J. (1970) Visibility of single atoms. *Science*, **168**, 1338–1340.

23. Crewe, A.V. (2009) in *Advances in Imaging and Electron Physics*, vol. 159 (ed. P.W. Hawkes), Elsevier, Amsterdam, pp. 1–61.
24. Wall, J., Langmore, J., Isaacson, M., and Crewe, A.V. (1974) Scanning-transmission electron-microscopy at high-resolution. *Proc. Natl. Acad. Sci. U.S.A.*, **71**, 1–5.
25. Rice, S.B., Koo, J.Y., Disko, M.M., and Treacy, M.M.J. (1990) On the imaging of Pt atoms in zeolite frameworks. *Ultramicroscopy*, **34**, 108–118.
26. Pennycook, S.J. (1989) Z-contrast STEM for materials science. *Ultramicroscopy*, **30**, 58–69.
27. Nellist, P.D. and Pennycook, S.J. (1996) Direct imaging of the atomic configuration of ultradispersed catalysts. *Science*, **274**, 413–415.
28. Sohlberg, K., Rashkeev, S., Borisevich, A.Y., Pennycook, S.J., and Pantelides, S.T. (2004) Origin of anomalous Pt-Pt distances in the Pt/alumina catalytic system. *ChemPhysChem*, **5**, 1893–1897.
29. Pennycook, S.J., Chisholm, M.F., Varela, M., Lupini, A.R., Borisevich, A., Peng, Y., van Benthem, K., Shibata, N., Dravid, V.P., Prabhumirashi, P., Findlay, S.D., Oxley, M.P., Allen, L.J., Dellby, N., Nellist, P.D., Szilagyi, Z.S., and Krivanek, O.L. (2004) Materials applications of aberration-corrected STEM. 2004 Focused Interest Group Pre-Congress Meeting "Materials Research in an Aberration Free Environment".
30. Pennycook, S.J., Chisholm, M.F., Lupini, A.R., Varela, M., Borisevich, A.Y., Oxley, M.P., Luo, W.D., van Benthem, K., Oh, S.H., Sales, D.L., Molina, S.I., Garcia-Barriocanal, J., Leon, C., Santamaria, J., Rashkeev, S.N., and Pantelides, S.T. (2009) Aberration-corrected scanning transmission electron microscopy: From atomic imaging and analysis to solving energy problems. *Philos. Trans. R. Soc. A.*, **367**, 3709–3733.
31. Krivanek, O.L., Chisholm, M.F., Nicolosi, V., Pennycook, T.J., Corbin, G.J., Dellby, N., Murfitt, M.F., Own, C.S., Szilagyi, Z.S., Oxley, M.P., Pantelides, S.T., and Pennycook, S.J. (2010) Atom-by-atom structural and chemical analysis by annular dark-field electron microscopy. *Nature*, **464**, 571–574.
32. Jin, C., Lin, F., Suenaga, K., and Iijima, S. (2009) Fabrication of a freestanding boron nitride single layer and its defect assignments. *Phys. Rev. Lett.*, **102**, 195505.
33. Alem, N., Erni, R., Kisielowski, C., Rossell, M., Gannett, W., and Zettl, A. (2009) Atomically thin hexagonal boron nitride probed by ultrahigh-resolution transmission electron microscopy. *Phys. Rev. B*, **80**, 155425.
34. Meyer, J., Chuvilin, A., Algara-Siller, G., Biskupek, J., and Kaiser, U. (2009) Selective sputtering and atomic resolution imaging of atomically thin boron nitride membranes. *Nano Lett.*, **9**, 2683–2689.
35. Okunishi, E., Ishikawa, I., Sawada, H., Hosokawa, F., Hori, M., and Kondo, Y. (2009) Visualization of light elements at ultrahigh resolution by STEM annular bright field microscopy. *Microsc. Microanal.*, **15**, 164–165.
36. Findlay, S.D., Shibata, N., Sawada, H., Okunishi, E., Kondo, Y., Yamamoto, T., and Ikuhara, Y. (2009) Robust atomic resolution imaging of light elements using scanning transmission electron microscopy. *Appl. Phys. Lett.*, **95**, 191913.
37. Oshima, Y., Sawada, H., Hosokawa, F., Okunishi, E., Kaneyama, T., Kondo, Y., Niitaka, S., Takagi, H., Tanishiro, Y., and Takayanagi, K. (2010) Direct imaging of lithium atoms in LiV_2O_4 by spherical aberration-corrected electron microscopy. *J. Electron Microsc.*, **59**, 457–461.
38. Rose, H. (1974) Phase-contrast in scanning-transmission electron-microscopy. *Optik*, **39**, 416–436.
39. Batson, P.E. (1993) Simultaneous STEM imaging and electron energy-loss spectroscopy with atomic-column sensitivity. *Nature*, **366**, 727–728.
40. Duscher, G., Browning, N.D., and Pennycook, S.J. (1998) Atomic column

resolved electron energy-loss spectroscopy. *Phys. Status Solidi A*, **166**, 327–342.

41. Browning, N.D., Chisholm, M.F., and Pennycook, S.J. (1993) Atomic-resolution chemical-analysis using a scanning-transmission electron-microscope. *Nature*, **366**, 143–146.

42. Varela, M., Lupini, A.R., van Benthem, K., Borisevich, A., Chisholm, M.F., Shibata, N., Abe, E., and Pennycook, S.J. (2005) Materials characterization in the aberration-corrected scanning transmission electron microscope. *Annu. Rev. Mater. Res.*, **35**, 539–569.

43. Beck, V. and Crewe, A.V. (1975) High-resolution imaging properties of STEM. *Ultramicroscopy*, **1**, 137–144.

44. Krivanek, O.L., Nellist, P.D., Dellby, N., Murfitt, M.F., and Szilagyi, Z. (2003) Towards sub-0.5 angstrom electron beams. *Ultramicroscopy*, **96**, 229–237.

45. Krivanek, O.L., Corbin, G.J., Dellby, N., Elston, B.F., Keyse, R.J., Murfitt, M.F., Own, C.S., Szilagyi, Z.S., and Woodruff, J.W. (2008) An electron microscope for the aberration-corrected era. *Ultramicroscopy*, **108**, 179–195.

46. Krivanek, O.L., Dellby, N., and Murfitt, M.F. (2009) in *Handbook of Charged Particle Optics* (ed. J. Orloff), CRC Press, Boca Raton, FL, pp. 601–640.

47. Krivanek, O.L., Dellby, N., Keyse, R.J., Murfitt, M.F., Own, C.S., and Szilagyi, Z.S. (2008) in *Aberration-Corrected Electron Microscopy* (ed. P.W. Hawkes), Academic Press, pp. 121–160.

48. Kisielowski, C., Freitag, B., Bischoff, M., van Lin, H., Lazar, S., Knippels, G., Tiemeijer, P., van der Stam, M., von Harrach, S., Stekelenburg, M., Haider, M., Uhlemann, S., Muller, H., Hartel, P., Kabius, B., Miller, D., Petrov, I., Olson, E.A., Donchev, T., Kenik, E.A., Lupini, A.R., Bentley, J., Pennycook, S.J., Anderson, I.M., Minor, A.M., Schmid, A.K., Duden, T., Radmilovic, V., Ramasse, Q.M., Watanabe, M., Erni, R., Stach, E.A., Denes, P., and Dahmen, U. (2008) Detection of single atoms and buried defects in three dimensions by aberration-corrected electron microscope with 0.5 Å information limit. *Microsc. Microanal.*, **14**, 469–477.

49. Nellist, P.D. and Pennycook, S.J. (1998) Subangstrom resolution by underfocused incoherent transmission electron microscopy. *Phys. Rev. Lett.*, **81**, 4156–4159.

50. Haider, M., Uhlemann, S., and Zach, J. (2000) Upper limits for the residual aberrations of a high-resolution aberration-corrected STEM. *Ultramicroscopy*, **81**, 163–175.

51. Haider, M., Müller, H., Uhlemann, S., Zach, J., Loebau, U., and Hoeschen, R. (2008) Prerequisites for a C_c/C_s-corrected ultrahigh-resolution TEM. *Ultramicroscopy*, **108**, 167–178.

52. Ronchi, V. (1964) Forty years of history of a grating interferometer. *Appl. Opt.*, **3**, 437–451.

53. Cowley, J.M. (1979) Adjustment of a STEM instrument by use of shadow images. *Ultramicroscopy*, **4**, 413–418.

54. Lin, J.A. and Cowley, J.M. (1986) Calibration of the operating parameters for an HB5 STEM instrument. *Ultramicroscopy*, **19**, 31–42.

55. James, E.M. and Browning, N.D. (1999) Practical aspects of atomic resolution imaging and analysis in STEM. *Ultramicroscopy*, **78**, 125–139.

56. Sawada, H., Sannomiya, T., Hosokawa, F., Nakamichi, T., Kaneyama, T., Tomita, T., Kondo, Y., Tanaka, T., Oshima, Y., Tanishiro, Y., and Takayanagi, K. (2008) Measurement method of aberration from Ronchigram by autocorrelation function. *Ultramicroscopy*, **108**, 1467–1475.

57. Lupini, A.R., Wang, P., Nellist, P.D., Kirkland, A.I., and Pennycook, S.J. (2010) Aberration measurement using the Ronchigram contrast transfer function. *Ultramicroscopy*, **110**, 891–898.

58. Lupini, A.R. and Pennycook, S.J. (2008) Rapid autotuning for crystalline specimens from an inline hologram. *J. Electron Microsc. (Tokyo)*, **57**, 195–201.

59. Glauber, R. and Schomaker, V. (1953) The theory of electron diffraction. *Phys. Rev.*, **89**, 667–671.

60. Cowley, J.M. (1969) Image contrast in a transmission scanning electron microscope. *Appl. Phys. Lett.*, **15**, 58–59.
61. Pogany, A.P. and Turner, P.S. (1968) Reciprocity in electron diffraction and microscopy. *Acta Crystallogr. A*, **24**, 103–109.
62. Pennycook, S.J. (2006) in *Encyclopedia of Condensed Matter Physics* (eds. F. Bassani, J. Liedl, and P. Wyder), Elsevier Science Ltd, Kidlington, Oxford, pp. 240–247.
63. Frank, J. (1973) Envelope of electron-microscopic transfer-functions for partially coherent illumination. *Optik*, **38**, 519–536.
64. Hawkes, P.W. (1978) in *Advances in Optical and Electron Microscopy*, vol. 7, (eds. V.E. Cosslett, and R. Barer), Academic Press, London, pp. 101–184.
65. Ritchie, R.H. and Howie, A. (1988) Inelastic-scattering probabilities in scanning-transmission electron-microscopy. *Philos. Mag. A*, **58**, 753–767.
66. Rose, H. (1976) Image formation by inelastically scattered electrons in electron microscopy. *Optik*, **45**, 139–158 and 87–208.
67. Kohl, H. and Rose, H. (1985) Theory of image-formation by inelastically scattered electrons in the electron-microscope. *Adv. Imaging Electron Phys.*, **65**, 173–227.
68. Oxley, M.P., Cosgriff, E.C., and Allen, L.J. (2005) Nonlocality in imaging. *Phys. Rev. Lett.*, **94**, 203906.
69. Oxley, M.P. and Pennycook, S.J. (2008) Image simulation for electron energy loss spectroscopy. *Micron*, **39**, 676–684.
70. Oxley, M.P., Varela, M., Pennycook, T.J., van Benthem, K., Findlay, S.D., D'Alfonso, A.J., Allen, L.J., and Pennycook, S.J. (2007) Interpreting atomic-resolution spectroscopic images. *Phys. Rev. B*, **76**, 064303.
71. Jesson, D.E. and Pennycook, S.J. (1993) Incoherent imaging of thin specimens using coherently scattered electrons. *Proc. R. Soc. Lond. A*, **441**, 261–281.
72. Jesson, D.E. and Pennycook, S.J. (1995) Incoherent imaging of crystals using thermally scattered electrons. *Proc. R. Soc. Lond. A*, **449**, 273–293.
73. Peng, Y., Oxley, M.P., Lupini, A.R., Chisholm, M.F., and Pennycook, S.J. (2008) Spatial resolution and information transfer in scanning transmission electron microscopy. *Microsc. Microanal.*, **14**, 36–47.
74. O'Keefe, M.A., Allard, L.F., and Blom, D.A. (2005) HRTEM imaging of atoms at sub-angstrom resolution. *J. Electron Microsc.*, **54**, 169–180.
75. Pennycook, S.J. and Jesson, D.E. (1990) High-resolution incoherent imaging of crystals. *Phys. Rev. Lett.*, **64**, 938–941.
76. Pennycook, S.J. and Jesson, D.E. (1991) High-resolution Z-contrast imaging of crystals. *Ultramicroscopy*, **37**, 14–38.
77. Pennycook, S.J. and Jesson, D.E. (1992) Atomic resolution Z-contrast imaging of interfaces. *Acta Metall. Mater.*, **40**, S149–SS59.
78. Nellist, P.D. and Pennycook, S.J. (1999) Incoherent imaging using dynamically scattered coherent electrons. *Ultramicroscopy*, **78**, 111–124.
79. Nellist, P.D. and Pennycook, S.J. (2000) in *Advances in Imaging and Electron Physics*, vol. 113 (eds. B. Kazan, T. Mulvey, and P.W Hawkes), Elsevier, New York, pp. 147–203.
80. Kambe, K. (1982) Visualization of Bloch waves of high-energy electrons in high-resolution electron-microscopy. *Ultramicroscopy*, **10**, 223–227.
81. Oxley, M.P. and Pennycook, S.J. (2010) The importance of detector geometry in STEM image formation. Proceedings of the International Microscopy Congress (IMC), p. 12_10.
82. Borisevich, A.Y., Lupini, A.R., and Pennycook, S.J. (2006) Depth sectioning with the aberration-corrected scanning transmission electron microscope. *Proc. Natl. Acad. Sci. U.S.A.*, **103**, 3044–3048.
83. van Benthem, K., Lupini, A.R., Kim, M., Baik, H.S., Doh, S., Lee, J.H., Oxley, M.P., Findlay, S.D., Allen, L.J., Luck, J.T., and Pennycook, S.J. (2005) Three-dimensional imaging of individual hafnium atoms inside a semiconductor device. *Appl. Phys. Lett.*, **87**, 034104.

84. van Benthem, K., Lupini, A.R., Oxley, M.P., Findlay, S.D., Allen, L.J., and Pennycook, S.J. (2006) Three-dimensional ADF imaging of individual atoms by through-focal series scanning transmission electron microscopy. *Ultramicroscopy*, **106**, 1062–1068.
85. Born, M. and Wolf, E. (1999) *Principles of Optics*, Cambridge University Press, Cambridge.
86. Behan, G., Cosgriff, E.C., Kirkland, A.I., and Nellist, P.D. (2009) Three-dimensional imaging by optical sectioning in the aberration-corrected scanning transmission electron microscope. *Philos. Trans. R. Soc. A*, **367**, 3825–3844.
87. Xin, H.L., and Muller, D.A. (2009) Aberration-corrected ADF-STEM depth sectioning and prospects for reliable 3D imaging in S/TEM. *J. Electron Microsc.*, **58**, 157–165.
88. De Jonge, N., Sougrat, R., Northan, B., and Pennycook, S. (2010) Three-dimensional scanning transmission electron microscopy of biological specimens. *Microsc. Microanal.*, **16**, 54–63.
89. Weyland, M., Midgley, P.A., and Thomas, J.M. (2001) Electron tomography of nanoparticle catalysts on porous supports: a new technique based on Rutherford scattering. *J. Phys. Chem. B*, **105**, 7882–7886.
90. Midgley, P.A., Weyland, M., Thomas, J.M., and Johnson, B.F.G. (2001) Z-contrast tomography: a technique in three-dimensional nanostructural analysis based on Rutherford scattering. *Chem. Commun.*, 907–908.
91. Arslan, I., Yates, T.J.V., Browning, N.D., and Midgley, P.A. (2005) Embedded nanostructures revealed in three dimensions. *Science*, **309**, 2195–2198.
92. Nellist, P.D., Behan, G., Kirkland, A.I., and Hetherington, C.J.D. (2006) Confocal operation of a transmission electron microscope with two aberration correctors. *Appl. Phys. Lett.*, **89**, 124105.
93. Nellist, P.D., Cosgriff, E.C., Behan, G., and Kirkland, A.I. (2008) Imaging modes for scanning confocal electron microscopy in a double aberration-corrected transmission electron microscope. *Microsc. Microanal.*, **14**, 82–88.
94. Cosgriff, E.C., D'Alfonso, A.J., Allen, L.J., Findlay, S.D., Kirkland, A.I., and Nellist, P.D. (2008) Three-dimensional imaging in double aberration-corrected scanning confocal electron microscopy, part I: elastic scattering. *Ultramicroscopy*, **108**, 1558–1566.
95. D'Alfonso, A.J., Cosgriff, E.C., Findlay, S.D., Behan, G., Kirkland, A.I., Nellist, P.D., and Allen, L.J. (2008) Three-dimensional imaging in double aberration-corrected scanning confocal electron microscopy, part II: inelastic scattering. *Ultramicroscopy*, **108**, 1567–1578.
96. Wang, P., Behan, G., Takeguchi, M., Hashimoto, A., Mitsuishi, K., Shimojo, M., Kirkland, A.I., and Nellist, P.D. (2010) Nanoscale energy-filtered scanning confocal electron microscopy using a double-aberration-corrected transmission electron microscope. *Phys. Rev. Lett.*, **104**, 200801.
97. Borisevich, A.Y., Lupini, A.R., Travaglini, S., and Pennycook, S.J. (2006) Depth sectioning of aligned crystals with the aberration-corrected scanning transmission electron microscope. *J. Electron Microsc.*, **55**, 7–12.
98. Oh, S.H., van Benthem, K., Molina, S.I., Borisevich, A.Y., Luo, W.D., Werner, P., Zakharov, N.D., Kumar, D., Pantelides, S.T., and Pennycook, S.J. (2008) Point defect configurations of supersaturated Au atoms inside Si nanowires. *Nano Lett.*, **8**, 1016–1019.
99. Lupini, A.R., Borisevich, A.Y., Idrobo, J.C., Christen, H.M., Biegalski, M., and Pennycook, S.J. (2009) Characterizing the two- and three-dimensional resolution of an improved aberration-corrected STEM. *Microsc. Microanal.*, **15**, 441–453.
100. Borisevich, A.Y., Chang, H.J., Huijben, M., Oxley, M.P., Okamoto, S., Niranjan, M.K., Burton, J.D., Tsymbal, E.Y., Chu, Y.H., Yu, P., Ramesh, R., Kalinin, S.V., and Pennycook, S.J. (2010) Suppression of octahedral tilts and associated changes in electronic

properties at epitaxial oxide heterostructure interfaces. *Phys. Rev. Lett.*, **105**, 087204.

101. Allen, L.J., Findlay, S.D., Oxley, M.P., and Rossouw, C.J. (2003) Lattice-resolution contrast from a focused coherent electron probe. Part I. *Ultramicroscopy*, **96**, 47–63.

102. Findlay, S.D., Allen, L.J., Oxley, M.P., and Rossouw, C.J. (2003) Lattice-resolution contrast from a focused coherent electron probe. Part II. *Ultramicroscopy*, **96**, 65–81.

103. James, R. (1962) *The Optical Principles of the Diffraction of X-rays*, Bell and Sons, London.

104. Hall, C.R. and Hirsch, P.B. (1965) Effect of thermal diffuse scattering on propagation of high energy electrons through crystals. *Proc. R. Soc. Lond. A*, **286**, 158–177.

105. Loane, R.F., Xu, P.R., and Silcox, J. (1991) Thermal vibrations in convergent-beam electron-diffraction. *Acta Crystallogr. A*, **47**, 267–278.

106. Muller, D.A., Edwards, B., Kirkland, E.J., and Silcox, J. (2001) Simulation of thermal diffuse scattering including a detailed phonon dispersion curve. *Ultramicroscopy*, **86**, 371–380.

107. Wang, Z. (1998) The 'frozen-lattice' approach for incoherent phonon excitation in electron scattering. How accurate is it? *Acta Crystallogr. A*, **54**, 460–467.

108. Wang, Z.L. (1998) An optical potential approach to incoherent multiple thermal diffuse scattering in quantitative HRTEM. *Ultramicroscopy*, **74**, 7–26.

109. Van Dyck, D. (2009) Is the frozen phonon model adequate to describe inelastic phonon scattering?. *Ultramicroscopy*, **109**, 677–682.

110. LeBeau, J.M., Findlay, S.D., Allen, L.J., and Stemmer, S. (2008) Quantitative atomic resolution scanning transmission electron microscopy. *Phys. Rev. Lett.*, **100**, 206101.

111. LeBeau, J.M., and Stemmer, S. (2008) Experimental quantification of annular dark-field images in scanning transmission electron microscopy. *Ultramicroscopy*, **108**, 1653–1658.

112. LeBeau, J.M., D'Alfonso, A.J., Findlay, S.D., Stemmer, S., and Allen, L.J. (2009) Quantitative comparisons of contrast in experimental and simulated bright-field scanning transmission electron microscopy images. *Phys. Rev. B*, **80**, 214110.

113. Thust, A. (2009) High-resolution transmission electron microscopy on an absolute contrast scale. *Phys. Rev. Lett.*, **102**, 220801.

114. Hytch, M.J. and Stobbs, W.M. (1994) Quantitative comparison of high-resolution TEM images with image simulations. *Ultramicroscopy*, **53**, 191–203.

115. Borisevich, A., Ovchinnikov, O.S., Chang, H.J., Oxley, M.P., Yu, P., Seidel, J., Eliseev, E.A., Morozovska, A.N., Ramesh, R., Pennycook, S.J., and Kalinin, S.V. (2010) Mapping octahedral tilts and polarization across a domain wall in $BiFeO_3$ from Z-contrast scanning transmission electron microscopy image atomic column shape analysis. *ACS Nano*, **4**, 6071–6079.

116. Oxley, M., Chang, H., Borisevich, A., Varela, M., and Pennycook, S. (2010) Imaging of light atoms in the presence of heavy atomic columns. *Microsc. Microanal.*, **16**, 92–93.

117. Pennycook, S.J., Varela, M., Lupini, A.R., Oxley, M.P., and Chisholm, M.F. (2009) Atomic-resolution spectroscopic imaging: past, present and future. *J. Electron Microsc.*, **58**, 87–97.

118. van Benthem, K. and Pennycook, S.J. (2009) Imaging and spectroscopy of defects in semiconductors using aberration-corrected STEM. *Appl. Phys. A*, **96**, 161–169.

119. Pennycook, S.J., Varela, M., Chisholm, M.F., Borisevich, A.Y., Lupini, A.R., van Benthem, K., Oxley, M.P., Luo, W., McBride, J.R., Rosenthal, S.J., Oh, S.H., Sales, D.L., Molina, S.I., Sohlberg, K., and Pantelides, S.T. (2010) in *The Oxford Handbook of Nanoscience and Nanotechnology*, vol. 2 (eds. A.V. Narlikar and Y.Y. Fu), Oxford University Press, Oxford, pp. 205–248.

120. Varela, M., Gazquez, J., Lupini, A.R., Luck, J.T., Torija, M.A., Sharma, M.,

Leighton, C., Biegalski, M.D., Christen, H.M., Murfitt, M., Dellby, N., Krivanek, O., and Pennycook, S.J. (2010) Applications of aberration corrected scanning transmission electron microscopy and electron energy loss spectroscopy to thin oxide films and interfaces. *Int. J. Mater. Res.*, **101**, 21–26.

121. Pennycook, S.J. and Nellist, P.D. (eds) (2011) *Scanning Transmission Electron Microscopy: Imaging and Analysis*, Springer.

122. Kimoto, K., Nakamura, K., Aizawa, S., Isakozawa, S., and Matsui, Y. (2007) Development of dedicated STEM with high stability. *J. Electron Microsc.*, **56**, 17–20.

123. Muller, D.A. and Grazul, J. (2001) Optimizing the environment for sub-0.2 nm scanning transmission electron microscopy. *J. Electron Microsc.*, **50**, 219–226.

124. Saito, M., Kimoto, K., Nagai, T., Fukushima, S., Akahoshi, D., Kuwahara, H., Matsui, Y., and Ishizuka, K. (2009) Local crystal structure analysis with 10-pm accuracy using scanning transmission electron microscopy. *J. Electron Microsc.*, **58**, 131–136.

125. Kimoto, K., Asaka, T., Yu, X., Nagai, T., Matsui, Y., and Ishizuka, K. (2010) Local crystal structure analysis with several picometer precision using scanning transmission electron microscopy. *Ultramicroscopy*, **110**, 778–782.

126. Intaraprasonk, V., Xin, H., and Muller, D. (2008) Analytic derivation of optimal imaging conditions for incoherent imaging in aberration-corrected electron microscopes. *Ultramicroscopy*, **108**, 1454–1466.

127. Haider, M., Hartel, P., and Müller, H. (2009) Current and future aberration correctors for the improvement of resolution in electron microscopy. *Philos. Trans. R. Soc. A*, **367**, 3665–3682.

128. Krivanek, O.L., Ursin, J.P., Bacon, N.J., Corbin, G.J., Dellby, N., Hrncirik, P., Murfitt, M.F., Own, C.S., and Szilagyi, Z.S. (2009) High-energy-resolution monochromator for aberration-corrected scanning transmission electron microscopy/electron energy-loss spectroscopy. *Philos. Trans. R. Soc. A*, **367**, 3683–3697.

129. Allard, L.F., Bigelow, W.C., Jose-Yacaman, M., Nackashi, D.P., Damiano, J., and Mick, S.E. (2009) A new MEMS-based system for ultra-high-resolution imaging at elevated temperatures. *Microsc. Res. Tech.*, **72**, 208–215.

130. Gai, P.L. and Boyes, E.D. (2009) Advances in atomic resolution in situ environmental transmission electron microscopy and 1 angstrom aberration corrected in situ electron microscopy. *Microsc. Res. Tech.*, **72**, 153–164.

131. de Jonge, N., Poirier-Demers, N., Demers, H., Peckys, D.B., and Drouin, D. (2010) Nanometer-resolution electron microscopy through micrometers-thick water layers. *Ultramicroscopy*, **110**, 1114–1119.

5
Electron Holography

Hannes Lichte

General Idea

The key problem of conventional intensity imaging is the loss of phase information of the image wave, in particular, with perfect aberration correction. This is healed by electron holography [1–3]: from holographic intensity data the wave is completely recovered, and hence one also has the phases available [4, 5] (Figure 5.1). Then the whole wave-optical arsenal of tools is applicable for propagating the wave at will through a "virtual" optical system, for example, for correction of residual aberrations, and for analyzing the wave completely by amplitude and phase, both in real space and in Fourier space. Therefore, most of the problems arising with pure intensity imaging would be solved. In a perfect holographic scheme we come up with the following:

- Perfect phase and amplitude contrast in separate phase and amplitude images over all reconstructed spatial frequencies out to the information limit of the transmission electron microscopy (TEM),
- A posteriori aberration correction including fine tuning by numerical wave processing, for obtaining the perfect object exit wave
- Wave-optical image processing for accessing and evaluating the object data
- No nonlinearities: waves are always linear
- All data intrinsically quantitative

All this is available at a lateral resolution limited, in principle, only by the information limit of the electron microscope. However, the decisive specific limitation of performance is presented by the restricted coherence of electrons, which essentially lowers the signal/noise properties in the recovered data. This is outlined in the section 5.2.9 about performance limits.

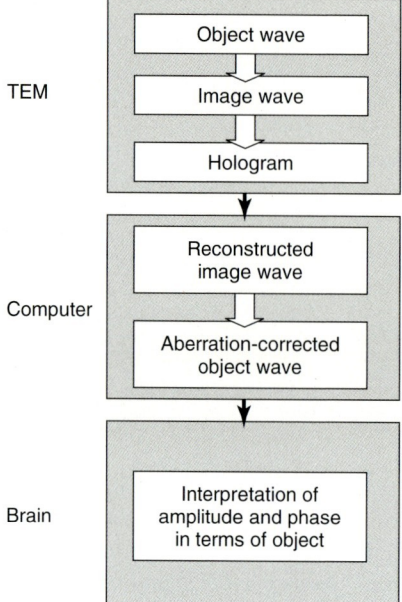

Figure 5.1 Holographic flux diagram of information. In the TEM, the object exit wave is distorted by residual lens aberrations, generally also in the case of an aberration-corrected microscope. The resulting image wave is recorded both by amplitude and phase in a hologram. This wave is reconstructed from the hologram, again by amplitude and phase, by means of numerical image processing; also, the wave is corrected from aberrations, and hence the object exit wave is at hand quantitatively; lateral resolution is given by the information limit of the TEM. Finally, the most difficult step is the interpretation of the findings, that is, of modulation of amplitude and phase, in terms of the object structure.

5.1
Image-Plane Off-Axis Holography Using the Electron Biprism

5.1.1
Recording a Hologram

Image-plane off-axis holograms [6] are recorded in the following way: the object wave is imaged by means of the objective lens into the image plane. With the help of an electron biprism [7] inserted between back focal plane and image plane, a plane reference wave adjacent to the object is coherently superimposed at an angle β to the image wave $\text{ima}(\vec{r}) = A(\vec{r}) \exp(i\phi(\vec{r}))$ (Figure 5.2).

The superposition gives rise to an interference pattern ("image-plane off-axis hologram")

$$I_{\text{hol}}(\vec{r}) = 1 + A^2(\vec{r}) + 2VA(\vec{r}) \cos(2\pi q_c x + \phi(\vec{r})) \tag{5.1}$$

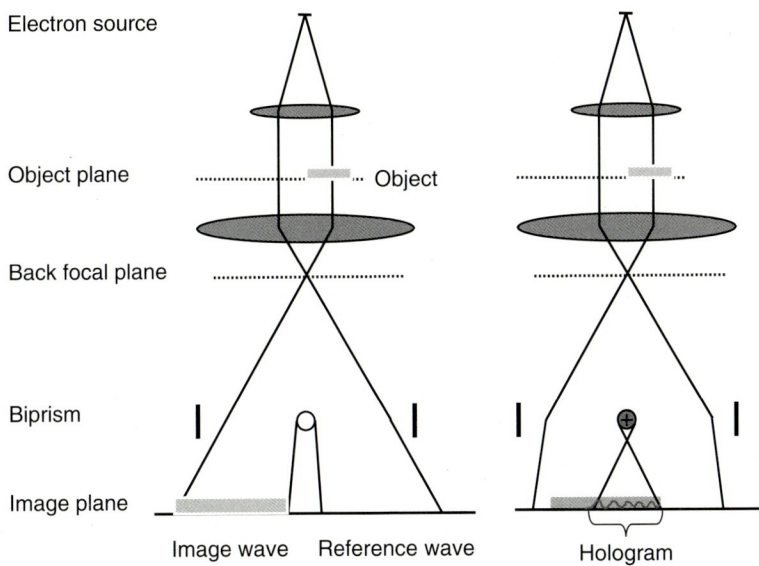

Figure 5.2 Principle of recording image-plane off-axis holograms in a TEM. In the image plane, the image wave configured by the objective lens is superimposed onto a structure-free reference wave. For given geometry, the biprism voltage determines both the field of view and fringe spacing (determining lateral resolution) by the angle of superposition. Usually, these are related to the object by division of the relations given in the figure by the magnification of the objective lens. The hologram is magnified to the level of the CCD camera and fed into a computer.

The contrast V ($0 \leq V \leq 1$) of the hologram fringes with spatial frequency $q_c = k\beta$ is determined by the degree of coherence $|\mu|$, by instabilities of the microscope V_{inst}, and by the modulation transfer function (MTF) of the detector V_{MTF}; furthermore, the fraction of inelastic interaction gives rise to object-related contrast damping V_{inel}. Therefore, $V = |\mu| V_{inst} V_{inel} V_{MTF}$ results. For convenience, $\vec{r} = (x, y)$ is related to the object exit plane. The hologram, suitably magnified by the projector lenses, is recorded by means of a charge-coupled device (CCD) camera. The magnification has to be chosen such that each hologram fringe is sampled by more than 4 pixels of the CCD camera [8]. Consequently, the pixel number of the camera limits the number of recordable fringes and hence the usable hologram width. The minimum size of a CCD camera should be 1k × 1k pixels.

Nowadays, the performance of commercial electron microscopes, equipped with high-brightness electron guns for coherent illumination, is so good that holograms can be recorded by anybody with due care. Besides the electron biprism, also available commercially, one needs the same accuracy for alignment and operation of a microscope, as needed for taking conventional electron micrographs at the atomic-resolution limit.

5.1.2
Reconstruction of the Electron Wave

From the hologram, the image wave is reconstructed with the help of numerical wave-optical image processing, which is very flexible in performing all thinkable mathematical steps of evaluating and displaying the data. Dedicated software for image processing is available, but one can perform the reconstruction also by writing the needed commands for more universal mathematics software, such as Mathematica®. The computing power of high-end desktopcomputer or even laptop computer is sufficient.

The principal way of reconstructing the image wave is the following (Figure 5.3): in the Fourier transform of the hologram,

$$\mathrm{FT}(I_{\mathrm{hol}}) =$$

$$\begin{array}{ll} \delta(\vec{q}) + \mathrm{FT}(A^2) & \text{Center band} \\ + V\mathrm{FT}(A\exp(i\phi)) \otimes \delta(\vec{q} - \vec{q}_c) & + \text{Sideband} \\ + V\mathrm{FT}(A\exp(-i\phi)) \otimes \delta(\vec{q} + \vec{q}_c) & - \text{Sideband} \end{array} \tag{5.2}$$

with spatial frequency $\vec{q}$ as Fourier coordinate, one finds three bands: the center band corresponds to the diffractogram of the conventional intensity image, which does not contain any image phase ϕ and hence is of no further interest here. The

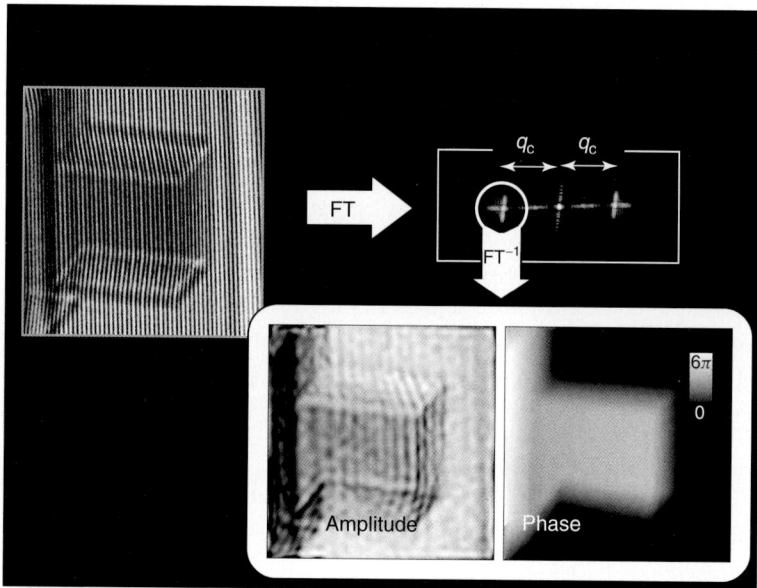

Figure 5.3 Reconstruction of the image wave from an image-plane off-axis hologram. In the Fourier spectrum of the hologram, one finds the three waves separated by the carrier spatial frequency $q_c = k\beta$ of the hologram fringes. The sideband with the Fourier transform of the image wave is selected, masked out, and centered. After inverse Fourier transform, the reconstructed image wave is displayed in real space by amplitude and phase.

two sidebands, however, represent the Fourier spectrum of the complete image wave and its conjugate, respectively. One of the two redundant sidebands is masked out and centered around $\vec{q} = \vec{0}$, so that after inverse Fourier transform the complete image wave

$$\text{ima}_{\text{rec}}(\vec{r}) = A_{\text{rec}}(\vec{r}) \exp(i\phi_{\text{rec}}(\vec{r})) \tag{5.3}$$

is found in real space; since the fringe contrast V damps the wave, special attention has to be paid to optimize all contributions to V, that is, coherence, stability of the microscope and its environment, and MTF of the camera. Furthermore, the inelastic contributions from the object have to be considered by selecting an optimum thickness optimizing the signal/noise properties of the reconstructed wave.

From the reconstructed wave given as a complex array of data $\Re + i\Im$, two images, that is, the amplitude image $A_{\text{rec}}(\vec{r})$ and the phase image $\phi_{\text{rec}}(\vec{r})$, can be extracted separately and, most essentially, quantitatively. The amplitude is calculated by means of

$$A_{\text{rec}}(\vec{r}) = \sqrt{\Re^2(\text{ima}_{\text{rec}}(\vec{r})) + \Im^2(\text{ima}_{\text{rec}}(\vec{r}))} \tag{5.4}$$

and the phase by

$$\phi_{\text{rec}}(\vec{r}) = \text{ArcTan}\left[\frac{\Im\left[\text{ima}_{\text{rec}}(\vec{r})\right]}{\Re\left[\text{ima}_{\text{rec}}(\vec{r})\right]}\right] \tag{5.5}$$

The data may be displayed as grayscale images, from which locally they may be read out for each pixel or as profile scans. Likewise, contour plots or arrow plots representing, for example, the phase gradient field are often very helpful for visual and quantitative interpretation. An example is shown in Figure 5.4.

5.2
Properties of the Reconstructed Wave

What is the fidelity of the reconstructed wave with the image wave present in the TEM? Ideally, $\text{ima}_{\text{rec}}(\vec{r}) \equiv \text{ima}(\vec{r})$ holds, that is, $V = 1$, $A_{\text{rec}}(\vec{r}) \equiv A(\vec{r})$, and $\phi_{\text{rec}}(\vec{r}) \equiv \phi(\vec{r})$. The properties of the really reconstructed image wave are outlined in the following.

5.2.1
Time Averaging

The reconstructed wave represents a wave, which is time averaged over exposure time. This comes about from collecting a large number of electrons in the hologram, where each may have "seen" a slightly different object during an extremely short interaction time of the order of 10^{-17} s; for example, with oscillating atoms, each electron experiences a seemingly frozen-in but different oscillation state of the atom. Therefore, the waves of subsequent electrons differ accordingly, and the

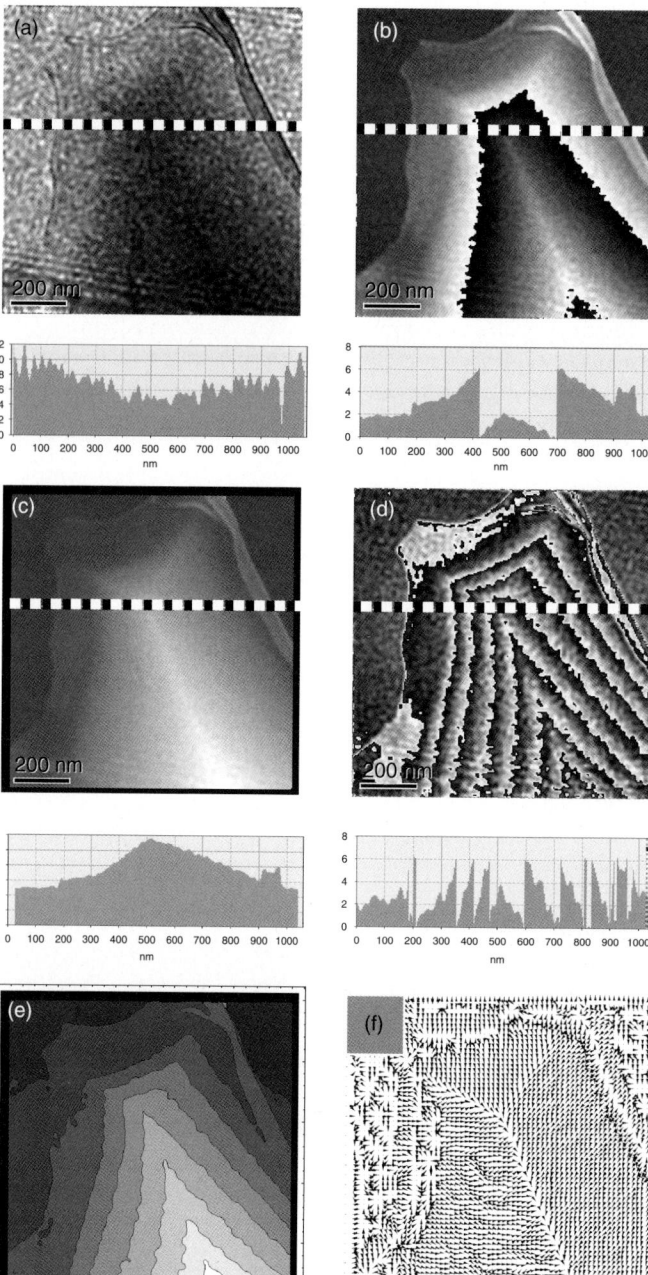

recorded hologram reads as

$$I_{\text{hol}}(\vec{r}) = \frac{1}{\tau}\int_0^\tau 1 + A^2(\vec{r};t)dt + \frac{2V}{\tau}\int_0^\tau A(\vec{r};t)\cos(2\pi q_c x + \phi(\vec{r};t))dt \qquad (5.6)$$

with the time-dependent amplitude $A(\vec{r},t)$ and phase $\phi(\vec{r},t)$. Consequently, the reconstructed wave results as

$$\text{ima}_{\text{rec}}(\vec{r}) = \frac{V}{\tau}\int_0^\tau A(\vec{r},t)\exp(i\phi(\vec{r},t))dt := A_{\text{rec}}(\vec{r})\exp(i\phi_{\text{rec}}(\vec{r})) \qquad (5.7)$$

integrated over exposure time τ of the hologram.

This time-averaged wave may have as such never existed. As an example, assume an object with a pure phase detail, which changes the phase shift from $\phi = \phi_0(\vec{r})$ to $\phi = 0$ halfway during exposure time; for example, this could happen by an atom hopping away under electron irradiation. The reconstructed phase will be halved in value $\phi_{\text{rec}}(\vec{r}) = \phi_0(\vec{r})/2$, and a reconstructed amplitude $A_{\text{rec}}(\vec{r}) = V\cos(\phi_0(\vec{r})/2)$ will show up, which was not present in the object. Of course, similar time-averaging effects exist with all other imaging techniques in the TEM; there, however, the averaging procedure is different [9].

5.2.2
Inelastic Filtering

In a usual intensity image, we find both elastic and inelastic contributions, which impede the interpretation. Therefore, a clear separation would be highly desirable.

As mentioned in the section 5.4.3 about inelastic coherence, we talk about inelastic interaction, if the total energy of the electron differs by more than 10^{-15} eV from its reference wave. Then we lose coherence with the purely elastic reference wave passing by the object, with the consequence that inelastically scattered electrons do not contribute to the hologram fringes nor to the reconstructed wave. This suggests a perfect filtering from inelastic contributions (not from the energy spread of illumination!) in the reconstructed wave. However, we found that at inelastic scattering also there is a certain extension of coherence *within* the inelastic wave field [10–12]; this amounts to a few 10 nm radius in the case of plasmons and surface plasmons. At the rim of an object, these inelastic coherent patches may reach out into the reference area and hence offer a coherent partner in the reference

Figure 5.4 Representation of the reconstructed wave. (a) Amplitude and (b) phase, as reconstructed. The grayscale increases from black to white with increasing phase values. The lines with strong black/white contrast are equal-phase lines. They show the transition $0 \leftrightarrow 2\pi$ arising from "phase wrapping" under evaluation of the arctan-function. As equiphasal lines, they give an intuitive idea about the phase distribution. Note from the line scans along the dotted lines that signal/noise is much better in phase than in amplitude. Furthermore, in kinematical approximation, the phase can directly be interpreted as projected potential. Differently processed phase images based on the same data set: (c) unwrapped by suitably adding 2π, (d) $4\times$ amplified, (e) contour map, (f) arrow-plot of phase gradient. The images b–f display the same phase distribution of the etched PZT-crystal looking like a bird's eye view of a hipped roof clearly revealing the etching structure of the crystal. Assuming homogeneous mean inner potential (MIP), the phase represents the thickness distribution $t(x,y) = \frac{\phi_{\text{rec}}(x,y)}{\sigma \text{MIP}}$.

wave for interference with the inelastic event in the object wave. Consequently, inelastic scattering may contribute to the hologram fringes, if the distance between points ("shear"), subsequently superimposed by the biprism, is smaller than the inelastic coherence extension. Therefore, filtering from inelastic interaction is not perfect, but at usual geometries, shear is so large that inelastic contributions to the reconstructed wave are negligibly small [13].

5.2.3
Basis for Recovering the Object Exit Wave

In particular, at high resolution, the image wave does not agree with the object exit wave because it is still distorted by aberrations

$$\text{ima}_{\text{rec}}(\vec{r}) = \text{obj}_{\text{rec}}(\vec{r}) \otimes \text{PSF}(\vec{r}) \tag{5.8}$$

However, it provides the basis for a posteriori correction and fine tuning of the coherent aberrations, in order to regain the (time-averaged) object exit wave by deconvolution from $\text{PSF}(\vec{r})$. This is best done in Fourier space by division of the image spectrum by the wave transfer function $\text{WTF}(\vec{q}) = \text{FT}(\text{PSF}(\vec{r}))$. Here, the most severe problem is the accurate determination of the parameters defining the wave aberration $\chi(\vec{q})$ active under recording the hologram [14]. An accuracy of $\pi/6$ is needed to suppress the crosstalk between amplitude and phase given by the $\sin \chi$-function and hence to clearly separate them in the reconstructed object exit wave.

5.2.4
Amplitude Image

The contrast V of the hologram fringes damps the reconstructed wave, for example, by the restricted spatial coherence, by instabilities, by the MTF of the CCD camera, and by inelastic interaction, that is, loss of coherent electrons. Contrast damping affects primarily the amplitude, and it should always be normalized with the amplitude reconstructed in vacuum set to $A_{\text{vac}} = 1$. Then, quantitative evaluation of the amplitude image is possible.

5.2.5
Phase Image

In addition to the wanted phase shift by the object, distortions of the hologram fringes also arise from imperfections of the biprism filament, from field aberrations of the optics, and from charging effects in the column. These unwanted artificial phase modulations are usually corrected by means of an empty reference hologram containing only the distortions.

Another problem with the phase arises, if the reference wave is not plane, instead modulated by far-reaching electric or magnetic fields from the object or from the surrounding outside the field of view. Then, an artificial phase is added to the wanted object-related phase with the risk of misinterpretation. In simple cases, the unwanted fields may be modeled and corrected for; however, this may be very cumbersome.

Owing to the periodicity of the ArcTan-function, phases larger than 2π appear "wrapped," which means that, surmounting 2π and multiples, the phase value starts at zero again. Correspondingly, in the phase image, lines of strong contrast appear at these phase jumps. For some evaluations, for example, in terms of electric potentials, phase wrapping is hindering. Also, before evaluation of the phase function by mathematical functions, phases must be unwrapped. Respective adding or subtracting 2π at these phase jumps allows unwrapping the phases for evaluation over the whole range of phase values. Corresponding algorithms are available [15]. Pathological phase jumps, where the wrapping line ends abruptly in the field of view, cannot be unwrapped easily. They arise because of undersampling, that is, because of a local steep phase gradient, the fringes are much closer than on average. These phase structures cannot be interpreted.

Fortunately, the phase values are not directly affected by the fringe contrast. However, as shown subsequently, the noise level increases, making the phase values less precise.

5.2.6
Field of View

The reconstructed field of view is given by the part of the hologram, which is captured by the camera. Optimum exploitation – also in the sense of lateral resolution – is given, if the hologram width just covers the CCD camera. Since more than 4 camera pixels have to sample one fringe, at the end the camera size determines the maximum recordable field of view.

5.2.7
Lateral Resolution

Lateral resolution is first of all given by the information content of the object exit wave, namely, the scattering properties of the object, and by the information limit of the TEM. Point resolution of the TEM does not play a role in lateral resolution, since the coherent aberrations can be corrected afterwards; however, it is very important for signal resolution, that is, for signal-to-noise ratio. To guarantee that the maximum spatial frequencies are resolved also in the reconstructed wave, the holographic procedure has to consider the following:

- Fringe spacing for sufficiently fine sampling
- Accuracy of a posteriori aberration correction better than $\pi/6$
- Signal/noise properties

5.2.7.1 Fringe Spacing
The radius q_{rec} of the aperture, used for cutting out the sideband in Fourier space, determines resolution in analogy to the Abbe theory of imaging; for highest

resolution, it should be as wide as possible. The maximum sensible radius q_{max}, however, is given by the distance q_c of the sideband from the centerband, that is, by fringe spacing $1/q_c$. Since, in general, the radius of the centerband (FT$\left[1 + A^2(\vec{r})\right]$) measures up to twice the one of the sideband (FT$\left[A\exp(i\phi)\right]$)

$$q_{max} \leq q_c/3 \tag{5.9}$$

has to be met for isolating the sideband from the centerband; this means that three hologram fringes have to cover 1 pixel to be reconstructed. For a weak pure phase object with $A = 1$ and hence FT$\left[A^2\right] = \delta(\vec{q})$, this relaxes to

$$q_{max} \leq q_c \tag{5.10}$$

Usually, an aperture with a radius between these two limits is applicable. In any case, the usable part of the sideband must not overlap with the centerband, to avoid disturbance from information situated in the centerband. Therefore, if for any reason q_c is limited, the Fourier spectrum of the image wave has to be restricted such that the above relations hold; this must be done under recording the hologram by means of an appropriate aperture in the back focal plane of the objective lens. In any case, only the part not overlapping with the centerband may be used for reconstruction. But, on the other side, it is disadvantageous to use finer fringes, that is, larger carrier frequencies than q_c given above, because then the very narrow fringes are more sensitive against instabilities, that is, V_{inst} drops; furthermore, one has to spend correspondingly more CCDpixels, making the field of view smaller.

5.2.7.2 Optimizing the Paths of Rays for Holography

Field of view and resolution (fringe spacing) found in the image plane with respect to the object depend on the optics and on the position of the biprism. Usually, the biprism is positioned in the selected area aperture, simply because there is a port that offers easy access. Of course, it depends on the paths of rays, which fields of view related to the object and can be realized with the biprism position fixed in the SA plane. Until recently, there were mainly two sets of parameters for operating our CM200 field emission gun (FEG) microscope: one allows a field of view of few micrometers in Lorentz mode using the Lorentz lens instead of the objective lens for medium-resolution applications; the other one in normal mode, using the objective lens, offers a field of some 10 nm for high-resolution holograms. Meanwhile, Sickmann et al. [16] have elaborated additional parameter sets varying the excitation of both the diffraction lens and the objective lens such that the gap between the two modes is bridged (Figure 5.5). This means that holography is now applicable in the same range of magnifications as conventional TEM imaging.

5.2.8
Digitization of the Image Wave

A comparably small number n_{rec} along one side w of the field of view of reconstructed pixels samples the reconstructed image wave; so the image measures $n_{rec} \cdot n_{rec}$ pixels. $n_{rec} = 2q_{max}w$ is given by the diameter $2q_{max}$ of the cutout sideband.

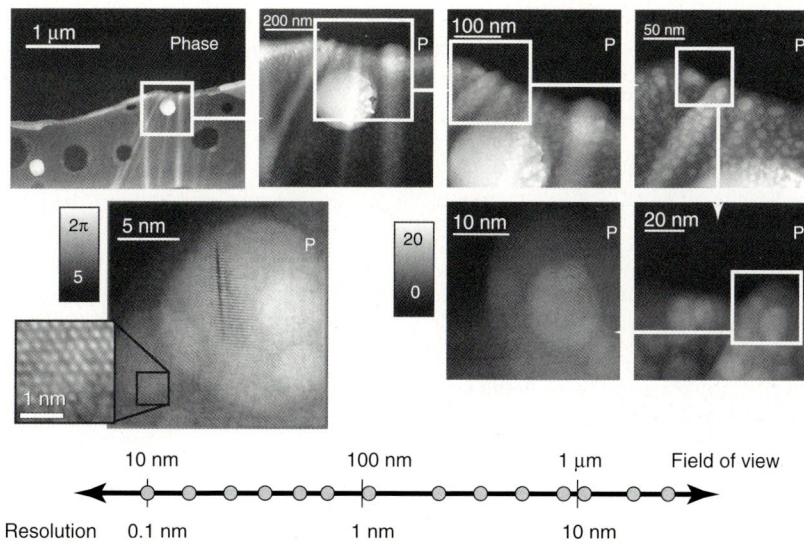

Figure 5.5 Range of field of view and resolution. By means of combined variation of imaging properties of objective and diffraction lens in our CM200 TEM, the field of view of the hologram may cover the range between several micrometers and some nanometers. Given the number of reconstructed pixels, resolution changes accordingly from some 10 nm to 0.1 nm. (From Sickmann et al. [16].)

For the case of recording the hologram by means of a CCD camera with n_{campix} camera pixels, n_{rec} can be derived as follows: At least 4 CCD pixels have to sample one hologram fringe, to avoid falsification of amplitude and phase. Since, because of $q_{max} \leq q_c/3$, three fringes make up 1 reconstructed pixel, one finds $n_{rec} \leq n_{campix}/12$ for a general object with strong amplitude $A(\vec{r})$, and $n_{rec} \leq n_{campix}/4$ for a pure weak phase object with $A(\vec{r}) \equiv 1$. Usually, however, the subimage with the cutout sideband is made larger than the minimum size n_{rec}. For this, a frame of zeros is arranged around the sideband ("zero padding"), to achieve a subimage with size of 2^n (n integer); this avoids aliasing and makes the subsequent Fourier transform easier. Of course, zero padding in Fourier space also correspondingly increases the number of pixels sampling the reconstructed wave in real space; however, there is no further gain in information, because zero padding simply generates interpolated values in the inserted pixels. This gives a nicer appearance, however, lateral resolution is still determined by n_{rec}.

5.2.9
Signal Resolution – Signal/Noise Properties

The hologram is built up by a finite number of electrons and hence always shows quantum noise. Therefore, contrast and position of the hologram fringes are not arbitrarily sharply defined; the same is true for the derived amplitude and phase of the reconstructed wave. Assume a wave reconstructed from a hologram with fringe contrast V, where N electrons have been collected in each resolved pixel. The error in phase may be estimated by means of the *phase detection limit*

$$\delta\varphi_{\lim} = \frac{\sqrt{2}\,\text{snr}}{V\sqrt{N}} \tag{5.11}$$

answering the question *which is the smallest phase difference detectable at a wanted signal/noise ratio* snr [8, 17]. This is considered in more detail in the following [18]:

As shown above, the contrast of the recorded hologram fringes is given by $V = |\mu|\,V_{\text{inel}}\,V_{\text{inst}}\,V_{\text{MTF}}$. The degree of coherence μ of illumination in the object entrance plane is practically given by spatial coherence μ^{sc}; temporal coherence need not be considered here, because, in usual holograms with less than 1000 fringes, it is always very close to unity. The average number N of electrons collected in a resolved pixel can be elaborated in the following way: for a Gaussian electron source, the total coherent current available at the degree of spatial coherence μ^{sc} is given as

$$I_{\text{coh}} = -\ln\left(|\mu^{\text{sc}}|\right)\frac{B}{k^2} \tag{5.12}$$

with brightness B and wave number k of illumination. With the hologram width w and the resolved pixel size $1/(2q_{\max})$ at resolution $q_{\max}$, the number of electrons per reconstructed pixel, reduced to an effective value by the detection quantum efficiency DQE of the CCD camera, reads as

$$N = -\ln\left(|\mu^{\text{sc}}|\right)\frac{B}{ek^2}\frac{1}{(2q_{\max}w)^2}\varepsilon\tau\,\text{DQE} \tag{5.13}$$

ε is ellipticity of illumination and τ exposure time.

Finally,

$$\delta\varphi_{\lim} = \frac{\sqrt{2\pi}\,\text{snr}\,n_{\text{rec}}}{|\mu^{\text{sc}}|\,V_{\text{inel}}\,V_{\text{inst}}\,V_{\text{MTF}}\sqrt{-\ln(|\mu^{\text{sc}}|)\frac{B}{ek^2}\varepsilon\tau\,\text{DQE}}} \tag{5.14}$$

results with the number $n_{\text{rec}} = 2q_{\max}w$ of reconstructed pixels. In effect, the degree of spatial coherence controls the term

$$\frac{1}{|\mu^{\text{sc}}|\sqrt{-\ln(|\mu^{\text{sc}}|)}} \tag{5.15}$$

which renders a minimal $\delta\varphi_{\lim}$ at an optimum $|\mu^{\text{sc,opt}}| = 0.61$

Sorting out the parameters given by the microscopic setup and by the disturbance level of the laboratory, one can define a figure of merit for the quality of a holographic system as

$$\text{NoiseFigure} = \frac{\sqrt{2\pi}}{|\mu^{\text{sc}}|\,V_{\text{inst}}\,V_{\text{MTF}}\sqrt{-\ln(|\mu^{\text{sc}}|)\frac{B}{ek^2}\varepsilon\tau\,\text{DQE}}} \tag{5.16}$$

It may depend on fringe spacing in that smaller values for V_{inst} result for smaller spacings at the same disturbance level. The above definition allows the abbreviated form

$$\delta\varphi_{\lim} = \text{NoiseFigure} \times \text{snr}\frac{n_{\text{rec}}}{V_{\text{inel}}} \tag{5.17}$$

which can be evaluated most easily for an object with given V_{inel}, once the NoiseFigure of a holographic system is determined. In a good laboratory, NoiseFigure $< 10^{-3}$ can be reached.

5.2.10
Amount of Information in the Reconstructed Wave

A suitable measure for the total information content ascertainable from a hologram can be defined as

$$\text{InfoCont} := n_{rec} \frac{2\pi}{\delta\varphi_{lim}} = \frac{2\pi\, V_{inel}}{\text{NoiseFigure snr}} \tag{5.18}$$

which means the number of reconstructed pixels multiplied by the number of phase values distinguishable in the phase range $[0, 2\pi]$ at a selected snr (Figure 5.6); evidently, InfoCont is a property of a given TEM setup used for recording an hologram. The only parameter depending on the object is the contrast reduction by inelastic interaction V_{inel}, which is related to object thickness t and mean free path for inelastic interaction λ_{inel} by means of $V_{inel} = \exp(-t/(2\lambda_{inel}))$. From the point of view of signal/noise ratio, there is an optimum thickness at $t_{opt} = 2\lambda_{inel}$, if the phase shift (signal) increases linearly with object thickness.

Evidently, at a given InfoCont, $\delta\varphi_{lim}$ can be adapted to the needs of measurement by means of the field of view w and lateral resolution q_{max}. For the needed $\delta\varphi_{lim}$, the resulting $n_{rec} = 2q_{max}w$ still allows recording a wide field of view at a correspondingly poor lateral resolution, or a small field of view at a good resolution. Field of view and lateral resolution are always concatenated by means of n_{rec}. For numeric examples, see Figure 5.7

For a given hologram width w related to the object, there is still the possibility of improving $\delta\varphi_{lim}$ a posteriori by selecting a smaller reconstruction aperture q_{max}, which increases the size of the reconstructed pixels hence reduces n_{rec}, alas, at

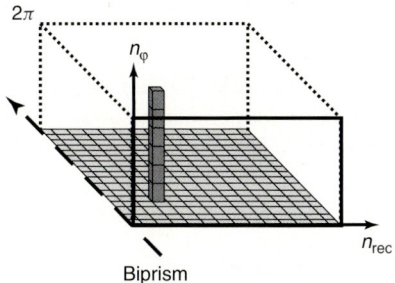

Figure 5.6 Information content in a hologram. The number of reconstructed pixels n_{rec} and the number of discernible phase values per 2π give InfoCont $= n_{rec}n_\varphi$, which is represented by the front face of the sketched information volume. Here, $n_{rec} = 14$ and $n_\varphi = 6$.

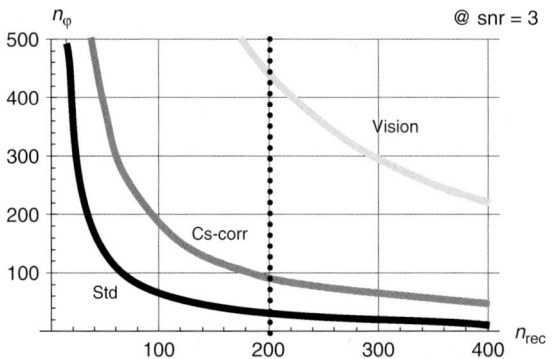

Figure 5.7 Phase resolution versus lateral resolution/field of view. Achievable phase resolution and number of reconstructed pixels for different information contents of holograms: 7.000 (std: CM200); 18.000 (special: Tecnai F20 Cs-corr); 88.000 (vision).

the cost of lateral resolution q_{max}. Specific details are discussed in the following sections for medium and high resolution.

5.3
Holographic Investigations

For electrons, the objects just consist of electric or magnetic fields. They modulate primarily the phase of the electron wave. The phase difference to field-free space is measured and evaluated in terms of the object.

5.3.1
Electric Fields

Figure 5.8 shows the general situation basic for the following. A part of the plane wave ψ_{ill} is crossing an object with an electric potential $V(x, y, z)$ and hence experiences a phase modulation $\varphi(x, y) = \sigma V_{proj}(x, y)$; $\sigma = 2\pi \frac{e}{hv}$ is the interaction constant with electron velocity v. $\varphi(x, y)$ is measured with respect to a reference part outside the object. Of course, for ease of interpretation, the reference wave should be selected from field-free space. The problems that otherwise arise with perturbed reference wave, for example, arising from long-range electrostatic fields, are investigated in [19].

5.3.1.1 Structure Potentials
Atoms present the Coulomb potential of the nucleus, which is screened and formed by the electron shell; then, the potential often is treated in free-atom approximation [20–22]. To understand findings at highest lateral resolution, the bond structure must also be taken into account, for example, by density functional

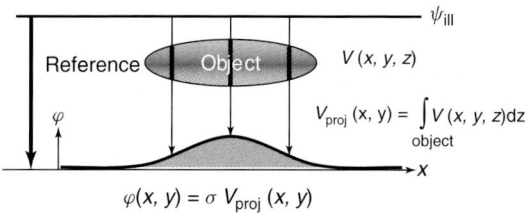

Figure 5.8 Electric phase shift. For the beam electrons, the object represents a potential distribution $V(x, y, z)$. In kinematical approximation, the phase shift relative to the reference wave is given by the projected potential $V_{proj}(x, y)$. By the projection, the structure information along beam direction is averaged out.

theory (DFT) calculation [23]. In any case, the atomic Coulomb potential is always positive with respect to field-free vacuum because there are no negative charge centers screened by positive shells. The beam electrons, propagating through, for example, a crystal, experience the potential distribution $V_{struct}(x, y, z)$. Therefore, their kinetic energy is changing along the path according to $e(U_a + V_{struct}(x, y, z))$, meaning that, by an atomic field, they are accelerated and decelerated back to the vacuum velocity, and hence modulated in phase; on the average, over the whole object they are faster in the solid than outside, and in between the atoms they have nearly vacuum velocity. According to the periodicity of the crystal, the potential may be expanded in a Fourier series as

$$V_{struct}(x, y, z) = V_{MIP} + \sum_{\vec{g} \neq 0} V_{\vec{g}} \exp(i2\pi \vec{g} \vec{r}) \tag{5.19}$$

with the potentials $V_{\vec{g}}$ as Fourier coefficients for the reciprocal lattice vectors $\vec{g}$ of the crystal. They depend on species and geometric arrangement of the constituent atoms.

$$V_{MIP} = \frac{1}{\Omega} \int_{uc} V_{struct}(x, y, z) dx dy dz \tag{5.20}$$

is the mean inner potential integrated about a unit cell of volume Ω. It is the zero component of the Fourier expansion, also valid for amorphous material. It is worth noting that these potentials do not leak out of the object, as they would do for a correspondingly charged object. In kinematic approximation, for example, tilted away from low-indexed zone axes, the influence of the $V_{\vec{g}}$ on the electron wave is very small, and hence the electron phase is given by the mean inner potential alone:

$$\varphi_{struct}(x, y) = \sigma \int_{thickness\ t} V_{struct}(x, y, z) dz \tag{5.21}$$
$$= \sigma V_{MIP} t$$

In case, local thickness t is known, V_{MIP} can be determined from the "projected potential" [24].

The contrast arising in a phase image gives the distribution of the constituting materials and thickness. This is shown in Figure 5.9 [25]: High impact polystyrene

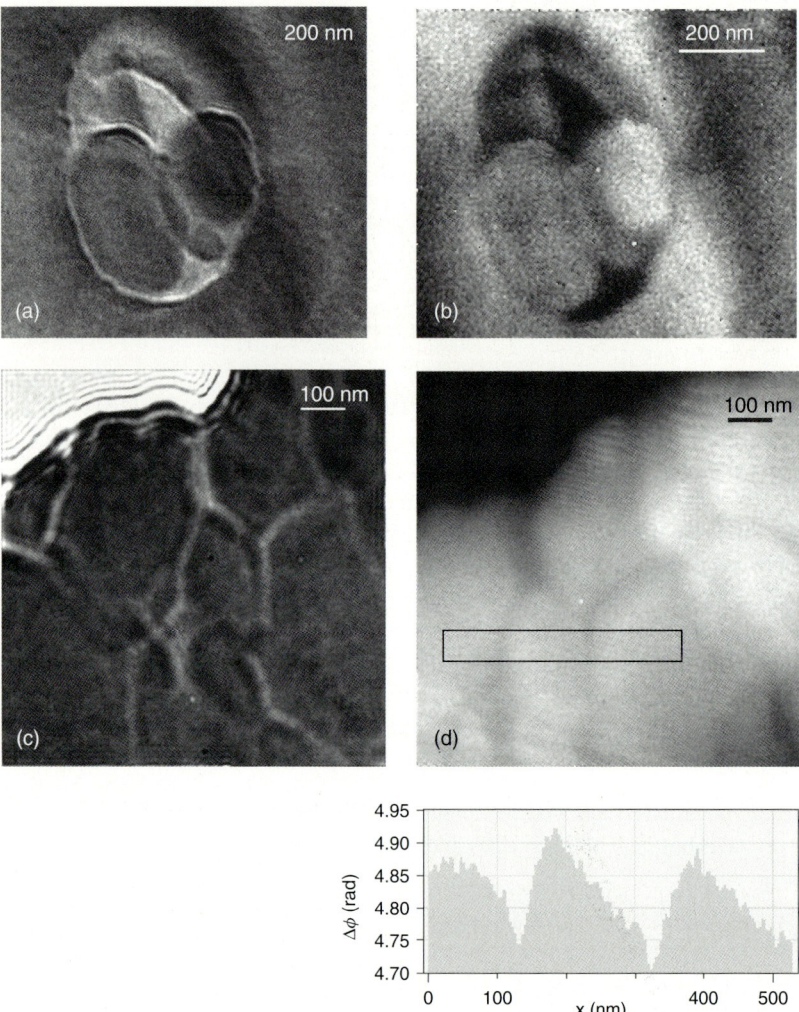

Figure 5.9 Holographic phase contrast. (a) Fresnel electron micrograph and (b) electron phase image reconstructed from a hologram of high-impact polystyrene (HIPS). In (a), strong Fresnel diffraction at the rims of the inclusions dominates, whereas in the phase image the large area structure of the inclusions is readily displayed by their projected potential. (c) Fresnel micrograph of different area of HIPS. Edge and interfaces of the sample show strong blurring caused by the Fresnel fringes at the large defocus. (d) Electron phase image of the same area as (c). The dark region at the top left corner is vacuum. The bright polystyrene inclusions separated by narrow darker regions of polybutadiene are clearly visible. Changes in topography along the thin cut are evidenced by the phase profile (below (d)) taken from the rectangular area as indicated. Notches between the wedge-shaped polystyrene particles indicate the polybutadiene rubber phase [25].

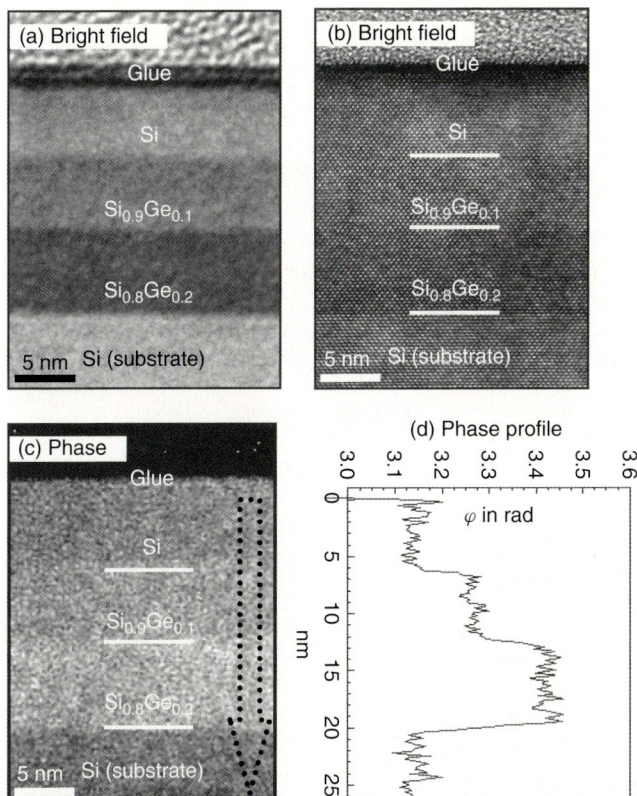

Figure 5.10 $Si_{1-x}Ge_x$ layers with different Ge-content x. (a) Bright field image exhibiting high contrast between the different layers, but only at medium resolution; the contrast is difficult to interpret quantitatively in terms of object composition. (b) High-resolution bright field image exhibiting atomic resolution but poor contrast difference between differently composed layers. (c) Phase image exhibiting high resolution (0.5 nm) as well as reasonable contrast. (d) Line scan from the phase image (c); averaged over 100 pixels, clearly shows the Ge concentration steps in the $Si_{1-x}Ge_x$ stack [26].

consists of the two constituents, polystyrene and polybutadiene. In the phase image, these may be localized and distinguished much better than in a usual strongly defocused Fresnel image. Also, the composition with changing contents and hence with changing V_{MIP} can quantitatively be determined from a phase image [26]. Figure 5.10 shows an example of a multilayer of $Si_{1-x}Ge_x$ with different x.

5.3.1.2 Intrinsic Electric Fields

Very essential for the function of many materials, in particular of the novel emerging materials, are intrinsic electric fields, which may arise, for example, from local enrichment or depletion of mobile charges also in a completely neutral solid. These fields are also difficult to image by conventional TEM; as large area phase structure, they are practically invisible. In analogy to the above, they produce an additional phase modulation

$$\varphi_{field}(x, y) = \sigma \int_{thickness} V_{field}(x, y, z) dz \tag{5.22}$$

and the total phase is just given by the sum $\varphi_{struct} + \varphi_{field}$.

Electric Fields in Doped Semiconductors Doping some areas with atoms of different valency controls the electric function of semiconductor devices; the dopants have to be on lattice position, to be electrically active. By doping, the solid remains electrically neutral; however, the kind and number of mobile carriers is enhanced. The dopant concentration is usually extremely small: at a moderate doping level of $10^{16}/cm^3$, the relative concentration is as small as 10^{-6}. Even at a comparably high doping of $10^{19}/cm^3$, the corresponding changes in the scattering properties of the object do not produce a visible contrast in a conventional TEM image; the mean inner potential would only change by about 0.1%, that is, by about some mV not yet measurable by electron holography either. However, by diffusion of the loosely bound extra electrons from n-dopants (or holes from p-dopants), a positive (negative) potential in the order of 1V builds up with respect to undoped regions. These potentials can be investigated by holography [27]. Figure 5.11 shows differently doped field effect transistors (FETs), which reveal the subsurface contacts of source and drain produced by ion implantation.

In comparison with holography, the disadvantage of materials analysis methods such as secondary ion mass spectrometer (SIMS) is that these measure, mostly at a poor lateral resolution, the absolute dopant concentration irrespective of whether the dopants are electrically active or not. It is a great advantage that holography measures the effectively reached potential distribution, which controls the function of the device.

The potential distribution produces a well-measurable phase shift (Figure 5.12). However, interpretation in terms of potentials is difficult, even if the object thickness is known. The reason is that often the potential is not homogeneous over thickness, because strong artifacts occur by the otherwise very advantageous preparation with the help of focused ion beam (FIB). By this preparation, at a targeted area of interest of a wafer, a thin lamella, some 100 nm thick, is cut out as a cross section for holographic investigation (Figure 5.13). The etching Ga-ions damage both surfaces of this lamella in an amorphized 20 nm thick layer, which is no more electrically active; these "dead-layers" contribute to the phase only by the nearly unchanged mean inner potential. Furthermore, in a transition layer underneath, also about 20 nm thick, crystallinity has survived, but it is heavily p-doped with Ga [29]. Only the inner core is pristine and hence represents the intact properties of the wafer. This is shown in Figure 5.14 at the example of pn-junctions in a Si-needle. Owing to the preparation-induced variation of potentials along the electron beam, the projected potential determined from the measured phase does not really represent the original wafer. For a more accurate evaluation, these artifacts have to be modeled and taken into account.

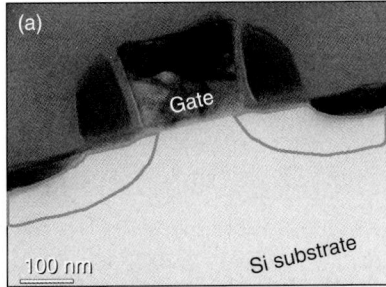
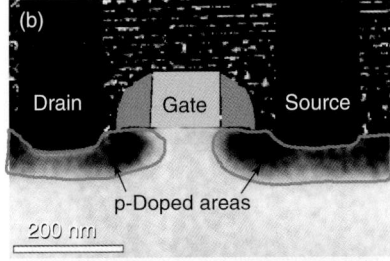
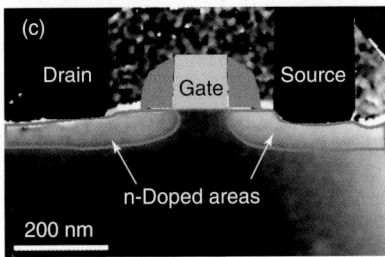

Figure 5.11 Dopant potentials in semiconductors. (a) Bright field TEM image of a MOS transistor. The frames indicate the supposedly doped areas around the source and drain that are not visible here. (b) Phase image of a p-type MOS transistor; the p-doping reduces the phase shift of the n-substrate: therefore, the doped areas appear darker. (c) Phase image of an n-type MOS transistor; the n-doping increases the phase shift with respect to the p-substrate; therefore, the doped area appears brighter. Compared to the conventional image in (a), implanted source/drain regions produce a substantial phase contrast. (From [28]).

Holographic tomography presented in section 5.4.1 offers a principle solution to this problem.

Electric Fields in Biological Objects Since electric fields do not show up in conventional TEM imaging, their existence is often unknown and surprising. An example is biomineralization, that is, the built-up of cartilage and ossicles from a fluoroapatite-gelatin system. The biomimetic fluoroapatite-gelatin system strongly resembles the hydroxyapatite collagen biosystem, which plays a decisive role in the human body as functional material in the form of bones [30] and teeth [31]. In both systems, the hierarchical and self-similar organization of nanocomposite

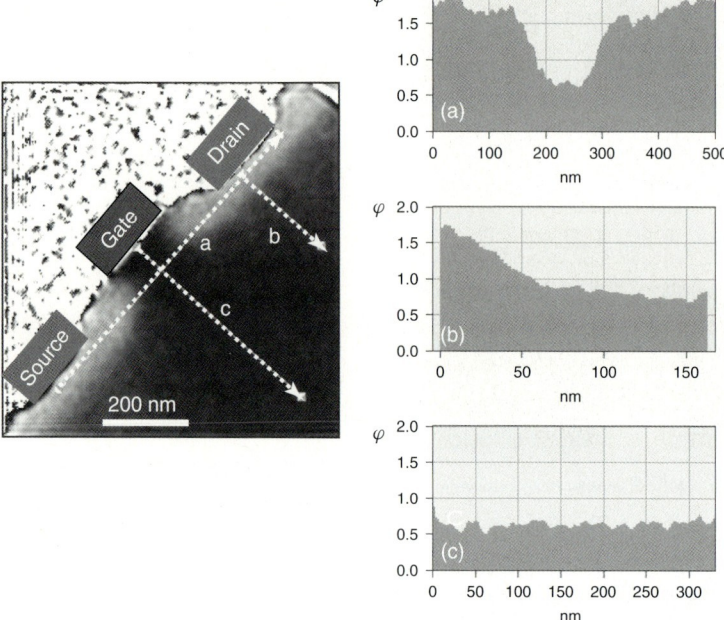

Figure 5.12 Potentials in semiconductors. The phase distribution produced by the positively charged n-doped regions with respect to the p-doped substrate can be measured by line scans (a) indentation in front of the gate along source-gate-drain, (b) from drain to substrate, (c) even potential in front of the gate. With known thickness, the potentials can be evaluated. (From [29].)

structures is of prominent relevance. The initial stages of the fractal growth of the fluoroapatite-gelatin composite comprise micrometer-sized, hexagonal, and prismatic composite seeds. These species develop in subsequent growth stages to dumbbell states and complete their development as closed, notched spheres. Mineralized macromolecules are present in the fluoroapatite-gelatin nanocomposites [32]. As early as 1999, the possible influence of intrinsic electric fields on the fractal morphogenesis of fluoroapatite-gelatin nanocomposite aggregates was discussed [33]. In fact, these fields were found by holography of central composite seeds, which represent the initial and hence the fundamental growth step during morphogenesis [34].

Figure 5.15 shows a conventional TEM micrograph of a hexagonal prismatic seed with a typical aspect ratio close to 3 : 1. Only the silhouette of the nontransparent seed is visible, whereas the phase image (eight times amplified) clearly shows the electric-potential distribution around the seed [34, 35]. This field is caused by a parallel orientation of triple-helical protein fibers of gelatin, which produce mesoscopic dipole fields. Consequently, the first clear evidence of a direct correlation between intrinsic electric fields and the self-organized growth of the fluoroapatite-gelatin biocomposite has been successfully provided.

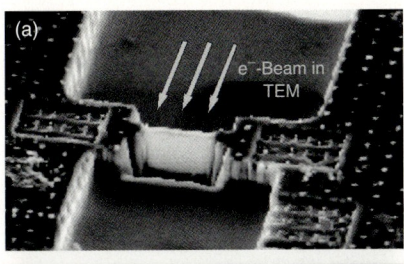

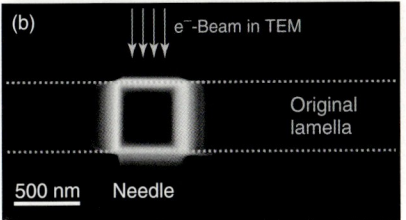

Figure 5.13 Focused ion beam (FIB) preparation for holography. (a) SEM image of a conventional FIB-prepared TEM lamella. (b) SEM top view of a needle prepared by cross-sectioning the TEM lamella in a second perpendicular etching process. Under FIB, severe damage by the ion bombardment occurs: amorphization of a surface layer and p-doping by the Ga-ions in a transition layer underneath; both layers are approximately 20 nm thick. These layers change the projected potential. (From [29]).

5.3.2
Magnetic Fields

Magnetic objects produce a magnetic phase shift

$$\varphi_{mag} = -2\pi \frac{e}{h} \Phi_{mag} \tag{5.23}$$

Contrary to the electric phase shift, it is nondispersive, meaning that it is, irrespective of the energy, the same for all electrons. Furthermore, since, at the end, the phase is given not by the magnetic field $\vec{B}$ but by the enclosed magnetic flux

$$\Phi_{mag} = \oint_{path\ loop} \vec{A}d\vec{s} = \int_{enclosed\ area} \vec{B}d\vec{f} \tag{5.24}$$

("Stokes theorem"), partial waves passing the object on the right and on the left are mutually shifted in phase, although they do not experience a Lorentz force from the $\vec{B}$-field. This is the famous Ehrenberg-Siday-Aharonov-Bohm-(ESAB-)effect [36, 37], experimentally proven by Moellenstedt and Bayh [38]. Later on it was confirmed by Tonomura et al. [39] by means of superconductors, and by Schmid et al. [40] using an encapsulated microcoil. Also, if the electrons run through the $\vec{B}$-field and hence experience a Lorentz force, the phase shift is always given by the respective enclosed magnetic flux. In general, the bending of the trajectories has to be taken into account for evaluation of the integral. With thin, electron-transparent magnetic films, however, one can mostly neglect the bending and integrate simply over z.

Basic relations between phase $\varphi_{mag}(x, y)$ and magnetic field $\vec{B}(x, y, z)$.

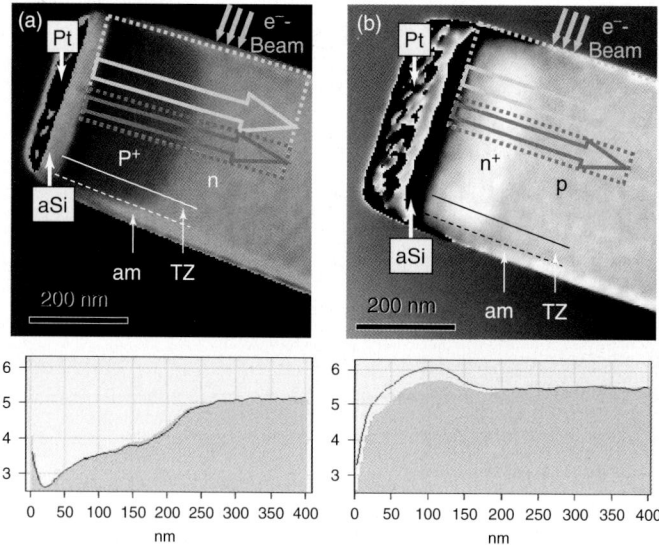

Figure 5.14 Phase images of (a) p–n and (b) n–p junction in a needle. The junctions are curved toward the edges because of changes in electrical activity at the lateral surfaces from ion bombardment. Amorphized layer (am) and transition zone (TZ) are indicated. Electrons propagate perpendicularly to the paper plane, but thanks to the square-shaped cross-section of the needle, the result is equivalent to the projection in the direction of the arrows. Line scans: Comparison of differently averaged phase profiles over the junction; top arrow: averaged from middle to edge, representing the projection with e-beam; bottom arrow: averaged over smaller core area, inaccessible by projection with the e-beam. Difference in behaviour for p- and n-doped structures is obvious. (From [29].)

The magnetic phase shift $\varphi_{\text{mag}} = -2\pi \frac{e}{h} \Phi_{\text{mag}}$ is always given by the magnetic flux Φ_{mag} enclosed between two respective trajectories superimposed in one point (x, y) of the interference pattern. Since, at the end, we want to measure the local distribution of the $\vec{B}(x, y, z)$-field from the integral quantity Φ_{mag}, we have to understand the relation between the electron phase $\varphi_{\text{mag}}(x, y)$ and $\vec{B}(x, y, z)$.

Assume a thin film homogeneously magnetized by $\vec{B} = (0, B_y, 0)$ and $\vec{B} = \vec{0}$ outside the film, as shown in Figure 5.16. With even film thickness t, the magnetic flux $\Phi_{\text{mag}}(x) = B_y t \cdot x$ linearly increases with x, and so the phase $\varphi_{\text{mag}}(x)$ decreases relative to the reference point in field-free space. The gradient of the phase wedge reads as

$$\frac{\partial \varphi_{\text{mag}}}{\partial x} = -2\pi \frac{e}{h} B_y t \qquad (5.25)$$

For x outside width b of the film, the phase remains constant, that is, $\frac{\partial \varphi_{\text{mag}}}{\partial x} = 0$, at the maximum phase difference $\Delta \varphi_{\text{mag}}$. The phase gradient indicates whether a respective electron would experience a Lorentz force or not. In Wentzel-Kramers-Brillouin (WKB) approximation, the corresponding deflection

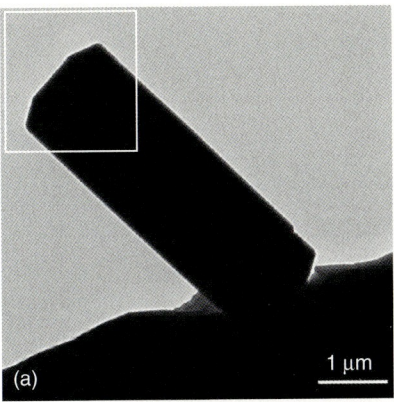 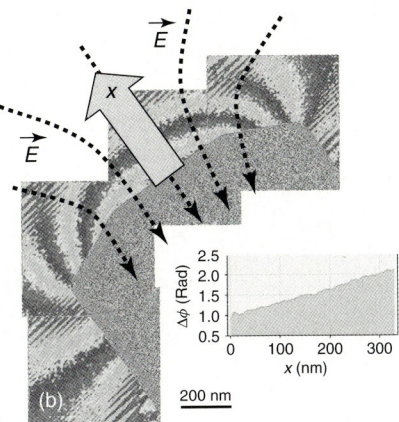

Figure 5.15 Electric field in biomineralization. (a) Conventional TEM micrograph of a free-standing hexagonal prismatic composite seed. (b) Reconstructed phase image (eight times amplified) of a hologram around the upper tip displays the stray field of a macroscopic electric dipole oriented parallel to the seed axis. The phase profile in the thick arrow gives a phase increase of about 1 rad per 300 nm corresponding to about 0.13 times the electric polarization observed for $BaTiO_3$. The hatched lines indicate the electric field $\vec{E}$ of the projected potential. These fields control growth in biomineralization. (From [34].)

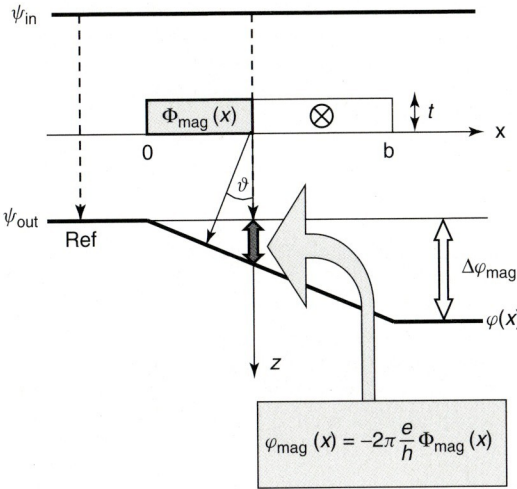

Figure 5.16 Illustration of magnetic phase shift. A plane electron wave ψ_{in} propagates through a thin film magnetized in the y-direction, that is, perpendicularly to the drawing plane. The magnetic layer is assumed infinitely extended in the y-direction to avoid a stray field outside. With respect to the indicated reference path, the phase of the object exit wave ψ_{out} decreases proportional to the respectively enclosed flux $\Phi_{mag}(x)$. Classically, a trajectory orthogonal to the wavefront would be deflected by the angle $\vartheta = \mathrm{grad}\varphi/(2\pi k)$ (wavenumber k). For large x-values beyond the thin film, the enclosed flux and hence the phase difference remains constant (ESAB-effect).

angle of the "classical" trajectories follows as

$$\vartheta(x) = \frac{1}{2\pi k} \frac{\partial \varphi_{\text{mag}}(x)}{\partial x} = -\frac{eB_y t}{p} \tag{5.26}$$

with kinetic momentum $p = \hbar k$.

The "projected" field

$$\vec{B}^{\text{p}} := \left(\int B_x dz, \int B_y dz \right) = \left(B_x^{\text{p}}, B_y^{\text{p}} \right) \tag{5.27}$$

integrated along the path through the object, that is, the z-coordinate, together with the one going through field-free space, determines the phase; only the in-plane field components B_x and B_y contribute.

The phase gradient

$$\nabla \varphi_{\text{mag}} = -2\pi \frac{e}{h} \left(B_y^{\text{p}}, -B_x^{\text{p}} \right) \tag{5.28}$$

with $|\nabla \varphi_{\text{mag}}| = 2\pi \frac{e}{h} |\vec{B}^{\text{p}}|$, and hence the projected field strength

$$|\vec{B}^{\text{p}}| = \frac{|\nabla \varphi_{\text{mag}}|}{2\pi} \frac{h}{e} \tag{5.29}$$

can be quantified with the number $\frac{h}{e} = 4.135 \cdot 10^3$ T nm^2. Between two successive iso-phasal lines mod 2π the flux $\Phi_{2\pi} = \frac{h}{e}$ is enclosed. Together with $\nabla \varphi_{\text{mag}} \cdot \vec{B}^{\text{p}} = 0$, saying that iso-phasal lines defined by $\varphi_{\text{mag}}(x, y) = \text{const}$ agree with the direction of the field lines $\vec{B}^{\text{p}}$, the projected magnetic field $\vec{B}^{\text{p}}$ is completely and quantitatively determined. However, for determination also of the z-component B_z^{p}, one would need a further hologram taken under such an object tilt that B_z^{p} contributes. Comparing two different points (x_0, y_0) and (x_1, y_1) in the phase image, one finds that the result is independent from the connection chosen. That is why the solution in the phase image is unambiguous. Examples are shown in Figures 5.17–5.19.

5.3.2.1 Distinction between Electric and Magnetic Phase Shift

Usually, objects consist of both electric and magnetic phase shifting components. Therefore, there is the need for distinguishing the two. This can be achieved by taking two holograms at different accelerating voltages because the electric phase shift is sensitive against the electron velocity, whereas the magnetic one is not. Another possibility [43] needs taking a second hologram with the object flipped over. Then the half sum of the two measured phase distributions reveals the electric component, whereas the half difference gives the magnetic one. Sometimes, symmetry aspects of the phase distribution allow deciding whether or not there is a substantial magnetic contribution. This is shown for a sphere in Figure 5.20.

Effects of Stray Fields If there is no perfect flux closure within a finite specimen, that is, $\vec{B} \neq \vec{0}$ also outside, two different effects may have to be considered:

- **Stray Field Above and Below the Object**: Often, a magnetic object does not show perfect flux closure so that a stray flux appears around it. The stray field

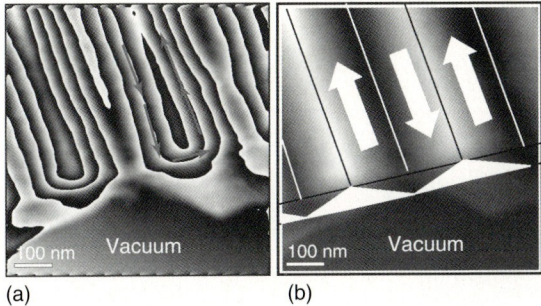

Figure 5.17 Phase images of magnetic inversion domains in thin cobalt films. (a) The phase-wrapping lines represent equiphasal lines mod 2π, hence the $\vec{B}$-field with direction indicated. From the phase gradient, the average strength of the projected in-plane $\vec{B}$-field results as 100 T nm. Assuming no stray field because of flux closure and a thickness of 100 nm, one obtains a magnetization equivalent of $B = 1$ T. The saturation magnetization of bulk cobalt is $B = 1.7$ T. (b) Phase unwrapping shows the whole phase range with magnetization indicated by white arrows. (In cooperation with Wolfgang Neumann, HU Berlin.)

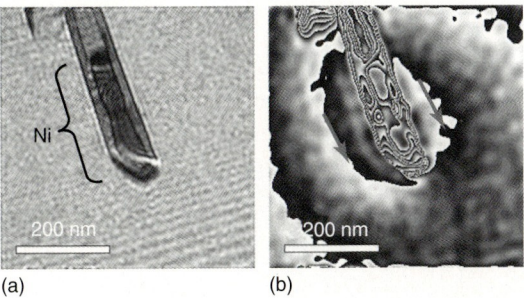

Figure 5.18 Phase image of a nickel rod encapsulated in a carbon nanotube. (a) Amplitude shows part of the nanotube filled with Ni. (b) Phase image reveals magnetic stray field around Ni. The undulations in the phase-wrapping lines stem from structural noise in the supporting carbon film [41].

around an arbitrary particle usually has a complicated structure, in particular in the close vicinity. Therefore, the principle effect is demonstrated again for a simple geometry. For example, for a single domain magnetized finite cylinder, the stray field in the middle is oriented opposite to the particle magnetization, as assumed in Figure 5.21. Evidently, the reverse outside flux reduces the total flux from inside the particle and hence also the measured phase shift. This may be quantitatively calculated and corrected, for example, for highly symmetric specimens such as cylindrical rods or spheres.

- **Stray Field Reaching into the Reference Area**: Usually we have to assume that the reference area is also not perfectly field-free. As illustrated in Figure 5.22, we see that the flux enclosed between superimposed "twin-rays" changes with

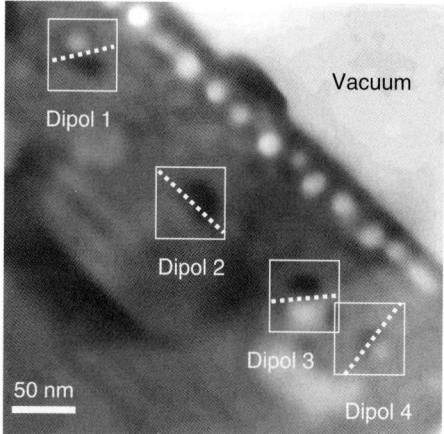

Figure 5.19 CoSm-particles in Si matrix. Some particles, for example, at the rim, produce a symmetric phase shift indicating changes in the electric inner potential. The framed areas contain antisymmetric phase shifts, which indicate magnetic stray fields and hence magnetic particles [42].

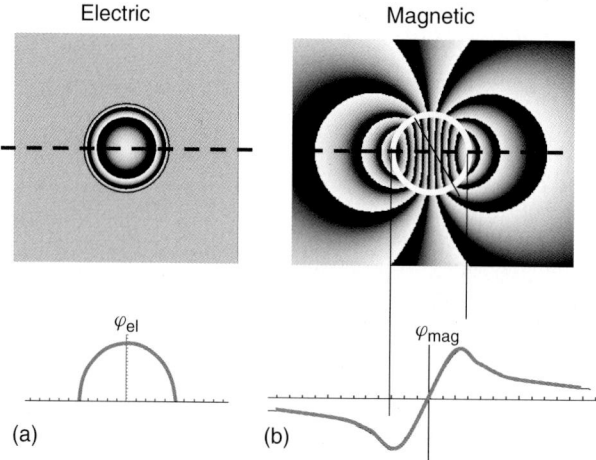

Figure 5.20 Comparison of electric and magnetic phase shift of a sphere. Electric and magnetic phase shift can clearly be distinguished by means of appearance and symmetry of the stray field with the help of the phase-wrapping lines. (a) A sphere with constant mean inner potential shows a rotational-symmetric appearance without the stray field; in case of a charged sphere, an electric stray field arises, which would also be rotational symmetric. (b) A homogeneously magnetized sphere exhibits a strict mirror-antisymmetric phase distribution, including the stray field outside the particle. Have in mind that in both cases the phase reveals the projected fields.

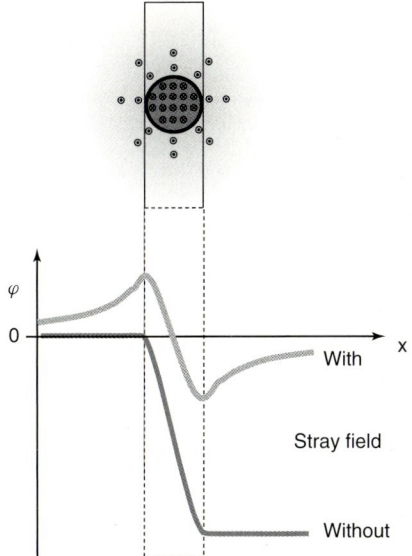

Figure 5.21 Role of the stray field in the object wave. Even the evaluation of magnetization (inner $\vec{B}$-field) of a simple, homogeneously magnetized cylinder from phase distribution is tedious. Without the stray field, the bottom curve applies. The extrema of the phase give the enclosed flux and hence, at known geometry, the $\vec{B}$-field. If a stray field is present, the parts projected from above and below the cylinder reduce the height difference between the extrema of the top curve. Furthermore, the curve approaches zero on both sides because of the lateral stray field. The lateral extension of the stray field depends on the cylinder length perpendicular to the drawing plane. In any case, for rotational symmetry, the reduction in the height difference of the extrema of the phase shift amounts to a factor of $2/\pi$, which has to be compensated for under evaluation, to obtain the magnetization correctly.

the considered x-coordinate because of both the change $\delta\Phi^{\text{obj}}$ in the object area and the change $\delta\Phi^{\text{ref}}$ in the reference area. Therefore, the total flux changes by $\delta\Phi = \delta\Phi^{\text{obj}} - \delta\Phi^{\text{ref}}$. Now the phase gradient reads as

$$\frac{\partial \varphi(x)}{\partial x} = -2\pi \frac{e}{h} \frac{\partial \Phi(x)}{\partial x} = -2\pi \frac{e}{h} \left(B_y^{\text{proj}}(x) - B_y^{\text{proj}}(x-d) \right) \quad (5.30)$$

with d the lateral distance ("shear") of the twin waves in the object plane. More generally, this reads

$$\nabla \varphi = -2\pi \frac{e}{h} \left(B_y^{\text{p}}(x), -B_x^{\text{p}}(x) \right) + 2\pi \frac{e}{h} \left(B_y^{\text{p}}(x-d), -B_x^{\text{p}}(x-d) \right)$$
$$= \nabla \varphi_{\text{obj}} - \nabla \varphi_{\text{ref}} \quad (5.31)$$

If the field $\vec{B}^{\text{p}}(x-d)$ in the reference area is not known, the object-related field $\vec{B}^{\text{p}}(x)$ cannot be determined. Often, however, the field in the reference area is comparably weak such that the object field can be determined, however, only at a correspondingly reduced accuracy.

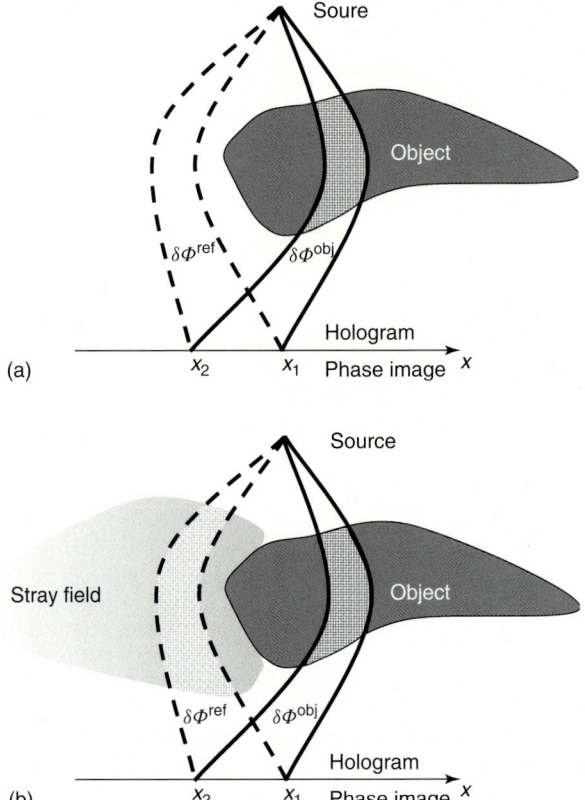

Figure 5.22 Role of lateral stray field in the reference wave. Assume a magnetized particle with and without a stray field. (a) No stray field. The phase difference between x_2 and x_1 is given by the object-related flux change $\delta\Phi_{obj}$. The $\vec{B}$-field of the object is determined correctly. (b) The inhomogeneous stray field in the reference area adds a related flux change $\delta\Phi_{ref}$, which, in general, also depends on x. Therefore, the phase difference cannot be uniquely interpreted in terms of the $\vec{B}$-field of the object.

Also, with magnetic objects there remains the projection problem, which makes the analysis of field structures along the z-direction impossible. Also, here, holographic tomography will be the method of choice. However, this will be more difficult than with electric potentials because here we deal with vector fields, which need tilt series about two independent directions.

5.3.3
Holography at Atomic Dimensions

Any object consists of atoms. Consequently, the elementary process is the interaction of the electron wave with a single atom given by the free-atom potential $V_{atom}(\vec{r})$ [20–23].

For the respective model, the object exit wave $\mathrm{obj}(x, y) = a(x, y) \exp(i\varphi(x, y))$ has to be calculated. Depending on the atomic number, there are different approximations: For very low atomic numbers, for example, carbon or oxygen, the linear approximation

$$\mathrm{obj}(x, y) = 1 + i\varphi(x, y) \tag{5.32}$$

holds with the phase

$$\varphi(x, y) = \sigma \int V_{\mathrm{atom}}(\vec{r}) dz \tag{5.33}$$

given by the projected potential. Stronger atoms require the phase grating approximation

$$\mathrm{obj}(x, y) = \exp(i\varphi(x, y)) \tag{5.34}$$

again with the phase given by Eq. (5.33).

Also, here, the so-called kinematic interaction gives the phase distribution simply by the projected potential: these atoms behave like pure phase objects. However, for even stronger atoms, the amplitude of the exit wave is also modulated. This results from the fact that one can no longer integrate simply along z; instead, one has to integrate along the paths orthogonal to the respective wave, already modulated under propagation through the previous part of the atomic field. This propagation effect results in a highly complicate modulation of the wave both in amplitude and phase ("dynamic interaction"). Equivalently in Fourier space, the scattering amplitude of the atom gets complex exhibiting a "scattering phase" [45].

The full dynamical calculation can, in most cases, only be done numerically by means of a multislice algorithm propagating the wave through the respective object model [46]. Figure 5.23 shows the peak phase shift of single atoms versus atomic number calculated by means of electron microscopy simulation (EMS) using multislice [47]. It reveals that different atom species may be distinguished in a phase image at atomic resolution.

In simple cases, one can configure an object consisting of many atoms by convolution of free-atom potentials into the respective positions known from crystallography. However, with improving resolution, tiny details also become essential and interesting, such as modification of the potential by bonds and intrinsic fields, or slight displacement of the atoms, for example, at surfaces, interfaces, and defects; for comprehensive understanding, *ab initio* calculations, including the relaxation of an atomic arrangement, have to be performed to obtain the detailed potential distribution $V_{\mathrm{struct}}(x, y, z)$ [49]. An example is shown in Figure 5.24.

Finally, the electron wave is propagated through the resulting potential distribution in order to obtain the object exit wave $\mathrm{obj}(x, y) = a(x, y) \exp(i\varphi(x, y))$ determined from a hologram. Unfortunately, dynamic results may be more difficult to interpret in terms of the object structure, the more the object thickness approaches extinction thickness. For example, then the phase shift does not increase linearly with thickness, or the phase shift from light atoms may become stronger than the

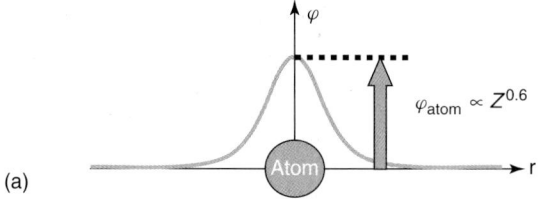

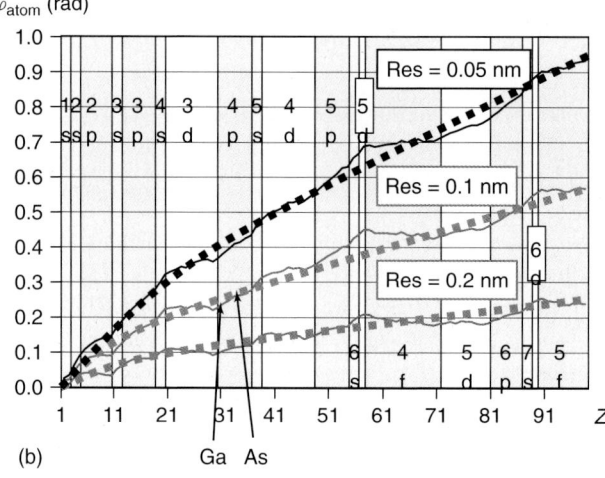

Figure 5.23 Phase shift by single atoms. (a) According to [48], the peak phase shift in the image wave of a single atom may be estimated as proportional to $Z^{0.6}$. (b) Peak phase shift of single atoms in the image plane at a lateral resolution Res versus atomic number Z at 300 kV, determined by EMS with one atom in a supercell 5 nm wide and 0.5 mm thick. The hatched lines show fitted proportionality to $Z^{0.6}$; deviations come from the orbital structure of the electron shell. It may happen that the phase shift decreases with increasing Z, in particular at poor lateral resolution. In general, the phase shift increases strongly with improved lateral resolution, which is a very essential benefit from the aberration corrector. (From [47].)

one from heavy atoms. A direct intuitive interpretation is hardly possible. To avoid this, objects for atomic resolution should be only a few nanometers thick. Alas, then the question arises, whether such a thin object still represents bulk properties or is dominated by surface effects such as surface reconstruction and distortions of the unit cell. Moreover, a further severe dynamic effect is the strong dependence of amplitude and phase modulation on slight mistilt out of zone axis, which has to be locally known for performing according simulation calculations.

Despite these problems, the object exit wave can in any case be written as $\text{obj}(x, y) = a(x, y) \exp(i\varphi(x, y))$ with amplitude $a(x, y)$ and phase $\varphi(x, y)$. The goal is the analysis of atomic structure as to the basic questions

- which atoms are where
- which bonds
- which fields or potentials are around

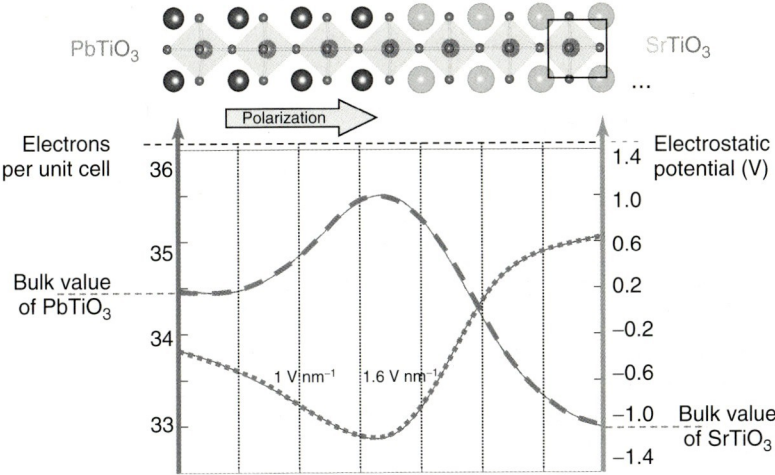

Figure 5.24 DFT simulation of a PbTiO$_3$/SrTiO$_3$ interface. By relaxation of the lattice, the titanium and oxygen atoms in PbTiO$_3$ shift to the left. This ferroelectric polarization of PbTiO$_3$ increases the number of electrons per unit cell and hence decreases the potential in the adjacent unit cells, also in SrTiO$_3$. Locally, the electric field strength may reach 1.6 V nm^{-1}. The potential variation of 1.6 V across the interface in a 10 nm thick object produces a phase shift of about $2\pi/60$. (Simulated by Axel Lubk and Sibylle Gemming.)

5.3.3.1 Special Aspects for Acquisition of Atomic Resolution Holograms
For analysis of the details of the atomic structure, one needs both lateral and signal resolution. These have been steadily improved since more than two decades [50].

5.3.3.2 Lateral Resolution: Fringe Spacing
Also for atomic resolution, the carrier frequency of the interference fringes has to be chosen as $q_c \geq 2\ldots 3q_{max}$. In fact, most of the high-resolution holograms are recorded with fringe spacings of about $s_{hol} = 1/q_c = 0.05$ nm or smaller, to reach a resolution of $q_{max} \approx 10$ nm^{-1}. The fine fringe spacing is a challenge because of the extreme sensitivity against disturbances and instabilities, in particular, with respect to the information content discussed in section 5.2.10.

5.3.3.3 Width of Hologram, Number of Fringes, and Pixel Number of CCD Camera
A prerequisite for quantitative analysis is the recording of the image wave as accurately as possible for all spatial frequencies transferred within the information limit of the objective lens into the image plane well above noise. The distortions from coherent aberrations can be corrected during reconstruction of the wave and hence can at first glance be neglected. Primarily, the following aspects have to be considered for acquisition of high-resolution electron holograms.

In order to catch all information spread out by the aberrations and hence needed for a posteriori correction of the coherent wave aberration, the field of view recorded by the CCD camera must be larger than

$$w_{min} = 4\,\text{PSF} \tag{5.35}$$

with the diameter PSF of the point-spread-function PSF $(\vec{r})$, which can be estimated by means of PSF $= \frac{1}{2\pi}\,\text{grad}\,\chi|_{max}$. This condition is equivalent to the well-known Rayleigh criterion. With the carrier frequency $q_c \geq 2\ldots 3q_{max}$, the minimum number of needed hologram fringes results as $n_{fringe} = 8\ldots 12q_{max}\text{PSF}$. Since each fringe has to be sampled by 4 CCD pixels, the pixel number of the CCD camera must be larger than $n_{pix} = 32\ldots 48q_{max}\text{PSF}$. This means that finally the objective lens and the CCD camera determine the resolution limit q_{max} achievable by high-resolution electron holography.

5.3.3.4 Adaptation of the Hologram Geometry

To control fringe spacing and width of hologram in terms of the object, they have to be related to the object plane. For this, one has to divide by the magnification $M = (a + b)/f$, with the distances a (back focal plane and biprism) and b (biprism and image plane), and f focal length of the objective lens. Related to the object plane, one obtains for the spatial frequency of the hologram fringes

$$q_c = \frac{2\gamma_0\,U_f a}{\lambda f} \tag{5.36}$$

and for the hologram width

$$w = \frac{2\gamma_0\,U_f b f}{a+b} - 2r_f\frac{f}{a} \tag{5.37}$$

Since $f \approx 1\,\text{mm}$ does not change much under the comparably small defocus values of some 100 nm, the only free parameters are the filament voltage U_f and the distance b, which can freely be selected by means of the excitation of the subsequent first intermediate lens. Because q_c does not depend on b, q_c and w can be controlled independently by means of U_f and b. A detailed analysis shows that there is an optimum position of the biprism in the path of rays, which would produce the optimum hologram in an image plane at a magnification of about 500 times [51].

5.3.3.5 Optimum Focus of Objective Lens

Since coherent aberrations have to be corrected anyway a posteriori, the focus can freely be optimized for taking holograms. It turns out that there is an optimum focus for holography [52] given as

$$D_z^{optimum} = -0.75\,C_s\left(\frac{q_{max}}{k}\right)^2 \tag{5.38}$$

where q_{max} denotes the highest desirable spatial frequency to be recorded. Generally speaking, this focus keeps the $\text{grad}\,\chi\,(q)$-function as low as possible over the whole range of spatial frequencies $[0, q_{max}]$. This offers the following advantages:

1) At optimum focus, the spatial coherence envelope is tuned such that the information limit q_{lim} of the electron microscope is increased to about $2q_{Scherz}$. Then, the information transfer is mainly limited by the temporal coherence

damping envelope, that is, by chromatic aberration and energy spread of illumination.

2) At optimum focus, the diameter of the point-spread-function is minimized as

$$\text{PSF} = \tfrac{1}{2} C_s \left(\tfrac{q_{max}}{k}\right)^3 \tag{5.39}$$

("disk of least confusion"). Therefore, the needed hologram width, which is a quarter of the one at Scherzer focus, is the smallest possible for a given q_{max}. This saves precious coherent electrons and pixel number of the CCD camera.

3) At optimum focus, the pixel number needed in Fourier space is also minimized, since regions with steep $\operatorname{grad}\chi(q)$ would need many pixels to avoid undersampling. Here, the combination of a low-C_s objective lens with the optimum focus for holography provides a wave aberration with a moderate gradient, which is less demanding in terms of sampling. Consequently, a CCD camera with 1k × 1k (2k × 2k) pixels allows the reconstruction of sidebands with 256 × 256 (512 × 512) pixels, and hence a resolution of 0.13 nm (0.1 nm) for a 300 kV electron microscope with a spherical aberration of about $C_s = 0.62$ mm can be achieved.

4) At optimum focus, the phase detection limit is improved, since the hologram width may be chosen smaller and the electron dose can be increased accordingly.

From the above considerations it is clear that atomic-resolution holography is possible and, after holographic correction of aberrations, the object exit wave can be reconstructed (Figure 5.25) [53, 54].

Therefore, in the strict sense, a TEM with an aberration corrector is not needed. Nevertheless, the restrictions encountered by the aberrations make it very difficult to reach highest performance: in particular, the signal resolution needed for detection of the very fine structure details, indispensable for a proper understanding of the object properties, is insufficient [18]. For example, the signal resolution reachable with our CM30 Special Tübingen TEM, specialized for atomic-resolution holography at about 0.1 nm, revealed a phase resolution of about $2\pi/20$; this is sufficient for detecting single heavy atoms such as gold with a peak phase shift $2\pi/12$ (Figure 5.26), but not oxygen with about $2\pi/50$.

Correspondingly, one can display amplitude and phase of heavy atoms such as gold and identify the atoms in the structure. In favorable cases, if the atomic phase shift difference is large enough according to Figure 5.23, one can also distinguish different species in a binary material such as GaAs, as shown in Figure 5.27. Evidently, holography allows materials analysis on atomic lateral resolution, if signal resolution is sufficient.

5.3.3.6 Demands on Signal Resolution

Electric Phase Shift by Elementary Charges According to $\varphi = \sigma V_{\text{proj}}$, in an object only 10 nm thick, potential variations ΔV would produce phase variations of about $2\pi/100 \cdot \Delta V$, and hence would require a signal resolution of $2\pi/1000$ for potential

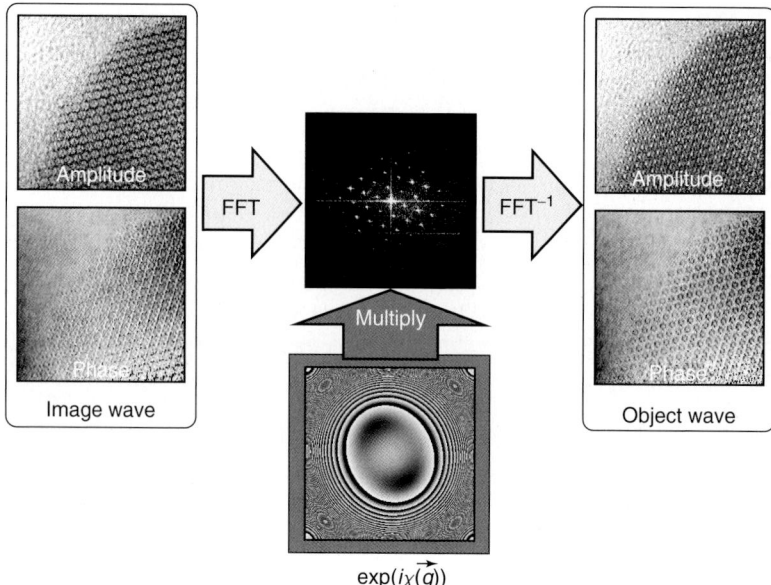

Figure 5.25 Holographic a posteriori correction of coherent aberrations. A phase plate conjugate to the one active in the TEM during the recording of the hologram is generated numerically and multiplied on the Fourier spectrum of the image wave. After inverse Fourier transform, the aberration-corrected object wave results. All coherent aberrations, also those higher than the third order, can be corrected if the aberration parameters are known. Accuracy of correction is given by accuracy of measurement of aberration parameters. Example: Philips CM30FEG ST/Special Tubingen TEM: $C_s = 1.2$ mm, $D_z = -45$ nm, $A_2 = -10$ nm, $a_{A2} = 30°$, $U_A = 300$ kV. [53].

details of $\Delta V = 0.1$ V. In solid state physics, depletion or enrichment by single elementary charges, and fractions thereof, are very essential for understanding the properties, for example, at an interface between $LaAlO_3$ and $SrTiO_3$ or a ferroelectric like $BaTiO_3$ grown on $SrTiO_3$. To get an idea about the phase shifting effect of a single charge, integration through the Coulomb field has been performed (Figure 5.28). While this integral always diverges for a single charge, it converges for a dipole, except at the singularities of the charge positions. For a pair of ±elementary charges at fixed positions, one obtains a peaked phase shift of $±2\pi/30$ at a lateral resolution of 0.1 nm, only very weakly depending on their mutual distance. If the charges are smeared about a unit cell, the width gets accordingly broader and the peak values shrink to less than $2\pi/100$. This shows that signal resolution significantly better than $2\pi/100$ would be needed to deal with these aspects of solids.

Magnetic Phase Shift by a Bohr Magneton The elementary quantity for magnetization is the Bohr Magneton (Figure 5.29). The magnetic moment of an atom is given as (noninteger) multiples, for example, in iron $|\vec{m}| = 2.2\mu_B$. Of course, it would be highly desirable to measure the magnetic moment of single atoms;

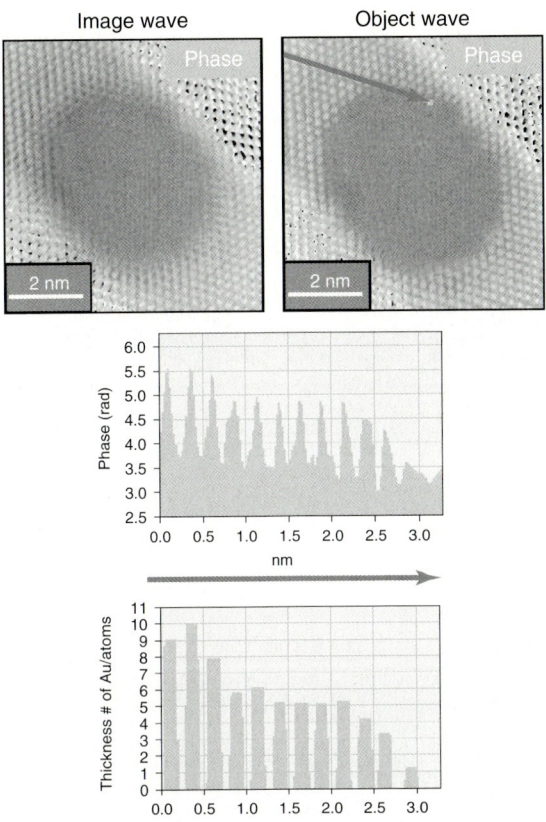

Figure 5.26 Improvement by a posteriori aberration correction. Reconstructed phase of the object exit wave from an [110]-oriented Au foil. In the image wave, by convolution with the point-spread function, the edge of the hole in the gold crystal is not atomically sharp. In the aberration-corrected object wave, the atoms clearly identify the edge. Both coherent and incoherent aberrations are corrected up to the information limit, yielding $q_{max} = 8.3\,\text{nm}^{-1}$. The indicated line scan shows the atomic phase shift, which, by comparison with EMS simulations, can be translated into a number of atoms in the respective column; the resulting numbers are very close to integers. Philips CM30FEG ST/Special Tübingen TEM [54]. (Christian Kisielowksi, NCEM Berkeley.)

however, the phase shift from a single Bohr Magneton in a unit cell with lattice constant $a = 0.4\,\text{nm}$ can be estimated as small as $\varphi_{\mu_B} \approx 2\pi \cdot 10^{-5}$. This is far beyond reach of the present signal resolution. For a substantial magnetic signal, one needs some $1000\mu_B$ such as present in a nanoparticle of some nanometer diameter (Figure 5.30) [55].

Improvement of Signal Resolution by Means of an Aberration Corrector Since the coherent aberrations can be corrected a posteriori, one would not need an aberration corrector for holography for achieving atomic lateral resolution. Nevertheless, it

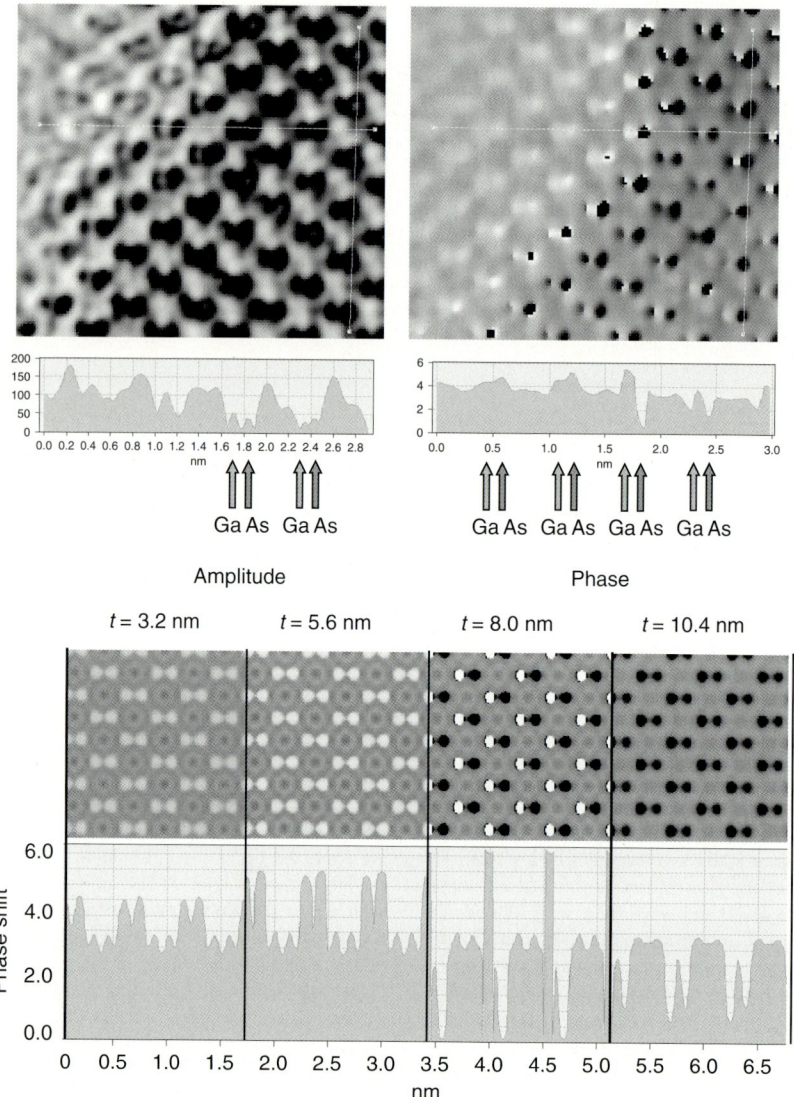

Figure 5.27 Which atom is where? In both amplitude and phase of the object wave of a wedge-shaped GaAs in [110]-orientation, the dumbbells are clearly resolved. The thickness increases to the right. While the amplitude image does not allow distinguishing between Ga and As, the phase image shows a phase shift at the Ga (Z = 31) weaker than at the As (Z = 33) positions. The phase shift increases from left to right with thickness; at 8 nm thickness, the As-phase shift exceeds 2π and hence is dark. For comparison, simulations by Karin Vogel are shown in the bottom. (See also Figure 5.37.)

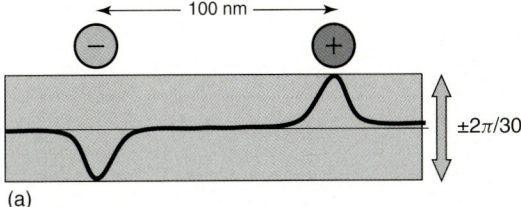

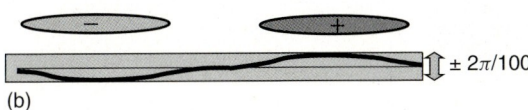

Figure 5.28 Phase shift by single charges. The projected potential of single charges diverges, but it converges for dipoles. (a) The phase shift of a dipole consisting of fixed elementary charges results as $\pm 2\pi/30$, only weakly depending on the distance between the charges. For this calculation, a lateral resolution of 0.1 nm is assumed, which smooths and restricts the peak values. (b) For charges smeared out over a unit cell, for example, in an interface, the phase shift gets broader and weaker.

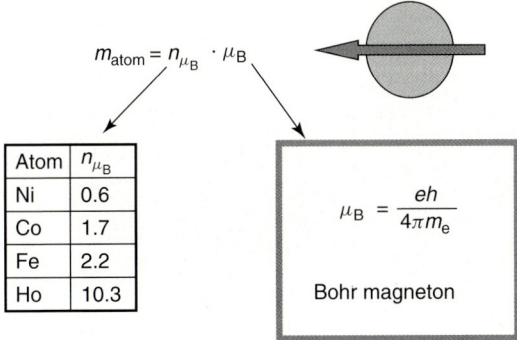

Figure 5.29 Phase shift by a Bohr Magneton. The elementary quantity for magnetization is the Bohr magneton, giving the magnetic moment m_{atom} of atoms as indicated in the table. Unfortunately, the phase shift $\varphi_{\mu_B} \approx 2\pi \cdot 10^{-5}$ by a Bohr magneton is very small. Therefore, to see a magnetic phase shift, one needs several thousand atoms in a magnetic domain.

turned out that an aberration corrector substantially improves signal resolution (Figure 5.31).

The following aspects summarize the main benefit of an aberration-corrected TEM for atomic holography.

- The diameter of the point-spread-function gets as small as the atoms, that is, about 0.1 nm in diameter. Consequently, the condition $w_{min} = 4\,PSF$ allows much smaller holograms, and hence the electron dose may significantly be increased. Therefore, the noise level is reduced and signal resolution enhanced.
- The information limit improves, and hence the contrast of high spatial frequency contributions to the wave strengthens the recorded signal.

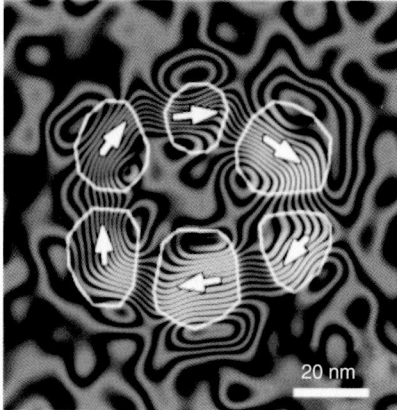

Figure 5.30 Magnetization in nanoparticles. The ring of cobalt particles provides nearly perfect flux closure and hence only weak stray fields. Therefore the magnetization gives the full magnetic signal (see Figure 5.35). The phase shift of a particle (Ø ≈ 20 nm, phase amplification 128) can be read as ≈ $2\pi/10$ [55]. Here, in a cross-section, we have approximately 2.500 unit cells, that is, 10,000 Co atoms. Evidently, the smallest detectable magnetization depends on signal resolution. From very good holograms, magnetism from Co particles with a diameter of 3–4 nm producing a phase shift of about $2\pi/160$ should be measurable.

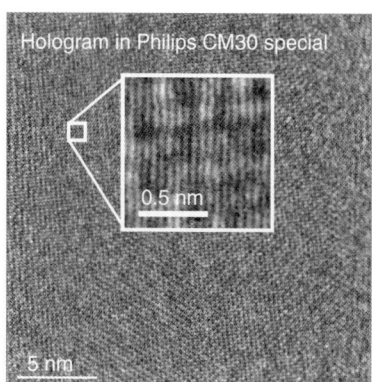 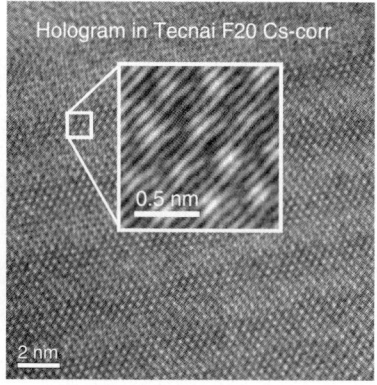

Figure 5.31 Improvement of hologram quality by an aberration corrector. With an aberration corrector, the point spread function shrinks considerably. Therefore, the width of the hologram may be made smaller. This improves the degree of coherence and the captured dose. Consequently, the contrast improves and quantum noise is reduced. In total, in our case, the phase resolution is improved by a factor of about 4 [56]. Object: GaAs/AlAs-multilayer and Recorded by Dorin Geiger.

- Opening the imaging aperture and hence the collection efficiency enhances the signal of the object in the recorded image wave.
- A lateral resolution makes correction easier: only the small values of the residual wave aberration have to be fine tuned.

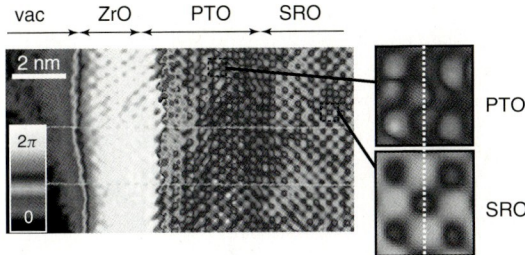

Figure 5.32 Phase image of PTO (PbTiO$_3$) grown on STO (SrRuO$_3$). In PTO, the unit cell is distorted in that the center atom is displaced, giving rise to the ferroelectric dipole. SRO is a conductor, centrosymmetric and hence not ferroelectric; ZrO is a capping layer. Very interesting is the transition from the noncentric unit cells in PTO to the centric ones in SRO. The phase shift increases in PTO toward the interface to SRO, whereas in SRO it is nearly flat; this flatness suggests that object thickness does not change much in this area. (In cooperation with Dr. Koichiro Honda, Fujitsu Laboratories Ltd., Japan.)

In fact, we found a substantial improvement in the performance of atomic-resolution holography with our Tecnai F20-Cscorr TEM [56]. Signal resolution improved to better than $2\pi/75$ at a lateral resolution close to 0.1 nm. However, this is not yet sufficient for the tiny details relevant for the properties of solids.

Which Fields Are Around? Intrinsic fields created by the atomic structure often determine properties of modern materials. These fields arise across grain boundaries or interfaces, or in polarized materials such as ferroelectrics. To understand materials properties, these fields have to be understood in relation to the atomic arrangement. We have made strong efforts on the investigation of ferroelectrics. As shown in the following, this topic is a highly demanding challenge. In fact one often finds – in addition to the atomic phase shift – a suggestive phase distribution (Figure 5.32). Also, the polarization of the unit cells is discernible from the asymmetric position of the center atoms in the unit cell, as predicted from DFT calculations (Figure 5.24).

The question arises how this atomic displacement develops across an interface from a ferroelectric to a nonferroelectric substance. It turns out that, instead of an abrupt change, there is a smooth transition in that also in the nonferroelectric material one finds a displacement induced by the ferroelectric neighbor (Figure 5.33). Our interpretation is that the ferroelectric field leaks out into the nonferroelectric and provokes there the displacement as ionic screening. This is supported by the finding that the width of the displaced zone in the nonferroelectric is narrower in the conducting SrRuO$_3$ than in isolating SrTiO$_3$, supposedly because the mobile carriers in SrRuO$_3$ strongly contribute by electronic screening.

Of course, it would be most interesting to see and measure the ferroelectric fields directly. There is, however, a row of severe problems involved. First of all, the voltages expected from the polarization values are of the order of only 1 V; consequently at an object thickness of 10 nm, suitable for atomic resolution, a

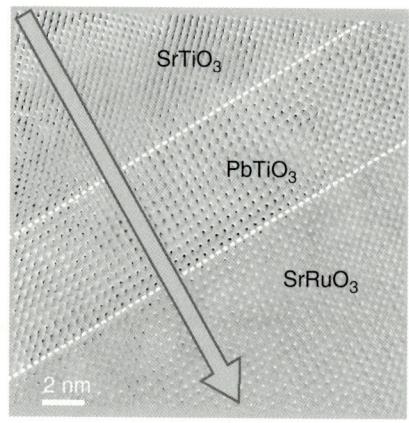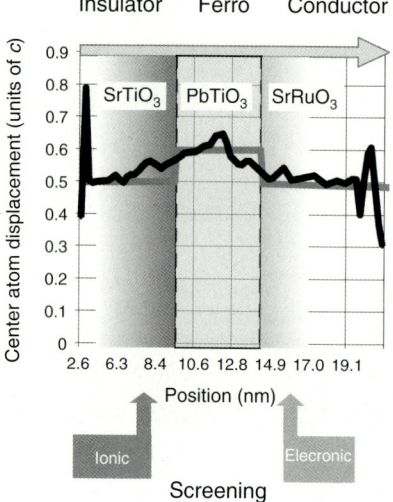

Figure 5.33 Displacement of the center atom in the unit cell across STO/PTO/SRO interfaces. In an abrupt model, the center atoms of STO and SRO would be centered, whereas in the ferroelectric PTO they would be displaced producing the polarization (gray line). However, as measured by Martin Linck and Axel Lubk, there is a displacement also in the adjacent zones in STO and SRO (black line). In STO, the screening zone by ionic displacement measures three unit cells, whereas in SRO it is only one unit cell wide; the reason is that in SRO the electronic screening by mobile carriers prevails and hence ionic screening is weak. In PTO, the displacement gradually builds up with distance from both interfaces. These measurements can favorably be performed in a phase image because the atoms are sharper than in the amplitude image. (In cooperation with Dr. Koichiro Honda, Fujitsu Laboratories Ltd., Japan.)

phase shift as small as $2\pi/100$ will result. Furthermore, for minimization of the total energy in the solid, compensation of the fields is expected, by either ionic or electronic screening. At the end, the question arises, whether at all there are appreciable fields left. Presumably, this strongly depends on the local properties of the object such as conductivity, and also on the local defect and interface structure, which may well vary on an atomic scale and hence give rise to different situations. This may be the reason why in our various findings there are examples without any significant phase shift attributable to electric fields (Figure 5.33); instead, in other examples (Figures 5.34 and 5.35) phase distributions appear, which seem to be characteristic of ferroelectric fields, but they are too large by roughly a factor of 10 compared to models. Presumably, in addition to ferroelectric origin, there are others – such as dynamic effects or differing charging effects – contributing to the phase. Unfortunately, a consistent interpretation of the findings is not yet possible with the present signal resolution and the present understanding of ferroelectricity.

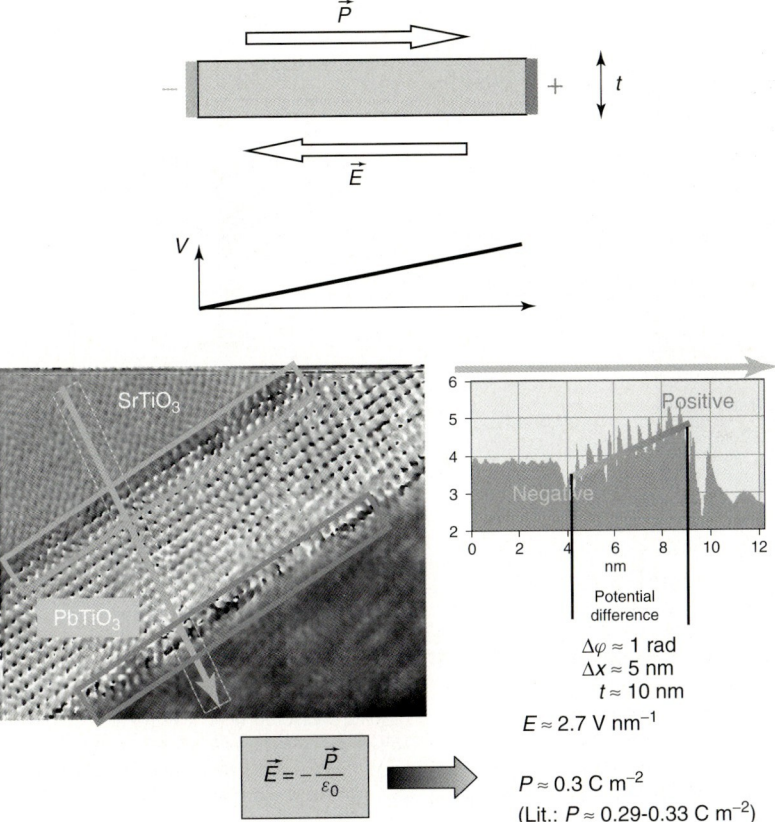

Figure 5.34 Phase distribution across the PTO layer. Scheme (top): The polarization $\vec{P}$ produces a surface charge distribution with an equivalent electric field $\vec{E} = -\vec{P}/\varepsilon_0$. Compensation is not accounted for, and hence the dielectric displacement $\vec{D} = \vec{0}$ is assumed. The projected electric potential would produce a phase shift with $\mathrm{grad}\varphi = \frac{\sigma t}{\varepsilon_0}\vec{P}$. Experimental findings (bottom): The phase shift increases strongly along the line scan across the 12 monolayers of the PTO. Trying to interpret this in terms of an intrinsic electric field, one finds an average value of 2.7 V nm^{-1}. In spite of the fact that the calculated polarization agrees quite well with literature values, we still refrain from this "simple" interpretation. Before that, the role of the electric displacement field $\vec{D} = \varepsilon_0\vec{E} + \vec{P}$, for example, from compensating charges, has to be considered in more detail. In any case, only the projected potential V derived from $\vec{E}$ contributes to the phase. Also, one has to ponder on the role of dynamic effects such as a possible polarization-induced bending of the PTO layer. (In cooperation with Dr. Koichiro Honda, Fujitsu Laboratories Ltd., Japan.)

Outlook: X-FEG and Optimum Position of Biprism After successful improvement of signal resolution by an aberration corrector, there is, besides the camera and optimizing position of the biprism, still the brightness of the electron source, which promises additional potential for further improvement. This has been verified recently by means of holograms [57] recorded in a Titan TEM equipped

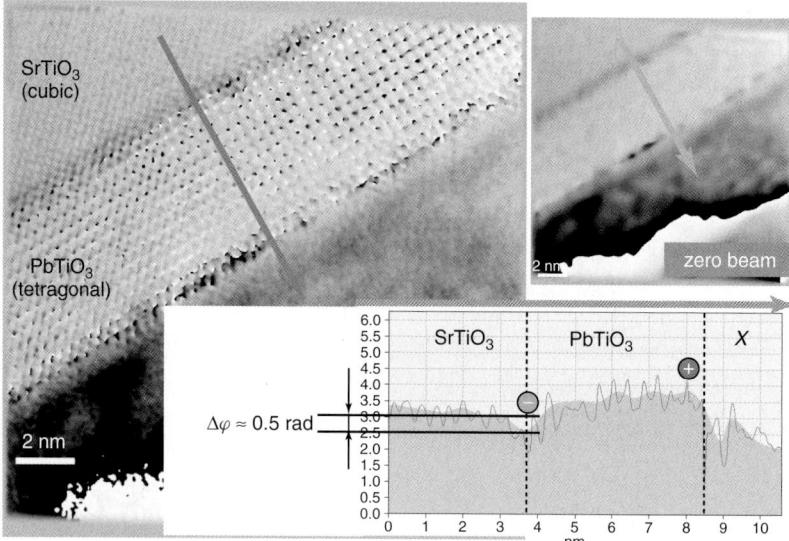

Figure 5.35 Charges in the interface and surface of PTO. In the phase images, both at atomic resolution and only reconstructed with the zero beam, the dark and bright seams at the interfaces suggest respective negative and positive charges at the edges of the PTO layer. This would, in principle, agree with the DFT calculation shown in Figure 5.38. However, the phase shift expected for the calculated voltage at an assumed thickness of 10 nm would only amount to about $2\pi/100$, instead of the $2\pi/12$ measured here. The reason for this discrepancy is not yet understood.

with an "X-FEG" electron gun developed by FEI [58]. An example is shown in Figure 5.36. We estimate the gain in brightness B by a factor of about 10. Correspondingly, the signal resolution is enhanced by another factor of $\sqrt{B} \approx 3$ (Figure 5.37). In fact, the best obtainable signal resolution of these holograms was determined as $2\pi/300$, which allows an astonishing quality of the results showing convincingly the future benefits of holography for materials characterization. The reconstructed wave clearly reveals the atom positions, in particular, after fine tuning of the aberrations (Figure 5.38).

Interestingly, the atoms can be spotted particularly well from the phase image, whereas the amplitude image of an atom seems to be intrinsically broader. Comparisons with simulations are also shown in [57]. The transition between the mutually rotated [110]-oriented gold crystals can clearly be analyzed lattice plane by lattice plane (Figure 5.39). Quantitative evaluation of the reconstructed wave again shows that the phase is superior to the amplitude in that it reveals more sophisticated details; furthermore, here the phase is less difficult to interpret, whereas the mechanisms forming the amplitude are more complicated. The phase distribution reveals the projected Coulomb potential of the nucleus screened by the electron shell, and hence the more atoms in z-direction, the higher the peak phase shift. By comparison with simulations, this allows counting the atoms precisely, as indicated in Figure 5.40.

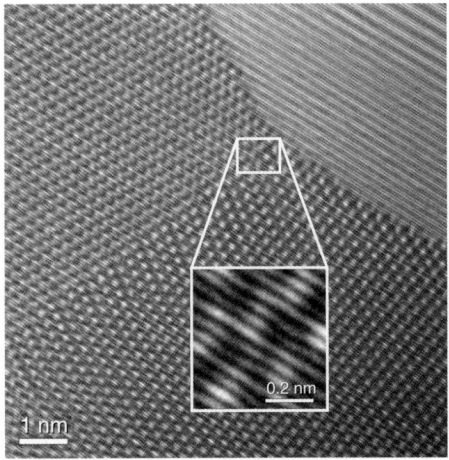

Figure 5.36 Atomic resolution hologram recorded with a Titan 300kV-X-FEG TEM. The combination of aberration corrector, X-FEG, with a brightness of about one order of magnitude higher, and acquisition at a magnification of 3.1 Mio by means of the GIF detector, allows recording best holograms. Biprism voltage 900 V, object: [110]-oriented gold crystal with a 90° tilt boundary. (From [57].)

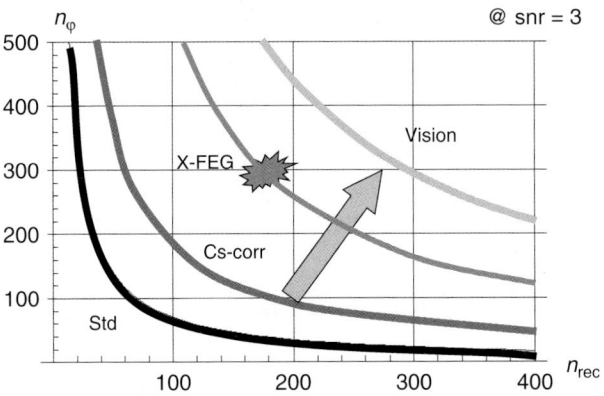

Figure 5.37 Information content of the Titan 300kV-X-FEG hologram. For reasonable n_{rec}, a phase resolution of $2\pi/200$ up to $2\pi/300$ can be reached. This is an improvement by the factor 5.75 with respect to our Tecnai F20 Cs-corr TEM, mainly due to the improved brightness of the X-FEG.

Instead, the amplitude distribution shows nearly evenly strong signals from the atom columns, irrespective of the number of atoms; furthermore, the atomic amplitudes are less sharply defined and hence give a strong signal also in the interatomic space, as seen from the correspondingly positioned line scans.

For a comprehensive interpretation of the findings, one also has to know the orientation of the object with respect to the electron beam. This is a difficult task

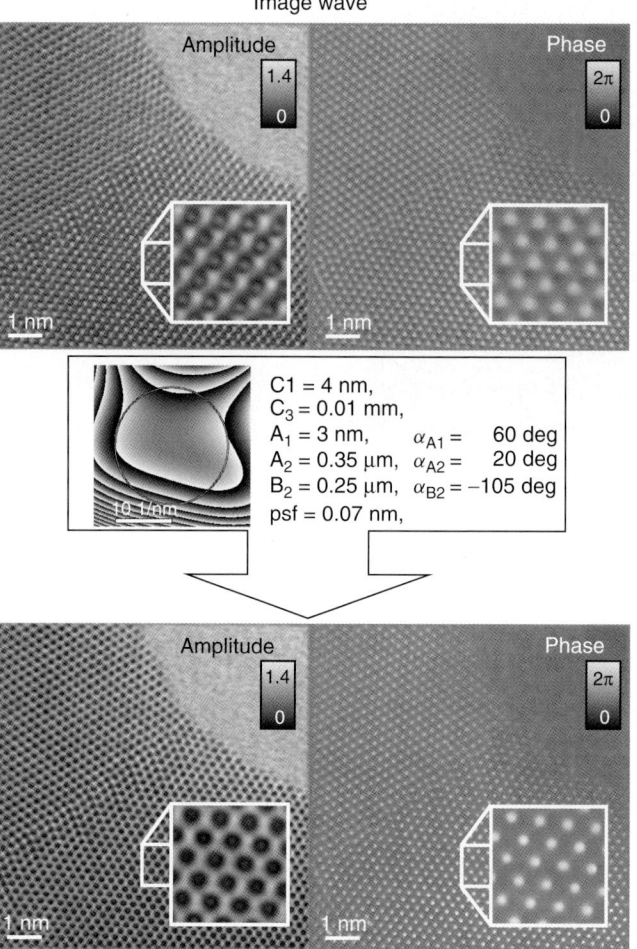

Figure 5.38 Reconstructed object wave after fine-tuning of aberrations. Because of restricted accuracy and time span of stability of the perfectly corrected state of the microscope, the careful a posteriori fine-tuning is indispensable to remove optical artifacts and to find the true fine structure of the object. In this case, the reconstructed image wave was still blurred by some defocus C_1 and spherical aberration C_3; furthermore, residual astigmatisms A_1 and A_2, as well as axial coma B_2, introduced anisotropic distortions, which might have given rise to misinterpretation in terms of the object. (From [57].)

in conventional imaging because toggling from imaging to diffraction in the TEM does not guarantee the same imaging conditions or exactly the same object area. Furthermore, from the Fourier transform of the image intensity ("diffractogram"), a mistilt of the object can hardly be detected because a diffractogram is always point symmetric. The Fourier transform of the reconstructed wave, however, shows the asymmetry of reflections, for example, from mistilt or noncentrosymmetry,

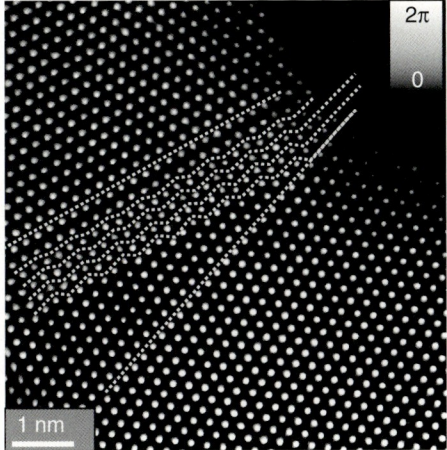

Figure 5.39 Phase image of transition zone between the [110] gold grains. In particular, the phase image with the comparably slim signals from atoms allows spotting of the atoms in the grain boundary at a high accuracy. It is a 90° tilt grain boundary described in detail in [59].

meaning that it behaves like an electron diffraction pattern. Therefore, holographic nanodiffraction is a powerful tool for comprehensive analysis of the findings: A small area containing only a few unit cells of the crystal is masked out and Fourier transformed; shifting the mask across the object wave, tilt variations can easily be recognized [44]. With the now available high-quality holograms, these methods also become increasingly powerful. Figure 5.41 shows that the gold crystal is nearly perfectly oriented in both areas adjacent to the boundary.

Additional information about the object can be extracted by means of reconstruction in the light of arbitrary reflections (Figure 5.42); this is a generalized form of dark-field imaging, which allows analysis of both amplitude and phase contributed from the reflections to the wave in real space. In fact, these phase images display the scattering phases of the diffracted waves such as the geometric phases, not the phase shift by the specimen. For example, such a phase image allows determining strain at the boundary.

5.4
Special Techniques

The hitherto-described results are achieved by furtherance of the standard method of off-axis image-plane holography. The image wave is recorded, reconstructed, and interpreted in terms of the object. In addition to the achievements, bringing holography increasingly to applications dealing with novel questions of materials science, new methods are also under development.

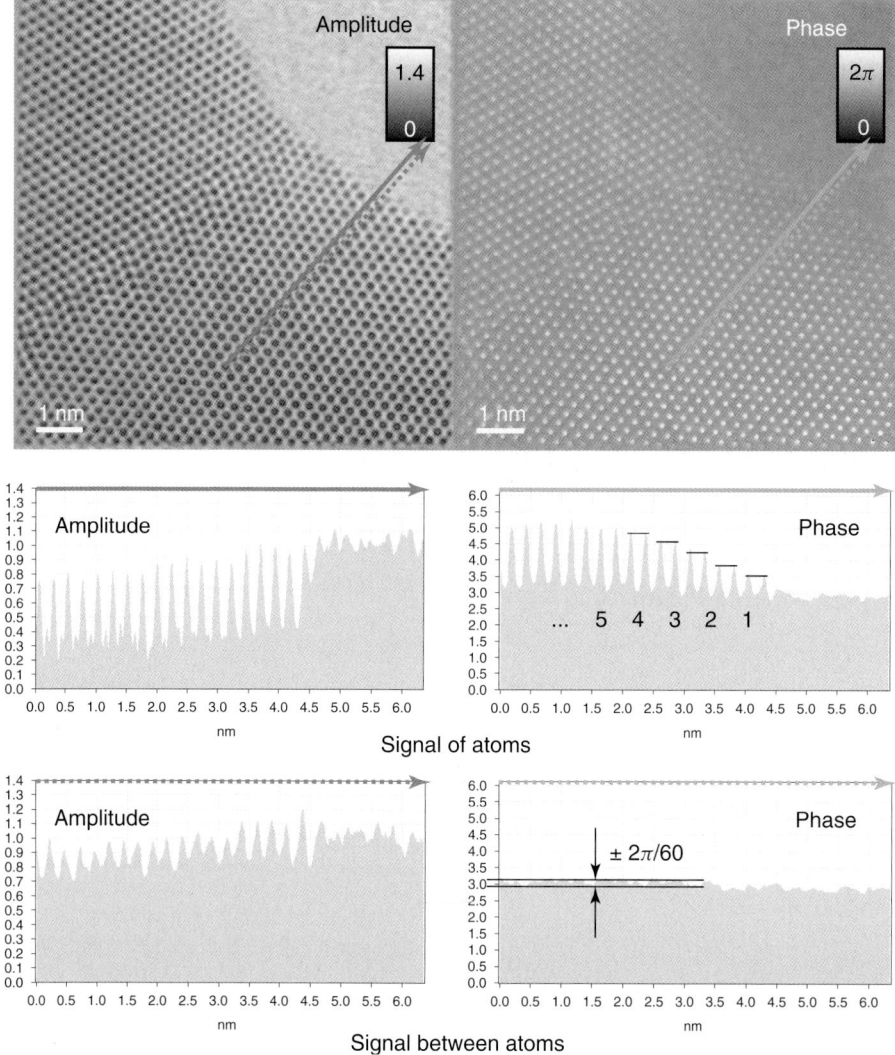

Figure 5.40 Evaluation of amplitude and phase image. (a) Amplitude and phase image with identical positions of line scans. (b) Line scans along a row of atom columns. In the amplitude, the height of indentations at the different columns is nearly the same, whereas the phase distinguishes very clearly between different numbers of atoms in the direction of the electron beam. Owing to the dynamical interaction, the phase step height created by one additional atom shrinks with the number of atoms. The phase signal for one atom approximately represents the projected atomic potential between the nucleus and the electron shell. (c) Line scans positioned in between two atom rows. In the amplitude, the signals of the atoms leak far out into the interatomic space. In the phase, there is no signal directly from the atoms, and hence small phase variations of $\pm 2\pi/60$ can easily be recognized; they are still a factor of about 5 stronger than noise and hence represent object information, however, not yet identified.

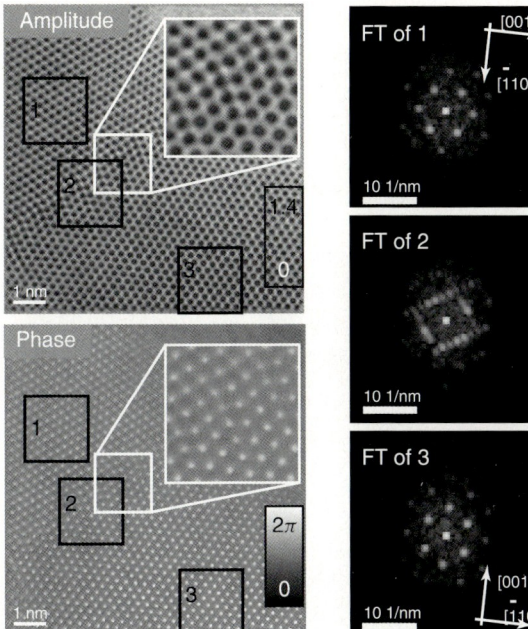

Figure 5.41 Holographic nanodiffraction. For each area selected by a numerical aperture, the respective Fourier transform shows a holographic diffraction pattern. Like an electron diffraction pattern, it reveals the nonsymmetrical excitation of reflections, for example, by noncentrosymmetry or mistilt. The lower grain area 3 of the gold crystal is nearly perfectly oriented along the [110] orientation, whereas grain area 1 is slightly twisted about [001]. The area 2 of the grain boundary is analyzed in detail in Figure 5.42.

5.4.1
Holographic Tomography

From light optics it is well known that LASER-holography allows 3D-imaging, for example, watching a checkerboard under changing perspectives by means of the parallax effect when moving the head. This is possible because – as in natural vision – light is scattered at the surfaces into a large angular range and hence every point of the object can be seen from every standpoint, unless shadowed by an obstacle. In particular, with Fresnel holography, light from every object point is collected in every pixel of the hologram and therefore every object point is visible from every direction of observation.

Things are, however, quite different with electron holography. First, Fresnel holography is poor because of the limited coherence. Second and most important is the fact that electron interaction produces a highly peaked forward scattering with an angular opening of less than 0.1rad. Therefore, the 2D projection effect is intrinsic to electron interaction with the object. Usual electron holography can hardly reconstruct the 3D structure (Figure 5.43).

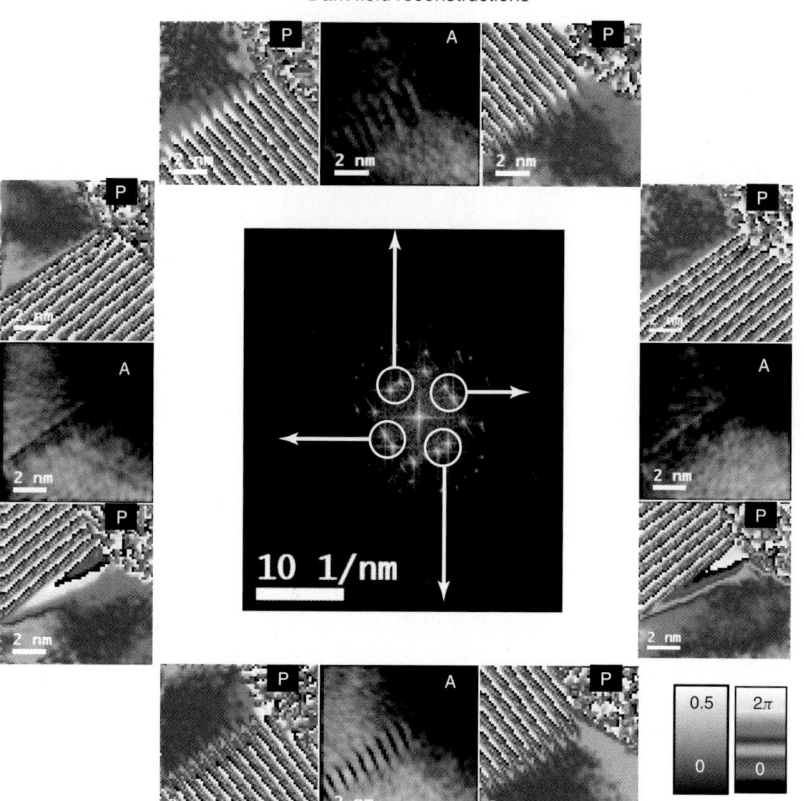

Figure 5.42 Holographic dark field reconstructions. Selecting reflections in Fourier space, after inverse Fourier transform, the respective contributions to amplitude and phase of the object wave are displayed. This is shown for the indicated pairs of {111} reflections for the two different grains. For each pair, amplitude and phase are reconstructed. The two phase images for each pair show the mutual geometric phases of the two grains; the phase in a grain is flat, if the respective reflection is centered in Fourier space before Fourier transform. Further details are found in [57].

Likewise, in any (S)TEM imaging method, the signal represented in one image represents the 2D projection of a 3D object property. Therefore, structure variations along the electron beam are averaged and not interpretable from a single 2D image. For true 3D imaging, tomography was established for various methods such as bright field or energy-filtered transmission electron microscopic (EFTEM) imaging in the (S)TEM. It widens the angular range by tilting the object: From tilt series out to ±90° at best, the 3D structure can be determined. A review is found in [60]. However, no phase-encoded information such as electromagnetic fields in semiconductors can be retrieved from conventional image intensities. Therefore,

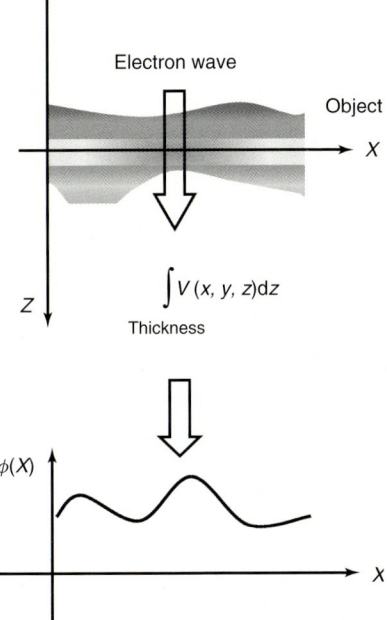

Figure 5.43 Problem of 2D holography. The phase represents the inner potential projected along the z-direction over thickness. Therefore, structure information along z is lost. This is not a problem as long as the object is homogeneous in z, and the thickness is known. But, for example, focused ion beam preparation, otherwise extremely useful for target preparation, damages surface layers by amorphization and doping, and the thickness may vary. Then, the projected potential cannot be interpreted properly.

the principle idea was adopted for developing holographic tomography [61, 62]. The scheme is shown in Figure 5.44.

Each tomography method must fulfill the projection theorem. It says that the signal obtained from a volume element of the object must be independent from the tilt; this is ideally true for the phase shift obtained from the projected potential, as long as there are no dynamic effects.

The method requires recording holograms at a large number of tilts with an angular increment of, say, 2° covering a tilt range of close to ±90°. The manual procedure applied in [61, 62] needs the admirable care and concentration of the operator over many hours. Therefore, despite the pioneering results shown, the manual method is not very conducive for application; furthermore, the results are only of limited reproducibility because of the many parameters, which have to be controlled by hand. In particular, the tilt-induced displacement of the specimen in all three directions makes huge problems.

Therefore, holographic tomography has made a big step forward by the recent development of the semiautomated software package Tomographic and Holographic Microscope Acquisition Software (THOMAS) by Daniel Wolf [63], which controls

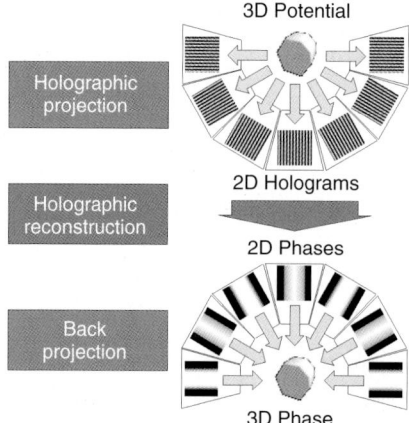

Figure 5.44 Scheme of holographic tomography. Tilting the object, a tilt series of 2D holograms is recorded. From the reconstructed 2D phases, the 3D phase distribution is built up by "back projection." The phases are ideally suited if kinematic, since projected potentials fulfill the projection theorem of tomography perfectly.

the whole tilt series. For each tilt, it adjusts the object positions, records holograms with and without object and, including normalization and phase unwrapping, it organizes the reconstruction of the 2D phase images. These are used for building up the 3D phase volume of the object by means of a weighted simultaneous iterative reconstruction technique (WSIRT)-algorithm. Most recent results are shown in Figures 5.45 and 5.46.

5.4.2
Dark-Field Holography

Already in 1986, Hanszen presented the method of dark-field holography [68] illustrated in Figure 5.47. For this, both object and reference waves are transmitted through a crystal. In the diffraction pattern in the back focal plane, a specific reflection is selected by means of a mask, and a hologram is recorded by superimposing different dark-field areas in the image plane of the crystal. If the object wave is from a nonperfect crystal area, for example, containing dislocations, whereas the reference wave is taken from a perfectly grown crystal area, the hologram allows investigating the imperfections by amplitude and phase.

Hÿtch et al. [69] very successfully adopted this method for measuring strain in semiconductors. With respect to an unstrained area, the displacement $\vec{u}(\vec{r}) = (u_x(\vec{r}), u_y(\vec{r}))$ of atoms produces a geometric phase distribution

$$\phi(\vec{r}) = -2\pi \vec{g} \vec{u}(\vec{r}) \qquad (5.40)$$

in the diffracted wave $\vec{g}$ conjugate to the lattice vector $\vec{a}$ in real space. The phases measured from two noncollinear diffraction waves $\vec{g}_{1,2}$ allow determining the

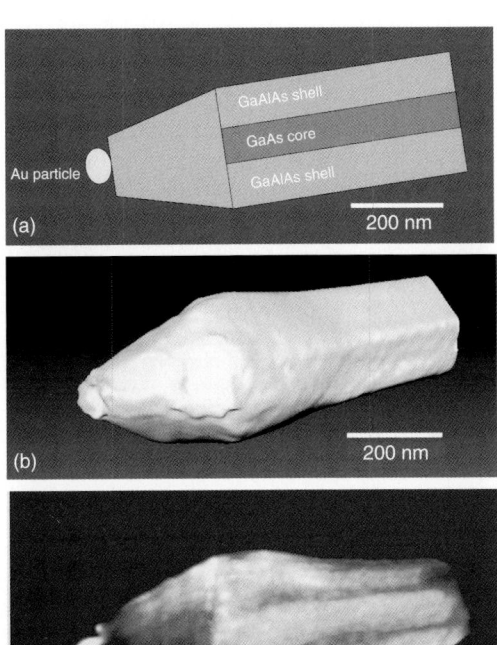

Figure 5.45 Figure 5.45 Holographic tomographic reconstruction of a GaAs/Al$_{0.33}$Ga$_{0.67}$As nanorod. (a): Schematic diagram of the object. The gold particle at the very tip controls growth during the metal oxide vapour phase epitaxy (MOVPE) process. Nanorod prepared by Lovergine and coworkers [64]. (b): Outer shape of the nanorod as displayed from the 3D iso-surface potential shows hexagonal growth along the shaft and transition to the tip area. For 3D observation, the reconstructed object can be freely rotated. (c): interior structure showing the GaAs-core and Al$_{0.33}$Ga$_{0.67}$As shell because of the different inner potentials, that is, the difference of mean inner potential (MIP), diffusion potentials, contact potentials, and so on. The gold particle with a very high MIP of 29 V produces a very strong phase shift; remarkable is the adjacent voltage drop in the particle. The difference of MIPs between GaAs and Al$_{0.33}$Ga$_{0.67}$As follows from DFT calculations [65] as 0.61 V. Again, for 3D observation of the interior structure, the reconstructed object can be freely rotated. (From [66].)

displacement field by means of

$$\vec{u}(\vec{r}) = -\frac{1}{2\pi}(\phi_1 \vec{a}_1 + \phi_2 \vec{a}_2) \tag{5.41}$$

The strain matrix is defined by the components $\varepsilon_{ij}(\overleftarrow{r}) := \partial u_i(\vec{r})/\partial j$, for $i = x, y$ and $j = x, y$, respectively, hence at the end given by $grad\phi(\vec{r})$ [70]. An example is shown in Figure 5.48.

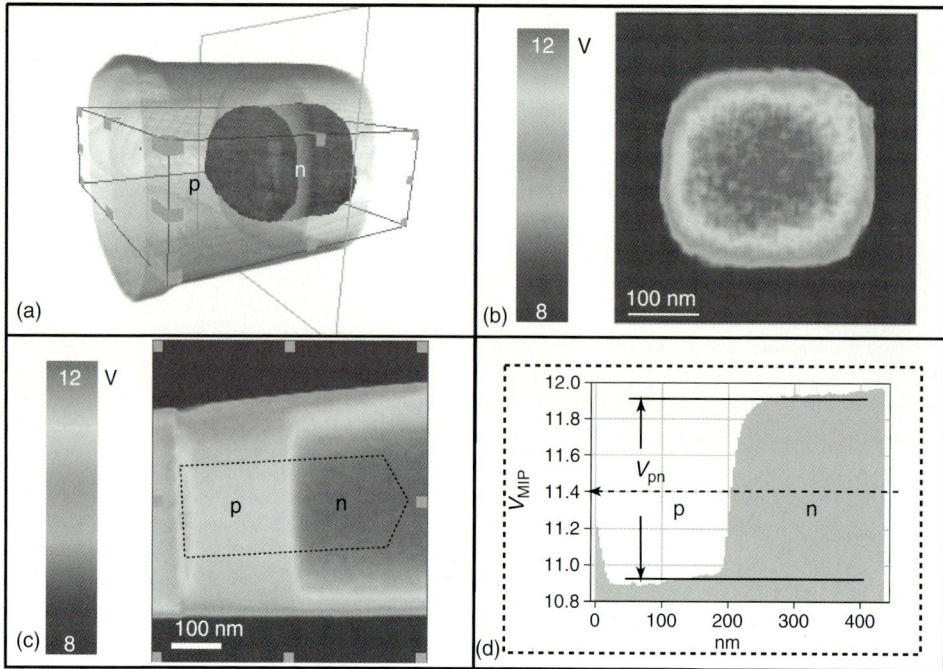

Figure 5.46 Holographic tomographic reconstruction of the potentials around a pn-junction in Si-needle. (a) 3D representation showing that the dopant-related potentials, for example, the n-doped volume (dark gray), do not reach out to the edges of the needle (light gray) because of the damage to the surface layers by the focused ion beam under preparation of the needle. (b) Cross-section through the needle at red frame in (a) shows potential distribution sliced through n-doped volume of the needle. (c) Center potential, averaged vertically over the core area as indicated in (b), is measured along needle axis through the intact center. (d) Potential across pn-junction from (c) reveals mean inner potential of silicon $V_{MIP} = 11.4$ V and pn-voltage $V_{pn} = 1.0$ V. These values are no more affected by FIB artifacts. (From [67].)

5.4.3
Inelastic Holography

As discussed in [75], inelastic interaction with energy transfer larger than about 10^{-15} eV destroys coherence with the elastic wave. The main question is whether the wave, "newborn" by an inelastic process, exhibits extended coherence at its origin (Figure 5.49). To answer this, the scheme shown in Figure 5.50 combining holography with energy filtering is used; note that the holograms are recorded in the image plane of the object [10–12, 71]. Increasing the biprism voltage, the distance ("shear") between subsequently superimposed inelastic electron patches is increased, which gives rise to a specific decrease in interference contrast (Figures 5.51 and 5.52); it is specific because, in addition to the general ensemble coherence of illumination, it reveals the state coherence given by the probability of

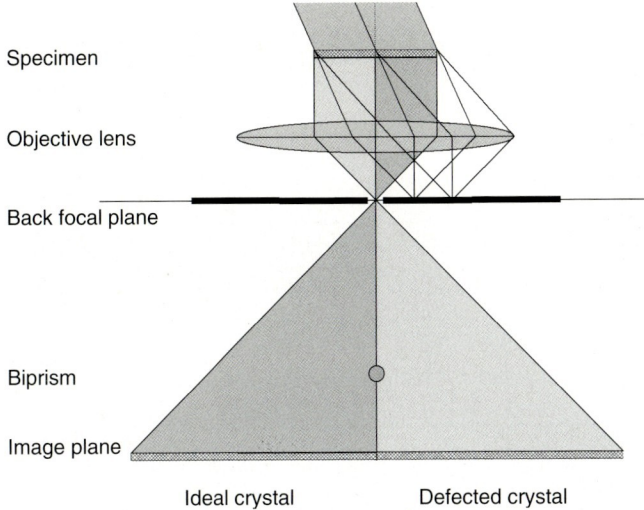

Figure 5.47 Scheme of dark field holography. The single-crystalline object is illuminated such that the selected diffraction wave is propagating along the optic axis. After applying a positive voltage, the biprism superimposes a partial wave from the ideal crystal with one from an area with defects. The reconstructed wave allows analyzing the defects by amplitude and phase in the light of the selected reflection [68].

excitation of different object states of same energy. In the case of Al-plasmons, the extension of state coherence was found reaching out to about 30 nm.

The interpretation is that there is an interaction volume of the measured size, where the incoming electrons have a certain probability for exciting the inelastic process. The extension of this volume measured by coherence can be described by the reduced density matrix for the fast electron after an inelastic scattering event with the energy transfer δE filtered out [72]

$$\rho_{out}(x, x'; \delta E) = \left(\frac{2\pi m e^2}{\varepsilon_0 \hbar^2 k}\right)^2 K_0(q_E |x|) K_0(q_E |x'|) \otimes S(x, x', \delta E) \quad (5.42)$$

$$\rho_{out}(x, x, \delta E'') = \left(\frac{2\pi m e^2}{\varepsilon_0 \hbar^2 k}\right)^2 K_0(q_E |x|) K_0(q_E |x'|) \otimes S(x, x', \delta E)$$

with $\otimes$ meaning convolution. It contributes to the intensity distribution of the hologram with the two waves superimposed under the shear $d = x - x'$ by means of [73]

$$I_{hol}(x) = \rho_{out}(x + d/2, x + d/2) + \rho_{out}(x - d/2, x - d/2) +$$
$$+ 2\Re \underbrace{\left[\mu(d) \rho_{out}(x - d/2, x + d/2)\right]}_{\text{off–diagonal element}} \quad (5.43)$$

The expression $\rho_{out}(x, x')$ combines the influence of the Coulomb delocalization of the inelastic interaction with the correlations in the excited quasi-particle field. The interaction sphere is described by the Bessel functions K_0 of second kind and

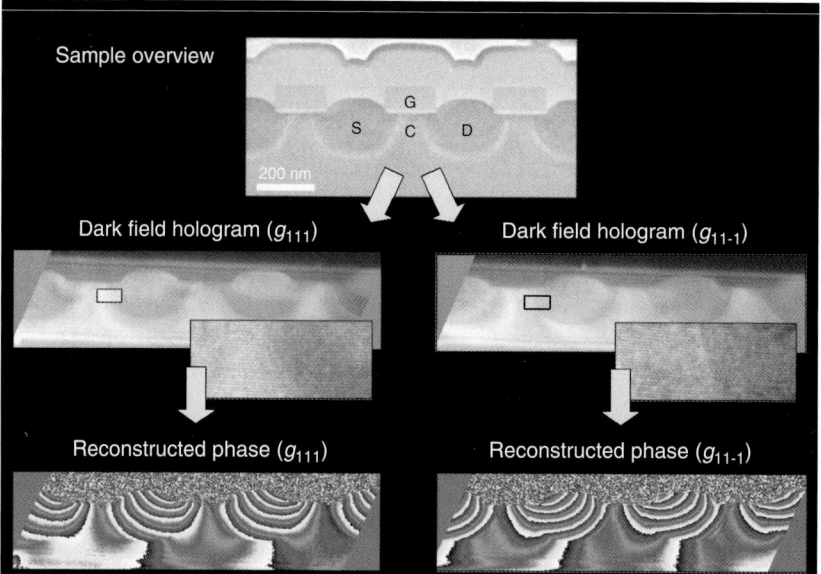
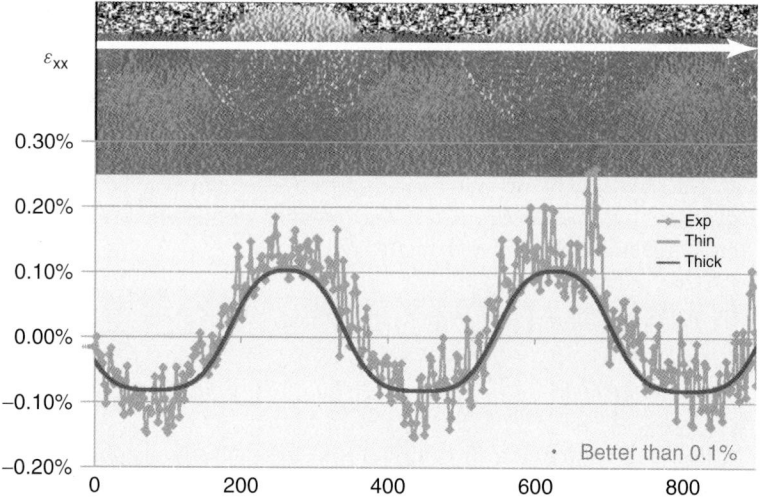

Figure 5.48 Strain measurement by dark field holography. The object consists of a row of FETs with gate G, source S, and drain D. Strain is produced by means of SiGe in S and D. Two dark field holograms are recorded with the g_{111} and $g_{11\bar{1}}$ waves, respectively. From the reconstructed phases, the displacement field $\vec{u}(\vec{r})$ is determined. The resulting strain ε_{xx} is shown in the bottom. The accuracy is better than 0.1%, which is equivalent to a strain of ± 50 MPa. (From [69].)

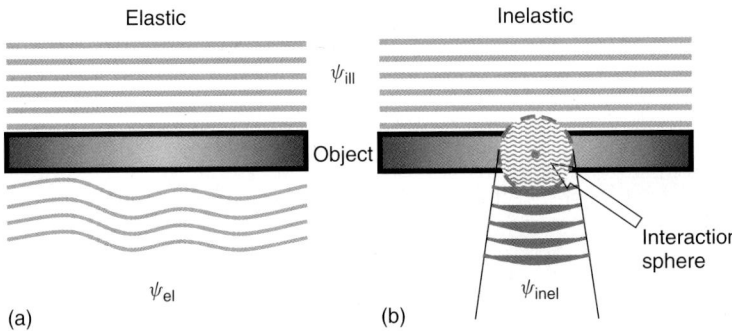

Figure 5.49 Elastic and inelastic interaction of the electron wave with an object. (a) At elastic interaction, the object exit wave ψ_{el} is phase modulated according to the electric and magnetic object potential under preservation of coherence properties of the wave. (b) At inelastic interaction, the illuminating wave ψ_{ill} collapses and gives rise to a "newborn" inelastic wave ψ_{inel} in the interaction sphere shown. The extension of the inelastic wave, that is, the interaction sphere, is determined by coherence measurements in the image plane of the object. At the edge of an object, the inelastic wave may also reach out laterally into vacuum.

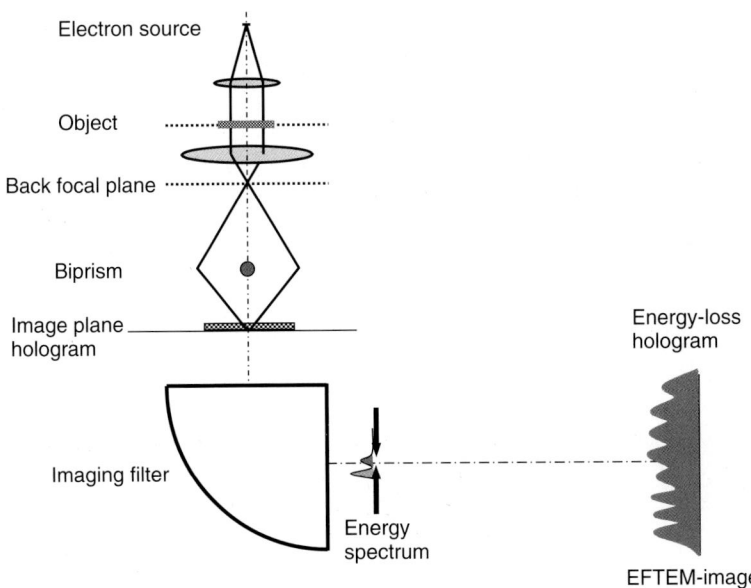

Figure 5.50 Setup for recording holograms with inelastically scattered electrons in an EFTEM. Both beams, subsequently superimposed, go through the object to make sure that both may suffer the same energy transfer at, for example, plasmon scattering. From the holograms built up by all electrons, elastic and inelastic, we select a certain window in the energy spectrum, to reveal the energy-loss hologram of interest. Please keep in mind that this is EFTEM imaging, meaning that we have an image-plane hologram of the inelastic process.

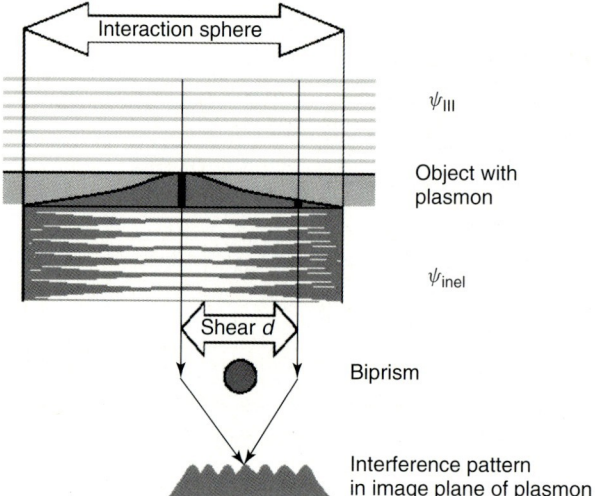

Figure 5.51 Scheme for measuring the extension of coherence in inelastic scattering. The plasmon is the source of the inelastic wave. We measure the extension of this source by measuring the coherent extension of the issued inelastic wave in the plasmon plane as follows: by means of the biprism voltage, the shear d is increased. The contrasts of the respective hologram fringes give the distribution of degree of coherence μ^{state} with shear d of the electrons exciting the plasmon state. The optics for imaging the plasmon into the interference plane is not drawn.

zeroth order, parameterized by the characteristic scattering vector with modulus $q_E = \delta E/(2E)k_0$. The correlation in the quasi-particle excitation is denoted by the density–density correlation function [72]

$$S(x, x', \delta E) = \int_{-\infty}^{+\infty} \int_{-\delta E}^{0} \rho_i(\vec{r}, \vec{r}', \varepsilon) \rho_f(\vec{r}', \vec{r}, \varepsilon + \delta E) d\varepsilon \, \exp(-iq_E(z - z')) dz dz' \tag{5.44}$$

with $\vec{r} = (x, z)$, ρ_i, and ρ_f the density matrices of initial and final state of the object with energy difference δE, and ε the integration variable in energy, respectively; S is the 2D-Fourier transform of the mixed dynamic form factor (MDFF) [74].

Recently, holograms have been recorded with surface plasmons in aluminum [75], which are outlined in Figure 5.53. The comparably strong interference contrast and wide extension of coherence of these processes is remarkable.

These experimental possibilities of inelastic holography and its findings have far-reaching consequences in that they show that also for imaging with inelastic electrons wave optics has to be applied, for example, with an appropriate, partial-coherent transfer theory. Furthermore, it opens the door for a thorough quantum mechanical analysis of the scattering processes by inelastic holography, including specific coherence properties and phases.

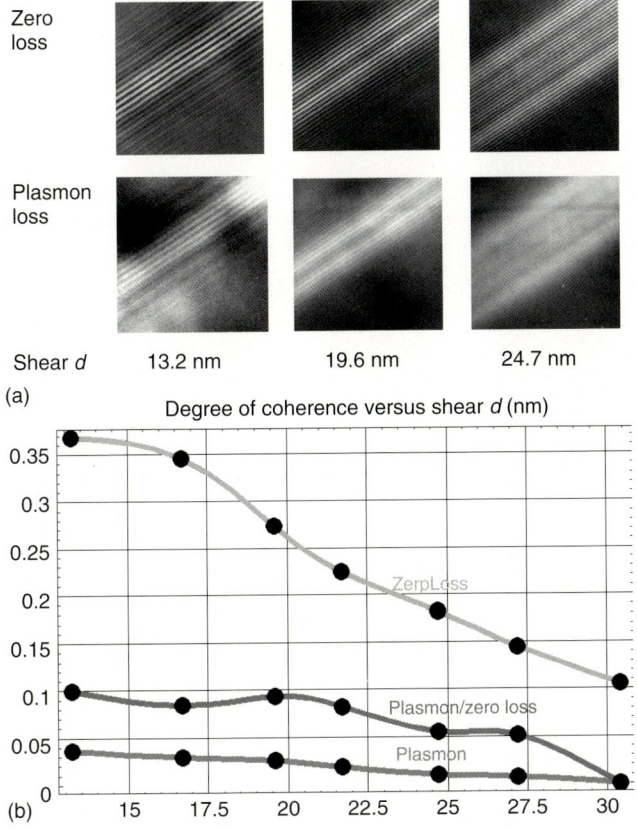

Figure 5.52 Electron holograms recorded with Al-plasmon-scattered electrons at increasing shear. (a) With increasing shear, fringe spacing decreases and overlapping width increases. For subsequent normalization with the "ensemble coherence" of illumination, the zero-loss holograms are also recorded. The result is a μ^{state} in the plasmon-scattered wave. Expectedly, the fringe contrasts in the plasmon-loss holograms are much lower than in the respective zero-loss holograms. For details, see [12]. (b) The curve plasmon/zeroLoss gives the μ^{state} of the plasmon-scattered electrons. Surprisingly, in spite of the generally poor coherence, coherent contributions show up in areas up to 30 nm wide at plasmon scattering. This is interpreted as the diameter of the newborn inelastic wave at the scatterer.

5.5
Summary

This review mainly describes the methodological background of image-plane off-axis electron holography. After more than three decades of development, it is now a unique and powerful method that is capable of investigating objects down to atomic resolution both laterally and in signal. Meanwhile, recording high-quality holograms and reconstructing the wave with highly sophisticated methods of numerical image processing allows investigating single atoms at 0.1 nm lateral

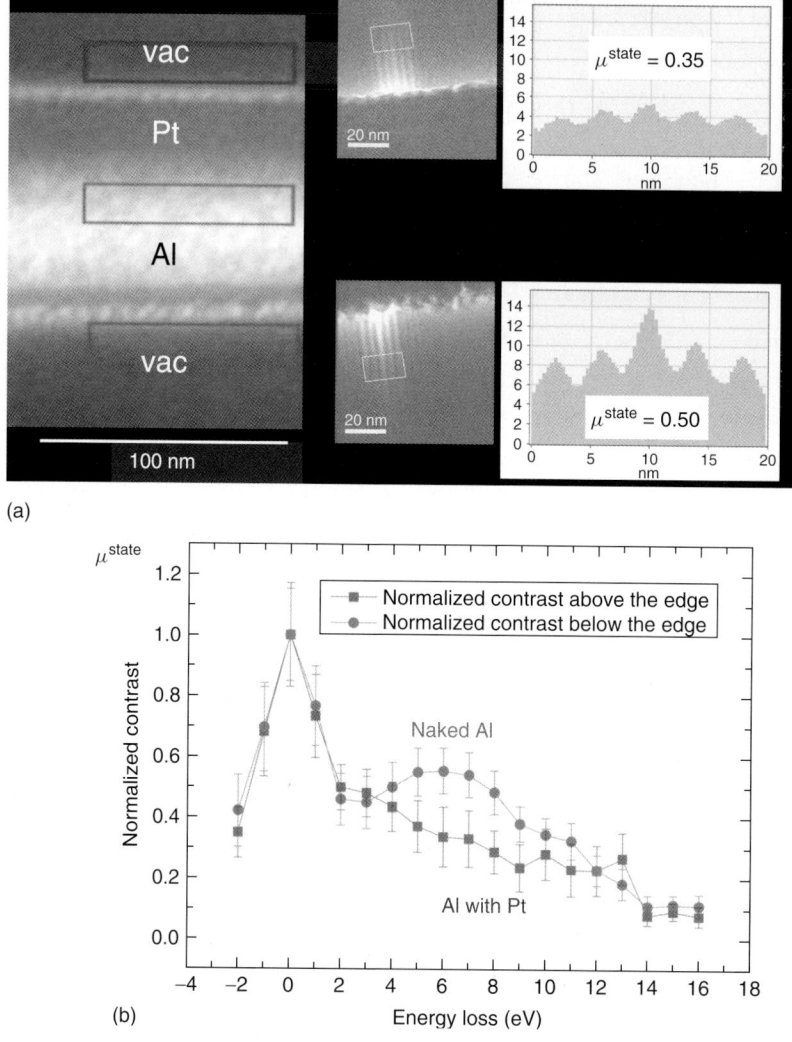

Figure 5.53 Surface plasmon holography. (a) Left: Aluminum plasmon energy-loss image (15 eV energy loss, 1 eV slit-width) of an aluminum bar with averaged spectra at three different positions: below, in the middle of, and above the bar (represented by rectangles). The filtered image shows a dark area in the upper part of the bar due to platinum remaining from FIB preparation (platinum plasmon loss: 35 eV [76]). The bright band shows more or less a pure aluminum part. (a) Right: Energy-loss filtered electron holograms (7 eV) of vacuum region and corresponding contrast profiles 25 nm above/below the edge of the bar. The averaged and zero-loss normalized (i.e., corrected for ensemble coherence) contrast above is about 0.35, the one below is about 0.50. (b) Normalized mean contrast of the fringes 25 nm above (square) and below (circle) the respective edge. The red curve shows a maximum around 6 eV, whereas the blue curve decreases monotonically. The contrast maximum is attributed to the Al surface plasmons. (From [75].)

resolution. The main advantage of holography is the availability of the complete object exit wave both in real space and in Fourier space; this gives a novel dimension for a posteriori exploiting ideally the object information contained in the wave by all numerical wave-optical algorithms one may think about. Typical benefits are as follows:

- Amplitude and phase of object exit wave separately
- Linearity: no interference effects between different reflections
- Quantitative results both in real space and in Fourier space
- Effective (not perfect) filtering out of inelastic information
- A posteriori fine tuning of aberrations
- Holographic nanodiffraction down to single unit cells
- Dark-field reconstruction in the light of arbitrary reflections

In addition to conventional imaging methods, the phase reveals specific information about the object structure, which is invisible in intensity images, such as electric-potential distributions up to atomic resolution and magnetic fields up to a nanometer scale; furthermore, scattering phases in Fourier space can also be measured. Improvement of signal resolution to better than $2\pi/300$ as standard at atomic lateral resolution is obtained with improved brightness; this is achievable already today with special instrumentation. Further improvement is expected from optimized paths of rays, as well as from improved modulation transfer and detective quantum efficiency of the detector.

However, with improving the signal resolution, not only do the wanted signals come up but also unwanted ones. The diversity of origins for occurrence and local variations of tiny phase shifting side effects is very large, as follows:

- Preparation artifacts
  - Isolating surface layers
  - Unwanted and unknown doping effects
  - Topography and thickness variations
  - Bending of the thin objects
  - Surface roughness
  - Differential etching and grooving at interfaces and defects
  - Ion implantation and amorphization by ion beams
- Effects of dynamic interaction
  - Effects of local tilt and thickness variation
  - Phase modulation with thickness non-linear
  - Phase jumps at extinction thicknesses.
- Electron beam induced phase shifts
  - Charging under the beam
  - Generation of electron–hole pairs and diffusion effects
  - Release of secondary electrons, even producing atomic contrast in scanning electron microscopic (SEM) images [77]
  - Different secondary yield at adjacent materials
  - Local charging and voltage drop by charge percolation giving artificial phase shifting pattern

- Changes in local conductivity along interfaces and at defects, which may be considered a property of the object.

These side effects may mimic sham properties of solids or obscure the true ones by artifacts. Therefore, with improvement of the holographic method, the most severe challenges arise with the interpretation of the findings; this is what we had to realize at the example of ferroelectrics at atomic resolution. Many of these effects, which mostly do not precipitate in conventional intensity images, give rise to phase shift and hence only show up and are particularly severe with holography. To make optimum use of the holographic benefits, we have to understand all these side effects.

Therefore, with the accessibility of smaller details of solids, the correct interpretation needs a deeper understanding of solids on an atomic scale, such as the mutual influence of different structures or materials at interfaces and defects. Consequently, more exact modeling of the objects at full relaxation is one of the future demands. Often, this cannot be extracted, for example, from textbooks; instead, a strong interaction with the respectively specialized communities in solid-state science is urgently needed. *Ab initio* calculations on large scales providing also the electronic properties on an atomic scale are being developed for applicability.

Besides the standard method, novel holographic methods are also under development.

- Holographic tomography allows measuring potential distributions in three dimensions without projection artifacts.
- By evaluation of the scattering phases, dark-field holography gives direct access to crystallographic details such as strain in large fields of view, because atomic resolution is not needed.
- Inelastic holography combines the methods of energy loss spectroscopy and holography, which will open new insight into the basic physics of inelastic processes.
- Finally, in future *in situ* experiments will help investigating real situations in solids, such as semiconductor devices in operation or the behavior of materials in extrinsic fields [78, 79]. Then, the measured change of the phases can uniquely be attributed to the varied parameter.

The final goal of all these efforts is collecting a comprehensive data set allowing the solution of the inverse problem, that is, finding the direct and unambiguous path leading from the findings back to the object structure – if possible at all.

Acknowledgments

Frequent discussions with Harald Rose TU Darmstadt/U Ulm and the joint efforts of the members of the Triebenberg-Laboratory Bernd Einenkel, Dorin Geiger, Andreas Lenk, Martin Linck (now at NCEM Berkeley), Axel Lubk (formerly Rother, now at CNRS Toulouse), Heide Müller, Marianne Reibold, Falk Roeder, John Sandino, Jan Sickmann, Sebastian Sturm, Karin Vogel, Daniel Wolf, and

Michael Lehmann, now TU Berlin, are indispensable for our progress. The fruitful cooperation with Bert Freitag at FEI is gratefully acknowledged. Also, the cooperation within the ESTEEM project substantially stimulates our work.

The Körber-Stiftung in Hamburg substantially promoted our research on atomic holography. The Francqui Foundation (Bruxelles) helped to establish the very fruitful cooperation with the group of D. Van Dyck, U Antwerp, on Inelastic Holography.

Our research was made possible by different grants obtained from the German Science Foundation (DFG), Federal Ministry of Education and Research (BMBF), German-Israeli Foundation, State of Saxony, and by the European Union (Framework 6, Integrated Infrastructure, Reference 026019 ESTEEM).

References

1. Gabor, D. (1948) A new microscopic principle. *Nature*, **161**, 563–564.
2. Leith, E.H. and Upatnieks, J. (1962) Reconstructed wave fronts and communication theory. *J. Opt. Soc. Am.*, **52**, 1123–1130.
3. Gabor, D. (1949) Microscopy by reconstructed wave fronts. *Proc. R. Soc. A*, **197**, 454–487.
4. Möllenstedt, G. and Wahl, H. (1968) Elektronenholographie und Rekonstruktion mit Laserlicht. *Naturwissenschaften*, **55**, 340–341.
5. Tonomura, A., Fukuhara, A., Watanabe, H., and Komoda, T. (1968) Optical reconstruction of image from Fraunhofer electron-hologram. *Jpn. J. Appl. Phys.*, **7**, 295.
6. Wahl, H. (1975) Bildebenenholographie mit Elektronen. Thesis. University of Tübingen.
7. Möllenstedt, G. and Düker, H. (1956) Beobachtungen und Messungen an Biprisma-Interferenzen mit Elektronenwellen. *Z. Phys.*, **145**, 377.
8. Lenz, F. (1988) Statistics of phase and contrast determination in electron holograms. *Optik*, **79**, 13–14.
9. Rother (Lubk), A., Gemming, T., and Lichte, H. (2009) The statistics of the thermal motion of the atoms during imaging process in transmission electron microscopy and related techniques. *Ultramicroscopy*, **109**, 139–146.
10. Harscher, A., Lichte, H., and Meyer, J. (1997) Interference experiments with energy filtered electrons. *Ultramicroscopy*, **69**, 201–209.
11. Lichte, H. and Freitag, B. (2000) Inelastic electron holography. *Ultramicroscopy*, **81**, 177–186.
12. Potapov, P.L., Lichte, H., Verbeeck, J., and van Dyck, D. (2006) Experiments on inelastic electron holography. *Ultramicroscopy*, **106**, 1012–1018.
13. Verbeeck, J., Bertoni, G., and Lichte, H. (2011) A holographic biprism as a perfect energy filter? *Ultramicroscopy*, **111**, 887–893.
14. Lehmann, M. and Lichte, H. (2002) Tutorial on off-axis electron holography. *Microsc. Microanal.*, **8**, 447–466.
15. Perkes, P. (2002) Phase unwrapping algorithm – plug in for digital micrograph©. Arizona State University (ASU).
16. Sickmann, J., Formánek, P., Linck, M., Mühle, U., and Lichte, H. (2011) Imaging modes for potential mapping in semiconductor devices by electron holography with improved lateral resolution. *Ultramicroscopy*, **111**, 290–302.
17. Voelkl, E. (2010) Noise in off-axis type holograms including reconstruction and CCD camera parameters. *Ultramicroscopy*, **110**, 199–210.
18. Lichte, H. (2008) Performance limits of electron holography. *Ultramicroscopy*, **108**, 256.
19. Matteucci, G., Missiroli, G.F., and Pozzi, G. (2002) Electron holography of long-range electrostatic fields, in *Advances in Imaging and Electron Physics*, vol. 122, Elsevier Science, USA, pp. 173–249.

20. Wentzel, G. (1926) Zwei Bemerkungen über die Zerstreuung korpuskularer Strahlen als Beugungserscheinung. *Z. Phys.*, **40**, 590–593.
21. Byatt, W.J. (1956) Analytical representation of Hartree potentials and electron scattering. *Phys. Rev.*, **104**, 1298.
22. Weickenmeier, A. and Kohl, H. (1991) Computation of absorptive form factors for high-energy electron diffraction. *Acta Crystallogr. A*, **47**, 590–597.
23. Schowalter, M., Rosenauer, A., Lamoen, D., Kruse, P., and Gerthsen, D. (2006) Ab-initio computation of the mean inner Coulomb potential of wurtzite-type semiconductors and gold. *Appl. Phys. Lett.*, **88**, 232108.
24. Jönsson, C., Hoffmann, H., and Möllenstedt, G. (1965) Messung des mittleren inneren Potentials von Be im Elektronen-Interferometer. *Phys. Kondens. Mater.*, **3**, 193.
25. Simon, P., Adhikari, R., Lichte, H., Micheler, G.H., and Langela, M. (2005) Electron holography and AFM studies on styrenic block copolymers and a high impact polystyrene. *J. Appl. Polym. Sci.*, **96**, 1573–1583.
26. Formánek, P. (2005) TEM-holography on device structures in microelectronics. PhD thesis. TU Cottbus.
27. Rau, W.-D., Schwander, P., Baumann, F.H., Höppner, W., and Ourmazd, A. (1999) Two-dimensional mapping of the electrostatic potential in transistors by electron holography. *Phys. Rev. Lett.*, **82**, 2614.
28. Lenk, A., Lichte, H., and Muehle, U. (2005) 2D-mapping of dopant distribution in deep sub micron CMOS devices by electron holography using adapted FIB-preparation. *J. Electron Microsc.*, **54**, 351–359A. doi: 10.1093/jmicro/dfi055
29. Lenk, A., Muehle, U., and Lichte, H. (2005) Why does a p-doped area show a higher contrast in electron holography than a n-doped area of same dopant concentration?, in *UK Springer Proceedings in Physics*, Springer, ISBN 978-3-540-31914-6 vol. 107 (eds A.G. Cullis and J.L. Hutchison), p. 205.
30. Bronner, F. and Farach-Carson, M. (eds) (2003) *Bone Formation*, Springer Publisher.
31. Teaford, M.F., Smith, M.M., and Ferguson M.W.J. (eds) (2000) *Development, Function and Evolution of Teeth*, Cambridge University Press.
32. Simon, P., Carrillo-Cabrera, W., Formánek, P., Göbel, C., Geiger, D., Ramlau, R., Tlatlik, H., Buder, J., and Kniep, R. (2004) On the real-structure of biomimetically grown hexagonal prismatic seeds of fluorapatite-gelatine-composites: TEM investigations along [001]. *J. Mater. Chem.*, **14**, 2218–2224.
33. Busch, S., Dolhaine, H., DuChesne, A., Heinz, S., Hochrein, O., Laeri, F., Podebrand, O., Vietze, U., Weiland, T., and Kniep, R. (1999) Biomimetic morphogenesis of fluorapatite-gelatin composites: fractal growth, the question of intrinsic electric fields, core/shell assemblies, hollow spheres and reorganization of denatured collagen. *Eur. J. Inorg. Chem.*, **10**, 1643–1653.
34. Simon, P., Zahn, D., Lichte, H., and Kniep, R. (2006) Electric dipole field induced morphogenesis of fluoroapatite-gelatine composites: a model system for general principles of biomineralisation? *Angew. Chem. Int. Ed.*, **45**, 1911–1915.
35. Simon, P., Lichte, H., Lehmann, M., Huhle, R., Formanek, P., and Ehrlich, H. (2008) Electron holography of biological samples: a review. *Micron*, **39**, 229.
36. Ehrenberg, W. and Siday, R.E. (1949) The refractive index in electron optics and the principles of dynamics. *Proc. Phys. Soc. (Lond.)*, **B62**, 8–21.
37. Aharonov, Y. and Bohm, D. (1959) Significance of electromagnetic potentials in quantum theory. *Phys. Rev.*, **115**, 485.
38. Moellenstedt, G. and Bayh, W. (1962) Kontinuierliche Phasenschiebung von Elektronenwellen im kraftfeldfreien Raum durch das magnetische Vektorpotential eines Solenoids. *Phys. Blaetter*, **18**, 299.
39. Tonomura, A., Osakabe, N., Matsuda, M., Kawasaki, T., Endo, J., Yano, S., and Yamada, H. (1986) Evidence for

the Aharonov-Bohm effect with magnetic field completely shielded from electron wave. *Phys. Rev. Lett.*, **56**, 792.
40. Möllenstedt, G., Schmid, H., and Lichte, H. (1982) Measurement of the phase shift between electron waves due to a magnetic flux enclosed in a metallic cylinder. Proceedings of the 10th International Congress on Electron Microscopy, Hamburg, Germany, vol. 1, p. 433.
41. Simon, P., Lichte, H., Drechsel, J., Formanek, P., Graff, A., Wahl, R., Mertig, M., Adhikari, R., and Michler, H.G. (2003) Electron holography of organic and biological materials. *Adv. Mater.*, **15** (17), 1475–1481.
42. Biskupek, J., Kaiser, U., Lichte, H., Lenk, A., Gemming, T., Pasold, G., and Witthuhn, W. (2005) TEM-characterization of magnetic samarium- and cobalt-rich-nanocrystals formed in hexagonal SiC. *J. Magn. Magn. Mater.*, **293**, 924.
43. Tonomura, A., Matsuda, T., Endo, J., Arii, T., and Mihama, K. (1986) Holographic interference electron microscopy for determining magnetic structure and thickness distribution. *Phys. Rev. B*, **34**, 3397–3402.
44. Orchowski, A., Rau, W.D., and Lichte, H. (1995) Electron holography surmounts resolution limit of electron microscopy. *Phys. Rev. Lett.*, **74** (3), 399
45. Reimer, L. (1989) *Transmission Electron Microscopy*, 2nd edn, Springer-Verlag.
46. Cowley, J.M. and Moodie, A.F. (1957) The scattering of electrons by atoms and crystals. I. A new theoretical approach. *Acta Crystallogr. A*, **10**, 609–619.
47. Linck, M., Lichte, H., and Lehmann, M. (2006) Electron holography: materials analysis at atomic resolution. *Int. J. Mater. Res* (formerly Z. Metallk.), **97**, 890.
48. Kirkland, J. (1998) *Advanced Computing in Electron Microscopy*, Plenum Press.
49. Lubk, A., Gemming, S., and Spaldin, N.A. (2009) First-principles study of ferroelectric domain walls in multiferroic bismuth ferrite. *Phys. Rev. B (Condens. Matter Mater. Phys.)*, APS, **80**, 104110.
50. Lichte, H. (1986) Electron holography approaching atomic resolution. *Ultramicroscopy*, **20**, 293.
51. Lichte, H. (1996) Electron holography: optimum position of the biprism in the electron microscope. *Ultramicroscopy*, **64**, 79–86.
52. Lichte, H. (1991) Optimum focus for taking electron holograms. *Ultramicroscopy*, **38**, 13–22.
53. Lehmann, M. (2004) Influence of the elliptical illumination on acquisition and correction of coherent aberrations in high-resolution electron holography. *Ultramicroscopy*, **100**, 9–23.
54. Lehmann, M. and Lichte, H. (2005) Electron holographic material analysis at atomic dimensions. *Cryst. Res. Technol.*, **40**, 149–160.
55. Dunin-Borkowski, R.E., Kasama, T., Wei, A., Tripp, S.L., Hÿtch, M.J., Snoeck, E., Harrison, R.J., and Putnis, A. (2004) Off-axis electron holography of magnetic nanowires and chains, rings, and planar arrays of magnetic nanoparticles. *Microsc. Res. Technol.*, **64**, 390.
56. Geiger, D., Lichte, H., Linck, M., and Lehmann, M. (2008) Electron holography with a Cs-corrected transmission electron microscope. *Microsc. Microanal.*, **14**, 68–81.
57. Linck, M., Freitag, B., Kujawa, S., Lehmann, M., and Niermann, T. State of the art in atomic resolution off-axis electron holography using high-brightness electron gun, *Ultramicroscopy*, in press.
58. Freitag, B., Knippels, G., Kujawa, S., Tiemeijer, P.C., Van der Stam, M., Hubert, D., Kisielowski, C., Denes, P., Minor, A., and Dahmen, U. (2008) *Proceedings of the EMC 2008*, Instrumentation and Methods, Springer, ISBN 978-3-540-85154-7 Vol. 1, pp. 55–56.
59. Medlin, D.L., Foiles, S.M., and Cohen, D. (2001) A dislocation-based description of grain boundary dissociation: Application to a 90° <110> tilt boundary in gold. *Acta Mater.*, **49**, 3689–3697.

60. Midgley, P.A. and Weyland, M. (2003) 3D electron microscopy in the physical sciences: the development of Z-contrast and EFTEM tomography. *Ultramicroscopy*, **96**, 413.
61. Lai, G., Hirayama, T., Ishizuka, K., and Tonomura, A. (1994) Three-dimensional reconstruction of electric-potential distribution in electron holographic interferometry. *J. Appl. Opt.*, **33**, 829–833.
62. Twitchett-Harrison, A.C., Yates, T.J.V., Nowcomb, S.B., Dunin-Borkowski, R.E., and Midgley, P.A. (2007) High-resolution three-dimensional mapping of semiconductor dopant profiles. *Nano Lett.*, **7**, 2020–2023.
63. Wolf, D., Lubk, A., Lichte, H., and Friedrich, H. (2010) Towards automated electron holographic tomography for 3D mapping of electrostatic potentials. *Ultramicroscopy*, **110**, 390–399.
64. Prete, P., Marzo, F., Paiano, P., Lovergine, N., Salvati, G., Lazzarine, L., and Sekiguchi, T. (2008) Luminescence of GaAs/AlGaAs core-shell nanowires grown by MOVPE using tertiary butylarsine. *J. Cryst. Growth*, **310**, 5114–5118.
65. Kruse, P., Schowalter, M., Lamoen, D., Rosenauer, A., and Gerthsen, D. (2006) Determination of the mean inner potential in III-V semiconductors, Si and Ge by density functional theory and electron holography. *Ultramicroscopy*, **106**, 105.
66. Wolf, D., Lichte, H., Pozzi, G., Prete, P., and Lovergine, N. (2011) Electron holographic tomography for mapping the three-dimensional distribution of the electrostatic potential in III-V semiconductor nanowires. *Appl. Phys. Lett.*, **98**, 264103
67. Wolf, D. (2010) Elektronenholographische Tomographie zur 3D-Abbildung von elektrostatischen Potentialen in Nanostrukturen. PhD-Thesis. Technische Universität Dresden, http://nbn-resolving.de/urn:nbn:de:bsz:14-qucosa-65125.
68. Hanszen, K.-J. (1986) Methods of off-axis electron holography and investigations of the phase structure in crystals. *J. Phys. D: Appl. Phys.*, **19**, 373–395.
69. Hÿtch, M.J., Houdellier, F., Hüe, F., and Snoeck, E. (2008) Nanoscale holographic interferometry for strain measurements in electronic devices. *Nature*, **453**, 1086.
70. Hÿtch, M.J., Snoeck, E., and Kilaas, R. (1998) Quantitative measurement of displacement and strain fields from HREM micrographs. *Ultramicroscopy*, **74**, 131–146.
71. Verbeeck, J., van Dyck, D., Lichte, H., Potapov, P., and Schattschneider, P. (2005) Plasmon holographic experiments: theoretical framework. *Ultramicroscopy*, **102**, 239–255.
72. Schattschneider, P., Nelhiebel, M., and Jouffrey, B. (1998) Density matrix of inelastically scattered fast electrons. *Phys. Rev. B*, **59**, 16.
73. Schattschneider, P. and Lichte, H. (2005) Correlation and the density-matrix approach to inelastic electron holography in solid-state plasmas. *Phys. Rev. B*, **71**, 045130.
74. Rose, H. (1976) Image formation by inelastically scattered electrons in electron microscopy. *Optik*, **45** (2), 139–158.
75. Röder, F. and Lichte, H. (2011) Inelastic electron holography – first results with surface plasmons. *Eur. Phys. J.: Appl. Phys.*, **54**, 33504.
76. Ravets, S., Rodier, J.C., Ea Kim, B., Hugonin, J.P., Jacubowiez, L., and Lalanne, P. (2009) Surface plasmons in the Young slit doublet experiment. *J. Opt. Soc. Am. B*, **26** (12), B28–B33.
77. Zhu, Y., Inada, H., Nakamura, K., and Wall, J. (2009) Imaging single atoms using secondary electrons with an aberration-corrected electron microscope. *Nat. Mater.*, **8**, 808. doi: 10.1038/NMAT2532
78. Twitchett, C., Dunin-Borkowski, R.E., and Midgley, P.A. (2002) Quantitative electron holography of biased semiconductor devices. *Phys. Rev. Lett.*, **88**, 238302.
79. Lenk, B., Einenkel, J., and Sandino, H., Lichte, A specially designed electric field holder for measurement of materials response to an external electric

field by electron holography. Conference Proceedings of ICM17, Rio de Janeiro, Brazil, 2010, 19.18.

Books

Hecht, E (1990) *Optics*, Addison Wesley. ISBN 0-201-11609-X.

Dunin-Borkowski, R.E., Kasama, T., Harrison, R.J., Kirkland, A.I., and Hutchison, J.L. (eds) (2007) Electron holography of nanostructured materials, in *Nanocharacterisation*, Royal Society of Chemistry, pp. 138–183. Chapter 5 in a volume entitled edited by ISBN-0854042418.

Dunin-Borkowski, R.E., Kasama, T., McCartney, M.R., Smith, D.J., Hawkes, P.W., and Spence, J.C.H. (eds) (2006) Electron holography, *Science of Microscopy*, Springer, pp. 1141–1195. Chapter 18 in a volume entitled edited by ISBN-0387252967.

Peng, L.M., Dudarev, S.L., and Whelan, M.J. (2010) *High-Energy Electron Diffraction and Microscopy*, Monographs on the physics and Chemistry of Materials, Oxford University Press. ISBN-978-0-19-850074-2.

Tonomura, A. (1999) *The Quantum World Unveiled by Electron Waves*, World Scientific. ISBN-981-02-2510-5.

Tonomura, A., Allard, L.F., Pozzi, G., Joy, D.C., and Ono, Y.A. (eds) (1995) *Electron Holography*, Elsevier. ISBN-0-444-82051-5.

Voelkl, E., Allard, L.F., and Joy, D.C. (eds) (1999) *Introduction to Electron Holography*, Kluwer Academics/Plenum Publishers. ISBN-0-306-44920-X.

Articles

Basics

Anaskin, F. and Stoyanova, I.G. (1965) The phase of an electron wave scattered over a semiplane. *Radio Eng. Electron. Phys.*, **13**, 1104–1110.

Marks, L.D. and Plass, R. (1994) Partially coherent and holographic contrast transfer theory. *Ultramicroscopy*, **55**, 165–170.

Scherzer, O. (1949) The theoretical resolution limit of the electron microscope. *J. Appl. Phys.*, **20**, 20.

Smith, D.J., Saxton, W.O., O'Keefe, M.A., Wood, G.J., and Stobbs, W.M. (1983) The importance of beam alignment and crystal tilt in high resolution electron microscopy. *Ultramicroscopy*, **11**, 263–282.

Method

Frost, B.G., Völkl, E., and Allard, L.F. (1996) An improved mode of operation of a transmission electron microscope for wide field off- axis holography. *Ultramicroscopy*, **63**, 15.

Harscher, A. and Lichte, H. (1996) Experimental study of amplitude and phase detection limits in electron holography. *Ultramicroscopy*, **64**, 57.

Heindl, E., Rau, W.D., and Lichte, H. (1996) The phase-shift method in electron off-axis holography: using neural network techniques. *Ultramicroscopy*, **64**, 87.

Ishizuka, K. (1993) Optimized sampling schemes for off- axis holography. *Ultramicroscopy*, **52**, 1.

Lehmann, M. (2000) Determination and correction of the coherent wave aberration from a single off-axis electron hologram by means of a genetic algorithm. *Ultramicroscopy*, **85**, 165.

Lehmann, M. (2004) Influence of the elliptical illumination on acquisition and correction of coherent aberrations in high-resolution electron holography. *Ultramicroscopy*, **100**, 9. doi: 10.1016/j.ultramic.2004.01.005

Leuthner, T., Lichte, H., and Herrmann, K.-H. (1989) STEM holography using the electron biprism. *Phys. Status Solidi A*, **116**, 113.

Linck, M. (2007) Optimum imaging conditions for recording high-resolution electron holograms in a Cs-corrected TEM. *Microsc. Microanal.*, **13** (Suppl. 3), 32.

Lichte, H., Geiger, D., Harscher, A., Heindl, E., Lehmann, M., Malamidis, D., Orchowski, A., and Rau, W.-D. (1996) Artefacts in electron holography. *Ultramicroscopy*, **64**, 67.

Lubk, A. (2010) Quantitative off-axis electron holography and (multi-)ferroic interfaces. PhD-Thesis. Universität Dresden, http://nbn-resolving.de/urn:nbn:de:bsz:14-qucosa-33452.

Lubk, A., Wolf, D., and Lichte, H. (2010) The effect of dynamical scattering in off-axis holographic mean inner potential and inelastic mean free path measurements. *Ultramicroscopy*, **110**, 438.

Mankos, M., Higgs, A.A., Scheinfein, M.R., and Cowley, J.M. (1995) Far- out- of- focus electron holography in dedicated FEG STEM. *Ultramicroscopy*, **58**, 87.

Rother (Lubk), A., Gemming, S., Chaplygin, I., Leisegang, T., and Lichte, H. (2008) Ab-initio simulation of the object exit wave of ferroelectrics. *AMTC Lett.*, **1**, 100.

Seidel, J., Martin, L.W., He, Q., Zhan, Q., Chu, Y., Rother (Lubk), A., Hawkridge, M.E., Maksymovych, P., Yu, P., Gajek, M., Balke, N., Kalinin, S.V., Gemming, S., Wang, F., Catalan, G., Scott, J.F., Spaldin, N.A., Orenstein, J., and Ramesh, R. (2009) Conduction at domain walls in oxide multiferroics. *Nat. Mater.*, **8**, 229.

Van Dyck, D., Lichte, H., and Spence, J.C.H. (2000) Inelastic scattering and holography. *Ultramicroscopy*, **81**, 187.

Weierstall, U. and Lichte, H. (1996) Electron holography with a superconducting objective lens. *Ultramicroscopy*, **65**, 13.

Interpretation

Koch, C.T. and Lubk, A. (2010) Off-axis and inline holography: a quantitative comparison. *Ultramicroscopy*, **110**, 460.

Latychevskaia, T., Formánek, P., Koch, C.T., and Lubk, A. (2010) Off-axis and inline electron holography: experimental comparison. *Ultramicroscopy*, **110**, 472.

Rose, H. (2010) Theoretical aspects of image formation in the aberration-corrected electron microscope. *Ultramicroscopy*, **110**, 488.

Rother (Lubk), A. and Scheerschmidt, K. (2009) Relativistic effects in elastic scattering of electrons in TEM. *Ultramicroscopy*, **109**, 154–160.

Voelkl, E. and Tang, D. (2010) Approaching routine $2\pi/1000$ phase resolution for off-axis type holography. *Ultramicroscopy*, **110**, 447.

Inverse Problem

Lentzen, M. (2010) Reconstruction of the projected electrostatic potential in high-resolution transmission electron microscopy including phenomenological absorption. *Ultramicroscopy*, **110**, 517.

Scheerschmidt, K. (2003) Parameter retrieval in electron microscopy by solving an inverse scattering problem, in *Proceedings of the 6th International Conference Mathematical and Numerical Aspects of Wave Propagation (WAVES 2003) (Germany, 2003)* (ed. G.C. Cohen), Springer, Berlin, p. 607 6ff.

Scheerschmidt, K. (2010) Electron microscope object reconstruction: Retrieval of local variations in mixed type potentials. Part I: theoretical preliminaries. *Ultramicroscopy*, **110**, 543.

Wang, A., Chen, F.R., Van Aert, S., and Van Dyck, D. (2010) Direct structure inversion from exit waves. Part I: theory and simulations. *Ultramicroscopy*, **110**, 527.

Applications

Electric Potentials

Beleggia, M. and Pozzi, G. (2010) Phase shift of charged metallic nanoparticles. *Ultramicroscopy*, **110**, 418.

Lichte, H., Reibold, M., Brand, K., and Lehmann, M. (2002) Ferroelectric electron holography. *Ultramicroscopy*, **93**, 199–212.

Simon, P., Huhle, R., Lehmann, M., Lichte, H., Mönter, D., Bieber, Th., Reschetilowski, W., Adhikari, R., and Michler, G.H. (2002) Electron holography on beam sensitive materials: organic polymers and mesoporous silica. *Chem. Mater.*, **14**, 1505–1514.

Tanji, T., Urata, K., Ishizuka, K., Ru, Q., and Tonomura, A. (1993) Observation of atomic surface potential by electron holography. *Ultramicroscopy*, **49**, 259–264.

Wanner, M., Bach, D., Gerthsen, D., Werner, R., and Tesche, B. (2006) Electron holography of thin amorphous carbon films: measurement of the mean inner, potential and a thickness-independent phase shift. *Ultramicroscopy*, **106**, 341.

Semiconductors

Beleggia, M., Capelli, R., and Pozzi, G. (2000) A model for the interpretation of holographic and Lorentz images of tilted reverse-biased p-n junctions in a finite specimen. *Philos. Mag. B*, **80**, 1071–1082.

Cherns, D. and Jiao, C.G. (2001) Electron holography studies of the charge on dislocations in GaN. *Phys. Rev. Lett.*, **87** (20), 205504.

Cooper, D., Ailliot, C., Barnes, J.-P., Hartmann, J.-M., Salles, P., Benassayag, G., and Dunin-Borkowski, R.E. (2010) Dopant profiling of focused ion beam milled semiconductors using off-axis electron holography; reducing artifacts, extending detection limits and reducing the effect of gallium implantation. *Ultramicroscopy*, **110**, 383.

McCartney, M.R., Gribelyuk, M.A., Li, J., Ronsheim, P., McMurray, J.S., and Smith, D.J. (2002) Quantitative analysis of one-dimensional dopant profile by electron holography. *Appl. Phys. Lett.*, **80**, 3213–3215.

Müller, E., Gerthsen, D., Brückner, P., Scholz, F., Kirchner, C., and Waag, A. (2006) Investigation of the electrical activity of dislocations in n-GaN and n-ZnO epilayers by transmission electron holography. *Phys. Rev. B*, **73**, 245316.

Orchowski, A., Rau, W.-D., Rücker, H., Heinemann, B., Schwander, P., Tillack, B., and Ourmazd, A. (2002) Local electrostatic potential and process-induced boron redistribution in patterned Si/SiGe/Si heterostructures. *Appl. Phys. Lett.*, **80**, 2556–2558.

Simon, P., Maennig, B., and Lichte, H. (2004) Conventional electron microscopy and electron holography of organic solar cells. *Adv. Funct. Mat.*, **14**, 669–676. doi: 10.1002/adfm.200304498

Li-Ion Batteries

Yamamoto, K., Iriyama, Y., Asaka, T., Hirayama, T., Fujita, H., Fisher, C.A.J., Nonaka, K., Sugita, Y., and Ogumi, Z. (2010) Dynamic visualization of the electric potential in an all-solid-state rechargeable lithium battery. *Angew. Chem. Int. Ed.*, **49**, 4414.

Holographic Tomography

Beanland, R., Sánchez, A., Hernandez-Garrido, J., Wolf, D., and Midgley, P. (2010) Electron tomography of III-V quantum dots using dark field 002 imaging conditions. *J. Micros.*, **237**, 148–154.

Magnetic

Beleggia, M., Kasama, T., and Dunin-Borkowski, R.E. (2010) The quantitative measurement of magnetic moments from phase images of nanoparticles and nanostructures, Ultramicroscopy. 110, 425–432.

Dunin-Borkowski, R.E., McCartney, M.R., Kardynal, B., and Smith, D.J. (1998) Magnetic interactions within patterned cobalt nanostructures using off-axis electron holography. *J. Appl. Phys.*, **84**, 374–378.

Lin, P.V., Camino, F.E., and Goldman, V.J. (2009) Electron Interferometry in the quantum hall regime: Aharonov-Bohm effect of interacting electrons. *Phys. Rev. B*, **80**, 125310.

McCartney, M.R., Agarwal, N., Chung, S., Cullen, D.A., Han, M.-G., He, K., Li, L., Wang, H., Zhou, L., and Smith, D.J. (2010) Quantitative phase imaging of nanoscale electrostatic and magnetic fields using off-axis electron holography. *Ultramicroscopy*, **110**, 375.

McCartney, M.R., Smith, David.J., Farrow, R.F.C., and Marks, R.F. (1997) Off-axis electron holography of epitaxial FePt films. *J. Appl. Phys.*, **82**, 2461–2465.

Ru, Q., Matsuda, T., Fukuhara, A., and Tonomura, A. (1991) Digital extraction of the magnetic- flux distribution from a electron interferogram. *J. Opt. Soc. Am. A*, **8**(11), 1739–1745.

Tonomura, A. (1999) *The Quantum World Unveiled by Electron Waves*, World Scientific. ISBN-981-02-2510-5.

Atomic Resolution

Jia, C.L., Houben, L., Thust, A., and Barthel, J. (2010) On the benefit of the negative spherical-aberration imaging technique for quantitative HRTEM. *Ultramicroscopy*, **110**, 500.

Lichte, H. Amelinckx, S., van Dyck, D., van Landuyt, J., van Tendeloo, G. (eds) (1997) Electron holography methods, *Handbook of Microscopy, Methods I*, VCH Verlagsgesellschaft, pp. 515–536. ISBN 3-527-29280-2.

Lichte, H. (1992) Electron holography I. Can electron holography reach 0.1nm resolution? *Ultramicroscopy*, **47**, 223.

Lichte, H. (1992) Electron holography II. First steps of high resolution electron holography into materials science. *Ultramicroscopy*, **47**, 231–240.

Lichte, H. and Rau, W.D. (1994) High-resolution electron holography with the CM30FEG-Special Tübingen. *Ultramicroscopy*, **54**, 310.

Rau, W.D. and Lichte, H. (1998) High-resolution off-axis electron holography, in *Introduction to Electron Holography* (eds , E. Völkl, , L.F. Allard, and , D.C. Joy), Kluwer Academic/Plenum Publishers, p. 201.

Röder, F., Lubk, A., Lichte, H., Bredow, T., Yu, W., and Mader, W. (2010) Long-range correlations in In2O3(ZnO)7 investigated by DFT calculations and electron holography. *Ultramicroscopy*, **110**, 400.

6
Lorentz Microscopy and Electron Holography of Magnetic Materials

Rafal E. Dunin-Borkowski, Takeshi Kasama, Marco Beleggia, and Giulio Pozzi

6.1
Introduction

The transmission electron microscope (TEM) is unparalleled in its ability to provide structural, chemical, and electronic information about materials over a wide range of length scales down to atomic dimensions [1]. However, it can also be used to quantitatively measure local variations in magnetic field with nanometer spatial resolution using Lorentz microscopy and electron holography. This chapter contains brief descriptions of the historical background and basis of the latter techniques, followed by examples taken from their application to the characterization of magnetic fields in nanoscale materials and devices. Each application emphasizes the sensitivity of the observed magnetic states and reversal mechanisms to local variations in microstructure and composition.

6.2
Lorentz Microscopy

6.2.1
Historical Background

The history of Lorentz microscopy in the TEM can be traced back to the 1940s, when magnetic fringing fields outside materials were characterized by using beam stops to form dark-field images [2, 3] and by analyzing distortions induced by magnetic fringing fields in shadow images of wire meshes [4]. In the 1950s, magnetic domains and domain walls inside thin films were imaged using shadow images [5] and out-of-focus images [6], as well as by using the edges of objective aperture strips to form images using only those electrons that had been refracted by the specimen in selected directions [7]. In the 1960s, the wave-optical description of the formation of Lorentz images of magnetic materials was used to interpret interference fringes in out-of-focus images of domain walls in Fe and NiFe films [8, 9] and low-angle electron diffraction patterns of periodic magnetic domains in Co films [10, 11]. This

Handbook of Nanoscopy, First Edition. Edited by Gustaaf Van Tendeloo, Dirk Van Dyck, and Stephen J. Pennycook.
© 2012 Wiley-VCH Verlag GmbH & Co. KGaA. Published 2012 by Wiley-VCH Verlag GmbH & Co. KGaA.

work was followed by a comparison of wave and geometrical optical approaches of Lorentz image interpretation [12], by the introduction of Zernike phase contrast and interference microscopy as two additional modes of Lorentz microscopy [13] and by the generation of magnetic phase contrast by using inhomogeneities in a thin C film inserted in the diffraction plane of the microscope [14].

It should be noted that each of these approaches for imaging magnetic fields in the TEM is sensitive only to the component of the magnetic induction within and around the specimen that is *perpendicular* to the electron beam. In addition, each technique is normally applied after switching off the conventional TEM objective lens, which would otherwise subject the specimen to a strong vertical magnetic field that could saturate its moment in the electron beam direction. Instead, an additional non-immersion Lorentz lens is often used as the primary imaging lens, with the specimen then situated either in a magnetic-field-free environment or in a precalibrated magnetic field that is applied to the specimen using dedicated magnetizing coils or by partially exciting the conventional microscope objective lens.

6.2.2
Imaging Modes

Lorentz microscopy is currently grouped primarily under the headings of Fresnel and Foucault imaging, as illustrated schematically in Figure 6.1, and differential phase contrast (DPC) microscopy [15]. All three techniques rely on the fact that magnetic specimens modulate the phase of the incident electron wave, with magnetic thin films, in particular, normally considered to be strong, albeit slowly varying, phase objects.

Fresnel imaging (Figure 6.1.a) involves the acquisition of highly defocused images of TEM specimens, in which magnetic domain walls appear as bright or dark lines depending on whether the images are acquired under- or overfocus. The technique is best suited to the characterization of extended thin films at low magnifications, as variations in specimen thickness at the edges of nanocrystals

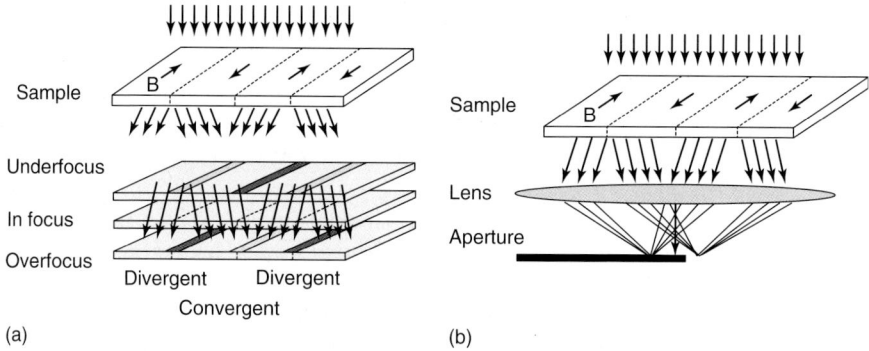

Figure 6.1 Schematic diagrams illustrating the formation of (a) Fresnel (out-of-focus) and (b) Foucault (magnetic dark-field) images in the TEM.

or patterned elements can introduce strong contrast that is non-magnetic in origin. The contrast in Fresnel images of magnetic materials is often interpreted simplistically in terms of the refraction of electrons toward or away from domain walls by the magnetization on either side of them, as shown schematically in Figure 6.1.a. The wave-optical nature of the image formation process becomes apparent when a sufficiently coherent electron source is used, as bright (convergent) wall contrast may then contain interference fringes that run parallel to the wall. The qualitative interpretation of Fresnel images is sometimes facilitated by the fact that "magnetization ripple" contrast, which arises from fluctuations in magnetization direction in polycrystalline specimens, usually lies perpendicular to the local magnetization direction [16]. On a more local scale, although the *spacing* of interference fringes with defocus in convergent domain wall images is constant with defocus, comparisons of fringe *contrast* with defocus between experiments and simulations can be used to determine domain wall widths [17]. Recently, the transport of intensity equation has been used to interpret defocus series of Fresnel images, in order to infer two-dimensional electron-optical phase variations semi-quantitatively [18]. However, such approaches are susceptible to artifacts, especially if the boundary conditions for the phase or its gradient at the edges of the field of view are unknown. The greatest advantage of the Fresnel mode of Lorentz microscopy remains its ease of application, especially for real-time studies of magnetization reversal processes [19].

Foucault imaging (Figure 6.1.b) involves the formation of a magnetic dark-field image of a specimen by selecting electrons that have been refracted by the magnetic field in the specimen using a displaced aperture placed either in the conventional diffraction plane of the microscope or in another conjugate plane. In contrast to Fresnel imaging, the recorded contrast arises from the magnetic domains themselves, rather than from domain walls, with bright contrast formed by electrons that have been allowed to pass through the aperture. If the relative direction of the aperture and the image are known, then the direction of the magnetic induction in the specimen can be determined. Classically, the Lorentz deflection angle is given by the expression

$$\beta_L = \frac{e\lambda t B_\perp}{h} \qquad (6.1)$$

where e is the electron charge, λ is the relativistic electron wavelength, $B_\perp$ is the average value of the component of the magnetic induction perpendicular to the electron beam direction within specimen thickness t, and h is Planck's constant. In practice, the technique can be difficult to implement experimentally because the scattering angles associated with magnetic fields in the specimen rarely exceed 100 μrad. As a result, the contrast is highly sensitive to the lateral position of the aperture and to its vertical placement in the back focal plane, and the technique is also not well suited to real-time studies of magnetization reversal processes if the position of the beam moves relative to the aperture when an external magnetic field is applied to the specimen. An alternative adaptation of the technique, which allows for more quantitative image interpretation, is referred to as *coherent Foucault imaging* [20]. The latter technique is normally applied in a microscope that is equipped with a field emission gun and involves replacing the blocking aperture by a phase-shifting aperture, whose thickness is chosen so that it provides a phase shift of close to π radians. It results in the formation of fringes that are analogous to those produced in an electron holographic interferogram (see below), with a separation between adjacent fringes that is equivalent to an enclosed magnetic flux of h/e. The advantage of the technique over electron holography is that no postprocessing is required. However, the technique is only applicable to regions that lie close to the specimen edge. In addition, it is often not readily applicable to the characterization of magnetic fields in continuous thin films, if the separation of adjacent fringes is large. The more direct approach of recording the fine structure in a low-angle diffraction pattern of a magnetic material is also sometimes implemented, although a highly coherent electron beam and a sufficiently large camera length are required, and the information in the diffraction pattern may originate from a relatively large illuminated specimen area.

DPC imaging [21–25] is usually performed in a scanning transmission electron microscope (STEM) using a rastered focused electron probe, with the local Lorentz deflection introduced by the specimen determined by recording difference signals between opposite segments of a quadrant detector as the probe is scanned across the specimen. The resolution of the technique is determined primarily by the probe size, which can be smaller than 10 nm with the specimen located in a magnetic-field-free environment. Real-time imaging of magnetization reversal processes is more difficult than using Fresnel imaging as a result of the scan rate. However, the total signal falling on the detector can also be used to record a standard incoherent bright-field image of the specimen, thereby providing structural and magnetic information about the specimen simultaneously. Variants of the technique include the use of an "annular quadrant" detector in a STEM [26] and the use of a "summed image DPC" approach in a conventional TEM to achieve contrast equivalent to that from STEM DPC images by recording summed series of Foucault images that are acquired using different aperture positions without the use of scanning coils or a quadrant detector [27, 28].

6.2.3
Applications

Over the past 40 years, each of the modes of Lorentz microscopy described above has been applied to the study of magnetic remanent states and reversal mechanisms in a wide range of materials, often in the presence of applied fields, currents, and reduced or elevated specimen temperatures in the TEM. *In situ* magnetizing experiments are complicated by the fact that the application of a horizontal magnetic field to a specimen results in deflection of the electron beam, requires the use of compensating coils to deflect the beam back to the optical axis of the microscope and usually limits the applied field range to a few hundred oersted. As a result, it is relatively common to tilt the specimen in a vertical magnetic field generated by the partially excited microscope objective lens. Such experiments also require care, as

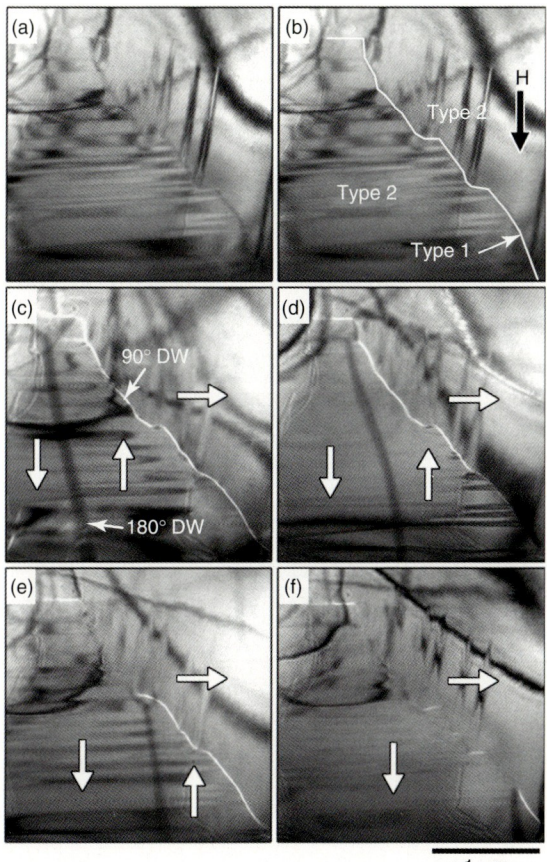

Figure 6.2 (a,b) Bright-field TEM images of a magnetite specimen that contains "Type 1" and "Type 2" ferroelastic twins. The images were recorded close to focus with the sample cooled below the Verwey transition. Diffraction contrast in differently oriented twins and curving bend contours are visible. A Type 1 twin boundary is marked with a white line in (b). The region to the left of this boundary has its [001] magnetic easy axis oriented top to bottom. The region to the right has its easy axis oriented left to right. (c–f) Lorentz (out-of-focus) TEM images recorded from the same region in different (downward) fields showing ferrimagnetic domain walls (DWs). The magnetic fields were applied by tilting the specimen in a 920 Oe vertical field generated by partially exciting the TEM objective lens. The in-plane components of the applied field are (c) 20, (d) 50, (e) 80, and (f) 150 Oe.

the out-of-plane (OOP) component of the magnetic field applied to the specimen can affect the details of the switching processes of interest.

Two recent examples of the use of Fresnel imaging to study materials in the presence of external stimuli are shown in Figures 6.2 and 6.3. Figure 6.2 shows a series of out-of-focus Lorentz (Fresnel mode) TEM images of a magnetite (Fe_3O_4) specimen [29, 30]. The images in Figure 6.2 were recorded using a liquid-nitrogen-cooled specimen holder with the specimen held below the Verwey

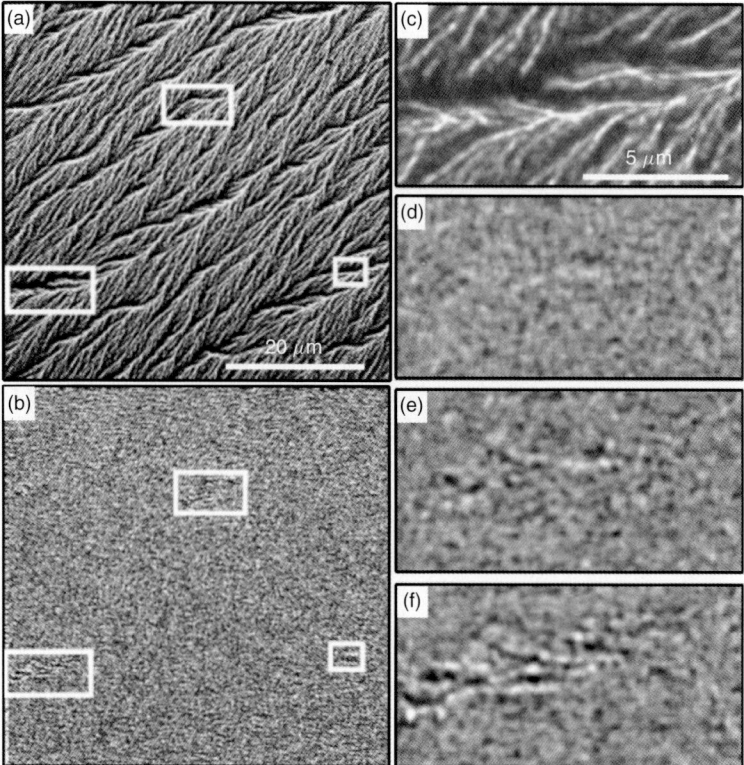

Figure 6.3 (a) Lorentz (out-of-focus) TEM image of a CoPtCr/alumina/Co magnetic tunnel junction recorded at remanence (in zero applied magnetic field). Domain walls and magnetization ripple contrast are due primarily to the soft Co layer. (b) Difference image generated between Lorentz images that had been recorded from the area shown in (a) after 0 and 1000 cycles of magnetization reversal of the soft layer, after initial magnetization of the hard layer using an in-plane field of ~10 kOe. (c) Enlargement of boxed area at bottom left of (a). (d–f) Difference images recorded from the same area after 5, 100, and 1000 cycles, respectively.

transition (~120 K), which is a first-order crystallographic phase transition below which the magnetocrystalline anisotropy of magnetite increases by an order of magnitude and the magnetic easy axis switches from the <111> directions of the parent cubic phase to the [001] direction of the low-temperature monoclinic phase. On cooling through the transition, the easy axis of the monoclinic phase can lie along any of the three <100> directions of the cubic phase, resulting in the formation of transformation twins. Although numerous studies have suggested that magnetic domain walls in magnetite can interact with the ferroelastic twin walls that form at low temperature [e.g., 26], no direct evidence for such interactions had been presented.

In the images shown in Figure 6.2, micrometer-scale monoclinic twin domains (referred to as *Type 1 twins*), which are separated from each other by jagged twin walls, are observed at low temperature. These twin domains are subdivided further by lamellar twins (referred to as *Type 2 twins*), which join at their ends to form needle twins that move in response to thermal or magnetoelastic stress by the advancement and retraction of their needle tips. Lateral motion of Type 1 and Type 2 twin walls is observed only within a few degrees of the phase transition temperature. Figure 6.2c–f shows a 180° magnetic domain wall being driven to the right by an increasing applied magnetic field, resulting in the conversion of a conventional 90° domain wall into a "divergent" 90° domain wall.

Figure 6.3 shows an example of the application of the Fresnel mode of Lorentz microscopy to the study, in plan-view geometry, of a layered magnetic structure that was deposited in the form of a tunnel junction based on a magnetically hard $Co_{75}Pt_{12}Cr_{13}$ layer (coercive field 1600 Oe) and a magnetically soft Co layer separated by a thin insulating alumina tunnel barrier [31]. In this sample, the average magnetic moment of the hard layer was observed to decay logarithmically as the adjacent soft layer was cycled magnetically, eventually becoming fully demagnetized. The magnetic stability of the hard layer was found to depend on the soft layer composition, the hard-layer magnetic moment, and the alumina tunnel barrier thickness. Figure 6.3 shows the application of Lorentz microscopy to study samples that had been prepared directly on electron-transparent silicon nitride windows, with the observed domain wall and magnetization ripple contrast originating primarily from the soft Co layer.

Difference images were generated between images that had been recorded before and after cycling the soft layer *in situ* in the TEM. The succession of enlargements (Figure 6.3c–f) demonstrates the growth of a demagnetized region in the CoPtCr film as a result of repeated field cycling. Through the examination of such images and subsequent comparisons with micromagnetic simulations, the hard-layer decay was found to result from the presence of magnetic fringing fields surrounding magnetic domain walls in the magnetically soft layer. The formation and motion of these walls caused statistical flipping of magnetic moments in randomly oriented grains in the hard CoPtCr layer, with a progressive trend toward disorder and eventual demagnetization. Random-axis anisotropy between adjacent grains in the CoPtCr layer, which gives this material desirable properties as a hard-disk medium, contributes to the mechanism of localized demagnetization. The strength of such short-range interactions between adjacent magnetic layers is important for device applications.

6.3
Off-Axis Electron Holography

6.3.1
Historical Background

The technique of electron holography, which was originally proposed by Denis Gabor as a means to correct for electron microscope lens aberrations [32, 33], is based on the formation of an interference pattern or "hologram" in the TEM, and provides the only direct means to measure the real-space phase shift of the electron wave that has passed through a specimen. Early applications of the technique were restricted by the limited brightness and coherence of the tungsten filaments that were used as electron sources [34]. It was not until the early 1970s that the off-axis mode of electron holography, making use of a Möllenstedt-Düker electron biprism [35], was used to study magnetic fields originating from thin films of CoFe [36] and single crystals of Ni [37]. In the latter study, the biprism edges were also observed to display zigzag or stepped shapes when they crossed domain walls. Subsequently, off-axis electron holograms of ferromagnetic domain walls were recorded successfully [38] and an off-axis Fresnel scheme was realized, which made use of a single crystal as an amplitude beam splitter, with the specimen inserted in the selected-area aperture plane [39].

In the 1980s, the research group of Akira Tonomura published a substantial body of work on electron holography of magnetic materials. Studies of magnetic fringing fields were carried out on microscopic NiFe horseshoe magnets [40], Fe wires containing domain walls [41] and individual γ-Fe_2O_3 [42] and barium ferrite [43] particles. Studies of magnetic microstructure within samples were carried out on in-plane and perpendicularly recorded high-density magnetic recording media [44, 45], individual fine Co particles [46] and arrays of NiFe particles [47]. One of the most elegant scientific contributions of electron holography was the experimental confirmation of the Aharonov–Bohm effect [48], which implies that potentials are more fundamental physical quantities in quantum mechanics than electric and magnetic fields and that such fields can influence electrons "without touching them." A series of electron holography experiments was performed on NiFe toroidal magnets covered with 300 nm thick layers of superconducting Nb. The Nb was deposited both to prevent electrons from penetrating the magnetic material and to confine the magnetic flux by exploiting the Meissner effect. It was estimated that a negligible fraction ($\sim 10^{-6}$) of the electron wave coherent extension should "touch" the magnetic field in such a toroid [49, 50]. The observations showed that the measured phase difference became quantized to a value of 0 or π when the specimen temperature was reduced below the superconducting critical temperature for Nb (5 K), that is, when a supercurrent was induced to circulate in the magnet. When the Nb was superconducting, an even number of flux quanta ($\phi_0 = h/2e = 2.07 \times 10^{-15}$ Tm^2) trapped in the toroid corresponded to a phase shift of 0, while an odd number corresponded to a phase shift of π. The observed quantization of magnetic flux and the measured phase differences with the magnetic field screened by the superconductor provided confirmation of the Aharonov–Bohm effect.

6.3.2
Basis and Governing Equations

A schematic diagram showing the typical electron-optical configuration used for the TEM off-axis mode of electron holography is shown in Figure 6.4. The region

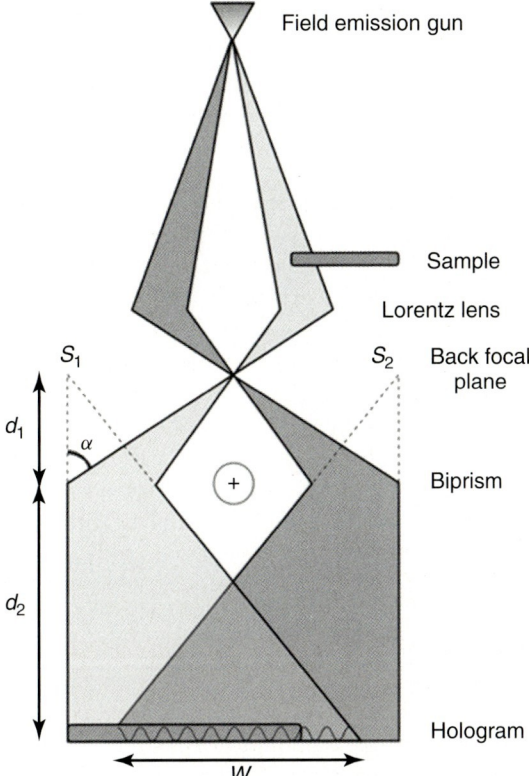

Figure 6.4 Schematic diagram of setup for off-axis electron holography. The symbols are defined in the text.

of interest on the specimen is positioned so that it covers approximately half the field of view. An electron biprism (a thin conducting wire), which is usually located close to the first image plane, has a positive voltage applied to it in order to overlap a "reference" electron wave that has passed through vacuum with the electron wave that has passed through the specimen.

The overlap region (of width W) marked in Figure 6.4 contains interference fringes and an image of the specimen, in addition to Fresnel fringes originating from the edge of the biprism wire, as shown in the form of experimental electron holograms in Figure 6.5. The electron-optical configuration in Figure 6.4 is equivalent to the use of two coherent electron sources S_1 and S_2 [51]. The application of a larger voltage to the biprism wire increases the separation of the two virtual sources and the width of the overlap region (Figure 6.5), placing a greater constraint on the spatial coherence of the source that is required to maintain sufficient interference fringe contrast.

The governing equations that describe image formation in electron holography can be written in terms of standard expressions for coherent image formation in the TEM. The electron wavefunction in the image plane is conventionally written

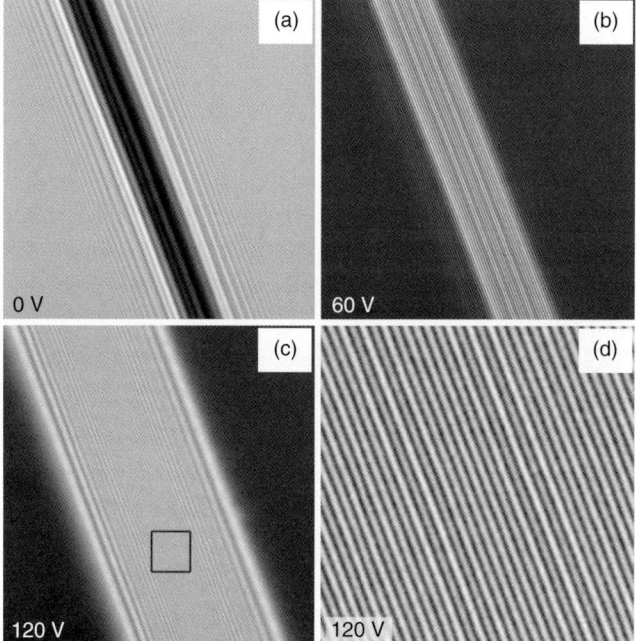

Figure 6.5 (a–d) Interference fringe patterns (electron holograms) recorded using biprism voltages of (a) 0, (b) 60, and (c,d) 120 V. In (a), only Fresnel fringes from the edges of the biprism wire are visible and (d) corresponds to the box marked in (c). The field of view in (a–c) is 916 nm.

in the form

$$\psi_i(\mathbf{r}) = A_i(\mathbf{r}) \exp[i\phi_i(\mathbf{r})] \quad (6.2)$$

where $\mathbf{r}$ is a two-dimensional position vector in the plane of the specimen and A and ϕ refer to amplitude and phase, respectively. In a bright-field TEM image, the intensity distribution is given by the expression

$$I(\mathbf{r}) = |A_i(\mathbf{r})|^2 \quad (6.3)$$

and all information that relates directly to the phase is lost. In contrast, the intensity distribution in an off-axis electron hologram can be obtained by considering the addition of a tilted plane reference wave to the complex specimen wave in the form

$$I_{\mathrm{hol}}(\mathbf{r}) = |\psi_i(\mathbf{r}) + \exp[2\pi i \mathbf{q}_c \cdot \mathbf{r}]|^2 \quad (6.4)$$

where the tilt of the reference wave is specified by a two-dimensional reciprocal space vector $\mathbf{q} = \mathbf{q}_c$. This expression can be rewritten in the form

$$I_{\mathrm{hol}}(\mathbf{r}) = 1 + A_i^2(\mathbf{r}) + 2A_i(\mathbf{r})\cos[\phi_i(\mathbf{r}) + 2\pi\mathbf{q}_c \cdot \mathbf{r}] \quad (6.5)$$

highlighting the fact that the intensity in the hologram consists of three separate contributions: the reference image intensity, the specimen image intensity, and a set of cosinusoidal fringes, whose local phase shifts and amplitudes are equivalent to the phase and amplitude, respectively, of the electron wavefunction in the image plane.

An experimental electron hologram of a specimen that contains a distribution of magnetic crystals is shown in Figure 6.6a, alongside a vacuum reference hologram (Figure 6.6b) and a magnified region of the specimen hologram (Figure 6.6c). The amplitude and phase shift of the specimen wave are recorded in the intensity and position, respectively, of the holographic interference fringes.

In order to extract phase and amplitude information from an electron hologram, it is usually first Fourier transformed. On the basis of Eq. (6.5), the Fourier transform of an electron hologram can be written in the form

$$\mathrm{FT}\left[I_{\mathrm{hol}}(\mathbf{r})\right] = \delta(\mathbf{q}) + \mathrm{FT}\left[A_i^2(\mathbf{r})\right] + \delta(\mathbf{q}+\mathbf{q}_c) \otimes \mathrm{FT}\left[A_i(\mathbf{r})\exp[i\phi_i(\mathbf{r})]\right] \quad (6.6)$$
$$+ \delta(\mathbf{q}-\mathbf{q}_c) \otimes \mathrm{FT}\left[A_i(\mathbf{r})\exp[-i\phi_i(\mathbf{r})]\right]$$

This equation describes a peak at the origin of reciprocal space corresponding to the Fourier transform of the vacuum reference image, a second peak centered at

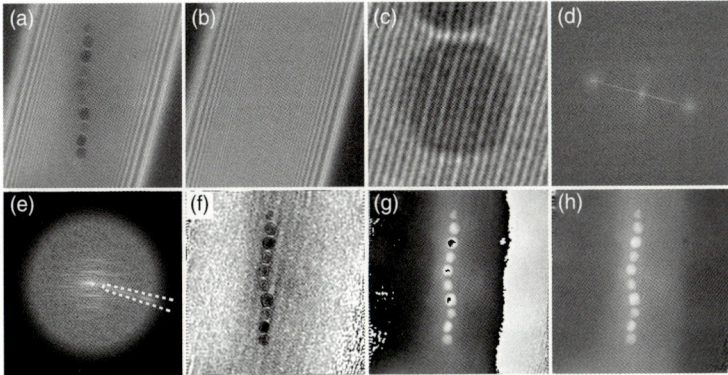

Figure 6.6 Sequence of processing steps required to convert a recorded electron hologram into an amplitude and phase image. (a) Experimentally acquired hologram of the region of interest (a chain of magnetite crystals). Coarser fringes, resulting from Fresnel diffraction at the edges of the biprism wire, are visible on either side of the overlap region. The overlap width and holographic interference fringe spacing are 650 and 3.3 nm, respectively. The field of view is 725 nm. (b) Vacuum reference hologram acquired immediately after the specimen hologram. (c) Magnified region of the specimen hologram, showing changes in the positions of interference fringes in a particle. (d) Fourier transform of the electron hologram in (a), containing a centerband, two sidebands, and a diagonal streak resulting from the presence of Fresnel fringes in the hologram. (e) One of the sidebands extracted from the Fourier transform, shown after applying a circular mask with smooth edges to reduce its intensity radially to zero. If required, the streak from the Fresnel fringes can be removed in a similar manner, by assigning a value of zero to pixels inside the region shown by the dashed line. Inverse Fourier transformation of the masked sideband is used to provide a complex image wave, which is then displayed in the form of (f) an amplitude image and (g) a modulo 2π phase image. Phase unwrapping algorithms are used to remove the 2π phase discontinuities from (g) to yield the final unwrapped phase image shown in (h).

the origin corresponding to the Fourier transform of a conventional bright-field TEM image of the specimen, a peak centered at $\mathbf{q} = -\mathbf{q}_c$ corresponding to the Fourier transform of the desired image wavefunction and a peak centered at $\mathbf{q} = +\mathbf{q}_c$ corresponding to the Fourier transform of the complex conjugate of the wavefunction. Figure 6.6d shows the Fourier transform of the hologram shown in Figure 6.6a. In order to recover the complex electron wavefunction, one of the two "sidebands" in the Fourier transform is selected (Figure 6.6e) and then inverse Fourier transformed. The amplitude and phase of the complex image wave are calculated using the expressions

$$A = \sqrt{\mathrm{Re}^2 + \mathrm{Im}^2} \tag{6.7}$$

$$\phi = \tan^{-1}\left(\frac{\mathrm{Im}}{\mathrm{Re}}\right) \tag{6.8}$$

where Re and Im are the real and imaginary parts of the real-space complex image wavefunction, respectively. Figure 6.6f,g shows the resulting amplitude and phase images, respectively. The amplitude image is similar to an energy-filtered bright-field TEM image, since the contribution of inelastic scattering to holographic interference fringe formation is negligible [52]. The phase image is initially calculated modulo 2π, meaning that 2π phase discontinuities appear at positions where the phase shift exceeds this amount (Figure 6.6g). If required, the phase image can then be "unwrapped" by using suitable algorithms, as shown in Figure 6.6h.

In Figure 6.6d, the streak from the "centerband" toward the sidebands is attributed to the presence of Fresnel fringes from the biprism wire that have a range of spacings. The contribution of this streak to the sideband can lead to artifacts in reconstructed amplitude and phase images and can be minimized by masking it from the Fourier transform before inverse Fourier transformation (Figure 6.6e). Alternatively, the Fresnel fringes can be eliminated by introducing a second biprism close to a different image plane in the microscope column [53, 54].

As phase information is stored in the lateral displacements of holographic interference fringes, long-range phase modulations arising from inhomogeneities in the charge and thickness of the biprism wire, as well as from lens distortions and charging effects (e.g., at apertures), can introduce artifacts into the reconstructed wavefunction. In order to take these effects into account, a reference electron hologram is usually obtained from vacuum alone by removing the specimen from the field of view without changing the optical parameters of the microscope (Figure 6.6b). Correction is then possible by performing a complex division of the specimen wavefunction by the vacuum wavefunction in real space and calculating the phase of the resulting complex wavefunction to obtain the distortion-free phase of the image wave.

6.3.3
Experimental Requirements

In order to generate a highly coherent electron beam for electron holography, a field emission electron gun is essential. Although no electron source is perfectly coherent, either spatially or temporally, the degree of coherence must be such that an interference fringe pattern of sufficient quality can be recorded within a reasonable acquisition time, during which specimen and/or beam drift must be negligible. It is common practice to adjust the condenser lens astigmatism to make the beam illumination elongated in a direction perpendicular to the biprism wire. Coherence is then maximized in the elongation direction, while the electron flux is maximized in the region of interest. The elongation direction of the illumination has to be aligned so that it is as close as possible to being perpendicular to the biprism, as a slight misalignment can lead to a dramatic decrease in interference fringe contrast. In order to maximize fringe visibility, a small spot size, small condenser aperture size, parallel illumination, and a low extraction voltage are usually required. In practice, a balance between coherence, intensity, and acquisition time must be achieved.

When characterizing magnetic materials using electron holography, the microscope objective lens is usually turned off because it creates a large magnetic field at the position of the specimen, just as for Lorentz microscopy. Instead, a non-immersion Lorentz lens is used as the primary imaging lens. The electron biprism, which is normally either a Au-coated glass fiber or a Pt wire with a diameter of ~1 µm, is usually positioned in place of one of the selected-area apertures close to the first image plane [55]. Lateral movement and rotation of the biprism is a useful feature, especially for studying magnetic materials.

Three coupled experimental parameters affect the quality of the final reconstructed phase images significantly: overlap width, interference fringe spacing, and interference fringe visibility. Generally, the application of a higher biprism voltage results in a larger overlap width, a finer interference fringe spacing, and a decrease in interference fringe contrast. The overlap width is given by the expression

$$W = 2\left(\frac{d_1 + d_2}{d_1}\right)\left(\alpha \frac{d_1 d_2}{d_1 + d_2} - R\right) \tag{6.9}$$

where d_1 is the distance between the focal plane and the biprism, d_2 is the distance between the biprism and image plane, R is the radius of the biprism wire, and α is the deflection angle defined in Figure 6.4. A thinner biprism wire results in a larger overlap width for a given interference fringe spacing, coupled with a better fringe visibility. The overlap width varies linearly with biprism voltage, as shown in Figure 6.7a,d for an FEI Titan TEM operated at 300 kV, and typically takes a value of between 0.5 and 1.5 µm when using the Lorentz lens as the primary imaging lens of the microscope. The interference fringe spacing is inversely proportional to the biprism voltage and is given by the expression

$$s = \lambda \left(\frac{d_1 + d_2}{2\alpha d_1}\right) \tag{6.10}$$

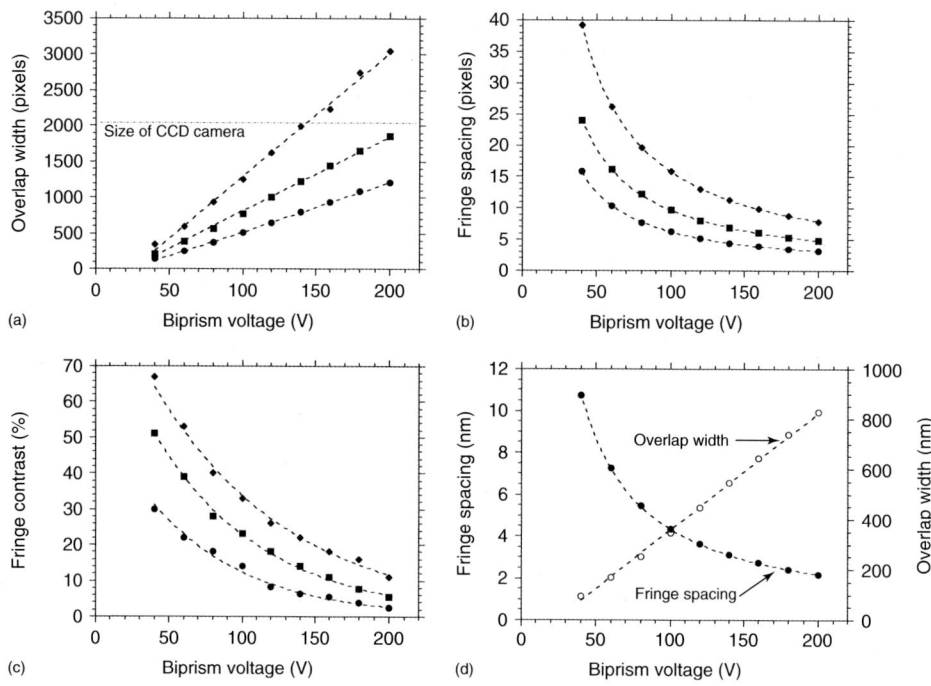

Figure 6.7 Measured dependence of (a) overlap width, (b) interference fringe spacing, and (c) interference fringe visibility recorded at indicated magnifications of × 1450 (circles), ×2250 (squares), and ×3700 (diamonds) on an FEI Titan 80-300ST TEM operated at 300 kV in Lorentz mode. A Gatan model 894 2k (2048 × 2048 pixel) UltraScan 1000FT camera mounted on a Gatan imaging filter was used. (d) Interference fringe spacing (filled circles) and overlap width (open circles) in nanometers plotted as a function of biprism voltage; CCD, charge-coupled device.

as shown in Figure 6.7b for an FEI Titan TEM. Interference fringe spacings of between 2 and 5 nm are typically used for examining magnetic materials, with a spatial resolution in the final phase image that is presently, at best, approximately three times the fringe spacing [56]. The interference fringe visibility, which is given by the expression

$$\mu = \left(\frac{I_{max} - I_{min}}{I_{max} + I_{min}} \right) \tag{6.11}$$

is one of the most important parameters for electron holography, as it is the primary factor that influences the phase resolution

$$\phi_{min} = \frac{SNR}{\mu} \sqrt{\frac{2}{N_{el}}} \tag{6.12}$$

where I_{max} and I_{min} are the maximum and minimum intensities, respectively, of the interference fringes, SNR is the signal-to-noise ratio in the hologram and N_{el} is the number of electrons collected per pixel [57, 58]. Figure 6.7c shows measurements of fringe visibility plotted as a function of biprism voltage for an FEI Titan TEM operated at different magnifications. The difference between the graphs is attributed to the modulation transfer function of the camera, suggesting that the use of a higher magnification can result in better phase resolution, although the field of view then becomes smaller. The number of electrons per pixel in a recorded electron hologram is usually at least 100–500 for an acquisition time of 2–8 s. If the biprism wire and the specimen are sufficiently stable, then a substantially longer acquisition time can be used [59].

6.3.4
Magnetic and Mean Inner Potential Contributions to the Phase

Off-axis electron holography can be divided into two modes: (i) high-resolution electron holography, in which the interpretable resolution in a high-resolution TEM image can be improved by the use of phase plates to correct for microscope lens aberrations [60, 61] and (ii) medium-resolution electron holography, in which magnetic and electrostatic fields in materials can be studied quantitatively, usually with nanometer spatial resolution [62]. Here, we concentrate on the medium-resolution application of electron holography to the study of magnetic fields in materials.

Neglecting dynamical diffraction (i.e., assuming that the specimen is thin and weakly diffracting), the phase shift recorded using electron holography can be written in the form

$$\phi(x, y) = \phi_e(x, y) + \phi_m(x, y) = C_E \int_{-\infty}^{+\infty} V(x, y, z) dz - \frac{e}{\hbar} \int_{-\infty}^{+\infty} A_z(x, y, z) dz \quad (6.13)$$

where V is the electrostatic potential, A_z is the component of the magnetic vector potential parallel to the electron beam direction,

$$C_E = \left(\frac{2\pi}{\lambda}\right)\left(\frac{E + E_0}{E(E + 2E_0)}\right) \quad (6.14)$$

is a constant that depends on the microscope accelerating voltage $U = E/e$ and E_0 is the rest mass energy of the electron. C_E takes values of 1.01×10^7, 8.64×10^6 and 6.53×10^6 rad m^{-1} V^{-1} at accelerating voltages of 80, 120, and 300 kV, respectively. If no external charge distributions or applied electric fields are present within or around the specimen, then the only electrostatic contribution to the phase shift originates from the mean inner potential V_0 of the material coupled with specimen thickness variations in the form

$$\phi_e(x, y) = C_E V_0 t(x, y) \quad (6.15)$$

where $t(x,y)$ is the specimen thickness. If the specimen has uniform composition, then the electrostatic contribution to the phase shift is proportional to the local specimen thickness at position (x,y). Alternatively, should the specimen thickness be known independently, ϕ_e can be used to provide a measurement of the local mean inner potential.

The difference between the value of the magnetic contribution to the phase shift at two arbitrary points with coordinates (x_1, y_1) and (x_2, y_2)

$$\Delta\phi_m = \phi_m(x_1, y_1) - \phi_m(x_2, y_2) = -\frac{e}{\hbar}\int_{-\infty}^{+\infty} A_z(x_1, y_1, z)dz + \frac{e}{\hbar}\int_{-\infty}^{+\infty} A_z(x_2, y_2, z)dz \quad (6.16)$$

can be written in the form of a loop integral in the form

$$\Delta\phi_m = -\frac{e}{\hbar}\oint \mathbf{A}\cdot d\mathbf{l} \quad (6.17)$$

for a rectangular loop formed by two parallel electron trajectories crossing the sample at coordinates (x_1, y_1) and (x_2, y_2) and joined, at infinity, by segments perpendicular to the trajectories. By virtue of Stokes' theorem,

$$\Delta\phi_m = \frac{e}{\hbar}\iint \mathbf{B}\cdot\hat{\mathbf{n}}\, dS = \frac{\pi}{\phi_0}\Phi(S) \quad (6.18)$$

so that a pictorial representation of the magnetic flux distribution throughout the specimen can be obtained by adding contours to a recorded phase image. A phase difference of 2π in such an image corresponds to an enclosed magnetic flux of 4.14×10^{-15} Tm^2. The relationship between the magnetic contribution to the phase shift and the magnetic induction can be established by considering the gradient of ϕ_m, which takes the form

$$\vec{\nabla}\phi_m(x, y) = \frac{e}{\hbar}\left[B_y^p(x, y), -B_x^p(x, y)\right] \quad (6.19)$$

where

$$B_j^p(x, y) = \int_{-\infty}^{+\infty} B_j(x, y, z)dz \quad (6.20)$$

are the components of the magnetic induction perpendicular to the incident electron beam direction projected in the electron beam direction.

In the special case when stray fields surrounding the specimen can be neglected, the specimen has a constant thickness and the magnetic induction does not vary with z within the specimen, Eq. (6.19) can be rewritten in the simplified form

$$\vec{\nabla}\phi_m(x, y) = \frac{et}{\hbar}\left[B_y(x, y), -B_x(x, y)\right] \quad (6.21)$$

in which there is a direct proportionality between the magnetic contribution to the phase gradient and the magnetic induction.

Equation (6.21) must, in general, be used with caution if the aim is to quantify the magnetic field strength and its direction on a local scale. However, for visualization purposes, it can be useful to relate the spacing and direction of phase contours to the direction of the projected magnetic field within the field of view. The few instances when the separation of electrostatic and magnetic contributions to the phase shift may be avoided include the special case of magnetic domains in a thin film of constant thickness (far from the specimen edge) and a measurement scheme for quantifying magnetic moments that is described below.

The contribution to the phase shift from local variations in mean inner potential is generally detrimental to studies of magnetic materials using electron holography. In particular, it can be much greater than the magnetic contribution for small (sub-50 nm) magnetic nanocrystals. A particular advantage of electron holography for the analysis of magnetic or electric fields in nanostructured materials is that unwanted contributions to the contrast from local variations in mean inner potential (i.e., composition and specimen thickness) can be removed from a holographic phase image more easily than from images recorded using alternative phase contrast techniques based on Lorentz microscopy. Figure 6.8 shows simulations of the mean inner potential and magnetic contributions to the phase shift for a 50 nm diameter Co crystal that is uniformly magnetized vertically. Although the particle is in a single magnetic domain state (Figure 6.8c), the mean inner potential dominates the recorded phase image. Separation of the desired magnetic contribution from the total phase shift is then essential.

In principle, the most accurate way of achieving this separation of the mean inner potential and magnetic contributions to the phase involves turning the specimen over after acquiring a hologram and acquiring a second hologram from the same region, as shown in Figures 6.9 and 6.10.

The two holograms are aligned after flipping one of them over digitally (Figure 6.9). Their sum and difference are then used to determine twice the mean inner potential and twice the magnetic contribution to the phase shift, respectively. The mean inner potential contribution can be subtracted from all subsequent phase images acquired from the same region of the specimen (Figure 6.10), followed by adding phase contours and colors to form final magnetic induction maps. A degree of smoothing of the final phase images is often used to remove statistical noise and artifacts resulting from misalignment of the pairs of phase images.

An alternative method, which is often more practical, involves performing a magnetization reversal experiment *in situ* in the microscope and then selecting pairs of holograms that differ only in the magnetization direction in the specimen. The addition of two phase images, in which the specimen is oppositely magnetized,

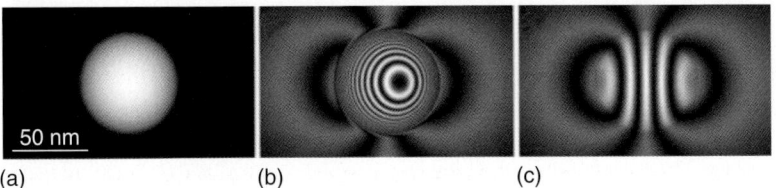

(a) (b) (c)

Figure 6.8 Simulations of the phase shift of a 50 nm diameter Co ($B = 1.8$ T; $V_0 = 17.8$ V) spherical particle, which is magnetized uniformly in the vertical direction. (a) Thickness map. (b) Total phase shift. (c) Magnetic contribution to the phase shift alone. In (b) and (c), the cosine of 12 times the phase is shown.

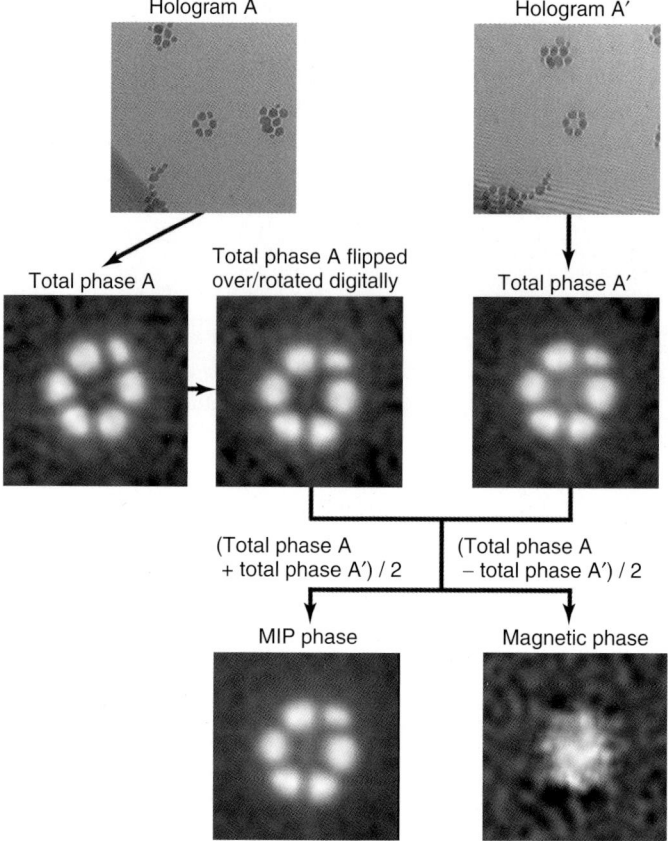

Figure 6.9 Experimental procedure used to obtain the mean inner potential (MIP) contribution to the phase shift for a self-assembled ring of 25 nm diameter Co crystals.

provides twice the mean inner potential contribution to the phase. If the magnetization in the specimen does not reverse perfectly, then reversal experiments may need to be repeated multiple times so that nonsystematic differences between reversed images are averaged out.

If the specimen cannot be turned over during an experiment and the magnetization in the specimen does not reverse exactly, then it may be possible to generate an "artificial" representation of the mean inner potential contribution to the phase either from a t/λ map (where t is the specimen thickness and λ is the inelastic mean free path) [63] or from a high-angle annular dark-field image. The constant of proportionality between the thickness map and the true mean inner potential contribution to the phase can be determined by least squares fitting to data collected near the specimen edge, where the magnetic contribution to the phase shift is smallest. This method may not be applicable in the presence of strong diffraction contrast or if the specimen contains regions that have substantially different

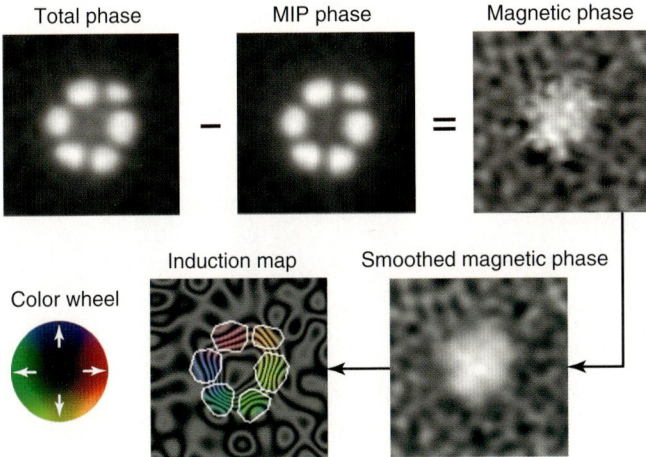

Figure 6.10 Experimental procedure used to obtain the magnetic contribution to the phase shift and final induction maps for a self-assembled ring of 25 nm diameter Co crystals. The induction map shows the cosine of 96 times the magnetic contribution to the phase. The color wheel shows the direction of the projected induction in each crystal.

compositions. In this situation, the use of two different microscope accelerating voltages or a phase image acquired above the Curie or Néel temperature of the specimen [64] may be used.

6.3.5 Applications

Figures 6.11–6.15 illustrate the application of off-axis electron holography to the characterization of magnetic remanent states and switching processes in nanoscale materials, to which it would be difficult to apply Lorentz microscopy. The off-axis electron holograms that are presented below were acquired at accelerating voltages of either 200 or 300 kV using field emission gun TEMs equipped with a Lorentz lens and an electron biprism located close to the conventional selected-area aperture plane of the microscope. The holograms were recorded digitally with the conventional microscope objective lens switched off and the specimen located in magnetic-field-free conditions (i.e., at remanence), after using the objective lens to apply a chosen field to the specimen. Vacuum reference holograms were always used to remove distortions associated with the imaging and recording system of the microscope.

Figure 6.11 illustrates the application of electron holography to the study of switching processes in nanoscale ring-shaped magnetic elements, which are of interest for magnetic recording and storage applications because they can support flux-closed (FC) magnetic states that would not be stable in disk-shaped elements of similar size, in the form of a montage of magnetic contributions to the phase shift

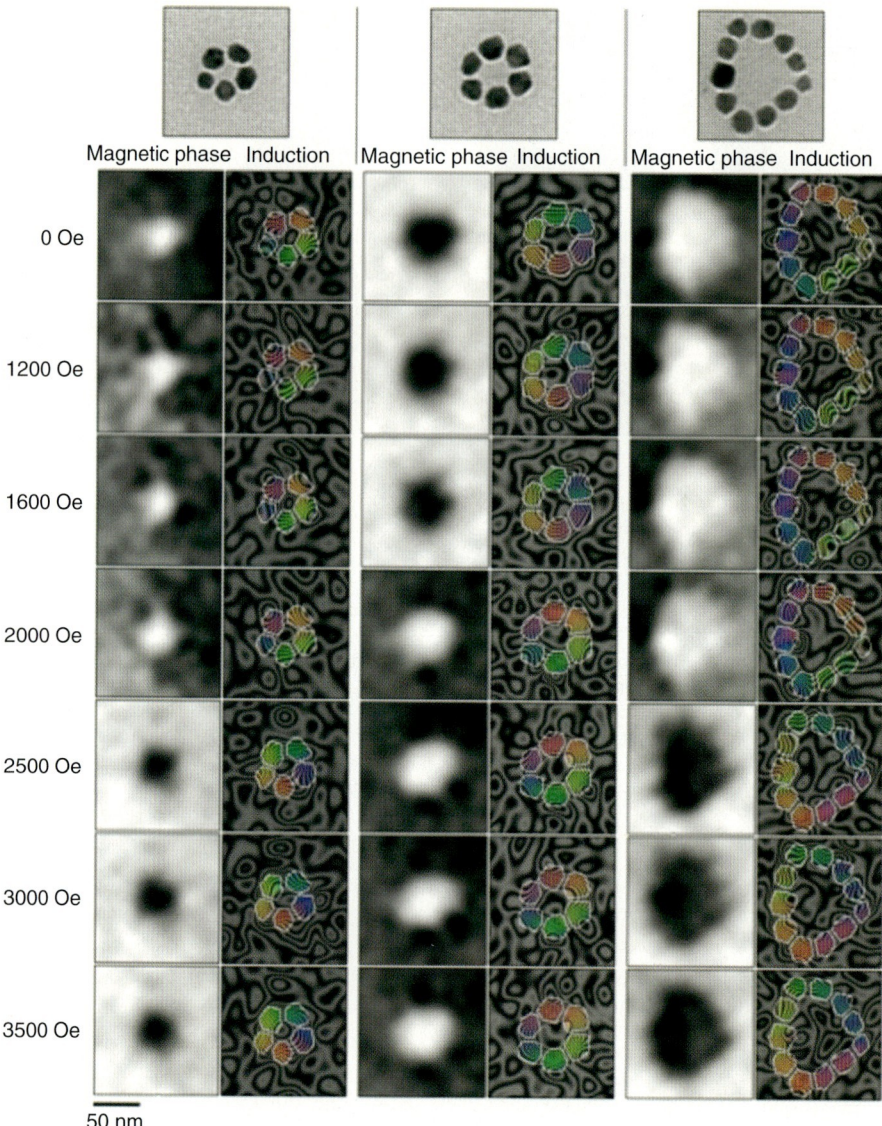

Figure 6.11 Top row: Defocused bright-field TEM images of rings of polycrystalline Co nanoparticles. Subsequent rows: Magnetic contributions to the phase shift and induction maps recorded at remanence using off-axis electron holography after initially saturating the rings using a large (−20 000 Oe) out-of-plane magnetic field and then applying the indicated out-of-plane fields in succession. The direction of the projected induction is shown using both contours and colors (red = right; blue = up; green = left; yellow = down). The contour spacing is 0.065 rad, corresponding to 96 times phase amplification.

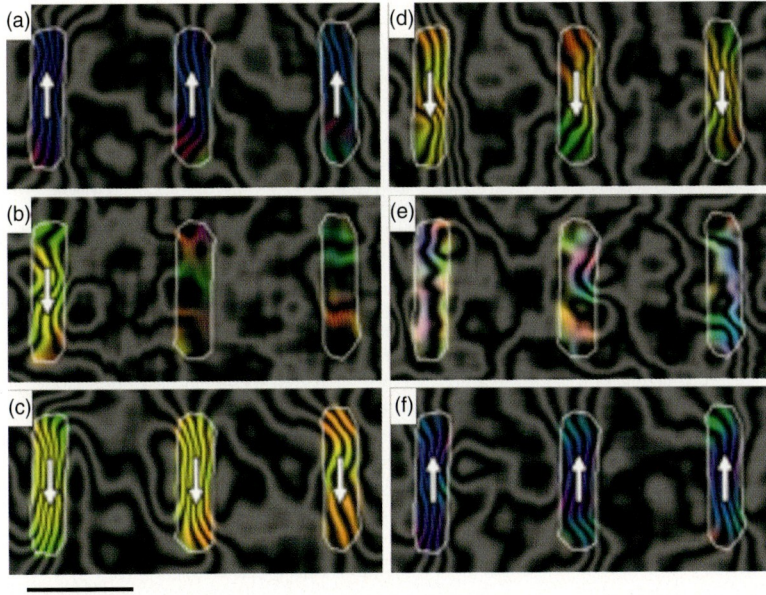

Figure 6.12 Magnetic induction maps showing six remanent magnetic states recorded using electron holography from three pseudo-spin-valve elements. Eighteen images similar to these were recorded in total. The outlines of the elements are shown in white. The contour spacing is $2\pi/64 = 0.098$ rad, such that the magnetic flux enclosed between any two adjacent black contours is $h/64e = 6.25 \times 10^{-17}$ Wb. The direction of the induction is shown using arrows and colors (red = right, yellow = down, green = left, and blue = up).

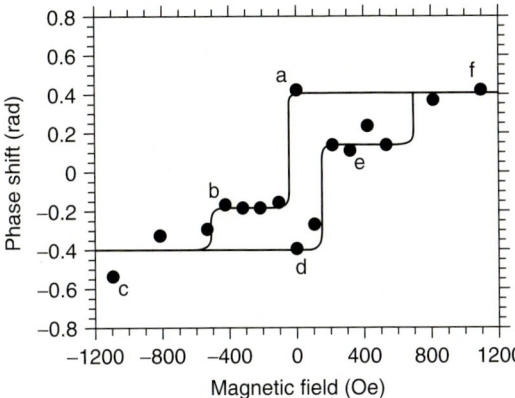

Figure 6.13 Remanent hysteresis loop measured directly from the experimental electron holographic phase images used to generate Figure 6.12. The graph shows the average of the magnetic contribution to the phase shift across the three elements, plotted as a function of the in-plane component of the magnetic field applied to the elements before recording the remanent states. The letters correspond to the six magnetic induction maps shown in Figure 6.12.

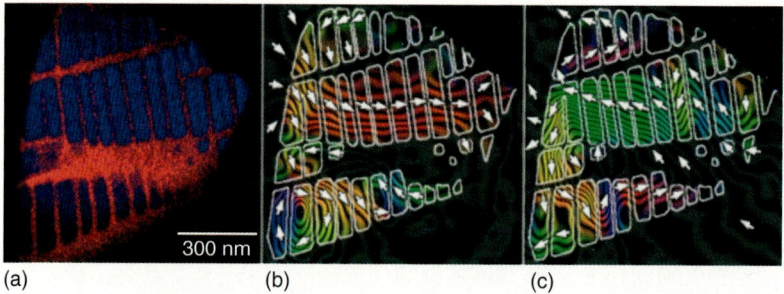

Figure 6.14 (a) Chemical map of naturally occurring titanomagnetite, obtained using energy-selected imaging in a Gatan imaging filter. Blue regions (corresponding to a background-subtracted Fe signal) show magnetite (Fe_3O_4) blocks, which are separated from each other by red regions (generated from a background-subtracted Ti signal) corresponding to paramagnetic ulvöspinel (Fe_2TiO_4). (b,c) show magnetic phase contours measured using electron holography from the same region, acquired after magnetizing the sample in different directions using the magnetic field of the microscope objective lens. The colors show the directions of the magnetic induction.

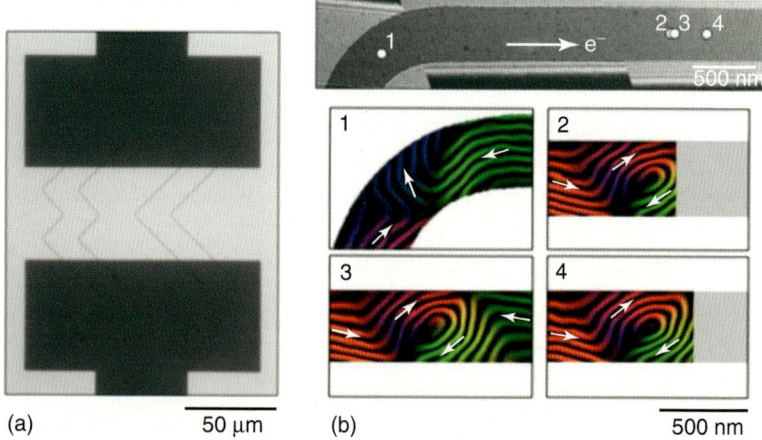

Figure 6.15 (a) Low-magnification TEM image of four permalloy zigzag wires (width: 430 nm, thickness: 11 nm) patterned onto a Si_3N_4 window using electron beam lithography. A kink in one wire was examined using electron holography. (b) Higher magnification TEM image of a permalloy wire, showing the motion of a magnetic domain wall after the injection of 10 μs pulses with a current density of 3.14×10^{11} A m^{-2}, showing magnetic induction maps obtained from a domain wall at each position. The long arrow in the top image indicates the direction of electron flow. The phase contour spacing is 0.785 rad.

and induction maps acquired from three self-assembled rings of 25 nm diameter polycrystalline Co nanoparticles. Each particle comprises a polycrystalline core of hexagonal-close-packed crystals of Co and a shell of ∼3 nm of CoO [65].

Holograms were recorded in magnetic-field-free conditions after applying chosen OOP magnetic fields *in situ* in the TEM by partially exciting the conventional microscope objective lens. An OOP field of −20 000 Oe was initially applied perpendicular to the plane of the specimen and reduced to zero. The sample was then taken out of the TEM in zero field, turned over, and placed back into the microscope. Chosen OOP fields of up to +20 000 Oe were then applied to the specimen in the TEM in succession by changing the current in the TEM objective lens. The applied field was reduced to zero before recording each hologram. The mean inner potential contribution to the phase shift was calculated from phase images that had been acquired before and after turning the specimen over. Although the initial chiralities (directions of magnetization) of the FC states in the individual rings are determined by the shapes, sizes, and positions of the constituent nanoparticles, reproducible magnetization reversal of most rings could be achieved by using OOP fields of between 1600 and 2500 Oe. The switching behavior was compared with micromagnetic simulations, which suggested that metastable states form at intermediate applied fields before FC reformation at remanence.

Figures 6.12 and 6.13 illustrate the application of electron holography to the characterization, in plan-view geometry, of a series of lithographically patterned elements that each contains a magnetically hard pinned layer and a magnetically soft free layer separated by a conducting spacer layer. The sample used in this study comprised a two-dimensional array of 75 × 280 nm pseudo-spin-valve elements, which were prepared from a polycrystalline $Ni_{79}Fe_{21}$ (4.1 nm)/Cu (3 nm)/Co (3.5 nm)/Cu (4 nm) film that had been sputtered onto an oxidized silicon substrate [66].

An approach based on focused ion-beam milling with Ga ions was used to minimize damage to the magnetic properties of the elements during preparation for TEM examination. Figure 6.12 shows six magnetic induction maps of three adjacent elements recorded at remanence after saturating the elements magnetically either parallel or antiparallel to their length using the magnetic field of the conventional microscope objective lens, followed by applying a reverse field. These images were used to identify separate switching of the Co and NiFe layers in the individual elements and to measure their switching fields, as shown in the form of a remanent hysteresis loop in Figure 6.13. A comparison of Figure 6.13 with micromagnetic simulations was used to infer that ∼20 nm of material at the edge of each element does not contribute to the magnetic signal, possibly as a result of oxidation during fabrication. In addition, the in-plane component of the induction in the Co layer is reduced by ∼40% from its nominal value. Differences in hysteresis between the elements are observed, and are thought to result from microstructural variations. Such switching variability must be minimized if similar elements are to be used in ultrahigh density magnetic recording applications.

Figure 6.14 illustrates the application of electron holography to the study of naturally occurring ferrimagnetic magnetite crystals that are arranged in the form of finely exsolved arrays of magnetic minerals in a naturally occurring rock [67]. The magnetic induction maps highlight the fact that the sizes, shapes, and arrangements of the crystals are optimized for use as stable recorders of the ancient geomagnetic field.

Figure 6.15 illustrates the use of Lorentz TEM, electron holography, and a TEM specimen holder that allows multiple electrical contacts [68] to be made to an electron-transparent specimen to study the competing effects of heating and spin torque on the current-induced motion of transverse and vortex-type domain walls in permalloy wires [69]. Here, the device comprised permalloy zigzag structures with a line width and thickness of 430 and 11 nm, respectively. The lines were fabricated directly onto electron-transparent Si_3N_4 windows using electron beam lithography, as shown in Figure 6.15a. Figure 6.15b shows the sequential positions of a magnetic domain wall that was subjected to 10 µs pulses with a current density of 3.14×10^{11} A m^{-2}. Electron holograms were acquired at each of these positions. It was observed that a transverse wall initially formed at a kinked region of the wire after the application of a magnetic field. After applying a current pulse, the domain wall moved by 2330 nm in the direction of electron flow and transformed into a vortex-type magnetic domain wall. After a second pulse, the vortex-type magnetic domain wall moved slightly in the same direction and became distorted (position 3 in Figure 6.15b), with the long axis of the vortex increasingly perpendicular to the wire length. This behavior may be associated with edge roughness or defects, which may restrict the movement of the wall. After a third pulse, the domain wall moved 260 nm further and retained its vortex state. More recent studies have involved the injection of smaller currents to influence thermally activated magnetic domain wall motion between closely adjacent pinning sites [70].

6.3.6
Quantitative Measurement of Magnetic Moments Using Electron Holography

The magnetic moment of a nanostructure is given by the expression

$$\mathbf{m} = \iiint \mathbf{M}(\mathbf{r}) d^3 \mathbf{r} \tag{6.22}$$

where $\mathbf{M}(\mathbf{r})$ is the position-dependent magnetization and $\mathbf{r}$ is a three-dimensional position vector. Unfortunately, an electron holographic phase image does not provide direct information about $\mathbf{M}(\mathbf{r})$. Instead, as detailed in the preceding sections, the recorded phase shift is proportional to the projection (in the electron beam direction) of the in-plane component of the magnetic induction $\mathbf{B}(\mathbf{r})$ both within and around the specimen. We have recently shown mathematically that the magnetic moment of a nanostructure can be measured quantitatively from a phase image or from its gradient components [71]. The basis of the algorithm is the relationship between the volume integral of the magnetic induction, a quantity that is referred to here as the "*inductive moment*" $\mathbf{m_B}$ and the *magnetic moment* $\mathbf{m}$. If a circular boundary around an isolated nanostructure of interest is chosen when integrating the phase shift in the form

$$\mathbf{m_B} = \frac{\hbar R^2}{e\mu_0} \int_0^{2\pi} d\theta [-\sin\theta, \cos\theta, 0] \phi(R\cos\theta, R\sin\theta) \tag{6.23}$$

where R is the radius of the integration circle, μ_0 is the vacuum permeability, and θ is the polar angle, then the following expression is found to be valid:

$$\mathbf{m} = 2\mathbf{m_B} \quad (6.24)$$

Furthermore, two orthogonal components of $\mathbf{m_B}$ can be extrapolated as a function of the radius of the integration circle to a circle of zero radius to yield a measurement that is free of most artifacts.

A measure of the magnetic moment carried out according to this scheme is model independent and does not rely on assumptions such as uniform magnetization of the particle or a priori knowledge such as the particle's morphology and/or composition. Furthermore, since the integration loop encloses the object and never crosses its boundaries, the procedure can be applied to the reconstructed phase image without the need for separating the mean inner potential and magnetic contributions to the phase. Finally, as the measurement is performed on the whole phase image and does not rely on pixel-by-pixel signal detection, the performance of the algorithm is affected by noise differently from that of conventional phase measurements. In particular, the algorithm appears to be particularly efficient at handling very noisy datasets.

Figure 6.16 illustrates one example of the application of the method to the measurement of the magnetic moment of a chain of three ferrimagnetic magnetite nanocrystals from a reconstructed phase image. Although the chain of particles is associated with a clear step in phase, the phase image in Figure 6.16 contains additional artifacts, including loops of contrast around two of the particles (possibly due to a slight defocus change between the two holograms that were used to create the magnetic induction map or changes in diffraction contrast), small changes in substrate morphology, and Fresnel fringes from the edges of the biprism wire, in addition to statistical noise.

In the example shown in Figure 6.16, the minimum radius of the integration circle that lies outside the physical boundary of the three crystals, about 100 nm, yields a magnetic moment of 2.29×10^6 μ_B oriented at 136° to the positive x axis (Figure 6.16a), where $\mu_B = 9.274 \times 10^{-24}$ J T^{-1} refers to one Bohr magneton. The maximum radius that is compatible with the available field of view and centered on the same position, 240 nm, yields a moment of 1.02×10^6 μ_B oriented at 126° (Figure 6.16b). By varying the radius in 1 nm increments between these values, a set of measurements for the two components of the magnetic moment that can be extrapolated quadratically to zero circle radius is obtained, as shown in Figure 6.16d.

The best-fitting values for the two components of the magnetic moment are $m_B^x = -(1.99 \pm 0.14)10^6$ μ_B and $m_B^y = (1.89 \pm 0.22)10^6$ μ_B, which, together, result in a vector of modulus $(2.74 \pm 0.18)10^6$ μ_B oriented at $(136 \pm 4)°$ to the positive x axis, approximately 20% higher in modulus (but oriented similarly) than that inferred when using the circle of smallest radius alone. The error bars are estimated from the square root of the statistical variance associated with the parabolic fit of each component and assigned to the modulus and angle according to standard error propagation theory.

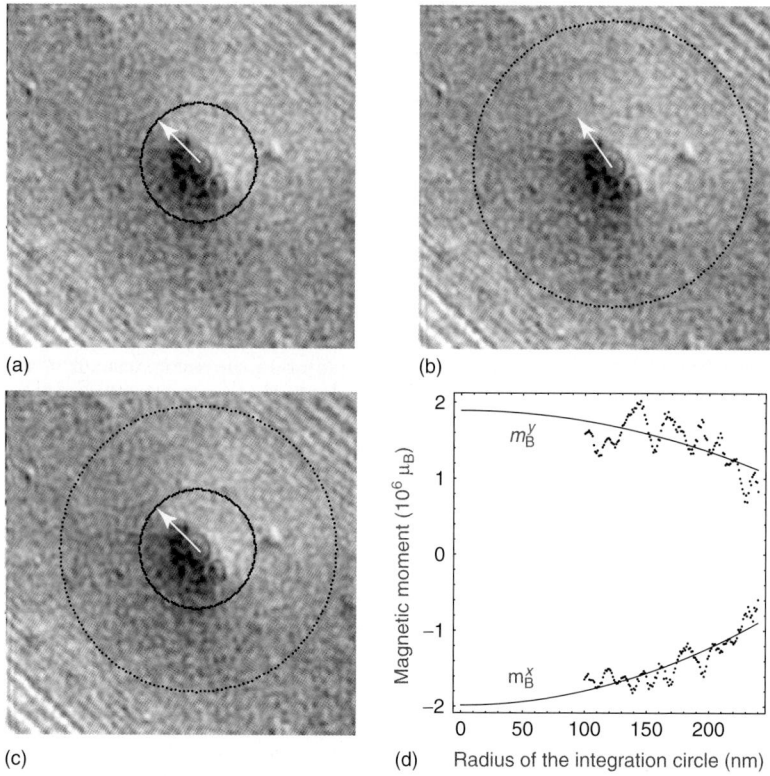

Figure 6.16 (a–c) Magnetic contribution to the recorded phase shift for a chain of three magnetite nanoparticles, each of which has a radius of ∼20 nm. The circles and arrows show the integration radius and the direction of the resulting inductive moments for radii of: (a) 100 nm, (b) 240 nm, and (c) when using integration radii between 100 and 240 nm to obtain a value for the magnetic moment extrapolated quadratically to zero integration radius. (d) Shows the parabolic fit of the two orthogonal components of the measured inductive moment.

In order to compare the results with predictions, the radii of the three magnetic particles were estimated to be approximately (20 ± 2) nm; dividing the magnetic moment (2 times the inductive moment, or $(5.5 \pm 0.4)10^6\ \mu_B$) by the total estimated volume of the particles $(1.0 \pm 0.2)10^5$ nm^3, an average magnetization of $(5.1 \pm 0.9)10^5$ A m^{-1} is obtained, corresponding to (0.64 ± 0.12) T. The expected magnetization of magnetite, as reported in the literature, corresponds to a magnetic induction of ∼0.6 T, confirming the soundness of our approach.

6.4
Discussion and Conclusions

The work that is described above is the result of many technical, experimental, and theoretical developments that we and others have pursued, involving the development of new approaches for the acquisition, analysis, and simulation of Lorentz images and electron holograms, the design and use of specimen holders that allow electrical contacts to be applied to magnetic devices *in situ* in the electron microscope, the comparison of recorded images with micromagnetic simulations and with information about the local compositions and morphologies of the same specimens, and the development of new strategies for the careful separation of the weak magnetic signal of primary interest from unwanted contributions to recorded holographic phase contrast. The resulting ability to study magnetic states and switching phenomena with nanometer spatial resolution is important for the development of a solid understanding of magnetic domain structures, reversal mechanisms, coercivities and interactions in thin films, isolated and closely-spaced magnetic nanoparticles and device structures, including the identification of the shapes and arrangements of magnetic elements that are stable, reversible, and reproducible at room temperature for future applications in magnetic recording and storage. Further work is required to optimize specimen preparation for magnetic characterization in the TEM, to increase the sensitivity of the technique for measuring weak fields and to improve its spatial and time resolution. The prospect of characterizing magnetic vector fields *inside* nanocrystals in three dimensions by combining electron tomography with electron holography is also of great interest [72, 73]. In the future, a phase sensitivity of below $2\pi/1000$ rad and subnanometer spatial resolution may be required for studies of materials such as diluted magnetic semiconductors. Such experiments will require the use of high-brightness electron sources, aberration correctors, highly sensitive recording media, drift-free specimen stages and holders, and sophisticated software for the automation of lengthy experiments.

Acknowledgments

We are grateful to A. Wei, N.S. Church, J.M. Feinberg, R.J. Harrison, M. Kläui, M. Pósfai, M.R. Scheinfein, M.R. McCartney, D.J. Smith, and many other colleagues for ongoing collaborations.

References

1. Smith, D.J. (2007) in *Nanocharacterisation* (eds A.I. Kirkland and J.L. Hutchison), RSC Publishing, Cambridge, pp. 1–27.
2. Marton, L. (1948) Electron optical "Schlieren" effect. *J. Appl. Phys.*, **19**, 687–688.
3. Marton, L. (1948) Electron optical observation of magnetic fields. *J. Appl. Phys.*, **19**, 863–864.
4. Marton, L. and Lachenbruch, S.H. (1949) Electron optical mapping of electromagnetic fields. *J. Appl. Phys.*, **20** (12), 1171–1182.
5. Hale, M.E., Fuller, H.W., and Rubinstein, H. (1959) Magnetic domain observations by electron microscopy. *J. Appl. Phys.*, **30**, 789–791.
6. Fuller, H.W. and Hale, M.E. (1960) Determination of magnetisation distribution in thin films using electron microscopy. *J. Appl. Phys.*, **31**, 238–248.
7. Fuller, H.W. and Hale, M.E. (1960) Domains in thin magnetic films observed by electron microscopy. *J. Appl. Phys.*, **31**, 1699–1705.

8. Boersch, H., Raith, H., and Wohlleben, D. (1960) Elektronenoptische untersuchungen weiβscher bezirke in dünnen eisenschichten. Z. Phys. A, 159, 388–396.
9. Boersch, H., Hamisch, H., Grohmann, K., and Wohlleben, D. (1962) Antiparallele weiβsche bereiche als biprisma für elektroneninterferenzen. II. Z. Phys. A, 167, 72–82.
10. Wade, R.H. (1967) Electric diffraction from a magnetic phase grating. Phys. Status Solidi (B), 19, 847–854.
11. Goringe, M.J. and Jakubovics, J.P. (1967) Electron diffraction from periodic magnetic fields. Philos. Mag., 15, 393–403.
12. Wohlleben, D. (1967) Diffraction effects in Lorentz microscopy. J. Appl. Phys., 38, 3341–3352.
13. Cohen, M.S. (1967) Wave-optical aspects of Lorentz microscopy. J. Appl. Phys., 38, 4966–4976.
14. Bowman, M.J. and Meyer, V.H. (1970) Magnetic phase contrast from thin ferromagnetic films in the transmission electron microscope. J. Phys. E: Sci. Instrum., 3, 927–929.
15. Chapman, J.N. (1984) The investigation of magnetic domain structures in thin foils by electron microscopy. J. Phys. D: Appl. Phys., 17, 623–647.
16. Chechenin, N.G., de Hosson, J.T.M., and Boerma, D.O. (2003) Effects of topography on the local variation in the magnetisation of ultrasoft magnetic films: a Lorentz microscopy study. Philos. Mag., 83, 2899–2913.
17. Lloyd, S.J., Mathur, N.D., Loudon, J.C., and Midgley, P.A. (2001) Magnetic domain wall width in $La_{0.7}Ca_{0.3}MnO_3$ thin films using Fresnel imaging. Phys. Rev. B, 64, 172407.
18. Volkov, V.V. and Zhu, Y. (2004) Lorentz phase microscopy of magnetic materials. Ultramicroscopy, 98, 271–281.
19. Uhlig, T. and Zweck, J. (2004) Direct observation of switching processes in permalloy rings with Lorentz microscopy. Phys. Rev. Lett., 93, 047203.
20. Johnston, A.B. and Chapman, J.N. (1995) The development of coherent Foucault imaging to investigate magnetic microstructure. J. Microsc., 179, 119–128.
21. Dekkers, N.H. and de Lang, H. (1974) Differential phase contrast in a STEM. Optik, 41, 452–456.
22. Chapman, J.N., Batson, P.E., Waddell, E.M., and Ferrier, R.P. (1978) The direct determination of magnetic domain wall profiles by differential phase contrast electron microscopy. Ultramicroscopy, 3, 203–214.
23. Morrison, G.R., Gong, H., Chapman, J.N., and Hrnciar, V. (1988) The measurement of narrow domain-wall widths in $SmCo_5$ using differential phase contrast electron microscopy. J. Appl. Phys., 64, 1338–1342.
24. Liu, Y. and Ferrier, R.P. (1995) Quantitative evaluation of a thin film recording head field using the DPC mode of Lorentz electron microscopy. IEEE Trans. Magn., 31, 3373–3375.
25. Kirk, K.J., McVitie, S., Chapman, J.N., and Wilkinson, C.D.W. (2001) Imaging magnetic domain structure in sub-500 nm thin film elements. J. Appl. Phys., 89, 7174–7176.
26. Chapman, J.N., McFadyen, I.R., and McVitie, S. (1990) Modified differential phase contrast Lorentz microscopy for improved imaging of magnetic structures. IEEE Trans. Magn., 26, 1506–1511.
27. Daykin, A.C. and Petford-Long, A.K. (1995) Quantitative mapping of the magnetic induction distribution using Foucault images formed in a transmission electron microscope. Ultramicroscopy, 58, 365–380.
28. Dooley, J. and De Graef, M. (1997) Energy filtered Lorentz microscopy. Ultramicroscopy, 67, 113–132.
29. Kasama, T., Church, N., Feinberg, J.M., Dunin-Borkowski, R.E., and Harrison, R.J. (2010) Direct observation of ferromagnetic/ferroelastic domain interactions in magnetite below the Verwey transition. Earth Planet. Sci. Lett., 297, 10–17.
30. Smirnov, A.V. and Tarduno, J.A. (2002) Magnetic field control of the low temperature magnetic properties of stoichiometric and cation-deficient

magnetite. *Earth Planet. Sci. Lett.*, **194**, 359–368.

31. McCartney, M.R., Dunin-Borkowski, R.E., Scheinfein, M.R., Smith, D.J., Gider, S., and Parkin, S.S.P. (1999) Origin of magnetisation decay in spin-dependent tunnel junctions. *Science*, **286**, 1337–1340.

32. Gabor, D. (1949) Microscopy by reconstructed wave-fronts. *Proc. R. Soc. London, Ser. A*, **197**, 454–487.

33. Gabor, D. (1951) Microscopy by reconstructed wavefronts II. *Proc. Phys. Soc. B*, **54**, 449–469.

34. Haine, M.E. and Mulvey, T. (1952) The formation of the diffraction image with electrons in the Gabor diffraction microscope. *J. Opt. Soc. Am.*, **42**, 763–769.

35. Möllenstedt, G. and Düker, H. (1956) Beobachtungen und messungen an biprisma-interferenzen mit electronenwellen. *Z. Phys.*, **145**, 377–397.

36. Tonomura, A. (1972) The electron interference method for magnetisation measurement of thin films. *Jpn. J. Appl. Phys.*, **11**, 493–502.

37. Pozzi, G. and Missiroli, G.F. (1973) Interference electron microscopy of magnetic domains. *J. Microsc.*, **18** (1), 103–108.

38. Lau, B. and Pozzi, G. (1978) Off-axis electron micro-holography of magnetic domain walls. *Optik*, **51**, 287–296.

39. Matteucci, G., Missiroli, G.F., and Pozzi, G. (1982) A new off-axis Fresnel holographic method in transmission electron microscopy: an application on the mapping of ferromagnetic domains. III. *Ultramicroscopy*, **8** (4), 403–408.

40. Matsuda, T., Tonomura, A., Suzuki, R., Endo, J., Osakabe, N., Umezaki, H., Tanabe, H., Sugita, Y., and Fujiwara, H. (1982) Observation of microscopic distribution of magnetic fields by electron holography. *J. Appl. Phys.*, **53**, 5444–5446.

41. Otani, Y., Fukamichi, K., Kitakami, O., Shimada, Y., Pannetier, B., Nozieres, J.-P., Matsuda, T., and Tonomura, A. (1997) in *Magnetic Ultrathin Films, Multilayers and Surfaces*, Materials Research Society Symposium Proceedings 475 (eds D.D. Chambliss, J.G. Tobin, D. Kubinski, K. Barmak, W.J.M. de Jonge, T. Katayama, A. Schuhl, and P.Dederichs), Materials Research Society, pp. 215–226.

42. Tsutsumi, M., Kugiya, F., Hasegawa, S., and Tonomura, A. (1989) A study on magnetisation model for particulate media. *IEEE Trans. Magn.*, **25**, 3665–3667.

43. Hirayama, T., Ru, Q., Tanji, T., and Tonomura, A. (1993) Observation of magnetic-domain states of barium ferrite particles by electron holography. *Appl. Phys. Lett.*, **63**, 418–420.

44. Osakabe, N., Yoshida, K., Horiuchi, Y., Matsuda, T., Tanabe, H., Okuwaki, T., Endo, J., Fujiwara, H., and Tonomura, A. (1983) Observation of recorded magnetisation pattern by electron holography. *Appl. Phys. Lett.*, **42**, 746–748.

45. Yoshida, K., Honda, Y., Kawasaki, T., Koizumi, M., Kugiya, F., Futamoto, M., and Tonomura, A. (1987) Measurement of intensity of recorded magnetisation on Co-Cr film by electron holography. *IEEE Trans. Magn.*, **23**, 2073–2075.

46. Tonomura, A., Matsuda, T., Endo, J., Arii, T., and Mihama, K. (1980) Direct observation of fine structure of magnetic domain walls by electron holography. *Phys. Rev. Lett.*, **44**, 1430–1433.

47. Runge, K., Nozaki, Y., Otani, Y., Miyajima, H., Pannetier, B., Matsuda, T., and Tonomura, A. (1996) High-resolution observation of magnetisation processes in $2\,\mu m \times 2\,\mu m \times 0.04\,\mu m$ permalloy particles. *J. Appl. Phys.*, **79**, 5075–5077.

48. Aharonov, Y. and Bohm, D. (1959) Significance of electromagnetic potentials in the quantum theory. *Phys. Rev.*, **115**, 485–491.

49. Osakabe, N., Matsuda, T., Kawasaki, T., Endo, J., Tonomura, A., Yano, S., and Yamada, H. (1986) Experimental confirmation of Aharonov-Bohm effect using a toroidal magnetic field confined by a superconductor. *Phys. Rev. A: At. Mol. Opt. Phys.*, **34**, 815–822.

50. Tonomura, A., Osakabe, N., Matsuda, T., Kawasaki, T., Endo, J., Yano, S., and Yamada, H. (1986) Evidence for Aharonov-Bohm effect with magnetic

field completely shielded from electron wave. *Phys. Rev. Lett.*, **56**, 792–795.

51. Dunin-Borkowski, R.E., McCartney, M.R., and Smith, D.J. (2004) in *Encyclopedia of Nanoscience and Nanotechnology*, vol. 3 (ed. H.S. Nalwa), American Scientific Publishers, Stevenson Ranch, pp. 41–100.

52. Verbeeck, J., Bertoni, G., and Lichte, H. (2011) A holographic biprism as a perfect energy filter? *Ultramicroscopy*, **111**, 887–893.

53. Yamamoto, K., Hirayama, T., and Tanji, T. (2004) Off-axis electron holography without Fresnel fringes. *Ultramicroscopy*, **101**, 265–269.

54. Harada, K., Akashi, T., Togawa, Y., Matsuda, T., and Tonomura, A. (2005) Optical system for double-biprism electron holography. *J. Electron Microsc.*, **54**, 19–27.

55. Lichte, H. (1996) Electron holography: optimum position of the biprism in the electron microscope. *Ultramicroscopy*, **64**, 79–86.

56. Völkl, E. and Lichte, H. (1990) Electron holograms for subångström point resolution. *Ultramicroscopy*, **32**, 177–180.

57. Lichte, H. (2008) Performance limits of electron holography. *Ultramicroscopy*, **108**, 256–262.

58. Völkl, E., Allard, L.F., Datye, A., and Frost, B.G. (1995) Advanced electron holography: a new algorithm for image processing and a standardized quality test for the FEG electron microscope. *Ultramicroscopy*, **58**, 97–103.

59. Cooper, D., Truche, R., Rivallin, P., Hartmann, J.-M., Laugier, F., Bertin, F., Chabli, A., and Rouviere, J.-L. (2007) Medium resolution off-axis electron holography with millivolt sensitivity. *Appl. Phys. Lett.*, **91**, 143501.

60. Lichte, H. (1991) Electron image plane off-axis electron holography of atomic structures. *Adv. Opt. Electron Microsc.*, **12**, 25–91.

61. Lehmann, M., Lichte, H., Geiger, D., Lang, G., and Schweda, E. (1999) Electron holography at atomic dimensions – present state. *Mater. Charact.*, **42**, 249–263.

62. Tonomura, A. (1992) Electron-holographic interference microscopy. *Adv. Phys.*, **41**, 59–103.

63. Egerton, R.F. (1996) *Electron Energy-Loss Spectroscopy in the Electron Microscope*, 2nd edn, Plenum, New York.

64. Loudon, J.C., Mathur, N.D., and Midgley, P.A. (2002) Charge-ordered ferromagnetic phase in $La_{0.5}Ca_{0.5}MnO_3$. *Nature*, **420**, 797–800.

65. Kasama, T., Dunin-Borkowski, R.E., Scheinfein, M.R., Tripp, S.L., Liu, J., and Wei, A. (2008) Reversal of flux closure states in cobalt nanoparticle rings with coaxial magnetic pulses. *Adv. Mater.*, **20**, 4248–4252.

66. Kasama, T., Barpanda, P., Dunin-Borkowski, R.E., Newcomb, S.B., McCartney, M.R., Castaño, F.J., and Ross, C.A. (2005) Off-axis electron holography of individual pseudo-spin-valve thin film magnetic elements. *J. Appl. Phys.*, **98**, 013903.

67. Feinberg, J.M., Harrison, R.J., Kasama, T., Dunin-Borkowski, R.E., Scott, G.R., and Renne, P.R. (2006) Effects of internal mineral structures on the magnetic remanence of silicate-hosted titanomagnetite inclusions: an electron holography study. *J. Geophys. Res.*, **111**, B12S15.

68. Kasama, T., Dunin-Borkowski, R.E., Matsuya, L., Broom, R.F., Twitchett, A.C., Midgley, P.A., Newcomb, S.B., Robins, A.C., Smith, D.W., Gronsky, J.J., Thomas, C.A., and Fischione, P.E. (2005) in *In-Situ Electron Microscopy of Materials*, Materials Research Society Sumposium Proceedings 907E (eds P.J. Ferreira, I.M. Robertson, G. Dehm, and H.Saka), Materials Research Society, pp. MM13-02–MM131-6.

69. Junginger, F., Kläui, M., Backes, D., Rüdiger, U., Kasama, T., Dunin-Borkowski, R.E., Heyderman, L.J., Vaz, C.A.F., and Bland, J.A.C. (2007) Spin torque and heating effects in current-induced domain wall motion probed by transmission electron microscopy. *Appl. Phys. Lett.*, **88**, 212510.

70. Eltschka, M., Wötzel, M., Rhensius, J., Krzyk, S., Nowak, U., Kläui, M., Kasama, T., Dunin-Borkowski, R.E., Heyderman, L.J., van Driel, H.J., and

Duine, R.A. (2010) Nonadiabatic spin torque investigated using thermally activated magnetic domain wall dynamics. *Phys. Rev. Lett.*, **105**, 056601.

71. Beleggia, M., Kasama, T., and Dunin-Borkowski, R.E. (2010) The quantitative measurement of magnetic moments from phase images of nanoparticles and nanostructures - I. Fundamentals. *Ultramicroscopy*, **110**, 425–432.

72. Lai, G., Hirayama, T., Ishizuka, K., and Tonomura, A. (1994) Three-dimensional reconstruction of electric-potential distribution in electron-holographic interferometry. *J. Appl. Opt.*, **33**, 829–833.

73. Phatak, C., Petford-Long, A.K., and De Graef, M. (2010) Three-dimensional study of the vector potential of magnetic structures. *Phys. Rev. Lett.*, **104**, 253901.

7
Electron Tomography

Paul Anthony Midgley and Sara Bals

In this chapter, we review the fundamentals, recent developments, and future possibilities of electron tomography. The different imaging modes that can be used for electron tomography are described and some applications within the field of nanomaterials are presented. We will also discuss (future) challenges for electron tomography, such as quantification and improvement of the resolution.

7.1
History and Background

7.1.1
Introduction to Nanoscale Systems

Nanoscale systems investigated within the fields of physics, biology, and chemistry are becoming ever smaller and more complex from a structural and chemical point of view. As described in other chapters, new developments within the field of transmission electron microscopy (TEM) enable the investigation of these systems at the atomic scale, not only to derive structural details but also chemical and electronic information. However, the majority of these techniques provide only a two-dimensional (2D) projection of what is a three-dimensional (3D) object. In order to understand the 3D morphology, composition, and so on, of a nanoscale object, it is necessary to turn to tomographic imaging using electron microscopy, that is, electron tomography.

7.1.2
Tomography

The history of tomography starts with the pioneering work of the mathematician Johan Radon who published a paper in 1917 on the projections of objects into a lower-dimensional space [1]. This projection or transform, known now as the *Radon transform*, and its inverse, form the mathematical basis of tomographic

Handbook of Nanoscopy, First Edition. Edited by Gustaaf Van Tendeloo, Dirk Van Dyck, and Stephen J. Pennycook.
© 2012 Wiley-VCH Verlag GmbH & Co. KGaA. Published 2012 by Wiley-VCH Verlag GmbH & Co. KGaA.

techniques used today. However, it took nearly 40 years for the ideas of Radon to be adopted, first by Bracewell [2], who proposed a method to reconstruct solar emission measured by a radio telescope. In 1963, the idea of an X-ray scanner [3] for medical imaging was proposed by Cormack, and built by Houndsfield in 1971 [4] allowing 3D reconstructions of the human body; Cormack and Hounsfield were jointly awarded the Nobel prize for Medicine in 1979. Since that time, many other forms of radiation have been used for life science tomography, including positron emission tomography (PET) [5], ultrasound CT [6], and zeugmatography (reconstruction from NMR imaging) [7]. In materials science, the development of high-resolution lab-based and synchrotron X-ray tomography [8–10] has been remarkable and the development of atom probe tomography offers true atomic resolution [11].

The first use of tomography in electron microscopy is seen in three key papers published in 1968. One, by de Rosier and Klug [12], determined the structure of a biological macromolecule whose helical symmetry enabled a 3D reconstruction to be made from a single projection (micrograph). A second by Hoppe *et al.* [13] showed how with a sufficient number of projections (images), it was possible to reconstruct fully asymmetrical systems and a third, by Hart [14], demonstrated how the signal-to-noise ratio in a projection could be improved by using an "average" re-projected image calculated from a tilt series of micrographs. Over the succeeding decades, the technique was adopted by electron microscopists, especially in the life sciences, pushing the theoretical understanding of the technique [15–18] and applying it, using BF images, to reconstruct a number of 3D biological structures.

The adoption of tomography by those practising electron microscopy in the materials science field took considerably longer. The first application of electron tomography to materials was the study of polymeric materials in the late 1980s adapting BF TEM techniques optimized for life science imaging to study the complex morphology of block copolymers [19], which can assemble into a range of complex nanostructures including lamellae and gyroid networks, and the porous nature of zeolites [20]. The use of BF images in the study of such materials is possible because of the noncrystalline nature of the specimen and their weak scattering potential.

Around the turn of the century however, it was realized that for strongly scattering, highly crystalline, objects, such as found predominantly in materials science, a different form of imaging would be necessary to enable tomography to work successfully. As we discuss later, the problem is that conventional BF images of crystalline materials made using (partially) coherent illumination are often dominated by contrast which cannot be used easily for tomographic reconstruction. Scanning transmission electron microscopy high-angle annular dark-field (STEM HAADF) images are far "simpler" to interpret and are now the basis for most materials-based electron tomography. Many examples of STEM HAADF imaging will be found in this book.

Electron tomography has now been applied to many classes of materials [21–24] and in addition to STEM tomography there has been development of many other tomographic imaging modes, for example, using energy-filtered transmission electron microscopy (EFTEM) for 3D compositional mapping [25, 26] and holography for mapping electric and magnetic fields, which are discussed later in this chapter.

The resolution of electron tomographic reconstructions is generally far higher in materials science than in the life sciences because the sample is normally far less beam sensitive, allowing a greater electron dose and greater signal at high spatial frequencies. Atomic resolution 2D imaging is commonplace (see examples in this book) but there are now also developments that indicate 3D atomic resolution is possible. The improvement in the mechanical stability of microscope goniometers, the automation of the acquisition, and the speed with which reconstructions are achieved have all contributed to making electron tomography a technique of ever-growing importance.

Although not the subject of this chapter, we should not forget that 3D imaging in the SEM FIB is a powerful way of revealing the 3D structure of "mesoscale" systems. Modern "dual-beam" instruments, which have both electron optical and ion optical columns, enable sequential "slicing and viewing," revealing an exposed fresh surface after milling away a thin section. It is possible to achieve subnanometer resolution in-plane and sub-10 nm slice thickness. By sacrificing the resolution it is possible to scale up the procedure to allow large volumes of material to be investigated; when combined with EBSD and energy-dispersive X-ray (EDX) spectroscopy this becomes a very powerful technique at the submicron level.

7.2
Theory of Tomography

In Radon's original paper [1], a transform, known now as the *Radon transform*, R, is discussed which maps a function $f(x, y)$ by a projection, or line integral, through f along all possible lines L with unit length ds:

$$Rf = \int_L f(x, y) ds \tag{7.1}$$

The experimental interrogation of an object by projections, or some form of transmitted signal, is then a discrete sampling of the Radon transform. The object $f(x, y)$ can then be reconstructed from projections Rf using the inverse Radon transform and the majority of reconstruction algorithms used in tomography are based on this inverse transform. The Radon transform converts real space data into "Radon space" (l, θ), where l is the line perpendicular to the projection direction and θ is the angle of the projection. An "image" in Radon space is called a *"sinogram."* A single projection of the object is a line at constant θ in Radon space. With a sufficient number of projections, an inverse Radon transform should reconstruct the object. Experimentally, the sampling of Radon space is always limited and any inversion will therefore be imperfect. In practice, the key is to limit those imperfections by acquiring as many images as possible or by adding a priori information.

Alternatively, one can understand the reconstruction process by realizing that a projection of an object at a given angle in real space is a central section through the Fourier transform of that object; this is known as the *"central slice theorem,"* or *"projection slice theorem"* [27]. By acquiring images (projections) at many different tilt angles, many central sections of Fourier space will therefore be sampled, increasing the information in the 3D Fourier space of the object. A Fourier inversion should yield the 3D object, whose fidelity will be limited by the "gaps" between the central slices brought about by the finite tilt sampling. Tomographic reconstruction from an inverse Fourier transform of the ensemble of Fourier-transformed projections is known as *direct Fourier reconstruction* [16]. However, if the number of images in the tilt series is limited, Fourier reconstruction methods can be problematic and instead most now use real space *backprojection* methods [28].

7.2.1
Real Space Reconstruction Using Backprojection

As the name suggests, backprojection methods are based on taking the set of projections acquired in a tilt series and sequentially backprojecting them, along the angle of the original projection, into a 3D space. Using a sufficient number of images (projections), from different angles, the superposition of all the backprojected images will generate the original object: this is known as *direct backprojection* [13, 15], see Figure 7.1.

In general (exceptions shown later), reconstruction using backprojection relies on the fact that the intensity in the projection is a monotonic function of the physical quantity to be reconstructed [29]; this is known as the *"projection requirement."* Reconstructions by direct backprojection appear to be blurred because of a relative enhancement of low-frequency information; the uneven sampling of spatial frequencies in the set of original projections is illustrated in Figure 7.2.

Even with regular sampling of spatial frequencies in each image (the spacing of the "data points" in the Figure 7.2), this will still result in a relatively higher sampling near the center of Fourier space. This can be corrected by using a weighting filter in Fourier space, typically a radially linear function rising from zero at the center to a maximum and then apodized, such that the Fourier transform is zero at the Nyquist frequency [30]. Using such a filter results in a *weighted backprojection* [31] and this correction is used widely, especially in the biological community.

Blurring and other artifacts can be corrected by using a different approach based on a priori information. The simplest version of this is to use the original data to correct the reconstruction. If the (imperfect) reconstruction is re-projected back along the original projection angles, the re-projections, in general, will be different from the original projections (images). The difference is characteristic of the deficiency of the reconstruction and can be backprojected to generate a

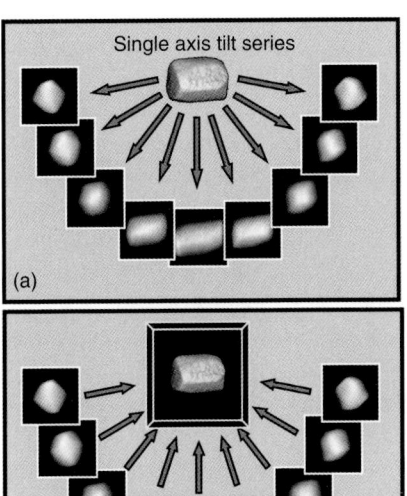

Figure 7.1 A schematic diagram of the tomographic reconstruction process using the backprojection method. In (a) a series of images are recorded at successive tilts. These images are backprojected in (b) along their original tilt directions into a 3D object space. The overlap of all the backprojections will define the reconstructed object.

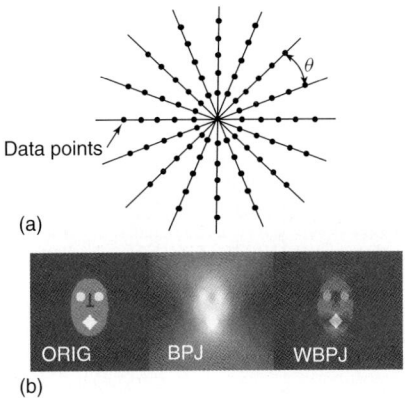

Figure 7.2 (a) Fourier space representation of the ensemble of projections, indicating the undersampling of high-spatial-frequency information. θ is the tilt increment between successive images. (b) An original phantom and its reconstruction by backprojection (BPJ) and weighted backprojection (WBPJ). In the WBPJ a weighting filter has reduced the blurring effect of the backprojection process.

"difference" reconstruction which is used to correct the imperfections in the reconstruction. This procedure is repeated iteratively until the reconstruction is constrained to best fit the original projections [32, 33]. There are a number of iterative reconstructions possible, the two most popular are the algebraic reconstruction technique (ART) [17], which compares the reconstruction with a single projection, correcting in a single direction, and then moving on to the next projection, and simultaneous iterative reconstruction technique (SIRT), which instead compares all the projections simultaneously. In practice, SIRT tends to be more stable computationally than ART when images are noisy [18]. The SIRT algorithm has been summarized in Figure 7.3.

If other a priori information is known, then the reconstruction can be further improved. "Discrete tomography" considers the object as discrete units, which can be spatially discrete (e.g., an atomic lattice) or discrete in terms of an object's

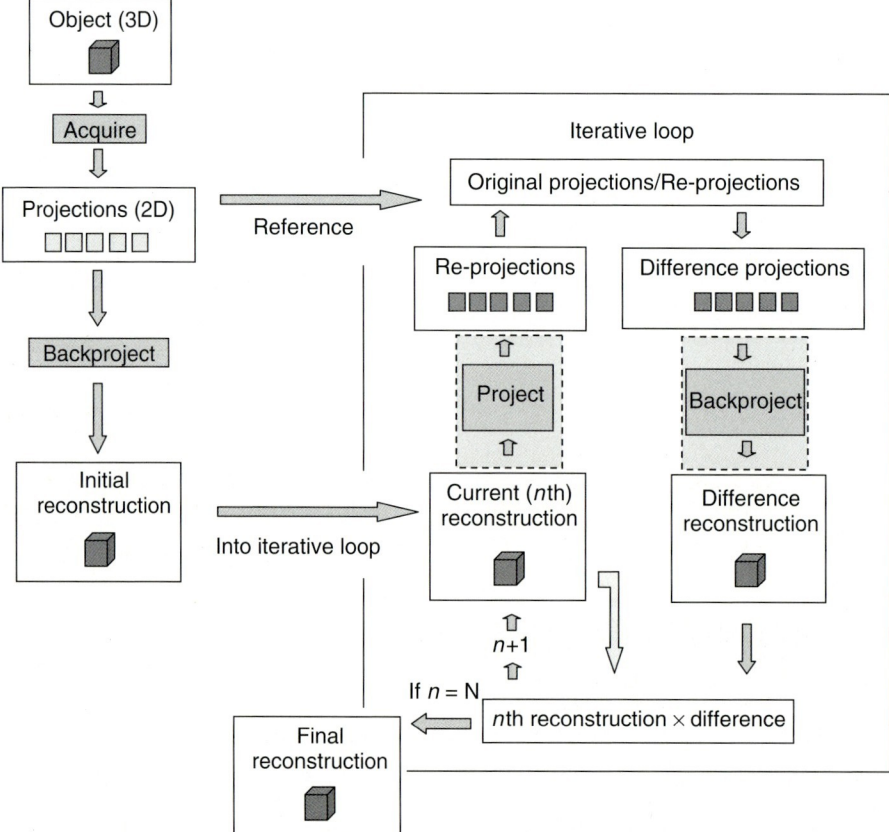

Figure 7.3 A schematic representation of the multiplicative SIRT algorithm, indicating the iterative loop in which re-projected data are compared to the original images.

density (i.e., the number of gray levels is discrete). We will comment more on this technique later.

There have also been attempts to use Bayesian techniques (maximum entropy methods) which try to find the simplest reconstruction taking into account the known projections, the noise in the data, sampling artifacts in the reconstruction, and the contrast limits of the original projections [34].

7.3 Electron Tomography, Missing Wedge, and Imaging Modes

As we discussed in the previous section, in general, the highest fidelity reconstructions will be when the number of images in the tilt series is maximized. Crowther et al. [15] established a simple equation relating the number of projections (N), the resolution (d), and the diameter of the reconstruction volume D:

$$d = \frac{\pi D}{N}$$

This assumes that the N projections are spread evenly over 180° but for tomography in an electron microscope there is almost always an upper limit to the tilt angle, leading to a loss of information in the least sampled direction (usually the optic axis), as seen in Figure 7.4. This "missing wedge" of information leads to an "elongation" of the reconstruction in that direction. An estimation of this elongation (e), as a function of the maximum tilt angle (α), is given by Radermacher and Hoppe [35]:

$$e = \sqrt{\frac{\alpha + \sin\alpha \cos\alpha}{\alpha - \sin\alpha \cos\alpha}}$$

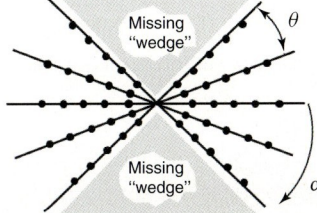

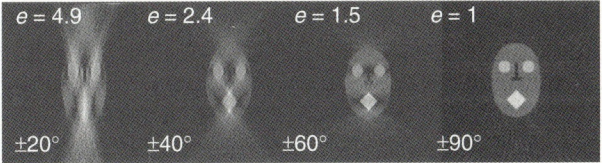

Figure 7.4 Illustration of the missing wedge in a Fourier space representation and in its effect on the reconstruction fidelity of a face phantom. The elongation factor, e, becomes quite pronounced as the tilt range is reduced.

TEM samples are often lamellae and this anisotropy can be modeled using a modified diameter (D) taking into account the lamella thickness (t) and the maximum tilt angle (α) [36]:

$$D = t \cos \alpha$$

This modification leads to an improved resolution prediction compared to the basic Crowther equation.

However, the Crowther equation is really only an estimate of the reconstruction resolution and does not take into account, for example, the noise present in the images and the limitations this will have on the reconstruction. An alternative approach is therefore to examine the intensity distribution of the object in Fourier space, and determine where this is above some threshold for noise. This is used in Fourier shell correlation (FSC) [37], the differential phase residual (DPR) [38], and the spectral signal-to-noise ratio (SSNR) method [39], methods developed originally for determining the resolution of single particle reconstruction [40, 41]. Recently, a more pragmatic method has been proposed, where no prior knowledge on image formation, alignment, and other experimental parameters is required. This method uses the intensity profile at the edge of, for example, a reconstructed nanoparticle to determine a measure for the resolution [42].

Clearly the need to acquire images at high tilt angles is paramount and, over the years, a number of dedicated TEM tomography holders have been built in-house and by commercial manufacturers, to allow specimens to be tilted to high angles even on conventional analytical or high-resolution instruments. In many cases, it is not practically possible to tilt to high angles but the missing information can be minimized by acquiring a second tilt series about an axis perpendicular to the first; this is known as *dual-axis tomography* and generates a "missing pyramid" of information as shown in Figure 7.5. The practical requirements for dual axis are a little more stringent than for conventional single-axis tomography and reconstruction algorithms have been developed to combine the benefits of dual-axis acquisition with iterative reconstructions [43, 44].

To achieve the "best" possible reconstruction, the missing wedge must be eliminated completely and specimen holders are available that allow complete rotation of the specimen using a separate rotation mechanism or a combination of an internal on-axis rotation coupled with the goniometer tilting mechanism. Typically, specimens used in such holders are needle-like or pillar-shaped and prepared using a FIB workstation. This type of holder was recently used to study a variety of samples including a zirconia/polymer nanocomposite and the complex 3D architecture of a transistor contact region [45–48]. Since this technique uses micropillars, comparable to those used for atom probe tomography, it is also interesting to think of a combination of both techniques; such an experiment was recently carried out by using the same pillar-shaped sample [49]. This experiment is very encouraging in a sense that electron tomography provides larger-scale morphological information and atom probe tomography yields atomic resolution, albeit for a relatively small volume, which can be up to $150 \times 150 \times 100 - 500$ nm^3.

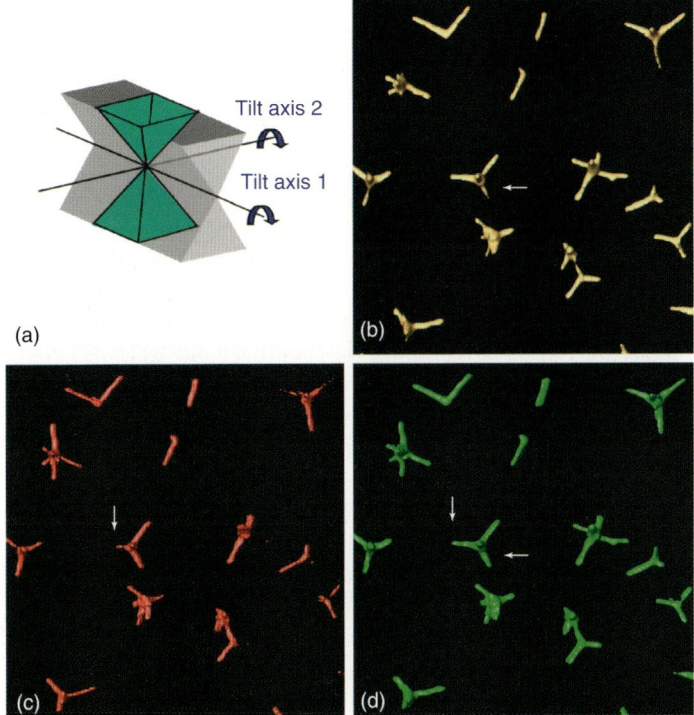

Figure 7.5 (a) Illustration showing how a dual-axis tilt series leads to a missing pyramid of information. (b–d) Reconstructions of cadmium telluride tetrapods from a dual-axis tilt series, reconstructed individually (b,c) and then as a dual-axis series (d). The arrows indicate a tetrapod where the missing wedge has had a particular effect on the individual reconstructions. Each "leg" of each tetrapod is better reconstructed in the dual-axis reconstruction. (Reproduced from Ref. [44], with permission from Elsevier.)

Even with the goniometer and sample holder optimized, there will still be a relative displacement of the object, in all three directions, after each tilt. Correcting for this, and finding the exact tilt axis position and angle, is crucial in achieving an "aligned" data set that will yield the best reconstruction. We will not discuss the process here but refer the reader to other texts for more information on this important practical step [22, 50]. Some practical issues related specifically to STEM acquisition are discussed in the next section.

7.4
STEM Tomography and Applications

As mentioned above, many materials are crystalline in nature and as such BF imaging leads to images that are riddled with unwanted contrast, such as bend contours and thickness fringes. For the projection requirement to be met and thus to use backprojection methods of reconstruction, an imaging mode is needed that minimizes that unwanted contrast and delivers an image intensity that changes in a monotonic manner with thickness, density, and so on. This need led to the development of STEM HAADF tomography, particularly for use in materials science [50] although it has also proven to be of use in some biological experiments [51].

As illustrated in chapter 4 of this book, images with minimal Bragg diffraction contrast can be formed in STEM mode by increasing the inner radius of an ADF detector to exclude Bragg-scattered beams [52]. The image intensity derived using this detector is a function of the projected thickness and the atomic number (Z) of the scattering atom, approaching a Z^2 relationship for a detector with a large inner radius. The lack of diffraction contrast in HAADF images allows them to be considered as projections of the structure in terms of thickness (up to a certain thickness value) and atomic number and thus to a good approximation meet the projection requirement for tomographic reconstruction.

The choice of the inner angle for the HAADF detector is critical and should be sufficiently large to ensure coherent effects are minimal. For thermal diffuse scattering (TDS) to dominate the image contrast, the inner angle should be $\geq \lambda/d_{th}$ [52] where λ is the electron wavelength and d_{th} is the amplitude of atomic thermal vibration.

Medium-resolution (~1 nm) STEM HAADF images are sensitive to changes in specimen composition with the intensity varying (for the most part) monotonically with composition and specimen thickness. However, HAADF images will depend on the excitation of Bloch states and associated channeling [53] and when a crystalline specimen is at or close to a major zone axis there is an increase in the STEM HAADF signal brought about by the localization of the beam onto atomic strings [54]. However, in general, this kind of strong channeling will occur infrequently during a tilt series and will have little effect on the overall intensity distribution in the reconstruction. One must also be careful regarding the contrast in images recorded from thick and heavy specimens. For example, in the case of tungsten contacts, the scattering is so strong that the scattered electrons are scattered outside the outer annulus of the HAADF detector so that the image intensity falls as the contacts becomes thicker in projection [55]. Such problems led to the development of incoherent BF STEM imaging for electron tomography, whose image intensity follows approximately Beer's Law, falls off exponentially with thickness and can be used even on very thick specimens [56].

In general a STEM HAADF image will take longer to acquire than BF TEM as the number of electrons collected on the detector is only a small fraction of those that are incident on the specimen. However, in general, although the total signal may be low, the contrast is much higher than in BF images. The relatively short focal depth of STEM can be exploited for automated focusing with a defocus series showing a clear trend in the sharpness/contrast of the image, enabling the optimum focus to be attained. If the specimen is slablike and tilted to high angles, it is likely that only part of the specimen will be in focus. If the tilt axis of the specimen is made perpendicular to the direction of the scan line, it is possible to adjust the beam crossover to match the change in specimen height. A focal ramp can be applied across the image to minimize the problems associated with a short depth of field; such "dynamic focusing" has been used for many years in scanning electron microscopy (SEM) of tilted surfaces.

STEM tomography has been used widely over the past decade or so to investigate the 3D structure of many materials. Here we illustrate its application by considering a few recent examples.

One area that has benefited enormously from the application of electron tomography has been the field of heterogeneous catalysis where structurally complex structures are used to support catalytically active nanoparticles. There is a pressing need to understand the geometry of the support (e.g., porosity, connectivity, surface area) and the distribution of the nanoparticles [57, 58]. Some catalysts are composed of crystalline nanoparticles whose surfaces are active or promote catalytic activity. Here tomography can be used to identify individual facets and surfaces. In the example shown in Figure 7.6, we see a reconstruction of metal oxide catalysts based on $(Ce/Tb/Zr)O_{2-x}$ with Au nanoparticles decorating the surface [59].

Here it becomes clear that the Au nanoparticles are anchored preferentially in the "valley" between two {111} facets. The contribution of interface energy to the total energy of such nanoparticles is thought to be of importance and so maximizing the interaction between the Au and oxide surface could lead to further stability.

In another example, we show how individual CdSe nanocrystals adopt a highly asymmetric structure (a "bullet" morphology) when anchored on carbon nanotube supports. Indeed the basal plane of the CdSe crystals matches remarkably well the graphite basal plane such that near-epitaxy is established [60]. This leads to a rather ordered arrangement of the nano-bullets along the outer wall of the multiwalled nanotube, see Figure 7.7.

As the next example, we show in Figure 7.8 cross sections of a rutile nanotube in which three different crystal morphologies are seen, labeled A, B, C. The

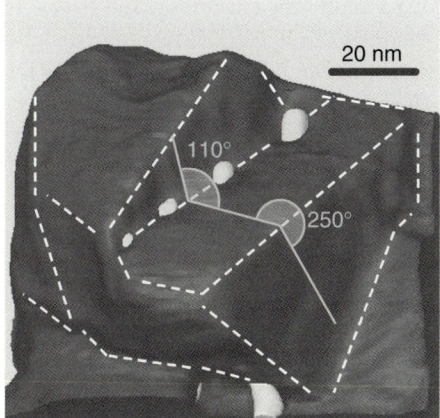

Figure 7.6 Reconstruction of a $(Ce/Tb/Zr)O_{2-x}$ catalyst with Au nanoparticles anchored preferentially in the crevice between two {111} facets. (Adapted from Ref. [59].)

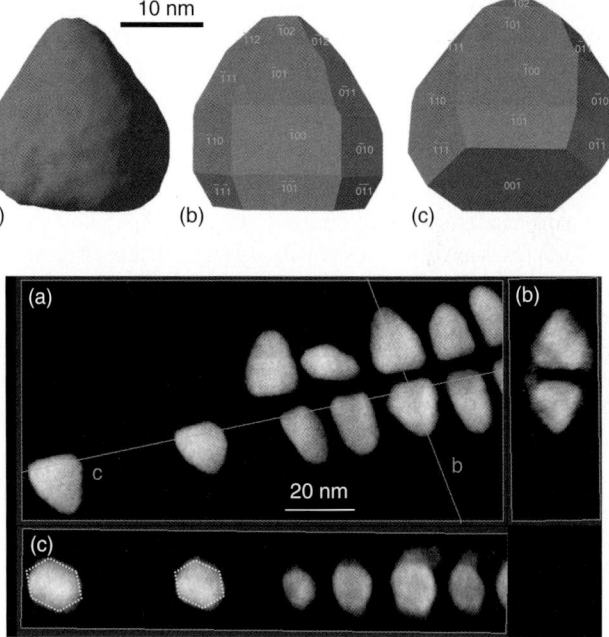

Figure 7.7 (a) Reconstruction of highly asymmetric CdSe "nano-bullets" and a model crystal showing the likely facets in the reconstruction. (b) The near-epitaxy between the CdSe basal plane and the graphite basal plane leads to a well-aligned array of CdSe on a multiwalled nanotube support. (Adapted from Ref. [60].)

different morphologies arise because of the thermal history. The nanotube is processed from an anatase sol–gel covering a multiwalled nanotube. The anatase is converted to rutile and then the carbon nanotube burned off. The two-step thermal processing leads to the unusual morphology that can be seen only when slices are taken through the 3D reconstruction [61]. A single image (projection) reveals very little!

7.5
Hollow-Cone DF Tomography

As discussed above, the projection requirement states that in order to obtain reliable 3D reconstructions, individual images should correspond to a projection through the structure, so that there is a monotonic variation with, for example, mass or density. Unfortunately, this is not in general the case for BF TEM images of crystalline samples, in which the image contrast will be dominated by Bragg diffraction. Despite the constraints of the projection requirement, in materials applications, 3D reconstructions based on BF TEM have been obtained in several

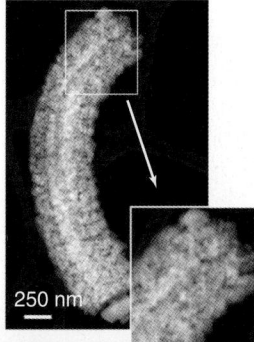

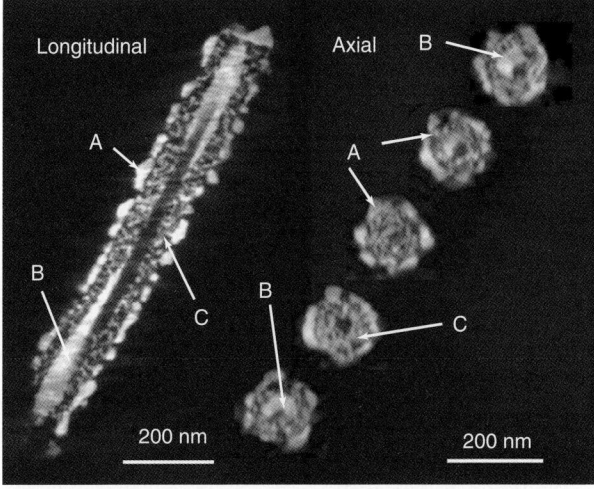

Figure 7.8 HAADF image of a rutile nanotube showing a complex morphology that is revealed in longitudinal and axial slices of a reconstruction. Three different morphologies of rutile crystal are evident, labeled A, B, and C. (Adapted from Ref. [61], with permission from Elsevier.)

cases, not only for noncrystalline porous materials but also for nanoparticles [62]. Although the use of BF TEM does not necessarily hamper the morphological (external) reconstruction of nanoscale particles, artifacts might be expected, for example, apparent voids in Pt nanoparticles, due to contrast reversals arising from changes to the diffraction conditions or from dynamical scattering [62]. Here, the use of STEM HAADF or EFTEM/EDX is a good alternative to the use of BF TEM for tomography.

An alternative approach to avoid or minimize the presence of diffraction contrast directly in TEM mode is to use annular dark-field transmission electron microscopy (ADF TEM) [63], hollow cone illumination [64], or precession techniques [65]. In ADF TEM, an annular objective aperture is used that acts as a central beam stop in the back focal plane of the objective lens. In this manner, the central

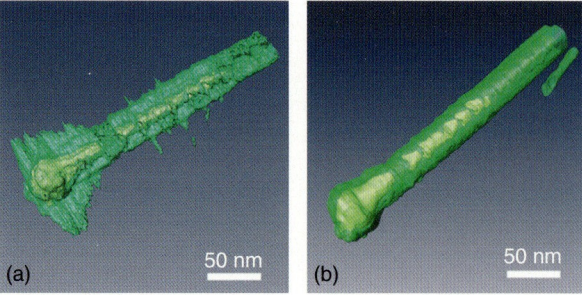

Figure 7.9 (a) Isosurface rendering of a 3D reconstruction based on STEM HAADF imaging. The Cu nanoparticle can be distinguished from the C nanotube, and smaller Cu parts are also present inside the nanotube. Reconstruction artifacts can be seen in the nanotube. (b) ADF TEM reconstruction of the same nanotube.

beam and all electrons scattered up to a certain angle are excluded from the imaging process. Annular apertures allow one to select electrons that have passed through given symmetric zones within the objective lens of the microscope. In this manner mass–thickness contrast is generated that can be interpreted quantitatively because it depends exponentially on the sample thickness [66]. In addition, ADF TEM meets the projection requirement for 3D reconstructions. The chemical sensitivity of ADF TEM imaging is similar to STEM HAADF imaging where the image follows a dependence close to Z^n with $n = 1.5 - 2$. It is therefore instructive to compare a 3D reconstruction obtained using ADF TEM to one using STEM HAADF. In Figure 7.9, this comparison is carried out for a sample that consists of carbon nanotubes filled with Cu nanoparticles. It should be noted that the ADF TEM images are recorded with a single exposure of 1 s, whereas the STEM HAADF images require a scan of approximately 12 s. The isosurface rendering of the STEM HAADF reconstruction is shown in Figure 7.9a: artifacts are present, which are related to a combination of the missing wedge and a high contrast that is present between the C tube and the Cu nanoparticles. The corresponding ADF TEM reconstruction, shown in Figure 7.9b appears smoother in comparison to the previous reconstruction, which is related to the fact that the ADF aperture acts as an *in situ* low-pass filter. To date, the spatial resolution of ADF TEM has been inferior to that of STEM HAADF, with the presence of spherical aberration acting as the main limitation. Therefore, the optimization of ADF TEM in aberration-corrected microscopes is very promising.

7.6
Diffraction Contrast Tomography

We stated above that in general for crystalline strongly scattering materials, the diffraction contrast seen in BF TEM images (and therefore also conventional DF

TEM images) does not satisfy the projection requirement. However, if great care is taken, it is possible to use diffraction contrast TEM images in a tomographic manner if the diffraction conditions do not change significantly over the range of the tilt series. Practically, this means either having a specialized high-tilt double-tilt rotate holder, aligning the sample very carefully before insertion into the microscope or using the DF tilts to correct for any changes in the diffraction condition along the tilt range.

An early example of this type of tomography was from a Ni-based superalloy, in which a superlattice reflection was used to image the cuboidal $L1_2$-ordered γ' precipitates in the A1-disordered γ matrix [67]. As the γ/γ' two-phase structure is annealed, γ particles appear in the γ' precipitates and coarsen into plate- or rodlike shapes. The 3D morphology and distribution of the γ particles were reconstructed to show the three {100} orientations of platelike γ particles, see Figure 7.10. In this case, the superlattice reflection is weak with a long extinction length and thus even at a relatively strong diffraction condition the image intensity varied reasonably monotonically with thickness.

Stereo microscopy has been used for many years to provide a 3D "image" of an object and in electron microscopy this has often been a network of dislocations. However, the precision with which the depth of the dislocation in the foil can be determined is limited by the small angle $(5 - 10°)$ that the sample must be tilted for the stereo effect to work. In order to achieve a true 3D representation of a dislocation network, a tomographic reconstruction is required. Barnard *et al.*, 2006 used a tilt series of weak-beam dark-field (WBDF) images to form a high-magnification tomographic reconstruction of a network of dislocations in a GaN epilayer [68]. The dislocations can be seen to interact with crack stresses and other dislocations nearby with some threading dislocations turning over to become in-plane dislocations. Of course, here we are not imaging the dislocation *per se* but the strain field around the core of the dislocation, which, so long as the diffraction conditions do not change significantly and the imaging reflection is weak, will give a faithful and accurate representation of the location of the dislocation in the epilayer.

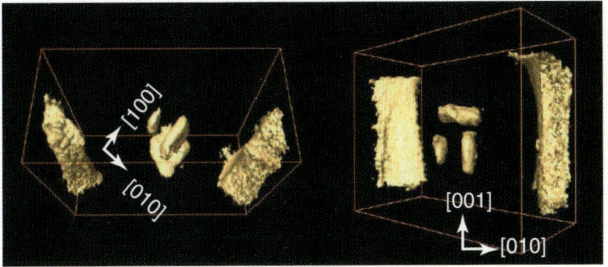

Figure 7.10 Surface rendered reconstruction of γ' precipitates in a Ni-based superalloy. The tomographic series was taken with a DF superlattice reflection. (Adapted from Ref. [67].)

This type of diffraction contrast tomography is practically very difficult and suffers from the fact that WBDF images will likely contain contrast from thickness variations and long-range sample bending that can disrupt the dislocation image contrast required. Thus for imaging dislocations, this method has been superseded by a STEM-based version using multiple scattered reflections that offers a "cleaner" image signal.

In the STEM method, we allow many diffracted beams to be recorded by a medium-angle annular dark-field (ADF) detector. Typically, the best DF contrast seems to be found by excluding the lowest-order diffracted beams with an angular range of say 10–40 mrad. These STEM MAADF images can be thought of as being equivalent to the summation of multiple DF TEM images with different excitation conditions. The net effect is to average out the extraneous thickness fringe and bend contour contrast and enhance the contrast associated with the dislocation, albeit with a slight loss of spatial resolution.

Figure 7.11a shows a comparison of a WBDF image of dislocations in Si with the equivalent STEM MAADF image of the same area. The lack of thickness fringe contrast is quite striking, as is the dislocation visibility. There are also important practical advantages of this STEM MAADF mode for dislocation imaging, which include the dynamic focusing possibility offered by STEM and the availability of auto-focusing and auto-tracking routines developed for STEM HAADF tomography. Figure 7.11b shows the STEM MAADF reconstruction [69] of dislocations in deformed silicon.

7.7
Electron Holographic Tomography

In a separate chapter 5 in this volume, electron holography has been shown to be able to reveal, through the reconstruction of the phase of the exit wavefunction, the presence of electrostatic potentials and magnetic fields. To a reasonable approximation, the phase of the wavefunction can be thought of as a projection of the potential (the scalar potential for electrostatics, a component of the vector potential for magnetic fields) and thus the phase may be used in a tomographic tilt series to reconstruct the 3D potential associated with the specimen.

In very specific cases, the phase can be simply related to a product of thickness and projected potential. The potential across p–n junctions is a relatively simple case where the change in the phase across the junction can be related directly to the "built-in potential" of the junction, provided diffraction contrast is weak (i.e., we are far from a strong scattering condition (zone axis, systematic row) and there are no magnetic fields). By acquiring a tilt series keeping that diffraction contrast weak, it is possible to reconstruct the 3D potential near a p–n junction. The result reveals how the potential change across the junction is reduced dramatically as the surfaces of the sample are approached. In this case, the Si p–n device was prepared by FIB, the Ga ions had probably caused some kind of knock-on damage, creating vacancies that then pinned dopant atoms near the mid-gap, depleting

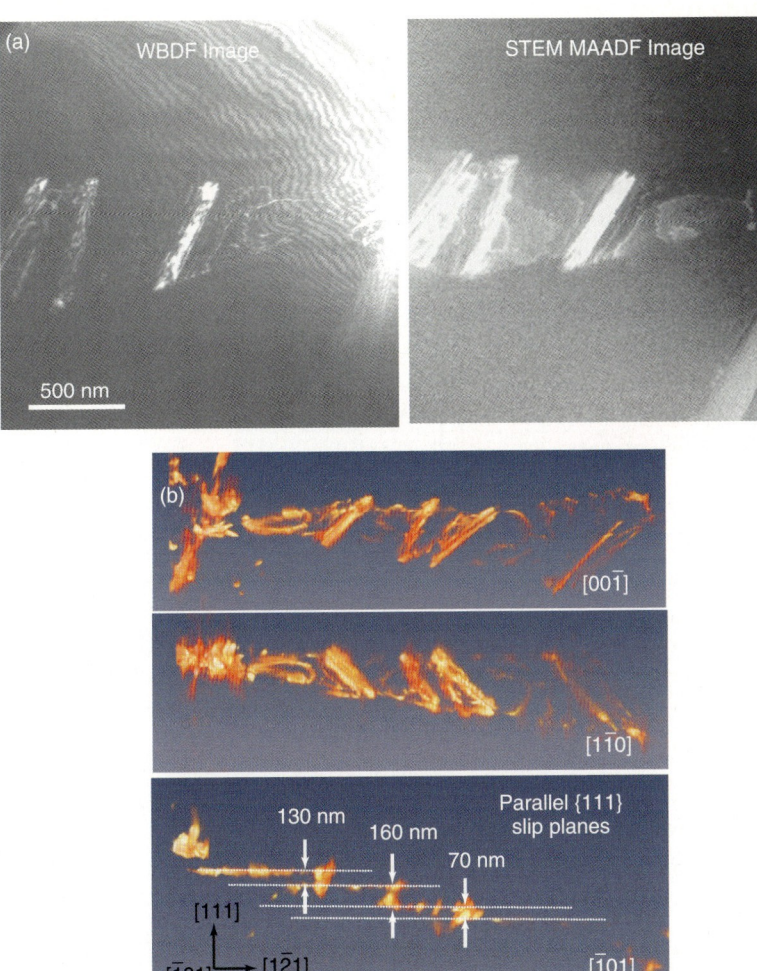

Figure 7.11 (a) Comparison of a weak-beam dark-field image (approximately g3g) and a corresponding STEM MAADF image of a dislocation array induced by indentation and annealing silicon. (b) Views of a STEM MAADF reconstruction showing parallel slip planes with approximately 100 nm separation. (Adapted from Ref. [69].)

carriers and reducing the built-in potential [70]. This had been postulated from previous 2D holography results but this was the first concrete 3D evidence of this happening. More recent work by the Dresden group, in which the acquisition of tilt series of holograms has been automated [71, 72], has led to the reconstruction of p–n junctions in a number of needle-shaped devices revealing with great accuracy the 3D potential distribution across the built-in junction, as shown in Figure 7.12.

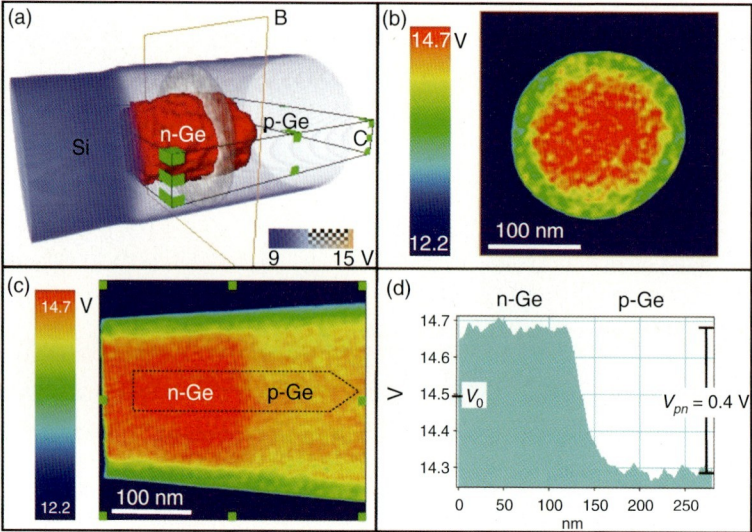

Figure 7.12 3D potential of a FIB-prepared germanium needle with p–n junction. (a) 3D Rendering with 14.42 V isopotential surface (red). (b) Cross section at the position shown in (a). (c) Longitudinal slice averaged over a thickness of 50 nm. The region for averaging is indicated in (a). (d) Line profile across the p–n junction along the arrow in (c) averaged over a width of 33 nm. (Courtesy of D. Wolf, adapted from Ref. [72].)

7.8
Inelastic Electron Tomography

Inelastic signals, such as those detected using electron energy-loss spectroscopy (EELS) and EDX spectroscopy can be used to map compositions not only in two but also in three dimensions. Early attempts to map chemical information in three dimensions using EDX were complicated by the directionality and inefficiency of the sample–detector geometry, by the need to tilt away from the detector, and by the consequent shadowing in half of the tilt series [26]. Recent work has taken advantage of needle-shaped specimens, where shadowing is eliminated and the detector geometry is not such a problem [73]. Also EFTEM can be used to acquire tilt series for electron tomography. EFTEM elemental maps are commonly based on three energy-filtered images acquired near an ionization edge. These maps however, are prone to diffraction effects through the coupling of elastic and inelastic signals. Diffraction effects can be minimized by forming jump-ratio images or by dividing elemental maps by low-loss (or zero-loss) images. However, care must be taken if such images are used for tomography as the resultant signal may not satisfy the projection requirement. This is related to the fact that most of the thickness information is lost by dividing two energy-filtered images in the jump ratio method. Both the ionization-edge signal and the background signal depend linearly on the thickness of the specimen. Because the energy-loss background intensity rises faster as a function of thickness than the ionization-edge signal, thicker areas may even give rise to a lower intensity in the jump-ratio map. EFTEM tomography based on elemental maps was first used in studies by Mobus and Inkson [23, 26] and Midgley and Weyland [22, 25]. A simpler form of EFTEM tomography using the post-edge contrast from a single energy loss enabled different phases in a complex polymer to be reconstructed [74]. However, in most experiments of this type, two or three images at each energy loss edge are acquired, if possible in an automated manner. This means that the acquisition of such a tilt series takes a very long time, resulting in a high total dose. Consequently, it is not straightforward to apply EFTEM tomography for beam sensitive materials. Nevertheless, it was shown by Leapman *et al.* that it is possible to map the three dimensional distribution of elements in biological samples by energy-filtered electron tomography [75]. At each energy loss the positions of the energy windows will closely control the SNR of the resultant maps, and as the quality of the reconstruction will depend on the quality of the projections, these should be chosen with some care. Over the last decades, a lot of effort has been devoted to an optimization of the acquisition and off-line processing in order to achieve nanoscale resolution and EFTEM tomography has been used to study a broad range of materials [76–79].

Image spectroscopy has also been extended to volume spectroscopy by recording a large energy series at every tilt angle [80]. A low-loss series of a nanocomposite composed of a multiwalled carbon nanotube encased in nylon was recorded every 3 eV over a wide range of tilts. The different plasmon excitation energies of the nylon ($\sim$22 eV) and the nanotube ($\sim$27 eV) enabled the two components of the composite to be distinguished. A similar approach was used to visualize the distribution of Si nanoparticles embedded in a Si matrix [81]. By reconstructing tomograms at individual energy losses, it becomes possible to identify a voxel or subvolume common to all of the energy-loss tomograms and, by plotting the intensity of the voxel as a function of energy loss, to extract spectral information from within the tomogram. In conventional EELS, spectral information is always projected through the structure, but now it is possible to extract spectral information from a subvolume without any projection artifact.

7.9
Advanced Reconstruction Techniques

Recently, a new reconstruction algorithm, called the *discrete algebraic reconstruction technique (DART)*, was introduced as a method to minimize reconstruction artifacts. DART is an iterative algebraic reconstruction algorithm, which alternates between steps of the SIRT algorithm from continuous tomography and certain discretization steps. The discretization is done by exploiting prior knowledge based on the fact that the studied structure consistently is composed of regions with homogeneous density, resulting in well-defined local contrast [82]. It has been shown recently that this technique leads to reconstructions with better quality for bright-field TEM [83], and STEM HAADF [84]. An example is shown in Figure 7.13 where orthogonal

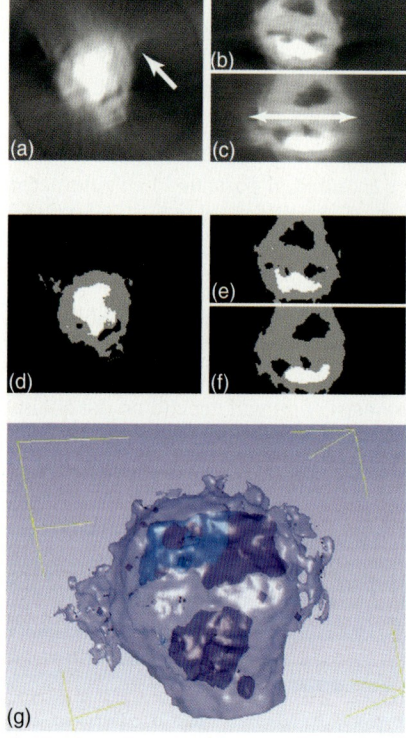

Figure 7.13 (a–c) Three orthogonal slices through a 3D reconstructed nanoparticle, based on conventional reconstruction techniques. Several artifacts, due to the missing wedge are indicated by white arrows. (d–f) Corresponding orthoslices based on discrete tomography. It can be seen that artifacts are drastically reduced. Different gray levels correspond to Cu, Cu_2O, and cavities respectively. (g) 3D rendering of the nanoparticle. (Adapted from Ref. [84].)

slices through a 3D reconstructed nanoparticle are shown. The orthoslices based on the reconstruction using SIRT are shown in Figure 7.13a–c. Artifacts (indicated by arrows in Figure 7.13a,c) are absent in the corresponding orthogonal slices of the discrete reconstructions and sharp boundaries between different parts of the nanoparticle are obtained (Figure 7.13d–f). It must be noted that the discrete reconstruction is segmented during the reconstruction. This is not the case for the conventional reconstruction. In practice, this means that quantitative results based on the conventional reconstruction algorithms may be subjective. Recent results have shown that the discrete reconstruction on the other hand results in a reliable segmentation and that quantitative results can be obtained in 3D [84]. In this case, 85% of the nanoparticle corresponds to Cu_2O (transparent area in (g)), 10% corresponds to cavities (dark purple area in (g)) and 5% corresponds to Cu (blue area in (g)). Quantitative electron tomography will be further discussed in the next section.

7.10
Quantification and Atomic Resolution Tomography

Often, the final goal when applying electron tomography is to obtain quantitative information in 3D. There are a number of examples of quantitative volumetric analysis in electron tomography. Jinnai *et al.*, demonstrated that characteristic lengths in block copolymers, measured from electron tomograms, agreed very closely with those derived from small-angle X-ray scattering (SAXS) and yielded insights into the physics behind their growth [85]. Given a segmented volume, it is relatively straightforward to measure internal structures, such as the size and distribution of silica inclusions in carbonaceous materials, the latter correlating well with the bulk resistivity [86]. A 3D analysis of a styrene network measured the mean curvature of a gyroidlike structure which indicated the effects of "packing frustration" [85]. In heterogeneous catalysis, electron tomography is the only technique that can provide direct measurement of the loading (mass of catalyst per unit surface area) and the local porosity [87].

In order to extract quantitative data from a 3D reconstruction, a segmentation step is required to determine the correspondence between different grayscales in the reconstruction and different compositions in the original structure. Segmentation is generally performed manually but this approach is very time consuming and even more important, can be subjective. Automatic segmentation by using a threshold at different gray levels might be a method to overcome this problem, but (missing wedge) artifacts can hamper this procedure. The influence of the missing wedge was studied by Kawase *et al.* [45] and Biermans *et al.* [88] in a quantitative manner and it was found that samples have to be tilted to at least $\pm 80°$ in other to obtain a reliable quantification. Even if projection images could be acquired from a full range of angles (e.g., using an on-axis rotation holder, see Section 7.3), several other types of artifacts will hamper the segmentation step. A quantitative interpretation based on the conventional reconstruction algorithms is therefore quite difficult. In this respect, the use of discrete tomography is very promising. It has been shown recently that this segmentation is indeed reliable and that quantification using discrete tomography is possible even for a tilt range of $\pm 60°$ [88].

Atomic resolution in 3D has been the ultimate goal in the field of electron tomography during the past few years [89, 90]. Recent advances have indeed pushed the resolution limit from the nanometer range to the atomic range. Aberration-corrected TEM enables one to increase the numerical aperture to high angles (30 mrad or more) leading to a reduction of the depth of focus in STEM Using this technique, which is called *"depth-sectioning tomography"* only very thin slices of a sample come into focus at the same time [91]. This effect can be used to obtain 3D information on a sample, by obtaining image sequences using different defocus values. Because of the high signal-to-noise ratio of an aberration-corrected microscope, it became possible to image individual impurity atoms inside the volume of a TEM sample. This depth-sectioning technique is clearly very powerful to study samples with a slablike geometry as tilting to high angles is no longer required. A disadvantage however is that the resolution in the depth direction is limited and rapidly becomes worse for laterally extended objects. A solution might be to use incoherent scanning confocal electron microscopy, but this greatly reduces the dose efficiency of the method [92]. Another option might be to combine the confocal approach with a limited tilt series and use iterative constraints.

Another possible route toward atomic resolution electron tomography is again the use of discrete electron tomography. In this case, the technique exploits the fact that crystals are discrete assemblies of atoms. Recently, it has been shown theoretically that discrete electron tomography is able to recover the shape of the particle as well as the position of its atoms from a limited number of 2D projections (10 or less) [93]. The reason that experimental results have been lacking until recently is that in all of the 2D projections, one should be able to determine the positions of the atom columns with great precision and in addition, one should be able to quantify the intensity in the images (TEM or STEM) in order to count the number of atoms in a column from a 2D projection. On the basis of the availability of new aberration-corrected microscopes in combination with advanced quantification methods, it was demonstrated recently that 3D reconstruction at the atomic scale is indeed feasible [94, 95]. This will open up a new level of characterization at the (sub) nano scale.

Acknowledgments

PAM would like to thank the colleagues who have contributed to this work over the years, including M. Weyland, J. M. Thomas, I. Arslan, J. S. Barnard, A. Hungria, J. C. Hernandez, Z. Saghi, J. Tong, J. Sharp, T. J. V. Yates and E. Ward. SB would like to express her gratitude for valuable discussions and fruitful collaborations to K. J. Batenburg, E. Sourty, X. Ke, E. Biermans, H. Heidari, B. Goris, C. Kisielowski and G. Van Tendeloo.

The authors thank the EC contract 026019 ESTEEM. PAM thanks the ERC for financial support under contract 291522 3DIMAGE.

References

1. Radon, J. (1917) Über die Bestimmung von Funktionen durch ihre Integralwerte längs gewisser Mannigfaltigkeiten. Ber. Verh. K. Sachs. Ges. Wiss. Leipzig, Math.-Phys. Kl., 69, 262–277.
2. Bracewell, R.N. (1956) Two dimensional aerial smoothing in radio astronomy. Aust. J. Phys., 9, 297–314.
3. Cormack, A.M. (1963) Representation of a function by its line integrals with some radiological applications. J. Appl. Phys., 34, 2722–2727.
4. Hounsfield, G.N. (1972) A Method and Apparatus for Examination of a Body by Radiation Such as X or Gamma Radiation, The Patent Office, London.
5. Brownell, G.L., Burnham, C.A., Hoop, B., and Bohning D.E. (1971) Proceedings of the Symposium on Dynamic Studies with Radioisotopes in Medicine, Rotterdam, 190, IAEA, Vienna.
6. Baba, K., Satoh, K., Sakamoto, S., Okai, T., and Ishii, S. (1989) Development of an ultrasonic system for three-dimensional reconstruction of the fetus. J. Perinat. Med., 17, 19–24.

7. Hoult, D.I. (1969) Rotating frame zeugmatography. *J. Magn. Reson.*, **33**, 183–197.
8. Banhart, J. (2001) Manufacture, characterisation and application of cellular metals and metal foams. *Prog. Mater. Sci.*, **46**, 559–632.
9. Marrow, T.J., Buffiere, J.-Y., Withers, P.J., Johnson, G., and Engelberg, D. (2004) High resolution X-ray tomography of short fatigue crack nucleation in austempered ductile cast iron. *Int. J. Fatigue*, **26**, 717–725.
10. Weierstall, U., Chen, Q., Spence, J.C.H., Howells, M.R., Isaacson, M., and Panepucci, R.R. (2002) Image reconstruction from electron and X-ray diffraction patterns using iterative algorithms: experiment and simulation. *Ultramicroscopy*, **90**, 171–195.
11. Miller, M.K. (2000) *Atom-Probe Tomography: Analysis at the Atomic Level*, Kluwer Academic/Plenum Press, New York.
12. de Rosier, D.J. and Klug, A. (1968) Reconstruction of three dimensional structures from electron micrographs. *Nature*, **217**, 130–134.
13. Hoppe, W., Langer, R., Knesch, G., and Poppe, C. (1968) Protein-kristallstrukturanalyse mit Elektronestrahlen. *Naturwissenschaften*, **55**, 333–336.
14. Hart, R.G. (1968) Electron microscopy of unstained biological material: the polytropic montage. *Science*, **159**, 1464–1467.
15. Crowther, R.A., de Rosier, D.J., and Klug, A. (1970) The reconstruction of a three-dimensional structure from projections and its application to electron microscopy. *Proc. R. Soc. Lond. A.*, **317**, 319–340.
16. Ramachandran, G.N. and Lakshminarayanan, A.V. (1971) Three-dimensional reconstruction from radiographs and electron micrographs: application of convolutions instead of fourier transforms. *Proc. Natl. Acad. Sci.*, **68**, 2236–2240.
17. Gordon, R., Bender, R., and Herman, G.T. (1970) Algebraic Reconstruction Techniques (ART) for three dimensional electron microscopy and xray photography. *J. Theor. Biol.*, **29**, 471–481.
18. Gilbert, P. (1972) Iterative methods for the three-dimensional reconstruction of an object from projections. *J. Theor. Biol.*, **36**, 105–117.
19. Spontak, R.J., Williams, M.C., and Agard, D.A. (1988) Three-dimensional study of cylindrical morphology in a styrene-butadiene-styrene block copolymer. *Polymer*, **29**, 387–395.
20. Koster, A.J., Ziese, U., Verkleij, A.J., Janssen, A.H., and de Jong, K.P. (2000) Three-dimensional transmission electron microscopy:∼ a novel imaging and characterization technique with nanometer scale resolution for materials science. *J Phys. Chem. B*, **104**, 9368–9370.
21. Spontak, R.J., Fung, J.C., Braunfeld, M.B., Sedat, J.W., Agard, D.A., Ashraf, A., and Smith, S.D. (1996) Architecture-induced phase immiscibility in a diblock/multiblock copolymer blend. *Macromolecules*, **29**, 2850–2856.
22. Midgley, P.A. and Weyland, M. (2003) 3D electron microscopy in the physical sciences: the development of Z-contrast and EFTEM tomography. *Ultramicroscopy*, **96**, 413–431.
23. Mobus, G. and Inkson, B.J. (2001) Three-dimensional reconstruction of buried nanoparticles by element-sensitive tomography based on inelastically scattered electrons. *Appl. Phys. Lett.*, **79**, 1369–1372.
24. Stegmann, H., Engelmann, H.H., and Zschech, E. (2003) Characterization of barrier/seed layer stacks of Cu interconnects by electron tomographic three-dimensional object reconstruction. *Microelectron. Eng.*, **65**, 171–183.
25. Weyland, M. and Midgley, P.A. (2003) Extending energy-filtered transmission electron microscopy (EFTEM) into three dimensions using electron tomography. *Microsc. Microanal.*, **9**, 542–555.
26. Mobus, G., Doole, R.C., and Inkson, B.J. (2003) Spectroscopic electron tomography. *Ultramicroscopy*, **96**, 433–451.
27. Deans, S.R. (1983) *The Radon Transform and Some of Its Applications*, John Wiley & Sons, Inc., New York, Chichester.
28. Vainshtein, B.K. (1970) *Sov. Phys. Crystallogr.*, **15**, 781.
29. Hawkes, P.W. (1992) The electron microscope as a structure projector, in

Electron Tomography: Three dimensional Imaging with the Transmission Electron Microscope (ed. J. Frank), Plenum Press, New York, London, 17–38.

30. Nyquist, H. (1928) Certain topics in telegraph transmission. *Trans. AIEE*, **47**, 617–644.

31. Gilbert, P.F.C. (1972) The reconstruction of a three-dimensional structure from projections and its application to electron microscopy. *Proc. R. Soc. Lond. B.*, **182**, 89–102.

32. Crowther, R.A. and Klug, A. (1971) ART and science, or, conditions for 3-D reconstruction from electron microscope images. *J. Theor. Biol.*, **32**, 199–203.

33. Bellman, S.H., Bender, R., Gordon, R., and Rowe, J.E. (1971) ART is science, being a defense of algebraic reconstruction techniques for three-dimensional electron microscopy. *J. Theor. Biol.*, **32**, 205–216.

34. Skoglund, U., Ofverstedt, L., Burnett, R.M., and Bricogne, G. (1996) Maximum-entropy three-dimensional reconstruction with deconvolution of the contrast transfer function: a test application with adenovirus. *J. Struct. Biol.*, **117**, 173–188.

35. Radermacher, M. and Hoppe, W. (1980) Properties of 3-D reconstructions from projections by conical tilting compared to single axis tilting. Proceedings of the 7th European Congress Electron Microscopy, Den Haag, 1980.

36. Radermacher, M. (1992) in *Electron Tomography: Three-dimensional Imaging with the Transmission Electron Microscope* (ed. J. Frank), Plenum Press, New York, London, p. 91.

37. Van Heel, M. and Harauz, G. (1986) Exact filters for general geometry three dimensional reconstruction. *Optik*, **73**, 146–156.

38. Frank, J., Verschoor, A., and Boublik, M. (1981) Computer averaging of electron micrographs of 40S ribosomal subunits. *Science*, **214**, 1353–1355.

39. Unser, M., Trus, B.L., and Steven, A.C. (1987) A new resolution criterion based on spectral signal-to-noise ratios. *Ultramicroscopy*, **23**, 39–51.

40. Henderson, R., Baldwin, J.M., Ceska, T.A., Zemlin, F., Beckmann, E., and Downing, K.H. (1990) Model for the structure of bacteriorhodopsin based on high-resolution electron cryo-microscopy. *J. Mol. Biol.*, **213**, 899–929.

41. Penczek, P.A. (2002) Three-dimensional spectral signal-to-noise ratio for a class of reconstruction algorithms. *J. Struct. Biol.*, **138**, 34–46.

42. Heidari Mezerji, H., Van den Broek, W., and Bals, S. (2011) A practical method to determine the effective resolution in incoherent experimental electron tomography. *Ultramicroscopy*, **111**, 330–336.

43. Tong, J., Arslan, I., and Midgley, P.A. (2006) A novel dual-axis iterative algorithm for electron tomography. *J. Struct. Biol.*, **153**, 55–63.

44. Arslan, I., Tong, J.R., and Midgley, P.A. (2006) Reducing the missing wedge: high-resolution dual axis tomography of inorganic materials. *Ultramicroscopy*, **106**, 994–1000.

45. Kawase, N., Kato, M., Nishioka, H., and Jinnai, H. (2007) Transmission electron microtomography without the "missing wedge" for quantitative structural analysis. *Ultramicroscopy*, **107**, 8–15.

46. Yaguchi, T., Konno, M., Kamino, T., and Watanabe, M. (2008) Observation of three-dimensional elemental distributions of a Si device using a 360°-tilt FIB and the cold field-emission STEM system. *Ultramicroscopy*, **108**, 1603–1615.

47. Jarausch, K., Thomas, P., Leonard, D.N., Twesten, R., and Booth, C.R. (2009) Four-dimensional STEM-EELS: enabling nano-scale chemical tomography. *Ultramicroscopy*, **109**, 326–337.

48. Ke, X., Bals, S., Cott, D., Hantshel, T., Bender, H., and Van Tendeloo, G. (2009) Three-dimensional analysis of carbon nanotube networks in interconnects by electron tomography without missing wedge artifacts. *Microsc. Microanal.*, **16**, 210–217.

49. Arslan, I., Marquis, E.A., Homer, M., Hekmaty, M.A., and Bartelt, N.C. (2008) Towards better 3D reconstructions by combining electron tomography and atom probe tomography. *Ultramicroscopy*, **108**, 1579–1585.

50. Frank J. (ed.) (2006) *Electron Tomography: Three-dimensional Imaging with the*

Transmission Electron Microscope, 2nd edn, Springer.

51. Porter, A.E., Gass, M.H., Muller, K., Skepper, J.N., Midgley, P.A., and Welland, M. (2007) Direct imaging of single-walled carbon nanotubes in cells. *Nat. Nanotechnol.*, **2**, 713–717.

52. Howie, A. (1979) Image contrast and localised signal selection techniques. *J. Microsc.*, **177**, 11.

53. Kirkland, E.J., Loane, R.F., and Silcox, J. (1987) Simulation of annular dark field stem images using a modified multislice method. *Ultramicroscopy*, **23**, 77–96.

54. Pennycook, S.J. and Nellist, P.D. (1999) in *Impact of Electron and Scanning Probe Microscopy on Materials Research* (ed. D.G. Rickerby), Kluwer, p. 161.

55. Dunin-Borkowski, R.E., Newcomb, S.B., Kasami, T., McCartney, M.R., Weyland, M., and Midgley, P.A. (2005) Conventional and back-side focused ion beam milling for off-axis electron holography of electrostatic potentials in transistors. *Ultramicroscopy*, **103**, 67–81.

56. Ercius, P., Weyland, M., Muller, D.A., and Gignac, L.M. (2006) Three-dimensional imaging of nanovoids in copper interconnects using incoherent bright field tomography. *Appl. Phys. Lett.*, **88**, 243116.

57. Ward, E.P.W., Yates, T.J.V., Fernández, J.-J., Vaughan, D.E.W., and Midgley, P.A. (2007) Three-dimensional nanoparticle distribution and local curvature of heterogeneous catalysts revealed by electron tomography. *J. Phys. Chem. C*, **111**, 11501–11505.

58. Beyers, E., Biermans, E., Ribbens, S., De Witte, K., Mertens, M., Meynen, V., Bals, S., Van Tendeloo, G., Vansant, E.F., and Cool, P. (2009) Combined TiO_2/SiO_2 mesoporous photocatalysts with location phase controllable TiO_2 nanoparticles. *Appl. Catal. B: Environ.*, **88**, 515–524.

59. Gonzalez, J.C., Hernandez, J.C., Lopez-Haro, M., del Rio, E., Delgado, J.J., Hungria, A.B., Trasobares, S., Bernal, S., Midgley, P.A., and Calvino, J.J. (2009) 3D characterization of gold nanoparticles supported on heavy metal oxide catalysts by STEM-HAADF electron tomography. *Angew. Chem. Int. Ed.*, **48**, 5313–5315.

60. Hungria, A.B., Juarez, B.H., Klinke, C., Weller, H., and Midgley, P.A. (2008) 3-D characterization of CdSe nanoparticles attached to carbon nanotubes. *Nano Res.*, **1**, 89–97.

61. Hungria, A.B., Eder, D., Windle, A.H., and Midgley, P.A. (2009) Visualization of the 3-dimensional microstructure of TiO_2 nanotubes by electron tomography. *Catal. Today*, **143**, 225–229.

62. Bals, S., Kisielowski, C., Croituru, M., and Van Tendeloo, G. (2005) Tomography using annular dark field imaging in TEM. *Microsc. Microanal.*, **11**, 2218–2219.

63. Bals, S., Van Tendeloo, G., and Kisielowski, C. (2006) A new approach for electron tomography: annular dark-field transmission electron microscopy. *Adv. Mater.*, **18**, 892–895.

64. Kaiser, U. and Chuvilin, A. (2003) Z-contrast imaging in a conventional TEM. *Microsc. Microanal.*, **9**, 78–79.

65. Rebled, J.M., Yedra, L., Estrade, S., Portillo, J., and Peiró, F. (2011) A new approach for 3D reconstruction from bright field TEM imaging: Beam precession assisted electron tomography. *Ultramicroscopy*, **111**, 1504–1511.

66. Bals, S., Kabius, B., Haider, M., Radmilovic, V., and Kisielowski, C. (2004) Annular dark field imaging in a TEM. *Solid State Commun.*, **130**, 675–680.

67. Hata, S., Kimura, K., Gao, H., Matsumura, S., Doi, M., Moritani, T., Barnard, J.S., Tong, J.R., Sharp, J.H., and Midgley, P.A. (2008) Electron tomography imaging and analysis of γ' and γ domains in Ni-based superalloys. *Adv. Mater.*, **20**, 1905–1909.

68. Barnard, J.S., Sharp, J., Tong, J.R., and Midgley, P.A. (2006) High-resolution three-dimensional imaging of dislocations. *Science*, **303**, 319.

69. Barnard, J.S., Eggeman, A.S., Sharp, J., White, T.A., and Midgley, P.A.

70. Twitchett, A.C., Dunin-Borkowski, R.E., and Midgley, P.A. (2002) Quantitative electron holography of biased semiconductor devices. *Phys. Rev. Lett.*, **88**, 238302.
71. Wolf, D., Lubk, A., Lichte, H., and Friedrich, H. (2010) Towards automated electron holographic tomography for 3D mapping of electrostatic potentials. *Ultramicroscopy*, **110**, 390–399.
72. Wolf, D. (2010) Elektronenholographische Tomographie zur 3D-Abbildung von elektrostatischen Potentialen in Nanostrukturen. PhD Thesis, University of Dresden, Germany.
73. Yaguchi, T., Konno, M., Kamino, T., Hashimoto, T., Ohnishi, T., and Watanabe, M. (2004) 3D elemental mapping using a dedicated FIB/STEM system. *Microsc. Microanal.*, **10**, 1030–1031.
74. Yamauchi, K., Takahashi, K., Hasegawa, H., Iatrou, H., Hadjichristidis, N., Kaneko, T., Nishikawa, Y., Jinnai, H., Matsui, T., Nishioka, H., Shimizu, M., and Fukukawa, H. (2003) Microdomain morphology in an ABC 3-miktoarm star terpolymer: a study by energy-filtering TEM and 3D electron tomography. *Macromolecules*, **36**, 6962–6966.
75. Leapman, R.D., Kocsis, E., Zhang, G., Talbot, T.L., and Laquerriere, P. (2004) Three-dimensional distributions of elements in biological samples by energy-filtered electron tomography. *Ultramicroscopy*, **100**, 115–125.
76. Weyland, M., Yates, T.J.V., Dunin-Borkowski, R.E., Laffont, L., and Midgley, P.A. (2006) Nanoscale analysis of three-dimensional structures by electron tomography. *Scr. Mater.*, **55**, 29–33.
77. Mobus, G. and Inkson, B.J. (2007) Nanoscale tomography in materials science. *Mater. Today*, **10**, 18–25.
78. Jin-Philipp, N.Y., Koch, C.T., and Van Aken, P.A. (2010) 3D elemental mapping in nanomaterials by core-loss EFTEM. *Microsc. Microanal.*, **16**, 1842–1843.
79. Goris, B., Bals, S., Van den Broek, W., Verbeeck, J., and Van Tendeloo, G. (2011) Exploring different inelastic projection mechanisms for electron tomography. *Ultramicroscopy*, **111**, 1262–1267.
80. Gass, M.H., Koziol, K.K.K., Windlem, A.H., and Midgley, P.A. (2006) Four-dimensional spectral tomography of carbonaceous nanocomposites. *Nano Lett.*, **6**, 376–379.
81. Yurtsever, A., Weyland, M., and Muller, D. (2006) 3-D imaging of non-spherical silicon nanoparticles embedded in silicon oxide by plasmon tomography. *Appl. Phys. Lett.*, **89**, 151920.
82. Batenburg, K.J., Bals, S., Sijbers, J., Kübel, C., Kaiser, U., Coronado, E.A., Midgley, P.A., Hernandez, J.C., and Van Tendeloo, G. (2009) 3D imaging of nanomaterials by discrete tomography. *Ultramicroscopy*, **109**, 730–740.
83. Bals, S., Batenburg, K.J., Liang, D., Lebedev, O., Van Tendeloo, G., Aerts, A., Martens, J., and Kirschhock, C. (2009) Quantitative three-dimensional modeling of zeotile through discrete electron tomography. *J. Am. Chem. Soc.*, **131**, 4769–4773.
84. Bals, S., Batenburg, K.J., Verbeeck, J., Sijbers, J., and Van Tendeloo, G. (2007) Quantitative three-dimensional reconstruction of catalyst particles for bamboo-like carbon nanotubes. *Nano Lett.*, **7**, 3669–3674.
85. Jinnai, H., Nishikawa, Y., Spontak, R.J., Smith, S.D., Agard, D.A., and Hashimoto, T. (2000) Direct measurement of interfacial curvature distributions in a bicontinuous block copolymer morphology. *Phys. Rev. Lett.*, **84**, 518–521.
86. Ikeda, Y., Katoh, A., Shimanuki, J., and Kohjiya, S. (2004) Nano-structural observation of in situ silica in natural rubber matrix by three dimensional transmission electron microscopy. *Macromol. Rapid Commun.*, **25**, 1186–1190.
87. Midgley, P.A., Thomas, J.M., Laffont, L., Weyland, M., Raja, R., Johnson, B.F.G., and Khimyak, T. (2004) High-resolution

scanning transmission electron tomography and elemental analysis of zeptogram quantities of heterogeneous catalyst. *J. Phys. Chem. B*, **108**, 4590–4592.
88. Biermans, E., Molina, L., Batenburg, K.J., Bals, S., and Van Tendeloo, G. (2010) Measuring porosity at the nanoscale by quantitative electron tomography. *Nano Lett.*, **10**, 5014–5019.
89. Bar Sadan, M., Houben, L., Wolf, S.G., Enyashin, A., Seifert, G., Tenne, R., and Urban, K. (2008) Bright-field electron tomography of inorganic fullerenes approaching atomic resolution. *Nano Lett.*, **8**, 891–896.
90. Saghi, Z., Xu, X., and Mobus, G. (2009) Model based atomic resolution tomography. *J. Appl. Phys.*, **106**, 024304.
91. Van Benthem, K., Lupini, A.R., Kim, M., Suck Baik, H., Doh, S., Lee, J.-H., Oxley, M.P., Findlay, S.D., Allen, L.J., Luck, J.T., and Pennycook, S.J. (2005) Three-dimensional imaging of individual hafnium atoms inside a semiconductor device. *Appl. Phys. Lett.*, **87**, 034104.
92. Xin, H.L. and Muller, D. (2009) Aberration-corrected ADF-STEM depth sectioning and prospects for reliable 3D imaging in S/TEM. *J. Electron Microsc.*, **58**, 157–165.
93. Jinschek, J.R., Batenburg, K.J., Calderon, H.A., Kilaas, R., Radmilovic, V.R., and Kisielowski, C. (2008) 3-D reconstruction of the atomic positions in a simulated gold nanocrystal based on discrete tomography: prospects of atomic resolution electron tomography. *Ultramicroscopy*, **108**, 589–604.
94. Van Aert, S., Batenburg, K.J., Rossell, M.D., Erni, R., and Van Tendeloo, G. (2010) Three-dimensional atomic imaging of crystalline nanoparticles. *Nature*, **470**, 374–377.
95. Bals, S., Casavola, M., Van Huis, M.A.,, Van Aert, S., Batenburg, K.J., Van Tendeloo, G., and Vanmaekelbergh, D. (2011) Three-dimensional atomic imaging of colloidal core-shell nanocrystals. *Nanoletters*, **11**, 3420–3424.

8
Statistical Parameter Estimation Theory – A Tool for Quantitative Electron Microscopy

Sandra Van Aert

8.1
Introduction

In materials science, the past decades are characterized by an evolution from macro- to micro- and, more recently, to nanotechnology. In nanotechnology, nanomaterials play an important role. The interesting properties of these materials are related to their structure. Therefore, one of the central issues in materials science is to understand the relations between the properties of a given material on the one hand and its structure on the other hand. A complete understanding of this relation, combined with the recent progress in building nanomaterials atom by atom, will enable materials science to evolve into materials design, that is, from describing and understanding toward predicting materials with interesting properties [1–5]. In order to understand the structure–properties relation, experimental and theoretical studies are needed [6–8]. Essentially, theoretical studies allow one to calculate the properties of materials with known structure, whereas experimental studies allow one to characterize materials in terms of structure. The combination of both approaches requires experimental characterization methods that can locally determine the unknown structure parameters with sufficient precision [3]. For example, a precision of the order of 0.01–0.1 Å is needed for the atom positions [9, 10].

The increasing need for precise determination of the atomic arrangement of nonperiodic structures in materials design and control of nanostructures explains the growing interest in quantitative analysis tools. Unlike qualitative materials characterization methods, which are based on a visual inspection of the observations, quantitative methods allow the extraction of local structure information at the subangstrom level [11]. Examples of observations in the field of electron microscopy are high-resolution transmission electron microscopy (HRTEM) images, electron energy loss spectra or high-angle annular dark field (HAADF) scanning transmission electron microscopy (STEM) images. The starting point of a quantitative analysis is the notion that one is not so much interested in the observations as such, but rather in the structure of the object under study. In other words, the observations are only regarded as a means toward this information. Throughout this chapter, it is shown that statistical parameter estimation theory provides useful

Handbook of Nanoscopy, First Edition. Edited by Gustaaf Van Tendeloo, Dirk Van Dyck, and Stephen J. Pennycook.
© 2012 Wiley-VCH Verlag GmbH & Co. KGaA. Published 2012 by Wiley-VCH Verlag GmbH & Co. KGaA.

methods to extract structure information with very high precision. When using this theory, observations are purely considered as data planes, from which structure parameters have to be determined as precisely as possible.

The purpose of this chapter is twofold. First, methods for the solution of a general type of parameter estimation problem often met in materials characterization or applied science and engineering in general is summarized. This will be the subject of Section 8.2. Second, applications of these methods in the field of transmission electron microscopy will be discussed in Section 8.3. In these applications, the goal is to determine unknown structure parameters, including atom positions, chemical concentrations, and atomic numbers, as precisely as possible from experimentally recorded images or spectra.

8.2
Methodology

8.2.1
Aim of Statistical Parameter Estimation Theory

In general, the aim of statistical parameter estimation theory is to determine, or more correctly, to estimate, unknown physical quantities or parameters on the basis of observations that are acquired experimentally [12]. In many scientific disciplines, observations are usually not the quantities to be measured themselves but are related to the quantities of interest. Often, this relation is a known mathematical function derived from physical laws. The quantities to be determined are parameters of this function. Parameter estimation then is the computation of numerical values for the parameters from the available observations. For example, if electron microscopy observations are made of a specific object, this function describes the electron–object interaction, the transfer of the electrons through the microscope, and the image detection. The parameters are the atom positions and atom types. The parameter estimation problem then becomes: computing the atom positions and atom types from the observations. As any experimenter will readily admit, observations contain "noise." This means that if a particular experiment is repeated under the same conditions, the resulting observations will fluctuate from experiment to experiment, and hence the estimated parameters, which are computed on the basis of these observations, will be different. The amount of variation in these estimated parameters quantifies the *precision* [13]. Moreover, modeling errors in the parameterized mathematical function may affect the outcome as well. Indeed, an incorrect or incomplete function will limit the *accuracy*, resulting in bias, that is, a systematic deviation of the estimated parameters from the true parameters. Ultimately, the unknown parameters need to be estimated as accurately and precisely as possible from the available observations. In the following sections, a framework will be outlined to reach this goal, including a model to describe statistical fluctuations in the observations, the limits to precision, the derivation of estimation procedures, model assessment, and the construction of confidence regions and intervals. For a detailed overview of statistical parameter estimation theory, the reader is referred to [11–17].

8.2.2
Parametric Statistical Model of the Observations

Experimental observations are prone to noise. Hence, the description of observations by means of a mathematical function alone is incomplete, since it does not account for these random fluctuations. An effective way to describe this behavior is by means of statistics. This implies that observations are modeled as stochastic variables. The values set by the mathematical function define the expectations of these stochastic variables. In a sense, these expectations would be the observations in the absence of noise. The fluctuation is the deviation of an observation from its expectation. In general, this description of observations as stochastic variables requires detailed knowledge about the disturbances that are acting on the observations [12]. However, as will be shown in Section 8.2.5, incorporating this prior information in the estimation procedure may significantly enhance its performance.

The observations are generally subject to statistical fluctuations. The usual way to describe the fluctuating behavior in the presence of noise is modeling the observations, as stochastic variables. Consider a set of observations $\{w_n, n = 1, \ldots, N\}$ and let w define the $N \times 1$ vector of observations:

$$w = (w_1, \ldots, w_N)^T \tag{8.1}$$

where the superscript T denotes transposition. By definition, a stochastic variable is characterized by its probability density function (distribution). If the observations are counting results, they are nonnegative integers and the probability density function defines the probability of occurrence of each of these integer outcomes. If the observations are continuous, the probability density function describes the probability of occurrence of an observation on a particular interval [13]. A set of observations is defined by its joint probability density function. The expectation value (mean value) of each observation, $\mathbb{E}[w_n]$, is defined by its probability density function. In statistical parameter estimation theory, it is assumed that these expectation values $\mathbb{E}[w_n]$ can be described by a functional model $f_n(\tau)$, which is parametric in the quantities τ to be measured, such as the locations of atom columns or atom column types:

$$\mathbb{E}[w_n] = f_n(\tau) \tag{8.2}$$

with $\tau = (\tau_1, \ldots, \tau_R)^T$ the $R \times 1$ vector of unknown parameters to be measured. The availability of such a model makes it possible to parameterize the probability density function of the observations, which is of vital importance to precise quantitative structure determination. An example is given below.

Example 8.1.

Let us consider statistically independent Poisson-distributed observations $w_n, n = 1, \ldots, N$. For example, these observations could be electron counting results detected using a CCD camera with a quantum efficiency equal to 1. The probability that the observation w_n is equal to ω_n is then equal to

$$\frac{(\lambda_n)^{\omega_n}}{\omega_n!} \exp(-\lambda_n) \tag{8.3}$$

where the parameter λ_n is equal to the expectation $\mathbb{E}[w_n]$. Furthermore, since the observations are independent, the probability $p(\omega)$ of a set of observations $\omega = (\omega_1, \ldots, \omega_N)^T$ is the product of all probabilities described by Eq. (8.3):

$$p(\omega) = \prod_{n=1}^{N} \frac{(\lambda_n)^{\omega_n}}{\omega_n!} \exp(-\lambda_n). \tag{8.4}$$

Next, suppose that the expectations λ_n can be described by a functional model $f_n(\tau)$:

$$\mathbb{E}[w_n] \equiv \lambda_n = f_n(\tau) \tag{8.5}$$

with $\tau = (\tau_1, \ldots, \tau_R)^T$ being the $R \times 1$ vector of unknown parameters to be measured. For example, in quantitative HRTEM, the expectation model of a crystal in zone-axis orientation can be described as

$$f_n(\tau) = f_{kl}(\tau) = \frac{N}{I_{norm}} |\psi(\mathbf{r}_{kl}; \tau) * t(\mathbf{r}_{kl}; \varepsilon, C_s)|^2 \tag{8.6}$$

with N being the total number of detected electrons in an image, $t(\mathbf{r}; \varepsilon, C_s)$ the point spread function of the electron microscope depending on microscope settings such as the defocus ε and the spherical aberration constant C_s, $\mathbf{r}_{kl} = (x_k \, y_l)^T$ the position of the pixel (k, l), and I_{norm} a normalization factor so that the integral of the function $|\psi(\mathbf{r}_{kl}; \tau) * t(\mathbf{r}_{kl}; \varepsilon, C_s)|^2 / I_{norm}$ is equal to 1. The parameterized function $\psi(\mathbf{r}_{kl}; \tau)$ represents the exit wave, which is formed directly beneath the specimen. The simplified channeling theory provides an expression for the exit wave [18, 19]:

$$\psi(\mathbf{r}; \tau) = 1 + \sum_{n=1}^{n_c} c_n \phi_{1s,n}(\mathbf{r} - \boldsymbol{\beta}_n) \left[\exp\left(-i\pi \frac{E_{1s,n}}{E_0} \frac{1}{\lambda} z\right) - 1 \right] \tag{8.7}$$

with z the object thickness, E_0 the incident electron energy, λ the electron wavelength, and n_c the total number of atom columns. The function $\phi_{1s,n}(\mathbf{r} - \boldsymbol{\beta}_n)$ is the lowest energy bound state of the nth atom column located at position $\boldsymbol{\beta}_n = (\beta_{xn} \, \beta_{yn})^T$, $E_{1s,n}$ is its energy, and c_n its excitation coefficient. The unknown parameters, represented by the parameter vector $\tau = (\beta_{x1}, \ldots, \beta_{xn_c} \, \beta_{y1}, \ldots, \beta_{yn_c})^T$ have to be estimated from the observations, in this case, an image. The quantities $w_n - \lambda_n$ are the so-called nonsystematic errors in the observations. If the expectation model (8.6) is correct, their expectations are equal to zero.

Example (8.1) illustrates that the expectation of the observations is an accurate description of what experimenters often call the model underlying the observations. In quantitative electron microscopy, this model corresponds to the model that is fitted to the images or spectra in a refinement procedure. It includes all ingredients needed to perform a computer simulation of this image or spectrum. In general, different estimation procedures, called *estimators*, can be used to estimate the unknown parameters of this model, such as least squares, least absolute values, or maximum likelihood (ML) estimators. Different estimators will have different properties, among which precision and accuracy are the most important ones. They will be discussed in Section 8.2.3. Furthermore, substitution of λ_n, as described by Eq. (8.6) in Eq. (8.4), shows how the probability density function depends on the unknown parameters to be measured. Probability density functions thus parameterized can be used for two purposes [13]. First, they may be used to define the concept of *Fisher information* and to compute the *Cramér–Rao lower bound* (CRLB), which is a lower bound on the variance of any unbiased estimator of the parameters. This is the subject of Section 8.2.4. Second, from the probability density function of the observations, the *ML estimator* of the parameters may be derived, which is the subject of Section 8.2.5.

8.2.3
Properties of Estimators

The performance of an estimator is often characterized by means of its accuracy and precision [12]. Accuracy concerns a systematic deviation of the estimated parameters from the true parameters. Suppose that

$$\widehat{\tau} = (\widehat{\tau}_1, \ldots, \widehat{\tau}_R)^T \tag{8.8}$$

is an estimator of the vector of parameters

$$\tau = (\tau_1, \ldots, \tau_R)^T \tag{8.9}$$

In general, an estimator is a function of the observations that are used to compute the parameters. Thus, every element $\widehat{\tau}_r$ is, like the observations, a stochastic variable. The accuracy of an estimator can be quantified by means of its *bias*, which is defined as the deviation of the expectation of the estimator from the hypothetical true value of the parameter. The bias of the estimator $\widehat{\tau}_r$ is thus defined as

$$\text{bias}(\widehat{\tau}_r) = \mathbb{E}[\widehat{\tau}_r] - \tau_r. \tag{8.10}$$

There may be different sources of bias. It may result from the choice of estimation procedure or from modeling errors. The first type of bias may require the use of a different estimator or may be cured by increasing the number of observations. It may often be detected by means of numerically simulated observations. The second type of bias is caused by inadequacies in the parametric model used to represent the expectations of the observations. An example is the absence of the point spread function of the electron microscope in the model used in Example (8.1). It is clear that modeling errors may corrupt the results of any estimator used. In Section 8.2.6, a discussion of the methods to test the validity of the parametric model is presented.

Precision concerns the spread of the estimated parameters among their mean values when the experiment is repeated under the same conditions. This is a result of the unavoidable presence of noise in the observations. If a particular experiment is repeated, the resulting observations will be different and therefore also the estimated parameters that are computed on the basis of these observations. The amount of variation in these estimated parameters quantifies the precision. The precision of an estimator can therefore be quantified by means of its *variance*, or equivalently, by means of its *standard deviation* being the square root of the variance. The variance of the rth element of $\hat{\tau}$ is defined as

$$\operatorname{var}(\hat{\tau}_r) = \mathbb{E}\left[(\hat{\tau}_r - \mathbb{E}[\hat{\tau}_r])^2\right] \tag{8.11}$$

Generally, one is interested in knowing the deviation of an estimator $\hat{\tau}_r$ from the true value τ_r of the parameter. This leads to the definition of the *mean squared error*:

$$\operatorname{mse}(\hat{\tau}_r) = \mathbb{E}\left[(\hat{\tau}_r - \tau_r)^2\right] \tag{8.12}$$

It can be shown that

$$\operatorname{mse}(\hat{\tau}_r) = \operatorname{var}(\hat{\tau}_r) + \operatorname{bias}^2(\hat{\tau}_r) \tag{8.13}$$

expressing that the mean squared error equals the sum of the variance and the square of the bias.

8.2.4
Attainable Precision

Different estimators of the same parameters from the same observations generally have different precisions. The question then arises what precision may be achieved ultimately from a particular set of observations. For the class of unbiased estimators (bias equals zero), this answer is given in the form of a lower bound on their variance, the so-called Cramér–Rao lower bound [13, 20–22].

Let $p(w; \tau)$ be the joint probability density function of a set of observations $\{w_n, n = 1, \ldots, N\}$. Example (8.1) makes clear how the dependence of $p(w; \tau)$ on the parameters τ to be measured can be established. This dependence can now be used to define the so-called *Fisher information matrix*. This Fisher information matrix, with respect to the elements of the $R \times 1$ parameter vector $\tau = (\tau_1, \ldots, \tau_R)^T$, is defined as the $R \times R$ matrix

$$F = -\mathbb{E}\left[\frac{\partial^2 \ln p(w; \tau)}{\partial \tau \, \partial \tau^T}\right] \tag{8.14}$$

where $p(w; \tau)$ is the joint probability density function of the observations $w = (w_1, \ldots, w_N)^T$. The expression between square brackets represents the Hessian matrix of $\ln P$, for which the (r, s)th element is defined by $\partial^2 \ln p(w; \tau)/\partial \tau_r \partial \tau_s$. The Fisher information is a measure of the physical fluctuations in the observations. In a sense, the Fisher information specifies the quality of the measurements. It expresses the 'inability to know' a measured quantity [22]. Indeed, use of the concept of Fisher information allows one to determine the highest precision, that is, the lowest variance, with which a parameter can be estimated unbiasedly. Suppose that $\hat{\tau}$ is any unbiased estimator of τ, that is, $\mathbb{E}[\hat{\tau}] = \tau$. Then, it can be shown that under general conditions, the covariance matrix $cov(\hat{\tau})$ of $\hat{\tau}$ satisfies [14]

$$\operatorname{cov}(\hat{\tau}) \geq F^{-1} \tag{8.15}$$

This inequality expresses that the difference between the left-hand and right-hand member is positive semidefinite. A property of a positive semidefinite matrix is that its diagonal elements cannot be negative. This means that the diagonal elements of cov $(\hat{\tau})$, that is, the variances of $\hat{\tau}_1, \ldots, \hat{\tau}_R$ are larger than or equal to the corresponding diagonal elements of F^{-1}:

$$\text{var}(\hat{\tau}_r) \geq \left[F^{-1}\right]_{rr} \tag{8.16}$$

where $r = 1, \ldots, T$ and $[F^{-1}]_{rr}$ is the (r, r)th element of the inverse of the Fisher information matrix. In this sense, F^{-1} represents a lower bound to the variances of all unbiased $\hat{\tau}$. The matrix F^{-1} is called the Cramér–Rao lower bound (CRLB) on the variance of $\hat{\tau}$. The derivation of the CRLB is illustrated in Example (8.2) [11].

Example 8.2.

(Poisson-distributed observations) The joint probability density function of a set of independent Poisson-distributed observations $\{w_n, n = 1, \ldots, N\}$ is given by

$$p(\omega) = \prod_n \frac{(\lambda_n)^{\omega_n}}{\omega_n!} \exp(-\lambda_n) \tag{8.17}$$

with $\omega = (\omega_1, \ldots, \omega_N)^T$. The logarithm of Eq. (8.17) is then described by

$$\ln p(\omega) = \sum_n \left(-\lambda_n + \omega_n \ln \lambda_n - \ln \omega_n!\right) \tag{8.18}$$

Suppose that the expectation values $\mathbb{E}[w_n] \equiv \lambda_n$ can be described by a functional model, which is parametric in the parameters to be measured: $\mathbb{E}[w_n] = f_n(\tau)$, with $\tau = (\tau_1, \ldots, \tau_R)^T$ the vector of parameters. Then, it follows from Eqs. (8.5), (8.14), and (8.18) that the elements of the $R \times R$ Fisher information matrix can be described as

$$\begin{aligned}
F_{rl} &= -\mathbb{E}\left[\sum_n \frac{\partial^2}{\partial \tau_r \partial \tau_l}\left(-f_n(\tau) + w_n \ln f_n(\tau) - \ln w_n!\right)\right] \\
&= -\mathbb{E}\left[\sum_n \left(\frac{-\partial^2 f_n(\tau)}{\partial \tau_r \partial \tau_l} + w_n \left(-\frac{1}{f_n^2(\tau)} \frac{\partial f_n(\tau)}{\partial \tau_r} \frac{\partial f_n(\tau)}{\partial \tau_l} + \frac{1}{f_n(\tau)} \frac{\partial^2 f_n(\tau)}{\partial \tau_r \partial \tau_l}\right)\right)\right] \\
&= \sum_n \left(\frac{\partial^2 f_n(\tau)}{\partial \tau_r \partial \tau_l} - \mathbb{E}[w_n]\left(-\frac{1}{f_n^2(\tau)} \frac{\partial f_n(\tau)}{\partial \tau_r} \frac{\partial f_n(\tau)}{\partial \tau_l} + \frac{1}{f_n(\tau)} \frac{\partial^2 f_n(\tau)}{\partial \tau_r \partial \tau_l}\right)\right) \\
&= \sum_n \frac{1}{f_n(\tau)} \frac{\partial f_n(\tau)}{\partial \tau_r} \frac{\partial f_n(\tau)}{\partial \tau_l}
\end{aligned} \tag{8.19}$$

The CRLB can now be found by taking the inverse of the Fisher information matrix.

Notice that for a scalar valued parameter τ, it follows from Eqs. (8.15) and (8.19) that

$$\text{var}(\hat{\tau}) \geq \frac{1}{\sum_n \frac{1}{f_n(\tau)} \left(\frac{\partial f_n(\tau)}{\partial \tau}\right)^2} \tag{8.20}$$

Notice that the CRLB is not related to a particular estimation method. It depends on the statistical properties of the observations, the measurement points, and, in most cases, the hypothetical true values of the parameters. This dependence on the true values looks, at first sight, to be a serious impediment to the practical use of the bound. However, the expressions for the bound provide the experimenter with the means to compute numerical values for it using nominal values of the parameters. This provides the experimenter with quantitative insight as to what precision might be achieved from the available observations. In addition, it provides an insight into the sensitivity of the precision of the parameter values [15].

8.2.5
Maximum Likelihood Estimation

As demonstrated above, every set of observations contains a certain amount of Fisher information. It is up to the experimenter to extract this information from the observations. For this purpose, different estimators may be used. Generally, different estimators have different precisions. The most precise estimator is the one that attains the CRLB, that is, the one that extracts all Fisher information [11]. It is known that there exists an estimator that achieves the CLRB at least asymptotically, that is, for an increasing number of observations. This estimator is the ML estimator. It is easily derived and has a number of favorable statistical properties [17].

Derivation of the Maximum Likelihood Estimator The derivation of the ML estimator, which requires the probability density function of the observations and its dependence on the unknown parameters to be known, consists of the following steps [13]:

- First, the available observations $w = (w_1, \ldots, w_N)^T$ are substituted for the corresponding variables $\omega = (\omega_1, \ldots, \omega_N)^T$ in the probability density function of the observations. Since the observations are numbers, the resulting expression only depends on the elements of $\tau = (\tau_1, \ldots, \tau_R)^T$.
- Next, the elements of $\tau = (\tau_1, \ldots, \tau_R)^T$, which are the hypothetical true parameters, are considered to be variables. To express this, they are replaced by $t = (t_1, \ldots, t_R)^T$. The resulting function $p(t)$ is called the *likelihood function* of the parameters of t for the observations w.
- Finally, the ML estimates $\hat{\tau}_{ML}$ of the parameters τ are computed. These are defined as the values of the elements of t that maximize $p(t)$:

$$\hat{\tau}_{ML} = \arg\max_t p(t) \qquad (8.21)$$

or, equivalently,

$$\hat{\tau}_{ML} = \arg\max_t \ln p(t) \qquad (8.22)$$

in which $\ln p(t)$ is called the *log-likelihood function*.

Example 8.3.

(Poisson-distributed observations) Suppose that a set of independent Poisson-distributed observations $\{w_n, n = 1, \ldots, N\}$ is available. Their parameterized joint probability density function is described by substituting Eq. (8.5) in Eq. (8.4). Substituting the available observations $w = (w_1, \ldots, w_N)^T$ for the corresponding variables $\omega = (\omega_1, \ldots, \omega_N)^T$ and replacing the true parameters $\tau = (\tau_1, \ldots, \tau_R)^T$ by $t = (t_1, \ldots, t_R)^T$ yields the likelihood function [11]:

$$p(t) = \prod_n \frac{(f_n(t))^{w_n}}{w_n!} \exp\left(-f_n(t)\right) \tag{8.23}$$

Taking the logarithm yields:

$$\ln p(t) = \sum_n \left[-f_n(t) + w_n \ln f_n(t) - \ln w_n!\right] \tag{8.24}$$

The ML estimates $\hat{\tau}_{ML}$ of the parameters τ are then given by

$$\hat{\tau}_{ML} = \arg\max_t \sum_n \left[-f_n(t) + w_n \ln f_n(t)\right] \tag{8.25}$$

Least Squares Estimation as a Special Case of Maximum Likelihood Estimation
For independent normally distributed observations $\{w_n, n = 1, \ldots, N\}$ with equal variance, the joint probability density function is given by

$$p(\omega) = \prod_n \frac{1}{\sqrt{2\pi}\sigma} \exp\left(-\frac{1}{2}\left(\frac{\omega_n - f_n(\tau)}{\sigma}\right)^2\right) \tag{8.26}$$

The likelihood function can now be obtained by substituting the available observations $w = (w_1, \ldots, w_N)^T$ for the variables $\omega = (\omega_1, \ldots, \omega_N)^T$ and replacing the true parameters $\tau = (\tau_1, \ldots, \tau_R)^T$ by $t = (t_1, \ldots, t_R)^T$. The logarithm of the likelihood function is then given by

$$\ln p(t) = -N \ln\left(\sqrt{2\pi}\sigma\right) - \frac{1}{2\sigma^2} \sum_n \left(w_n - f_n(t)\right)^2 \tag{8.27}$$

The ML estimates $\hat{\tau}_{ML}$ of the parameters τ are then given by

$$\hat{\tau}_{LS} = \arg\min_t \sum_n \left(w_n - f_n(t)\right)^2 \tag{8.28}$$

with $f_n(\tau)$ the expectation model. This example thus shows that the ML estimator is equal to the well-known *uniformly weighted least squares estimator* for independent normally distributed observations with equal variance [11].

Properties of the Maximum Likelihood Estimator The ML estimator has a number of favorable statistical properties [11, 17]. First, it can be shown that this estimator achieves the CRLB asymptotically, that is, for an infinite number of observations. Therefore, it is asymptotically most precise. Second, it can be shown that the

ML estimator is consistent, which means that it converges to the true value of the parameter in a statistically well-defined way if the number of observations increases. Third, the ML estimator is asymptotically normally distributed, with a mean equal to the true value of the parameter and a variance–covariance matrix equal to the CRLB. In electron microscopy, the number of observations is generally large enough for these asymptotic properties to apply. This can be assessed by estimating from artificial, simulated observations. Given the advantages of the ML estimator, it is often used in practice.

8.2.5.1 The Need for a Good Starting Model

The relative ease with which the likelihood function can be derived does not mean that its maximum can be found easily [11]. Finding the ML estimate corresponds to finding the global maximum in an R-dimensional parameter space, where R is the number of parameters to be estimated. In quantitative structure determination, each possible combination of the R parameters corresponds directly to an atomic structure. It is represented by a point in an R-dimensional parameter space. The search for the maximum of the likelihood function is usually an iterative numerical procedure. In electron microscopy applications, the dimension of the parameter space is usually very high. Consequently, it is quite possible that the optimization procedure ends up at a local maximum instead of at the global maximum of the likelihood function, so that the wrong structure model is suggested, which introduces a systematic error (bias). To solve this dimensionality problem, that is, to find a pathway to the maximum in the parameter space, good starting values for the parameters are required. In other words, the structure has to be resolved. This corresponds to X-ray crystallography, where one has to first resolve the structure and afterward refine it. Resolving the structure is not trivial. It is known that the details in HRTEM images do not necessarily correspond to features in the atomic structure. This is not only due to the unavoidable presence of noise but also to the dynamic scattering of the electrons on their way through the object, and the image formation in the electron microscope, both of which have a blurring effect. As a consequence, the structure information of the object may be strongly delocalized, which makes it very difficult to find good starting values for the structure parameters. However, it has been shown that starting values for the parameters, and hence a good starting structure of the object, may be found by using so-called direct methods. Direct methods, in a sense, invert the imaging process and the dynamic scattering process using some prior knowledge which is generally valid, irrespective of the (unknown) structure parameters of the object. The starting structure obtained with such a direct method can be obtained in different ways [11]. Examples of such methods are high-voltage electron microscopy, aberration correction in the electron microscope, HAADF STEM, focal-series reconstruction, and off-axis holography. A common goal of these methods is to improve the interpretability of the experimental images in terms of the structure and may as such yield an approximate structure model that can be used as a starting structure for ML estimation from the original images.

8.2.6
Model Assessment

As mentioned above, ML estimation requires the probability density function of the observations and its dependence on the unknown parameters to be known. It has been shown that this dependence is often established by the availability of a parametric model that describes the expectations of the observations. Substitution of this so-called *expectation model* in the expression for the probability density function provides a parameterized probability density function from which the ML estimator can be derived. Obviously, such an estimation procedure is valid if and only if the expectation model is an accurate description of the true expectations [11, 13]. As discussed earlier, an inaccurate expectation model may indeed lead to biased estimates. Therefore, it is important to test the validity of the expectation model before attaching confidence to the parameter estimates obtained. If the model is inadequate, the estimated parameters could be biased. Therefore, it must be modified and the analysis continued until a satisfactory result is obtained [23]. Many statistical model assessment methods are based on specific distributional assumptions. In particular, the assumption that the deviations of the observations from their expectations (i.e., the noise contributions) are independent, zero mean normally distributed with common variance, is common in classical statistical tests. The normality assumption is often justified by appealing to the central limit theorem, which states that the resultant of many disturbances, none of which is dominant, will tend to be normally distributed [23]. In [11], some model assessment methods based on the normality assumption have been reviewed.

A very effective test that is applicable to any probability density function of the observations is the so-called likelihood ratio test. This test is discussed in detail in den Dekker *et al.* [11]. The aim of this test is to conclude whether there is reason to reject the expectation model or not. In comparison to other tests, the likelihood ratio test has the advantage that replications of the observations are not required. Here, the likelihood ratio test will be presented for Poisson-distributed observations [11].

Example 8.4.

(Likelihood ratio test for Poisson-distributed observations) Suppose that a set of Poisson-distributed observations $\{w_n, n = 1, \ldots, N\}$ is available and that we wish to test the so-called null hypothesis (H_0) that the expectations of these observations can be described by the expectation model $f_n(\tau)$ against the alternative hypothesis (H_1) that the expectations of these observations cannot be described by $f_n(\tau)$. It can be shown that, when H_0 is true, the statistic

$$\text{LR} = 2 \sum_{n=1}^{N} \left(-w_n + w_n \ln \frac{w_n}{f_n(\widehat{\tau}_{\text{ML}})} + f_n(\widehat{\tau}_{\text{ML}}) \right) \tag{8.29}$$

is approximately chi-square distributed with $N - R$ degrees of freedom with R the number of unknown parameters. Furthermore, $f_n(\widehat{\tau}_{\text{ML}})$ in Eq. (8.29) represents the expectation model evaluated at the ML estimates $\widehat{\tau}_{\text{ML}}$. The likelihood ratio test uses LR as the test statistic. The hypothesis H_0 is accepted at the significance level α (which can be freely chosen) if:

$$\text{LR} \leq \chi^2_{N-R,1-\alpha} \tag{8.30}$$

or, alternatively,

$$\chi^2_{N-R,\alpha/2} \leq \text{LR} \leq \chi^2_{N-R,1-\alpha/2} \tag{8.31}$$

with $\chi^2_{N-R,q}$ the qth quantile of a chi-square distribution with $N-R$ degrees of freedom. The meaning of the significance level α is that if H_0 is true, the probability of rejecting H_0 is α and the probability of accepting H_0, making the correct decision, is $1 - \alpha$.

In Section 8.3.3, this likelihood ratio test will be applied to test the expectation model used in a quantitative analysis of electron energy loss spectra.

8.2.7
Confidence Regions and Intervals

In order to evaluate the level of confidence to be attached to the obtained ML estimates, confidence regions and intervals associated with these estimates are required. A summary of existing methods to compute such regions and intervals is presented in [11]. One of these methods is based on the asymptotic normality of the ML estimator. It is preferred especially if the experiment cannot be replicated. For example, an approximate $100(1-\alpha)\%$ confidence interval for the rth element of τ, τ_r, is given by

$$[\widehat{\tau}_{ML}]_r \pm \lambda_{(1-\alpha/2)} \sqrt{\left[F^{-1}\right]_{rr}} \tag{8.32}$$

with $[\widehat{\tau}_{ML}]_r$ the rth element of the parameter vector $\widehat{\tau}_{ML}$, $\left[F^{-1}\right]_{rr}$ the (r,r)th element of F^{-1} and $\lambda_{(1-\alpha/2)}$ the $(1-\alpha/2)$ quantile of the standard normal distribution. The meaning of a $100(1-\alpha)\%$ confidence interval is that it covers the true element τ_r of τ with probability $1-\alpha$. Usually, F^{-1} is a function of the parameters to be estimated. The confidence intervals (8.32) are then derived by using approximations of F^{-1}. A useful approximation $\widehat{F}^{-1}$ of F^{-1} may be obtained by substituting $\widehat{\tau}_{ML}$ for τ in the expression for the CRLB, yielding:

$$\widehat{F}^{-1} = F^{-1}\big|_{\tau=\widehat{\tau}_{ML}} \tag{8.33}$$

8.3
Electron Microscopy Applications

In the following sections, applications of statistical parameter estimation theory are outlined. The notions resolution and precision are discussed in 8.3.1. In 8.3.2, 8.3.3, and 8.3.4, applications to reconstructed exit waves, electron energy loss spectra, and HAADF STEM images are shown, respectively. Finally, in 8.3.5, its use to optimize the design of electron microscopy experiments is discussed.

8.3.1
Resolution versus Precision

The performance of electron microscopy experiments is usually expressed in terms of two-point resolution. One of the earliest and most famous criteria for two-point resolution is that of Rayleigh. This criterion defines the possibility of perceiving separately two point sources in an image formed by a diffraction-limited imaging system or objective lens. According to Rayleigh, two point sources of equal brightness are just resolvable when the maximum of the intensity pattern produced by the first point source falls on the first zero of the intensity pattern produced by the second point source [24]. Consequently, the Rayleigh resolution limit is given by the distance between the central maximum and the first zero of the diffraction-limited intensity point spread function of the imaging system concerned. Rayleigh's choice of resolution limit is based on presumed resolving capabilities of the human visual system when it is used to detect differences in intensity at various points of the composite intensity distribution. Since Rayleigh's days, several other resolution criteria similar to Rayleigh's have been proposed. These criteria include the well-known point resolution often used in electron microscopy [25].

In fact, classical resolution criteria do not concern detected images but calculated images, that is, noise-free images exactly describable by a known parametric two-component mathematical model. The shape of the component function is assumed to be exactly known. However, classical resolution criteria disregard the possibility of using this a priori knowledge [26]. Obviously, in the absence of noise, numerically fitting the known two-component model to the images with respect to the unknown parameters, for example, the positions of the two components, would result in a perfect fit. The resulting solution for the positions and, therefore, for the distance would be exact, and no limit to resolution would exist, no matter how closely spaced the two components. In reality, however, calculated images do not occur. Instead, one must deal with detected images [27]. In this case, the model will never be known exactly and systematic errors are introduced. Furthermore, the images will never be noise-free, so that nonsystematic, or statistical, errors will also be present. These are the differences of each observation and its expectation. As a result of these nonsystematic errors, the estimated positions will fluctuate from experiment to experiment. The amount of variation in these estimated positions quantifies the precision. As shown in Section 8.2.4, the attainable precision can be quantified using the so-called Cramér–Rao lower bound.

The difference between resolution and precision can be illustrated as follows [28–30]. Suppose that the microscope is able to visualize individual atoms or atom columns so that the structure can be resolved. In other words, neighboring atoms or atom columns can be discriminated in the images. Furthermore, it will be assumed that the expectations of the Poisson-distributed image pixel values can be modeled as a summation of Gaussian peaks centered at the atom or atom column positions. The position coordinates of these atoms or atom columns are the unknown parameters in the expectation model. It then follows that the precision, expressed in terms of the lower bound on the standard deviation of these coordinate estimates, that is, the square root of the CRLB, is approximately equal to [29, 30]:

$$\sigma_{CR} \approx \frac{\rho}{\sqrt{N}} \tag{8.34}$$

where ρ represents the width of the Gaussian peaks and N represents the number of detected electrons per atom or atom column. The width of the peaks can be shown to be proportional to the Rayleigh resolution [31]. It is now clear that it is not only the resolution that matters but the electron dose as well. If the resolution can only be improved at the expense of a decrease in the number of detected electrons, both effects have to be balanced under the existing physical constraints so as to produce the highest precision [32, 33]. Section 8.3.5, further discusses how to optimize the microscope settings in terms of the attainable precision with which unknown structure parameters can be estimated.

8.3.2
Atom Column Position Measurement

HRTEM has evolved toward an established method to determine the internal atomic structure of materials at a local scale [6]. Over the years, different so-called *direct methods* have been developed, improving the direct interpretability of the images down to a resolution of the order of 0.5 Å. Examples of such methods are aberration-corrected transmission electron microscopy [34–36] and exit wave reconstruction [37]. These methods are often used to measure shifts in the atom positions. Although they greatly enhance the visual interpretability of the images in terms of the atomic structure, a combination with quantitative methods is required in order to measure atom column positions with a precision of the order of a few picometers [38–45]. As an example, we focus here on the measurement of the atom column positions of a compound, $Bi_4 W_{2/3} Mn_{1/3} O_8 Cl$, with a precision in the picometer range [41]. Therefore, exit wave reconstruction combined with statistical parameter estimation has been used. Figure 8.1a shows a part of the phase of the reconstructed exit wave. This phase reveals the light oxygen atom columns in the presence of heavier atom columns. Although the phase of the exit wave is

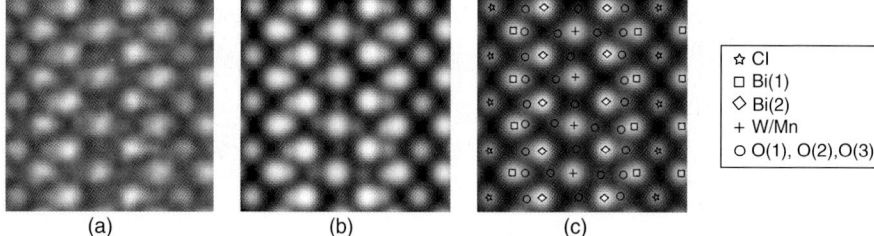

Figure 8.1 (a) Phase of the exit wave. (b) Model evaluated at the estimated parameters. (c) Overlay indicating the estimated positions of the different atom column types [41].

often considered as the final result, it has been used here as a starting point for quantitative refinement of the atom column positions. Therefore, a model has been proposed to describe the expectations of the pixel values in the reconstructed phase. The unknown parameters of this model, including the atom column positions, have been estimated in the least squares sense. Figure 8.1b,c shows the model evaluated at the estimated parameters. Next, interatomic distances have been computed from the estimated atom column positions. From sets of equivalent distances, mean interatomic distances and their corresponding standard deviations have been calculated. The standard deviation, being a measure of the precision, ranges from 0.03 Å to 0.1 Å. Furthermore, a good agreement has been found when comparing these results with X-ray powder diffraction data. It is also found that projected distances beyond the information limit of the microscope (1.1 Å) can be determined with picometer range precision [41].

In order to evaluate the limitations of the proposed method when applied to reconstructed exit waves, simulation studies have been carried out [46]. More specifically, the effects of modeling errors, noise, the reconstruction method, and amorphous layers have been quantified. The presence of amorphous layers, which is hard to avoid under realistic experimental conditions, has been found to be the most important factor limiting the reliability of atom column position estimates. Figure 8.2a,b shows the phase of a simulated exit wave of $SrTiO_3$ [001] without amorphous layers and with amorphous layers both on the top and bottom of the sample, respectively. The symbols indicate the estimated atom column positions. Table 8.1 shows the corresponding root mean square errors, given by the square root of Eq. (8.12) or (8.13), measuring the spread of the estimated positions about the true positions, which have been used as an input for the simulations. These numbers clearly show the strong influence of amorphous layers on the estimated atom column positions. Moreover, amorphous layers seem to have a greater impact on the reliability of the position measurement of light element atom columns (O) than on that of heavier element atom columns (Sr and TiO). Furthermore, it can be shown that the root mean square errors increase with the increasing thickness of the amorphous layers and that they are independent of the restoration method. However, despite the strong influence of the amorphous layers, errors in the picometer range are still feasible.

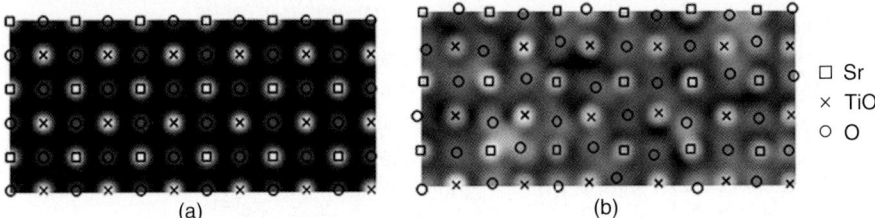

Figure 8.2 Phase of a simulated exit wave of $SrTiO_3$ [001] (a) without amorphous layers and (b) with amorphous layers both on the top and bottom of the sample. The symbols indicate the estimated atom column positions [46].

Table 8.1 Estimated root mean square errors (RMSE) for the Sr, TiO, and O atom columns obtained from the phase of a simulated exit wave (a) without amorphous layers and (b) with amorphous layers on the top and bottom of the crystalline sample of 2 nm thickness [46].

RMSE (pm)	Sr	TiO	O
a	0.05	0.05	0.05
b	5.3	5.7	21.8

8.3.3
Model-Based Quantification of Electron Energy Loss Spectra

For a long time, the interpretation of electron energy loss spectra has mostly been done visually or by using conventional quantification methods [47–52]. More recently, Verbeeck et al. [53–56] have introduced the use of the ML method to determine elemental concentrations. As introduced in Section 8.2.2, the starting point of this method is the availability of a parametric statistical model of the observations. This model describes both the expectations of the observations and the fluctuations of the observations about the expectations. The physical model describing the expectations contains the parameters of interest, which, in most cases, can be directly related to chemical concentrations in the sample. Next, quantitative chemical concentration measurements can be obtained by fitting this model to an experimental spectrum. Therefore, a criterion of goodness of fit is used, which is optimized with respect to the unknown parameters. In contrast to most conventional quantification methods, a physics-based parametric model is used, which mimics all processes involved in the recording of an EELS spectrum.

As an example, we will consider a set of 200 EELS spectra recorded from the same area of an h-BN sample [54]. These spectra can be regarded as a set of replications of the same experiment, which ideally only differ by the noise in the spectra. In Verbeeck et al. [54], a combination of a power law background, two Hartree Slater edges [57] for the B and N K-edge, and an empirical model for the fine structure has been proposed to model the expectations of the experimental EELS observations. Furthermore, the observations have been assumed to be statistically independent and Poisson distributed about their expectations. Given this parametric statistical model of the observations, the unknown parameters of this model, including the concentrations of B and N, can be estimated using the ML estimator as discussed in Section 8.2.5. Because of its optimal properties, this estimator is generally preferred. In Figure 8.3, the refined model is represented by the gray full line together with the underlying experimental EELS spectrum which is represented by black dots. This model has been evaluated at the estimated parameters. Although this figure shows an excellent visual match between the experiment and the model, the likelihood ratio test, as discussed in Section 8.2.6, can be used to quantitatively

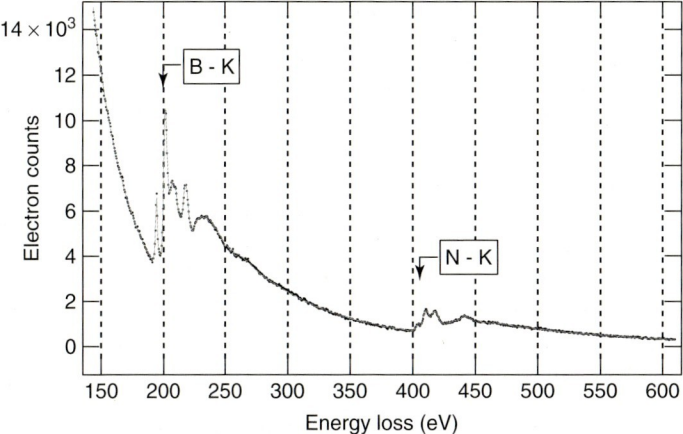

Figure 8.3 EELS model evaluated at the ML estimates (gray full line) and the corresponding experimental spectrum (black dots) of an h-BN sample showing a good visual agreement [54].

conclude whether there is reason to reject the expectation model or not. The significance level α has been chosen equal to 0.05 as is common practice in statistical hypothesis testing. The meaning of this significance level is that there is a probability of 5% to reject the expectation model when it is true. For the given experiment, the test statistic LR, given by Eq. (8.29), which can be computed given the experimental observations and the refined model, falls within the two critical values given in (8.31). Therefore, the refined model has been accepted. Next, from the set of estimated parameters the B/N ratio can be computed. This estimation procedure has been repeated for all 200 replicated spectra. From the thus obtained 200 estimated B/N ratios, 95% confidence intervals for the sample mean and standard deviation have been computed. These are shown in Table 8.2. The proportion of accepted models could be shown to be equal to 91% for the chosen significance level of 5%. Given this high level of confidence that can be attached to the refined models, it is expected that the B/N ratios have been estimated very accurately. Furthermore, the advantage of having repetitive measurements of the same experiment is the ability to compare the thus obtained precision with the attainable precision as discussed in Section 8.2.4. For each experiment, the lower bound on the standard deviation with which the B/N ratio can be measured has been estimated using the square root of Eq. (8.33). The 95% confidence interval on the mean of these lower bounds is given in Table 8.2. Since there is no overlapping with the confidence interval for the standard deviation, it should be concluded that the highest attainable precision has not been reached experimentally. However, from this comparison it is directly clear that the experimental measurements are approaching the lower bound. In [56] it has been shown that this discrepancy is caused by correlation in the recording process. The highest attainable precision can indeed be obtained experimentally by taking correlation into account in the

Table 8.2 Summary of the quantitative results obtained from 200 experimental EELS spectra of h-BN [54].

Mean B/N ratio	[1.0447–1.0473]
Standard deviation of B/N ratio	[0.0101–0.0123]
Estimated lower bound on the standard deviation	[0.00844–0.00847]
Accepted LR tests at significance level $\alpha = 0.05$	91%

parametric statistical model of the observations. Finally, in [55], it has been shown that by using this model-based measurement of the concentrations, the precision improved by a factor of 3 as compared to conventional quantification methods. As expected, the estimates are independent of the thickness, although this is not the case using conventional methods.

8.3.4
Quantitative Atomic Resolution Mapping using High-Angle Annular Dark Field Scanning Transmission Electron Microscopy

It is generally expected that local chemical information can be retrieved from HAADF STEM images. It is well known that HAADF STEM images show Z-contrast. One of the advantages is therefore the possibility to visually distinguish between chemically different atom column types. However, if the difference in atomic number of distinct atom column types is small or if the signal-to-noise ratio is poor, direct interpretation of HAADF STEM images is inadequate. Moreover, detailed probe characteristics need to be taken into account for correct interpretation of the images [58–60]. For example, if the tails of the electron probe have contributions on neighboring atom columns, intensity transfer from one atom column to neighboring atom columns may occur. Such effects also hamper the direct interpretation of the images, especially when studying interfaces. For these reasons, quantitative analysis of HAADF STEM images is required. Over the years, several approaches have been proposed for quantitative HAADF STEM [61–67].

Recently, a new quantitative HAADF STEM method has been developed [68] showing that here also the use of statistical parameter estimation theory is very helpful in order to extract local chemical information with good accuracy and precision. The method is based on a quantification of the total intensity of the scattered electrons for the individual atom columns using statistical parameter estimation theory. In order to apply this theory, a model is required to describe the image contrast of the HAADF STEM images. Therefore, a simple, effective incoherent model has been assumed: a parameterized object function, which is peaked at the atom column positions, convoluted with the probe intensity profile. Next, the scattered intensities, corresponding to the volumes under the peaks of the object function, are estimated by fitting this model to experimental HAADF STEM images. These estimates can then be used as a performance measure to distinguish between different atom column types and to identify the nature of unknown columns using statistical hypothesis testing. The reliability of the method is supported by means of simulated HAADF STEM images as well as a combination of experimental images and electron energy loss spectra [68].

It has experimentally been shown that statistically meaningful information on the composition of individual columns can be obtained even if the difference in averaged atomic number Z is only 3. Figure 8.4a shows an enlarged area from an experimental HAADF STEM image of an $La_{0.7}Sr_{0.3}MnO_3$–$SrTiO_3$ multilayer structure close to the $SrTiO_3$ substrate using an FEI TITAN 80–300 microscope operated at 300 kV. Note that no visual conclusions could be drawn concerning the sequence of the atomic planes at the interfaces. The refined model is shown in Figure 8.4b. Visually comparing Figure 8.4b with Figure 8.4a does not reveal any significant systematic deviations of the refined model from the experimental observations. This is illustrated more clearly in Figure 8.4c, where the data of the experiment (Figure 8.4a) and refined model (Figure 8.4b) are shown together after averaging along the horizontal direction. From this figure, it is also clear that the impression of a varying background is very well described by the proposed expectation model. Clearly, the background trend may equally well be due to the overlap of neighboring columns at the interface. This shows that the simple incoherent model is able to describe the image contrast adequately. Assuming that the model is correct, Figure 8.4b may be regarded as an optimal image reconstruction. Figure 8.4d shows the experimental observations together with an overlay, indicating the estimated positions of the columns together with their atom column types. The composition of the columns away from the interfaces is assumed to be known, whereas the composition of the columns in the planes close to the interface (indicated by the symbol "X") is unknown. Histograms of the

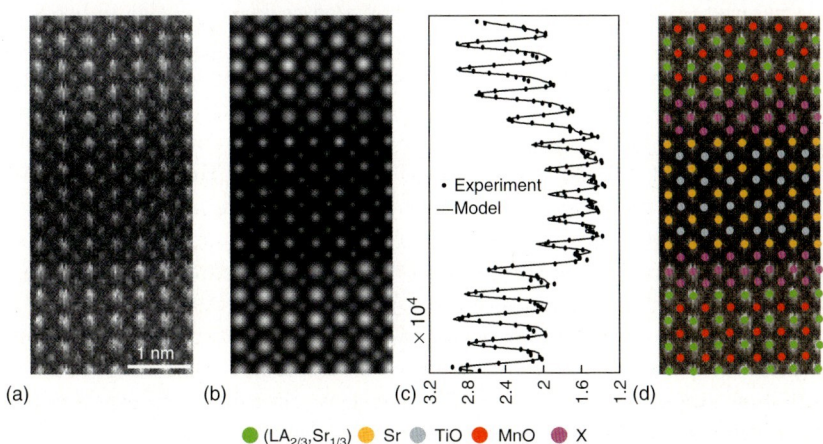

Figure 8.4 (a) Enlarged section of an experimental HAADF STEM image of an $La_{0.7}Sr_{0.3}MnO_3$–$SrTiO_3$ multilayer structure close to the $SrTiO_3$ substrate. (b) Refined model. (c) Experimental data (a) and refined model (b) averaged along the horizontal direction. (d) Overlay indicating the estimated positions of the columns together with their atom column types. The columns whose composition is unknown are indicated by the symbol "X" [68].

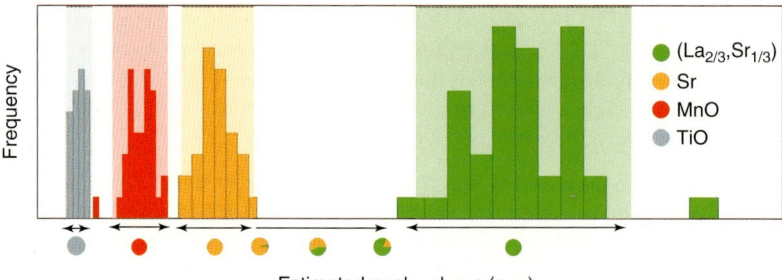

Figure 8.5 Histograms of the estimated peak volumes of the columns in Figure 8.4, whose composition is assumed to be known. The colored vertical bands represent the corresponding tolerance intervals. The unknown columns can then be identified by comparing their estimated peak volumes with these tolerance intervals. Single-colored dots will be used to indicate columns whose estimated peak volume falls inside a tolerance interval, whereas pie charts will be used to characterize columns whose estimated peak volume falls outside the tolerance intervals [68].

estimated peak volumes of the known columns are presented in Figure 8.5 and show the statistical nature of the result. The colored vertical bands correspond to 90% tolerance intervals. It is important to note that these tolerance intervals are not overlapping, meaning that columns, even for which the difference in averaged atomic number is only 3 (TiO and MnO), can clearly be distinguished. Next, in order to identify an unknown column, its estimated peak volume is compared with the colored confidence intervals. Its column type is equal to the column type corresponding to the tolerance interval containing the estimated peak volume of the unknown column. A column whose estimated peak volume falls in between the tolerance intervals corresponding to Z_i and Z_j will have an averaged atomic number in between Z_i and Z_j. For this multilayer structure, such effects are the result of intermixing or diffusion of different types of atoms [69]. Single-colored dots are used to indicate columns whose estimated peak volume falls inside a tolerance interval, whereas pie charts are used otherwise. The sizes of the two segments are an indication of the relative position of the estimated peak volume with respect to the nearest tolerance intervals. The results of this quantification are shown in Figure 8.6. The lower planes of the quantified areas consist of purely TiO and MnO columns, respectively. However, intermixing of La and Sr and of Ti and Mn is observed in the upper two planes. The degree of intermixing is more important for the interface on top of the STO layer than for the interface at the bottom of the STO layer. These results have been confirmed by an experimental EELS line scan, proving that the method works well in practice [68].

8.3.5
Statistical Experimental Design

An important purpose for which the expressions for the CRLB can be used is the optimization of the experimental design. Experimental design can be defined

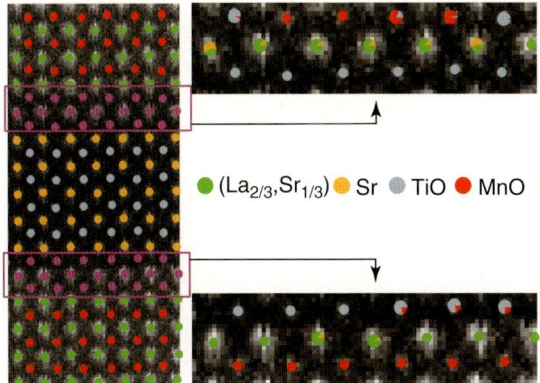

Figure 8.6 Results of the quantification of the unknown column types of the areas on the top and at the bottom of the STO layer [68].

as the selection of free variables in an experiment to improve the precision of the measured parameters. By calculating the CRLB, the experimenter gets an impression whether for a given experimental design the precision attainable is sufficient for the purposes concerned. If not, the experimental design has to be changed. If this is not possible, it has to be concluded that the observations are not suitable for the purposes of the measurement procedure. In this way, the experimental design can be optimized so as to attain the highest precision [15]. In [28, 31–33, 70–73], it has been shown how optimizing the design of quantitative electron microscopy experiments may enhance the precision with which structure parameters can be estimated.

The optimal experimental design of HRTEM experiments has been reconsidered in terms of the attainable precision with which atom column positions can be estimated [31, 32, 72, 73]. Optimization of the experimental design requires knowledge of the joint probability density function of the observations. For HRTEM applications, this function has been introduced in Example (8.1) of Section 8.2.2, where the observations have been assumed to be a set of independent Poisson-distributed observations. In this case, it follows from Example (8.2) of Section 8.2.4 that the elements of the Fisher information matrix are given by Eq. (8.19). The CRLB, giving an expression for the attainable precision, can then be found by taking the inverse of the Fisher information matrix. From the evaluation of the lower bound on the variance with which atom column positions can be estimated, it follows that for microscopes operating at intermediate accelerating voltages of the order of 300 kV, the potential merit of the use of a spherical aberration corrector or a chromatic aberration corrector is only of a limited value and that the use of a monochromator usually does not pay off in terms of precision, presuming that specimen drift puts a practical constraint on the experiment. These results follow from Figure 8.7, where the lower bound on the standard deviation of the position coordinates of a gold atom column is evaluated as a function of the spherical aberration constant.

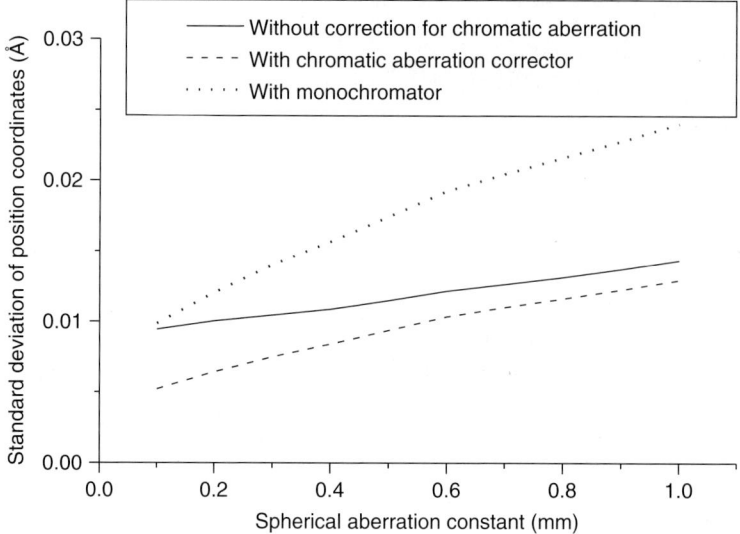

Figure 8.7 The lower bound on the standard deviation of the position coordinates of an atom column of a 50 Å thick gold [100] crystal as a function of the spherical aberration constant for a microscope operating at 300 kV equipped with or without chromatic aberration corrector or monochromator. In this evaluation, the recording time is kept constant [72, 73].

The solid curve corresponds to a microscope without correction for chromatic aberration, that is, a microscope without a chromatic aberration corrector and a monochromator. The dashed curve corresponds to a microscope with a chromatic aberration corrector, for which the chromatic aberration constant is equal to 0 mm. The dotted curve corresponds to a microscope with a monochromator, for which the energy spread corresponds to a full width at half maximum height of 200 meV. In this figure, it is assumed that the specimen drift is the relevant physical constraint. Hence, the recording time is kept constant. Consequently, the total number of detected electrons is smaller in the presence of a monochromator. Therefore, it follows from Figure 8.7 that it is, in principle, possible to obtain a precision of the order of 0.01 Å, even with a microscope that is not corrected for spherical and chromatic aberration. Notice, however, that for noncorrected microscopes, the interpretation of the images is much more difficult, complicating the derivation of a good starting model. As discussed in Section 8.2.5.1, good starting models are necessary in order to avoid that the optimization ends up at a local optimum of the criterion of goodness of fit instead of at the global optimum introducing a systematic error (bias).

For amorphous instead of crystalline structures, the conclusions are different, since amorphous structures are very sensitive to radiation damage so that radiation sensitivity rather than specimen drift will be the limiting factor in the experiment. Although the precision improves by increasing the incident dose, it has to be taken into account here that every incident electron has a finite probability to

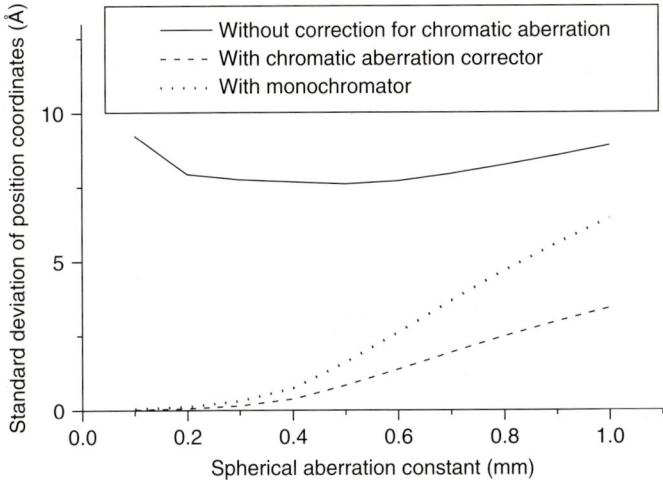

Figure 8.8 The lower bound on the standard deviation of the position coordinates of an atom column of a 50 Å thick silicon [100] crystal as a function of the spherical aberration constant for a microscope operating at 50 kV equipped with or without chromatic aberration corrector or monochromator. In this evaluation, the recording time is kept constant [72, 73].

damage the structure. The structure can be damaged either by displacing an atom from its position or by changing chemical bonds because of ionization. Therefore, a compromise between precision and radiation damage has to be made, which turns out to be optimal when the accelerating voltage is reduced. However, at low accelerating voltages, the instrumental aberrations become important. For this reason, correction of both the spherical aberration and the chromatic aberration by either a chromatic aberration corrector or a monochromator will be essential. By means of statistical experimental design, it has been shown that a substantial improvement of the precision may be obtained if both the spherical and chromatic aberrations are corrected [31, 72, 73]. This is shown in Figure 8.8.

8.4 Conclusions

In this chapter, it has been shown how quantitative transmission electron microscopy may greatly benefit from statistical parameter estimation theory to estimate unknown structure parameters. Ultimately, these parameters need to be estimated as accurately and precisely as possible from the available observations including HRTEM images, EELS spectra, or HAADF STEM images. In Section 8.1, a framework has been outlined to reach this goal. The parameterized joint probability density function of the observations has been introduced as a model to describe statistical fluctuations in the observations. This description requires a (usually physics-based) model describing the expectation values of the observations. Furthermore, it requires detailed knowledge about the disturbances that are acting on the observations. A thus parameterized probability density function can then be used for two purposes. First, it can be used to define the concept of Fisher information and to compute the CRLB. This is a theoretical lower bound on the variance of any unbiased estimator of the parameters. Second, from the probability density function of the observations the ML estimator of the parameters can be derived. The use of the ML estimator is motivated by the fact that this estimator has favorable statistical properties. However, there is no use in attaching confidence to ML parameter estimates as long as the model has not been accepted. A very effective technique for model assessment is the likelihood ratio test, which has been presented for Poisson-distributed observations. Finally, it has been shown how to construct confidence intervals for ML parameter estimates.

In Section 8.2, statistical parameter estimation theory has been applied to various kinds of observations acquired by means of transmission electron microscopy. Using aberration-corrected HRTEM images or reconstructed exit waves in combination with model-based parameter estimation theory, positions of atom columns can be measured with a precision of the order of a few picometers even though the resolution is still two orders of magnitude larger. Recent studies show that the presence of amorphous layers, which is hard to avoid experimentally, is an important factor limiting the reliability of the atom column position estimates. In the field of electron energy loss spectroscopy, the application of the ML method has been shown to be very successful in determining elemental concentrations. As compared to existing quantification methods, the precision improved by a factor of 3. Furthermore, by taking the effect of correlation in the recording process into account, the highest attainable precision, given by the CRLB, can be attained experimentally. Local chemical information can also be retrieved from HAADF STEM images. Although such images show Z-contrast, small differences in atomic number cannot visually be detected. Moreover, detailed probe characteristics need to be taken into account for a correct interpretation of the images. Therefore, quantitative analysis methods are required. Also here, statistical parameter estimation theory is very helpful. By taking the probe characteristics into account, very small differences in averaged atomic number can be distinguished. Finally, it has been illustrated how the expressions for the CRLB can be used for the optimization of the experimental design so as to attain the highest precision.

Acknowledgments

Sincere thanks to A. van den Bos, who unfortunately passed away too soon, for his enthusiastic and expert guidance in my understanding of statistical parameter estimation theory. I would like to acknowledge A. J. den Dekker for many fruitful and enlightening theoretical discussions. Many thanks to S. Bals and J. Verbeeck without whose experimental expertise and knowledge the practical examples shown in this chapter would not have been possible.

References

1. Wada, Y. (1996) Atom electronics: a proposal of nano-scale devices based on atom/molecule switching. *Microelectron. Eng.*, **30**, 375–382.
2. Olson, G.B. (1997) Computational design of hierarchically structured materials. *Science*, **277**, 1237–1242.
3. Olson, G.B. (2000) Designing a new material world. *Science*, **288**, 993–998.
4. Reed, M.A. and Tour, J.M. (2000) Computing with molecules. *Sci. Am.*, **282**, 68–75.
5. Browning, N.D., Arslan, I., Moeck, P., and Topuria, T. (2001) Atomic resolution scanning transmission electron microscopy. *Phys. Status Solidi B*, **227**, 229–245.
6. Spence, J.C.H. (1999) The future of atomic resolution electron microscopy for materials science. *Mater. Sci. Eng. R*, **26**, 1–49.
7. Muller, D.A. and Mills, M.J. (1999) Electron microscopy: probing the atomic structure and chemistry of grain boundaries, interfaces and defects. *Mater. Sci. Eng., A*, **260**, 12–28.
8. Springborg, M. (2000) *Methods of Electronic-structure Calculations: From Molecules to Solids*, John Wiley & Sons, Ltd, Chichester.
9. Muller, D.A. (1999) Why changes in bond lengths and cohesion lead to core-level shifts in metals, and consequences for the spatial difference method. *Ultramicroscopy*, **78**, 163–174.
10. Kisielowski, C., Principe, E., Freitag, B., and Hubert, D. (2001) Benefits of microscopy with super resolution. *Physica B*, **308–310**, 1090–1096.
11. den Dekker, A.J., Van Aert, S., van den Bos, A., and Van Dyck, D. (2005) Maximum likelihood estimation of structure parameters from high resolution electron microscopy images. Part I: a theoretical framework. *Ultramicroscopy*, **104**, 83–106.
12. van den Bos, A. (2007) *Parameter Estimation for Scientists and Engineers*, John Wiley & Sons, Inc, New Jersey.
13. van den Bos, A. and den Dekker, A.J. (2001) Resolution reconsidered - conventional approaches and an alternative, in *Advances in Imaging and Electron Physics*, vol. 117, (ed. P.W. Hawkes), Academic Press, San Diego (CA), pp. 241–360.
14. van den Bos, A. (1982) Parameter estimation, in *Handbook of Measurement Science*, vol. 1, (ed. P. Sydenham), John Wiley & Sons, Ltd, Chicester, pp. 331–377.
15. van den Bos, A. (1999) Measurement errors, in *Encyclopedia of Electrical and Electronics Engineering*, vol. 12, (ed. J.G. Webster), John Wiley & Sons, Inc, New York, pp. 448–459.
16. Seber, G.A.F. and Wild, C.J. (1989) *Nonlinear Regression*, John Wiley & Sons, Inc, New York.
17. Cramér, H. (1946) *Mathematical Methods of Statistics*, Princeton University Press, Princeton, NJ.
18. Van Dyck, D. and Op de Beeck, M. (1996) A simple intuitive theory for electron diffraction. *Ultramicroscopy*, **64**, 99–107.
19. Geuens, P. and Van Dyck, D. (2002) The S-state model: a work horse for HRTEM. *Ultramicroscopy*, **93**, 179–198.
20. Jennrich, R.I. (1995) *An Introduction to Computational Statistics - Regression Analysis*, Prentice Hall, Englewood Cliffs, NJ.
21. Stuart, A. and Ord, K. (1994) *Kendall's Advanced Theory of Statistics*, Arnold, London.
22. Frieden, B.R. (1998) *Physics from Fisher Information - A Unification*, Cambridge University Press, Cambridge.
23. Bates, D. and Watts, D. (1988) *Nonlinear Regression Analysis and its Applications*, John Wiley & Sons, Inc, New York.
24. Rayleigh, L. (1902) Wave theory of light, in *Scientific Papers by John William Strutt, Baron Rayleigh*, vol. 3, Cambridge University Press, Cambridge, pp. 47–189.
25. O'Keefe, M.A. (1992) 'Resolution' in high-resolution electron microscopy. *Ultramicroscopy*, **47**, 282–297.
26. den Dekker, A.J. and van den Bos, A. (1997) Resolution: A survey. *J. Opt. Soc. Am. A*, **14**, 547–557.

27. Ronchi, V. (1961) Resolving power of calculated and detected images. *J. Opt. Soc. Am. A*, **51**, 458–460.
28. Van Aert, S., Van Dyck, D., and den Dekker, A.J. (2006) Resolution of coherent and incoherent imaging systems reconsidered - Classical criteria and a statistical alternative. *Opt. Express*, **14**, 3830–3839.
29. Bettens, E., Van Dyck, D., den Dekker, A.J., Sijbers, J., and van den Bos, A. (1999) Model-based two-object resolution from observations having counting statistics. *Ultramicroscopy*, **77**, 37–48.
30. Van Aert, S., den Dekker, A.J., Van Dyck, D., and van den Bos, A. (2002) High-resolution electron microscopy and electron tomography: resolution versus precision. *J. Struct. Biol.*, **138**, 21–33.
31. Van Dyck, D., Van Aert, S., den Dekker, A.J., and van den Bos, A. (2003) Is atomic resolution transmission electron microscopy able to resolve and refine amorphous structures. *Ultramicroscopy*, **98**, 27–42.
32. den Dekker, A., Van Aert, S., Van Dyck, D., van den Bos, A., and Geuens, P. (2001) Does a monochromator improve the precision in quantitative HRTEM. *Ultramicroscopy*, **89**, 275–290.
33. Van Aert, S., den Dekker, A.J., Van Dyck, D., and van den Bos, A. (2002) Optimal experimental design of STEM measurement of atom column positions. *Ultramicroscopy*, **90**, 273–289.
34. Rose, H. (1990) Outline of a spherically corrected semiaplanatic medium-voltage transmission electron microscope. *Optik*, **85**, 19–24.
35. Haider, M., Uhlemann, S., Schwan, E., Rose, H., Kabius, B., and Urban, K. (1998) Electron microscopy image enhanced. *Nature*, **392**, 768–769.
36. Batson, P.E., Dellby, N., and Krivanek, O.L. (2002) Sub-ångstrom resolution using aberration corrected electron optics. *Nature*, **418**, 617–620.
37. Coene, W.M.J., Thust, A., Op de Beeck, M., and Van Dyck, D. (1996) Maximum-likelihood method for focus-variation image reconstruction in high resolution transmission electron microscopy. *Ultramicroscopy*, **64**, 109–135.
38. Jia, C.L., Lentzen, M., and Urban, K. (2003) Atomic-resolution imaging of oxygen in perovskite ceramics. *Science*, **299**, 870–873.
39. Van Aert, S., den Dekker, A.J., van den Bos, A., Van Dyck, D., and Chen, J. (2005) Maximum likelihood estimation of structure parameters from high resolution electron microscopy images. Part II: a practical example. *Ultramicroscopy*, **104**, 107–125.
40. Ayache, J., Kisielowski, C., Kilaas, R., Passerieux, G., and Lartigue-Korinek, S. (2005) Determination of the atomic structure of a $\Sigma 13 SrTiO_3$ grain boundary. *J. Mater. Sci.*, **40**, 3091–3100.
41. Bals, S., Van Aert, S., Van Tendeloo, G., and Ávila Brande, D. (2006) Statistical estimation of atomic positions from exit wave reconstruction with a precision in the picometer range. *Phys. Rev. Lett.*, **96**, 096106.
42. Houben, L., Thust, A., and Urban, K. (2006) Atomic-precision determination of the reconstruction of a 90° tilt boundary in $YBa_2Cu_3O_{7-\delta}$ by aberration corrected HRTEM. *Ultramicroscopy*, **106**, 200–214.
43. Urban, K. (2008) Studying atomic structures by aberration-corrected transmission electron microscopy. *Science*, **321**, 506–510.
44. Jia, C.L., Mi, S.B., Urban, K., Vrejoiu, I., Alexe, M., and Hesse, D. (2008) Atomic-scale study of electric dipoles near charged and uncharged domain walls in ferroelectric films. *Nat. Mater.*, **7**, 57–61.
45. Jia, C.L., Mi, S.B., Faley, M., Poppe, U., Schubert, J., and Urban, K. (2009) Oxygen octahedron reconstruction in the $SrTiO_3/LaAlO_3$ heterointerfaces investigated using aberration-corrected ultrahigh-resolution transmission electron microscopy. *Phys. Rev. B*, **79**, 081405.
46. Van Aert, S., Chang, L.Y., Bals, S., Kirkland, A.I., and Van Tendeloo, G. (2009) Effect of amorphous layers on the interpretation of restored exit waves. *Ultramicroscopy*, **109**, 237–246.
47. Egerton, R.F. and Malac, M. (2002) Improved background-fitting algorithms for

48. Su, D.S. and Zeitler, E. (1993) Background problem in electron-energy-loss spectroscopy. *Phys. Rev. B*, **47**, 14734–14740.
49. Pun, T., Ellis, J.R., and Eden, M. (1984) Optimized acquisition parameters and statistical detection limit in quantitative EELS. *J. Microsc.*, **135**, 295–316.
50. Pun, T., Ellis, J.R., and Eden, M. (1985) Weighted least squares estimation of background in EELS imaging. *J. Microsc.*, **137**, 93–100.
51. Egerton, R.F. (1982) A revised expression for signal/noise ratio in EELS. *Ultramicroscopy*, **9**, 387–390.
52. Unser, M., Ellis, J.R., Pun, T., and Eden, M. (1987) Optimal background estimation in EELS. *J. Microsc.*, **145**, 245–256.
53. Verbeeck, J. and Van Aert, S. (2004) Model based quantification of EELS spectra. *Ultramicroscopy*, **101**, 207–224.
54. Verbeeck, J., Van Aert, S., and Bertoni, G. (2006) Model-based quantification of EELS spectra: including the fine structure. *Ultramicroscopy*, **106**, 976–980.
55. Verbeeck, J. and Bertoni, G. (2008) Accuracy and precision in model based EELS quantification. *Ultramicroscopy*, **108**, 782–790.
56. Verbeeck, J. and Bertoni, G. (2008) Model-based quantification of EELS spectra: treating the effect of correlated noise. *Ultramicroscopy*, **108**, 74–83.
57. Rez, P. (1982) Cross-sections for energy loss spectrometry. *Ultramicroscopy*, **9**, 283–287.
58. Hillyard, S. and Silcox, J. (1995) Detector geometry, thermal diffuse scattering and strain effects in ADF STEM imaging. *Ultramicroscopy*, **58**, 6–17.
59. Klenov, D.O., Zide, J.M., Zimmerman, J.D., Gossard, A.C., and Stemmer, S. (2005) Interface atomic structure of epitaxial ErAs layers on (001)In$_{0.53}$Ga$_{0.47}$As and GaAs. *Appl. Phys. Lett.*, **86**, 241901.
60. Dwyer, C., Erni, R., and Etheridge, J. (2010) Measurement of effective source distribution and its importance for quantitative interpretation of STEM images. *Ultramicroscopy*, **110**, 952–957.
61. Ourmazd, A., Taylor, D.W., Cunningham, J., and Tu, C.W. (1989) Chemical mapping of semiconductor interfaces at near-atomic resolution. *Phys. Rev. Lett.*, **62**, 933–936.
62. Anderson, S.C., Birkeland, C.R., Anstis, G.R., and Cockayne, D.J.H. (1997) An approach to quantitative compositional profiling at near-atomic resolution using high-angle annular dark field imaging. *Ultramicroscopy*, **69**, 83–103.
63. Voyles, P.M., Muller, D.A., Grazul, J.L., Citrin, P.H., and Gossmann, H.J.L. (2002) Atomic-scale imaging of individual dopant atoms and clusters in highly n-type bulk Si. *Nature*, **416**, 826–829.
64. Erni, R., Heinrich, H., and Kostorz, G. (2003) Quantitative characterisation of chemical inhomogeneities in Al-Ag using high-resolution Z-contrast STEM. *Ultramicroscopy*, **94**, 125–133.
65. Klenov, D.O. and Stemmer, S. (2006) Contributions to the contrast in experimental high-angle annular dark-field images. *Ultramicroscopy*, **106**, 889–901.
66. LeBeau, J.M., Findlay, S.D., Allen, L.J., and Stemmer, S. (2008) Quantitative atomic resolution scanning transmission electron microscopy. *Phys. Rev. Lett.*, **100**, 206101.
67. LeBeau, J.M., Findlay, S.D., Wang, X., Jacobson, A., Allen, L.J., and Stemmer, S. (2009) High-angle scattering of fast electrons from crystals containing heavy elements: simulation and experiment. *Phys. Rev. B*, **79**, 214110.
68. Van Aert, S., Verbeeck, J., Erni, R., Bals, S., Luysberg, M., Van Dyck, D., and Van Tendeloo, G. (2009) Quantitative atomic resolution mapping using high-angle annular dark field scanning transmission electron microscopy. *Ultramicroscopy*, **109**, 1236–1244.
69. Muller, D.A., Kourkoutis, L.F., Murfitt, M., Song, J.H., Hwang, H.Y., Silcox, J., Dellby, N., and Krivanek, O.L. (2008) Atomic-scale chemical imaging of composition and bonding by aberration-corrected microscopy. *Science*, **319**, 1073–1076.
70. den Dekker, A.J., Sijbers, J., and Van Dyck, D. (1999) How to optimize the design of a quantitative HREM experiment

so as to attain the highest precision. *J. Microsc.*, **194**, 95–104.

71. Van Aert, S. and Van Dyck, D. (2001) Do smaller probes in a scanning tranmission electron microscope result in more precise measurement of the distances between atom columns. *Philos. Mag. B*, **81**, 1833–1846.

72. Van Aert, S., den Dekker, A.J., and Van Dyck, D. (2004) How to optimize the experimental design of quantitative atomic resolution TEM experiments. *Micron*, **35**, 425–429.

73. Van Aert, S., den Dekker, A.J., van den Bos, A., and Van Dyck, D. (2004) Statistical experimental design for quantitative atomic resolution transmission electron microscopy, in *Advances in Imaging and Electron Physics*, vol. 130, (ed. P.W. Hawkes), Academic Press, San Diego, CA, pp. 1–164.

9
Dynamic Transmission Electron Microscopy

Nigel D. Browning, Geoffrey H. Campbell, James E. Evans, Thomas B. LaGrange, Katherine L. Jungjohann, Judy S. Kim, Daniel J. Masiel, and Bryan W. Reed

9.1
Introduction

Transmission electron microscopy (TEM) has long played a key role in driving our scientific understanding of extended defects and their control of the properties of materials – from the earliest TEM observations of dislocations [1] through to the current use of aberration-corrected TEMs to determine the atomic structure of grain boundaries [2]. With the current generation of aberration-corrected and monochromated TEMs, we can now obtain images with a spatial resolution approaching 0.05 nm in both the plane-wave, phase contrast TEM and the focused probe, Z-contrast scanning transmission electron microscopy (STEM) modes of operation [3–5]. In addition to the increase in the spatial resolution, aberration correctors also provide an increase in the beam current and subsequently the signal-to-noise levels (contrast) in the acquired images. This means that small differences in structure and composition can be more readily observed and, for example, in the STEM mode of operation, complete 2-D atomic resolution elemental maps can be generated using electron energy loss spectroscopy (EELS) [6, 7]. Furthermore, the EEL spectra that are obtained using a monochromated microscope also show vast improvements over the spectra that could be obtained a few years ago – allowing bonding state changes to be observed from core-loss spectra with high precision [8] and the low-loss region of the spectrum to be used to map fluctuations in optical properties [9–11]. Taken all together, these newly developed capabilities for (S)TEM provide a comprehensive set of tools to measure, quantify, and understand the atomic scale properties of nanoscale materials, interfaces, and defects.

However, for all of the high precision in the measurements described above, a key feature of the experiments is that the object being studied has to remain stationary. For most high-resolution observations, there is a constant battle to overcome drift, charging, mechanical instabilities, stray fields, beam damage, and so on, to achieve this stability criterion. The closest that aberration-corrected imaging has come to studying a dynamic event at high spatial resolution, is when the beam itself

Handbook of Nanoscopy, First Edition. Edited by Gustaaf Van Tendeloo, Dirk Van Dyck, and Stephen J. Pennycook.
© 2012 Wiley-VCH Verlag GmbH & Co. KGaA. Published 2012 by Wiley-VCH Verlag GmbH & Co. KGaA.

causes a structural change to occur and sequential images show a progression of atomic movements [12]. Of course, dynamic events have been studied in an electron microscope before, with a whole subfield of *in situ* microscopy, with either complete microscopes or specimen stages, being used to study the effects of mechanical deformation, the effect of gas pressure on catalytic activity and nanostructure nucleation and growth, beam damage, and even reactions taking place in liquids [13–32]. In all of these cases, spatial resolution is sacrificed for *in situ* conditions and temporal resolution, and the temporal resolution of the observation is limited by the signal reaching the camera used to record the observation, that is, the number of electrons needed to form an interpretable image. For typical beam currents in commercial thermionic and field emission electron microscopes, this places a practical limit on temporal resolution of ∼1 ms. In many cases, this means that the imaging process itself is just too slow to see the critical details of the phenomenon being studied – we may as well just perform static measurements.

As shown in Figure 9.1, there are a wide range of dynamic phenomena that occur in both inorganic and organic structures on timescales well below 1 ms. In some cases, as with dislocation motion, it is not necessary to achieve atomic spatial resolution to observe the phenomenon, while in others, such as atomic diffusion, it is. There is therefore a range of length- and timescales that it would be ideal to access by experimental techniques – roughly 1 μm to 1 Å spatial resolution coupled with 1 μs to 1 fs temporal resolution. This desire to achieve high temporal/spatial resolution is not new, and dynamic observations have been shown by optical and

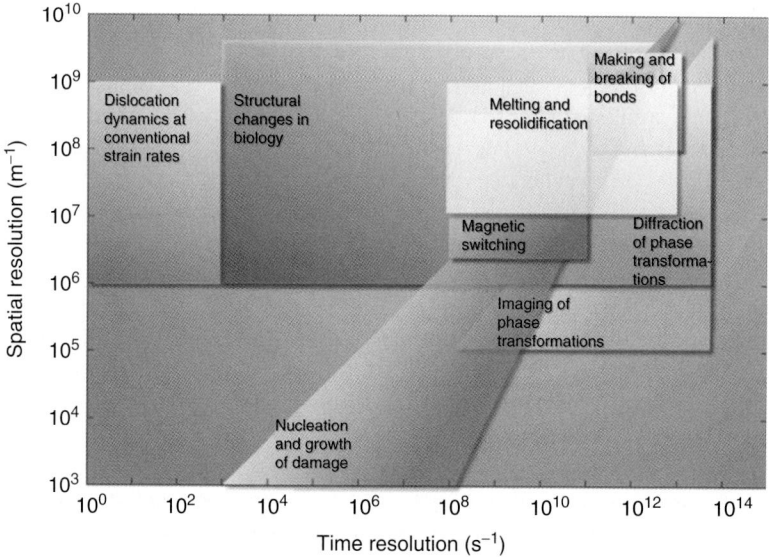

Figure 9.1 Properties of materials and biological systems and the length/timescales needed to observe them.

X-ray means [33–37]. However, while femtosecond spectroscopy and pump-probe experiments have proved to be very successful, they are typically limited in spatial resolution and often require the image to be inverted from a diffraction pattern – there is no direct image of the process taking place. These limitations can be overcome by using electron pulses on the same timescale to analyze materials [38]. The immediate benefit of using electrons is that the interaction of the electrons with the material being analyzed is much stronger, producing more signal. This advantage has been used for many years in the field of ultrafast electron diffraction (UED) [39–42]. If the electron pulses can be combined with the TEM methods described above, then the beam can be directed and focused, images, spectra, and diffraction patterns can be obtained from localized areas, and direct high-resolution images of dynamic events can also be obtained – avoiding the need to invert the diffraction pattern.

The ability to study dynamic processes in materials on a timescale approaching 1 ns is the main driving force behind the development of the dynamic transmission electron microscope (DTEM) at Lawrence Livermore National Laboratory (LLNL) [43–52]. To achieve this temporal resolution while still maintaining the direct high-resolution imaging capability of a TEM, required the modification of a conventional TEM to create and control large electron bunches – this development follows the groundbreaking research of Bostonjoglo and co-workers in this area [53–55]. In the remainder of this chapter, the basic physical principles behind the creation of electron bunches, their control, and the expected TEM image resolution are defined. The modifications to the microscope electron optics necessary to turn a conventional TEM into a DTEM are also described in detail for both the current DTEM and for new instruments that are on the horizon. It should be noted here that the goal of this particular development is to obtain complete images from single-shot experiments, making the approach radically different from the ultrafast, or stroboscopic, TEM that has been developed by Zewail and co-workers [56–60].

9.2
Time-Resolved Studies Using Electrons

The engineering of a single-shot nanosecond TEM imaging capability starts with a physical understanding of the electron pulses themselves. Because of the need to extract information from the sample at an extraordinary rate (measured in bits of information per square nanometer per nanosecond) compared with conventional TEM, DTEM has to operate at extremely high current densities. Yet it must do so with minimal sacrifice of spatial and temporal coherence. This is because of the physics of TEM image contrast, including the electron–sample interaction and the electron–optical properties (including aberrations) of the lens system.

Large local convergence angles α at the sample plane (which are equivalent to small spatial coherence lengths; see Figure 9.2a) tend to wash out the contrast. This is important because α plays a central role in the law of conservation of

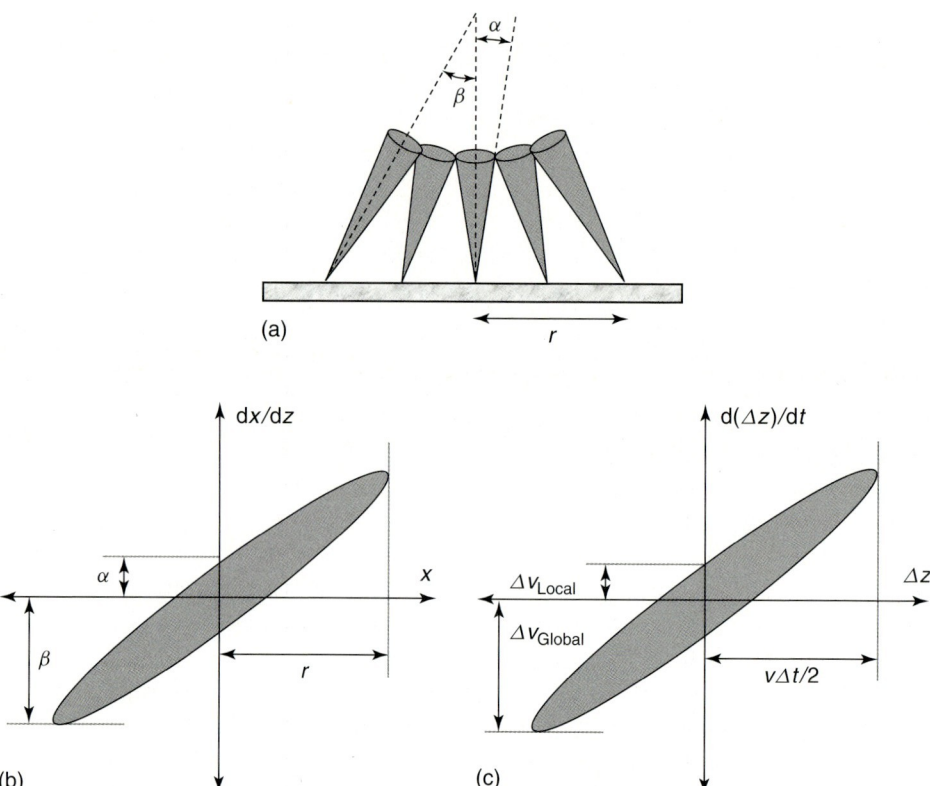

Figure 9.2 (a) Illustrating local (α) versus global (β) measures of beam convergence/divergence. While α is a measure of lateral spatial coherence at a given point, and the product $\varepsilon = r\alpha$ is a nearly conserved quantity in the absence of apertures, β can be changed rather arbitrarily if enough transfer lenses are available. (b) The same beam profile as a lateral phase space ellipse. The emittance ε is equal to the area of this ellipse times a factor of order unity. (c) A similar phase space plot for the longitudinal (z) direction, where Δz is the axial distance of an electron from the pulse center of mass and v is the nominal pulse velocity. In this case, Δv_{Global} is very important, as it couples with the chromatic aberration of the objective lens.

brightness (discussed below); increasing current density at the sample generally comes at the cost of increasing α and thus reducing the spatial coherence. The range of acceptable convergence angles may vary greatly with the contrast mechanism and spatial frequency range of interest. For example, coherent diffractive imaging may require $\alpha \ll 1$ mrad, while for high-angle annular dark-field (HAADF) TEM imaging $\alpha \sim 10$ mrad may be acceptable. No matter what the mechanism is, a reduction in spatial coherence eventually reduces the relevant image contrast to the point where the features of interest are indistinguishable from artifacts and noise. A textbook example of this is the manner in which the convergence angle cuts off the contrast-transfer function for high-resolution phase contrast imaging, and we will work within this formalism for the examples shown below. Temporal incoherence (usually expressed in terms of energy spread), by coupling to the lens system's chromatic aberration, also reduces contrast in a similar manner and must be kept to a minimum in any TEM.

Such "conventional" limits on spatiotemporal coherence apply to all TEM imaging. Single-shot DTEM adds an additional constraint, namely, that a single nanosecond-scale pulse contains enough electrons – typically $\sim 10^7 - 10^8$ – to provide an image of acceptable contrast and resolution. This eliminates the option of simply increasing spatial coherence (by introducing apertures and adjusting condenser lenses) and acquiring for a longer time, and it puts stringent requirements on the brightness and current of the electron pulse. Specifically, the brightness must be at least $\sim 10^7$ A cm$^{-2}\cdot$ sr^{-1} (and preferably orders of magnitude higher), while the beam current must be on the milliampere scale [61]. Add to this the requirement that the energy spread be kept below 5 eV, and this greatly restricts the kinds of electron sources that may be used. At the time of writing, the field is dominated by a technology shown to reliably produce such electron beam parameters over a very large number of shots – namely, linear photoemission produced by ultraviolet pulsed lasers directed onto large-area metallic cathodes. However, there are some experimental electron sources and source concepts that may eventually replace this technology [62–66].

In addition to these conventional effects, DTEM must contend with additional nonlinear effects specifically related to the extremely high current densities. This arises from one basic physical problem: electrons are charged fermions. Thus they will repel each other over long distances, collide with each other at short distances, and refuse to fill phase space (the six-dimensional space combining three-dimensional position with three-dimensional momentum) more densely than one electron per quantum state. As a result, electron beams are subject to space charge effects, inhomogeneous scattering effects (following the terminology of [67]; these are also called *statistical Coulomb effects* by some authors and are considered a subset of space charge effects by others), and fundamental limits on spectral brightness (i.e., brightness for a given energy spread).

Space charge effects limit the current density extractable from an electron gun geometry and (together with the energy spread) are an important factor in determining the maximum brightness attainable with a given gun [68]. They also create lens aberrations [69] and defocus electron beams (forcing slight adjustments of lens strengths as the beam current changes). For very short pulses on the picosecond scale or faster, space charge effects can also dramatically broaden both the duration and the energy spread of the pulse [70–73], thus degrading both spatial and temporal resolution. Resolution is further degraded by inhomogeneous effects in the post-sample crossovers. Analysis of this effect is computationally challenging [50, 69] but quite important since it implies rather stringent limits on resolution, and since this effect cannot be reversed with aberration correction or any other technology.

9.2.1
Brightness, Emittance, and Coherence

Let us examine the physical properties that limit the performance of pulsed electron instruments. Ultimately, the performance is governed by the brightness of the electron source. While this is true for any TEM, it is especially true for a single-shot DTEM, which demands a very high fluence (electrons per unit area) be delivered in a very short time. We will define brightness B as

$$B = \frac{Ne}{(\pi r^2)(\pi \alpha^2) \Delta t} \quad (9.1)$$

where N is the number of electrons per pulse, e is the electronic charge, r is the radius of the electron beam, α is the local convergence semi-angle, and Δt is the time duration of a pulse. B is very nearly conserved for a beam propagating at a fixed accelerating voltage. By "nearly conserved," we mean that B is a constant in an ideal electron column, but in real systems, a number of physical effects (e.g., Boersch effects [67], aberrations, and space charge effects [69]) can cause the effective brightness to degrade with propagation distance. If the voltage is not constant, then the normalized brightness (equal to fundamental constants times $\lambda^2 B$, with λ the electron wavelength ([68], and using the relativistic de Broglie relation $\lambda = h/(\beta \gamma mc)$)) is the relevant conserved quantity that includes all relativistic corrections. The quantities r, α, and Δt are finite and the brightness as we have defined it is an effective average over a finite area. This definition has the advantage of direct experimental relevance, but as a result, our B is not a precisely conserved quantity in a freely propagating beam; lens aberrations, space charge, and statistical Coulomb effects can all reduce our effective average B as a beam propagates through a column. This is particularly true for the subpicosecond regime, in which pulses will expand longitudinally so that Δt is no longer constant [70–73].

We also define the related quantity emittance as $\varepsilon = r\alpha$, which is itself a nearly conserved quantity just as B is, provided no electrons are blocked at apertures. A smaller ε implies a higher-quality beam, with higher spatial coherence for a given spot size. The phase space filled by electrons (Figure 9.2b) can be transformed by lenses, so long as its area does not change. If the lenses are aberration-free, then a phase space ellipse remains an ellipse throughout the column, although its tilt and aspect ratio can change. Multipole lens elements such as those in aberration correctors can perform more complex operations, for example, turning an aberrated sigmoidal shape into an approximate ellipse. In principle, a sufficiently complex lens system could allocate this phase space area in just about any desired manner (which, for most experiments, would be an ellipse at the sample, although some results from light optics suggest other possibilities [74]). In time-resolved electron microscopy, we also need to consider the longitudinal phase space (Figure 9.2c), which shows the distribution of electron speeds (or energies) as a function of arrival time (or of longitudinal position) relative to the center of mass of the bunch. The combination of this longitudinal phase space with the transverse phase spaces in the x- and y-directions (Figure 9.2b) comprise the six-dimensional phase space mentioned above. Longitudinal phase space area is also approximately conserved and can be manipulated with various combinations of space charge expansion, ballistic propagation, and pulse compressors (which we may think of as temporal lenses). This is the basis of a proposal to improve the temporal resolution of UED while still minimizing the energy spread [75]. These authors have shown a calculation of the relevant phase space at the sample (Figure 9.4a of [66]), which includes a strong sigmoidal component arising from nonlinearities in the system, which are essentially temporal lens aberrations. In principle, this could be corrected with the appropriate nonlinear temporal focusing, but it remains to be seen whether the necessary control could be achieved in practice.

The longitudinal space charge effects that cause Δt to increase as a pulse propagates are only very weakly coupled to the aberrations that cause the emittance ε to degrade (although they do affect the rate of evolution of the lateral phase space ellipse [70–73]). Thus the product $B\Delta t$ (which is proportional to the number of electrons in a pulse) is also nearly conserved. Combining this effect with the variation of brightness with accelerating voltage, we find that the dimensionless quantity

$$N_C \equiv \frac{\pi^2 \lambda^2 B \Delta t}{e} \tag{9.2}$$

(which we will call the *coherent fluence*) is a convenient figure of merit that is very nearly constant as a function of propagation distance for any electron pulse we would likely be using. This definition is motivated by a recognition that the lateral coherence length is given by $r_C = \lambda/\alpha$, so that our quantity N_C is essentially the number of electrons per lateral coherence area, per pulse (to within definition-dependent factors of order unity). No amount of lensing, aperturing, acceleration, or space charge dynamics will allow a user to significantly improve this value once the pulse has left the gun. As we shall show, the coherent fluence plays a central role in the theory of resolution limits for single-shot pulsed imaging. For example, if N_C is not much more than 1, then coherent single-shot imaging is impossible, no matter how good the lens system is.

The coherent fluence is related to very fundamental quantum mechanical properties of the electrons. Consider the Pauli exclusion limit, such that only two spin-1/2 electrons may occupy a single spatial wave function (which claims a volume h^3 of six-dimensional phase space, h being Planck's constant). Suppose the phase space volume is cylindrical in both real space (with radius r and height Δz) and momentum space (with radius Δp_r and height Δp_z). We then define the maximum number of electrons in the pulse as

$$N_{\text{Degenerate}} = 2 \frac{(\pi r^2 \Delta z)(\pi \Delta p_r^2 \Delta p_z^2)}{h^3} \tag{9.3}$$

But electrons must also obey the Heisenberg's uncertainty relation as well as basic kinematics of massive particles, thus $\Delta p_r = \alpha h/\lambda$; $\Delta p_z = \Delta E/v$; and $\Delta z = v\Delta t$ (with v the average speed of an electron in the pulse). Combining these, we find that the coherent fluence can be expressed as

$$N_C = \frac{\pi}{2} \frac{N}{N_{\text{Degenerate}}} \frac{\Delta E \Delta t}{\hbar/2} \equiv f_0 \, f_{\text{Pauli}} \, f_{\text{Heisenberg}} \tag{9.4}$$

where f_0 is a factor of order unity that depends on our particular choice of phase space profiles and definitions of distributions widths (strictly speaking, the rms widths should be used in the Heisenberg uncertainty product); $f_{\text{Pauli}} \leq 1$ is the degeneracy (i.e., the degree to which the pulse satisfies the Pauli exclusion principle); and $f_{\text{Heisenberg}} \geq 1$ is the amount of excess energy spread beyond that required by the time duration and the uncertainty principle. Since coherent imaging requires $N_C \gg 1$, this shows that the *only* way to achieve coherent single-shot imaging is to compensate for the low Fermi degeneracy (10^{-4} or less for typical pulses) by increasing the energy spread far beyond the transform limit. In practice, this means that picosecond-scale single-shot coherent electron imaging is inherently challenging and will almost certainly require chromatic aberration correction. This result is so firmly rooted in the fundamental physics that it is difficult to see how it could ever be bypassed. Notice that the result does not even depend on the spatial resolution.

In summary, the emittance is a measure of the phase space size and the brightness is a measure of how densely this phase space is filled, and a sufficiently flexible illumination system could allocate this phase space in just about any manner that would be desired, provided certain combinations of parameters are conserved. Assuming, then, that we have such a system and an electron gun of some specified coherent fluence N_C, we would like to determine the beam conditions and imaging system capabilities that would enable a specified level of performance.

9.2.2
Single-Shot Space–Time Resolution Trade-Offs

To do this, we will consider the interplay of brightness, coherence, shot noise, and the instrumental contrast-transfer function, with the goal of determining the spatial resolution limit as a function of time resolution Δt. These calculations are in the spirit of an order-of-magnitude estimate for the performance as limited by the microscope itself; we assume an ideal sample with 100% contrast arising from the electron–sample interaction, although modifying the results for finite sample contrast would not be difficult. Space charge and stochastic blurring effects are not included. We will use standard textbook formulae for a conventional TEM's partially coherent contrast-transfer functions ($T(r)$) for both phase and amplitude contrast [76], including spatial and temporal incoherence effects through the usual envelope approximations. We will assume that the user is interested in getting the best possible contrast at a given spatial frequency, with a fixed brightness and single-shot time resolution, which means (i) adjusting the objective lens defocus to maximize $|T|$ for the desired spatial frequency band (this is done implicitly for all the results shown here) and (ii) increasing the current density by converging the beam until α is large enough that the spatial incoherence starts to significantly reduce $|T|$. We define a coherence factor $f_{coherence} = \lambda/r\alpha$, with λ the electron wavelength, r the radius of the smallest feature to be resolved, and α the half-width of the angular distribution function. In other words, we have scaled the pixel size to the lateral coherence length via the dimensionless factor $f_{coherence}$. Employing the above definition of brightness, we find that the number of electrons per pixel area πr^2 is

$$N = \frac{\pi^2 \lambda^2 B \Delta t}{e} \cdot \frac{1}{f_{coherence}^2} = \frac{N_C}{f_{coherence}^2} \qquad (9.5)$$

The coherent fluence has come up naturally from the imposition of our condition (ii).

We then apply the Rose criterion [77], which specifies the minimum number of particles that need to be detected before a pixel can be said to be resolved in the presence of shot noise. The governing formula is $N^{-1/2} = |T|/f_{Rose}$, with f_{Rose} typically set to ~ 5 and T the contrast-transfer function. Combining equations yields the formula

$$N_C = \left(\frac{f_{coherence} f_{Rose}}{T} \right)^2 \qquad (9.6)$$

where the coherent fluence is set equal to a combination of dimensionless parameters. This defines a curve (actually two curves, one for phase and one for amplitude contrast, which have different values of $T(r)$ which can be plotted to show the trade-off of spatial and temporal resolution. This is done in Figure 9.3. The absolute contrast-transfer function $|T|$ is maximized with respect to defocus at the spatial frequency of one cycle per $2r$ (where r is the spatial resolution). We also sought to minimize the required N_C as a function of α, but the inverse-linear dependence of $f_{coherence}$ on α more than balanced the relatively weak α dependence of our approximate transfer function T, suggesting that the optimum was at extremely large convergence angles. However, the standard formulae for the spatial coherence envelope come from a first-order approximation, and they break down at large α, so that the experimentally relevant optimum is probably closer to $f_{coherence} = 1$. In concrete terms, converging the electron beam increases the signal-to-noise ratio very quickly, but after a certain point, the spatial incoherence rapidly gets so bad (in a way that is not captured by the basic equations) that no imaging is possible. As a practical compromise, we match the coherence length to the target resolution, that is, we set $f_{coherence}$ to 1, for all coherent imaging modes.

Figure 9.3 also includes a simpler curve that estimates the resolution limit for incoherent HAADF-TEM imaging. Since this is an incoherent imaging mode, we no longer vary α; rather we set it to a fairly large value (10 mrad) that is on the order of the inner radius of an HAADF-TEM aperture [78]. The relevant formula becomes

$$N_C = \left(\frac{\lambda f_{Rose}}{r\alpha} \right)^2 f_{collected}^{-1} \qquad (9.7)$$

where $f_{collected}$ is the fraction of electrons that scatter into the annular hole in the HAADF-TEM aperture. This quantity varies a great deal with the aperture size and

9 Dynamic Transmission Electron Microscopy

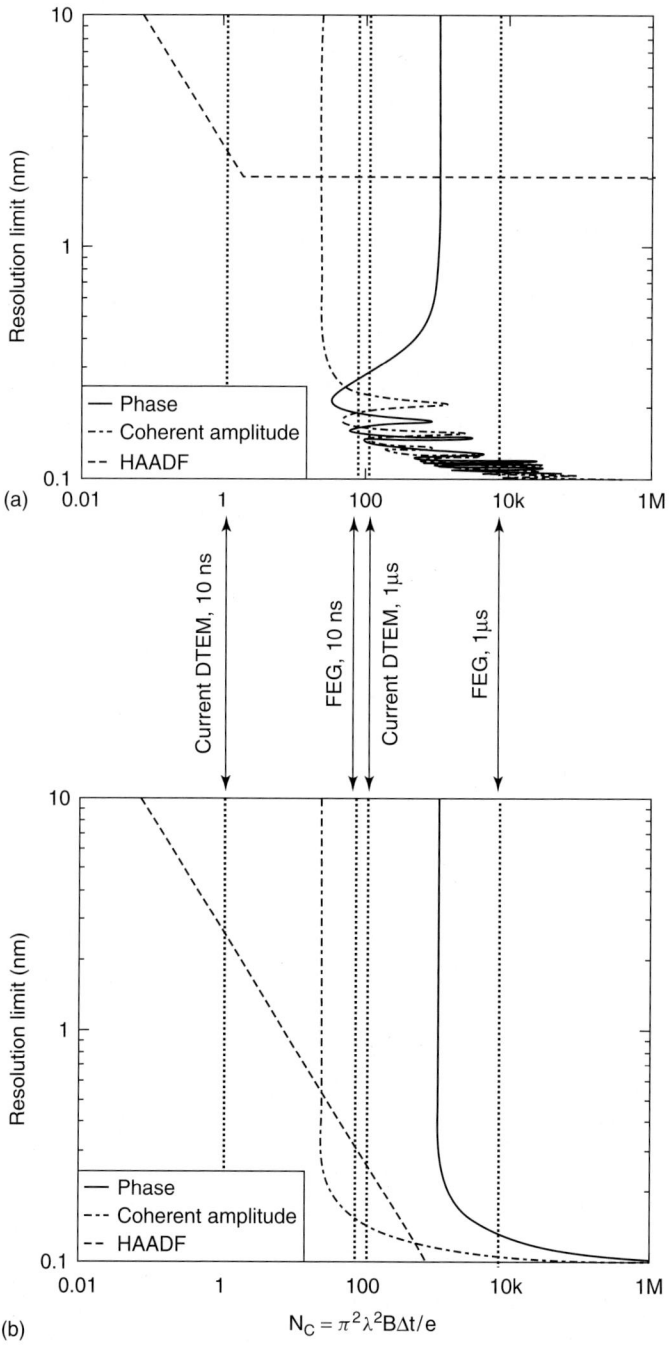

the mass thickness of the sample; we have arbitrarily set it to 0.2 for purposes of rough calculation. The resolution in HAADF-TEM mode will be limited by spherical aberration to $r_{min} \sim C_S \alpha^3$, and we have cut off the spatial resolution below this point for the HAADF curves in Figure 9.3.

From Figure 9.3, we may conclude a number of things, keeping in mind that the models employed are intended for rough estimates only. First, the current DTEM at LLNL should be capable of incoherent imaging on the scale of ~10 ns and a few nanometers, with the use of an HAADF aperture and an ideal 100%-contrast sample. At present, the DTEM reaches better than 10 nm resolution in 15 ns exposures in conventional mass-thickness (incoherent) bright-field imaging [79], with real samples that inevitably have less than 100% contrast, so the calculations seem to be reasonably close to reality in this indirect comparison. Extending the DTEM's pulse duration into the microsecond regime should enable coherent imaging modes, including some phase contrast imaging at resolutions near 0.3 nm. Increasing the brightness to be comparable to that of a field emission gun (FEG) would enable nearly the same performance at 10 ns as would be possible with the current brightness at 1 µs, while the FEG brightness at 1 µs should be capable of atomic resolution imaging over a wide spatial bandwidth. C_S correction (Figure 9.3b) would allow all three imaging modes to push down to angstrom-scale resolution, where the chromatic aberration becomes the limit according to the present models. Values below 0.1 nm were not calculated owing to the breakdown of the approximations in the spatial coherence envelope calculation, since the convergence angles at the coherence-matched condition become very large near this point. Also, we have neglected electron–electron scattering effects in the imaging lens system, which previous calculations [50] indicate may be the dominant resolution limit in this regime.

Addition of a phase plate (which, for small convergence angles, allows the user to swap the coherent amplitude and phase contrast-transfer functions) to an aberration-corrected, high-brightness system could, in principle, enable atomic resolution at the scale of 10 ns, provided the electron–electron scatter can be minimized. Electron–electron scatter may turn out to be a more serious problem than chromatic aberration, which can be either minimized at the source (by reducing the energy spread of the photoemitted electrons) or corrected in the imaging system. It may be that a polarized electron source [67] could help with the electron–electron scattering problem, but at present this concept is quite speculative. A more direct method is to go to much higher accelerating voltages [50]. This may have disadvantages in terms of radiation damage to the sample and

Figure 9.3 (a) C_S-limited and (bottom) C_S-corrected resolution limits as a function of the scaled product of brightness and pulse duration (or coherent fluence) for single-shot imaging. Vertical dashed lines are N_C values for four scenarios, as indicated. The parameters are 200 keV kinetic energy, $C_S = 2$ mm (a), 5 µm (b), $C_C = 2$ mm, $\Delta E = 3$ eV, DTEM brightness 3×10^7 A cm$^{-2} \cdot$ sr^{-1}, FEG brightness 2×10^9 A cm$^{-2} \cdot$ sr^{-1}, $f_{Rose} = 5$, $f_{coherence} = 1$, $\alpha_{HAADF} = 0.01$, $f_{collected} = 0.2$. These curves are for ideal samples with 100% contrast; the curves for real samples will be shifted somewhat to the right.

the difficulty of lens engineering, but it may be just about the only way to achieve single-shot near-atomic resolution in subnanosecond electron pulses. Annular dark-field TEM imaging may also help to reduce this effect.

9.3
Building a DTEM

The ongoing effort at LLNL has focused on the single-shot approach described above, developing a DTEM with this capability [43–52]. The single-shot approach was pioneered by Bostonjoglo and co-workers [53–55] who demonstrated a spatial resolution of ~200 nm with pulse duration <10 ns. Since all the information is obtained from a single specimen drive event, this technique is able to measure irreversible and unique material events that cannot be studied stroboscopically [56–60]. The obvious caveats to the single-shot approach are that it requires a high-brightness source ($>10^7$ A cm^{-2}sr at the very least, and preferably much more) and that electron–electron interactions in high current electron pulses (i.e., nanosecond pulses) will limit spatial resolution. In addition, modern TEMs are not designed for high current operation necessary for single-shot technique, and thus, the instrument must be modified and optimized to accommodate high current pulsed electron probes, making it a much more technically challenging approach. Yet despite its difficulties, the single-shot approach is worth pursuing because it dramatically widens the range of phenomena that can be studied, since most processes of interest in materials science are irreversible. In this section, the design and construction of the LLNL DTEM for single-shot operation, which currently operates with spatial resolution better than 10 nm using 15 ns electron pulses, will be described. This section will discuss the challenges and limitations of the single-shot approach.

9.3.1
The Base Microscope and Experimental Method

The DTEM is built on the JEOL 2000FX microscope platform (Figure 9.4). The electron optical column has been modified to provide laser access to the photocathode and specimen. A brass drift section has been added between the gun alignment coils and condenser optic that contains a 1 in. laser port and a 45° Mo mirror, which directs an on-axis 211 nm laser pulse toward an 825 μm Ta disk photocathode.[1] The 12 ns UV laser pulse photoexcites a 15 ns FWHM electron pulse from the cathode. This pulse is then accelerated through the electron gun and passes through a hole in the Mo[2] laser mirror and into the electron optics of the TEM column. The

---

1) The Ta disk photocathode can also be used as a thermionic source in nominal TEM operational mode.
2) Solid molybdenum mirrors have a high reflectivity (>60%) in the UV range, and molybdenum can be easily polished to produce high-quality, uniform surfaces (low wavefront distortion) needed for laser applications.

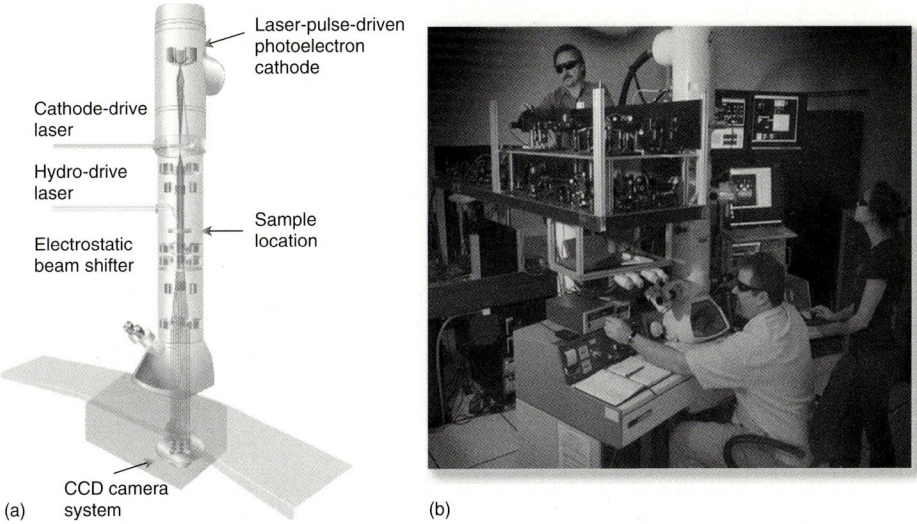

Figure 9.4 (a) Schematic of the LLNL DTEM and (b) installed and operating DTEM at LLNL.

electron pulse is aligned and illuminates specimens as in standard TEM operation, and thus all imaging modes can be utilized, for example, bright-field, dark-field, and selected-area electron diffraction. A critical step in reengineering the TEM for obtaining high current electron pulses in the single-shot mode was adding electron optics and column sections between the accelerator and condenser sections to better couple the photoemitted electron pulse into the condenser electron optics. Specifically, a weak lens has been installed above the Mo laser mirror and brass drift section (Figure 9.5). The lens provides increased current by focusing the spatially broad pulsed electron beam through the hole in the laser mirror and condenser system entrance apertures. In prior configurations without the coupling lens, it was found that the hole in the Mo laser mirror was the limiting aperture and reduced the electron current by a factor of 20. The coupling lens combined with appropriate condenser lens settings and imaging conditions preserves the brightness of the gun and improves beam quality by reducing the aberrations that can result from a spatially broad electron pulse and high-angle, off-axis electrons.

Time-resolved experiments in the DTEM are conducted by first initiating a transient state in the sample and then taking a snapshot of the transient process with the 15 ns electron pulse at some preferred time delay after the initiation. In most DTEM experiments, the transient process is initiated with a second laser pulse that enters the TEM column through a modified high-angle X-ray port. For nanosecond timescale experiments, neodymium-doped YAG lasers with pulse duration from 3 to 25 ns are used that can produce fluences up to 1500 J cm^{-2} on the specimen, which is high enough to turn most specimens into a plasma. Thus, by controlling the laser energy and spot on the sample, wide ranges of temperatures in the sample, and sample heating rates can be produced. The fundamental wavelength

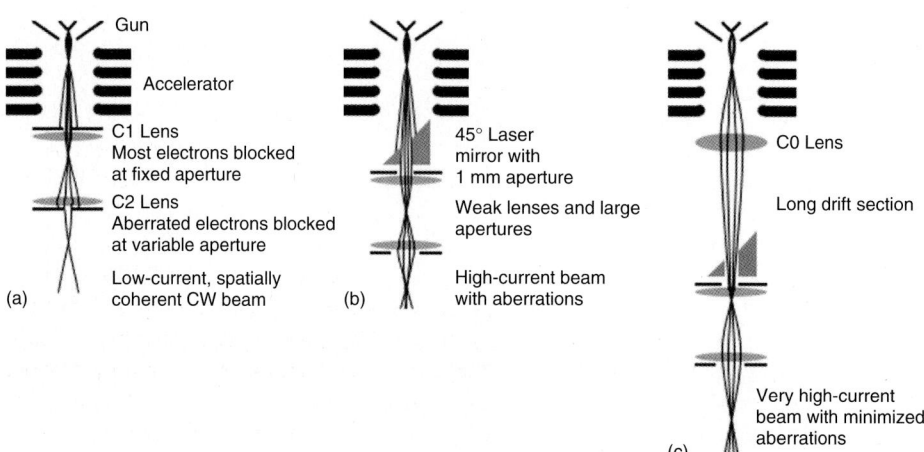

Figure 9.5 Schematic beam paths in the gun, accelerator, and condenser lenses for (a) conventional CW TEM, (b) the first version DTEM, and (c) the current DTEM with the added C0 lens. While the conventional design is an excellent optimization for few second exposure times, many DTEM experiments call for a broader ability to sacrifice spatial coherence for increased signal without introducing excessive aberrations. The weak C0 lens solves this problem.

(1064 nm) of these lasers can be frequency converted using nonlinear harmonic generation crystals, for example, doubled (532 nm), tripled (355 nm), or quadrupled (266 nm), as dictated by the absorption characteristics of the sample and the desired experimental conditions. For instance, metals have broadband absorption and, thus, all of these wavelengths can be used, while certain semiconductors only absorb sufficient amounts of laser energy in the visible or UV range and may require frequency-doubled or frequency-tripled laser pulses.

In a typical DTEM experiment, the instrument and specimen are first aligned in standard, thermionic mode, and then alignment is optimized further in pulsed mode, which requires both precise positioning of the laser on the photocathode and different lens settings due to homogeneous space charge effects and pulse expansion. Since DTEM pump-probe experiments usually permanently alter the sample, the sample is moved to a new location between shots or an entirely new sample may need to be introduced into the microscope. Therefore, the on-average evolution of an irreversible materials process is studied through a series of pump-probe experiments with different time delays. For ultrafast optical pump-probe experiments, the time delay is often generated using an optical delay line. This is impractical for the wide range of time delays used in the DTEM that are sometimes on the microsecond time scale, and it would require 300 m of distance per microsecond of delay time. Instead, the DTEM controls the time delays electronically, triggering each of the two lasers independently, enabling a precision of ~1 ns even with time delays of many microseconds. To enable precise calibration of pulse arrival times, laser energy, and spot sizes, each laser passes through a beam splitter shortly before entering the TEM column, and the split-off beam is directed to cameras, photodiodes, and energy meters placed at a position optically equivalent to the sample position (for the sample drive laser) and the cathode (for the cathode laser).

9.3.2
Current Performance of Single-Shot DTEM

In its current configuration, the DTEM can acquire images with better than 10 nm spatial resolution with 15 ns electron pulses. An example of the current spatial resolution and image capabilities is given in Figure 9.6, which shows a cross-sectional view of gold (dark layers) and carbon (light contrast layers) multilayer foil. The individual layer thicknesses are less than 10 nm, and the layers are clearly resolved in the single-shot image (Figure 9.6a). To better illustrate the resolution limits in the single-shot image, the pixel intensity across the multilayer was measured and plotted (Figure 9.6b). The intensity from 9 nm thick layers is clearly visible above the background, indicating that the resolution is at least 9 nm, if not better. A critical advancement of LLNL DTEM over prior instruments developed by the Bostanjoglo group is the ability to capture dynamical contrast images, previously thought impossible owing to lack of electron beam coherence in the TU Berlin instruments. The unique electron source and lens system of the LLNL DTEM allows higher electron beam current and coherence, enabling the use of objective apertures to obtain bright-field and dark-field images of defects such as grain boundaries, dislocations, and stacking faults. Figure 9.7 shows an example of a single-shot, 15 ns exposure DTEM image that clearly shows the line defects (dislocations) and planar defects (stacking faults) in this stainless steel material. Dynamical image contrast formation is an important capability, since these defects

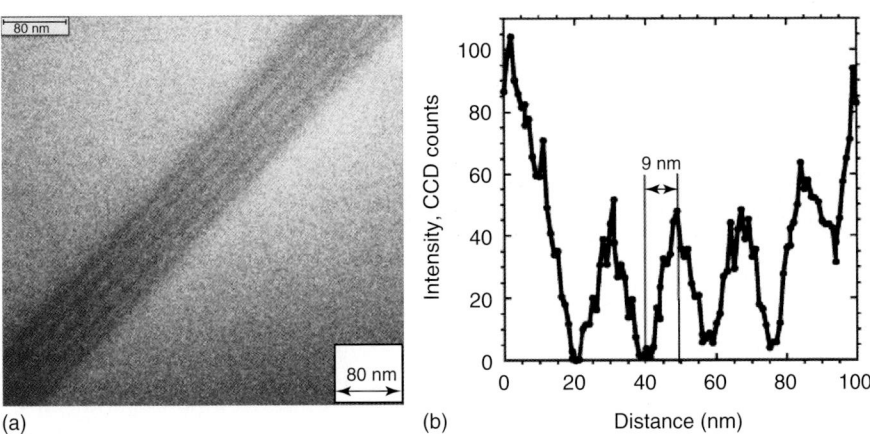

Figure 9.6 (a) Single-shot 15 ns image of a Au/C grating showing resolution less than 10 nm. (b) Plot of the pixel intensity across the multilayer.

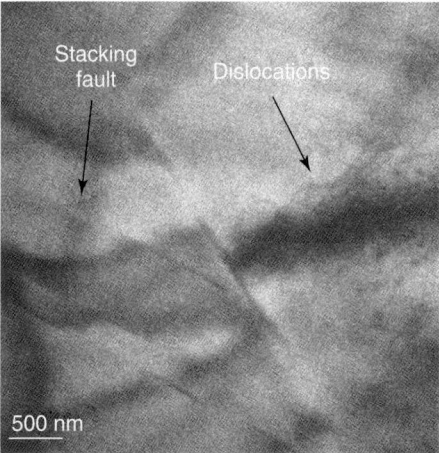

Figure 9.7 Single-shot pulsed image of a stainless steel sample, which clearly shows that DTEM can image material defects such as dislocations and stacking faults with a single 15 ns exposure.

play an important role in the mechanical properties of materials, and capturing their dynamics under a transient thermal or mechanical load can provide a better understanding of how the mechanical behavior of the material evolves. Figure 9.8 shows the effect of including a HAADF detector aperture in the beam path. This form of imaging shows higher contrast and resolution than the bright-field image (as expected from the discussion in Section 9.2).

9.4
Applications of DTEM

There are many applications of single-shot DTEM, ranging from phase transitions to nanoscale catalysis all the way to biological molecules and systems. In this section, we will focus on one particular application of the DTEM – reactive nanolaminate films – which highlights the main techniques and capabilities of the instrument.

9.4.1
Reactive Nanolaminate Films

Reactive nanolaminate films attract interest because of their useful ability to store energy and then release it rapidly. The stored energy is used for various applications requiring localized heating such as soldering temperature-sensitive devices [80, 81], igniting other reactions, or neutralizing biological materials [82]. Reactive nanolaminates also provide an opportunity to study rapid, self-propagating

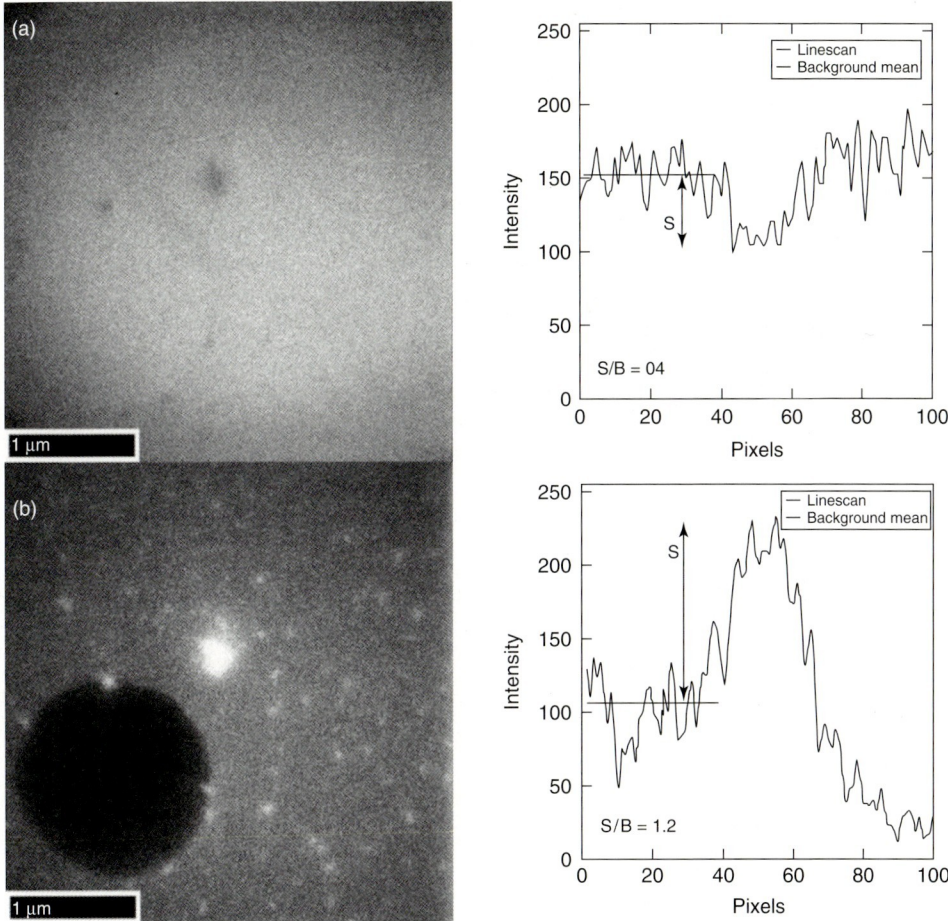

Figure 9.8 (a) Bright-field DTEM, 12 ns pulse duration and (b) ADF DTEM, 12 ns pulse duration. The contrast improvement provided by ADF TEM enables dynamics to be imaged at improved spatial and temporal resolution.

reactions that present a scientifically interesting case of coupled nonequilibrium processes (atomic diffusion, heat diffusion, chemical reactions, and phase transformations) that occur over a wide range of length- and timescales, and produce unexpected transient structures that relax in microseconds [49].

The self-propagating reactions in nanolaminates are initiated in a localized area by an external stimulus and are characterized by the layers interdiffusing, heat being released, and a reaction front propagating parallel to the layers of the reactive multilayer foil (RMLF). The reaction front can have extremely high heating rates of $\sim 10^6 - 10^9$ K s^{-1} [83]. In addition, reaction front speed and temperature can be manipulated by tailoring the layer thickness, foil chemistry, and the extent of pre-mixing [81, 84–86]. However, little is known about how the reactants mix and transform into the final reaction products. Increasing our knowledge of the transient behavior provides insights into the nature of rapid formation reactions and could provide improved methods for tailoring reactive laminate performance for specific applications.

The combination of fast reaction speeds and submicron structure in these self-propagating reactions makes direct characterization of phase transformations, morphological progressions, and rate-limiting mechanisms challenging. Recent studies that use high-speed imaging or *in situ* X-ray diffraction of self-propagating reactions in nanolaminates have increased our understanding of these reactions [83, 87, 88], but these studies still lack nanoscale temporal and spatial resolutions. The DTEM is particularly well suited to address this opportunity in characterization capability. Using the DTEM capabilities, the moving reaction front of reactive nanolaminates has been observed *in situ* [49]. The DTEM images show a transient cellular morphology that disappears during cooling. Such studies provide an opportunity to gain fundamental insights into the mechanisms that control the self-propagation of exothermic reactions. Here we show how the DTEM can identify intermetallic products and phase morphologies as exothermic formation reactions self-propagate in Al/Ni nanolaminate films of Al-rich, Ni-rich, and equiatomic chemistries.

9.4.2
Experimental Methods

The RMLF samples studied in the DTEM have one of three Al–Ni atomic compositions: Al-rich, Ni-rich, or balanced, with Al to $Ni_{0.91}V_{0.09}$ atomic ratios of 3 : 2, 2 : 3, and 1 : 1. The vanadium is added to the Ni to stop its magnetism and thereby simplify sputter deposition. V is a substitutional alloying element in Ni and continues to be substitutional in the intermetallic reaction products. The films were deposited onto polished NaCl crystals for liftoff onto clamping Cu TEM grids using deionized water. The films were capped on top and bottom with half-thickness Ni layers to prevent Al oxidation in air and during the actual experiment, while still maintaining the desired composition. The films had an as-grown total thickness of 125 nm and five to six bilayers, measured by profilometer and TEM cross-section micrographs.

The orientation of the experimental setup inside of the DTEM is shown in the schematic in Figure 9.9. First, the pump laser pulse warms a small region of the RMLF, initiating the reaction at a position far enough from the observation area (hundreds of micrometers) for the reaction front to reach steady-state propagation conditions at the point of observation. The reaction front radiates in all directions from its initiation site in the freestanding RMLF at a constant speed. Next, the probing electron pulse is emitted from the photocathode at a user-defined time. The electron pulse intersects the foil at a location along the microscope optic axis (away from the initiation site) at a time to probe the reaction front as it arrives. As with other pump-probe experiments, it is necessary to define a time-zero reference point for the experiments. Here (with the exception of the low-magnification

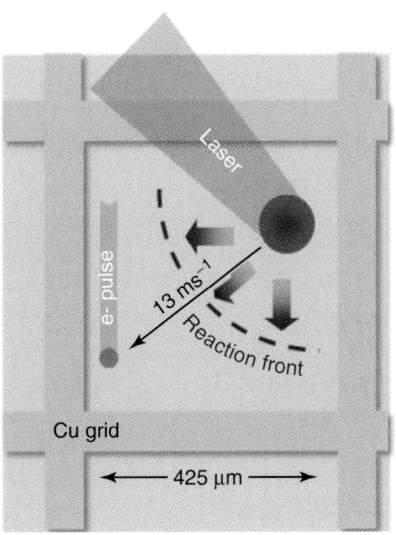

Figure 9.9 Schematic representation of a DTEM snapshot process. In a time-zero snapshot, the electron pulse probes the sample as the reaction front passes the center of the microscope optic axis and micrograph (or area used for diffraction). Plan-view observations are made so that the electron pulse transmits through all of the sample layers and the viewing z-direction is normal to the multilayers.

images described below), time-zero is defined by the arrival of the reaction front at the center of the viewing area. Time-zero is experimentally determined during each session through observations of the reaction front's position in multiple images. In diffraction, this definition is the same, however diffraction is collected from a comparatively large area (up to tens of microns) and the data at time-zero include information from regions both before and after the reaction front. The typical time-resolved diffraction experiments in this work are made up of a series ("before," "during," and "after") of snapshots. Unreacted and reacted diffraction patterns are collected from the foils in a static state and are used as references during experimental setup and for camera length calibration during data analysis. Diffraction experiments were collected using a 100 μm condenser aperture to remove the electrons that would intersect the sample at high angles (thus improving reciprocal-space resolution). Rather than using a selected area diffraction (SAD) aperture, the pulsed beam was spread to a limited diameter of illumination. This method was essential for collecting enough electrons for single-shot analysis.

9.4.3
Diffraction Results

Single-pulse diffraction studies of the reactive foils probe the reacting specimen at a specified time after the pump laser initiation. The collected diffraction ring data are rotationally averaged for an improved signal-to-noise ratio. They are plotted as

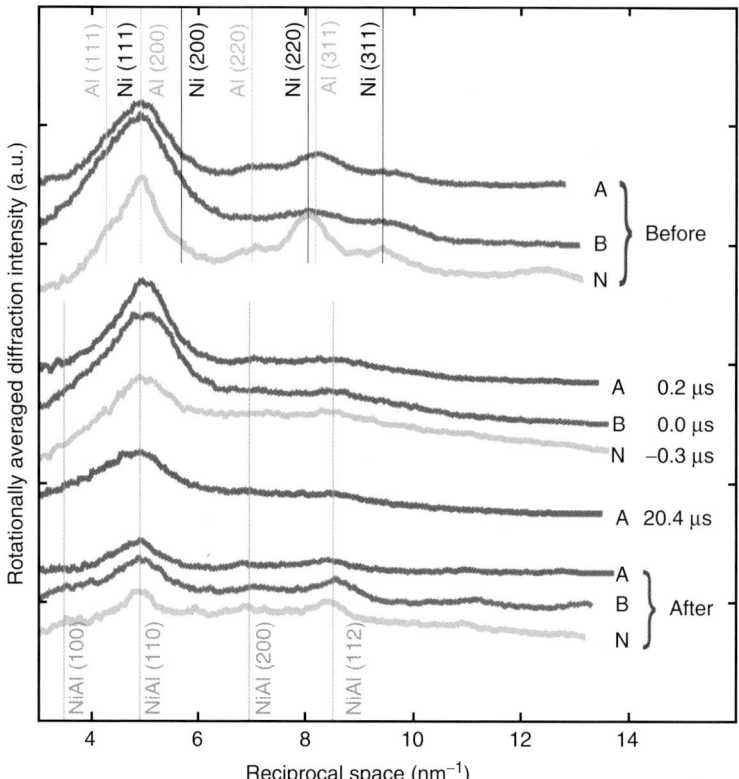

Figure 9.10 Snapshot ring diffraction patterns are displayed as the rotational average with respect to reciprocal spacing for clarity. A, B, and N correspond to foils of Al-rich (3Al:2Ni atomically), balanced (1Al:1Ni atomically), and Ni-rich (2Al:3Ni atomically), respectively. Timing of the single-shot data acquisition is as indicated with reference to the mixing reaction front arrival at the viewing optic axis, t_0 (see Figure 9.9).

intensity versus position in reciprocal space. The results in Figure 9.10 show that all three stoichiometries show only a single reaction product of the intermetallic, ordered B2 phase, NiAl. Diffraction snapshots acquired before the mixing front arrives clearly show the initial as-fabricated samples as a superposition of pure FCC Al and pure FCC Ni patterns. The data acquired as the reaction front arrives (at $t = 0\,\mu s$) contain peaks corresponding to diffraction rings of B2 NiAl. The diffraction peaks obtained after the reaction zone passes are entirely attributable to B2 NiAl. Crystalline peaks are present in all diffraction patterns covering many different time delays, indicating that the films remain crystalline, at least in part, throughout the dynamic reaction.

As the hot reaction front passes through the area probed by the electron pulse, the background intensity increases, and the diffraction ring patterns become more diffuse owing to increased thermal diffuse scattering (TDS). Seconds later, after the specimen is fully reacted and cooled, the samples have converted entirely to

NiAl without signs of transitions to other phases such as Al_3Ni_2 or $AlNi_3$, unlike other findings from differential scanning calorimetry and X-ray diffraction work of various material systems of nanolaminates [83, 89, 90]. The result that all three types of these far-from-equilibrium nanolaminates make a single transition to the NiAl phase should be noted since the material systems are predicted form various other intermetallic phases [91] for the 3:2 and 2:3 Al:Ni compositions, when in equilibrium. TDS increases the background, blurs the ring patterns, and reduces the intensity of the higher-order reflections for times at least to 20.4 μs, as seen in the Al-rich diffraction curve. Data at longer time delays were not acquired since the reacted foils would often tear and move far out of the field of view. By directly comparing 20.4 μs and "after" single-shot data (effectively collected when the thin foil is cooled to ambient temperature), the TDS effects are apparent by the blurring of peak details. This long-lasting TDS effect is also found in the atomically balanced and Ni-rich foils.

9.4.4
Imaging Results

The reaction front was directly imaged using single pulses to observe the layer mixing process and the reaction front morphology. A series of low-magnification images with a $\sim(500\,\mu m)^2$ field of view is shown in Figure 9.11 from the three film compositions. The reaction propagates radially from the laser initiation site. The mixing front appears in the "during" micrograph as a sharp change in contrast that has traveled a particular distance following reaction initiation. Again, the time delay is the difference between the ignition and the snapshot image. In these low-resolution images, the reaction front appears smooth and featureless for all three types of samples.

To examine morphological details at the reaction front at higher resolution, we begin with the Al-rich system. In bright-field imaging conditions, the reaction

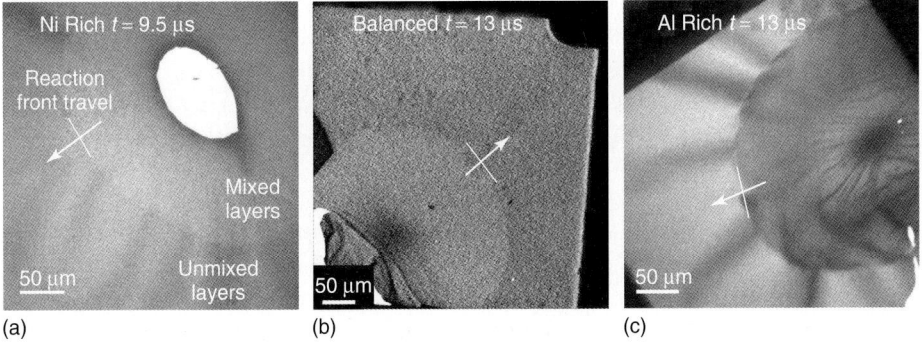

Figure 9.11 (a–c) Low-magnification images showing the propagation of the reaction front for Ni-rich, stoichiometric, and Al-rich samples.

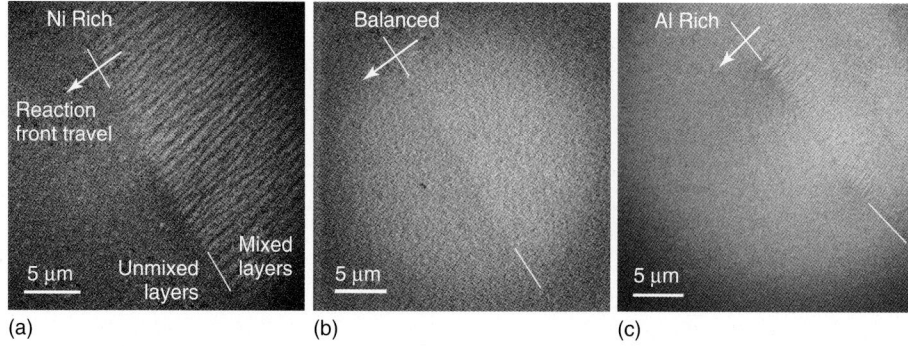

Figure 9.12 (a) Time-resolved image of the reaction front in Al-rich foils grown to a 3:2 Al:Ni ratio imaged with 15 ns temporal resolution. The dotted line approximately indicates the mixing front as it travels toward the lower left of the image. Cellular features are observed that are 3 μm long. (b) Time-resolved image of the stoichiometrically balanced foils show the passing of the reaction front in which no cellular structure is observed. (c) Time-resolved image of the Ni-rich mixing front showing the presence of a cellular structure, albeit existing over a time and length scale different from that of the Al-rich sample.

front is closely followed by an elongated, fingerlike cellular structure (Figure 9.12). Although these features are somewhat blurred at the front, they appear to have slightly scalloped edges. In this image, the transient cellular features are observed in their entirety, from start to finish, and are measured to be ~3.2 μm long with a periodicity of 390 ± 55 nm. On the basis of the length of the cellular features and the reaction front velocity, a straightforward calculation shows that a given point on the foil remains in a cellular morphology for roughly 230 ns. Immediately behind the reaction front, the material was found by diffraction to be nanocrystalline NiAl. The electron pulse interaction area was limited to a 3.3 μm diameter region on the foil, making the Al-rich diffraction work accurate to this dimension. As the metastable cellular features are quite short, there is a small possibility that other intermetallic phases may be forming temporarily in this cellular region. This, however, seems unlikely, because in over 20 Al-rich diffraction experiments near the reaction front we only observed conversion to the NiAl phase. No discernable intermetallic phases aside from NiAl were seen at any point in time. Thus if other crystalline phases are present, they must exist either very briefly or in very small quantities, or both.

Similar bright-field results for the Ni-rich system are also shown (Figure 9.12) for comparison with the Al-rich system. The cellular features are substantially longer than those observed in Al-rich films, having a length greater than 40 μm (equivalent to a 3 μs duration at a given point) and a periodicity of 790 ± 220 nm in width. The Ni-rich cellular features are also more uniform in morphology compared to the markedly tortuous features found in the Al-rich foils although they show a similar distinct line of formation and gradual termination, establishing their transient nature. The dark lines between the cellular features are 220 ± 80 nm wide.

Similar to the Al-rich films, it appears that the cellular features behind the reaction front are at least partially solid NiAl intermetallic. Diffraction data obtained for these Ni-rich foils (Figure 9.11) confirmed the almost immediate formation of NiAl at the reaction front with an error in timing of ± 0.3 µs. In diffraction work, a condenser aperture was used so that the beam interacts with a round sample area 11.2 µm in diameter, far smaller than the 40 µm width of the cellular formation region. Diffraction work of the previous section taken in synchronization with the arrival of the visible reaction front (from imaging) proves that crystalline NiAl intermetallic formation occurred within 300 ns of $t = 0$, which would be equivalent to a distance 3.9 µm behind the reaction front. The dark intensity between the "cells" fades away in the last micrograph of the series in Figure 9.14, at a point in time where the reaction should have completed some microseconds earlier. The solid solubility range of the NiAl B2 phase increases as the temperature drops [90], so that at room temperature nearly all of the excess nickel could be reabsorbed into a stable homogeneous B2 structure with a Ni : Al ratio near 3 : 2.

To complete the set of experiments, Figure 9.12 also shows the results of experiments where the reaction front is observed in the balanced composition foils. The front appears as a moderately straight, featureless variation in intensity over a distance less than 1 µm wide. No cellular structure is observed in this case.

9.4.5
Discussion

The as-deposited Al/Ni reactive multilayer films are in a state that is far from equilibrium. As the films react, the high reaction temperatures and nanoscale diffusion distances allow them to approach equilibrium rapidly. To understand the observations of morphological transients in the Al-rich and Ni-rich samples and the lack of such transients in the balanced samples, we consider regions of the equilibrium Al–Ni phase diagram where phase separation of the NiAl intermetallic and Al–Ni liquid is expected. Considering the reaction temperature and as-grown chemical compositions, the Ni-rich and Al-rich foils both fall into two-phase fields of NiAl + liquid on either side of the congruent melting point in the equilibrium phase diagram [91]. The stoichiometrically balanced foils reside in the solid NiAl region of the phase diagram throughout the reaction process and fall short of entering the liquid phase field. Thus, the equilibrium phase diagram suggests the Al-rich and Ni-rich samples could show phase separation while the balanced foils should not.

The diffraction patterns show that all of the foils are at least partially solid at all times, to within the time resolution of 15 ns, equivalent to 200 nm of reaction front travel. Additionally, time-resolved imaging data suggest that cellular features form in a time of <100 ns. Two types of phenomena may be proposed to explain the sudden formation of the cellular structure: *solid + liquid* or *solid + solid* phase separations. It is improbable that the phase separation is governed by a mechanism of purely solid-state diffusion [92], considering that features 390–790 nm wide form in ~100 ns. Intrinsic solid-state diffusion coefficients, D, are 1.9 and 0.65 μm^2/s at T_m and D_0, the pre-exponential factor is 1.7 and 1.9 m^2 s^{-1} for Al and Ni, respectively [93]. These diffusivities are too low to account for the features observed with these large separation distances. Using these numbers, we calculate a characteristic distance for solid-state diffusion, $d = (Dt)^{1/2}$, of about 0.3 nm. If we use the calculated diffusivity of Ni in *liquid* Al of 1.8×10^{-8} m^2/s [94] at 1770 K, then the characteristic length is about 40 nm, enough to ensure complete mixing of neighboring layers but not enough to explain formation of the cellular structure (390–790 nm) via liquid-state diffusion in ~100 ns. Thus the explanation must invoke a mass transport mechanism significantly faster than solid- or liquid-state diffusion, leaving bulk convective motion in the liquid phase as the most likely candidate.

Liquid is capable of rapid convective motion once it forms as a separate phase, including beading up on a surface under the influence of surface tension. Therefore, we hypothesize that the observed cellular structures arise from a complex evolution of mechanisms, as follows: heat from the reaction front arrives at a given point, causing solid-state diffusion to initiate interdiffusion. The atomic mixing releases heat to further promote the process. When the local temperature is high enough, the Al layers melt and atomic intermixing accelerates further. The liquid layers provide enhanced mobility through convective flow and permit rapid coalescence of the liquid layers into thicker structures that will evolve to minimize their interface energies. If the liquid volumes extend through the thickness of the films, they would be visible as cellular structures in plan view, as seen in Figure 9.12. As the liquid Al(Ni) layers are coalescing, atomic diffusion in the liquid (near the interfaces) can lead to chemical ordering and the nucleation of NiAl. Thus, we hypothesize that the cellular structures form in the Al-rich and Ni-rich samples through a coalescence of the liquid layers within the films and the nucleation of the B2 NiAl intermetallic. When these samples cool and reenter a single-phase region, the remaining liquid is absorbed back into the solid B2 NiAl intermetallic.

This hypothesis is consistent with the observations reported here and with the known thermodynamic properties of the materials. It is especially telling that the balanced 1:1 stoichiometric foils do not produce a visible transient cellular structure during reaction propagation. At 1700 K, they are expected to transition directly into NiAl, as is detected in the diffraction patterns. The calculated maximum temperature does not exceed the NiAl melting temperature of 1911 K. Thus, while the Al layers near the reaction front will melt and thereby enhance chemical mixing, the liquid phase is not an equilibrium phase during the rapid heating experiment. Only the B2 NiAl intermetallic is in equilibrium in the binary samples. Thus, the binary samples are unable to form liquid solutions that are stable enough to allow for coalescence and formation of a solid/liquid cellular structure. The lack of cellular structure in the stoichiometric film therefore supports the hypothesis that bulk motion of the liquid is essential to the formation of these structures.

In conclusion, in all three types of RMLFs studied, the NiAl intermetallic formed within nanoseconds after the mixing reaction front and it is the only phase present after the foils are reacted. Dynamic imaging revealed the presence of transient cellular structures only in off-stoichiometric foils. The cellular features are found to form as the material passes through a two-phase field of liquid + NiAl. Varying dimensions of the cellular features appear, which, we speculate, can be traced to differences in excess reaction energy and atomic mobility. Unlike in slow heating experiments, Al_3Ni_2 does not form in Al-rich foils owing to Al depletion during the self-propagating reactions; therefore, the material at the reaction zone does not travel through a purely thermal trajectory into the two-phase field of liquid + NiAl. Finally, this study confirms that new questions on very fine spatiotemporal scales can be answered using DTEM.

9.5
Future Developments for DTEM

DTEM as described in the previous sections is a work in progress. Given the advanced level of understanding of the electron optics requirements, there are some clear pathways forward to improve the performance of the instrument. Some of these developments are already underway – inclusion of an arbitrary waveform generator, movie mode, the use of aberration correctors, and the incorporation of *in situ* stages. Others are concepts that still require further development – novel electron sources, the use of pulse compression, and moving to relativistic energy beams (up to 5 MV). In this section, the use and benefits of these new additions to DTEM will be discussed in detail.

9.5.1
Arbitrary Waveform Generation Laser

At nanosecond timescales, the electron pulse duration corresponds closely with the laser pulse duration, that is, there is little temporal broadening of the pulse from space charge effects. Q-switched lasers can generate high-energy optical pulses with lengths from 1 to 100 ns. The current DTEM uses a neodymium-doped yttrium lithium fluoride laser that produces a pulse with a wavelength of 1053 nm and FWHM pulse duration of 25 ns. This pulse undergoes two nonlinear processes (frequency multiplication to the fifth harmonic and photoemission in the space-charge-limited regime), with the net effect that that electron pulse has a 15 ns duration.

Since not all experiments require nanosecond time resolution, a compromise can be made between the temporal and spatial resolutions (as described in detail in Section 9.2). For instance, microsecond time resolution is sufficient for some catalytic reactions and dislocation dynamics studies, and these studies can benefit from the added electron dose and increased spatial resolution possible with longer exposure times. For most Q-switched laser systems, the resonator cavity, lasing medium, and excitation source determine the pulse duration and repetition rate. Therefore these parameters are very difficult to change in routine operation. To enable a flexible laser system and tailor the laser parameters for a given experiment, an arbitrary waveform generation (AWG) laser can be used that can temporally shape laser pulses, thus allowing easy changes to the pulse duration.

For example, an AWG laser system being developed for the existing DTEM includes a waveform generator that drives a fiber-based electro-optical modulator, thus temporally shaping a continuous wave fiber laser seed. The modulated waveforms are then amplified using Yb-fiber-based and bulk, diode-pumped YAG amplifiers. The final IR pulses can have energies in excess of 1 J, and they can be frequency converted and delivered to the DTEM at appropriate energy for use as cathode or specimen drives. The AWG cathode drive laser enables continuously variable and controlled electron pulse durations from 250 µs down to 5 ns. Using an AWG laser to drive sample reactions enables precise control of pulse time and shape so that we can achieve better control of the drive conditions, thus providing a wide variety of temperature–time profiles on the sample. At present, the sample is rapidly heated with a pulse of ~10 ns duration (a number which we have very little ability to change), and it then cools according to its own thermal conductivity, usually on a timescales of 10–100 µs. By compensating the thermal conduction heat loss from the sample region of interest with an appropriately shaped pulse, one can maintain the sample temperature at a constant level for some microseconds before allowing it to cool. This will enable a broad class of experiments in catalysis and surface science. The AWG capability also enables the production of electron pulse trains, an essential element of the development of high time resolution movies.

9.5.2
Acquiring High Time Resolution Movies

The ability to acquire high time resolution movies (dubbed *Movie Mode DTEM*) expands the DTEM's science capabilities in single-shot mode by providing detailed histories of unique material events on the nanometer and nanosecond scale. The existing DTEM uses single-pump/single-probe operation, building up a process's typical time history by repeating an experiment with varying time delays at different sample locations. The Movie Mode DTEM upgrade currently being installed will enable single-pump/multiprobe operation. This will provide the ability to track the creation, motion, and interaction of individual defects, phase fronts, and chemical reaction fronts, providing invaluable information of the chemical, microstructural, and atomic level features that influence the dynamics and kinetics of rapid material processes. For example, the potency of a nucleation site is governed by many factors related to defects, local chemistry, and so on. While a single pump-probe snapshot provides statistical data about these factors, a multiframe movie of a unique event allows all of the factors to be identified and explored in detail. It provides unprecedented insight into the physics of rapid material processes from their early stages (e.g., nucleation) to completion, giving direct, unambiguous information regarding the dynamics of complex processes.

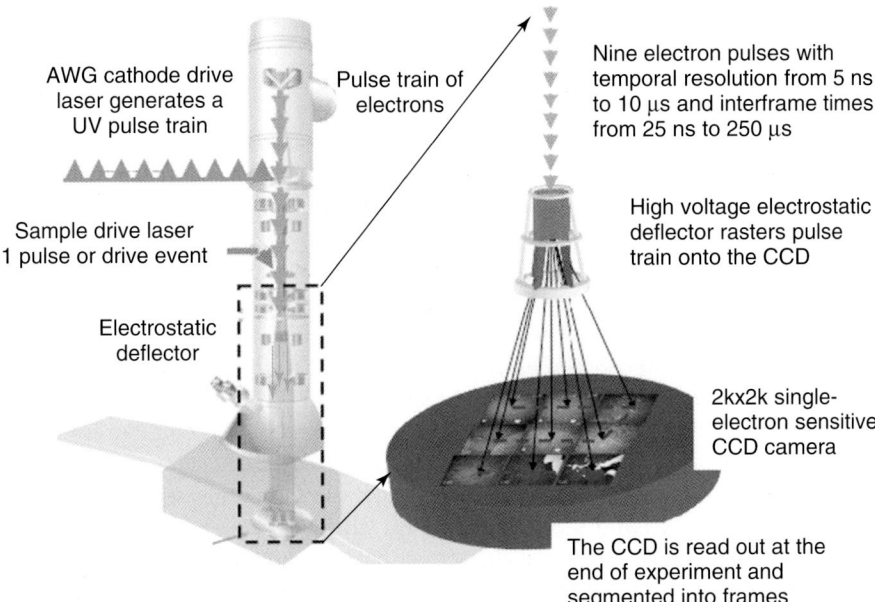

Figure 9.13 Schematic of the movie mode technology that enables single-pump/multiprobe operation and true, *in situ* microscopy capabilities in the DTEM in which multiframe movies of ultrafast material dynamics can be acquired.

A core component of the movie mode technology (Figure 9.13) is the AWG cathode laser system in which a series of laser pulses can be produced with user-defined pulse durations and delays that stimulates a defined photoemitted electron pulse train for a single sample drive event. Each pulse will capture an image of the sample at a specific time. A fast-switching electrostatic deflector located below the sample directs each image to a separate patch on a large, high-resolution CCD camera. At the end of the experiment, the entire CCD image is read out and segmented into a time-ordered series of images, that is, a movie. The current technology should enable up to 25-frame movies with interframe times as low as 25 ns. This frame rate is 6 orders of magnitude faster than modern video-rate *in situ* TEM. Future versions of movie mode may include fast-framing CCD technology, which can capture hundreds of frames within a few microseconds. The operating principle of these devices is that the photoelectron CCD data from multiple frames is stored in on-chip buffers that are read out at the end of the acquisition.

9.5.3
Aberration Correction

To further maximize the current and the spatial coherence in the illumination on the specimen, the pre-specimen objective lens should be corrected for spherical aberration. Aberration correctors have been employed on several commercial instruments, and work is underway to implement their use with the short electron pulses on a DTEM currently being installed at UC-Davis. Correcting for spherical aberration appears to reduce incoherent broadening of the contrast-transfer function and accentuate phase contrast from small signals, making them attractive, for example, for use in liquid cell biological imaging (see the next section). It also allows the use of larger aperture and convergence angles at the specimen, thereby increasing the dose without significant loss in beam coherence. To ensure that the maximum contrast is maintained after the beam has interacted with the specimen, the post-specimen objective lens also should be aberration corrected. In this case, the lens should be optimally corrected for both spherical and chromatic aberrations (the instrument being installed at UC-Davis is only spherical aberration corrected after the specimen). The spherical aberration correction has the same advantages as for the pre-specimen lens. Additionally, the chromatic aberration correction will reduce the energy spread associated with photoelectron generation and the global space charge effect, which will enable high-resolution images with even a large, initial energy spread. Current DTEMs have an energy spread in the beam that is estimated to be of ∼8–10 eV. With the use of the RF cavity for pulse compression (see later), this energy spread may increase by a factor of 10. Although chromatic aberration correctors are still being developed for conventional TEM use, the current designs should be able to cope with energy spreads up to ∼100 eV, making them viable for all DTEM applications.

9.5.4
In Situ Liquid Stages

The tunable drive laser described earlier provides a unique capability to initiate reactions for observation within the DTEM. However, observing reactions inside the high vacuum of a microscope does not yield information on how a reaction propagates in the material's native environment. To address this current limitation, *in situ* gas and fluid TEM stages have been designed to expand the experimental regime applicable to DTEM imaging. Recent progress in the field of *in situ* TEM research has focused on imaging interfaces between nanomaterials and surrounding media such as liquid and gas. The two main developments for *in situ* TEM research have been the modification of side-entry TEM stages housing sealed environmental chambers [95–98] and the modification of an entire microscope to accommodate low vapor pressure and liquid or gas injection [99]. In the case of the DTEM, the sealed environmental chamber provides the best opportunity to accurately control the gas/liquid environment around the specimen while maintaining the electron optics of the column (DTEM already changes the electron optical approach to the column, so using the sealed stages reduces the need for more complex arrangements).

The assembly of the environmental chambers for both *in situ* gas and fluid stages is nearly identical. Both types of stages confine the sample between two electron-transparent membranes (generally amorphous silicon nitride). These membranes create a vacuum seal to the outside environment (such as the high vacuum of a conventional TEM) while providing a fully enclosed chamber containing the gas or liquid solution of choice. Additionally, the fluid and gas path lengths of the stages can be varied to optimize experiments for attainable resolution and contrast or accommodate samples of different sizes. For example, the *in situ* fluid stages can be imaged at fluid thicknesses from 50 to 1000 nm while the *in situ* gas stages have a range of 50–250 μm.

Obviously, these *in situ* stages can be utilized with conventional TEMs to observe reactions that are regulated electrochemically or thermally but on timescales limited to fast-readout cameras (typically several milliseconds). Figure 9.14 depicts results from an *in situ* continuous-flow fluid stage. In this experiment, citrate-stabilized 10 nm gold nanoparticles were imaged at ambient temperature with a JEOL JEM-2100F TEM and a fluid path length of 500 nm. On parallel beam illumination of the sample, the gold nanoparticles show evidence of movement and flocculation over time as indicated by the white arrows in Figure 9.14a,d. While the mechanism of this electron-induced flocculation is currently being investigated, this example demonstrates the capability of imaging materials with *in situ* fluid stages. By combining these stages with the fast time resolution of the DTEM, mechanisms of nucleation, growth, assembly/disassembly, and morphological changes can be

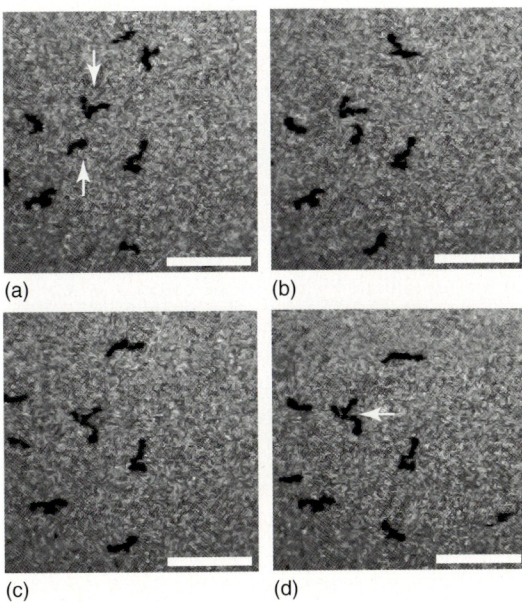

Figure 9.14 Time series depicting electron beam–induced movement and flocculation of 10 nm gold nanoparticles in water using the *in situ* fluid stage at UC-Davis. Two nanoparticle clusters in (a) combine to form a single cluster in (d) as indicated by white arrows. Panels (b–d) were acquired 30, 60, and 90 s respectively after (a). Scale bars represent 100 nm for all panels.

studied at high temporal and spatial resolution in fields ranging from materials science to biology.

9.5.5
Novel Electron Sources

To achieve higher spatial resolution and produce higher-quality bright-field images, the gun brightness in DTEM must be improved. Photoemission can achieve a gun brightness in excess of 10^{10} A cm^{-2} sr^{-1}. The photocathode brightness can be substantially improved by optimizing the source material, laser parameters, and high-voltage accelerator design to reduce emittance growth and increase the photoemitted electron current. Another essential element in the source design is to appropriately match the laser photon energy to the cathode work function in order to minimize the thermal energy of the photoemitted electrons and associated thermal emittance that reduces gun brightness. The optimal source material and laser parameters, therefore, must be chosen on the basis of the best compromise between quantum efficiency and thermal emittance and emittance growth effects. In practice, the range of available photocathode materials and laser wavelengths facilitates this optimization. However, to fully optimize the system and electron currents, the gun and accelerator section in the DTEM must be modified to increase the electric field at the cathode surface.

The Child–Langmuir effect limits the steady-state current that may be emitted from a cathode as a function of the cathode–anode separation and the potential difference. In general, the main approaches for increasing current in future DTEM designs are the same as those that have been developed for other high-brightness electron sources: increase the electric field at the surface and decrease the acceleration gap. The electron accelerators of modern TEMs are designed so as to provide for smooth acceleration of the beam to reduce emittance growth and increase spatial and spectroscopic imaging resolution, and thus, the electric field gradients are small. This is true except in the immediate vicinity of the cathode, where the Wehnelt (for a thermionic geometry) or the extraction and/or suppression electrodes (for a field emission geometry) can create highly inhomogeneous electric fields. An alternate solution is to use RF-based photoguns that generate strong (100 MV m^{-1}), spatially smooth electric fields and have high brightness. These guns also produce megaelectron volt electrons, which can reduce space charge effects and increase spatiotemporal resolution. Other solutions for gun designs may include the use of a cooled gas to emit an ultracold gas of electrons. Ultimately, the electron gun must generate both high peak current and high beam quality to obtain the brightness needed for subnanosecond single-shot images. As noted above, however, the fundamental properties of electrons will likely force the beam to have a large energy spread, so that chromatic aberration correction may be required for subnanosecond single-shot imaging.

This capability will enable observation of very fast, irreversible processes in materials occurring on the subnanosecond timescales, such as irreversible phase transformation induced by high-pressure impulse loading or microstructural evolution induced by energetic particle bombardment. Operating in the megaelectron volt range has the additional benefit of greatly increasing the maximum sample thickness (in transmission) from the tens of nanometers attainable with current instruments to the micrometer scale, which may reduce surface effects that may dominate processes in thin foils. Pulsed RF guns have been used for many years as electron sources for synchrotrons and can generate extremely bright, pulsed electron beams [100–105]. The bright beams can be generated because of the high RF fields (typically 100 MV m^{-1} or more) used to accelerate the beam over a short distance. The beams become relativistic in a few centimeters, quickly reducing the space–charge force repulsion, and the electrons are typically accelerated to ~5–6 MeV ranges, depending on the RF amplitude and the electron beam launch phase. In principle, electron optics can be engineered to operate with such a gun and electron energy ranges. However, megaelectron volt DTEM may look much more like an accelerator than an HVEM.

9.5.6
Pulse Compression

At present, DTEM time resolution is set by the electron pulse duration, which is essentially equal to the cathode laser pulse duration, and is dictated by the experiment's fluence requirements and the electron gun brightness. The size, shape, and temporal resolution of the electron pulse impacting the specimen is a function of the broadening and de-coherence effects in the column. In a next-generation DTEM, an RF cavity will conceivably be incorporated into the column and used to shape the pulse by applying a field, for example, to speed up the electrons at the back of the pulse and slow down electrons at the front of the pulse, thereby shortening the overall pulse duration. It may also have the flexibility to shape the pulse in more ways than just longitudinal compression. RF cavity compression devices are commonly used in acceleration facilities to compress multimegavolt pulses of nanocoulomb charge down to picosecond durations while maintaining normalized emittance values ~1μm rad and energy spread less than ~1%. The location of the RF cavity is likely to be between the condenser lens and the objective lens, allowing for the shortest distance (time) after the electron pulse shaping until the interaction with the specimen, while still allowing control of the pulse with a final lens. Pulse compression may be the only way to dig deeply into the subnanosecond regime with kilovolt accelerating voltages.

9.6
Conclusions

The use of photoemission sources has been shown to provide the ability to study the dynamics in materials and biological systems with nanometer and nanosecond resolution in single-shot mode. The DTEM that is currently operational uses the most basic electron optical components and can be readily upgraded to improve the overall combined spatiotemporal performance. By incorporating *in situ* stages into the microscope, dynamic processes under widely varying environmental conditions can also be studied. In this case, the ability to control the beam and the stimulus to the specimen through a laser will provide unprecedented control and reproducibility to experiments that are not afforded by conventional microscopes and heating stages.

Acknowledgments

Aspects of this work were performed under the auspices of the U.S. Department of Energy by Lawrence Livermore National Laboratory and supported by the Office of Science, Office of Basic Energy Sciences, Division of Materials Sciences and Engineering, of the U.S. Department of Energy under Contract DE-AC52-07NA27344. Aspects of this work were also supported by DOE NNSA-SSAA grant number DE-FG52-06NA26213, by NIH grant number RR025032-01, by NSF grant, and by ExxonMobil.

References

1. Hirsch, P.B., Horne, R.W., and Whelan, M.J. (1956) *Philos. Mag.*, **1**, 677.
2. Jia, C.L. and Urban, K. (2004) *Science*, **303**, 2001–2004.
3. Haider, M., Uhlemann, S., Schwan, E., Rose, H., Kabius, B., and Urban, K. (1998) *Nature*, **392**, 768–769.
4. Batson, P.E., Dellby, N., and Krivanek, O.L. (2002) *Nature*, **418**, 617–620.
5. Erni, R., Rossell, M.D., Kisielowski, C., and Dahmen, U. (2009) *Phys. Rev. Lett.*, **102**, 096101.
6. Muller, D.A., Kourkoutis, L.F., Murfitt, M., Song, J.H., Wang, H.Y., Silcox, J., Dellby, N., and Krivanek, O.L. (2008) *Science*, **319**, 1073–1076.
7. Kimoto, K., Asaka, T., Nagai, T., Saito, M., Matsui, Y., and Ishizuka, K. (2007) *Nature*, **450**, 702–704.
8. Lazar, S., Hebert, C., and Zandbergen, H.W. (2004) *Ultramicroscopy*, **98**, 249–257.
9. Mitterbauer, C., Kothleitner, G., Grogger, W., Zandbergen, H., Freitag, B., Tiemeijer, P., and Hofer, F. (2003) *Ultramicroscopy*, **96**, 469–480.
10. Nelayah, J., Kociak, M., Stephan, O., de Abajo, F.J.G., Tence, M., Henrard, L., Taverna, D., Pastoriza-Santos, I., Liz-Marzan, L.M., and Colliex, C. (2007) *Nat. Phys.*, **3**, 348–353.
11. Arslan, I., Hyun, J.K., Erni, R., Fairchild, M.N., Hersee, S.D., and Muller, D.A. (2009) *Nano Lett.*, **9**, 4073–4077.
12. Ortalan, V., Uzun, A., Gates, B.C., and Browning, N.D. (2010) Towards full structure determination of bimetallic nanoparticles with an aberration corrected electron microscope. *Nature Nanotechnology*, **5**, 843–847.
13. Espinosa, H.D., Zhu, Y., and Moldovan, N. (2007) *J. Microelectromech. Syst.*, **16**, 1219–1231.
14. Gai, P.L. (2002) *Top. Catal.*, **21**, 161–173.
15. Robach, J.S., Robertson, I.M., Lee, H.J., and Wirth, B.D. (2006) *Acta Mater.*, **54**, 1679–1690.
16. Deb, P., Rawat, V., Kim, H., Kim, S., Oliver, M., Marshall, M., Stach, E., and Sands, T. (2005) *Nano Lett.*, **5**, 1847–1851.
17. Xin, R.L., Leng, Y., and Wang, N. (2006) *J. Cryst. Growth*, **289**, 339–344.
18. Han, W.Q., Wu, L.J., Zhu, Y.M., and Strongin, M. (2005) *Nano Lett.*, **5**, 1419–1422.

19. Min, K.H., Sinclair, R., Park, I.S., Kim, S.T., and Chung, U.I. (2005) *Philos. Mag.*, **85**, 2049–2063.
20. Howe, J.M. and Saka, H. (2004) *MRS Bull.*, **29**, 951–957.
21. van der Veen, J.F. and Reichert, H. (2004) *MRS Bull.*, **29**, 958–962.
22. Oh, S.H., Kauffmann, Y., Scheu, C., Kaplan, W.D., and Rühle, M. (2005) *Science*, **310**, 661–663.
23. Radisic, A., Ross, F.M., and Searson, P.C. (2006) *J. Phys. Chem. B*, **110**, 7862–7868.
24. Radisic, A., Vereecken, P.M., Hannon, J.B., Searson, P.C., and Ross, F.M. (2006) *Nano Lett.*, **6**, 238–242.
25. Radisic, A., Vereecken, P.M., Searson, P.C., and Ross, F.M. (2006) *Surf. Sci.*, **600**, 1817–1826.
26. Minor, A.M., Lilleodden, E.T., Stach, E.A., and Morris, J.W. Jr. (2004) *J. Mater. Res.*, **19**, 176–182.
27. Zhou, G.W. and Yang, J.C. (2005) *J. Mater. Res.*, **20**, 1684.
28. Hofmann, S., Sharma, R., Ducati, C., Du, G., Mattevi, C., Cepek, C., Cantoro, M., Pisana, S., Parvez, A., Cervantes-Sodi, F., Ferrari, A.C., Dunin-Borkowski, R., Lizzit, S., Petaccia, L., Goldoni, A., and Robertson, J. (2007) *Nano Lett.*, **7**, 602.
29. Sharma, R. (2005) *J. Mater. Res.*, **20**, 1695.
30. Minor, A.M., Asif, S.A.S., Shan, Z.W., Stach, E.A., Cyrankowski, E., Wyrobek, T.J., and Warren, O.L. (2006) *Nat. Mater.*, **5**, 697.
31. Shan, Z.W., Mishra, R.K., Asif, S.A.S., Warren, O.L., and Minor, A.M. (2008) *Nat. Mater.*, **7**, 115.
32. Clark, B.G., Robertson, I.M., Dougherty, L.M., Ahn, D.C., and Sofronis, P. (2005) *J. Mater. Res.*, **20**, 1792.
33. Loveridge-Smith, A., Allen, A., Belak, J., Boehly, T., Hauer, A., Holian, B., Kalantar, D., Kyrala, G., Lee, R.W., Lomdahl, P., Meyers, M.A., Paisley, D., Pollaine, S., Remington, B., Swift, D.C., Weber, S., and Wark, J.S. (2001) *Phys. Rev. Lett.*, **86**, 2349.
34. Kalantar, D.H., Belak, J.F., Collins, G.W., Colvin, J.D., Davies, H.M., Eggert, J.H., Germann, T.C., Hawreliak, J., Holian, B.L., Kadau, K., Lomdahl, P.S., Lorenzana, H.E., Meyers, M.A., Rosolankova, K., Schneider, M.S., Sheppard, J., Stolken, J.S., and Wark, J.S. (2005) *Phys. Rev. Let.*, **95**, 075502 - 1–075502-4.
35. Hawreliak, J.A., Kalantar, D.H., Stolken, J.S., Remington, B.A., Lorenzana, H.E., and Wark, J.S. (2008) *Phys. Rev. B*, **78**, 220101.
36. Ng, A., Parfeniuk, D., and DaSilva, L. (1985) *Phys. Rev. Lett.*, **54**, 2604.
37. Kalantar, D.H., Remington, B.A., Colvin, J.D., Mikaelian, K.O., Weber, S.V., Wiley, L.G., Wark, J.S., Loveridge, A., Allen, A.M., and Hauer, A.A. (2000) *Phys. Plasmas*, **7**, 1999.
38. Spence, J. and Howells, M.R. (2002) *Ultramicroscopy*, **93**, 213–222.
39. Zewail, A.H. (2006) *Annu. Rev. Phys. Chem.*, **57**, 65–103.
40. Zewail, A.H. (2000) *J. Phys. Chem. A*, **104**, 5660–5694.
41. Siwick, B.J., Dwyer, J.R., Jordan, R.E., and Miller, R.J.D. (2003) *Science*, **302**, 1382–1385.
42. Cao, J., Hao, Z., Park, H., Tao, C., Kau, D., and Blaszczyk, L. (2003) *Appl. Phys. Lett.*, **83**, 1044–1046.
43. Masiel, D.J., LaGrange, T., Reed, B.W., Guo, T., and Browning, N.D. (2010) *ChemPhysChem*, **11**, 2088–2090.
44. Taheri, M.L., McGowan, S., Nikolova, L., Evans, J.E., Teslich, N., Lu, J.P., LaGrange, T., Rosei, F., Siwick, B.J., and Browning, N.D. (2010) *Appl. Phys. Lett.*, **97**, 032102.
45. Reed, B.W., LaGrange, T., Shuttlesworth, R.M., Gibson, D.J., Campbell, G.H., and Browning, N.D. (2010) *Rev. Sci. Instrum.*, **81**, 053706.
46. Reed, B.W., Armstrong, M.R., Browning, N.D., Campbell, G.H., Evans, J.E., LaGrange, T.B., and Masiel, D.J. (2009) *Microsc. Microanal.*, **15**, 272–281.
47. LaGrange, T., Grummon, D.S., Browning, N.D., King, W.E., and Campbell, G.H. (2009) *Appl. Phys. Lett.*, **94**, 184101.
48. Taheri, M.L., Reed, B.W., Lagrange, T.B., and Browning, N.D. (2008) *Small*, **4**, 2187–2190.

49. Kim, J.S., LaGrange, T.B., Reed, B.W., Browning, N.D., Taheri, M.L., Armstrong, M.R., King, W.E., and Campbell, G.H. (2008) *Science*, **321**, 1472–1475.
50. Armstrong, M.R., Browning, N.D., Reed, B.W., and Torralva, B.R. (2007) *Appl. Phys. Lett.*, **90**, 114101.
51. Armstrong, M., Boyden, K., Browning, N.D., Campbell, G.H., Colvin, J.D., DeHope, B., Frank, A.M., Gibson, D.J., Hartemann, F., Kim, J.S., King, W.E., LaGrange, T.B., Pyke, B.J., Reed, B.W., Shuttlesworth, R.M., Stuart, B.C., and Torralva, B.R. (2007) *Ultramicroscopy*, **107**, 356–367.
52. LaGrange, T.B., Armstrong, M., Boyden, K., Brown, C., Browning, N.D., Campbell, G.H., Colvin, J.D., DeHope, B., Frank, A.M., Gibson, D.J., Hartemann, F., Kim, J.S., King, W.E., Pyke, B.J., Reed, B.W., Shuttlesworth, R.M., Stuart, B.C., and Torralva, B.R. (2006) *Appl. Phys. Lett.*, **89**, 044105.
53. Bostonjoglo, O. and Leidtke, R. (1980) *Adv. Imaging Electron Phys.*, **60**, 451.
54. Bostonjoglo, O., Elschner, R., mao, Z., Nink, T., and Weingartner, M. (2000) *Ultramicroscopy*, **81**, 141.
55. Bostonjoglo, O. (2002) *Adv. Imaging Electron Phys.*, **121**, 1.
56. Barwick, B., Park, H.S., Kwon, O.H., Baskin, J.S., and Zewail, A.H. (2008) *Science*, **322**, 1227–1231.
57. Baum, P., Yang, D.S., and Zewail, A.H. (2007) *Science*, **318**, 788–792.
58. Yang, D.S., Lao, C., and Zewail, A.H. (2008) *Science*, **321**, 1660–1664.
59. Kwon, O.H., Barwick, B., Park, H.S., Baskin, J.S., and Zewail, A.H. (2008) *Nano Lett.*, **8**, 3557–3562.
60. Carbone, F., Kwon, O.H., and Zewail, A.H. (2009) *Science*, **325**, 181–184.
61. King, W.E., Campbell, G.H., Frank, A., Reed, B., Schmerge, J.F., Siwick, B.J., Stuart, B.C., and Weber, P.M. (2005) *J. Appl. Phys.*, **97**, 111101.
62. Boussoukaya, M., Bergeret, H., Chehab, R., Leblond, B., and Leduff, J. (1989) *Nucl. Inst. Meth. Phys. Res. A*, **279**, 405–409.
63. Garcia, C.H. and Brau, C.A. (2002) *Nucl. Inst. Meth. Phys. Res. A*, **483**, 273–276.
64. Hommelhoff, P., Sortais, Y., Adhajani-Talesh, A., and Kasevich, M.A. (2006) *Phys. Rev. Lett.*, **96**, 077401.
65. Claessens, B.J., van der Geer, T., Taban, G., Vredenbregt, E.J.D., and Luiten, O.J. (2005) *Phys. Rev. Lett.*, **95**, 164801.
66. Zolotorev, M., Commins, E.D., and Sannibale, F. (2007) *Phys. Rev. Lett.*, **98**, 184801.
67. Rose, H. and Spehr, R. (1983) *Adv Electron. Electron Phys. Suppl.*, **13C**, 475–530.
68. Reiser, M. (1994) *Theory and Design of Charged Particle Beams*, John Wiley & Sons, Inc., New York, p. 65.
69. Kruit, P. and Jansen, G.H. (1997) in *Handbook of Charged Particle Optics* (ed. J. Orloff), CRC Press, Boca Raton, FL, pp. 275–318.
70. Siwick, B.J., Dwyer, J.R., Jordan, R.E., and Miller, R.J.D. (2002) *J. Appl. Phys.*, **92**, 1643–1648.
71. Reed, B.W. (2006) *J. Appl. Phys.*, **100**, 034916.
72. Michalik, A.M. and Sipe, J.E. (2006) *J. Appl. Phys.*, **99**, 054908.
73. Gahlmann, A., Park, S.T., and Zewail, A.H. (2008) *Phys. Chem. Chem. Phys.*, **10**, 2894–2909.
74. Vellekoop, I.M., van Putten, E.G., Lagendijk, A., and Mosk, A.P. (2008) *Opt. Express*, **16**, 68–75.
75. van Oudheusden, T., de Jong, E.F., van der Geer, S.B., Op't Root, W.P.E.M., Luiten, O.J., and Siwick, B.J. (2007) *J. Appl. Phys.*, **102**, 093501.
76. Krivanek, O.L. (1992) in *High-Resolution Transmission Electron Microscopy and Associated Techniques* (eds P. Buseck, J. Cowley, and L. Eyring), Oxford University Press, New York, pp. 519–567.
77. Rose, A. (1948) in *Advances in Electronics and Electron Physics* (ed. Marston), Academic Press, New York.
78. Bals, S., Kabius, B., Haider, M., Radmilovic, V., and Kisielowski, C. (2004) *Solid State Commun.*, **130**, 675–680.
79. LaGrange, T., Campbell, G.H., Reed, B.W., Taheri, M., Pesavento, J.B., Kim, J.S., and Browning, N.D. (2008) *Ultramicroscopy*, **108**, 1441–1449.

80. Wang, J., Besnoin, E., Duckham, A., Spey, S.J., Reiss, M.E., Knio, O.M., Powers, M., Whitener, M., and Weihs, T.P. (2003) *Appl. Phys. Lett.*, **83** (19), 3987–3989.
81. Wang, J., Besnoin, E., Duckham, A., Spey, S.J., Reiss, M.E., Knio, O.M., and Weihs, T.P. (2004) *J. Appl. Phys.*, **95** (1), 248–256.
82. (a) Zhao, S.J., Germann, T.C., and Strachan, A. (2006) *J. Chem. Phys.*, **125** (16); (b) Kim, J.S., LaGrange, T., Reed, B.W., Taheri, M.L., Armstrong, M.R., King, W.E., Browning, N.D., and Campbell, G.H. (2008) *Science*, **321** (5895), 1472–1475.
83. Trenkle, J.C., Koerner, L.J., Tate, M.W., Gruner, S.M., Weihs, T.P., and Hufnagel, T.C. (2008) *Appl. Phys. Lett.*, **93** (8)
84. Floro, J.A. (1986) *J. Vac. Sci. Technol. A-Vac. Surf. Films*, **4** (3), 631–636.
85. Ma, E., Thompson, C.V., Clevenger, L.A., and Tu, K.N. (1990) *Appl. Phys. Lett.*, **57** (12), 1262–1264.
86. Mann, A.B., Gavens, A.J., Reiss, M.E., VanHeerden, D., Bao, G., and Weihs, T.P. (1997) *J. Appl. Phys.*, **82** (3), 1178–1188.
87. Adams, D.P., Rodriguez, M.A., McDonald, J.P., Bai, M.M., Jones, E., Brewer, L., and Moore, J.J. (2009) *J. Appl. Phys.*, **106** (9).
88. McDonald, J.P., Hodges, V.C., Jones, E.D., and Adams, D.P. (2009) *Appl. Phys. Lett.*, **94** (3).
89. Blobaum, K.J., Van Heerden, D., Gavens, A.J., and Weihs, T.P. (2003) *Acta Mater.*, **51** (13), 3871–3884.
90. Edelstein, A.S., Everett, R.K., Richardson, G.Y., Qadri, S.B., Altman, E.I., and Foley, J.C. (1994) *J. Appl. Phys.*, **76** (12), 7850–7859.
91. Singleton, M. (1990) in *Binary Alloy Phase Diagrams*, vol. 2 (ed. T. Massalski), ASM International, Metals Park, OH, pp. 181.
92. Wei, H., Sun, X.F., Zheng, Q., Guan, H.R., and Hu, Z.Q. (2004) *Acta Mater.*, **52** (9), 2645–2651.
93. Brown, A.M. and Ashby, M.F. (1980) *Acta Metall.*, **28** (8), 1085–1101.
94. Du, Y., Chang, Y.A., Huang, B.Y., Gong, W.P., Jin, Z.P., Xu, H.H., Yuan, Z., Liu, Y., He, Y., and Xie, F.Y. (2003) *Mater. Sci. Eng. A-Struct. Mater. Prop. Microstruct. Process.*, **363** (1–2), 140–151.
95. Williamson, M.J., Tromp, R.M., Vereecken, P.M., Hull, R., and Ross, F.M. (2003) *Nat. Mater.*, **2**, 532–536.
96. Creemer, J.F., Helveg, S., Hoveling, G.H., Ullmann, S., Molenbroed, A.M., Sarro, P.M., and Zandbergen, H.W. (2008) *Ultramicroscopy*, **108**, 993–998.
97. Zheng, H., Smith, R.K., Jun, Y.-W., Kisielowski, C., Dahmen, U., and Alivisatos, A.P. (2009) *Science*, **324**, 1309–1312.
98. de Jonge, N., Peckys, D.B., Kremers, G.J., and Piston, D.W. (2009) *Proc. Natl. Acad. Sci. U.S.A.*, **106**, 2159–2164.
99. Gai, P.L. (2002) *Microsc. Microanal.*, **8**, 21.
100. Musumeci, P., Moody, J.T., Scoby, C.M., Gutierrez, M.S., Bender, H.A., and Wilcox, N.S. (2010) *Rev. Sci. Instrum.*, **81**, Article no: 013306.
101. Bolton, P.R., Clendenin, J.E., Dowell, D.H., Ferrario, M., Fisher, A.S., Gierman, S.M., Kirby, R.E., Krejcik, P., Limborg, C.G., Mulhollan, G.A., Nguyen, D., Palmer, D.T., Rosenzweig, J.B., Schmerge, J.F., Serafini, L., and Wang, X.J. (2002) *Nucl. Instrum. Methods Phys. Res. Sect. A-Accel. Spectrom. Detect. Assoc. Equip.*, **483**, 296.
102. Dowell, D.H., Bolton, P.R., Clendenin, J.E., Gierman, S.M., Limborg, C.G., Murphy, B.F., Schmerge, J.F., and Shaftan, T. (2007) *Nucl. Instrum. Methods Phys. Res. Sect. A-Accel. Spectrom. Detect. Assoc. Equip.*, **507**, 331.
103. Schmerge, J.F., Dowell, D., and Hastings, J. (2004) First National Lab and University Alliance Workshop on Ultrafast Electron Microscopies.
104. Serafini, L. (1996) *IEEE Trans. Plasma Sci.*, **24**, 421.
105. Travier, C. (1994) *Nucl. Instrum. Methods Phys. Res. Sect. A-Accel. Spectrom. Detect. Assoc. Equip.*, **340**, 26.

10
Transmission Electron Microscopy as Nanolab
Frans D. Tichelaar, Marijn A. van Huis, and Henny W. Zandbergen

Apart from imaging and diffraction techniques, many methods exist in transmission electron microscopy (TEM) to characterize the structural, chemical, and physical properties of materials. Spectroscopy methods such as energy-dispersive X-ray (EDX) and electron energy loss spectroscopy (EELS) are universally applied to locally study material composition and structure; to a much lesser extent cathodoluminescence (CL) is employed. Furthermore, so-called *in situ* methods (experiments that are performed inside the TEM) such as heating, cooling, and straining using dedicated specimen holders are often used to investigate the relationship between structural and physical properties. Environmental transmission electron microscopes (ETEMs) are used to investigate changes in materials in contact with gases. Recently, much effort has been taken to perform such experiments inside a TEM with a high degree of control over the physical parameters involved at the nanoscale, while minimizing the negative effects on the observation conditions, such as scattering of the electron beam by a gas atmosphere, specimen drift, and specimen vibrations induced by temperature changes. These type of experiments are referred to when we speak of *nanolab* in this chapter. Often, dedicated specimen holders based on microelectronic mechanical system (MEMS) techniques are used for nanolab experiments. Of course, there is a diffuse boundary between "TEM as nanolab" and ETEM and *in situ* TEM, as seen in other chapters.

This chapter is mostly about MEMS-based devices and/or piezo-controlled contacting and about the TEM holders to support these MEMS devices, including all the interfacing between the MEMS devices and the equipment outside the TEM to control manipulations in the MEMS devices. Also, we describe recent developments in electron-generated CL. The following nanolab setups are discussed:

1) TEM and measuring the electrical properties
2) TEM with MEMS-based heaters
3) TEM with a gas nanoreactor
4) TEM with a liquid nanoreactor
5) TEM and measuring optical properties

Handbook of Nanoscopy, First Edition. Edited by Gustaaf Van Tendeloo, Dirk Van Dyck, and Stephen J. Pennycook.
© 2012 Wiley-VCH Verlag GmbH & Co. KGaA. Published 2012 by Wiley-VCH Verlag GmbH & Co. KGaA.

6) Sample preparation for nanolab experiments
   a. Electron beam sculpting
   b. Gallium ion beam sculpting
   c. Helium ion beam sculpting.

10.1
TEM and Measuring the Electrical Properties

The understanding of the details of nanostructural changes in nanodevices under the influence of an electrical current and vice versa, requires both imaging of the nanostructural changes and the results of the electrical measurements. Generally, when the effects of an electrical current through nanostructures on these structures are studied, electrical data and TEM data are acquired sequentially. However, it is not obvious at all that electrical properties over this experimental sequence can be obtained by interpolation of these data. Thus, a much better experimental setup is to do both measurements simultaneously and this can, in principle, be achieved with *in situ* TEM.

Two approaches are followed to perform *in situ* TEM with electrical measurements. The first one is to make devices, similar to those in "normal" nanoelectrical characterizations, whereby a nanostructure is in contact with two electrodes (often with a gate electrode nearby), but with the investigated nanostructure either on a very thin substrate such as SiN, or free standing [1]. The second approach is to use a piezo-driven needle that makes contact with a part of the nanostructure, whereby the support of this nanostructure is the counter electrode.

A setup for the first approach [2] is given in Figure 10.1. It consists of an MEMS type of sample (Figure 10.1), which can be quite complex, since such structures can be made using top-down nanofabrication techniques. For instance, the chip in

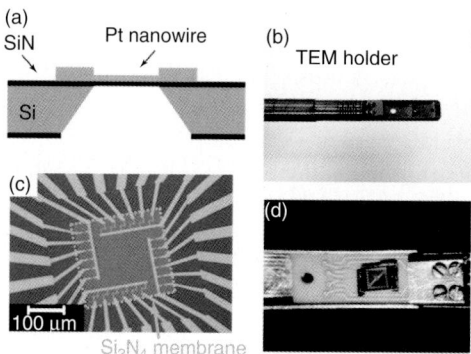

Figure 10.1 (a) Schematic drawing of a MEMS-based nanostructure. Sizes of the working electrode and the nanowire are exaggerated for clarity. (b) The tip of the TEM holder, which can operate at ~100–300 K. (c) Central part of the MEMS chip with the thin SiN membrane indicated, which supports 24 Pt nanolines. (d) Part of the specimen holder tip showing the chip and a chip carrier.

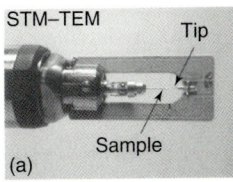 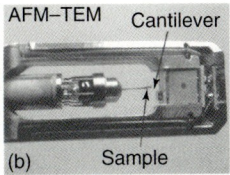

Figure 10.2 Experimental setups within (a) the STM–TEM and (b) the AFM–TEM holder. The locations of the samples, the STM tip, and the Si cantilever are indicated. (Reprinted with permission from Ref. [4], Copyright 2008 IEEE.)

Figure 10.1c contains 24 Pt nanobridges. Such a MEMS device requires an interface at the holder connecting to dedicated measurement equipment, which should be sufficiently advanced to measure currents in the nanoampere range. The tip of the holder shown in Figure 10.1b [3] has eight electric feedthroughs for carrying out the electrical measurements on chiplike samples connected to a measurement setup. It is important to mention that for any measurement setup, stray currents can easily destroy the nanodevice, and, for instance, a sudden release of built-up charge created by the electron beam can melt the nanodevice. Evidently, real-time video recording using a fast scan camera (up to 25 images per s) is also required to correlate features appearing on current–voltage ($I-V$) characteristics or changes in resistance of the sample with changes in its nanostructure.

A setup for the second approach is given in Figure 10.2. In this case, one part of a nanostructure is positioned on an electrode, whereas the counter electrode is the tip of a piezo stage [4]. The needle direction and position have to be adjustable in the X, Y, and Z directions to make contact with a given nanostructure, for instance, an entangled nanotube (NT) array.

10.1.1
TEM and Measuring Electrical Properties. Example 1: Electromigration

Using a sample holder that allows positioning of a MEMS device, Gao et al. [2] investigated the grain growth and electromigration processes in Pt polycrystalline metal nanolines at different electron currents (Figure 10.3). The MEMS device consists of a number of Pt nanolines on a 100 nm silicon nitride window. The dimensions of the Pt nanowire are $300 \times 200 \times 14\,nm^3$. The 100 nm thick SiN membrane is somewhat thick for high-resolution imaging, but if high resolution is required the SiN can be thinned with a plasma down to, for instance, 20 nm after the whole Pt deposition process has been performed. A direct correlation between the evolution of the grain size and the change in the resistance can be made. The sample was current annealed by increasing a bias voltage from 0 to 300 mV. Once the resistance increased, the voltage was reduced (feedback control) and next ramped up (which is visible in Figure 10.3a). This allows for a more careful formation of a nanogap. If the current increase is continued, an electromigration process is started (Figure 10.3c). On further continuation, a small gap with a

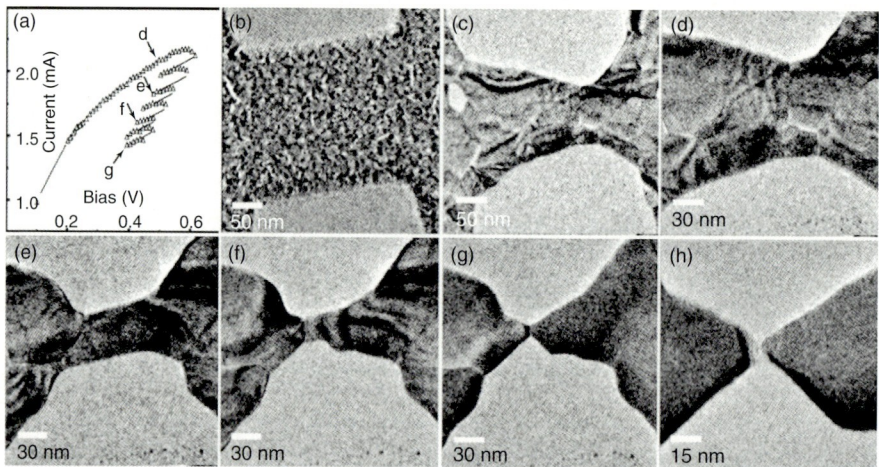

Figure 10.3 (a) A current–voltage curve of an electromigration process with feedback control. Triangles in the figure are the measured *I*–*V* data points; the black line indicates the feedback loop. The *I*–*V* curve corresponds to the electromigration from images (d) to (g), more precisely, the four locations indicated by the arrows and labels correspond to the images with the same labels. (b–h) *In situ* TEM imaging of the electromigration. (b) The original Pt nanowire. (c–h) Feedback control leads to symmetric electrodes on both sides of the final gap.

separation of a few nanometers between two electrodes is formed (Figure 10.3h). At each step in the process, the applied current can be correlated with the specimen shape. The formed gap shows typical *I*–*V* properties, as is shown in Figure 10.4.

10.1.2
TEM and Measuring Electrical Properties. Example 2: Carbon Nanotubes

Using a scanning tunneling microscopic (STM) unit fitted into a TEM specimen holder, Costa *et al.* [5] studied the electrical properties of N-doped carbon nanotubes (CNTs). The NTs were stuck onto a Cu-alloy panel fixed inside the TEM holder; a piezo-driven stage with a mechanically polished and etched tungsten needle of 50 μm diameter served as a moveable counterpart. The needle could be adjusted to make contact with a given NT protruding from the entangled NT array within a range of 5, 5, and 1 μm in the *X*, *Y*, and *Z* directions, respectively. The two terminal *I*–*V* curves were measured using a DC power supply. Typically, the current was measured for input voltages ranging from 0 to ∼10 V. A diagram of the electrical measurements and manipulation using the TEM–STM unit are shown in Figure 10.5. A series of measurements on the nonlinear *I*–*V* characteristics was done, in which a selected NT was first contacted and next measured (Figure 10.5). Next, a ∼250 nm gap was cut, using a quick sweep of −10 and 10 V, and the NT was recontacted. This sweep leads to momentous extreme currents, which induce fatal structure failure, allowing sections of 200–500 nm to be cut in sequence and with high reproducibility. Figures 10.5c,e show the details of the first cut and the

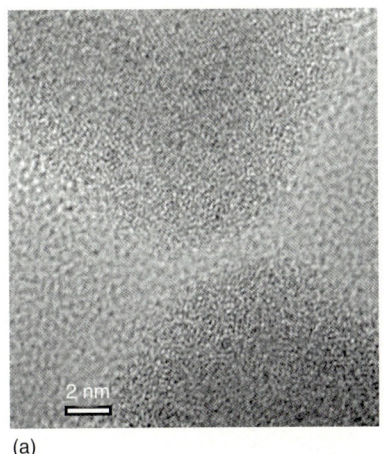
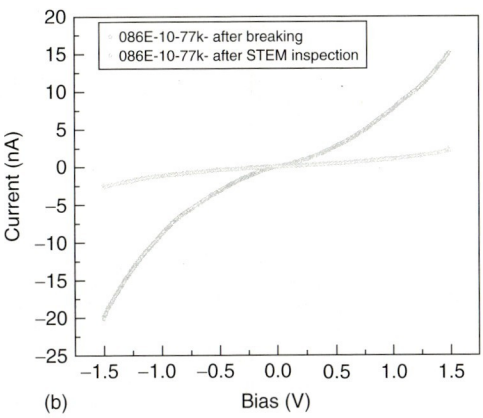

Figure 10.4 A typical *in situ* TEM result of a recent experiment in which a nanogap was created in a Pt bridge by electromigration. The Pt bridge is on a ~40 nm thick SiN window. A nanogap just after formation can be seen in (a). The corresponding *I–V* curve is shown as a dark gray line in (b). After performing a subsequent STEM experiment, the light gray *I–V* line was measured. It shows three problems: (i) the gap size is changing after formation (because of strain relaxation of the bridge caused by recrystallization during electromigration), (ii) the image is blurred by the SiN substrate, and (iii) the Pt surfaces show an unknown amorphous structure (possibly a Pt–Si alloy).

second contact, respectively. Linear *I–V* curves were observed for the recontacted NTs. Probing of the pure NT required a three-step procedure: clean, cut, and recontact. The analysis of hundreds of spectra showed that the shape transition from the nonlinear to the linear regimes is, in fact, a fairly common occurrence in the N-doped CNT system.

10.2
TEM with MEMS-Based Heaters

An example of a MEMS-based heater is shown in Figure 10.6. The principle is a ~1000 nm thick membrane in which a heater is embedded [6, 7]. Because of the small volume of the membrane, the required power to bring and keep a sample at a certain temperature is very small. In the example of the heater shown in Figure 10.6, the heater is connected to four electrical connections that allow heating on the one hand and measurement of the resistance for temperature determination on the other hand.

A MEMS-based heater holder has – compared to conventional heating holders – the big advantage that the heat produced is much lower. Consequently, the heat transfer to the holder and the resulting drift of the specimen are also much smaller (drifts below 0.5 nm/min are relatively easily obtained at 500 °C). Still, the drift can be significant, in particular, when a large change in temperature

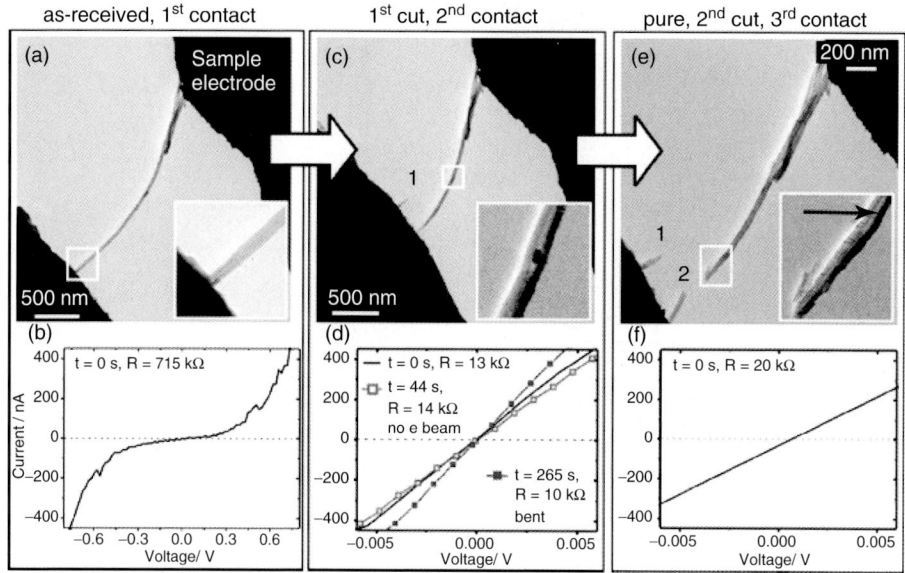

Figure 10.5 Different stages of N-doped CNT tailoring and corresponding spectra. (a,b) The as-received N-doped C nanostructure showed nonlinear I–V characteristics when contacted. (c,d) A ~250 nm section (indicated by 1) was cut, using a quick sweep of −10 and 10 V, and the NT was recontacted. Exclusively, linear relations were observed. The inset shows an encapsulated metal particle. (e,f) Probing of the pure NT required a three-step procedure: clean, cut, and recontact. Initially, the bias voltage was slowly increased to ~8 V, leaving the pristine tube structure. In the inset image of (e), the black arrow points to the void left by the particle in inset (c). Next, a −10 and 10 V sweep resulted in an additional cut ~500 nm long (indicated by 2). Finally, the pure NT was recontacted. The I–V kept its linear characteristics with the extracted resistance being in the same range as those obtained in (d). (Reprinted with permission from Ref. [5]. Copyright 2007 American Institute of Physics.)

is applied. In this respect, it is useful to consider the MEMS-heater/TEM-holder ensemble as a series of thermal resistors (Figure 10.7). When the heater is at an elevated temperature, the membrane between the heater and the Si part of the heater chip will transfer some heat to the Si part, depending on its thermal resistance. Likewise, all the components between the membrane and the outside tube of the holder, including heat radiation, will transfer heat to the outer tube of the holder. If the tube is heated by only 0.01 °C, the displacement of the tip can still be 20 nm. On the one hand, a nearly zero specimen drift without having to wait for too long can only be achieved when the coupling between the heater and holder is minimized. On the other hand, a certain mechanical stability of the holder is required for good high-resolution imaging, which implies a rather rigid fixing of the MEMS heater to the holder. The optimization of a low heat transfer, fast reduction of specimen drift after a temperature change, and maintaining the imaging resolution are technological challenges.

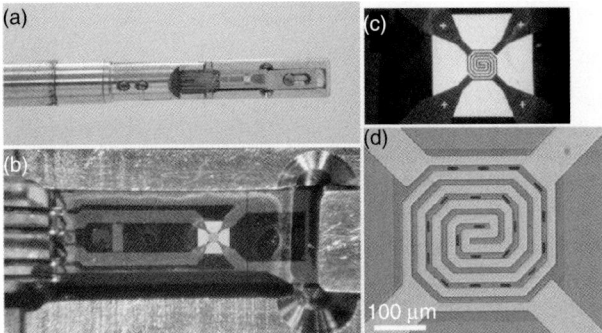

Figure 10.6 A heating specimen holder with MEMS microheater. (a,b) Optical images of the holder with a mounted MEMS microheater that is fabricated using silicon-based fabrication technology. (c,d) Optical images of the center of the microheater with an embedded, planar Pt wire for local heating. Between the windings (showing dark in (d)) are electron-transparent viewing windows (dimensions 5×20 µm) where the SiN is approximately 15 nm thick.

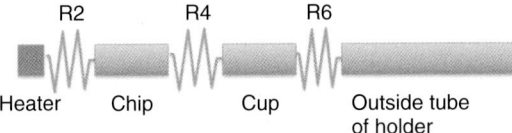

Figure 10.7 Schematic representation of the various thermal resistors (R2, R4, and R6) that determine the heat transfer of the heater to the outer tube of the TEM holder. Note that the outer tube is connected to several parts of the goniometer with components that each have their own thermal resistance and contacts with the final heat sink. Also, the specimen itself, its contact with the support of the heater, and the support-heater contact introduce thermal gradients and thus the real situation is more complicated than indicated here.

10.2.1
TEM with MEMS-Based Heaters. Example 1: Graphene at Various Temperatures

First isolated in 2004 [8], graphene has received tremendous scientific attention because of its unique electronic properties [9]. Graphene also features edge dynamics [10] and mechanical properties [11], opening up even more opportunities, such as its use in sequence genomic DNA using nanopores [12–14]. In order to harvest the many promising properties of graphene in applications, a technique is required to cut, shape, or sculpt the material on a nanoscale without damage to its atomic structure, as this drastically influences the electronic properties of the nanostructure. A temperature-dependent self-repair mechanism allows for damage-free atomic-scale sculpting of graphene using a focused electron beam. The temperature has a remarkable effect on the changes induced by 300 keV electrons [15]. At room temperature (RT), a rapid amorphization occurs (Figure 10.8a), which hampers detailed high-resolution electron microscopy imaging. At temperatures of 200 °C, electron beam irradiation leads to amorphization with only short

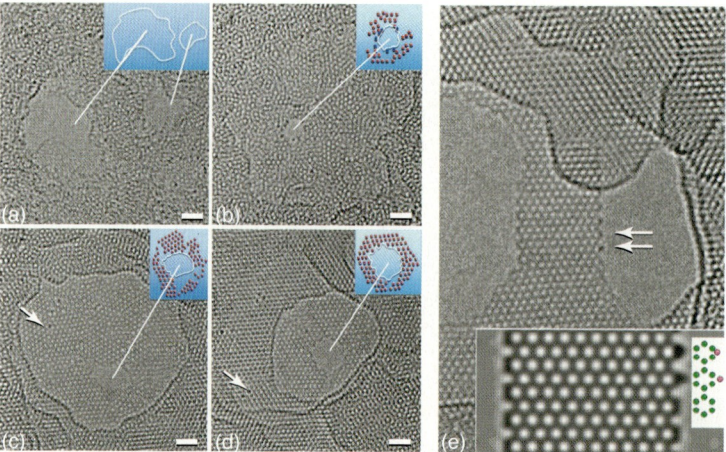

Figure 10.8 Influence of temperature on the sculpting of few-layer graphene by an electron beam. (a) at room temperature (RT), (b) at 200 °C, (c) at 500 °C, and (d) at 700 °C. Red arrows indicate some of the C ad-atoms trapped at defects. The insets in (b–d) show the positions of the identifiable hexagons (red dots) and the estimated position of the edge (white line). The blue dots in the inset at 200 °C are ad-atoms, (e) shows a nanoribbon formed at 600 °C, and the inset in (e) shows a calculated image for three configurations of the edge of which two have a C ad-atom. Typical exposure times are 8–12 s; scale bars, 1 nm.

range order (Figure 10.8b). At 500 °C, the electron beam results in the formation of polycrystalline monolayers. The single crystalline graphene transforms into polycrystalline graphene with clear grain boundaries that are straight but short. At 700 °C, remarkably, graphene conserves its full crystallinity even under a very intense electron beam, as shown in Figure 10.8d. These sculpting conditions can be used to make certain shapes of graphene [15], for instance, a nanopore as in Figure 10.8d or a nanoribbon as is shown in Figure 10.8e.

10.2.2
TEM with MEMS-Based Heaters. Example 2: Morphological Changes on Au Nanoparticles

The MEMS-based heater holder has a superior performance in imaging resolution and drift (1.0 Å and 0.02 nm s^{-1}, respectively, obtained in an aberration-corrected Titan microscope operating at 300 kV after stabilization for 20 min) [7]. This technological achievement enables monitoring of nanoscale transitions in real time and with atomic resolution. Using this holder, the thermal stability and structural transitions of metal nanoparticles, semiconductor nanocrystals, and heterogeneous nanostructures was investigated [6, 7, 16].

One important field of interest is catalysis, as it is known that the catalytic activity of nanoparticles is related to the presence of atomic edges and steps at the nanoparticle surface. The overall morphology of the nanoparticle and the presence

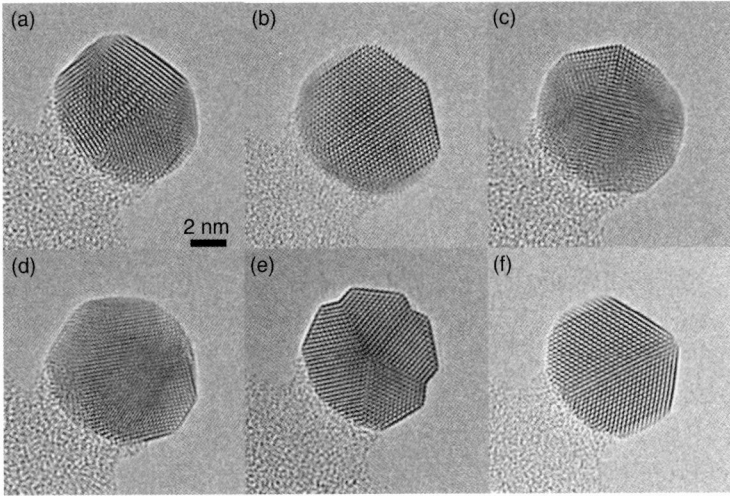

Figure 10.9 Morphological transformations of a 7 nm gold particle. The heater temperature is set to 415 °C in (a–d) and to 450 °C in (e) and (f). (a) Initial decahedral structure, (b) rotated decahedral structure, (c) distorted icosahedral structure, (d) cuboctahedral structure, and (e) highly faceted marks decahedral structure with fivefold twin axis parallel to illumination. (f) Again, Marks decahedral structure with fivefold axis inclined normal to the electron beam.

of atomic defects at the surface change drastically at elevated temperatures, as shown in Figure 10.9 where a single Au nanoparticle undergoes multiple and very drastic structural transformations [6].

10.3
TEM with Gas Nanoreactors

When investigating the reaction of a gas with a solid at relevant gas pressures, the ETEM approach can often not be employed, because the gas pressure in an ETEM is limited, in particular, if high-resolution information is needed. The solution to this pressure restriction is to confine the high-pressure gas in between two thin membranes that are sufficiently electron transparent. In the 1970s and earlier, this approach was already used up to 1 bar pressure, but imaging resolution was limited to a few nanometers because of the relatively thick windows and a large separation between them [17]. Nowadays, such a configuration can be made relatively easily using MEMS technology. In the setup used in Delft [18–21], the MEMS-based nanoreactor is composed of two Si chips with square-shaped 1 μm thick SiN membranes that contain very thin SiN windows about 20 nm thick (Figure 10.10a,b [20]). The chips contain a Pt heater and an inlet and outlet for the gas. With such a setup, 4 bar hydrogen pressures inside a TEM have been realized. The specimen holder includes O-rings to obtain leak-tight connections

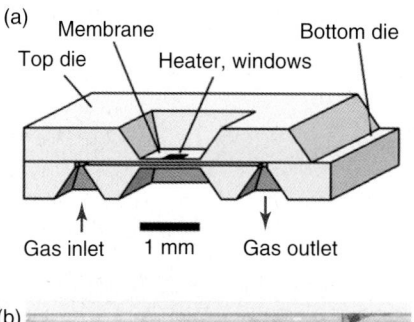
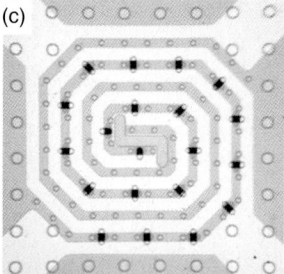

Figure 10.10 Illustration of a nanoreactor device. (a) Schematic cross section of the nanoreactor. (b) Optical image of the TEM holder with the integrated nanoreactor and the four electrical probe contacts. (c) Optical close-up of the nanoreactor membrane. The small ovaloids are the electron-transparent windows. The circles are the SiO$_2$ spacers that define the minimum height of the gas channel. (d) A low-magnification TEM image of a pair of adjacent 10 nm thick windows. Their alignment creates a highly electron-transparent (bright) square through which high-resolution TEM imaging can be performed.

between the holder and the nanoreactor (Figure 10.10b). We have developed a gas supply system to allow any gas pressure between 0 and 10 bar to pass via the TEM holder through the nanoreactor. Various measurements were taken to reduce contamination originating from hydrocarbons, which may, for instance, originate from the gas container. Others [21–25] have developed similar nanoreactors, all based on the use of thin electron transparent windows. The gap between these windows varies from 2 μm to 1 mm for the various reported devices. Evidently, the smaller the gap, the higher the pressure can be without significant loss in resolution. Note that these nanoreactors can also be used for other experimental techniques such as scanning X-ray transmission microscopy [19].

10.3.1
TEM with Gas Nanoreactors. Example 1: Hydrogen Storage Materials

In order to understand the loading and unloading mechanism of hydrogen storage materials, it is important to study this process at the atomic level. *In situ* TEM at application pressures (1–10 bar) of hydrogen gas and temperatures up to 500 °C are expected to provide the most direct information. Such a pressure, in combination with the high vacuum of the TEM, can be achieved using a MEMS-based nanoreactor as described above. Starting with a gap between the

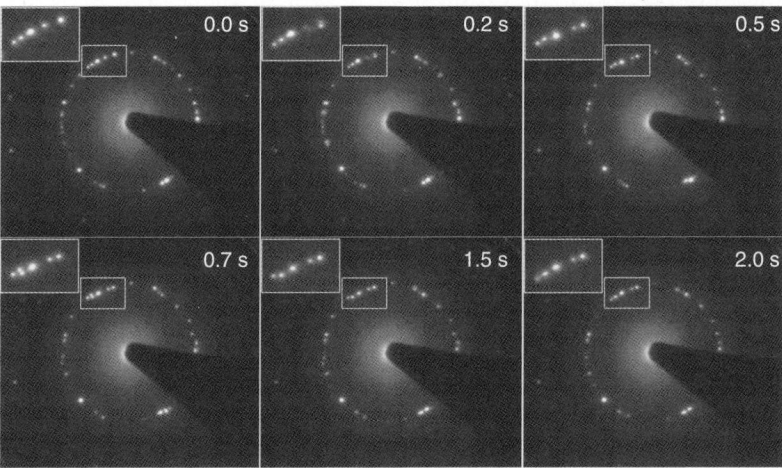

Figure 10.11 Frames from a movie taken in selected-area electron diffraction (SAED) mode of the hydrogen loading of palladium during an increase in the hydrogen pressure. One can see that the hydrogen loading occurs within 2 s, and that within this time some particles transform faster than others.

membranes of about 2 μm, a pressure and temperature increase results in a bulging of the membranes. For instance, if the membranes are 1 mm wide, the gap will become ∼40 μm for 2 bar and a temperature of 400 °C. A solution is to make the windows less wide or anchor the two membranes to each other [25]. The combination of high pressures and heating allows the study of all kinds of chemical reactions under industrially relevant gas pressures.

Figure 10.11 shows the sequence of diffraction patterns recorded during an increase in the hydrogen pressure [21]. In general, within 1–2 s all Pd particles are hydrogenated, but within this time frame one can observe that some particles are hydrogenated quicker than others. Upon loading with hydrogen at 200 mbar the lattice expands by about 3.4%, which is consistent with bulk values (3.3% change) [26]. Both heating at high temperature and lowering the pressure resulted in dehydrogenation.

10.3.2
TEM with Gas Nanoreactors. Example 2: STEM Imaging of a Layer of Gold Nanoparticles at 1 bar

De Jonge [27] has performed a scanning transmission electron microscopy (STEM) study on a layer of gold nanoparticles on top of a TiO_2 layer with a $CO/O_2/He$ gas mixture at atmospheric pressure. The gas column was 0.36 mm thick. Figure 10.12 shows a STEM image, which is noise reduced with a convolution filter. The STEM image shows gold islets of several different sizes. The background signal varied over the image, which can possibly be explained by a combination of thickness

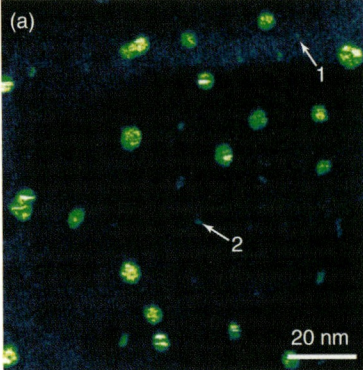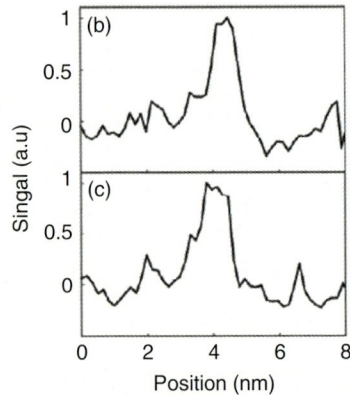

Figure 10.12 One bar STEM imaging of gold nanoparticles in 1% CO/5% O_2/He gas. (a) Image showing small gold islands on the top SiN window recorded at a magnification of 600 000, a pixel size of 0.17 nm, and a pixel-dwell time of 8 μs. For improved visibility of the nanoparticles, a convolution filter was applied and the signal intensity was color-coded. The thickness of the gas column was 360 μm. (b) Line-scan signal versus horizontal position, over the nanoparticle indicated with arrow #1 in (a). The background level was set to zero. (c) Line-scan signal versus horizontal position, over the nanoparticle indicated with arrow #2 in (a). (Reprinted with permission from Ref. [27]. Copyright 2010 American Chemical Society.)

variation of the TiO_2 layer and the formation of carbon contamination during imaging. To determine the spatial resolution achieved, line scans over six of the smallest nanoparticles were made. Two of these are indicated by the arrows in Figure 10.12a, and their line scans are shown in Figure 10.12b,c. The background level was determined from the average of the 20 pixels at the left side of the peak and then set to zero. The full width at half-maximum (fwhm) of the peak above the background level in Figure 10.12b is 0.8 nm, and in Figure 10.3c is 1.0 nm, giving an indication of the sizes of the nanoparticles. Thus, STEM imaging of 1 nm Au particles in the presence of a 360 μm CO/O_2/He gas column of 1 bar can be achieved.

10.4
TEM with Liquid Nanoreactors

For a long time, microscopists have tried to study processes involving water in a TEM. The difficulty is the incompatibility of vacuum with water. There are mainly two approaches to resolve this issue: the environmental chamber with small apertures and differential pumping, and a cell with windows that is completely sealed off from the vacuum. The second approach allows a much wider range of systems to be studied. Only a few studies have been published involving windowed cells; see [28–34]. A recent example is the study of muscle filament contraction by Sugi et al. [35]. They used a cell based on the design by Fukami et al. [36, 37] that has windows made out of 20 nm carbon over a copper grid that can stand pressure

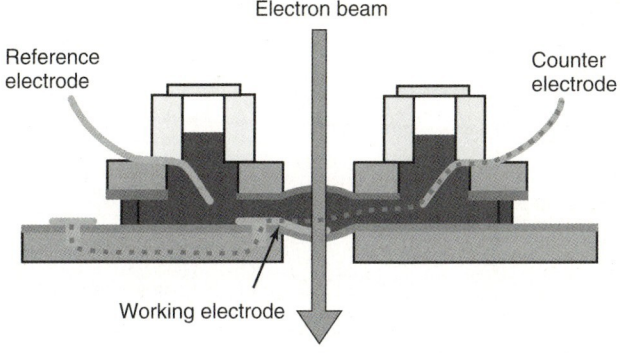

Figure 10.13 Schematic view of a TEM liquid cell. The sizes of the working electrode and viewing windows are exaggerated for clarity. The TEM imaging beam is indicated with a light gray arrow, and the electrical current path due to plating is indicated with a dark gray dotted line.

differences up to 1 atmosphere. The specimen is placed on the lower carbon film and it is kept wet by constantly circulating air saturated with water vapor through the cell.

In 2002, Ross and coworkers developed a liquid cell by means of silicon-based technology that is completely sealed [38]; see Figure 10.13 for a schematic drawing of a TEM liquid cell [39]. It has 100 nm thick SiN windows, and contains a liquid layer of approximately 1 μm thick, without the necessity (or possibility) to circulate vapor-saturated air to keep the specimen wet. Such cells were used for various *in situ* electrochemical nucleation and growth experiments [30, 31, 40]. The main limitation was reported to be the limiting height of the cell, which results in ∼1000 times slower growth rate of clusters. The growth kinetics of individual islands is characterized by two growth exponents as expected from models for diffusion limited growth with a transition from 3D diffusion to 1D diffusion [30].

10.4.1
TEM with Liquid Nanoreactors. Example 1: Cu Electrodeposition

Figure 10.14 shows several frames taken from two TEM movies of changes of Cu particles on Au electrodes, with the potential set to +0.15 V, whereby the clusters are stripped [39]. A cell similar to the one in Figure 10.13 is used. The liquids are 0.10 M $CuSO_4 \cdot 5\,H_2O$ + 1 vol.% H_2SO_4 + 50 ppm HCl and the additives polyethylene glycol (PEG) (left block) and 3-mercapto-1-propanesulfonic acid sodium salt (MPS) (right block). Prior to recording these data, the copper clusters were deposited outside the electron beam area and then moved inside the beam. After 2 s, the potential is set to +0.15 V and the clusters strip. In the experiment with PEG, the average brightness of the area without copper deposit is constant. Note that no bright area is left after the copper has been stripped, as is

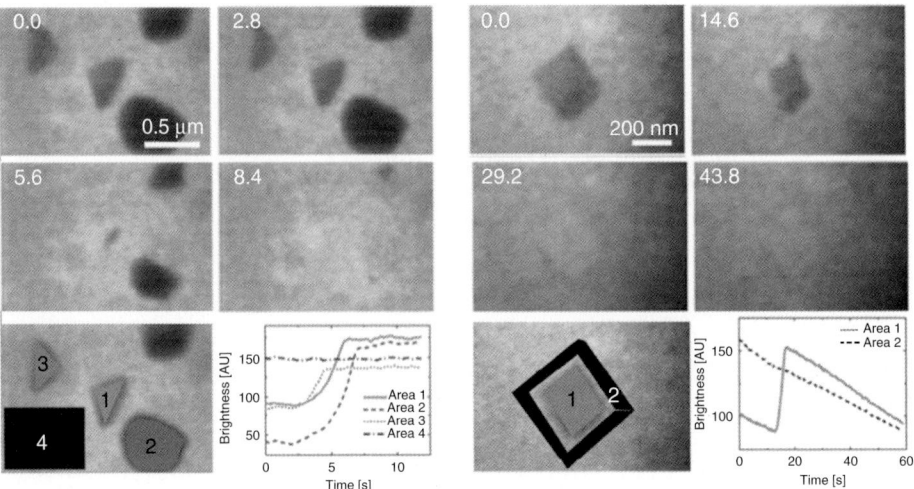

Figure 10.14 Left block: Cl + PEG on a Au electrode, Right block: Cl + MPS on a Au electrode. Stripping of clusters of Cu on a Au electrode and the changes in contrast due to the electron beam irradiation. The top four images in each block are movie frames showing the stripping of Cu particles. The time in seconds is given in the upper left corner. Each graph shows the average brightness in pixel values (which are arbitrary units) as a function of time, for each of the series. (Permission to reproduce from Ref. [39] is acknowledged.)

the case in the images for MPS. For the MPS, the average brightness of the area without copper deposit (area 2) decreases linearly. Note that on the positions where the copper is stripped, a bright area is left. The brightness difference between areas 1 and 2 reduces a factor 2.4 between $t = 18$ s and $t = 58$ s.

The primary processes of energy loss in inelastic scattering are molecular excitation, ionization, phonon production, and other collective excitations. These primary excitations can be converted to heat directly but they can also produce radicals, cause bond scission, and produce secondary electrons. In this way, PEG and MPS molecules can be cross-linked to form a polymerized carbon-rich layer. In water, the processes produce, for example, H^+ and OH^- radicals, H_2O_2 molecules and molecular H_2 [41]. These species are very reactive and also increase the rate of destruction of hydrocarbons. The differences in the results for PEG and MPS show that it is not just formation of carbon-rich species, since both molecules can be cracked by the electron beam if they would be present, for instance, as absorbed molecules on a conventional sample in the vacuum of a TEM. According to Reimer [42], the knowledge of the individual damage processes is very poor and, moreover, almost no one has looked at inorganic species in liquid water [29]. The chemical interaction of radiation with matter is the field of radiation chemistry, and constitutes a serious complication for the case of TEM with an incredibly high radiation dose. Results from radiation chemistry, where the doses are much lower, are therefore only applicable to the very early stages of contrast formation.

10.4.2
TEM with Liquid Nanoreactors. Example 2: Nucleation, Growth, and Motion of Small Particles

In order to observe colloidal nanocrystal growth, Zheng et al. [33] used a liquid nanoreactor based on the reactor developed by Williamson et al. [40], who made a self-contained liquid cell with an improved resolution in the subnanometer range and used these disposable liquid cells to image *in situ* platinum nanocrystal growth in solution using various TEMs. Pt particles were prepared by dissolving Pt (acetylacetonate) (10 mg ml^{-1}) in a mixture of *o*-dichlorobenzene and oleylamine (9 : 1 in volume ratio). After loading the cell with the growth solution, the cell was sealed and loaded into the microscope. Within the electron-transparent window, the volume with the reaction solution with a thickness of about 200 nm was confined between two silicon nitride membranes (25 nm thickness each). The growth of Pt nanocrystals in solution was initiated by electron beam irradiation. The relatively high beam intensity was briefly used to initiate the nucleation. Next, nanocrystals nucleated and grew. Figure 10.15a shows a sequence of video images recorded after 0.0, 12.1, 24.2, and 77.0 s durations of exposure to electron beam radiation. From the initial growth solution of Pt^{2+} precursor, a large number of Pt nanocrystals emerged, and new particles continued to appear. The nucleation under a constant electron beam irradiation spanned more than 10 s (see the number of particles as a function of time in Figure 10.15b). Particle growth and nucleation occurred in parallel (see particles highlighted by arrows in Figure 10.2a, indicating examples of growth). Along with the conventional particle growth by means of monomer addition from solution, frequent coalescence events between the particles were observed. At the early stage of the growth, the number of particles gradually increased and reached a maximum at 21.0 s. Subsequently, the number of particles dropped significantly and eventually settled at a constant value. Although some smaller particles were seen to dissolve completely, the decrease in the number of particles was mainly due to the coalescence events between individual particles (see the number of coalescence events as a function of time in Figure 10.15b).

10.5
TEM and Measuring Optical Properties

Cathodoluminescence (CL) is used to investigate the optical properties of materials by exposing a material to an electron beam, and collecting the ejected photons. This method has been applied many times in a scanning electron beam (SEM) [43, 44] and also in a TEM [45, 46]. The smaller the electron beam, the higher the spatial resolution of CL, and more importantly: the structure from which the signal comes can be investigated at the same time with a high spatial resolution. In a STEM, the spatial resolution can be very small (∼1 Å) and the CL signal for each beam position can be combined with an EELS signal. An advantage over CL in a TEM is the possibility to make a map of the CL signal, that is, optical

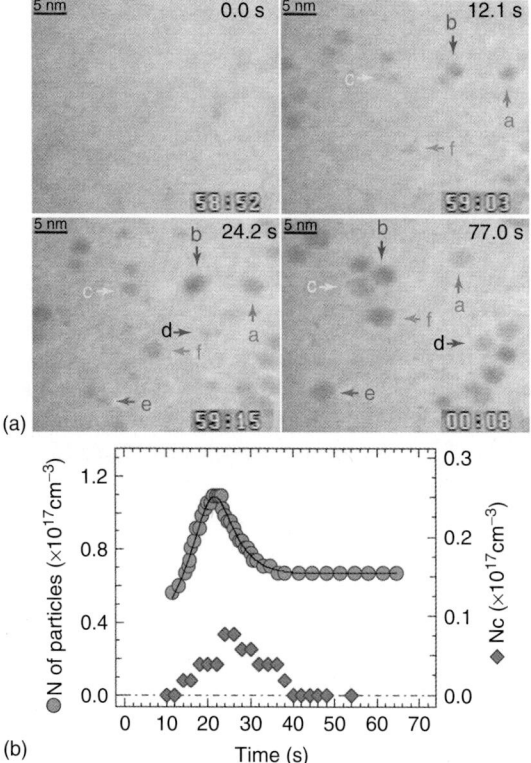

Figure 10.15 Growth and coalescence of Pt nanocrystals. (a) Video images acquired at 0.0, 12.1, 24.2, and 77.0 s of exposure to the electron beam. Specific particles are labeled with arrows. The growth trajectories of these individual particles reveal the multiple pathways leading to size focusing. (b) Number of particles (left axis) and number of coalescence events (Nc, right axis) during an interval of 2.0 s versus time. Particles nucleate and grow during the adjustment of focus for imaging (0–10 s). (From Ref. [33]; reprinted with permission from AAAS.)

imaging can be done at a resolution not much worse than the spatial resolution for atomic position imaging. The combination of the high-energy resolution of CL spectroscopy (10 meV) with a high spatial resolution of STEM opens new possibilities such as the measurement of the spectra of quantum emitters in relation to their size and structure (see below).

In Université Paris Sud Orsay, in the framework of ESTEEM,[1] an existing VG STEM was modified to accommodate a specimen cartridge designed to collect the light signal radiating from the sample where the electron beam interacts with the specimen [47]. The collection is achieved by a combination of a parabolic mirror, a lens system, and an optical fiber in the holder. The signal is subsequently measured on a charge-coupled device (CCD) chip and processed (Figure 10.16).

1) ESTEEM The European Union Framework fp6 program under a contract for an Integrated Infrastructure Initiative. Reference 026019 ESTEEM.

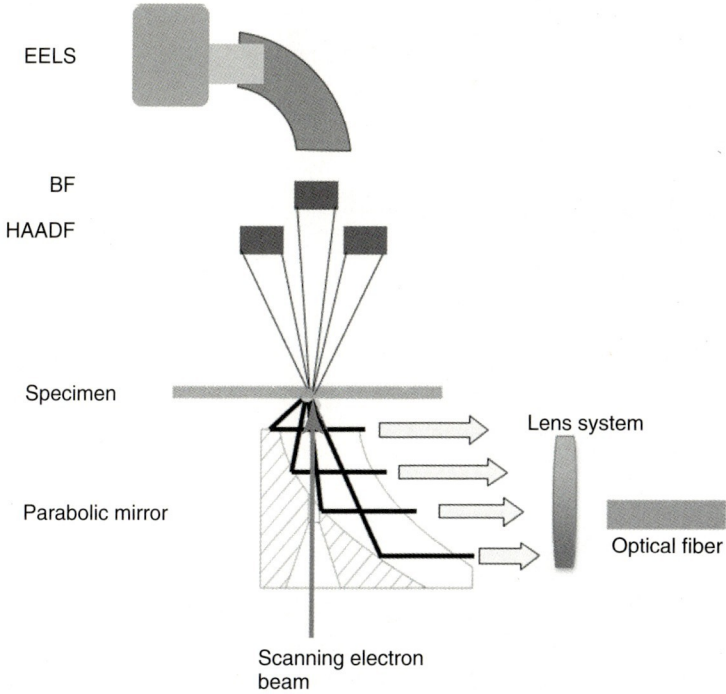

Figure 10.16 Scheme of the light gathering construction. The parabolic mirror and the lens system with the optical fiber have been built into a sample holder that is inserted into the STEM. (Courtesy of M. Kociak (Université Paris Sud, Orsay).)

The specimen stage is cooled to liquid nitrogen temperature to increase the signal and to avoid a decrease in luminescence efficiency over time (especially in semiconducting materials) caused by electron radiation damage. In addition, specimen contamination is less at lower temperature. A removable cartridge, on which the sample stays, is in mechanical and thermal contact with a metallic holder that is cooled through a metallic lead by liquid nitrogen and this metallic holder can be slightly heated, allowing control over the temperature.

The system is capable of 1 nm spatial resolution when recording in high angle annular dark field (HAADF) mode in combination with simultaneous CL spectrum collection with a few tens of millielectronvolt resolution. The CL spectrum collection time for each 1 nm × 1 nm pixel is 30–500 ms. The results first published refer to a GaN/AlN quantum disc system as depicted in Figure 10.17. In the system, the Si-doped GaN discs vary in thickness from 4 (1 nm) to 14 monolayers. The CL spectrum of such a nanowire is ambiguous; therefore, the local probing of the system was done with the CL STEM. Figure 10.17 shows three different spectra collected at three different positions, showing the CL dependency on the disc

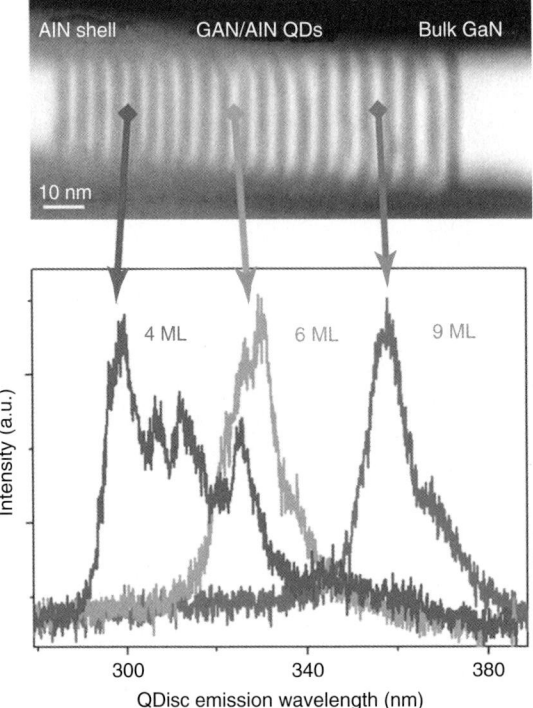

Figure 10.17 Three CL spectra from three differently sized GaN/AlN quantum discs. AlN appears dark in the upper HAADF image, GaN bright. The thickness of the discs is indicated in the number of atomic monolayers at the spectra. For comparison: 30 nm line position difference corresponds to 90 MeV energy difference. (Reprinted with permission from Ref. [47]. Copyright 2011 American Chemical Society.)

thickness. In Figure 10.18 the peak position, the intensity, and fwhm are mapped for the complete nanowire. The peak position in the spectrum from the pure GaN on the right of the nanowire corresponds closely to the earlier reported value of the near band edge line of GaN nanowires. The emission wavelength values of the individual discs are plotted in Figure 10.19. An increase in the emission wavelength with the disc thickness is observed, as expected for a quantum confinement effect. The larger wavelengths for the thicker discs compared to bulk GaN (smaller band gap) are expressions of the quantum-confined Stark effect (see also the simulated values). The scattering is caused by strain variations along the nanowire.

These experiments demonstrate the capability of spectral imaging with the CL STEM with a spatial resolution of about 1 nm and an energy resolution of about 10 meV. Especially the correlation of spectral details at high energy resolution with structural features (quantum emitters) on the nm level could not be achieved before. For example, evidence was found for the quantum-confined Stark effect leading to an emission below the bulk GaN band gap for disks thicker than 2.6 nm.

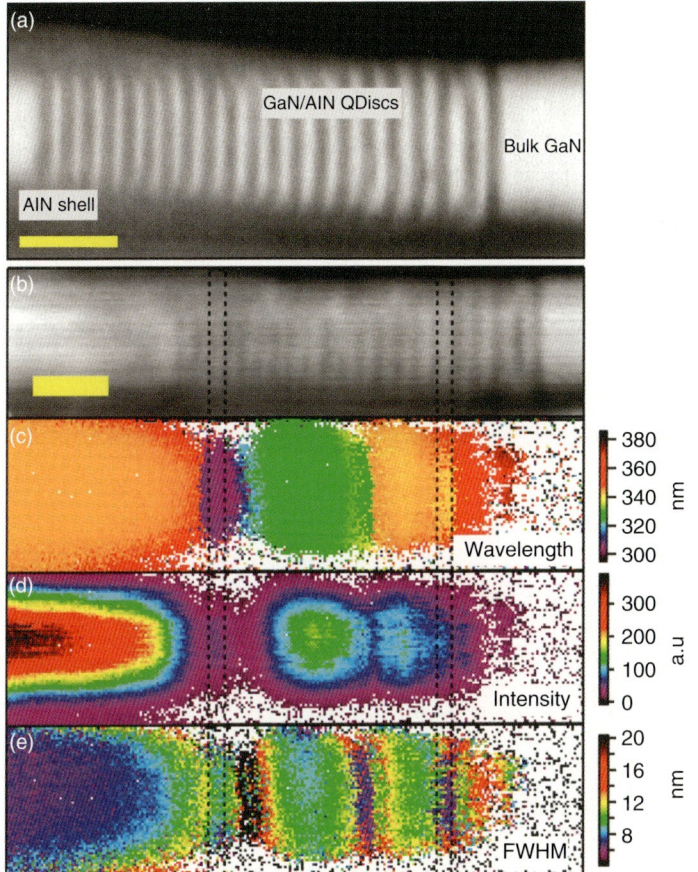

Figure 10.18 Spectral imaging of a GaN/AlN nanowire (NW) with 20 QDisks. It was acquired with 256 per 64 pixels of spatial sampling, 0.6 nm/pixel, and with 256 pixels of spectral sampling, 2 nm/pixel (wavelength), within 6 min. (a) HAADF image of a GaN/AlN nanowire (GaN appears whiter, AlN darker). Scale bar is 20 nm. (b) HAADF of the NW, acquired simultaneously with the CL. Scale bar is 20 nm. (c) Wavelength position of the most intense peak. (d) Intensity of the most intense peak. (e) Full width at half-maximum of the most intense peak. Some individual spectral features can be correlated to some individual quantum disks, as emphasized by the black dashed rectangles. (Reprinted with permission from Ref. [47]. Copyright 2011 American Chemical Society.)

10.6
Sample Preparation for Nanolab Experiments

This section focuses on the realization of the mounting and shaping of nanostructures, by "sculpting matter on a nanoscale," and on preventing artifacts, in particular the presence of C contamination. The first step in the process is the mounting of a sample on a device that enables the *in situ* experiments. Either

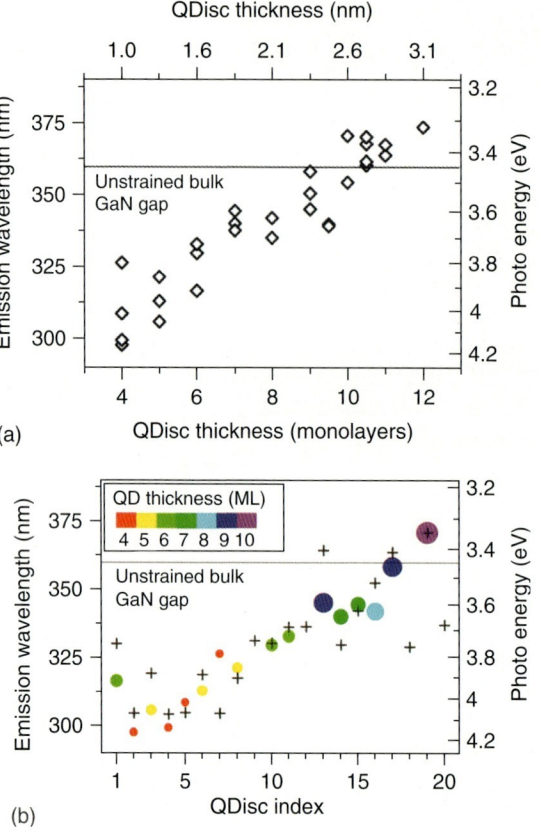

Figure 10.19 (a) Emission wavelength as a function of QDisc thickness measured with monolayer resolution of the individual Qdiscs in nanowires, as in Figure 10.18. For a given thickness, the dispersion in emission wavelength values is caused by varying positions within the nanowires (NWs). (b) CL energy of individual QDiscs shown as they are distributed inside the nanowire. The experimental values are given by full symbols whose color and size indicate the thickness in MLs of each QDisk according to the legend. The results of the simulations in the assumption that only a pyroelectric field is present are given by the black crosses. The emission of each QDisc is clearly affected not only by its size but also by its exact position along the nanowire. For a given QDisc size (same marker size and color) the emission wavelength systematically increases from the left to the right of the nanowire. (Reprinted with permission from Ref. [47]. Copyright 2011 American Chemical Society.)

the sample has to be positioned precisely on a given location (for instance, two electrodes), or the contact with the measurement equipment has to be made on an already existing sample (e.g., top-down evaporation of Au contacts on the CNT). For the first approach, the wedging technique [14] is very useful, in which a selected area from a sample is collected on a polymer film floated on a water surface and position on the measurement device by slowly lowering the water level. For the

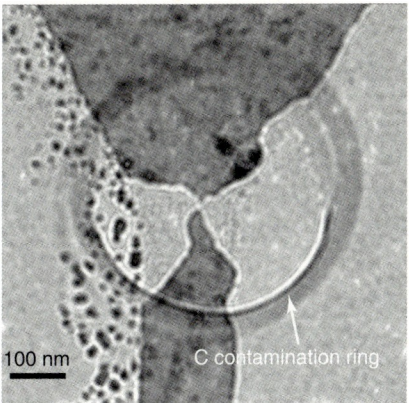

Figure 10.20 Example of the formation of a C ring, due to cracking of hydrocarbons at the edge of the electron beam, which was used for the imaging of gap formation in a Au bridge. This experiment was done at room temperature and with a polymer chip carrier. The source of the C contamination is mainly the polymer of the chip carrier.

latter approach, standard top-down electron beam lithography techniques can be used. Next, depending on the material and the application, the nanostructure that one wants to characterize, e.g., by electrical measurements, has to be shaped in the desired shape. For instance, in the case of graphene, one may want to create and measure a nanoribbon. The shaping can be done with focused beams: Ga ions, He ions, or electrons. The He beam and electron beam are mostly used for nanostructures. Irrespective of which method of (re)shaping is used, some C deposition is likely to occur. Depending on the type of *in situ* measurement, this C contamination is just annoying (leading to, e.g., loss of resolution) or completely detrimental, for example, for the case of electrical measurements where shortcuts run over the C contamination (see Figure 10.20), and for the case of catalyst particles where surface properties are lost. This C contamination has to be either completely prevented (as for catalysts) or at least removed (for electrical measurements).

10.6.1
Sculpting with the Electron Beam

The damaging electron beam/matter interaction can be taken advantage of when changing the specimen in a desired manner: specimens can be recrystalized or sculpted with the electron beam on a nanometer scale to study nanometer-sized atomic arrangements [48–54]. Sculpting a specimen with the electron beam means that material is removed: the atoms in the path of the electrons migrate out of their position by ionization processes or knockout. Usually, an electron current

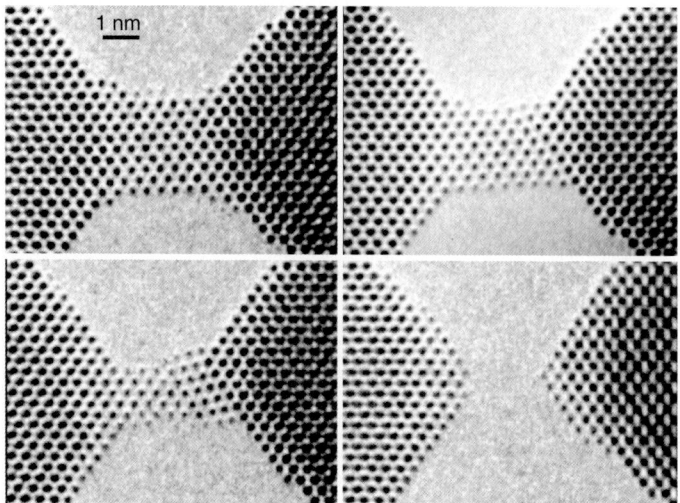

Figure 10.21 Selection of images from a movie of the formation of a gap, starting with a bridge in a freestanding Au film. The holes in the top and bottom were made by electron beam irradiation before the recording of this series. In the end, a 2 nm gap is created.

density of 10–100 pA/(nm)^2 is used with a beam size of 10–100 (nm)^2. Sculpting at a nanometer scale has, for instance, been done to make nanoelectrodes for measuring electrical properties of single molecules or nano particles inside the electron microscope [55–59].

Figure 10.21 shows a sequence of movie stills showing the sculpting of a pair of nanoelectrodes out of a freestanding gold foil using a 2–10 nm wide electron beam in a FEI Titan. First, a 20 nm wide bridge was made by drilling away material with a Ga ion beam in a focused ion beam (FIB), and subsequently with the electron beam in the TEM. The material forming the bridge was gradually removed. Once a gap was formed, the material retracted on both sides to form a gap 2 nm wide, as a result of probably strain relaxation and surface energy minimization. In Pt, the atomic mobility is much lower, resulting in less retraction once the gap is formed: the distance was only 0.6 nm in a Pt gap formation experiment. Such a gap can, for instance, be used to measure the transport properties of a small Au nanoparticle trapped in between two TEM made electrodes [55, 60]. The electrical measurements show a Coulomb blockade suppression of the current even at RT, in good agreement with the single-electron tunneling behavior expected for this device.

Electron beam sculpting is also used for the *in situ* formation of a nanopore [11] of a desired size in freestanding membranes of SiO_2 [58], Si_3N_4 [61], SiO_2/SiN [62], $SiO_2/SiN/SiO_2$ [56], or graphene [14]. Figure 10.22 shows one of the first experiments of nanopore formation in SiO_2 [58] by means of an electron beam.

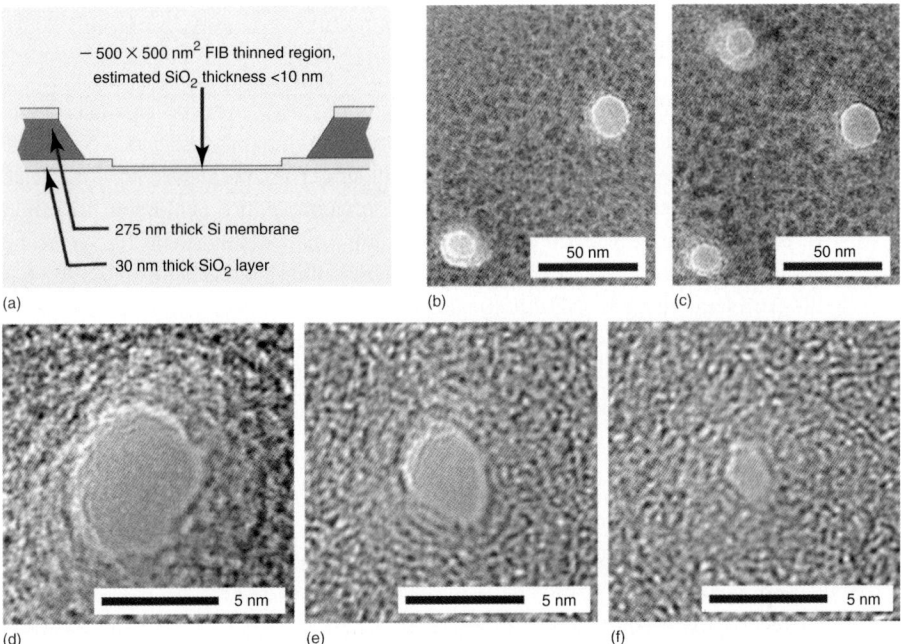

Figure 10.22 (a) Cross section of a thin SiO$_2$ membrane within a silicon-based membrane. Fabrication of this structure starts with a bare silicon membrane, which is oxidized with about 30 nm of SiO$_2$ on both sides. Using electron beam lithography and reactive-ion etching, we open up 1 × 1 μm^2 in the oxide layer. After a KOH wet etch, we obtain 30 nm thick SiO$_2$ membranes. Subsequently, these were thinned further in a focused ion beam (FIB) microscope to a final estimated thickness of less than 10 nm. (b) TEM micrograph of a part of a membrane with two holes that were drilled by a focused electron beam inside the TEM microscope. (c) TEM micrograph after drilling a third hole in the membrane depicted in (b). (d–f) Sequence of TEM images obtained on a shrinking nanopore with an initial diameter of about 6 nm and a final diameter of only 2 nm. (Reprinted with permission from Macmillan Publishers Ltd: Nature Materials [58], Copyright 2003.)

Interestingly, increased viscous flow in the membrane induced by the electron beam enables the flow of material [63]. A critical diameter d_c was calculated from the difference in surface energy of a closed and open pore. It was found that $d_c = h$, with h the thickness of the membrane. At diameters below d_c the pore will shrink, otherwise it will grow. The estimated membrane thickness of 40 nm was in good agreement with the observed critical size of 50–80 nm.

Such nanopores are used to detect biopolymers such as DNA and RNA by translocation: individual molecules are pulled, one by one, through the pore by an electric field [64–66]. Important factors in the analysis of DNA translocations are sidewall abruptness of the pores and the chemical composition of the wall surface [67]. The details of the chemistry involved in these nanopore wall/polymer molecule interactions need further study.

10.6.2
Sculpting with the Ga Ion Beam

Ga ions (and several other heavy ions) have been used over the past decades to remove material, and by focusing the ions one can remove material at a local scale. With an FIB one can also obtain images, but since the ions also remove material, one is actually changing the sample during imaging, which is a major limitation in particular for very small structures. An important development has been the combination of an FIB and an SEM in one machine: the Dual Beam, which has the advantage that most of the imaging can be done with the electron beam, and thus that the imaging does not change the sample significantly. Three-dimensional structures such as SQUIDS can be made with an FIB [68, 69]; see Figure 10.23 for an example.

10.6.3
Sculpting with the Helium Ion Beam

The beam of a Helium ion microscope (HIM) can be used very efficiently as a cutting and sculpting tool for nanosamples; for a review see [71], for examples, see [70–73]. It can also be used for the preparation of thin electron microscopy samples, for example, for $Cu_xBi_2Se_3$ [74]. The operation principle of the HIM is similar to that of an FIB with the difference that helium ions are used instead of gallium ions. In Figure 10.24, the interaction of He ions with Si is demonstrated by Monte Carlo calculations and compared to the same when using Ga ions. The unique

(a)

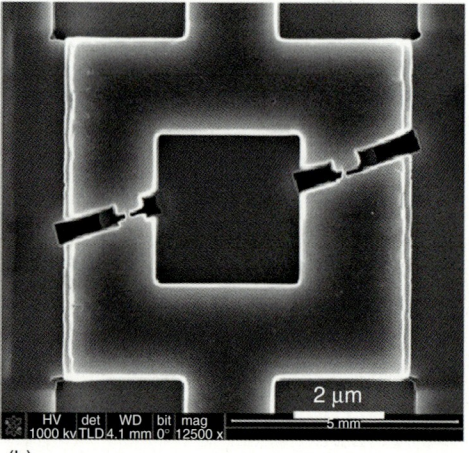
(b)

Figure 10.23 SEM images of the MgB2 nanobridge SQUID fabricated by the focused ion beam method: (a) bird's-eye view and (b) top view. (Reprinted with permission from Ref. [68]. Copyright 2009, IOP Publishing Ltd.)

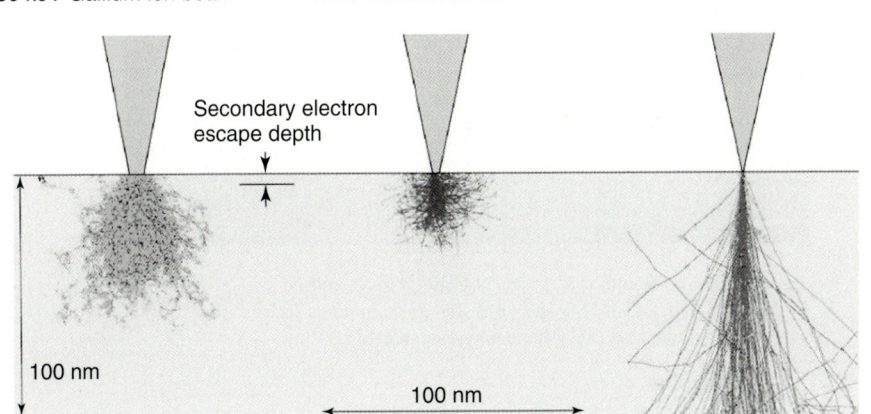

Figure 10.24 Monte Carlo calculations on the trajectories of Ga and He ions in silicon. The He ions penetrate much deeper into the Si. On the other hand, the spread in the first 20 nm is much smaller. The effect of the deeper penetration is disadvantageous for Si, because the trapped He can amorphisize the Si or form bubbles, but for some other materials such as Bi_2Se_3 (planar structure with van der Waals bonding between Se–Se planes) one can obtain thin areas by milling even with a 90° incidence angle of the He ions, indicating that the possibility of out-diffusion of He plays an important role. (Reprinted with permission from Ref. [75]. Copyright 2007, American Institute of Physics.)

interaction of the helium ions with the sample material allows for high-resolution scanning imaging possibilities as well as very precise nanofabrication.

As an example, we show the sculpting of a Pt nanobridge with a HIM Orion ZEISS operated at an acceleration voltage of 25 kV. Figure 10.25 shows the sample modification of a polycrystalline Pt nanobridge (300 nm × 200 nm × 14 nm) positioned on a 100 nm thick Si_3N_4 membrane. As a first step, the membrane near the bridge was removed. Then, a desired shape of the bridge was defined in a

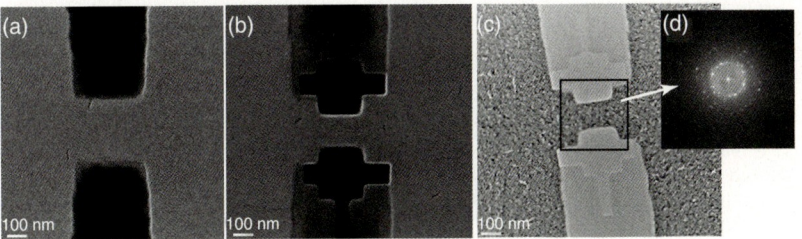

Figure 10.25 The result of cutting of the polycrystalline bridges with HIM. (a) A HIM image of the initial state of sample, (b) a HIM image of the bridge after sculpting, (c) a TEM image of the bridge after sculpting, and (d) the Fourier transform of the area outlined in (c).

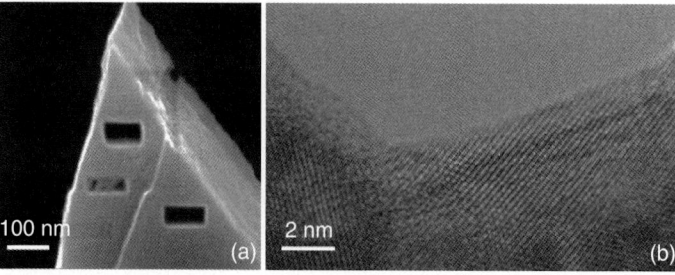

Figure 10.26 Example of a cut in $Cu_xBi_2Se_3$ crystal imaged by (a) HIM and (b) HREM. The $Cu_xBi_2Se_3$ is continuous up to the edge, indicating that little damage is created by the He ion beam.

pattern generator program and nanosculpting of the sample was performed. The developed technique allows for fast and efficient cutting of both Pt samples and Si_3N_4 membranes (Figure 10.25a,b) [76]. Depending on the size of the modified area, cutting takes from a few seconds to a few minutes. The sculpted samples were examined with the TEM (Figure 10.25c,d). It can be seen that the sample remains crystalline up to the edge and that no significant defects have been created. Figure 10.26 shows that after He sculpting, HREM imaging in a TEM is possible [74].

References

1. Heersche, H.B., Lientschnig, G., O'Neill, K., van der Zant, H.S.J., and Zandbergen, H.W. (2007) In situ imaging of electromigration-induced nanogap formation by transmission electron microscopy. *Appl. Phys. Lett.*, **91** (7), 072107.
2. Gao, B., Rudneva, M., McGarrity, K.S., Xu, Q., Prins, F., Thijssen, J.M., Zandbergen, H.W., and van der Zant, H.S.J.(2011) In-situ transmission electron microscopy imaging of grain growth in a platinum nanobridge induced by electric current annealing. *Nanotechnology*, **22**, 205705.
3. Rudneva, M., Gao, B., van der Zant, H.S.J., and Zandbergen, H.W. (2010) In-situ electrical characterization combined with simultaneous TEM observation. Proceedings International Microscopy Congress 17, Rio de Janeiro, Brazil, September 19–24, 2010 (eds G. Solórzano and W. de Souza), p. M19.14.
4. Golberg, D., Costa, P.M.F.J., Mitome, M., and Bando, Y. (2008) In-situ TEM electrical and mechanical properties measurements of one-dimensional inorganic nanomaterials. 2nd IEEE International Nanoelectronics Conference, 2008, Vols. 1–3, 1127–1131.
5. Costa, P., Golberg, D., Mitome, M., and Bando, Y. (2007) Nitrogen-doped carbon nanotube structure tailoring and time-resolved transport measurements in a transmission electron microscope. *Appl. Phys. Lett.*, **91** (22), 223108.
6. Young, N.P., van Huis, M.A., Zandbergen, H.W., Xu, H., and Kirkland, A.I. (2010) Transformations of gold nanoparticles investigated using variable temperature high-resolution transmission electron microscopy. *Ultramicroscopy*, **110** (5), 506–516.
7. van Huis, M.A., Young, N.P., Pandraud, G., Creemer, J.F., Vanmaekelbergh, D., Kirkland, A.I., and Zandbergen, H.W.

(2009) Atomic imaging of phase transitions and morphology transformations in nanocrystals. *Adv. Mater.*, **21** (48), 4992–4995.
8. Novoselov, K.S., Geim, A.K., Morozov, S.V., Jiang, D., Zhang, Y., Dubonos, S.V., Grigorieva, I.V., and Firsov, A.A. (2004) Electric field effect in atomically thin carbon films. *Science*, **306** (5296), 666–669.
9. Castro Neto, A.H., Guinea, F., Peres, N.M.R., Novoselov, K.S., and Geim, A.K. (2009) The electronic properties of graphene. *Rev. Mod. Phys.*, **81** (1), 109–162.
10. Girit, C.O., Meyer, J.C., Erni, R., Rossell, M.D., Kisielowski, C., Yang, L., Park, C.H., Crommie, M.F., Cohen, M.L., Louie, S.G. et al. (2009) Graphene at the edge: stability and dynamics. *Science*, **323** (5922), 1705–1708.
11. Lee, C., Wei, X.D., Kysar, J.W., and Hone, J. (2008) Measurement of the elastic properties and intrinsic strength of monolayer graphene. *Science*, **321** (5887), 385–388.
12. Garaj, S., Hubbard, W., Reina, A., Kong, J., Branton, D., and Golovchenko, J.A. (2010) Graphene as a subnanometre trans-electrode membrane. *Nature*, **467** (7312), 190–193.
13. Postma, H.W.C. (2010) Rapid sequencing of individual DNA molecules in graphene nanogaps. *Nano Lett.*, **10** (2), 420–425.
14. Schneider, G.F., Kowalczyk, S.W., Calado, V.E., Pandraud, G., Zandbergen, H.W., Vandersypen, L.M.K., and Dekker, C. (2010) DNA translocation through graphene nanopores. *Nano Lett.*, **10** (8), 3163–3167.
15. Song, B., Schneider, G.F., Xu, Q., Pandraud, G., Dekker, C., and Zandbergen, H.W. (2011) Atomic-scale electron-beam sculpting of defect-free graphene nanostructures. *Nano Lett.*, **11** (6), 2247–2250.
16. Figuerola, A., van Huis, M., Zanella, M., Genovese, A., Marras, S., Falqui, A., Zandbergen, H.W., Cingolani, R., and Manna, L. (2010) Epitaxial CdSe-Au nanocrystal heterostructures by thermal annealing. *Nano Lett.*, **10** (8), 3028–3036.
17. Baker, R.T.K., Barber, M.A., Waite, R.J., Harris, P.S., and Feates, F.S. (1972) Nucleation and growth of carbon deposits from nickel catalyzed decomposition of acetylene. *J. Catal.*, **26** (1), 51–62.
18. Creemer, J.F., Helveg, S., Kooyman, P.J., Molenbroek, A.M., Zandbergen, H.W., and Sarro, P.M. (2010) A MEMS reactor for atomic-scale microscopy of nanomaterials under industrially relevant conditions. *J. Microelectromech. Syst.*, **19** (2), 254–264.
19. de Smit, E., Swart, I., Creemer, J.F., Hoveling, G.H., Gilles, M.K., Tyliszczak, T., Kooyman, P.J., Zandbergen, H.W., Morin, C., Weckhuysen, B.M. et al. (2008) Nanoscale chemical imaging of a working catalyst by scanning transmission X-ray microscopy. *Nature*, **456** (7219), 222–225.
20. Creemer, J.F., Helveg, S., Hoveling, G.H., Ullmann, S., Molenbroek, A.M., Sarro, P.M., and Zandbergen, H.W. (2008) Atomic-scale electron microscopy at ambient pressure. *Ultramicroscopy*, **108** (9), 993–998.
21. Yokosawa, T., Alan, T., Pandraud, G., de Kruijff, T.R., Dam, B., and zandbergen, H.W. (2011) In-situ TEM at 1-4 bar hydrogen pressure on hydrogen storage materials. *Ultramicroscopy*, **112** (1), 47–52.
22. Parkinson, G.M. (1989) High-resolution in situ controlled-atmosphere transmission electron microscopy (CATEM) of heterogeneous catalysts. *Catal. Lett.*, **2** (5), 303–307.
23. Alan, T., Gaspar, J., Paul, O., Zandbergen, H.W., Creemer, F., and Sarro, P.M. (2010) Characterization of ultrathin membranes to enable tem observation of gas reactions at high pressures. IMECE 2009: Proceedings of the Asme International Mechanical Engineering Congress and Exposition, Vol. 11, pp. 327–331.
24. http://www.hummingbirdscientific.com (accessed December 30 2011).
25. Mele, L., Santagata, F., Pandraud, G., Morana, B., Tichelaar, F.D., Creemer, J.F., and Sarro, P.M. (2010) Wafer-level assembly and sealing of a MEMS nanoreactor for in situ microscopy. *J. Micromech. Microeng.*, **20** (8), 085040.

26. Maeland, A.J., and Gibb, T.R.P. (1961) X-Ray diffraction observations of Pd-H2 system through critical region. *J. Phys. Chem.*, **65** (7), 1270.
27. de Jonge, N., Bigelow, W.C., and Veith, G.M. (2010) Atmospheric pressure scanning transmission electron microscopy. *Nano Lett.*, **10** (3), 1028–1031.
28. Swift, J.A., and Brown, A.C. (1970) Environmental cell for examination of wet biological specimens at atmospheric pressure by transmission scanning electron microscopy. *J. Phys. E: Sci. Instrum.*, **3** (11), 924.
29. Butler, E.P. and Hale, K.F. (1981) *Dynamic Experiments in the Electron Microscope*, North-Holland (Elsevier), Amsterdam.
30. Radisic, A., Ross, F.M., and Searson, P.C. (2006) In situ study of the growth kinetics of individual island electrodeposition of copper. *J. Phys. Chem. B*, **110** (15), 7862–7868.
31. Radisic, A., Vereecken, P.M., Hannon, J.B., Searson, P.C., and Ross, F.M. (2006) Quantifying electrochemical nucleation and growth of nanoscale clusters using real-time kinetic data. *Nano Lett.*, **6** (2), 238–242.
32. Ross, F.M. (2006) in *Science of Microscopy* (eds P.W. Hawkes and J.C. Spence), Kluwer Publishers, pp. 445–534.
33. Zheng, H.M., Smith, R.K., Jun, Y.W., Kisielowski, C., Dahmen, U., and Alivisatos, A.P. (2009) Observation of single colloidal platinum nanocrystal growth trajectories. *Science*, **324** (5932), 1309–1312.
34. Grogan, J.M. and Bau, H.H. (2010) The nanoaquarium: a platform for in situ transmission electron microscopy in liquid media. *J. Microelectromech. Syst.*, **19** (4), 885–894.
35. Sugi, H., Akimoto, T., Sutoh, K., Chaen, S., Oishi, N., and Suzuki, S. (1997) Dynamic electron microscopy of ATP-induced myosin head movement in living muscle thick filaments. *Proc. Natl. Acad. Sci. U.S.A.*, **94** (9), 4378–4382.
36. Fukami, A. and Murakami, S. (1979) Progress of electron-microscopy of wet specimens using film-sealed environmental cell – its serious problems for biological-materials. *J. Electron Microsc.*, **28**, S41–S48.
37. Kohyama, N., Fukushima, K., and Fukami, A. (1979) Observation for hydrate form of clay-minerals by means of environmental cell method. *J. Electron Microsc.*, **28** (3), 258–259.
38. Ross, F.M., Williamson, M.J., Tromp, R.M., Hull, R., and Vereecken, P.M. (2002) In situ transmission electron microscopy of copper electrodeposition. *Microsc. Microanal.*, **8** (S02), 420–421.
39. den Heijer, M. (2008) In-situ transmission electron microscopy of electrodeposition. MSc thesis, Delft University of Technology, Kavli Institute of Nanoscience, pp. 1–118.
40. Williamson, M.J., Tromp, R.M., Vereecken, P.M., Hull, R., and Ross, F.M. (2003) Dynamic microscopy of nanoscale cluster growth at the solid-liquid interface. *Nat. Mater.*, **2** (8), 532–536.
41. Garrett, B.C., Dixon, D.A., Camaioni, D.M., Chipman, D.M., Johnson, M.A., Jonah, C.D., Kimmel, G.A., Miller, J.H., Rescigno, T.N., Rossky, P.J. et al. (2005) Role of water in electron-initiated processes and radical chemistry: Issues and scientific advances. *Chem. Rev.*, **105** (1), 355–389.
42. Reimer, L. (1997) *Transmission Electron Microscopy; Physics of Image Formation and Microanalysis*, Springer, Berlin.
43. de Abajo, F.J.G. (2010) Optical excitations in electron microscopy. *Rev. Mod. Phys.*, **82** (1), 209–275.
44. Dierre, B., Yuan, X.L., and Sekiguchi, T. (2010) Low-energy cathodoluminescence microscopy for the characterization of nanostructures. *Sci. Technol. Adv. Mater.*, **11** (4), 043001.
45. Yamamoto, N., Araya, K., and de Abajo, F.J.G. (2001) Photon emission from silver particles induced by a high-energy electron beam. *Phys. Rev. B*, **6420** (20), 9.
46. Strunk, H.P., Albrecht, M., and Scheel, H. (2006) Cathodoluminescence in transmission electron microscopy. *J. Microsc. (Oxford)*, **224**, 79–85.
47. Zagonel, L.F., Mazzucco, S., Tence, M., March, K., Bernard, R., Laslier, B., Jacopin, G., Tchernycheva, M., Rigutti,

L., Julien, F.H. et al. (2011) Nanometer scale spectral imaging of quantum emitters in nanowires and its correlation to their atomically resolved structure. *Nano Letters*, **11**, 568.

48. Troiani, H.E., Miki-Yoshida, M., Camacho-Bragado, G.A., Marques, M.A.L., Rubio, A., Ascencio, J.A., and Jose-Yacaman, M. (2003) Direct observation of the mechanical properties of single-walled carbon nanotubes and their junctions at the atomic level. *Nano Lett.*, **3** (6), 751–755.

49. Takai, Y., Kawasaki, T., Kimura, Y., Ikuta, T., and Shimizu, R. (2001) Dynamic observation of an atom-sized gold wire by phase electron microscopy. *Phys. Rev. Lett.*, **8710** (10), 106105.

50. Rodrigues, V., Fuhrer, T., and Ugarte, D. (2000) Signature of atomic structure in the quantum conductance of gold nanowires. *Phys. Rev. Lett.*, **85** (19), 4124–4127.

51. Remeika, M. and Bezryadin, A. (2005) Sub-10 nanometre fabrication: molecular templating, electron-beam sculpting and crystallization of metallic nanowires. *Nanotechnology*, **16** (8), 1172–1176.

52. Oshima, Y., Kondo, Y., and Takayanagi, K. (2003) High-resolution ultrahigh-vacuum electron microscopy of helical gold nanowires: junction and thinning process. *J. Electron Microsc.*, **52** (1), 49–55.

53. Kondo, Y. and Takayanagi, K. (1997) Gold nanobridge stabilized by surface structure. *Phys. Rev. Lett.*, **79** (18), 3455–3458.

54. Kizuka, T. (1998) Atomic process of point contact in gold studied by time-resolved high-resolution transmission electron microscopy. *Phys. Rev. Lett.*, **81** (20), 4448–4451.

55. Zandbergen, H.W., van Duuren, R.J.H.A., Alkemade, P.F.A., Lientschnig, G., Vasquez, O., Dekker, C., and Tichelaar, F.D. (2005) Sculpting nanoelectrodes with a transmission electron beam for electrical and geometrical characterization of nanoparticles. *Nano Lett.*, **5** (3), 549–553.

56. Wu, M.Y., Smeets, R.M.M., Zandbergen, M., Ziese, U., Krapf, D., Batson, P.E., Dekker, N.H., Dekker, C., and Zandbergen, H.W. (2009) Control of shape and material composition of solid-state nanopores. *Nano Lett.*, **9** (1), 479–484.

57. van den Hout, M., Hall, A.R., Wu, M.Y., Zandbergen, H.W., Dekker, C., and Dekker, N.H. (2010) Controlling nanopore size, shape and stability. *Nanotechnology*, **21** (11), 115304.

58. Storm, A.J., Chen, J.H., Ling, X.S., Zandbergen, H.W., and Dekker, C. (2003) Fabrication of solid-state nanopores with single-nanometre precision. *Nat. Mater.*, **2** (8), 537–540.

59. Krapf, D., Wu, M.Y., Smeets, R.M.M., Zandbergen, H.W., Dekker, C., and Lemay, S.G. (2006) Fabrication and characterization of nanopore-based electrodes with radii down to 2 nm. *Nano Lett.*, **6** (1), 105–109.

60. Bezryadin, A., Dekker, C., and Schmid, G. (1997) Electrostatic trapping of single conducting nanoparticles between nanoelectrodes. *Appl. Phys. Lett.*, **71** (9), 1273–1275.

61. Kim, M.J., Wanunu, M., Bell, D.C., and Meller, A. (2006) Rapid fabrication of uniformly sized nanopores and nanopore arrays for parallel DNA analysis. *Adv. Mater.*, **18** (23), 3149–3153.

62. Wu, M.Y., Krapf, D., Zandbergen, M., Zandbergen, H., and Batson, P.E. (2005) Formation of nanopores in a SiN/SiO2 membrane with an electron beam. *Appl. Phys. Lett.*, **87** (11), 113106.

63. Ajayan, P.M. and Iijima, S. (1992) Electron-beam-enhanced flow and instability in amorphous silica fibers and tips. *Philos. Mag. Lett.*, **65** (1), 43–48.

64. Storm, A.J., Chen, J.H., Zandbergen, H.W., and Dekker, C. (2005) Translocation of double-strand DNA through a silicon oxide nanopore. *Phys. Rev. E*, **71** (5), 051903.

65. Li, J.L., McMullan, C., Stein, D., Branton, D., and Golovckenko, J. (2001) Solid state nanopores for single molecule detection. *Biophys. J.*, **80** (1), 339A–339A.

66. Kasianowicz, J.J., Brandin, E., Branton, D., and Deamer, D.W. (1996) Characterization of individual polynucleotide molecules using a membrane channel.

Proc. Natl. Acad. Sci. U.S.A., **93** (24), 13770–13773.

67. Smeets, R.M.M., Keyser, U.F., Krapf, D., Wu, M.Y., Dekker, N.H., and Dekker, C. (2006) Salt dependence of ion transport and DNA translocation through solid-state nanopores. *Nano Lett.*, **6** (1), 89–95.

68. Lee, S.G., Hong, S.H., Seong, W.K., and Kang, W.N. (2009) All focused ion beam fabricated MgB2 inter-grain nanobridge dc SQUIDs. *Supercond. Sci. Technol.*, **22** (6), 064009.

69. Kim, S.J., Hatano, T., Kim, G.S., Kim, H.Y., Nagao, M., Inomata, K., Yun, K.S., Takano, Y., Arisawa, S., Ishii, A. et al. (2004) Characteristics of two-stacked intrinsic Josephson junctions with a submicron loop on a $Bi_2Sr_2CaCu_2O_{8+d}$ (Bi-2212) single crystal whisker. *Physica C*, **412–414**, 1401–1405.

70. Postek, M.T., Vladar, A., Dagata, J., Farkas, N., Ming, B., Wagner, R., Raman, A., Moon, R.J., Sabo, R., Wegner, T.H. et al. (2011) Development of the metrology and imaging of cellulose nanocrystals. *Meas. Sci. Technol.*, **22** (2), 024005.

71. Ogawa, S., Thompson, W., Stern, L., Scipioni, L., Notte, J., Farkas, L., and Barriss, L. (2010) Helium ion secondary electron mode microscopy for interconnect material imaging. *Jpn. J. Appl. Phys.*, **49** (4), 04DB12.

72. Postek, M.T., Vladar, A.E., and Ming, B. (2009) in *Frontiers of Characterization and Metrology for Nanoelectronics: 2009* (eds Seiler D.G., Diebold, A.C. McDonald, R. Garner, C.M. Herr, D. Khosla, R.P. and Secula, E.M.) American Institute of Physics, Melville, pp. 249–260.

73. van den Boom, R.J.J., Parvaneh, H., Voci, D., Huynh, C., Stern, L., Dunn, K.A., Lifshin, E., Seiler, D.G., Diebold, A.C., McDonald, R., Garner, C.M., Herr, D., Khosla, R.P., and Secula, E.M. (eds) (2009) in *Frontiers of Characterization and Metrology for Nanoelectronics: 2009*, American Institute of Physics, Melville, pp. 309–313.

74. Rudneva, M., Veldhoven, Ev., Shu, M.S., Maas, D., and Zandbergen, H.W. (2010) Preparation of electron transparent samples of pre-selected areas using a helium ion microscope. Conference Proceedings of International Microscopy Congress 17, Rio de Janeiro, Brazil, September 19–24 (eds G. Solórzano and W. de Souza), p. I18.18.

75. Notte, J., Ward, B., Economou, N., Hill, R., Percival, R., Farkas, L., and McVey, S. (2007) in *Frontiers of Characterization and Metrology for Nanoelectronics: 2007* (eds Seiler D.G., Diebold, A.C. McDonald, R. Garner, C.M. Herr, D. Khosla, R.P. and Secula, E.M.) American Institute of Physics, Melville, pp. 489–496.

76. Rudneva, M., van Veldhoven, E., Maas, D., and Zandbergen, H. (2011) Helium ion microscope as a sculpting tool for nanosamples. Conference proceedings of International Union of Microbeam Analysis Societies – V, Seoul, Korea, 22–27 May. (to be published).

11
Atomic-Resolution Environmental Transmission Electron Microscopy

Pratibha L. Gai and Edward D. Boyes

11.1
Introduction

Many dynamic processes occur at gas–solid or liquid–solid interfaces and greatly impact the solid's properties, either in nature or in technology [1–5]. Direct observation of microstructural evolution under dynamic reaction conditions is a powerful scientific tool. *In situ* electron microscopy (EM) under controlled reaction conditions provides dynamic information on processes, which cannot be obtained directly and readily by other techniques. We are particularly interested in properties of solids, including catalytic materials and nanomaterials, under reaction conditions of gaseous environments and operating temperatures, and in exploring structure–property relationships associated with them. In heterogeneous catalysis, for instance, it is well known that the state of the catalyst and the catalyst's properties are inextricably dependent on the reaction environment. Direct observations of solid catalysts and their structural evolution under dynamic reaction conditions are key to obtaining insights into the relationship between the nanostructure of the solid and reaction mechanisms [1–8], because it is often not possible to infer the dynamic state of the material from postmortem examinations of the static material.

Although high-resolution transmission electron microscopy (TEM) in high vacuum has advanced significantly over the preceding decades, providing a versatile means for directly studying the structure of a solid with real-space resolution down to the atomic scale, there are important applications in materials science in which the role of the environment on a sample is critical. This is because the natural environment for reacting materials is not high vacuum and many reactions occur in gas environments and at elevated temperatures, at the atomic scale. However, applying high-vacuum TEM for *in situ* studies of gas–solid interactions is extremely demanding, in part because of the very short path length of electrons in dense media (solids, fluids) and the need for high-vacuum conditions to avoid compromising the atomic-resolution imaging and analytical capabilities. The key challenge for establishing such extreme reaction conditions inside an EM vacuum column is to confine gas environments in the close vicinity of a sample operating at elevated temperatures.

Handbook of Nanoscopy, First Edition. Edited by Gustaaf Van Tendeloo, Dirk Van Dyck, and Stephen J. Pennycook.
© 2012 Wiley-VCH Verlag GmbH & Co. KGaA. Published 2012 by Wiley-VCH Verlag GmbH & Co. KGaA.

A number of notable *in situ* experiments have relied on modifications to the standard TEM operations. In some of the earlier applications [4], the sample was sandwiched between thin windows to contain the saturated water vapor environment necessary for both preservation and reaction of the sample. Electron-transparent windows have been used to contain gases and solvents in EMs, but there are problems in reliably sustaining a large pressure difference across a window that is thin enough to permit electron penetration suitable for high-resolution studies. The additional diffuse scattering in the windows may obscure the already limited image contrast and window cells are generally not compatible with heating or tilting systems.

The complications and potential for catastrophic failure of windows can be avoided by substituting small apertures above and below the sample to restrict the diffusion of gas molecules, but to allow the penetration of the electron beam. Aperture systems are robust and they can easily be made compatible with regular sample heating or cooling stages. In practice, a series of apertures, with differential pumping systems connected between them, are needed above and below the sample to get a minimum useful vapor pressure around the sample, while maintaining high vacuum in the rest of the EM. Typically, pairs of apertures are added above and below the sample with pumping lines attached between them.

To realize environmental transmission electron microscopic (ETEM) studies, different approaches have been pursued to confine gas (or liquid) environments inside a TEM's vacuum column. In early ETEM systems, environmental cell (ECELL, or microreactor) jigs were designed [1, 4, 6, 9]; and references therein] to be inserted between the pole pieces of the objective lens of the EM, necessitating the frequent rebuilding of the EM to effect the changeover in functionality and presenting difficulties with reliability, as well as the alignment of the electron beam through the series of apertures.

11.2
Atomic-Resolution ETEM

To address these challenges, an atomic-resolution ETEM has been pioneered by the authors of this chapter for probing of gas–solid reactions directly at the atomic level under controlled atmosphere and temperature conditions [10, 11]. This development is now the basis of atomic-resolution ETEMs across the world and has opened up the unique possibility to monitor, with atomic resolution, solid materials (such as heterogeneous catalysts and nanomaterials), during exposure to reactive gas environments and elevated temperatures on practical materials. In this chapter, we provide a review of some recent applications that exploit atomic-resolution ETEM for understanding the role of gas–surface interactions in nanocatalysts and nanomaterials in their functioning state. Temperature-, time-, and pressure-resolved studies on real systems are possible, bridging for the first time the material and pressure gaps at operating temperatures. Although the use of surface science techniques in very high vacuum with extended single crystal surfaces to model and understand the fundamentals of heterogeneous catalysis has been invaluable, these studies were limited by the "pressure and materials gap" that exists between these studies and real systems at reaction temperatures.

In addition to providing direct and unparalleled insights into the dynamic evolution of structural changes at the atomic level, the ETEM allows the detection, in real time, of surface as well as subsurface structural phenomena including the defect evolution, access to metastable states, reaction mechanisms, changes in the chemical composition, monolayer segregation, and active sites on specific catalyst surfaces involved in binding gas molecules. These can be explored by closely mimicking the working conditions of practical reactors, allowing us a look at the real atomic world. Entire patterns of key catalyst activation, operation, and aging can be studied in a range of gases and temperature regimes that provide a fundamental understanding of kinetics and reaction processes. These studies enable optimization of synthesis, processing, and stability of materials. The atomic-resolution ETEM development is therefore a major change from the conventional methods. Below, we describe the development of the method that has opened up a new field for studying gas–solid reactions at the atomic level.

11.3
Development of Atomic-Resolution ETEM

In this chapter, we outline recent developments in ETEM.

Recently, we have taken a new approach to organize and design instruments that are dedicated to ECELL operations and are integral to the EM [11]. In our case, the modifications of the instrument are considered permanent. First, it is based on an excellent, modern, and computer-controlled EM (Philips CM30T TEM/STEM) system with a proven high-resolution lattice imaging performance. Second, the whole column, and not just the region around the sample, has been completely redesigned for the ECELL functionality. Third, a custom set of pole pieces incorporating radial holes for the first stage of differential pumping, but with acceptably low astigmatism and no measurable deleterious effect on imaging, has been constructed for the instrument.

The basic geometry is a four-aperture system, in pairs above and below the sample (Figure 11.1), but the apertures are now mounted inside the bores of the objective lens pole pieces rather than between them as in previous designs. This approach allows unrestricted use of regular sample holders in a relatively narrow gap lens ($S = 9$ mm) with much lower aberration coefficients ($Cs = Cc = 2$ mm) than have been possible with previous ECELL designs. Regular microscope apertures are centered in precision bushes in each pole piece. The apertures have direct metal-to-metal seals around the rim. In the manufacture of electron lenses, the priority is to minimize misalignment. It was necessary to measure each bore optically and to construct custom bushes for each position.

Another important feature is that the controlled environment ECELL volume is the regular sample chamber of the EM. Because the regular sample chamber in the

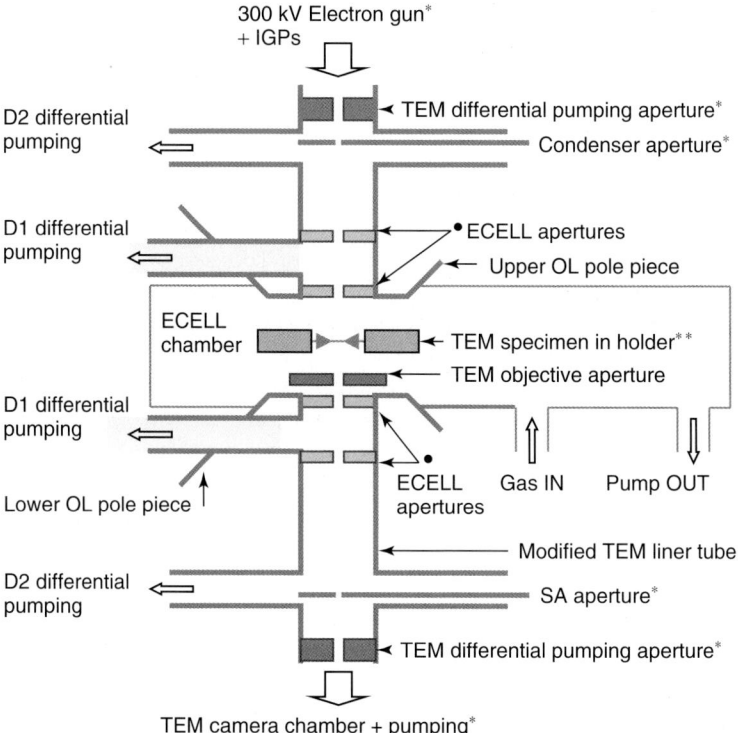

Figure 11.1 A schematic of the basic geometry of the aperture system in the pioneering development of atomic-resolution ETEM by Gai and Boyes to probe gas–solid reactions at the atomic level. Radial holes through objective pole pieces (OL) are for gas lines at D. The OL pole pieces are above and below the specimen holder and the lower OL pole piece is indicated. Gas inlet, the first stage of differential pumping lines (D1) between the environmental cell (ECELL) apertures, condenser aperture, a second stage of pumping (D2) at the condenser lens, selected area (SA) diffraction aperture, parallel electron energy loss spectroscopy (PEELS) and TEM camera vacuum are indicated. The sample chamber is the gas reaction cell. The novel ETEM design of Gai and Boyes has been adopted by TEM manufacturers (FEI) and the development has been highlighted by the American Chemical Society's C&E News (1995, 2002). Ion getter pump (IGP).

ETEM is the ECELL volume, it is permanently mounted and integrated. Therefore, once the modifications are made to the EM column, the ETEM can be operated either in gas environments or in vacuum (as a conventional TEM), without compromising the atomic resolution. Differential pumping systems connected between the apertures consist of molecular drag pumps (MDP) and turbomolecular pumps (TMP). This system permits high gas pressures in the ECELL sample region while maintaining high vacuum in the rest of the ETEM. A conventional reactor-type gas-manifold system constructed with stainless steel enables the inlet of flowing gases into the ECELL of the ETEM, and is compatible with the ultrahigh voltage (UHV) and gas pressure conditions. Operating pressures are measured using multiple pirani gauges calibrated for the appropriate gases and disposed around the system to ensure adequate engineering diagnostics. The sample is in a dynamic flowing gas environment similar to conditions in technological reactors, with a positive flow rate of gas. A hot stage allows samples to be heated routinely to about 1000 °C. A mass spectrometer is included for gas analysis. For dynamic atomic resolution studies, a few millibars (mbar) up to 30 mbar of gas pressure are typically used in the ECELL. Higher gas pressures are possible, but this compromises the resolution (because of multiple scattering effects of the electron beam through denser gas layers). Electronic image shift and drift compensation help to stabilize high-resolution images for data recording, with a time resolution of the order of 1/30 s or better on charge-coupled device (CCD), with real-time digitally processed low light TV (LLTV) camera video system, or on movie film; and minimally invasive low-dose electron beam techniques are used throughout. For example, a video system connected to the ETEM apparatus can facilitate digital image processing and real-time recording of dynamic events at a time scale of milliseconds, consistent with contact times of gas molecules in chemisorption reactions in technological reactors [11–13].

In our development of atomic-resolution ETEM, atomic-resolution, scanning transmission electron microscopy (STEM), hot stage, and PEELS/GIF (parallel electron energy loss spectroscopic/Gatan image filtering) functionalities have been combined in a single instrument. The combination is required to aid simultaneous determination of the dynamic structure and composition of the reactor contents. The ETEM system is used as a nanolaboratory with multiprobe measurements. Designing of novel reactions and nanosynthesis are possible. The structure and chemistry of dynamic catalysts are revealed by atomic imaging, electron diffraction, and chemical analysis while the sample is immersed in controlled gas atmospheres and at the operating temperature. For chemical microanalyses, a commercial Gatan PEELS/GIF system is fitted to provide elemental analysis during *in situ* experiments with low gas pressures [11, 14], and fast and minimally invasive high-resolution chemical mapping with filtered TEM images. The analyses of the oxidation state in intermediate phases of the reaction are possible, as illustrated in the following sections. In principle, extended energy loss fine structure (EXELFS) studies are also possible. In many applications, the determination of the size and subsurface location of particles requires the use of the dynamic STEM system (integrated with ETEM), with complementary methods for chemical and crystallographic analyses. The first atomic-resolution ETEM and the schematic of accessories are illustrated in Figure 11.2a,b, respectively.

Because of the small amounts of solid reactant in the microscope sample, measurement of reaction products are carried out on larger samples in a microreactor operating under similar conditions and used for the nanostructural correlation. The correlation between the dynamic nanostructure and reactivity is crucial for a better understanding of gas–solid reactions and the development of improved materials and processes.

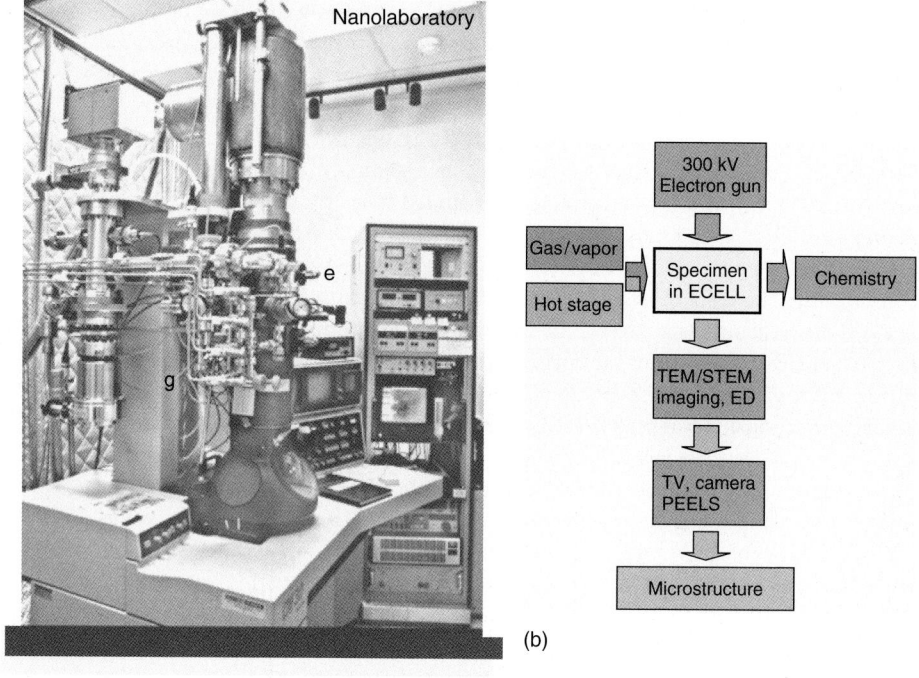

Figure 11.2 (a) The first atomic-resolution E(S)TEM (CM30 ETEM/STEM with LaB$_6$ gun) and (b) schematic of the accessories for probing gas–solid reactions [11].

11.4
Experimental Procedures

The impact of the charged particle electron beam in the ETEM on chemical reactions must always be minimized or eliminated by the use of appropriate very low electron dose techniques (with doses well below the threshold for damage). The signal is amplified by a LLTV camera. These procedures also reduce any local temperature rise under the beam. The effect of the conditions are checked in parallel "blank" calibration experiments (where dynamic experiments are conducted without the beam and the beam switched on only to record the final reaction end point and then compared to *in situ* data under low-dose imaging). These precautions are taken to ensure that there are no deleterious effects of the electron beam [1–3, 5, 12, 13]. The aim is completely noninvasive characterization under benign conditions without any contamination. Under carefully simulated conditions, close to those in practical reactors, data from *in situ* ETEM can be directly related to structure–property relationships in technological processes [3].

Systematic hydrocarbon oxidation experiments on oxide surfaces and nanomaterial transformations have shown that generally gas pressures of a few millibars are sufficient for obtaining insights into structural changes in gas–solid reactions and for the technological correlation. Generally, the coverage of the catalyst surface and the gas-catalyst interface are observed to be important irrespective of higher pressures around the sample in the oxidation reactions studied [1, 15, 16]. This is confirmed by the fact that the same defects are observed in molybdenum trioxide in reducing gases of hydrogen and propylene at a pressure of 130 mbar [1, 15], and at 13 mbar, at the same operating temperatures [15, 16]. This is illustrated in Figure 11.3a: (a) reduced domain defect structures and (b) extended crystallographic shear (CS) plane defect structures at 300 and 400 °C respectively, in

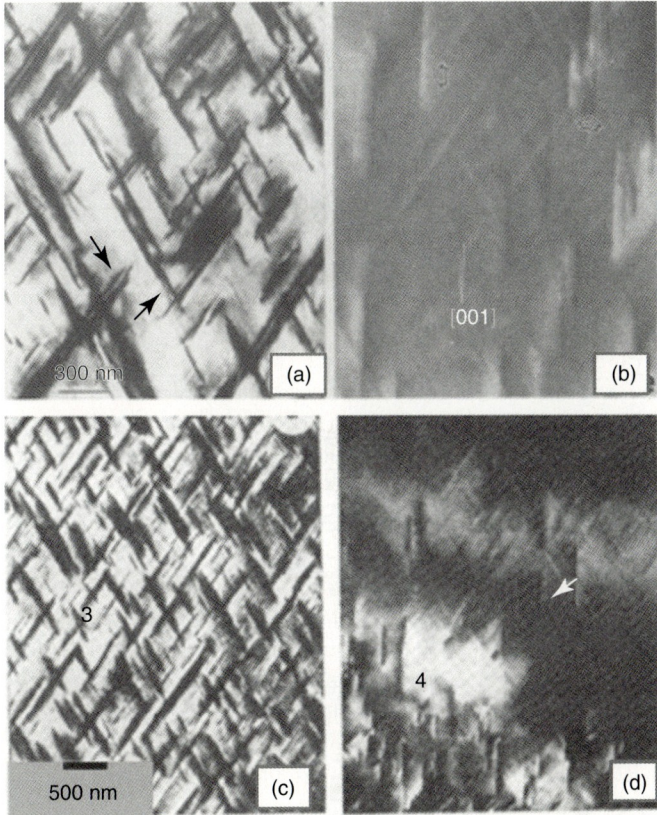

Figure 11.3 The effect of gas pressures on surface modifications in dynamic reactions in real-time, in MoO_3 oxidation catalysts: (a) reduced domain defect structures and (b) extended crystallographic shear plane defect structures at 300 and 400 °C, respectively, in hydrogen (or propylene) at a gas pressure of 130 mbar. ((a) and (b) are at the same magnification.) The same defects are observed at 13 mbar pressure of the gas at the same temperatures shown in (c) and (d). ((c) and (d) are at the same magnification.) The same defects were also observed in ~1 bar of hydrogen gas pressure. The data indicate that the surface coverage of the catalyst and the gas-catalyst interface are of great importance in the oxidation reactions.

hydrogen (or propylene) at a gas pressure of 130 mbar. ((a) and (b) are at the same magnification.) The same defect structures are observed at 13 mbar pressure of the gas at the same temperatures, shown in (c) and (d). ((c) and (d) are at the same magnification.) The same defects are also present in mixtures of 13 mbar of C_3H_6 and oxygen gases [15] and the same surface modifications were observed in ~1 bar of hydrogen gas pressure. There are further confirmations including the correlation of ETEM studies with data from technological reactors in alkane oxidation [3, 12, 17, 18]. However, some Pd-based or combustion catalysts may require higher gas pressures. Examples of *in situ* ETEM studies reported on MoO_3, V_2O_5, and mixed-metal oxides in hydrocarbon–oxygen (or air) mixtures have demonstrated that the surface structure of the catalyst is continuously changing as a function of the temperature, reducing up to about 400 °C even in the presence of oxygen (because of the strong driving force of the reducing gas), leading to the formation of different defect structures at different temperatures [1, 15]. The defects form by breaking different structural bonds in the catalyst because of the surface reaction and only a fraction of a percent of defects can control an entire catalytic process [3]. It is therefore important to analyze the nature of defects to obtain a better understanding of the difference between catalytically beneficial defects and catalytically detrimental defects. It is then possible to operate in the temperature regime where surface defects are beneficial to catalytic reactions. In general, surface structural transformations in catalytic oxides in reduction–oxidation reactions are applicable to oxide crystallography.

In addition to calibration experiments, several conditions are required for successful studies using the ETEM. Electron-transparent samples are necessary and most practical catalyst powders and ceramic oxides meet this requirement. Ultrahigh-purity heater materials and sample grids capable of withstanding elevated temperatures and gases are required (such as stainless steel, Ti, or Mo). The complex nature of catalysis with gas environments and temperature requires a stable design of the ETEM instrument to simulate the real world at the atomic resolution. For high temperature studies in gas environments, normally samples are supported on finely meshed grids of high-purity Ti or Mo with experiments carried out on crystals protruding from the edge of grid bars. (Carbon support films are not suitable for high temperature experiments.)

The novel design of the *in situ* atomic-resolution ETEM [11] has been adopted by TEM manufacturers FEI for commercial production and later versions of our atomic-resolution ETEM instrument (later versions including the CM 200-300 series, Tecnai, and Titan (including the aberration-corrected (AC) versions) have been installed in laboratories around the world, resulting in research opportunities for scientists as well as benefits to technology and generating substantial revenues. The novel atomic-resolution ETEM's capabilities have been featured by the American Chemical Society's Chemical and Engineering News [19, 20] and Materials Research Society (MRS) bulletins) [5, 21].

The first atomic-resolution ETEM has operated with a LaB_6 electron emission source. With the advent of field emission gun (FEG)-ETEMs, electron beams are very strong and extra care (including very low dose imaging and calibration experiments) is essential in performing ETEM experiments to prevent beam damage to samples. For example, in electron energy loss spectroscopy (EELS) under regular illumination conditions, gas ionization in FEG-ETEM can result in peaks due to inelastic scattering of electrons and damage to samples.

11.5
Applications with Examples

In the following sections, we describe, primarily from our research, examples that demonstrate how atomic-resolution ETEM can be applied to a variety of practical problems to advance our understanding of the role of gas–surface interactions in nanomaterials and catalysts in their functioning state. For a more detailed presentation of ETEM applications in the literature, we refer the reader to other publications (e.g., [9, 21–36]). We also describe some recent developments with double AC *in situ* E(S)TEM.

11.6
Nanoparticles and Catalytic Materials

11.6.1
Dynamic Nanoparticle Shape Modifications, Electronic Structures of Promoted Systems, and Dynamic Oxidation States

The extreme nature of catalysis with the complex combination of temperature and gas environment presents a formidable challenge in catalytic studies. Visualizing live catalysts at work at the atomic level is critical to understanding their action and the development of improved catalysts and processes. The finding that nanoclusters can undergo dynamic shape changes in response to changes in the gas environment [6, 11, 25] has the general implication that the distribution of catalytic active sites, and hence the catalytic activity, can be regulated by reaction conditions. Moreover, atomic-resolution ETEM can provide information about the gas-induced changes in the distribution of surface sites in catalyst materials.

ETEM is playing a key role in understanding gas–nanocatalyst interface interactions in real time at the atomic level. Nanoparticles of Pt, Pd, and Au, of only a few nanometers (nm) in size, are important in technological processes such as fuel cells, catalytic reforming, and hydrogenation processes. However, sintering results in the loss of active surface area of nanoparticles, and can impact the selectivity of their performance. One clear example of atomic scale interfacial reactions is demonstrated by Pt nanoparticles supported on titania (rutile), which is of interest in reforming catalysis. Figure 11.4 shows atomically resolved dynamic ETEM images of the same Pt particle, denoted by P and an arrow, and showing the 0.23 nm (111) atomic lattice spacings, when the catalyst is activated in hydrogen gas at 300 and 450 °C, respectively. At the lower temperature, the Pt nanoparticle surfaces

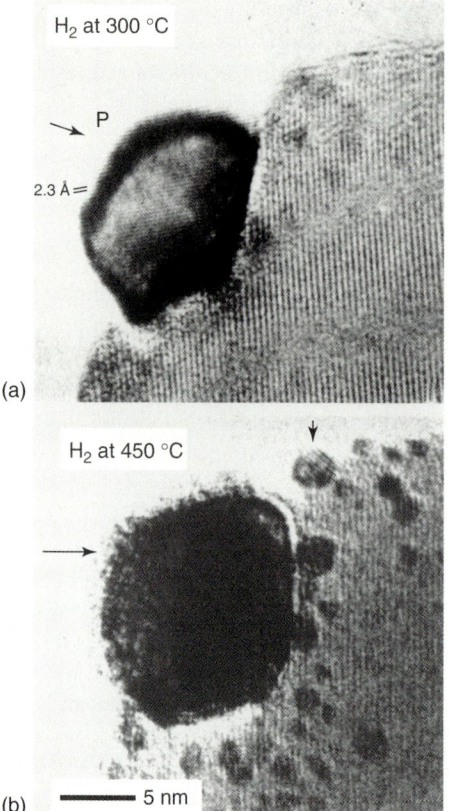

Figure 11.4 Atomically resolved real-time ETEM studies of metal–support interactions and particle shape changes in nanophase Pt/TiO$_2$ catalysts from the same area and the same particle. (a) and (b) are at the same magnification. (a) *In situ* dynamic catalyst activation in hydrogen (at 3 mbar gas pressure), images at 300 °C. The (111) lattice atomic spacings (0.23 nm) are clearly resolved in the Pt metal particle (P) under controlled conditions. (b) The same Pt particle imaged at 450 °C, also in hydrogen. Catalyst deactivation with growth of the support oxide monolayer indicated by the arrow and the development of nanoscale single-crystal clusters.

are clean and faceted, whereas at higher temperatures strong metal–support interactions are induced, leading to the growth of a passivating Ti-rich subnanometer overlayer [5, 11]. The results are important in understanding the Pt dispersion, metal–support interactions, the catalyst deactivation, and the fraction of the active metal available for gas chemisorption and catalysis.

ETEM studies have been invaluable in understanding gas–nanoparticle interactions and electronic structural changes. For example, electronic structures of the promoted catalytic materials following gas–solid reactions can be usefully determined using the cathodoluminecence (CL) method in the EM. Development of CL in the EM enables the determination of electronic structural properties, band

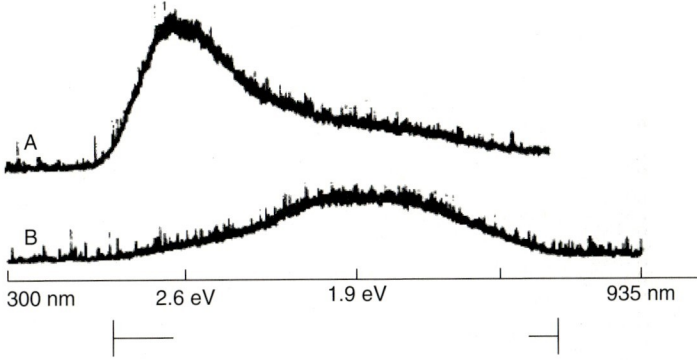

Figure 11.5 Cathodoluminescence spectra of Sb-promoted tin oxide nanocatalysts: A, the reacted Sn/Sb oxide nanocatalyst and B, the unreacted catalyst, showing a peak shift indicative of chemical or structural change and surface segregation of the promoter Sb.

gap energies, and promoter distributions in nanoparticle catalysts [7]. In this work, the wavelength of light or other radiation emitted by electrons in electron-probe stimulation and electron–hole pair combination is directly related to the local band gap energies of nanoparticles. Sb-promoted tin oxide nanocatalysts, of interest in catalysis and sensors, show marked differences in band gap energies between the fresh sample and the reacted sample following *in situ* hydrogen reduction [7]. Figure 11.5 shows the change in band gap energy with band gaps of 2.6 eV for the reacted sample and 1.9 eV for the unreacted sample. Quantitative chemical analyses indicate possible surface segregation of Sb [7]. The CL studies with *in situ* experiments have important implications in understanding the promoter distribution and electronic structural changes in the nanoscale particle systems.

ETEM studies have been used to explore the role of gas environments on nanoparticles at higher pressures, which illustrate that reduction–oxidation (redox) behavior plays an important role in catalytic processes. *In situ* studies at high gas pressures of ∼260 mbar in a high-voltage ETEM (operating at 1 MeV) have led to the discovery of new interaction mechanisms in Cu metal nanoparticles supported on nonwetting ceramic oxides such as alumina. The system is of interest in catalytic and electronics technologies [6]. The dynamic studies show the bulk diffusion of the metal nanoparticles through the ceramic oxide substrate in oxygen environments. This is illustrated in Figure 11.6 [6]. Figure 11.6a shows nanoparticles of Cu on alumina in a CO environment at ∼200 °C and (b) shows the same particles in an oxygen gas environment at 200 °C, with both gases at a pressure of 260 mbar. The bulk diffusion of Cu is confirmed by electron spectroscopy for chemical analysis (ESCA or XPS) depth profile of Cu diffusion through alumina substrate following oxidation at ∼200 °C, shown in (c). The sample is etched from the alumina edge (i, indicated at the origin on the horizontal axis), toward the alumina–Cu interface indicated by i_s (on the horizontal axis) showing diffusion through the alumina.

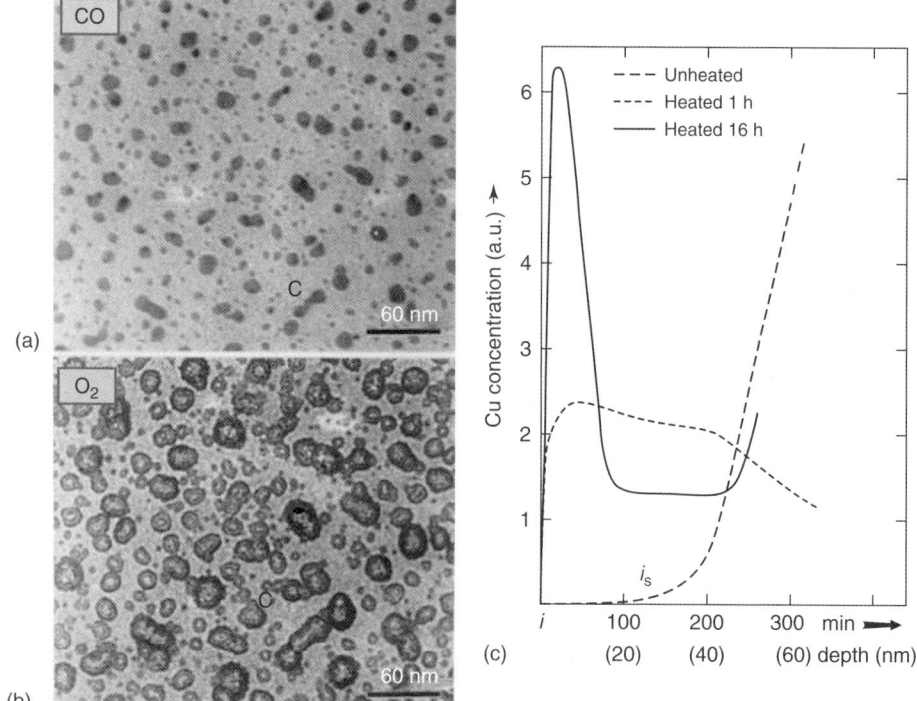

Figure 11.6 Effect of different gas environments: *in situ* real-time ETEM observations of dynamic Cu/alumina in different gas environments. Dynamic images are recorded from the same area of the sample in 260 mbar of gas pressures of CO and of oxygen at ~200 °C. (a) CO gas; (b) oxygen gas. Complex wetting and diffusion of Cu through the bulk of alumina is observed in (b). (c) ESCA depth profile of Cu diffusion through alumina substrate following oxidation at ~200 °C. The sample is etched from the alumina edge **i**, toward the alumina/Cu interface indicated by i_s (on the horizontal axis) showing diffusion through alumina. The ESCA data confirm the *in situ* observations. Cu/alumina is important in catalysis an electronics applications.

This ETEM research has impacted the design of metal nanocatalysts, processes of regeneration of nanoparticle catalysts, and metal conductors deposited on ceramic substrates in micro- and nanoelectronics.

Supported bimetallic (two-metal) nanosystems are important in chemical processing and petrochemical technologies. Their enhanced selectivity to a particular product is due to changes in their surface and electronic structures introduced by the second element. The catalytic properties such as gas chemisorption on nanoparticle surfaces and reactions are largely determined by the composition of their surfaces. Copper–ruthenium (Cu–Ru) nanoparticles are of interest in catalytic hydrogenolysis of ethane gas to methane and their behavior in ethane is therefore important. *In situ* ETEM of Cu–Ru supported on carbon have been investigated in ethane gas [3]. The catalyst nanostructure is shown in Figure 11.7a.

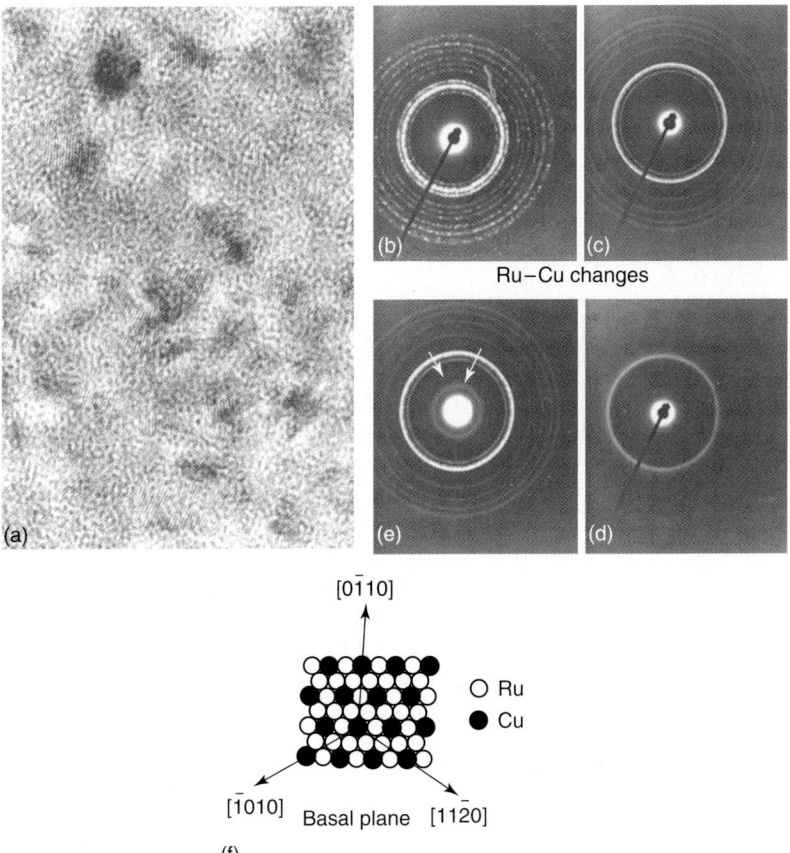

Figure 11.7 Dynamic ETEM studies of bimetallic Cu–Ru nanocatalysts supported on carbon in ethane gas in real time. (a) The nanostructure of the catalyst showing smaller Ru nanoparticles (with lattice spacings of 0.27 nm) and larger Cu particles. Dynamic electron diffraction in ethane gas from RT to ~300 °C. (b) Real time pattern with rings from Cu and hexagonal Ru. The first three rings correspond to d-spacings of 0.234, 0.214, and 0.206 nm. (c) and (d) show gradual modifications. (e) Intermediate phase with extra rings forming a superlattice at ~250 °C. (f) Proposed $CuRu_3$ superstructure of the intermediate alloy phase.

Dynamic electron diffraction in ethane from room temperature (RT) to ~300 °C shows the formation of an intermediate metastable phase, which could be indexed as an alloy phase $CuRu_3$. This intermediate nanophase was hitherto unknown, as the Cu–Ru system is not known to form alloy phases in the bulk. Figure 11.7b–e shows dynamic electron diffraction observations in ethane: the first three rings in (b) correspond to d-spacings of 0.234, 0.214, and 0.206 nm, respectively; (e) shows the formation of an intermediate phase with extra rings (arrowed) forming a superlattice at ~250 °C. The extra rings are not from Cu or from their oxides, and can be indexed as 1/3 (0 1 −1 2) and 1/3 (0 0 0 4) reflections. A proposed model of the new $CuRu_3$ superlattice structure is shown in Figure 11.7f. Rings due to this alloy formation are also observed in the reduction in H_2. The formation of the alloy phase has implications in the hydrogenolysis reaction [3].

11.7 Oxides

ETEM has played a vital role in understanding structural modifications leading to defects on oxide surfaces during selective hydrocarbon oxidation reactions. Metal oxides are used as catalysts because of their ability to provide lattice oxygen in the oxidation of hydrocarbons to a variety of products, including energy sources, polymers, and agricultural products. The ways in which oxides release structural oxygen and accommodate deviations in oxygen stoichiometry, without changing the overall structure under operating conditions, are critical to the science and technology of oxide catalysis.

In situ ETEM studies have provided key insights into the nature and the role of point and extended defects, and of active sites on oxide surfaces in hydrocarbon oxidation reactions [1, 22, 34]. Atomic-resolution ETEM studies of butane oxidation over vanadyl pyrophosphate catalysts ($(VO)_2P_2O_7$, or abbreviated to VPO) with parallel catalytic reactivity testing, have led to the discovery of a novel, reversible anion-vacancy-induced crystal glide shear reaction mechanism [10]. The glide shear mechanism leads to the formation of partial screw dislocations bounding stacking faults (glide shear plane defects), accommodating the shape misfit between the reduced surface layer and the subsurface. This mechanism explains the release of oxide structural oxygen, the accommodation of anion nonstoichiometry and the preservation of anion vacancies associated with active Lewis acid sites at the surface without collapsing (changing) the overall structure of the catalyst. This structural integrity is essential to the long life of the catalyst. The positively charged anion-vacancy sites are important for the hydrocarbon activation (they can accept electrons from the hydrocarbon molecule) and in the replenishment of the lost catalyst oxygen from the gas phase oxygen. In situ ETEM has revealed that the glide shear is a fundamental and crucial transformation mechanism in several important oxide materials and the key to an efficient oxide catalyst [12]. The studies have provided the basis for the understanding of the solid-state selective oxidation process [10, 12] and have been described in recent reviews [2, 23, 28, 31].

These fundamental ETEM studies have further shown earlier long-standing (~40 years) hypotheses in the literature to be incorrect: these hypotheses suggested that the elimination of anion-vacancy sites and irreversible catalyst structural collapse, referred to as crystallographic shear mechanism (forming partial edge dislocations bounding stacking faults, referred to as CS plane defects), could lead to active sites. The ETEM work further showed that collapsed CS defects are a consequence of the gas–solid reaction and catalytically detrimental. The ETEM work on VPO has led to the development of improved catalyst materials for *n*-butane oxidation in the manufacture of polymers [17, 18].

11.8
In situ Atomic Scale Twinning Transformations in Metal Carbides

The dynamic atomic scale reactions in transition metal carbides, including WC, are of considerable interest in fuel cell technologies and nanoelectronics, and therefore nanostructural transformations are of great importance [37]. WC powder exhibits extensive atomic scale twin boundary defect structures. WC has a hexagonal structure, with $a = 0.2906$ nm and $c = 0.2838$ nm). Atomic-resolution ETEM studies of the same thin area of (100) WC sample in 20% hydrogen/helium at a gas pressure of 3 mbar are illustrated in Figure 11.8. The twin structures are along the [001] direction with lattice modulations along [010]. Figure 11.8a shows twin defects at RT, with the corresponding electron diffraction (ED) in (c). Figure 11.8b,d shows the dynamic image and ED, respectively, of the same area at 450 °C, reacted for 15 min. Figure 11.8c shows the elimination of almost all of the atomic twins and the transformation to W nanometal and carbon nanostructures [38]. The presence of the metal was confirmed by the emergence of [100] reflections of cubic α-W with a lattice parameter of $a = 0.32$ nm. The structure of WC in [100] and the possible mechanism for the transformation leading to W are shown in (e) and (f), respectively. Similar atomic scale interactions are important in other transition metal carbides and nitrides.

11.9
Dynamic Electron Energy Loss Spectroscopy

Dynamic EELS in the atomic-resolution E(S)TEM can be usefully employed to monitor changes in the chemical oxidation state of reacting catalysts of interest in emission control [39–41] and other nanomaterials. A dynamic EELS study of the Rh(1%)/Ce$_{0.8}$Pr$_{0.2}$O$_{2-x}$ catalyst is shown in Figure 11.9, where spectra are recorded under H$_2$ gas pressure (4 mbar) at temperatures ranging from RT to about 900 °C [39]. The peak positions for some of the characteristic features of Ln^{4+} and Ln^{3+} species are marked by the solid and dashed vertical lines, respectively. Remarkably, the data illustrate that the evolution of the lanthanides (Ce and Pr) can be followed separately. The data show the Ce reduction from 4$^+$ to 3$^+$ oxidation state at elevated temperatures. The cerium M$_{4,5}$ ratio (referred to as *"white-line"* ratio) changes as Ce is reduced from 4$^+$ to 3$^+$ state at high temperatures in the reducing environment. The change in the "white-line" ratio can be used to monitor the extent of reduction in individual nanoparticles with temperature and time [39]. *In situ* ETEM studies have also been vital in understanding the nucleation and growth of one-dimensional nanomaterials, including carbon nanotube reactions and growth kinetics in gas atmospheres [2, 26, 42, 43]. Other ETEM studies include nanoparticle systems [21, 44].

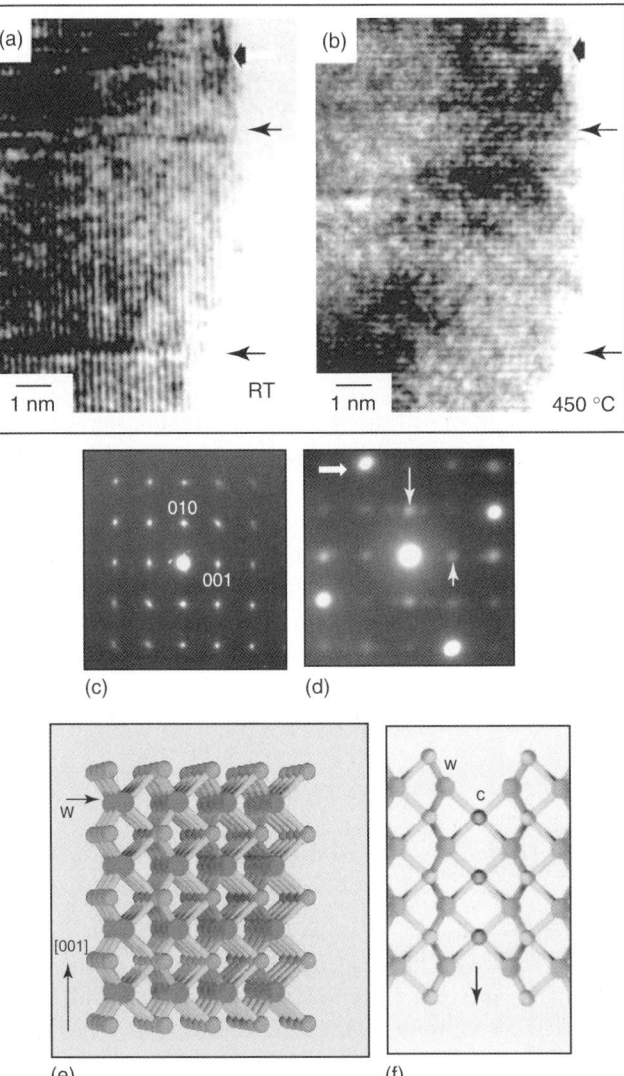

Figure 11.8 *Atomic-resolution* ETEM studies of real-time reduction of the same area of WC powder sample in hydrogen/He gas: (a) Room temperature (RT); (b) dynamic reaction at 450 °C, indicating the elimination of almost all of the twins; (c) the corresponding electron diffraction; (d) dynamic electron diffraction pattern of (b) indicating the emergence of {100} cubic tungsten reflections (arrowed); (e) structural schematic of hexagonal WC in (100) orientation; W atoms (arrowed at W; in green) are surrounded by C atoms; and (f) proposed mechanism for the formation of W and carbon nanostructures: showing a twin boundary structure along [001] (arrowed) and carbon atoms gathered in the defects react with hydrogen and diffuse out. This is accompanied by readjustments of the atoms to create the metal structure.

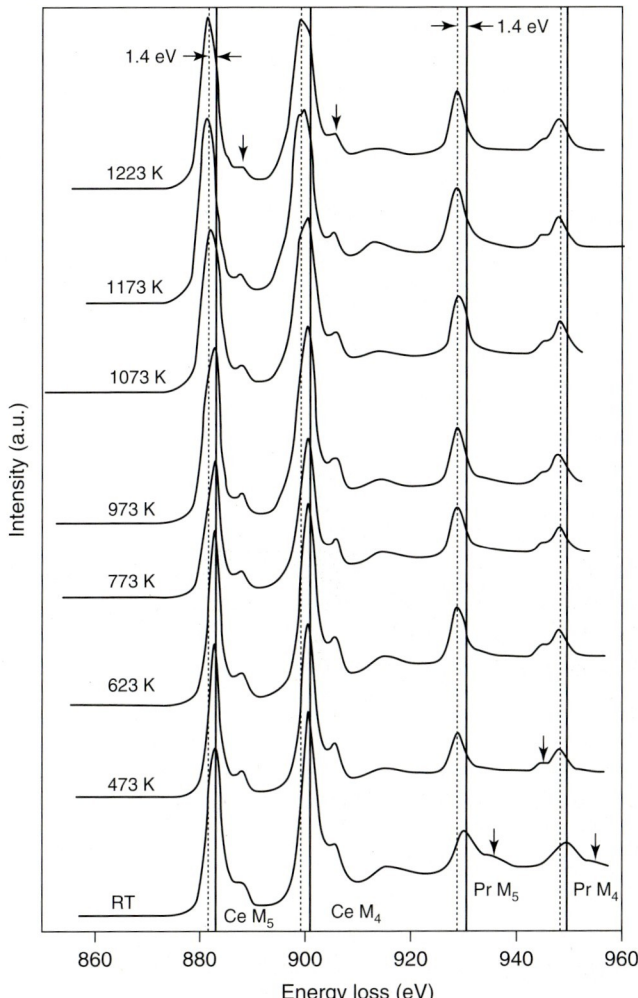

Figure 11.9 Dynamic EELS study of Rh/Ce$_{0.8}$Pr$_{0.2}$O$_{2-x}$ in reaction environments of gas pressure and temperatures: spectra recorded under hydrogen gas pressure (4 mbar) at the indicated temperature. The peak positions for some of the characteristic features of Ln^{4+} and Ln^{3+} species marked by solid and dashed vertical lines, respectively. Changes in fine structure of M$_{4,5}$ lines: 4$^+$ to 3$^+$ state is observed. The figure is reproduced by permission of the Royal Society of Chemistry: http://www.rsc.org.

11.10
Technological Benefits of Atomic-Resolution ETEM

In addition to being adopted for commercial production by TEM manufacturers and used by researchers internationally, our atomic-resolution ETEM has contributed to the invention of a catalytically controlled, breakthrough *in situ* nanocoating process for titania pigments for high-strength polymer applications [45]. The patented invention has been commercialized by DuPont, and has resulted in substantial revenues. Other ETEM contributions include the development of improved alkane oxidation catalysts [17, 18]. *In situ* EM studies have contributed to the development of stable ceramic oxide supports (for catalysts and in electronics packaging applications), which yield properties of their high temperature analogs at RT without undergoing phase transformations [46, 47]. Developments of additional materials and reactions have been reported [9, 21–36].

11.11
Other Advances

As described in the preceding sections, atomic-resolution ETEM with a differential pumping system for flowing gas represents a significant advance for *in situ* studies under controlled reaction environments with operating gas pressures up to 30 mbar and in selected cases, higher. Atomic-resolution *in situ* images of molecular sieves at higher gas pressures of 0.5 bar of hydrogen have also been reported [48].

Other important developments include sample holder-based developments. One approach for *in situ* environmental studies is to employ a thin-window cell, consisting of a TEM sample holder that encloses the sample and gas between two electron-transparent thin films such as carbon [49, 50]. Such reaction cells with electron-transparent windows separate the fluid phase from the EM vacuum and can, in principle, be applied in a regular EM with exchangeable objective lens pole pieces for atomic resolution studies. Different window cells for studies at high temperature and RT are possible. Other developments include the application of microelectromechanical systems (MEMS) technology to construct specimen holders with gas containment and high gas pressures (~1 bar) [51]. With window cell systems, it would be desirable to achieve accurate temperature measurements at the sample and tilting to crystallographic zone axes for orienting samples and analyzing defect structures. *In situ* studies with femtosecond time resolution have been reported [52].

11.12
Reactions in the Liquid Phase

Many commercial hydrogenation and polymerization processes are derived from liquids at operating temperatures and the associated chemical reactions occur on the subnanometer scale. Recent advances in biological catalysis and molecular electronics also take place in liquid environments. Probing reactions in solutions is therefore of great importance in the development of advanced catalytic and bioelectronics technologies. We review the progress of wet-ETEM in the following sections.

As described in the previous sections, "window cells," where a sample is sandwiched between electron-transparent windows, have been used to study hydrated or biological samples. Several research groups have been active in the area of wet-ETEM using window cells or liquid injection directly into an ECELL [49, 53]. However, in these developments, liquid–solid reactions have been studied with limited resolution at RT, with nonheating and nontilting sample stages and the results tend to be averaged over large areas.

The development of wet-cell methods in the ETEM for *in situ* direct probing of catalyst-liquid reactions in gas environments at operating temperatures at the subnanometer level has been reported [54, 55]. Atomic-resolution ETEM studies have contributed to the development of a low temperature heterogeneous process for the liquid phase hydrogenation of adiponitrile (ADN) to hexamethylene diamine (HMD), a key component in the manufacture of linear polyamide, nylon (6,6), and for the polymer synthesis in the liquid phase [55]. Wet-ETEM has opened up opportunities of dynamic imaging of wet reactions in polymer processes [55].

For example, novel cobalt-promoted ruthenium nanocatalysts on rutile titania nanosupport have been developed for the hydrogenation of low vapor pressure ADN ($NC(CH_2)_4CN$) to HMD ($H_2N(CH_2)_6NH_2$) at a low temperature of ∼100 °C. A liquid-feed heating holder [54] enables *in situ* nanosynthesis of organic HMD fibers and polymers in liquid–gas–solid environments in the ETEM. For the hydrogenation of ADN in the liquid phase, Co–Ru/titania nanocatalysts are immersed in the ADN liquid (in methanol and 0.15 wt% NaOH solvent) in the liquid holder. Nanoliters to microliters of the liquid are injected (using procedures similar to those employed in chromatographic methods) over the catalysts, with the catalyst immersed in flowing liquids. Flowing hydrogen gas is inserted using the gas-manifold system in the ETEM simultaneously and the sample is heated to ∼100 °C. Figure 11.10 illustrates liquid–solid–gas reactions: (a) Co–Ru nanocatalyst on nanoscale titania; (b) reaction of ADN liquid in hydrogen over the nanocatalyst at 100 °C, showing the first dynamic imaging of desorbed organic product, HMD fibers on the nanoscale [55]. The *in situ* studies in wet environments at the nanoscale have elucidated novel hydrogenation and polymerization reactions using ADN and hitherto unknown structures of the organic products. Scale-up laboratory experiments confirm the *in situ* wet-ETEM studies with high hydrogenation selectivity and activity of the catalyst [55]. HMD solution is then added to adipic acid solutions over the catalyst and dynamic polymer formation is observed at 188 °C [55]. The development of wet-ETEM is thus powerful in developing polymer and related catalytic reactions in the liquid phase.

11.13
In situ Studies with Aberration Correction

Aberration correction (AC) is particularly beneficial in dynamic *in situ* experiments because there is rarely, if ever, the opportunity to record from a scene that may be continuously changing, a systematic through focal series of images for subsequent

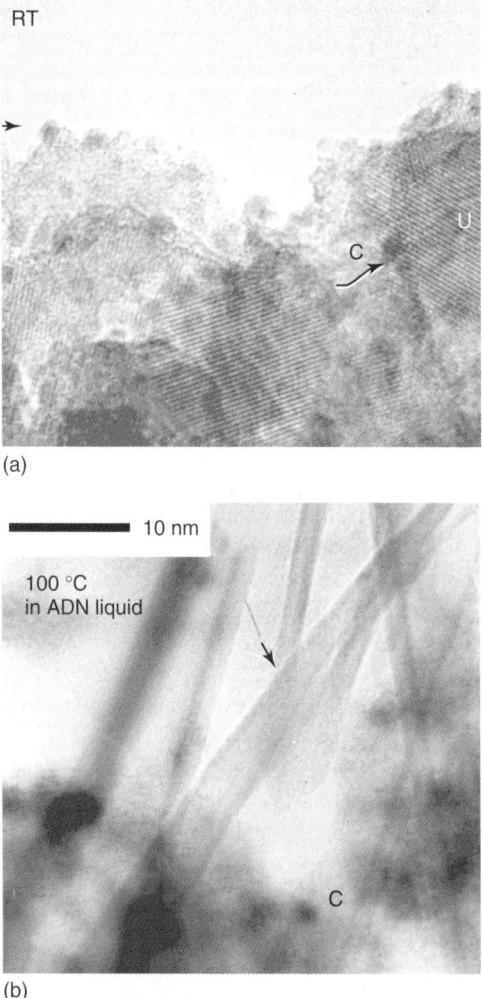

Figure 11.10 Wet-ETEM studies in liquid–solid–gas reactions: (a) Co–Ru nanocatalyst on nanoscale titania and (b) reaction of adiponitrile liquid in hydrogen over the nanocatalyst at 100 °C showing desorption of HMD nono fibers.

data reconstruction. In these applications, it is necessary to extract the maximum possible information from each single image in a continuously changing sequence. It is also desirable to limit the electron dose exposure to ensure minimally invasive conditions and to avoid secondary effects such as contamination. On the basis of these considerations, we proposed AC dynamic *in situ* EM [56]. our in-house AC E(S)TEM development and its applications to nanoparticles and in the development of new energy-efficient nanocatalysts for biofuels are described below.

To observe changes in the nanostructure as a function of operating temperature under controlled *in situ* conditions in an AC environment, the following procedures have been employed by the authors [38, 57]. The need to accommodate special specimen holders in an AC instrument has been one of the more important criteria driving the specification of a double aberration-corrected (2AC) EM, namely, the JEOL 2200FS (2AC) FEG TEM/STEM operating at 200 kV, in the recently established Nanocentre at the University of York. (Currently all the AC-ETEMs use the earlier atomic-resolution ETEM design of Gai and Boyes [10, 11] described in the preceding sections. Since aligning the sample into a zone axis orientation is a prerequisite for atomic-resolution EM of crystalline materials, an increased specimen tilt range is also desirable. Both these conditions benefit from the larger gap (HRP/Midi) objective lens pole piece [38, 57], shown in Table 11.1 [38, 57] below. The Cs (C3) aberration correctors [58] on both the STEM probe forming and TEM image sides of the instrument were used to provide the desired expanded specimen geometry with minimal effect on the 1 Å and below imaging performance of the system. The effect of the increased pole piece gap on Cs in this range is much more evident ($\times 2$) than on Cc ($\times 1.2$); and of course Cs is now corrected (Cs ~ 0).

The advantages of the configuration we have adopted include promoting a contrast transfer function (CTF) extending to higher spatial frequencies and resolution in the data; allowing image recording at close to zero defocus, including to strengthen and simplify interpretation of information at internal interfaces and external nanoparticle surfaces; analyzing small (<2 nm) nanoparticles and clusters on supports, using high-resolution TEM as well as high-resolution STEM; facilitating high-angle annular dark-field (HAADF) STEM and extending HAADF STEM resolution to 1 Å (0.1 nm) and below. (HAADF STEM resolution at <1 Å (0.1 nm) in RT studies have been reported [59, 60]). As well as benefiting from improved resolution at 1Å and below, it becomes important to be able to set the conditions to avoid the previously intrusive CTF and defocus sensitive oscillations in image contrast in the spatial resolution range from 1 to 3Å. This is where atomic neighborhoods in crystals lie in low index projections and it is especially important in studying surfaces of ultrasmall nanoparticles, atomic scale defects,

Table 11.1 Advantages of the wider gap polepiece lens employed in our development of double AC ETEM/ESTEM. Aberration correction at higher Cs values allows the wider gap (Midi) objective lenses to accommodate higher sample tilts and a hot stage without significant performance penalty.

Pole piece type	Gap range (mm)	Uncorrected Cs (mm)	Uncorrected Cc (mm)	Cc with Cs AC (mm)	Standard hot stage fits?
UHR	2.2–2.5	0.5–0.6	1.0–1.2	1.4	No
Midi	4.3–5.4	1.0–1.2	1.2–1.4	1.6	Yes

and internal interfaces. It is a key topic of interest in nanoparticle studies, for example, in considering the possible origin of sintering mechanisms important in heterogeneous catalyst design including control of processing, activation, operation, and deactivation mechanisms. (Calibration experiments were performed according to procedures described in Section 11.1.)

11.14
Examples and Discussion

We show in both theory (*www.maxsidorov.com* [61]), with contrasting CTFs shown in Figure 11.11, and in practice, it is possible with Cs aberration correction to combine demonstrated spatial resolution around 1Å with the HRP lens pole piece with the large ~4.3 mm gap [38, 57] needed to accommodate a standard hot stage (Gatan model 628); and with it in operation (using Gatan Digital Micrograph). This is an example of using aberration correction to combine the limited added space required for *in situ* experiments with a high level of imaging performance with which such facilities were previously incompatible; and thereby to extend considerably application-specific and relevant TEM and STEM experimental capabilities.

The system is in practice stability limited (CTF envelope terms) and some of the practical steps necessary to deliver this powerful combination of capabilities will be covered elsewhere with further examples of the new tool in action. These considerations quickly set a limit to how far the lens gap can be stretched without beginning to compromise performance too seriously; taking into consideration realistically attainable stabilities in internal electronics and mechanics, and in key external environmental factors.

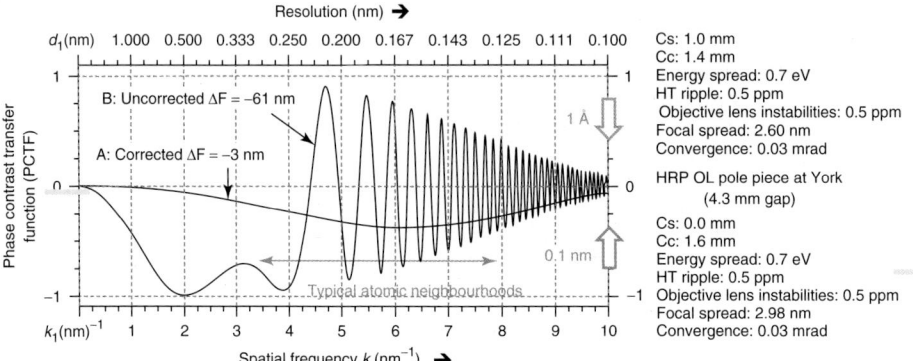

Figure 11.11 Calculated contrast transfer functions (CTFs) for uncorrected (B) and aberration-corrected (A) imaging conditions of the HRP version of the double aberration-corrected JEOL 2200FS FEG TEM/STEM at the University of York (UK).

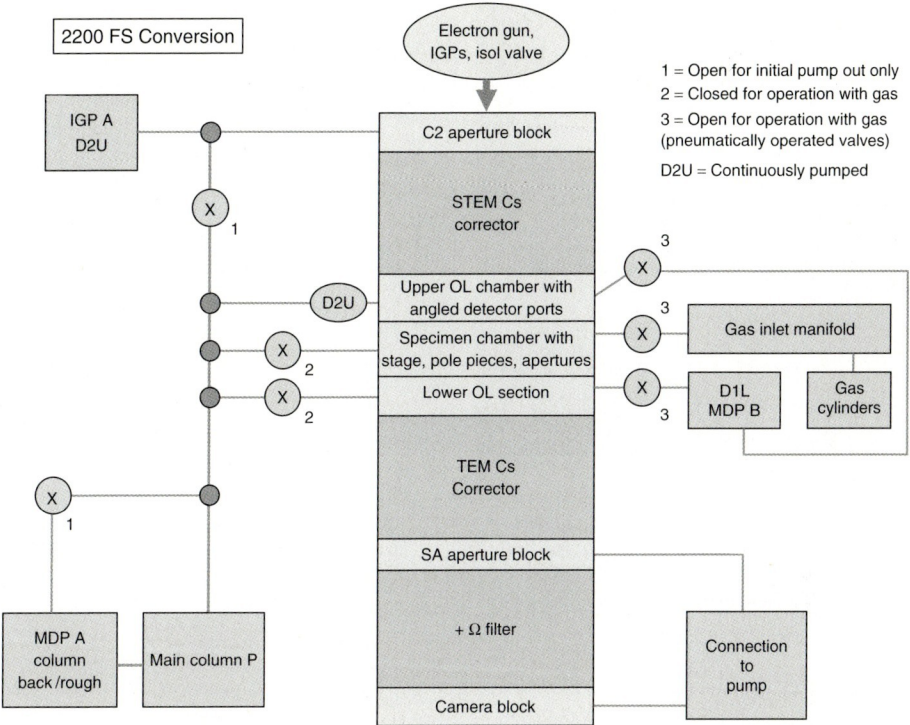

Figure 11.12 A schematic of the double aberration-corrected JEOL 2200FS conversion carried out in-house at the University of York.

Our in-house 2AC E(S)TEM development is further configured to be tolerant to out-gassing hot samples, and as a key foundation for controlled gas environment experiments [38, 57]. A schematic of the JEOL 2AC 2200FS conversion carried out in-house at York is shown in Figure 11.12 [38, 57]. The instrument configuration is considered to be beneficial for *in situ* developments with additional capabilities [11]. In our case, both capabilities are combined on the remotely controlled JEOL 2200 FS platform. In the 2AC E(S)TEM, single-atom imaging has been possible in practical catalysts: Figure 11.13(a) illustrates gold single atoms on titania substrate. Au nanocatalysts supported on titania, containing atomically dispersed gold and nanoclusters of gold, are of interest in the selective oxidation of carbon monoxide. Figure 11.13(b) shows nano-ceria with atomic resolution in hydrogen gas.

TEM correction montage procedures (Zemlin tableaux) and image resolution to better than 1 Å have been maintained with the hot stage in the *in situ* 2AC-EM, including with power connected and the samples held at elevated temperatures; so far to ~900 °C [57]. Figure 11.14a shows an *in situ* AC-TEM image of Pt–Pd nanoparticles on carbon support at RT and (b) shows the corresponding fast

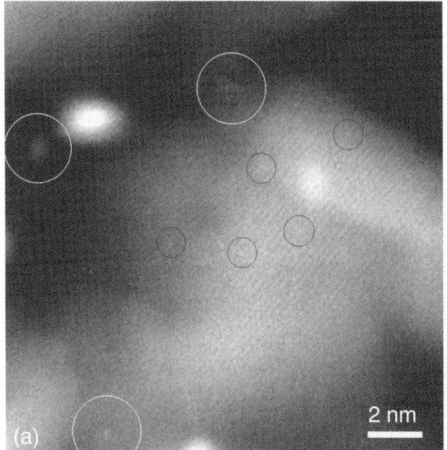 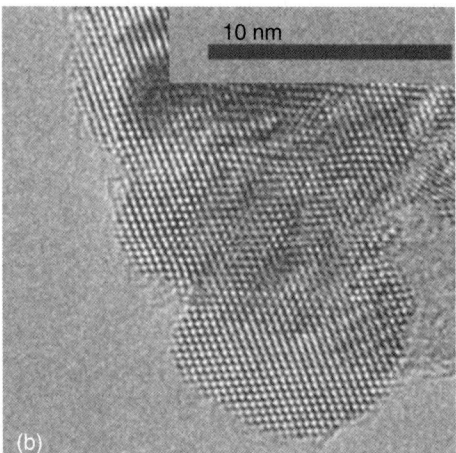

Figure 11.13 Single atoms of gold (black circles) and subnanometer clusters of gold (white circles) in Au/titania practical catalyst, imaged (a) imaged in HAADF-STEM in modified *in situ* double aberration-corrected E(S)TEM; (b) In-situ double AC-ETEM: atomic resolution image of nano-ceria in hydrogen gas initially at about 0.05 mbar gas pressure measured at the sample.

Fourier transform/optical diffractogram (FFT/OD). The particles are in the size range of ≤1–2 nm, with the carbon support contributing strongly to the diffraction. Figure 11.13c,d show *in situ* images of a selected Pt–Pd nanoparticle on carbon at 500 °C, and the corresponding FFT/OD with <0.11 nm resolution, respectively. *In situ* HAADF-STEM studies have shown atomic resolution [62].

11.15
Applications to Biofuels

Alternatives to fossil fuels are being sought in order to reduce the world's dependence on nonrenewable resources. Biofuels can be derived from plant oil feedstocks and biological materials [63]. Many countries have set targets for partial replacement of fossil fuels with renewable biofuels. Biodiesel is one such clean and renewable biofuel. Solid catalysts (e.g., alkaline earth oxides) offer advantages, especially pertaining to biodiesel separation and the opportunity for continuous process operation. Although solid catalysts have great promise in plant oil triglyceride transesterification to biodiesel, the identification of active sites and operating surface nanostructures during processing is required for the development of energy-efficient heterogeneous catalysts for biofuel synthesis [64]. Alkaline earth oxide, such as MgO, is of technological interest as a strong candidate for use in the solid base catalyzed plant triglyceride transesterification. It adopts a cubic structure with $a = 0.4212$ nm.

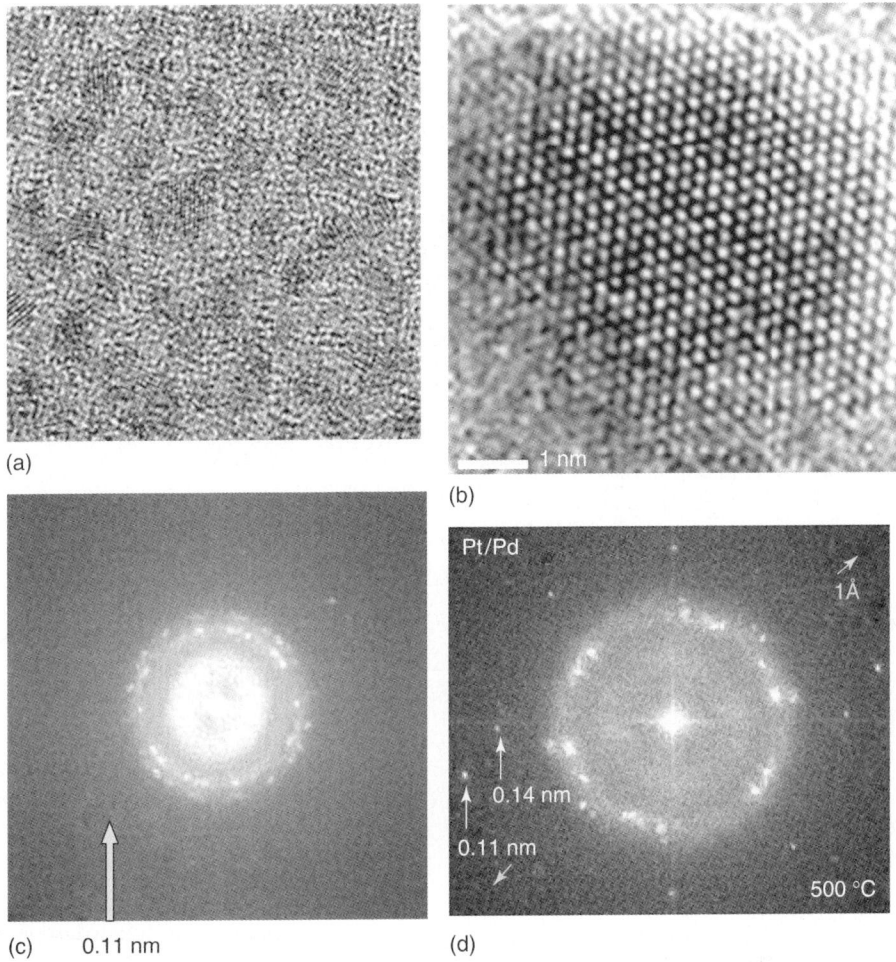

Figure 11.14 *In situ* aberration-corrected electron microscopy observations: (a) Pt–Pd nanoparticles on carbon at room temperature (RT); (b) the corresponding FFT/optical diffractogram (OD). The particles are in the size range ≤1–2 nm with the carbon support contributing strongly to the diffraction; (c) *in situ* image of a selected Pt–Pd nanoparticle on carbon at 500 °C; and (d) the corresponding FFT/OD with <0.11 nm resolution.

High-quality nanocrystalline-MgO (nano-MgO) practical powder catalysts can be prepared using magnesium hydroxide methoxide precursor ($Mg(OH)(OCH_3)$) [64]. The dynamic transformation of the precursor can be followed under controlled calcination/activation conditions using *in situ* AC-TEM at the 0.1 nm level. The dynamic observations of the catalyst nanostructure are quantified with parallel studies of the catalyst performance (which is the catalytic turnover frequency (TOF_{cat})) and physicochemical studies [64]. The dynamic transformation is illustrated in

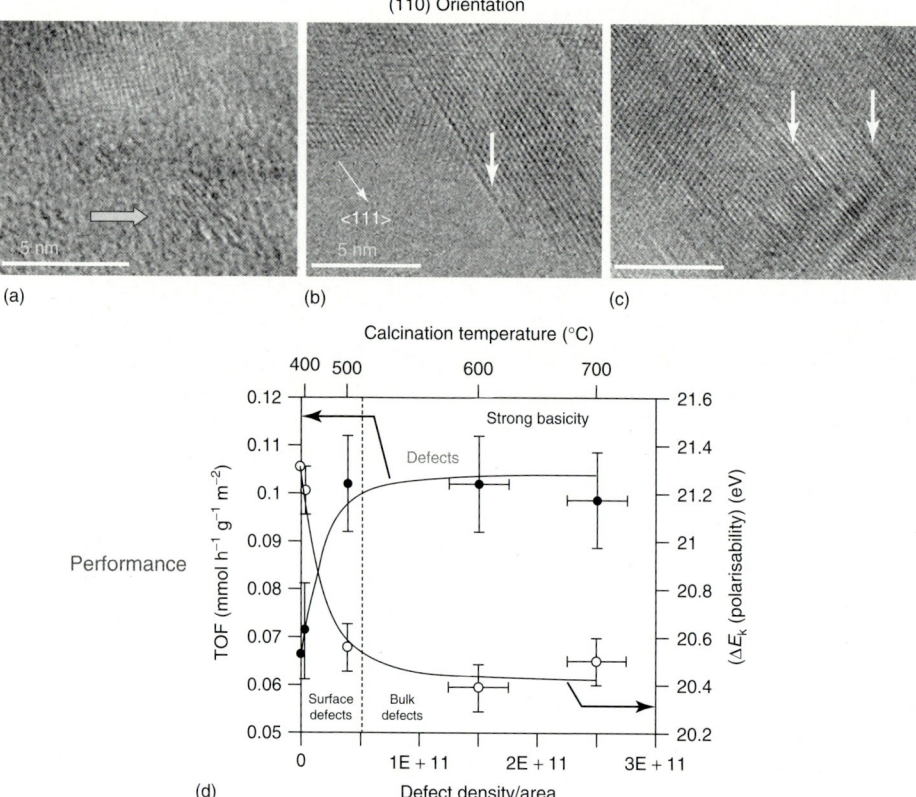

Figure 11.15 Real-time dynamic in situ AC-TEM studies: (a) single and overlapping nanocrystals of only about 3–5 nm (arrowed) are observed at room temperature (RT); (b) nucleation of defects at 500 °C in calcinations in surface profile imaging. The defects are at or near the surface (e.g., indicated by arrows). They are viewed in (110) projection and are along <111> directions; (c) profile image (at the same magnification as in (b)) at 700 °C shows the defects have increased in concentrations and have grown along their length extending into the bulk (some are indicated by arrows). (d) Structure–activity (TOF$_{cat}$) relationship between the (TOF$_{cat}$) and defect densities (solid circles), determined by in situ aberration-corrected electron microscopy, and surface polarizability determined by auger electron spectroscopy (AES) for trygliceride conversion over nanoscale MgO. Regions where surface defects predominate are indicated at ~500 °C. An excellent correlation is observed between the surface defect density and superior catalytic activity leading to new, energy-efficient nanocatalysts for biofuel synthesis.

Figure 11.15. Figure 11.15a shows randomly orientated, overlapping single crystallites of only 3–4 nm mean width, embedded within an amorphous medium. The crystallites are primarily in (110)- and (001)-type orientations at RT; (b) extended defects are observed at the catalyst surface at 500 °C; (c) the growth of the defects is observed at 700 °C. The defect analyses have indicated that they are partial screw dislocations formed by glide shear mechanism along <111> direction and the shear or the displacement vector **R** of the type, $\pm a/2$ <111> lies in the plane of shear. The defects implicate coplanar anion vacancies in active sites of the catalyst in the transesterification of plant triglycerides to biodiesel.

TOFs$_{cat}$ are calculated from the rate of triglyceride conversion. The resulting rates are normalized to the total catalyst surface area to provide an activity per unit area.

$$\text{TOF}_{cat}(\text{mmol h}^{-1} \text{ m}^{-2}) = \text{RTG}(\text{mmol h}^{-1})/(\text{mcat(g)} \times \text{SA}(\text{m}^2\text{g}^{-1})) \quad (11.1)$$

where RTG is the rate of triglyceride conversion, mcat is the mass of catalyst used, and SA is the BET surface area of catalyst. This assumes that all surface sites are equivalent, and therefore will represent an average over the different facets exposed following any given calcination. An excellent linear correlation between the surface atomic scale glide shear defect density from *in situ* EM and TOF$_{cat}$ is observed as shown in Figure 11.15d, with optimum performance at 500 °C. The linear correlation between the surface defect density (and therefore polarizability) and activity facilitates process optimization, leading to new, energy-efficient nanocatalysts for biofuel synthesis.

11.16
Conclusions

Dynamic *in situ* experiments can be very efficient, and productive of data, in scanning a range of experimental conditions in a single session, and in mechanistic studies. Moreover, if as is often the case in catalysis, the key data for rational designs of new materials can only be obtained from dynamic *in situ* studies. The limits on the systems are a maximum gas pressure in the millibar range dictated by electron scattering and finally leakage into the gun. The emphasis in ETEM is on dynamic *in situ* studies of reaction and other change mechanisms under controlled, near real-world conditions of environment and temperature; with minimally invasive EM for analysis. Any compromises in TEM performance with the ETEM functionality added are now very limited. On a personal note, we were happy to find that although our custom prototype atomic-resolution-ETEM took significant time to carry out research, design and develop it in our laboratory, the first lattice resolution images were achieved within a few minutes of first switching on the beam! The atomic-resolution ETEM has led to an advanced understanding of gas–solid interactions and the development of improved materials and processes. Another important outcome is that our ETEM development has been adopted by leading TEM manufacturers for commercial production, being used by researchers worldwide and has resulted in considerable scientific output and revenues. Other important developments are based on sample holders and include thin window cells consisting of a TEM sample holder that encloses the sample and the gas phase between two gas-tight electron-transparent thin films and the application of MEMS technology to construct specimen holders with gas containment and high gas pressures (~1 bar). Aberration correction is particularly beneficial to *in situ* studies. Its benefits include being able to capture in a single image a full range of spatial frequencies without Cs-induced contrast reversals and minima in the CTF of an uncorrected EM and minimization of electron delocalization effects in TEM images. These effects are most important in structural discontinuities such as defects, which play a key role in materials properties and chemical activity.

Acknowledgments

We are grateful for the assistance in developing our ETEM, from L. Hanna (DuPont). We thank the University of York, Yorkshire Forward (the regional development agency), the European Union through the European Regional Development Fund and JEOL (UK) Ltd. for the sponsorship of the Nanocentre at the University of York, K. Yoshida for collaboration and Ian Wright for assistance in sample preparation. We thank the Royal Society of Chemistry (UK) and J.J. Calvino (University of Cadiz) for permission to reproduce Figure 11.9.

References

1. Gai-Boyes, P.L. (1992) *Catal. Rev. Sci. Eng.*, **34**, 1, and references therein.
2. Gai, P.L. (1999) *Top. Catal.*, **9**, 19.
3. Gai, P.L. and Boyes, E.D. (2003) *Electron Microscopy in Heterogeneous Catalysis*, Institute of Physics Publishing, London.
4. Butler, E.P. and Hale, K.F. (1981) *Dynamic Experiments*, North Holland, Amsterdam.
5. Gai, P.L., Boyes, E.D., Hansen, P., Helveg, S., Giorgio, S., and Henry, C. (2007) *Mater. Res. Soc. (MRS) Bull.*, **32** (12), 1044–1048.
6. Gai, P.L., Smith, B.C., and Gowen, G. (1990) *Nature*, **348**, 430.
7. Boyes, E.D., Gai, P.L., and Warwick, C. (1985) *Nature*, **313**, 666.
8. Salmon, F., Park, C., and Baker, R. (1999) *Catal. Today*, **53**, 305.
9. Sharma, R. and Weiss, K. (1998) *Microsc. Res. Technol.*, **42**, 270.
10. Gai, P.L. et al. (1995) *Science*, **267**, 661.
11. Boyes, E.D. and Gai, P.L. (1997) *Ultramicroscopy*, **67**, 219.
12. Gai, P.L. (1997) *Acta Crystallogr.*, **B53**, 346.
13. Gai, P.L. (1998) *Adv. Mater.*, **10**, 1259.
14. (a) Boyes, E.D. and Gai, P.L. (1998) in *Electron Microcopy* (eds H. Claderon Benavides and M.J. Yacaman) IOP Publication, p. 511; (b) Gai, P.L. and Boyes, E.D. (1997) *In-situ EM in Materials Research*, Kluwer.
15. Gai, P.L. (1981) *Philos. Mag. (London)*, **43** (4), 841.
16. Gai, P.L. and Labun, P.A. (1985) *J. Catal.*, **94** (79), 94.
17. Gai, P.L., Kourtakis, K., Coulson, R., and Sonnichsen, G. (1997) *J. Phys. Chem. B*, **101**, 9916.
18. Kourtakis, K. and Gai, P.L. (2004) *J. Mol. Catal.*, **220**, 93.
19. Haggin, J. (1995) *Chem. Eng. News*, **73** (30), 39.
20. Jacoby, M. (2002) *Chem. Eng. News*, **80** (31), 26.
21. Gai, P.L., Sharma, R., and Ross, F. (2008) *Mater. Res. Soc. MRS Bull.*, **33**, 107.
22. Gai, P.L. (1983) *J. Solid State Chem.*, **49**, 25.
23. Gai, P.L. (2002) *Top. Catal.*, **21**, 161.
24. Sharma., R. (2001) *Microsc. Microanal.*, **7**, 494.
25. Hansen, P.L. et al. (2002) *Science*, **295**, 2053.
26. Boyes, E.D. and Gai, P.L. (2004) *Microsc. Today*, **12**, 24.
27. Thomas, J.M. and Gai, P.L. (2004) *Adv. Catal.*, **48**, 171.
28. Gai, P.L. and Calvino, J.J. (2005) *Annu. Rev. Mater. Res.*, **35**, 465.

29. Datye, A., Hansen, P.L., and Helveg, S. (2006) *Adv. Catal.*, **50**, 77.
30. Stach, E. (2006) *Microsc. Microanal.*, **12**, 2 (In situ Tutorials).
31. Gai, P.L. (2007) in *Nanocharacterisation (RSC Nanoscience and Nanotechnology)*, Chapter 7 (eds A. Kirkland and J. Hutchison), Royal Society of Chemistry, p. 268.
32. Gai, P.L., Torardi, C.C., and Boyes, E.D. (2007) *Turning Points in Solid State Chemistry*, Chapter 45, Royal Society of Chemistry, p. 745.
33. Laursen, A.B. et al. (2010) *Angew. Chem.*, **49**, 3504.
34. Gai, P.L., Bart, J.C.J., and Boyes, E.D. (1982) *Philos. Mag. (London)*, **A45**, 531.
35. Helveg, S. and Hansen, P.L. (2005) *Catal. Today*, **111**, 68.
36. Hernandez, J.C., Yoshida, K., Boyes, E.D., Midgley, P., Christensen, C.H., and Gai, P.L. (2011) *Catal. Today*, **160**, 165.
37. Oyama, T. (1996) *Chemistry of Transition Metal Oxides*, Blackie International, London.
38. Gai, P.L. and Boyes, E.D. (2009) *Microsc. Res. Technol. (Wiley)*, **72**, 153.
39. Lopez-Cartes, C., Bernal, S., Calvino, J.J., Cauqui, M.A., Blanco, G., Perez-Omil, J.A., Hansen, P.L. et al. (2003) *Chem. Commun.*, **5**, 644.
40. Sharma, R., Crozier, P.A., Kang, Z., and Eyring, L. (2004) *Philos. Mag.*, **84**, 2731.
41. Sharma, R. and Crozier, P.A. (2005) in *TEM for Nanotechnology* (ed. Z. Wang), Springer and Tsingua UniversityPress, p. 531.
42. Helveg, S., Lopez-Cartes, C., Sehetsed, J., Hansen, P.L., Clausen, B., Rostrup-Nielsen, J., Abild-Pederson, F., and Norekov, J. (2004) *Nature*, **427**, 426.
43. Hoffmann, S.R., Sharma, R., Ducati, C., Robertson, J. et al. (2007) *Nano Lett.*, **7**, 602.
44. Liu, R.J., Crozier, P., Hucul, D., Blackson, J., and Salaita, G. (2005) *Appl. Catal. A.*, **282**, 111.
45. Subra, N.S., Diemer, R.B., and Gai, P.L. (2005) U.S. Patent 6,852,306 B2.
46. Gai, P.L. (2006) U.S. Patent 7,101,820 B2.
47. Gai, P.L. et al. (1993) *J. Solid State Chem.*, **106**, 35.
48. Thomas, M., Terasaki, O., Zhou, W., Gonzalez-Calbet, A., and Gai, P.L. (2001) *Acc. Chem. Res.*, **34** (7), 583.
49. Daulton, T., Little, B., Lowe, K., and JonesMeehan, J. (2001) *Microsc. Microanal.*, **7**, 470.
50. Giorgio, S., Sao Joao, S., Nitsche, S., Sitja, G., and Henry, C.R. (2006) *Ultramicroscopy*, **106**, 503.
51. Creemer, J. et al. (2008) *Ultramicroscopy*, **108**, 993.
52. Zewail, A.H. (2010) *Science*, **328**, 187.
53. Fukushima, K., Ishikawa, A., and Fukami, A. (1985) *J. Electron Microsc.*, **34**, 47.
54. Gai, P.L. (2002) *Microsc. Microanal.*, **8**, 21.
55. Gai, P.L., Kourtakis, K., and Boyes, E.D. (2005) *Catal. Lett.*, **102** (1–2), 1.
56. (a) Gai, P.L. and Boyes, E.D. (2005) *Microsc. Microanal.*, **11**, 1526, and references therein; (b) Gai, P.L., and Boyes, E.D. (2006) In-situ electron microscopy. Proceedings of the International IMC 16, Vol. 1, p. 53.
57. Gai, P.L. and Boyes, E.D. (2010) *J. Phys.Conf. Ser.*, **241**, 012055.
58. Haider, M. et al. (1998) *Nature*, **392**, 768.
59. Allen, L.J., Findlay, S., Lupini, A.R., Oxley, M.P., and Pennycook, S.J. (2003) *Phys. Rev. Lett.*, **91**, 105503.
60. Batson, P. et al. (2002) *Nature*, **418**, 617.
61. www.maxsidorov.com
62. Gai, P.L., Yoshida, K., Shute, C., Jia, X., Walsh, M.J., Ward, M.R., Dresselhaus, M.S., Weertman, J.R., and Boyes, E.D. (2011) *Microsc. Res. Technol.*, **74**, 664.
63. Graedel, T.E. (2002) in *Handbook of Green Chemistry and Technology*, (eds J.H Clark and D. Macquarrie), Blackwell, Oxford.
64. Gai, P.L., Montero, J., Lee, A., Wilson, K., and Boyes, E.D. (2009) *Catal. Lett.*, **132**, 182.

12
Speckles in Images and Diffraction Patterns
Michael M. J. Treacy

12.1
Introduction

Speckle is a topic that is not often discussed in electron microscopy. It has perhaps been examined more extensively in coherent X-ray and laser scattering contexts [1, 2]. It is also a topic of importance in astronomy, where speckled star images provide the information needed to correct ground-based telescope images for atmospheric irregularities [1, 3]. However, there are not many examples of the application of speckle to the study of materials in electron microscopy. Fluctuation electron microscopy (FEM) is one such application [4]. Speckle in electron scattering data from thin samples is generally associated with noise, which can arise from low illumination fluence, weak scattering from the sample, detector inefficiency, and structural irregularities in the sample. As a rule, and usually with good reason, speckle in our experimental electron scattering data tends to be viewed as a nuisance phenomenon that lowers the signal-to-noise ratio and the resolution of our experiments. Speckle is usually something to be avoided, or ignored.

However, speckle arising from the structure of our sample is not necessarily information-free. Diffraction patterns and micrographs of real materials tend to appear noisy with respect to an "ideal," perfectly homogeneous specimen.[1] Disordered materials, such as nanocrystalline compacts, glasses, and amorphous materials will exhibit strong speckle at length scales comparable to the length scale of the heterogeneities, and so speckle analysis offers a way to probe structural correlations within disordered materials [5, 6]. An important concept here is that it is the fluctuations in scattering within the sample that inform us about the underlying structure; the scattering "noise" is not information-free.

The term *fluctuation microscopy* has evolved since the introduction of the method in the mid-1990s [5]. It refers to experimental methods that examine spatial fluctuations in scattering from small sample volumes for the purposes of identifying the presence of medium-range order, or (MRO). Generally, MRO refers to structural

---

1) Strictly speaking, if perfectly homogeneous samples existed they would defeat the purpose of high-resolution electron microscopy, which is to explore structural and chemical heterogeneities down to atomic resolution.

Handbook of Nanoscopy, First Edition. Edited by Gustaaf Van Tendeloo, Dirk Van Dyck, and Stephen J. Pennycook.
© 2012 Wiley-VCH Verlag GmbH & Co. KGaA. Published 2012 by Wiley-VCH Verlag GmbH & Co. KGaA.

correlations at length scales between about 0.5 and 3 nm, although there is no firm definition. Below 1 nm correlations can be referred to as *short range*, and above 2.5 nm as long range. Perhaps wisely, there is no sharp delineation between the categories. There are now a number of reviews of fluctuation microscopy available [4, 7–9]. As a technique, fluctuation microscopy is evolving rapidly, and even the most recent reviews are already showing their age.

This chapter is based on a set of lecture notes that were presented to graduate students in the Department of Materials at Oxford University, as part of The Frontiers of Microscopy and Microanalysis course. Consequently, the style is that of a tutorial rather than a review. The aim is to provide a basic introduction to the origins and uses of electron speckle – a topic that is not often discussed in the standard electron microscopy textbooks. It is not intended as an up-to-date review of FEM, although there will be a general description of this technique. Also, it is not intended as a review of electron microscopy of amorphous materials, even though these materials are the primary application of electron speckle analysis. Instead, my goal is to explain this somewhat esoteric topic in simple terms. Hopefully, by demystifying the topic a little, this chapter provides the reader with the basis for a better physical understanding of the causes of speckle in image and diffraction data. It will show how an analysis of speckle can provide useful information about disorder in amorphous samples. Although the scattering theory underpinning FEM tends to be more complex than that for plane-wave diffraction, the mathematical treatment will be kept as simple as possible.

Speckle analysis of amorphous materials is still an emerging area of research. Perhaps, if I succeed in adumbrating the murky areas of the topic, this chapter will inspire future discoveries.

12.2
What Is Speckle?

Speckle is a term used to describe the seemingly random patterns of bright and dark spots that occur in images and diffraction patterns of an object. I emphasize dark spots because speckle patterns are fluctuations in intensity, which can be either positive or negative, about a mean value. In day-to-day usage, a fluctuation tends to mean a variation as time changes. In our context, we are also interested in variations as position changes. In physics, space and time are both important field parameters across which field variables can change. In principle, we are interested in both types of fluctuation, but here the focus is on just the spatial variations, and assume that structures are not changing with time. The human eye (mine particularly) tends to place undue emphasis on the bright spots in a pattern. Presumably, this is because a bright spot represents an event, such as the arrival of an electron (or lots of electrons), whereas dark regions represent a relative absence. In fact, the absence of intensity at a point can be just as important a signal as the presence of intensity.

There are two types of speckle that we will distinguish here. Although their physical origins appear to be different, both types are inherently properties of electrons.

1) **Shot noise.** The first type of speckle is associated with the illumination, and is not related to the object. When the irradiation fluence rate (electrons per unit area per unit time) is low, the inherently random arrival of electrons at the detector (similar to the arrival of raindrops on a puddle) becomes easy to see at the detector time frame rates usually encountered. For modern charge coupled device (CCD) cameras, exposure times are ≥ 1 ms. Some researchers like to subdivide this noise category further into two types: Poisson noise, when the fluence (electrons per unit area) is so low, approximately ≤ 10 counts per pixel that the pixel intensity distribution follows Poisson statistics; and Gaussian noise, when the fluence is higher, approximately >10 counts per pixel. The only difference between these two subcategories are the mathematical approximations used to describe them. The underlying physics describing them is identical. This speckle from noise, although important, is not the focus of this chapter.

2) **Coherent interference.** The second type of speckle is associated with the specimen. This type potentially tells us something about the atomic arrangements, and this is the speckle that interests us here. In a sample comprising randomly placed scatterers, the speckle arises from the random phases of the scattered waves. Such waves, when they interfere in the image or diffraction plane, generate a random intensity pattern that is reproducible, unlike the noise pattern.

Recall, the wave-particle duality of quantum mechanics tells us that photons and electrons behave as particles at the instants of their creation and destruction (i.e., emission and detection), but behave as waves as they travel between those two events. Thus, electrons scatter as waves, but are detected as particles. Shot noise is associated with the particulate behavior of electrons, whereas coherent interference is associated with their wavelike behavior. The two types inevitably coexist in all scattering experiments.

There is potentially a third type of speckle that can come from the illumination source. If the source is equivalent to an assembly of smaller coherent sources (e.g., a filament with many sharp points), then speckled coherent interference patterns can occur in the illumination itself. This speckle is closely related to the second type listed above, but is independent of the scattering in the sample. However, the intensity patterns in the illumination over the sample will, in turn, influence the scattering contribution from each sample region.

Additional noise can arise in images and diffraction patterns because of inefficiencies in the detector. This noise is unrelated to the sample itself, but can depend on the intensity being detected, so it can be an important factor when quantitative measurements are required. This latter noise generally does not contribute speckle to images, but instead modifies the speckle inherent to the arriving signal.

12.3
What Causes Speckle?

Speckle is the interference pattern from random, or quasi-random, sources. It is rather like the patterns of bright light that one sees at the bottom of outdoor swimming pools. The ripples at the water surface bend (phase shift) sunlight and an interference pattern appears at the bottom of the pool. In the pool, the pattern is continuously changing because the water surface is continuously changing. If the water were frozen suddenly, the speckle pattern would remain approximately constant, changing slowly because the sun moves across the sky. In electron microscopy, speckle is usually constant in time, unless the atomic arrangements are changing with time. However, as with the sun's motion, speckle patterns will change when the illumination conditions are changed.

Speckle is not unique to electron microscopy, and so the discussion here applies to optical scattering too provided we keep in mind that polarization is an additional consideration in both X-ray and visible optics. To help us understand the origin of coherent speckle, we shall remind ourselves of the essential optics that govern imaging and diffraction. In Figure 12.1, a ray diagram of a simple microscope is represented. It shows plane-wave illumination arriving at a sample, depicted as parallel rays. Although we are describing an interference phenomenon, which is a property of waves, for simplicity the diagram represents the wave motion as rays that travel perpendicularly to the wavefront.[2] The scattering radiates from each point of the sample as spherical waves.

We shall assume here that the electrons are scattered elastically off the potential $\phi(r)$ surrounding each atom nucleus. r is the radial distance from the atomic nucleus. Consequently, since the potential extends over a short distance from the nucleus before being screened by the orbital electrons, each atom has a short physical extent, and is not a true point scatterer. The finite width of the atom is equivalent to the scattering from a point, but with the spherical wave amplitude being modulated by a scattering form factor. This describes how the scattered spherical wave amplitude, from the projected atom potential, decreases with scattering angle θ. A simple way to express this is in terms of the phase change $\varphi(x, y)$ of a plane wave as it passes through the atom along the z direction. This can be expressed as

$$\varphi(x, y) = -\frac{2\pi}{\lambda} \frac{1}{2E_0} \int \phi(r) dz \equiv -\sigma \phi(x, y) \qquad (12.1)$$

where E_0 is the energy of the electron beam, λ the electron wavelength, and we have set $\sigma = \pi/\lambda E_0$. The integral is along the incident beam direction. The minus sign arises because the negatively charged electrons are pulled toward the positively charged nucleus, thereby advancing their phase relative to the unscattered electron.

---

[2] Although it makes little difference to the discussion here, we should remember that rays are inherently particulate in nature.

12.3 What Causes Speckle?

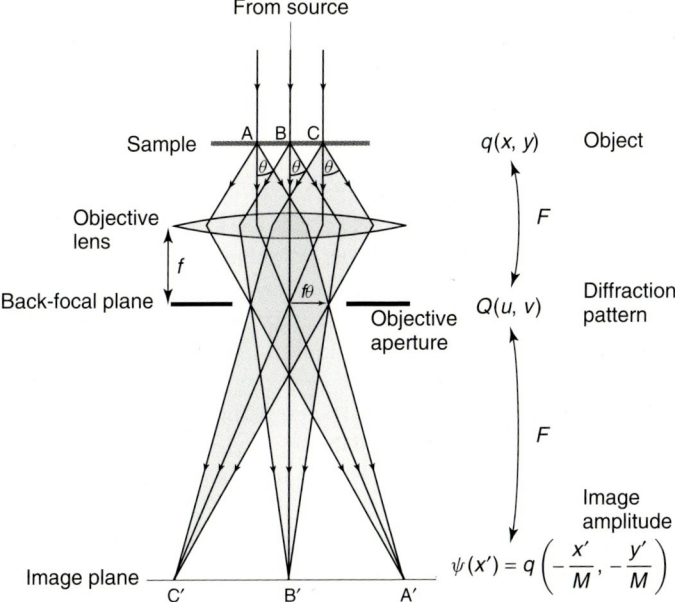

Figure 12.1 Abbé theory. Ray diagram depicting the passage of electron illumination through the sample to the image on the detector plane. There are three regions where the rays can be thought of as representing spherical waves originating from, or arriving at, points. (i) The object plane where scattered waves originate. (ii) The image plane where the same rays converge. (iii) A crossover in the back-focal plane of the objective lens, where the diffraction pattern is formed. The latter two (image and diffraction) are strongly dependent on the coherence of the illumination wavefront.

The specimen exit wave becomes

$$q(x, y) = \exp\{-i\sigma\phi(x, y)\} \quad (12.2)$$

This constitutes the phase object approximation (POA) and applies to thin samples with many atoms, and $\phi(x, y)$ is the combined projected potential distribution of all atoms.

The weak phase object approximation (WPOA) goes one step further by assuming that $\sigma\phi(x, y) \ll 1$ everywhere. This then allows us to expand the exponential as a series and keep only the first-order term:

$$q(x, y) \approx 1 - i\sigma\phi(x, y) \quad (12.3)$$

The diffraction wave function, $Q(u, v)$, is related to the exit wave function $q(x, y)$ by a Fourier transform

$$Q(u, v) = \iint q(x, y) \exp\{2\pi i(ux + vy)\}\,dx\,dy \equiv F[q(x, y)] \quad (12.4)$$

where u and v are the reciprocal lattice wave vectors parallel to the Cartesian real-space x and y directions. The symbol F denotes a two-dimensional Fourier transform. Provided the exit wave function $q(x, y)$ is known, then the function $Q(u, v)$ will describe the scattered amplitude in the diffraction plane.

12.4
Diffuse Scattering

The alert reader will notice that Eq. (12.4) is exactly the same as the equation describing the diffracted wave function from a periodic object (a crystal), or from a single atom. In the case of a periodic object, the bright intensities in the diffraction pattern are the lattice reflections.[3] If u and v are small compared to $1/\lambda$, these bright diffraction spots will be distributed periodically in the diffraction plane. The sharp, periodic lattice reflections are the result of strong, resonant constructive interferences between scattered waves from each atom. These are not commonly referred to as "*speckle*" because there is no randomness in the relative phases of the waves. However, what we usually refer to as speckle has the same origin, coming from Eq. (12.4).

In real samples, at ordinary temperatures, there is thermal motion that introduces a small aperiodicity. Each atom vibrates as part of a spectrum of phonon waves, which displace the atom slightly off the true lattice point at any instant in time, t. Such displacements detune the resonance conditions slightly, causing some scattering to appear in between the sharp peaks. If such thermal motion were frozen, so that each atom was stationary at its instantaneous position, speckle would appear between the sharp peaks. This speckle would tell us something about the thermal displacements at that instant. However, typical infrared phonon frequencies are of the order of $\sim 10^{14}$ Hz. The rapid motion means that the diffuse scattering observed in most experiments is the incoherent sum of many different displaced configurations. In turn, the speckle is averaged away leaving us with just a smoothly varying diffuse background. The delicate speckle that might have told us instantaneous structural details is gone.

Essentially, thermal diffuse scattering is the time-averaged speckle pattern associated with scattering from atoms that are displaced stochastically, by thermal phonons, from their lattice positions.

Not all atomic displacements in crystals come from phonons. Atomic impurity atoms, vacancies, and defects create potentially static displacements (strains) of atoms. In principle, a careful analysis of speckle in the diffuse scattering could reveal some of the details of these strain fields when there is a dilute distribution of defects. Of course, a single defect will not create significant speckle in between lattice peaks; speckle arises from multiple waves interfering at a point.

Weak-beam imaging is an example of how the static distribution of diffuse scattering associated with a defect can be exploited to learn important details of a single defect structure.

---

3) Strictly, in the WPOA, these reflections should not be called *Bragg reflections* since there are no lattice planes as such. The potential is projected onto a flat sheet, and only $(hk0)$ indices can be generated from the projected lattice lines.

12.5
From Bragg Reflections to Speckle

Let us now look more closely at how speckle emerges in scattering patterns from disordered samples. A good place to start is to inspect the scattering from a single atom. In Figure 12.2a, the projected potential of a hypothetical single point atom is represented. It is the tiny, barely discernible, white dot at the center of the panel. The scattering pattern from such a hypothetical point atom is shown in Figure 12.2b as a uniform intensity everywhere. I have ignored the intense bright zero-order peak that would appear at the center of Figure 12.2b, which represents the unscattered transmitted beam, and have shown only the intensity scattered by the point atom.

Real atoms are not true points. The potential outside a neutral atom is weak because the strong positive potential associated with the nucleus is cancelled, or

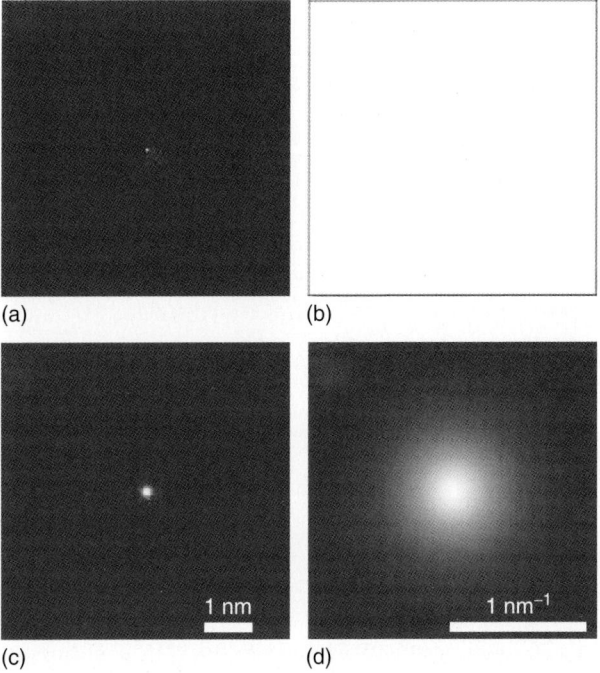

Figure 12.2 Scattering from a single atom. (a) Projected potential of a single point atom. (b) Scattering (i.e., diffraction pattern) from a point atom. The intensity is uniform with no interference, no speckle. (c) Projected potential of a single Si atom. There is a small, nonzero extent to the atom. (d) Scattering pattern from a single Si atom. The scattering intensity falls off with scattering angle. The width of this distribution is inversely proportional to the effective atom width. The function that describes the fall-off in wave amplitude is referred to as the *atomic scattering factor*, $f(u, v)$. The intensity here falls off as $|f(u, v)|^2$.

screened, by the orbital electrons whose total charge is equal and opposite to that of the nucleus. Inside the atom, however, the screening is partial and the projected potential is strongly positive through the nucleus. Figure 12.3c depicts the projected potential of a single Si atom, after allowing for partial screening near the nucleus. The bright region at the center is still small, but it is no longer a true point. The scattering pattern (d) is no longer uniform. The scattered intensity falls off with increasing scattering angle (or increasing scattering vector). The amplitude of this fall-off is described by the atomic form factor, $f(u,v)$ which applies to the scattered wave function. Thus, the intensity distribution in (d) is the square of this function, $|f(u,v)|^2$.

There is no speckle visible in the scattering patterns (Figure 12.2a,c). Technically, the projected potentials show one bright speckle, the atom itself, at the center. However, speckle can arise because of the statistics of the scattering from the atom. As each incident electron is scattered, it will be detected at a point. It is only after many electrons have been scattered that the smooth intensity distribution shown here will emerge.

A similar calculation is shown in Figure 12.3 for two silicon atoms that are separated by 1.5 nm. Panel (a) shows the projected potential of the two atoms. Panel (b) shows the scattered intensity pattern that results if a very large number of electrons are scattered by the pair, assuming that the atoms do not move during the experiment.[4]

The intensity distribution falls off in an identical manner to that for a single Si atom, shown in Figure 12.2d. However, the pattern is modulated by interference fringes. The fringes are perpendicular to the line connecting the two atoms. We can see how these fringes arise by writing the projected potential explicitly as the sum of the potentials of two single Si atoms centered at (x_1, y_1) and (x_2, y_2)

$$\phi(x, y) = \phi_{Si}(x - x_1, y - y_1) + \phi_{Si}(x - x_2, y - y_2) \tag{12.5}$$

It is helpful to rewrite this as

$$\phi(x, y) = \phi_{Si}(x - x', y - y') \times \left[\delta(x_1 - x')\delta(y_1 - y') + \delta(x_2 - x')\delta(y_2 - y')\right] \tag{12.6}$$

The delta functions $\delta(x_1 - x')$, and so on, are zero everywhere except at $x' = x_1$, and so on, where they are infinitely large, but in such a way that when integrated over all x they evaluate to 1. If we put this into Eq. (12.3) for the weak phase object exit wave function $q(x, y)$, and then insert that into Eq. (12.4) for the diffracted wave function we get the (momentarily) complicated-looking expression for $Q(u, v)$ for

---

[4] Newton's third law of motion tells us that the atoms will recoil after each scattering event in order to conserve momentum. We are ignoring that complicating issue here to keep the discussion simple.

12.5 From Bragg Reflections to Speckle

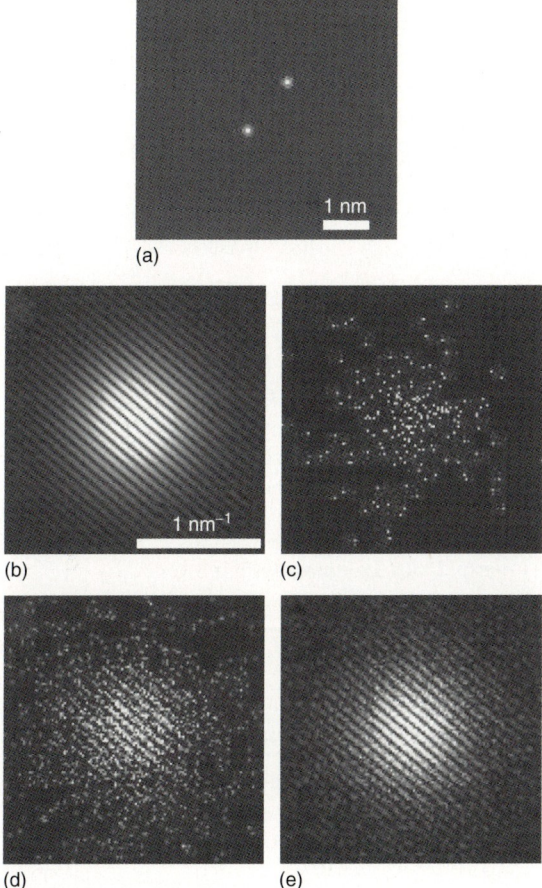

Figure 12.3 Speckle arising from shot noise. The scattering calculated for two Si atoms separated by 1.5 nm. (a) Projected potential of the two Si atoms. (b) Diffraction pattern showing Young-type interference fringes. The fringe intensity fades because of the fall-off of the atomic scattering form factor with increasing scattering angle. (c) Diffraction pattern when there are only 200 two Si atoms electron counts total. Random speckle due to Poisson-type shot noise dominates the pattern, and the interference fringes are not easily discerned. (d) 2000 counts total. The fringes in the bright central region are now clearer. (e) 20 000 counts total. The shot noise is now mostly Gaussian-type and most of the important features of the interference pattern are now visible.

$$Q_2(u,v) = i\sigma \iint \phi_{Si}(x-x', y-y') \left[\delta(x_1-x')\delta(y_1-y') + \delta(x_2-x')\delta(y_2-y') \right]$$
$$\times \exp\{2\pi i(ux + vy)\} \, dx \, dy \quad (12.7)$$

To make the math a bit simpler, let us say that atom 1 is at the origin, that is $(x_1, y_1) = (0, 0)$. This then makes $(x_2, y_2) \equiv (\Delta x, \Delta y)$, the displacement of atom 2 relative to atom 1. Thus,

$$\begin{aligned}
Q_2(u,v) &= -i\sigma \iint \phi_{Si}(x-x', y-y')[\delta(x')\delta(y') + \delta(\Delta x - x')\delta(\Delta y - y')] \\
&\quad \times \exp\{2\pi i(ux+vy)\}dx\,dy \\
&= -i\sigma \iint \phi_{Si}(x,y) \exp\{2\pi i(ux+vy)\}dx\,dy - i\sigma \iint \phi_{Si}(x-\Delta x, y-\Delta y) \\
&\quad \times \exp\{2\pi i(ux+vy)\}dx\,dy \\
&= -i\sigma \int \phi_{Si}(x,y) \exp\{2\pi i(ux+vy)\}dx\,dy \times [1 + \exp\{2\pi i(u\Delta x + v\Delta y)\}]
\end{aligned}$$
(12.8)

The last step is accomplished by substituting $x'' = x - \Delta x$ into the second integral in the middle line, and then noticing that the two integrals are the same, except for a phase factor. The integral represents the wave function scattered from a single atom, $Q_1(u,v)$. The second term inside the square brackets in the last line of Eq. (12.8) represents the identical scattering from an atom that has been displaced by $(\Delta x, \Delta y)$ relative to the first atom. Notice that the square brackets are outside the integral. Replacing the integral by $Q_1(u,v)$, we can simplify things further to get

$$\begin{aligned}
Q_2(u,v) &= Q_1(u,v) \times [1 + \exp\{2\pi i(u\Delta x + v\Delta y)\}] \\
&= 2Q_1(u,v) \times \left[\frac{\exp\{-\pi i(u\Delta x + v\Delta y)\} + \exp\{\pi i(u\Delta x + v\Delta y)\}}{2}\right] \\
&\quad \times \exp\{\pi i(u\Delta x + v\Delta y)\} \\
&= 2Q_1(u,v) \times \cos\{\pi(u\Delta x + v\Delta y)\} \times \exp\{\pi i(u\Delta x + v\Delta y)\}
\end{aligned}$$
(12.9)

The intensity from two atoms is therefore related to the intensity from one atom via

$$I_2(u,v) = |Q_2(u,v)|^2 = 4I_1(u,v) \times \cos^2\{\pi(u\Delta x + v\Delta y)\}$$
(12.10)

Here, I have set $I_1(u,v) = |Q_1(u,v)|^2$. Since

$$\cos^2(x) = \frac{1}{2}[1 + \cos(2x)]$$
(12.11)

Equation (12.10) becomes

$$I_2(u,v) = 2I_1(u,v)\left[1 + \cos\{2\pi(u\Delta x + v\Delta y)\}\right]$$
(12.12)

This tells us that whenever $(u\Delta x + v\Delta y)$ changes by a full integer, the phase changes by 2π and we get a new fringe. These are the Young's fringes seen in Figure 12.3b. If we place both atoms on the x-axis by setting $\Delta y = 0$, we see more clearly that the fringe spacing in reciprocal space corresponds to a change in u of $1/\Delta x$. The fringes move further apart when the atoms move closer together, and vice versa.

Equation (12.10) shows that the peak fringe intensity for two atoms is four times greater than that for one atom. However, the total scattering is only twice as much, approximately, because the dark fringes between the bright ones dip down to zero intensity. This nonlinear scaling of coherent intensity with the number of scatterers is an important feature of speckle.

The role of shot noise in generating speckle is shown in Figure 12.3c–e. Panel (c) shows the same fringe pattern as in (b), except there are only 200 electrons detected over the whole pattern. This noise generates strong Poisson-type speckle that overwhelms the fringe pattern. When the counts are increased by a factor of ten, so that there are 2000 counts over the whole pattern, as shown in (d), the fringe pattern is now evident near the central bright zone. A further factor of ten (e), improves the fringe visibility further and the speckliness of the pattern diminishes. Although this type 1 speckle is important, it is related primarily to noise in the illumination and detection system, and is not derived from the sample.

The evolution of diffraction patterns as a function of the number of atoms and their distribution are shown in Figure 12.4. We have already seen the patterns for one atom and two atoms, which are reproduced in Figure 12.4a,b. There is nothing particularly speckled about these noise-free time-averaged patterns. As the number of atoms increases from 1 to 10, the diffraction pattern develops additional sets of fringes such that by 10 atoms, there is already a lot of fine-scale structure in the diffraction pattern (Figure 12.4f). The pattern is already quite speckled.

Why does this fine-scale structure appear so quickly in the pattern? Already, by four atoms, the simple fringiness of the pattern is gone. The clue to this complexity becomes apparent when we remind ourselves of the two-atom example. Two atoms are sufficient to generate interference fringes. Every pair of atoms in the sample contributes its own fringe pattern. For N atoms in a sample, where $N > 1$, there are $N(N-1)/2$ distinct atom pairs, and hence (potentially) an equal number of distinct fringe patterns. For 10 atoms, this means $10 \times 9/2 = 45$ distinct fringe patterns. That is a lot of Fourier components, and is certainly sufficient to represent a complicated distribution.

Amorphous models of silicon clusters containing 1000 atoms show complex speckle patterns (Figure 12.4g). The cubic shape of the cluster generates a set of fringes close to the zero-order beam. These represent the Fourier transform of a cube that, in this model, is 2.72 nm on a side. However, at higher scattering vectors, pronounced near-random speckle is evident with no particularly strong bright spots. Other models, containing just a hint of MRO (Figure 12.4h), show similar speckle patterns, but with stronger bright peaks. In different orientations, these peaks twinkle on and off, with bright spots appearing in different locations (Figure 12.4i). Finally, a perfect crystalline particle generates almost no random speckle (Figure 12.4j). Instead, a few sharp and intense bright spots appear. These are the lattice spots, which occur when the sample contains many atom pairs with the same relative separation vector. In a crystalline cluster, most of the $1/2 \times 1000 \times 999 \simeq 500\,000$ Fourier components are identical, which is why the pattern is simpler. These spots represent strong resonant constructive interference. Although the physical process that generates the lattice spots is identical to that which generates the speckle, the lack of randomness means that these spots are not usually referred to as *speckle*. This is why analysis of crystalline materials is

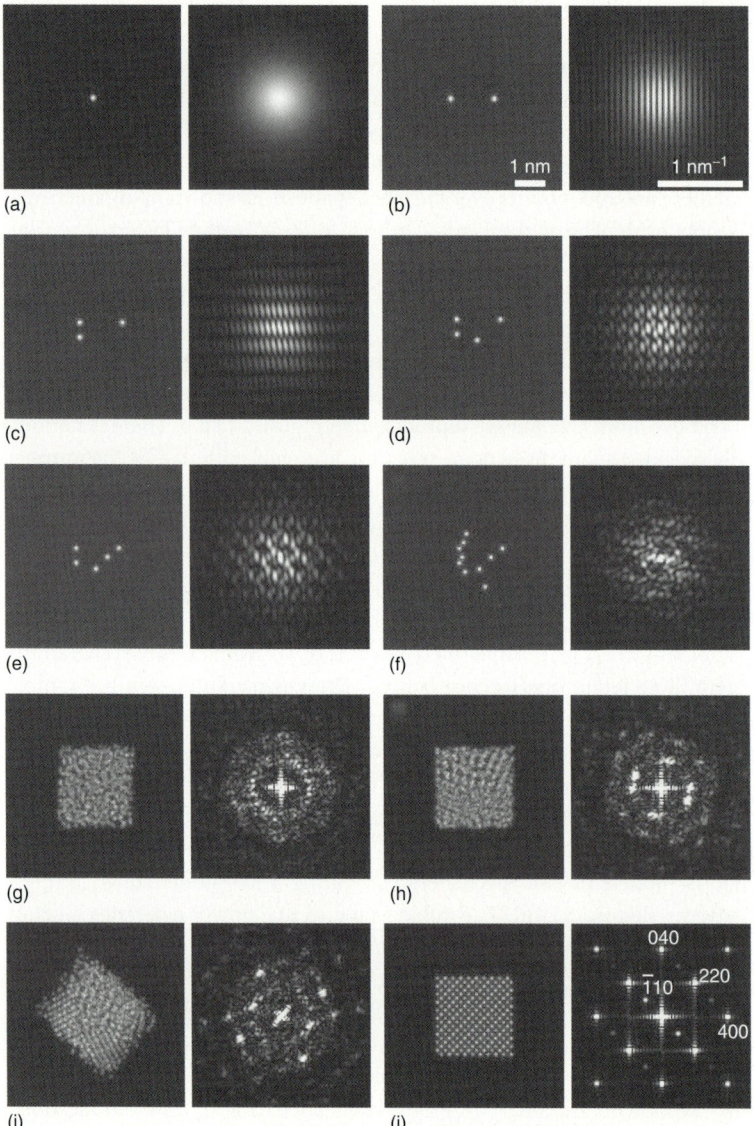

Figure 12.4 Evolution of speckle in calculated diffraction patterns from various models of silicon structures. Each image pair represents the projected potential on the left, and the corresponding diffracted intensity on the right. (a) 1 atom. (b) 2 atoms. (c) 3 atoms. (d) 4 atoms. (e) 5 atoms. (f) 10 atoms. (g) 1000 atoms arranged as a continuous random network inside a cube measuring ∼2.72 nm on a side. (h) 1000 atoms in a cube measuring 2.72 nm on a side containing ∼1 nm diameter paracrystalline grains. (i) Same model as (h) but in a random orientation. (j) 1000 atoms arranged as a cubic crystallite of diamond-cubic Si, 2.71 nm on a side. (Crystalline Si is denser than amorphous Si.)

much easier; there are way fewer Fourier components to identify. The indices of the major spots are given in Figure 12.4j.[5]

Clearly, if the diffraction wave function were known, as it is for these computations, then it would be a simple matter to perform an inverse Fourier transform, $F^{-1}(Q(u, v))$, to reconstruct the projected potential of the sample, exactly as shown to the left of each diffraction pattern. However, since we measure only the diffracted intensities, we know the wave function amplitudes, but not the phases. Sophisticated phase retrieval methods are being developed to overcome this shortcoming. They work for model systems, but do not yet work reliably for real materials. Nevertheless, phase retrieval is an exciting area of research to be in right now.

12.6
Coherence

Real experimental diffraction and imaging data are rarely as straightforward as the simple, idealized models presented in the previous section. Specimen drift, contamination, beam damage, detector inefficiencies, and inelastic scattering are important considerations for almost all microscopy studies. Many of them can be mitigated with patience, persistence, anticontaminators, energy-filtering spectrometers, and judicious choice of operating voltage.

For the quantitative study of speckle, illumination coherence is also important, and is worth discussing in more detail here. Wolf presents an excellent discussion of coherence in his recent book [10]. First, we will remind ourselves of the basics.

We know that a pure plane wave can be represented by a function of the type

$$\psi(\mathbf{r}, t) = \psi_0 \exp\{i(\mathbf{k}\cdot\mathbf{r} - \omega t\} \tag{12.13}$$

ψ_0 is the amplitude of the wave. $\mathbf{k}$ is the three-dimensional wave vector, and $\mathbf{r}$ is a point in three-dimensional space. t is time, and ω is the angular frequency of the wave, where $\omega = 2\pi v$ with v being the frequency. The wave travels in the direction $\mathbf{k}$. Ideally, $\mathbf{k}$ is a constant vector, and ω is a constant value.

In practice, a wave describing illumination from a source contains a spread of k values, and a spread in ω values. For both photons and electrons, ω can be a single value (i.e., a laser source), which then dictates that the amplitude $\mathbf{k} = |\mathbf{k}|$ also must be single valued. However, the vector direction of $\mathbf{k}$ need not be single valued. This situation allows interference between the different waves to occur. An example is the focusing of a probe where the set of plane waves in the probe, arriving from slightly different directions, interfere constructively at the focal point to produce an

---

[5] A note to the observant reader: The $\{\bar{1}10\}$ reflections are normally forbidden in an infinite, cubic-diamond-structure, Si crystal, but become allowed in small particles when the destructive interferences are not 100% complete, as has happened in the simulation for the crystalline cube in Figure 12.4j.

intense spot. This same situation can be viewed alternatively as a set of spherical waves emerging from a point source, and then converging onto a point through the focusing action of a lens. Although the illumination intensity is focused onto a small region, all of the component plane waves have a fixed phase relationship between themselves. Such illumination is fully coherent, even though it is spatially delimited at the focal point and at the source.

12.6.1
Temporal Coherence

If there is an energy spread in the illumination such that there is a mean value of ω, $\bar{\omega}$, and an effective bandwidth $\Delta\omega$, where each individual value of ω is generated randomly, then over a time Δt the wave will lose coherence with itself. In simple terms, the wave oscillations will drift by more than π in phase relative to a perfect reference wave of frequency $\bar{\omega}$, and with no spread in ω. This will occur in time

$$\Delta t \approx \frac{2\pi}{\Delta\omega} \tag{12.14}$$

As $\Delta\omega \to 0$, $\Delta t \to \infty$, as we would expect for full temporal coherence.

12.6.2
Coherence Length

The corresponding length Δl of the wave train that remains coherent is the distance that the wave travels in the time Δt:

$$\Delta l = v\Delta t = \frac{2\pi v}{\Delta\omega} \approx \left[\frac{\bar{\lambda}}{|\Delta\lambda|}\right]\bar{\lambda} \tag{12.15}$$

where v is the wave velocity, which equals the electron velocity, $\Delta\lambda$ is the spread in wavelength of the electron corresponding to the frequency spread $\Delta\omega$, and $\bar{\lambda}$ is the mean wavelength. (The above expression is accurate for photons, but is approximate for electrons because their speed v also changes when ω and λ change; the dispersion relation for electrons is more complicated, but the correct answer is not that much different.) The typical coherence length for a 200 kV electron, which has an energy spread $\Delta E = 0.5$ V can be found by first looking up the electron velocity at 200 kV, $v = 0.695c = 2.09 \times 10^8$ m/s. $\Delta\omega$ is found from $\Delta\omega = 2\pi e\Delta E/h$, where e is the charge on the electron and h is Planck's constant. Inserting the numbers, we find that the coherence length $\Delta l \approx 1.7$ µm. This is almost two orders of magnitude longer than the thickness of a typical thin sample, ~0.02 µm, and so the coherence length is not usually an important consideration in electron imaging or diffraction from plane waves.

12.6.3
Spatial Coherence

As we have just seen, temporal coherence relates to the extent of coherence along the beam direction. Spatial coherence relates to the perpendicular direction. It tells us the lateral extent along the nominal wavefront before we lose coherence.

Curiously, if a point source emits uncorrelated electrons, the illumination remains fully coherent. This counterintuitive result comes about because each electron plane wave resembles that given by Eq. (12.13), apart from a phase factor, φ, giving $\psi(\mathbf{r}, t) = \psi_0 \exp\{i(\mathbf{k} \cdot \mathbf{r} - \omega t + \varphi)\}$. Adding the contribution from all uncorrelated electrons is the same as adding identical, randomly phase-shifted, sine waves. The outcome is just another simple sine wave with an averaged phase factor. The resultant illumination is still a perfectly coherent plane wave, a result that is rather surprising. It is the final absolute phase over the whole wave that is indeterminate. But it is the same indeterminate phase everywhere, and does not affect the coherence between pairs of points illuminated by the wave.

A loss of coherence will arise when our source subtends a finite solid angle $\Delta\Omega$ at the sample. The sine waves are no longer identical. Radiation is emitted incoherently from points over the source, which emit plane waves with slightly different tilt angles with respect to a fixed point on the sample. Each point emits radiation that is uncorrelated with respect to the radiation from neighboring points. At any instant, t, a fully coherent interference pattern appears at the sample. The time-averaged illumination, however, remains coherent (on average) in little patches of area

$$\Delta A = \frac{\bar{\lambda}^2}{\Delta\Omega} \qquad (12.16)$$

at the sample (Figure 12.5). For example, the sun subtends an angular diameter at the Earth of about $2\theta_s = 0.5\,°$. Light from the sun originates at points near the sun's surface. Clearly, light coming from one side of the sun's limb (edge) is completely uncorrelated with respect to the light coming from the diametrically opposite side. Nevertheless, the illumination from the sun exhibits some coherence as it arrives on Earth. The coherence area is found by setting $\bar{\lambda} = 550\,\text{nm}$ (green). The sun subtends a solid angle at Earth of $\Delta\Omega \approx \pi\theta_s^2 \approx \pi\,(0.25 \times \pi/180)^2 \approx 6 \times 10^{-5}$ sr. This yields a coherence area of $\Delta A \approx 5000\,\mu\text{m}$. Assuming that ΔA is a circular patch, the coherence width is thus

$$\Delta W = \sqrt{\frac{4\Delta A}{\pi}} \qquad (12.17)$$

which gives us $\Delta W \approx 80\,\mu\text{m}$ for sunlight. This means that we could use sunlight for a Young's slits experiment, and would get interference fringes provided the slits were closer than $80\,\mu\text{m}$; the illumination arriving at the two slits would have partial mutual coherence. However, the coherence length of visible sunlight is short. The visible range of wavelengths is about 400 nm (violet) to about 700 nm (deep red). This will give $\Delta\lambda = 300\,\text{nm}$. Inserting into Eq. (12.15) we find that the coherence length is about two wavelengths, $\Delta l \approx 2\bar{\lambda} \approx 1\,\mu\text{m}$. This will impose

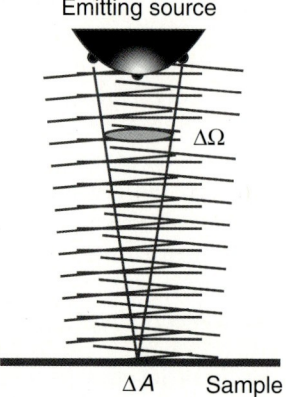

Figure 12.5 Coherence area. An extended source of irradiation is at the top, and subtends a solid angle $\Delta\Omega$ with respect to the sample plane. Radiation from points across the source arrive at the sample from slightly different directions and interfere. Because emitting points on the source are mutually uncorrelated, the time-averaged illumination is consistently coherent only within little patches of area ΔA.

exacting demands on the slit plane; each slit must be within 1 μm of the plane that is perpendicular to the sun's direction. This tolerance seems to be too demanding for any type of kitchen-table experiment to show Young's fringes with sunlight.[6]

The above discussion of sunlight is clearly of importance to those who wish to study speckle due to wave interference at the bottom of swimming pools! It is now clear that the analogy offered earlier in these notes is limited in scope. Nevertheless, it remains a valid example of speckle.

Let us now consider a star. Stars are excellent point sources, and the light arriving on Earth from them should be very good plane waves, ignoring atmospheric density variations. Nevertheless, like the sun, light from different points on the surface of the star will be statistically uncorrelated. Optical interferometry on Earth has been used to determine the coherence area of starlight and in turn determine the star's angular diameter. For example, the orange star Betelgeuse in the Orion constellation has been measured to produce a coherence area of $\sim 6\,m^2$ on Earth, giving a solid angle subtended by the star of 4.1×10^{-14} sr as viewed from Earth.

Similar issues arise in electron microscopy. Our electron sources are not true points, even for the sharp tungsten field emitters, and consequently the coherence width of the illumination at the sample can be small if we are not careful. Keep in mind that the crucial parameter in Eq. (12.16) is the solid angle of the source as viewed from the sample. This is controlled by the projector lenses, which can move the image of the source closer, or further from the sample.

As a final point, in the scanning transmission electron microscope (STEM), the cold field emitter is sufficiently small that it is usually assumed that electron waves

---

6) Nevertheless, the author intends to try it out one day.

are emitted coherently across the (small) tip, as if they originated from the focal point below the curved surface. In practice, such sources are partially coherent.

12.6.4
Coherence Volume

The coherence volume is that volume around each scatterer within which the time-averaged scattering from neighbors exhibits significant coherence strength (Figure 12.6). The coherence volume is just the product $\Delta V = \Delta A \times \Delta l$, which gives us

$$\Delta V \approx \frac{1}{\Delta \Omega} \left(\frac{\bar{\lambda}}{\Delta \lambda} \right) \bar{\lambda}^3 \tag{12.18}$$

When such illumination scatters from a crystal of volume V, the coherent diffracted intensity is equivalent to the sum of diffraction patterns from $V/\Delta V$ coherent subvolumes, and the spot broadening is equivalent to that from a smaller probe of width ΔW, rather than the (wider) width of the full illumination.

When pursuing quantitative diffraction studies, such as in fluctuation microscopy (discussed later), knowledge of the extent and shape of the coherence volume can be important. Hollow-cone illumination is a convenient way to control the coherence volume in the transmission electron microscope (TEM) implementation of fluctuation microscopy [11]. Typically, the illumination is tilted with respect to the optic axis so as to form a tilted dark-field image. The illumination is then precessed electronically so as to maintain the same tilt angle while changing the azimuthal angle steadily [12]. Clearly, the illumination is incoherent as a function of the azimuthal angle since the illumination at each azimuth is emitted at different times. Since the azimuthal frequency is usually less than 100 Hz, neighboring emission points in the cone will be completely outside each other's coherence volume. From the above, we know that for 200 kV illumination, $v = 2.09 \times 10^8$ m/s and the coherence length is $\Delta l = 1.7 \times 10^{-6}$ m. It takes the electron 8×10^{-15} s to traverse this length, in which time hollow-cone illumination precessing at 100 Hz will traverse an angle of 3×10^{-10} degrees. Clearly, the hollow-cone illumination source is highly uncorrelated. The coherence volume of hollow-cone illumination is narrow in extent, and has a sinusoidal oscillation along the beam axis (Figure 12.7).

12.7
Fluctuation Electron Microscopy

FEM is a technique that examines the coherent speckle originating in the scattering from disordered materials. With the knowledge of the coherence volume, the speckle statistics can be probed to obtain clues about any underlying ordering that may be present.

The method is a hybrid imaging/diffraction technique. It examines the scattering from small volumes in the sample, whose lateral extent is defined by the microscope

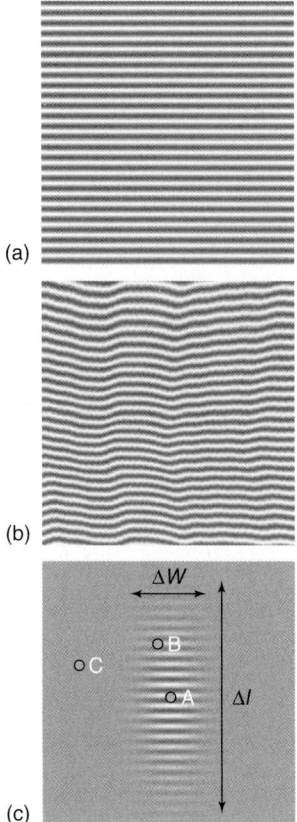

Figure 12.6 Coherence volume. Panel (a) is a depiction of the instantaneous amplitude of a perfect plane wave traveling down the page. Bright represents wave peaks, black the troughs. Panel (b) depicts the instantaneous wave amplitude of a wave field that is partially coherent. Its frequency and wavelength at any point are changing with time. The wavefronts are no longer flat, and the relative location of neighboring peaks varies with time. Despite the nonplane-wave nature of the wavefront, this instantaneous wave distribution is 100% coherent since there is a definite phase relationship between any pair of points across the whole wave. It is the time average of this wave that exhibits partial coherence. Panel (c) depicts the time-averaged wave with respect to a point at the peak of a wave. The coherence length is a characteristic length along the propagation direction (vertical) within which the sine wave stays within about $\pm\pi$ phase of a reference sine wave. The coherence width is the lateral extent of the wavefront that stays within $\pm\pi$ phase of the reference wave. The volume delineated by these dimensions is the coherence volume. For a scatterer A placed at the center of this distribution, the maximum time-averaged strength of interference with a second scatterer is given by the amplitude of coherence strength at that second scatterer. Thus, the time-averaged scattered wave from B exhibits significant interference with that from A, whereas the time-averaged interference between A and C is washed out. The coherence length Δl and coherence width ΔW are indicated. There is intensity everywhere in this wave field. However, constructive interference can only occur between points that lie within the coherence volume.

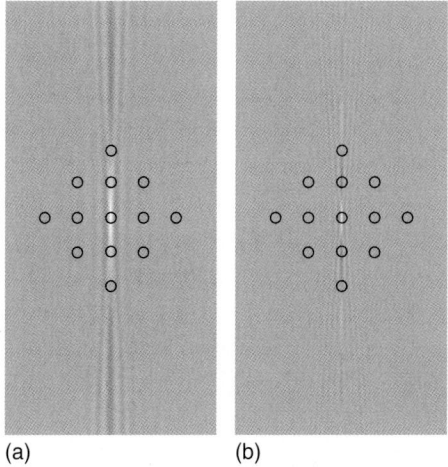

Figure 12.7 Coherence volume for hollow-cone illumination. In these views, the hollow-cone illumination is traveling down from the top of the page. The bright regions represent that volume around the central atom wherein scattering will produce constructive interference. The dark regions are where destructive interference will occur. A [100] view of a Si crystallite is superimposed as a scale reference. As the hollow-cone angle is increased (from a to b), the coherence volume gets narrower, and the sinusoidal oscillations along the optic axis get shorter in period. Atoms that had been scattering constructively are now scattering destructively. This allows the internal structure of the cluster to be probed. (From Ref. [11]).

resolution, and whose length is essentially delimited by the sample thickness. The "fluctuations" referred to in this context are *not* fluctuations in time, but rather fluctuations in position. FEM can be conducted in either a TEM or a STEM; the two methods are equivalent via the reciprocity principle. A conceptually easy way to think of FEM is in terms of a focused STEM probe that scans the sample. The microdiffraction pattern then changes (or twinkles) with time as the probe is scanned (Figure 12.8). Nevertheless, the fluctuations are still with respect to position.

A surprising, and important, feature of the FEM technique is that it is a low-resolution technique. For maximum sensitivity, the resolution (or probe size) needs to be comparable to the length scale of the structural ordering being probed. For MRO, this is the ~1.0 to ~2.5 nm length scale. Ironically, in this age of aberration-corrected electron microscopes, high resolution is a hindrance in FEM studies because, in an almost literal sense, "one cannot see the wood for the trees."

The equivalent experiment can also be conducted in a TEM. Either tilted dark-field images, or hollow-cone dark-field images, are collected as a function of tilt angle. In the STEM experiments, it is the diffraction patterns that are speckled. In the TEM images, it is the dark-field images that are speckled. An example of a TEM dark-field image of amorphous carbon is shown in Figure 12.9.

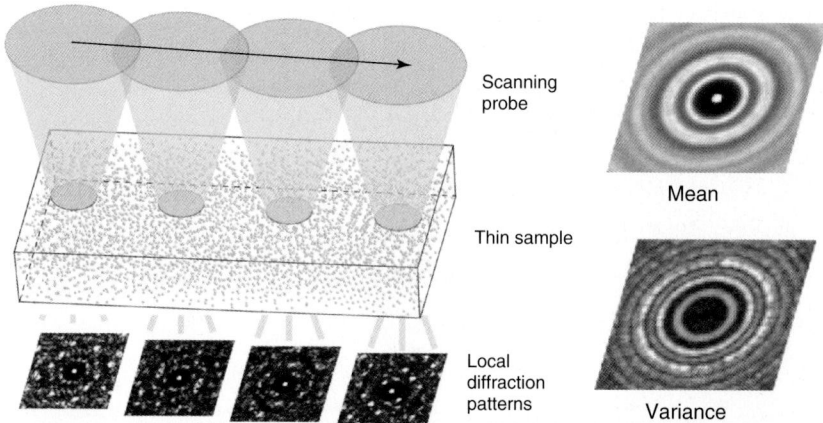

Figure 12.8 Depiction of an idealized FEM experiment conducted in a STEM on a disordered material. As the probe is scanned across the sample, the microdiffraction pattern (which is speckled) changes, as shown at the bottom. A measure of the speckliness is the intensity variance of all the microdiffraction patterns collected from the sample. The variance reveals those diffraction vectors (structural length scales) that fluctuate the most. (From Ref. [8]).

12.7.1
Measuring Speckle

As we learned in the earlier sections, a random assembly of atoms (such as the carbon film in Figure 12.9) is going to generate a speckled image (or diffraction pattern). We know that an ordered crystal generates sharp diffraction peaks, and high-resolution images will show an ordered projection of the lattice. At low resolution, the image of a uniformly thick, unbent, defect-free crystal will show no contrast. It will be speckle-free, except for shot noise. One can easily imagine all the gradations between perfect order and complete disorder. The speckle will evolve in a manner that is consistent with the "rules" governing the order/disorder.

The materials of most interest here are the amorphous materials that contain traces of MRO. All amorphous materials must possess a short-range order. This is dictated by the bond lengths and angles imposed by chemical considerations. Even gas molecules evince traces of short-range ordering in that the hard sphere radius of each molecule imposes an exclusionary volume to all other molecules; no two molecules can get closer than their combined radii.

How do we distinguish between speckle arising from short-range order alone from that arising from MRO? First, we need to define what we mean by "medium-range order," which is frequently referred to as *MRO*. There is no hard definition of these ordering length scales. It is impractical to impose any hard definition since the length scales of interest depend on the material. Correlations in amorphous materials are topologically different from linear correlations in

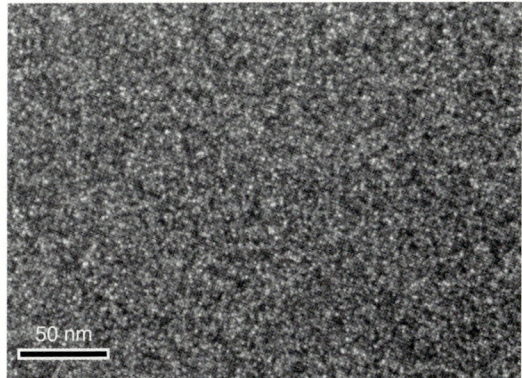

Figure 12.9 Tilted dark-field image of a ~10 nm thick film of amorphous carbon that was prepared by sputtering. The resolution of the image is about 1 nm. Low resolution is essential for the FEM technique. At high resolution, the probed volumes are too small. The speckle seen here is due in part to shot noise and in part to random structural alignments. There is no medium-range order in this sample. One cannot just point to a bright (or dark) speckle in the image and claim it is an ordered region.

polymers, for example. As alluded to earlier, for atomically glassy materials, approximate definitions would be

- short-range order: ≤1.0 nm
- medium-range order: 1.0–2.5 nm
- long-range order: ≥2.5 nm.

These are not hard-and-fast definitions. For example, depending on the context, the range for MRO can be extended to 0.5–3.0 nm.

It is clear that a quantitative measure of the speckle is needed. Visually, we know what it is at a qualitative level. However, we cannot get away with qualitative phrases such as "more speckled than" or "less speckled than" based on eyeballing images or diffraction patterns.

The speckle represents a variation of a signal about some mean value. That mean value will tell us little about the variations. One of the simplest statistical measures of variation is the variance. The standard deviation of a distribution is the square root of the variance. In statistical terms, the mean value of a variable r, $<r>$, that is distributed according to a normalized probability distribution $P(r)$, is

$$\langle r \rangle = \int rP(r)dr \tag{12.19}$$

where the integral is over the range of r-values allowed. The mean is also referred to as the *first moment of the distribution*. The normalization condition is

$$1 = \int P(r)dr \tag{12.20}$$

which is the zeroth moment of the distribution; it is constant, and here defined to be 1.

The variability of the distribution is measured by the second moment,

$$\langle r^2 \rangle = \int r^2 P(r) dr \tag{12.21}$$

Of course, higher moments exist, and are found simply by increasing the power of r in the integral. For our purposes, the second moment is sufficient for now. It tells us about the width or variability of the distribution.

Suppose our distribution were uniform, and everywhere has the value r_0. Then our distribution is $P(r) = \delta(r - r_0)$. The first moment is $\langle r \rangle = r_0$, and our second moment is $\langle r^2 \rangle = r_0^2$. The variance Var of the distribution is then found by subtracting the square of the mean from the second moment; thus,

$$\text{Var} = \langle r^2 \rangle - \langle r \rangle^2 \tag{12.22}$$

So, for our constant distribution, the variance is zero, $V = 0$. This confirms that there is no variability, as we already know.

In FEM experiments, we are interested in the variation in intensity, so in all of the above equations we replace r by I. For an image, such as that shown in Figure 12.9, the intensity is distributed as a function of position (x, y) in the image as $I(x, y)$. For simplicity, let us suppose that this intensity distribution has been normalized by dividing the raw data by the integral $\iint I(x, y) dx\, dy$. This appears as a double integral since we are averaging over both the x and y position coordinates. The intensity variance is then

$$\text{Var} = \langle I(x, y)^2 \rangle_{xy} - \langle I(x, y) \rangle_{xy}^2 \tag{12.23}$$

The subscript xy on the angular braces is reminding us that the average is over position, and not some other variable such as time.

Equation (12.23) is very easy to implement in a computer program, or in DigitalMicrograph^{TM}. One just takes the mean intensity in the image, then one squares the image everywhere (replacing every pixel with the square of its value) and then takes the mean of that "squared" image. The variance is just the mean-of-the-square minus the square-of-the-mean. Note that the variance can never be negative.

In FEM experiments, the idea is to explore the variance as a function of scattering vector (u, v). In the TEM experiment, this corresponds to the illumination tilt vector. For hollow-cone illumination, this would be replaced by the cone radius vector amplitude, k. Since the intensity falls off naturally as a function of scattering angle, it is useful to remove this variation as a function of (u, v). The normalized variance nVar is then obtained by dividing the variance by the square of the mean,

$$\begin{aligned} \text{nVar}(u, v) &= \frac{\langle I(x, y, u, v)^2 \rangle_{xy} - \langle I(x, y, u, v) \rangle_{xy}^2}{\langle I(x, y, u, v)^2 \rangle_{xy}} \\ &= \frac{\langle I(x, y, u, v)^2 \rangle_{xy}}{\langle I(x, y, u, v) \rangle_{xy}^2} - 1 \end{aligned} \tag{12.24}$$

Again, this is easily computed. Notice that the explicit dependence on the scattering vector is included. There is no averaging over the scattering vector, only over sample position. Here, $k = \sqrt{(u^2 + v^2)}$.

In STEM FEM experiments (STFEM), the equations are identical. In this case, it is the average and variance of all the diffraction patterns that is being computed. It is not the mean and variance within the patterns, as it is for the image intensities. The averaging and variance is still being computed with respect to (x, y) as given by the different probe positions for each pattern (Figure 12.8).

Figure 12.10 gives an example of how speckle reveals underlying MRO in a sample of 14.4 nm thick amorphous evaporated germanium. In Figure 12.10(A) are the hollow-cone dark-field images as a function of scattering vector k, which is related to the hollow-cone angle. Visually, the speckliness changes with k. This change is clear in the normalized variance plots (B), which show that the normalized variance has peaks. The presence of peaks tells us that the underlying structure has extended regions exhibiting periodicities corresponding to the reciprocal of the lattice vectors at the peaks.

The length scale of these regions is controlled by the microscope resolution. In the above image, a 7 μm diameter objective aperture was used, giving an angular diameter of $\alpha = 2.0$ mrad at the sample, and thus a resolution of $R = 1.22\lambda/\alpha = 1.5$ nm.

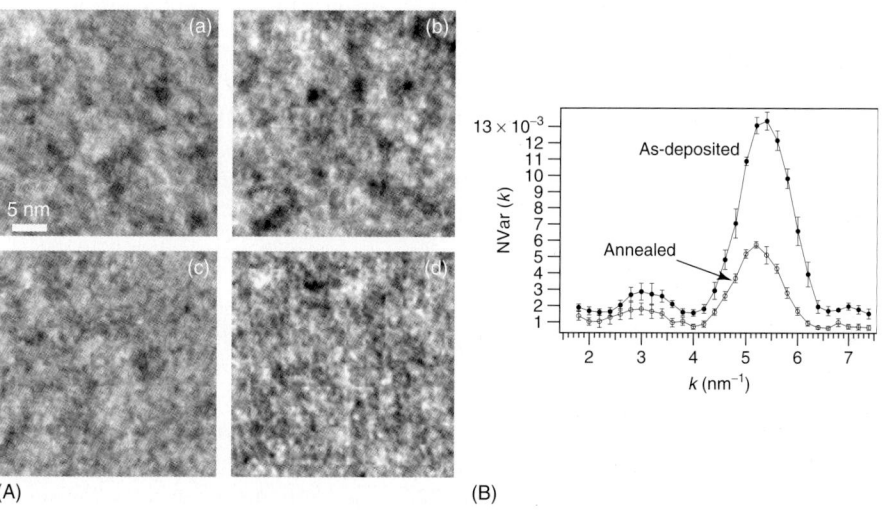

Figure 12.10 (A) A series of hollow-cone dark-field images of a 14.4 nm thick amorphous Ge sample as a function of scattering vector amplitude k. (a) $k = 2.5$ nm^{-1}, (b) $k = 3.0$ nm^{-1}, (c) $k = 4.2$ nm^{-1}, (d) $k = 5.7$ nm^{-1}. The speckle varies as a function of k, being most pronounced at $k = 5.7$ nm^{-1}, revealing that the underlying atomic structure is not random. (B): The normalized variance plot as a function of k. Pronounced peaks are present. A random sample would show no peaks in the speckle variance, since there should be no special k vectors within the structure. When the sample is annealed at a temperature below the recrystallization temperature, the normalized variance of the scattering decreases. The material becomes more disordered on annealing. (From Ref. [13]).

This is the length scale being probed by this data. In STEM, it is still the objective (i.e., the probe-forming) aperture that controls the probe size. The resolution, or probe size, is an additional variable that can be used to explore the speckle statistics.

FEM is a statistical technique. As a rule, one cannot point at a white dot on the image and claim that it is a region with MRO. Speckle also arises from random atom alignments. The width of such speckle in images is governed by the microscope resolution, and so even random speckle will appear as 1.5 nm dots in the image. An embedded crystallite will show up as an extended region, with width much larger than the resolution. This spatial correlation of the speckle can be interpreted as ordering on length scales larger than the resolution.

In diffraction patterns, the speckle size is governed by the reciprocal of the coherence width at the specimen, $1/\Delta W$, and by the reciprocal width of the length scale of the ordering, $1/L$. It is the larger of these two values that dominates. The width of the diffraction speckle decreases when the objective lens is defocused by an amount Δf. Then, the speckle width is governed by the angle subtended by ΔW or L at the focal plane, that is, the smaller of the two values $\Delta W/(\lambda |\Delta f|)$ or $L/(\lambda |\Delta f|)$. Here, I assume that $|\Delta f| > \Delta W$ and L.

A cartoon view of how FEM works is presented in Figure 12.11.

12.8
Variance versus Mean

Why does FEM exhibit such sensitivity to MRO? Is it really more sensitive than diffraction alone? To address these questions, we need to examine a little more closely what the intensity variance is measuring. The normalized variance of a set of atoms illuminated under tilted dark-field conditions is [5]

$$n\text{Var}(k) = \frac{\sum_i \sum_j \sum_m \sum_n \left(N_0 A_{im} A_{mj} - A_{ij}\right) A_{mn} e^{2\pi i k \cdot (r_{ij} + r_{mn})}}{\sum_i \sum_j \sum_m \sum_n A_{ij} A_{mn} e^{2\pi i k \cdot (r_{ij} + r_{mn})}} \quad (12.25)$$

$r_{ij} = r_j - r_i$ is the vector separating atoms j and atom i (pointing toward j). The A_{ij} terms represent the amplitude (not intensity) of the point spread function at atom j when the function is centered on atom i. For a circular objective aperture, this function is the Airy disc amplitude. In this notation, the mean intensity in an image is simply:

$$\langle I(k)\rangle_{xy} = \sum_i \sum_j A_{ij} e^{2\pi i k \cdot (r_{ij} + r_{mn})} \quad (12.26)$$

Clearly, the mean diffraction is a much simpler function. It is a two-body function, and so only depends on atom pairs. Conversely, the variance is a four-body function, and thus depends on atom pair–pair combinations. It is these higher order pair–pair correlations that give the variance such sensitivity to MRO compared to the mean diffraction data. The speckle is telling us about the four-body correlations.

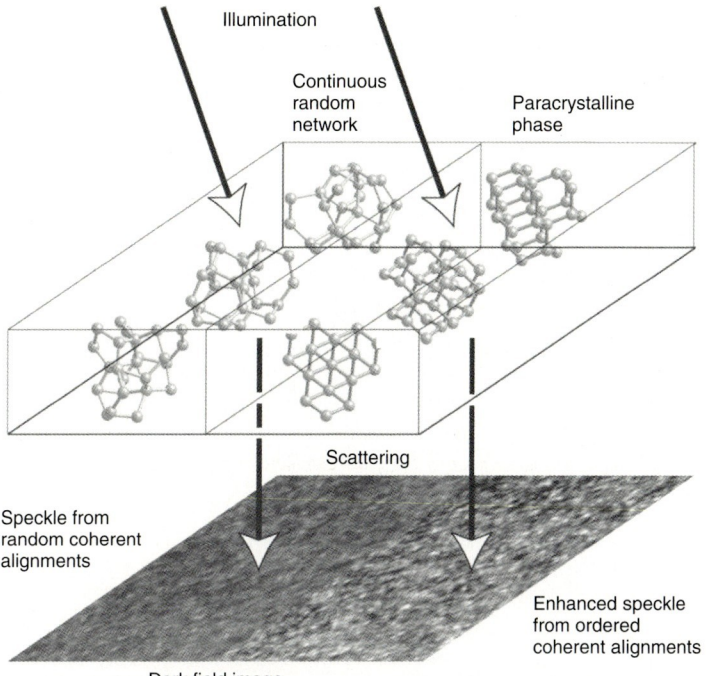

Figure 12.11 A cartoon view of how FEM works. A hypothetical Si sample is irradiated by tilted illumination to form a tilted dark-field image, which appears at the bottom. For simplicity, no apertures or image-forming lenses are shown. On the left, the sample is a continuous random network. Small, ~1 nm diameter isolated Si clusters are shown. Through random alignments, a speckle pattern is generated in the image (bottom left). On the right, the sample contains small crystalline regions of similar size. The coherent scattering from such regions is strong, since if three noncollinear atoms are aligned into a pseudo-Bragg reflection, then the other atoms in the cluster will also be aligned automatically. The amplitude of the speckle increases, and the variance increases as the Q vectors corresponding to the strongly coherent scattering.

Specimens with almost identical diffraction patterns can give strikingly different normalized variances.

It should be evident that the mean diffraction data (Eq. (12.26)) is a much easier function to invert than is the normalized variance (Eq. (12.25)). Thus, despite the sensitivity of the speckle to MRO, diffraction remains a powerful tool for exploring the short-range order in glassy materials. Radial distribution functions are fairly straightforwardly obtained from electron diffraction data [14–17]. However, the complexity of the four-body distribution function makes similar analysis for speckle a practical impossibility.

However, it is not a hopeless cause. With diligence, a researcher can explore a set of conjectured models and then compute the normalized variance expected. The model that matches the experimental data the best will give useful clues about the type of MRO present in the sample (see [18, 19]).

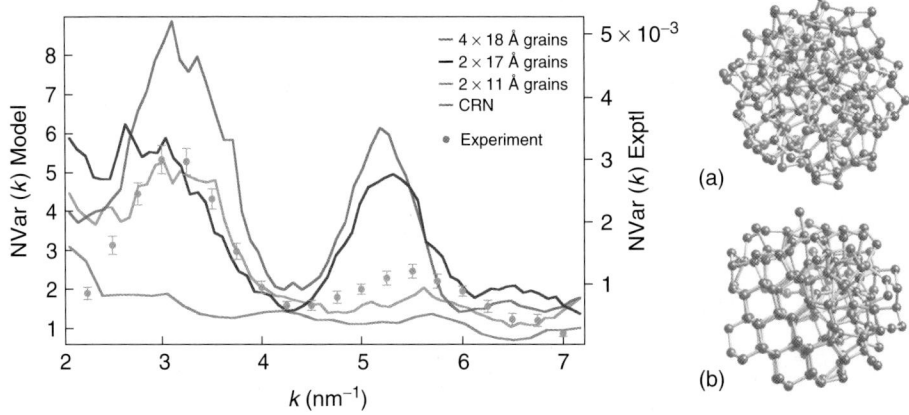

Figure 12.12 Left: Plots showing the normalized variance for evaporated amorphous Si (points) and for several models, including the continuous random network (CRN). The data is fitted best by a paracrystalline model, which contains small grains of strained crystalline Si. The paracrystalline model is shown in (b). The CRN model is shown in (a). (Reproduced from [20]).

Figure 12.12 presents the experimental normalized variance for evaporated amorphous silicon. The calculated normalized variance is superimposed for several models, including the continuous random network (CRN). The diffraction patterns for these models are almost indistinguishable, indicating that they all have a similar short-range order. However, the normalized variances are pronouncedly different. The data is most consistent with the presence of paracrystalline Si grains about 1.2 nm in diameter. These grains are similar to nanocrystallites of cubic Si, but have large shear strain gradients. Notice that the CRN model shows no pronounced variance peaks. This is because it contains no extended ordering at the resolution of the data (1.5 nm).

12.9
Speckle Statistics

Evident in the plot in Figure 12.12 is the fact that the experimental variance is a factor of 2000 or so less than that for the models. If FEM is to be a quantitative tool, we need to understand this huge difference. The reason for this reduced variance in experimental data is to do with coherence. First, let us explore what we would expect from a perfectly random speckled signal.

Suppose that the exit wave amplitude at any given point is $a + ib$, where a and b are independently Gaussian-distributed with the same standard deviation, so that:

$$P(a,b) da db = \frac{1}{2\pi\sigma^2} \exp\left(\frac{-(a^2+b^2)}{2\sigma^2}\right) da\, db \qquad (12.27)$$

We rewrite $a + ib = c \exp(i\phi)$, and note that $da\, db$ is equivalent to $c\, dc\, d\phi$. Thus, $P(c, \phi) dc d\phi \equiv P(a, b) c\, dc d\phi$. Since intensity is given by $I = a^2 + b^2 \equiv c^2$, we have for the mean value $\langle I \rangle$

$$\langle I \rangle = \frac{1}{2\pi\sigma^2} \int_{c=0}^{\infty} \int_{\phi=-\pi}^{\pi} c^3 \exp\left(\frac{-c^2}{2\sigma^2}\right) dc\, d\phi$$
$$= 2\sigma^2 \tag{12.28}$$

Therefore,

$$P(c)dc = \frac{1}{\sigma^2} c \exp\left(\frac{-c^2}{2\sigma^2}\right) dc \tag{12.29}$$

Noting that $I = c^2$, $\langle I \rangle = \sigma^2$ and that $2\, c\, dc = dI$, we get finally

$$P(I)dI = \frac{1}{\langle I \rangle} \exp\left(\frac{-I}{\langle I \rangle}\right) dI \tag{12.30}$$

This is a negative exponential, which is quite a remarkable result.

The second moment of intensity $\langle I^2 \rangle \equiv \langle c^4 \rangle$ is found similarly,

$$\langle I^2 \rangle = \frac{1}{2\pi\sigma^2} \int_{c=0}^{\infty} \int_{\phi=-\pi}^{\pi} c^5 \exp\left(\frac{-c^2}{2\sigma^2}\right) dc d\phi$$
$$= 8\sigma^4$$
$$= 2 \langle I \rangle^2 \tag{12.31}$$

The normalized variance for such a random distribution is then:

$$\text{nVar} = \frac{\langle I^2 \rangle}{\langle I \rangle^2} - 1$$
$$= \frac{2 \langle I \rangle^2}{\langle I \rangle^2} - 1$$
$$= 1 \tag{12.32}$$

Simulations for the CRN model in Figure 12.12 give an approximately constant line near nVar ≈ 1. The non-zero value is because of sampling noise in this model.

The mystery of the suppressed variance in our data is resolved when we include the effects of spatial incoherence. If we model the source as comprising m independent (i.e., mutually uncorrelated) sources, then the intensity distribution is no longer a simple negative exponential, but instead is the convolution of m independent negative exponentials. The result is the so-called gamma distribution,

$$P(I) = \frac{m^m}{(m-1)!} \frac{I^{m-1}}{\langle I \rangle^m} \exp\left(\frac{-Im}{\langle I \rangle}\right) \tag{12.33}$$

The normalized variance is nVar $= 1/m$ for $m \geq 1$. This indicates that experimental hollow-cone data is acquired with an illumination that is equivalent to as many as $m = 2000$ mutually incoherent sources. The true number is likely to be less than this, since inelastic scattering, and multiple scattering in thicker samples, will also play a role in suppressing the variance.

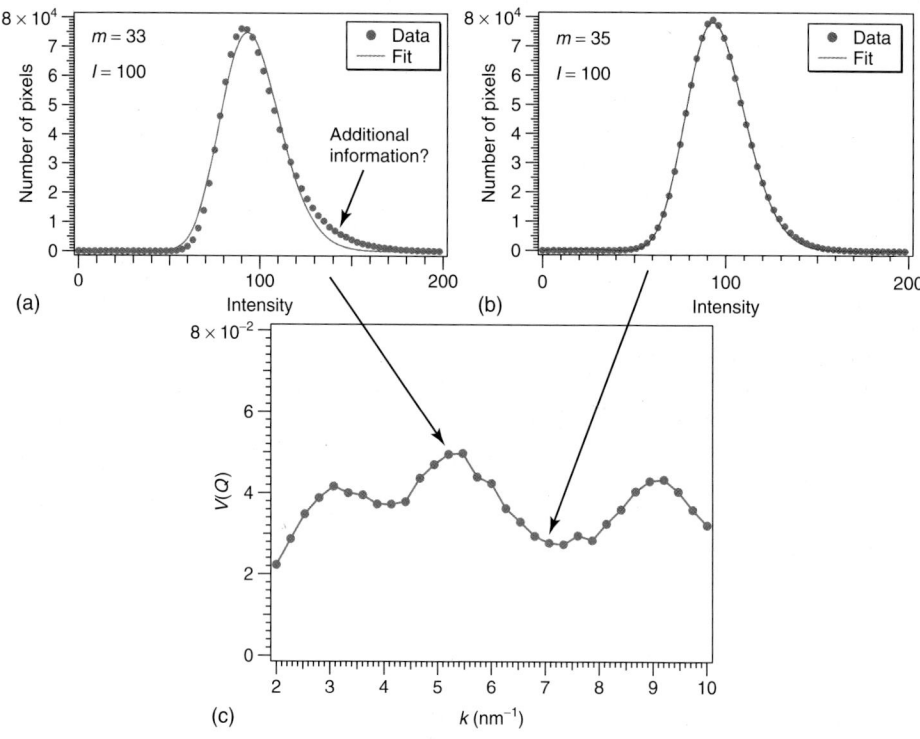

Figure 12.13 (a) Histogram of the speckle intensity distribution in the amorphous carbon film shown in Figure 12.9 at scattering vector $k = 5.5$ nm^{-1}. (b) At scattering vector $k = 7.0$ nm^{-1}. The data fits the gamma distribution very well for $m = 35$. (c) The normalized variance plot $V(k)$ for amorphous carbon. The speckle intensity distribution deviates slightly from the gamma distribution at the peaks in the $V(k)$ because additional structural information is present.

However, in tilted dark field, low values of m can be obtained even with fully coherent illumination. The statistics for the amorphous carbon speckle shown in Figure 12.9 are presented in Figure 12.13 for two experimental tilt conditions. The fit to a gamma distribution is impressive, finding a value for m in the range 33–35. For tilt vector $k = 7.0$ nm^{-1}, there is a dip in the normalized variance curve, and the fit to the gamma distribution is excellent, suggesting that the amorphous carbon speckle under these scattering conditions is indistinguishable from that from a random structure (i.e., a CRN). However, at a tilt vector $k = 5.5$ nm^{-1} the fit to a gamma distribution shows small differences. This is consistent with the presence of nonrandom structural information. We do not yet know how to extract this additional information directly from such plots.

The gamma distribution has the peculiar property that, for a given m, and maintaining I constant, it maximizes the Shannon informational entropy $-\int P \ln(P) dI$.

The excellent fit at $k = 7.0 \text{ nm}^{-1}$ is telling us that the speckle distribution under those conditions is free of additional information – that is, no significant MRO is present in this particular sample.

The mean intensity in these images is $\langle I \rangle = 100$, which suggests that shot noise is a strong contributor to the speckle in these images, in addition to the structural speckle. The statistics of the shot noise will resemble the gamma distribution. However, shot noise cannot be the source of the shoulders in these plots, which appears to be structural in origin.

It is not yet clear what the parameter m means in real experimental data, particularly since fully coherent illumination should give $m = 1$. In an ideal experiment with fully coherent illumination, we should have $m = 1$. Instead, most such experiments give $m > 10$. In an ideal experiment, m is equivalent to the number of incoherent sources that illuminate the sample. More likely, m is a measure of decoherence in the scattering, arising from beam damage and energy-loss events. However, in practice it includes a contribution from shot noise and possibly beam damage. This is an important issue to resolve if FEM is to become a quantitative technique.

For those interested in finding out more about fluctuation microscopy, there is an excellent online bibliography maintained by Prof. Paul M. Voyles at the University of Madison-Wisconsin [21].

12.10
Possible Future Directions for Electron Speckle Analysis

This chapter presents a tutorial-style overview of the subject of speckle in electron microscopy. It is not a comprehensive review, and I have presented it in such a way that fluctuation microscopy is the "climax," since this technique actively exploits speckle for the study of amorphous materials. This is not a closed subject, and is still an actively researched field. It is clear that important developments are yet to come, particularly in the area of quantification. I present here a short summary of areas that still need work.

- **The inversion problem.** There is no mathematical or algorithmic tool at present that can invert the four-body data in the normalized variance. However, the ability to invert the two-body diffraction data is now well established. A promising approach is to use Monte Carlo-type methods, which involve random jostling of the model's atomic coordinates, and selecting those moves that improve the fit with data. This works well with diffraction data. By incorporating variance data, in conjunction with diffraction data (i.e., the mean data), such Monte Carlo methods may be steered toward models that exhibit medium-range order. This approach offers much promise and is being actively pursued by several groups at present [22, 23]. As with all Monte Carlo methods, the data constraints (the number of points in the plots) are usually vastly outweighed by the degrees of freedom (i.e., the $3N$ atomic x_i; y_i; z_i coordinates for the $i = 1$; N atoms in a

model). Solutions tend not to be unique, and caution is needed when interpreting data. Solutions can be greatly improved by adding a realistic atomic potential to the constraints, limiting the space of solutions to only those structures that make sense chemically.
- **Data blindness**. By focusing on the mean and variance only, much of the data available is being ignored. For example, the third and fourth moments of the intensity distribution (the skew and kurtosis) tell us about the outlying distribution of intensities. In addition, spatial correlations (that is, broad speckle patches) are also ignored. What could the statistics of speckle dimensions tell us?
- **Could all this be done more simply?** Is diffraction intensity variance the best way to access the higher order correlations?
- **How to display multidimensional data.** We know what a two-body radial distribution function looks like. What does the four-body distribution function look like? Do we need to know the full details? Is a pair–pair function, in separation and relative orientation, sufficient? Is it still too complicated?
- **The eternal promises of high resolution.** Will exit wavefront reconstruction in aberration-corrected microscopes be sufficient to resolve MRO? How thick can the samples be before the projection becomes indecipherable?
- **Tomography.** Will atomic-resolution electron tomography offer the breakthrough for structural studies of amorphous materials? For example, can we map the distribution of ordered regions within a sample? Will the coherent speckle confuse the back-projection algorithms?
- **Phase-retrieval methods.** Potentially, phase retrieval offers even higher resolution than aberration-corrected images. This is because no image is formed, and the (axial) aberrations only modify the phase, and not the position of the diffracted rays. (Off-axis aberrations such as barrel distortion and coma do shift the diffraction data.) Shot noise will be an important limiting factor here. However, despite the difficulties, this is a very exciting area to work in these days.
- **Speckle intensity histograms**. Why do gamma distributions fit the speckle intensity histograms so well for CRN-type materials? Is there a deeper meaning to the parameter m? Intensity histograms must contain more information than the mean and variance alone (they are just two numbers). Is the additional information in the histogram useful? Does beam damage, and the associated temporal twinkling, contribute to an effective m-value? Is an ideal $m = 1$ experiment possible?

The study of amorphous materials and disordered systems is still a "hot" topic, albeit a difficult one. It is not usually possible to interpret micrographs and diffraction patterns of disordered materials directly. Instead, there are additional layers of statistical interpretation necessary. Amorphous materials are intrinsically much harder to study than crystals because, unlike crystals, every atom has a unique structural environment. Their study by diffraction inevitably includes a study of speckle. It is hoped that this tutorial has given the reader insights into this interesting area of research.

References

1. Dainty, J.C. (1975) in *Laser Speckle and Related Phenomena* (ed. J.C. Dainty), Springer-Verlag, New York, pp. 255–280.
2. Goodman, J.W. (1975) in *Laser Speckle and Related Phenomena* (ed. J.C. Dainty), Springer-Verlag, New York, pp. 60–68.
3. McAlister, H.A. (1985) High angular resolution measurements of stellar properties. *Ann. Rev. Astron. Astrophys.*, **23**, 59–87.
4. Treacy, M.M.J., Gibson, J.M., Fan, L., Paterson, D.J., and McNulty, I. (2005) Fluctuation microscopy: a probe of medium range order. *Rep. Prog. Phys.*, **68** (12), 2899–2944.
5. Treacy, M.M.J. and Gibson, J.M. (1996) Variable coherence microscopy: a rich source of structural information from disordered materials. *Acta Crystallogr. A*, **52**, 212.
6. Gibson, J.M., Treacy, M.M.J., and Voyles, P.M. (2000) Atom pair persistence in disordered materials from fluctuation microscopy. *Ultramicroscopy*, **83**, 169–178.
7. Voyles, P.M. and Abelson, J.R. (2003) Medium-range order in amorphous silicon measured by fluctuation electron microscopy. *Sol. Energy Mater. Sol. Cells*, **78**, 85–113.
8. Treacy, M.M.J. (2005) What is fluctuation microscopy? *Microsc. Today*, **13** (5), 20–21.
9. Bogle, S.N., Voyles, P.M., Khare, S.V., and Abelson, J.R. (2007) Quantifying nanoscale order in amorphous materials: simulating fluctuation electron microscopy of amorphous silicon. *J. Phys.-Condens. Matter*, **19** (45), 455204.
10. Wolf, E. (2007) *Introduction to the Theory of Coherence and Polarization of Light*, Cambridge University Press, Cambridge.
11. Treacy, M.M.J. and Gibson, J.M. (1993) Coherence and multiple scattering in "Z"-contrast images. *Ultramicroscopy*, **52**, 31.
12. Krakow, W. and Howland, L.A. (1976) A method for producing hollow cone illumination electronically in the conventional transmission microscope. *Ultramicroscopy*, **2**, 53–67.
13. Gibson, J.M. and Treacy, M.M.J. (1997) Diminished medium-range order observed in annealed amorphous germanium. *Phys. Rev. Lett.*, **78**, 1074.
14. Graczyk, J.F. and Moss, S.C. (1969) Scanning electron diffraction attachment with electron energy filtering. *Rev. Sci. Instrum.*, **40**, 424433.
15. Moss, S.C. and Graczyk, J.F. (1969) Evidence of voids within as-deposited structure of glassy silicon. *Phys. Rev. Lett.*, **23**, 1167–1171.
16. Cockayne, D.J.H. and McKenzie, D.R. (1988) Electron- diffraction analysis of polycrystalline and amorphous thin-films. *Acta Crystallogr.*, **A44**, 870878.
17. Cockayne, D.J.H. (2007) The study of nanovolumes of amorphous materials using electron scattering. *Annu. Rev. Mater. Res.*, **37**, 159–187.
18. Zhao, G., Buseck, P.R., Rougée, A., and Treacy, M.M.J. (2009) Medium-range order in molecular materials: fluctuation electron microscopy for detecting fullerenes in disordered carbons. *Ultramicroscopy*, **109**, 177–188.
19. Zhao, G., Treacy, M.M.J., and Buseck, P.R. (2010) Fluctuation electron microscopy of medium- range order in ion-irradiated zircon. *Philos. Mag.*, **90**, 4661–4677.
20. Gibson, J.M. and Treacy, M.M.J. (1998) Paracystallites found in evaporated amorphous tetrahedral semiconductors. *J. Non-Cryst. Solids*, **231**, 99–110.
21. Voyles, P.M. Introduction to Fluctuation Microscopy, http://tem.msae.wisc.edu/FEM/index.html (accessed 6th Jan, 2010).
22. Biswas, P., Tafen, D., and Drabold, D.A. (2005) Experimentally constrained molecular relaxation: the case of GeSe$_2$. *Phys. Rev. B*, **71**, 054204.
23. Hwang, J., Clausen, A.M., Cao, H., and Voyles, P.M. (2009) Reverse monte carlo structural model for a zirconium-based metallic glass incorporating fluctuation microscopy medium-range order data. *J. Mater. Res.*, **24**, 3121–3129.

13
Coherent Electron Diffractive Imaging
J.M. Zuo and Weijie Huang

13.1
Introduction

This chapter describes the principles of coherent electron diffractive imaging and its realizations. In the broadest definition, diffractive imaging refers to the use of diffraction intensity for imaging without the benefit of interference that can occur after the sample as in high-resolution electron microscopy (HREM) or electron holography. In both these cases, interference of diffracted beams or different parts of a beam is used to detect internal atomic structure or potential. It should be said that diffractive imaging is not a new concept. Crystallographers routinely image atoms in 3D molecules by solving the inversion problem of crystal diffraction as long as they can be crystallized. In scanning transmission electron microscopy (STEM), diffraction intensity is integrated and mapped based on the probe position using the bright-field and annular dark-field (ADF) detectors. Another successful example of diffractive imaging is the electron backscattering diffraction (EBSD) used for microstructure mapping in scanning electron microscopes. The technique employed in EBSD involves diffraction pattern processing and recognition that goes beyond recording of diffraction intensities. Scanning electron nanodiffraction (SEND) is the TEM version of EBSD, but with much higher spatial resolution. SEND has been reviewed before [1]. The focus of this chapter is on electron coherent diffractive imaging (CDI) based on inversion of electron diffraction patterns recorded from isolated objects or areas of continuous sample isolated by nanometer-sized apertures.

Electrons have short wavelengths, interact strongly with matter, and can be focused using electromagnetic lenses. These properties make electrons a useful complementary probe to X-rays or neutrons. Compared with electron imaging, electron diffraction provides reciprocal space information and can be used to overcome limitations of electron direct imaging. The use of HREM imaging for quantitative structure determination is complicated by the interpretation of image contrast. Only in extremely thin samples of light elements can the HREM image contrast be related directly to the sample's projected potential. For samples of reasonable thicknesses, the image contrast is a mixture of complex exit wave function and phases introduced by the imaging lens; image interpretation in general requires modeling of the electron scattering process and the properties of the electron imaging lens [2]. In STEM, the image contrast is less sensitive to focus when an ADF detector is used. Nonetheless, a proper interpretation of ADF-STEM image contrast also requires modeling of the electron scattering process (Chapter 2). The resolution of STEM imaging is limited by the probe size, scan distortions, and scan noises in a STEM. For organic materials susceptible to radiation damage, both HREM and STEM imaging are often not an option because the amount of electron dose required to produce a sufficient image contrast can be larger than the material's radiation damage threshold. Electron diffraction, on the other hand, can work at low-dose situations by averaging over many unit cells for crystals. Certain structural information, such as lattice spacing and crystal orientations, is also easier to obtain from diffraction patterns than from images. Information about the crystal orientation, unit cell dimensions, and sample thickness can be obtained from convergent beam electron diffraction (CBED) patterns using the well-established techniques [3].

Imaging, in principle, can be achieved by inverting the recorded diffraction patterns. The benefits for doing this will become clear, as this chapter shows. Inversion of electron diffraction patterns requires the solution of the so-called phase problem. The diffraction pattern records the intensity of the Fourier transform of the exit wave function, not the phase. Inverse Fourier transform thus requires the phase missing in the diffraction pattern. This is known as the *phase problem in diffraction*. Critical to the inversion of diffraction patterns is to find the phases of the diffracted waves. In crystallography, the phase problem is solved based on a *priori* information about the crystal structure. The a *priori* information includes the sharply peaked atomic charge density and the periodicity of the crystal. An attempt to solve the phase problem directly in electron diffraction was to use interference between diffraction discs in electron ptychography [4]. The concept of ptychography was first proposed by Hoppe [5] and then further developed by Rodenburg [4]. In the original ptychography, electron diffraction patterns are recorded over an area of a crystal using a coherent probe with a diameter less than the size of the crystal unit cell, and the diffraction intensity at the middle of the overlapping disks is processed as a function of the probe position to form atomic resolution images [4]. For imaging, we must consider objects that are not perfect, infinite crystals.

The ability to invert diffraction patterns to form images has attracted considerable interest recently in the X-ray diffraction community, where the lack of high-resolution imaging lens has been a major obstacle toward X-ray imaging. In electron diffraction, the additional phase introduced by the imaging lens aberrations does not affect the diffraction intensity, and diffractive imaging by solving the phase problem provides atomic-resolution imaging at diffraction-limited resolution. The inversion of electron diffraction patterns of nanometer-sized objects is helped by the fact that the small object leads to broadened diffraction peaks; in the case of coherent electron diffraction, the broadening gives additional diffraction information and under not so restricted conditions can lead to inversion of diffraction patterns [6].

This chapter is organized in five sections plus a conclusion. After this introduction, Section 13.2 covers coherent electron diffraction techniques for nanometer-sized objects. Section 13.2 is followed by a description of the noncrystallographic phase problem in Section 13.3, which this chapter aims to solve. Section 13.4 describes different iterative transformation algorithms (ITAs) for solving the noncrystallographic phase problem, their requirements, and testing. Section 13.5 uses quantum dots as an example to demonstrate the phasing of experimental electron diffraction patterns and introduces techniques for achieving this.

13.2
Coherent Nanoarea Electron Diffraction

There are two approaches to coherent nanoarea electron diffraction (NED): one is based on a nanometer-sized coherent electron probe where the probe size controls the volume of the sample for diffraction and the other is based on the use of an aperture placed at the image plane of the objective lens (OL) to select diffracted beams from a small sample area for selected NED.

A setup for a small coherent electron beam is shown in Figure 13.1. It consists of a field emission electron gun, an illumination system with four condenser lenses, including the objective prefield and a third condenser lens or minilens. The field emission gun (FEG) provides the source brightness and lateral coherence required for coherent diffraction. The three magnetic lenses setup (condenser I, II, and the objective prefield) for the illumination system is common in modern transmision electron microscopes (TEMs). Condenser I is a demagnifying lens that is used to reduce the effective electron source size for improved lateral coherence. The small, parallel beam is achieved by reducing the convergence angle of the condenser II

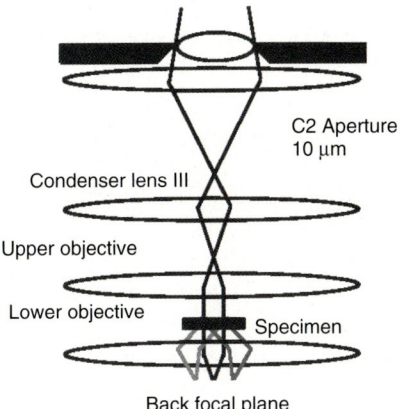

Figure 13.1 Schematic diagram of an electron illumination system for coherent nanoarea electron diffraction.

crossover using the additional condenser III and placing the crossover at the focal plane of the objective prefield, whereby it forms a parallel beam illumination on the sample. For a condenser aperture of 10 μm in diameter, the probe diameter is ~50 nm with an overall magnification factor of 1/200 in the JEOL 2010F electron microscope (JEOL, USA). The beam size can be further reduced with additional magnetic lenses in the illumination system, as demonstrated by Wen et al. using a microscope equipped with a probe C_s corrector [7]. All electrons illuminating the sample in NED are recorded in the diffraction pattern. NED in an FEG microscope thus can be used to provide higher beam intensity for electron nanodiffraction (the probe current intensity using a 10 μm condenser II aperture in JEOL 2010F is ~10^5 e s^{-1} nm^2) [8].

The electron probe formed using the above-described illumination system can be described using the same principle of the STEM probe formation. In STEM, the electron beam crossover from the last condenser lens is imaged by the OL. The difference between the OLs' focal length and the focal length required to image the crossover onto the sample is defined as the defocus. For a convergent beam of electrons, the lens aberrations introduce angle-dependent phases, $\chi(\mathbf{K})$, with $\mathbf{K}$ standing for the part of the incident beam wave vector perpendicular to the optical axis. The phase $\chi(\mathbf{K})$ from the OL aberrations and its relation to the focused electron probe used in STEM is described in Chapter 4. For coherent electron nanodiffraction using a defocused probe, we must also consider the electron source wave function $\phi_S(\mathbf{R})$ formed by the condenser lens and its contribution to the electron probe. According to the image formation theory, the electron probe on the sample is an image of $\phi_S(\mathbf{R})$ magnified by the lens magnification M. The image is a convolution of $\phi_S(\mathbf{R})$ and the OL resolution function $T(\mathbf{R})$:

$$\phi_P(\mathbf{R}) = \phi_S(-\mathbf{R}/M) \otimes T(\mathbf{R})$$
$$= \int_{-\infty}^{\infty} \phi_S(-M\mathbf{K}) A(\mathbf{K}) \exp[i\chi(\mathbf{K})] \exp(2\pi i \mathbf{K} \cdot \mathbf{R}) d\mathbf{K} \quad (13.1)$$

where $A(\mathbf{K})$ is the aperture function with a value of 1 for $|\mathbf{K}| < \Theta/\lambda$ and 0 beyond with Θ standing for the beam convergence angle. The electron beam energy spread and the chromatic aberration are neglected in Eq. (13.1). Equation (13.1) also assumes that the illuminating electron wave is perfectly coherent across the condenser aperture.

In the NED mode, the electron beam crossover is placed close to, or at, the front focal plane. The electron source in this case is magnified ($M \gg 1$). The sample is also placed away from the electron source image after the OL (the image plane) near the back focal plane of the OL. This large underfocus must be included as a part of the lens aberration function in Eq. (13.1) in order to simulate the electron probe in NED [8]. The probe magnification is used to reduce the electron beam convergence angle for the parallel beam diffraction. To demonstrate this, we assume a Gaussian distribution for the electron source:

$$\phi_S(\mathbf{R}/M) = A \exp(-a^2 R^2/M^2) \quad (13.2)$$

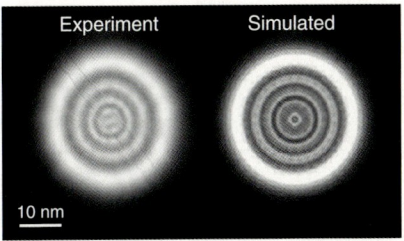

Figure 13.2 Experimental and simulated electron nanoprobe used in nanoarea electron diffraction (NED). The simulation used $C_s = 1$ mm and $\Delta f = -360$ nm.

where a is one over the probe half width at A/e. The Fourier transform of this Gaussian probe after OL is

$$\phi_S(\mathbf{K}) = \frac{A\sqrt{\pi}}{a} \exp\left[-K^2/(a/M\pi)^2\right] \quad (13.3)$$

The width of the beam in the reciprocal space is reduced by a factor of $1/M$. The source function in NED with its large probe magnification leads to a reduced electron beam convergence angle. The Gaussian half width of the defocused electron beam formed using a 10 μm condenser aperture is ~0.05 mrad in the JEOL2010F TEM. We note that the real space probe in NED is a convolution of the magnified source with $T(\mathbf{R})$. The dominant probe features come from $T(\mathbf{R})$, as shown in Figure 13.2, for a comparison between an experimental probe and simulation based on $T(\mathbf{R})$ alone [8].

A selected area aperture is used to define a small area of diffraction in selected-area electron diffraction (SAED) (Figure 13.3). The electron illumination is spread out over a large area of sample for a parallel beam in this case. In conventional TEM with a large OL spherical aberration (C_s), the area defined by the aperture shifts at the sample proportional to C_s and diffraction angle at a power of 3, as illustrated in Figure 13.3. For $C_s = 1$ mm and a diffraction angle of 50 mrad, the displacement is 125 nm. This makes SAED not so useful for electron nanodiffraction in conventional TEM. Recently, Yamasaki and his colleagues at Nagoya University demonstrated that with a TEM aberration corrector, electron diffraction patterns can be recorded from areas of ~10 nm.

The area of diffraction in SAED performed in an aberration-corrected TEM is defined by the selected area aperture. Using this method, a small area can be selected from a continuous sample with well-defined boundaries using a sharp aperture at the resolution of an aberration-corrected TEM. An electron beam with sharp edges can also be formed by imaging the condenser aperture on to the sample using electron lenses in the illumination system. This has been demonstrated in TEM equipped with a probe corrector using the transfer lenses in the probe corrector for aperture imaging [7, 9]. The electron beam, as shown in Figure 13.2, provides a parallel beam for high resolution in electron diffraction but lacks the edge sharpness, which works better with isolated nanostructures. Figure 13.4 shows an example of electron diffraction of a single carbon nanotube of 1.49 nm

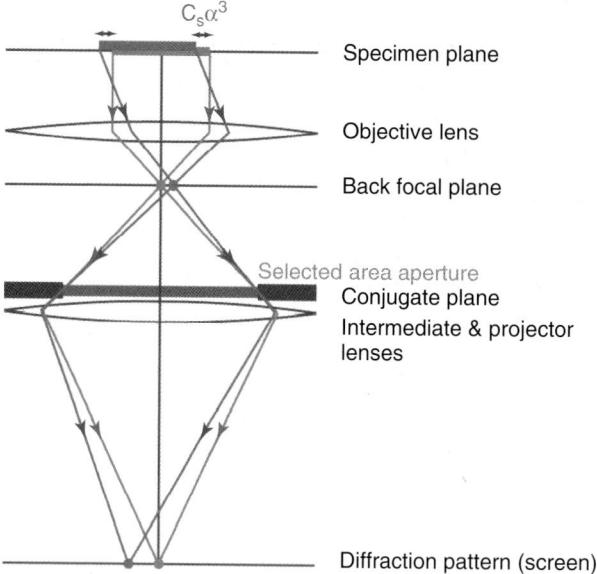

Figure 13.3 A schematic illustration of the displacement of the selected area on the sample caused by the spherical aberration of the objective lens. (The figure was provided by Jun Yamasaki of Nagoya University, Japan).

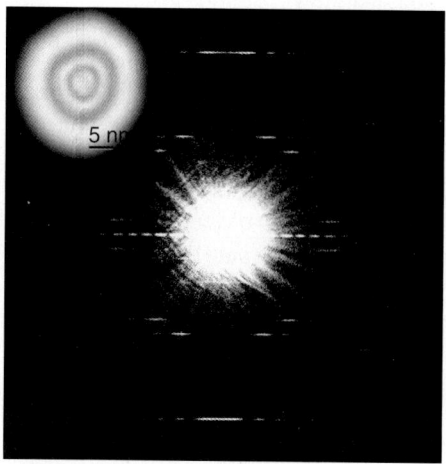

Figure 13.4 Nanoarea electron diffraction pattern from a single-walled carbon nanotube 1.49 nm in diameter with encapsulated C60 molecules. The pattern was recorded using 80 keV electrons. The inset at the top left corner is the electron beam used to record the diffraction pattern. The nanotube is visible in the beam image. The small probe size was used to isolate a single tube for diffraction. (The figure was provided by Ke Ran of University of Illinois, USA).

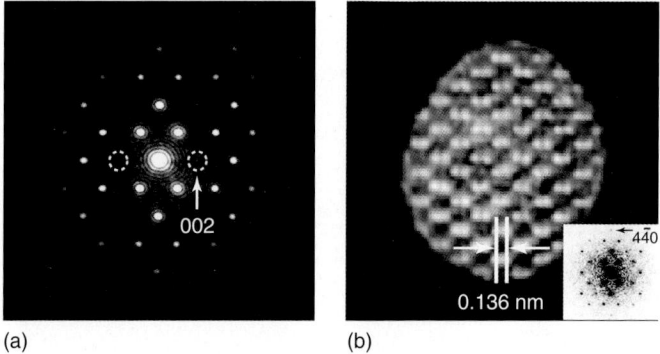

Figure 13.5 Selected area nanodiffraction using aberration-corrected TEM, (a) recorded diffraction pattern and (b) reconstructed image (amplitude). (Figure from Ref. [10]).

in diameter with a chiral vector of (13,8). The nanotube was supported on holey carbon films. The diffraction pattern was recorded from a section of the tube over a hole in the carbon film. For comparison, Figure 13.5 shows the electron diffraction pattern and the selected sample area obtained using an aberration-corrected TEM by Morishita et al. [10].

13.3
The Noncrystallographic Phase Problem

The scattered waves at the back focal plane of the OL satisfy the so-called Fraunhofer diffraction condition [11] and can be described as the Fourier transform of the exit wave function $\psi_{\text{exit}}(\mathbf{r})$:

$$\Psi(\mathbf{k}) = \int \psi_{\text{exit}}(\mathbf{r}) \exp(-i\mathbf{k} \cdot \mathbf{r}) d^3\mathbf{r} = |\Psi(\mathbf{k})| \exp(-i\varphi(\mathbf{k})) \qquad (13.4)$$

where $\mathbf{k}$ is the wave vector of the scattered wave and $\varphi(\mathbf{k})$ is the phase of the complex amplitude of $\Psi(\mathbf{k})$. For small nanostructures, such as carbon nanotubes, electron diffraction is well described by the kinematical approximation. At a given scattering angle, the scattered electron wave is a sum of the scattered waves over the volume of the structure:

$$\psi_{\text{exit}}(\mathbf{k}) \approx \int \left[1 + i\pi\lambda U(\mathbf{r}')\right] e^{-2\pi i \mathbf{k} \cdot \mathbf{r}'} \varphi_0(\mathbf{r}') d\mathbf{r}'$$

$$= \varphi_0(\mathbf{k}) + i\pi\lambda \int U(\mathbf{r}') e^{-2\pi i \mathbf{k} \cdot \mathbf{r}'} \varphi_0(\mathbf{r}') d\mathbf{r}' \qquad (13.5)$$

Here $U(\mathbf{r}) = 2m|e|V(\mathbf{r})/h^2$ with $V(\mathbf{r})$ as the Coulomb potential of the sample, $\mathbf{k}$ is the scattered wave vector. The illuminating electron wave function $\varphi_0(\mathbf{r})$ is formed by the electron lens, as described in the previous section.

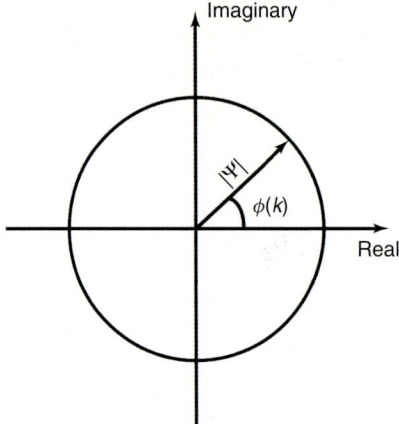

Figure 13.6 The complex structure factor $\Psi(\mathbf{k})$ can be represented on an Argand diagram by a vector in the complex plane of length $|\Psi(\mathbf{k})|$ and having an angle with the real axis of $\varphi(\mathbf{k})$.

In the Argand diagram of Figure 13.6, the complex wave is represented by a vector in the complex plane with a length of $|\Psi(\mathbf{k})|$ and at an angle of $\varphi(\mathbf{k})$ to the real axis. The exit wave function $\psi_{exit}(\mathbf{r})$ can be reconstructed by inverse Fourier transform of $\Psi(\mathbf{k})$. Experimentally, however, one can only measure the length of the complex vector ($|\Psi(\mathbf{k})|$), while the phase angle $\varphi(\mathbf{k})$ cannot be measured directly from the diffraction pattern. This is known as the *phase problem in crystallography*. The phase problem thus prevents one from direct inversion of diffraction using Fourier transformation.

The phase, in principle, can be measured by holography. In holography, a reference wave is used to interfere with the object wave, which gives a set of interference fringes. The maxima of the fringes are locations where the phases of the scattered wave match that of the reference wave; therefore, the phase of the scattered wave can be measured from the intensities recorded in the interference pattern. When the hologram is illuminated with the same reference wave, one can reconstruct the original object that yielded the scattered wave by a backward propagation. The reillumination stage is equivalent to the inverse Fourier transformation of the hologram. Lichte *et al.* have shown experimentally that in off-axis electron holography, the complete information about amplitude and the phase of the electron exit wave can be reconstructed numerically from a single hologram [12, 13]. The reference wave is created by splitting the illumination using an electron biprism in imaging. The same experiment in diffraction requires diffractive waves from an aperture. This has been demonstrated in the case of soft X-ray diffraction [14]. In electron diffraction, such experiments can be done using selected area diffraction.

There are a number of established crystallographic methods to solve the phase problems for crystals. For inorganic crystals and organic molecules with a small number of atoms, direct methods [15, 16] are widely used. Direct methods are a group of *ab initio* phase determination techniques based on mathematical procedures that compare structure factor amplitudes derived from a single crystal. For example, by using a statistically correct phase relation proposed by Karle and Karle [15, 16], $\varphi(\mathbf{h}) \approx \varphi(\mathbf{k}) + \varphi(\mathbf{h} - \mathbf{k})$, one can obtain the phase of **h** from the phases of **k** and **h-k**. For macromolecules such as proteins, the large number of atoms (of the order of $10^2 - 10^5$) makes deriving structures using direct methods computationally prohibitive. Alternatively, methods have been developed based on atomic replacement using chemically modified molecules, which are more efficient for macromolecular phasing.

Solving crystal structure by direct methods or other crystallographic methods requires the preparation of a crystalline specimen. Many biologically important macromolecules, such as viruses and cells, cannot be crystallized. Unlike crystals, diffraction of noncrystalline materials gives broad peaks and a continuous background. The difficulty in crystallizing nonperiodic structures prevents the structure determination at the atomic resolution through conventional crystallography methods. Overcoming this difficulty requires the solution of the phase problem for nonperiodic structures or the so-called noncrystallographic phase problem.

13.4
Coherent Diffractive Imaging of Finite Objects

The motivation to realize diffractive imaging using electrons is different from X-rays. It is difficult to obtain a direct image with X-rays because of the lack of the high-resolution X-ray lens. In an electron microscope, direct images can be readily formed. However, the imaging resolution is limited by the aberrations of the magnetic OL. In conventional TEM operated at 200 kV with a high-resolution pole piece, the resolution is typically ~2 Å. Aberration correction has led to significant improvement in image resolution [17–19]. However, information transfer in aberration-corrected TEM is still limited by the chromatic aberration and the beam convergence, which in combination impose a damping envelope function on the contrast transfer function (CTF) for the image formation. The CTF is also low for most spatial frequencies (Figure 13.7) compared to the 100% amplitude information transfer in diffraction except in diffraction intensities below the background noise level. Information transfer in the diffraction pattern, thus, is much more efficient. If one can utilize the high-frequency information present in the diffraction pattern through diffractive imaging, much higher resolution can be achieved for imaging.

A number of experimental methods have been proposed to image nonperiodic structures from diffraction patterns. Gabor [21] first proposed a two-stage imaging process. In the first stage, a diffraction pattern of the specimen is recorded on a photographic plate using a divergent electron beam emerging from a point source. The diffraction pattern recorded this way is essentially a hologram (since scattered and unscattered beams overlap and interfere) and thus carries both the phase and amplitude of the electron wavefront. In the second stage, the plate is

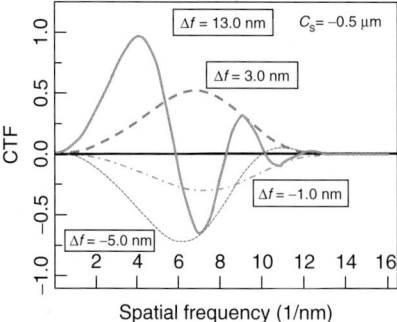

Figure 13.7 CTF of aberration-corrected TEM. Parameter used for the CTF calculations are wavelength 1.97 pm, $C_s = -0.5$ mm, and semi-angle of beam convergence 0.2 mrad. (After Zaoli Zhang and Ute Kaiser, *Ultramicroscopy*, **109**, 1114 (2009), Ref. [20]).

illuminated using visible light and the electron wavefront that emerges from the specimen is reconstructed using optical lenses. Since the spherical aberrations of the optical lenses are easier to correct than electron lenses, the image resolution could be improved in the second stage. However, atomic resolution has never been achieved using Gabor's idea because of a number of technical difficulties: first, an ideal point source is very hard to realize; second, the electron beam that Gabor used had limited coherence; therefore, not enough inference was formed to carry the phase information. More recently, Spence and his coworkers experimented with a field emission point source for point projection microscopy [22, 23].

Recent breakthroughs in CDI come from the convergence of several ideas that has led to a working solution to the noncrystallographic phase problem.

13.4.1
Brief History of Coherent Diffractive Imaging

Sayre [24] observed in 1952 that Bragg diffraction under-samples diffracted intensity relative to Shannon's theorem. In the early 1980s, the development of ITAs, especially the introduction of feedback by Fienup [25], led to a remarkably search-based method for the inversion of diffraction intensity data. The parallel development was the iterative real-space algorithm in protein crystallography using solvent flattening and noncrystallographic symmetry for phase recovery or phase improvements [26]. In 1998, Sayre et al. [27] took the important step of combining the Fienup-type iterative algorithm with the idea of reconstructing the diffraction pattern. In 1999, Miao et al. [28] reported the first successful reconstruction of a microfabricated object. In their work, they achieved a reconstruction of an array of gold dots at a resolution of 75 nm using 1.7 nm wavelength X-rays. Their success was followed by a number of theoretical insights and successful experimental efforts from a number of research groups. Robinson et al. [29] used hard X-ray (1.035 Å wavelength) from the advanced photon source in the Argonne National Laboratory to record coherent diffraction patterns from Au nanocrystals of a few hundreds of nanometers in diameter. From the diffraction data, they successfully reconstructed the shape of the nanocrystal. Later on, Williams et al. extended Robinson's work to mapping the three-dimensional (3D) intensity distribution of a Bragg peak, and successfully phased the 3D diffraction data to reveal the internal strain of the Au nanocrystal. Theoretical insight into ITAs was obtained through the realization that iterations can be viewed as Bregman projections in Hilbert space; this insight has led to the systematic categorization of different iterative algorithms and the development of new algorithms [30, 31].

13.4.2
Oversampling

The idea of oversampling is based on the information theory for sampling a continuous, but finite object, which was first formulated by Harry Nyquist in 1928 [32] and further developed by Claude E. Shannon in 1949 [33]. The Nyquist–Shannon theorem states that [33]: "if a function $f(x)$ vanishes outside the points $x = \pm a/2$, then its Fourier transform $F(k)$ is completely specified by the values that it assumes at the points $k = 0, \pm 1/a, \pm 2/a, \ldots$" The minimum sampling frequency of $1/a$ is called the *Nyquist frequency*.

Discrete sampling is used in digital representation of the object or diffraction patterns, where a continuous object of dimension a is approximated by discrete points denoted by $x = 0, \ldots, N-1$ and the Fourier transform is carried out via summation for a set of discrete frequencies denoted by k:

$$F(k) = \sum_{x=0}^{N-1} f(x) \exp(2\pi i k \cdot x/N) \tag{13.6}$$

$$f(x) = \frac{1}{N^2} \sum_{k=0}^{N-1} F(k) \exp(2\pi i k \cdot x/N) \tag{13.7}$$

where Eqs. (13.6) and (13.7) denote the forward and inverse Fourier transform, respectively. The smallest frequency in Fourier transform is $1/a$. The sampling relation is illustrated in Figure 13.8.

In a diffraction experiment, what is measured is $|F(k)|^2$ instead of $F(k)$. Inverse Fourier transforming $|F(k)|^2$ gives the autocorrelation function of $f(x)$:

$$f(x) \otimes f(x) = \text{FT}^{-1}\left\{|F(k)|^2\right\} \tag{13.8}$$

where FT^{-1} denotes inverse Fourier transform and $\otimes$ denotes convolution. The autocorrelation function $f(x) \otimes f(x)$ has exactly twice the dimension of the original function $f(x)$; for example, if $f(x)$ has a dimension of a, its autocorrelation function then has a dimension of $2a$. To obtain the complete autocorrelation function, one must sample the diffraction pattern $|F(k)|^2$ at an interval of $1/2a$. According to the nature of discrete Fourier transformation (Eqs. (13.6) and (13.7)), the corresponding object $f(x)$ should have a physical dimension of $2a$, meaning that the total field of

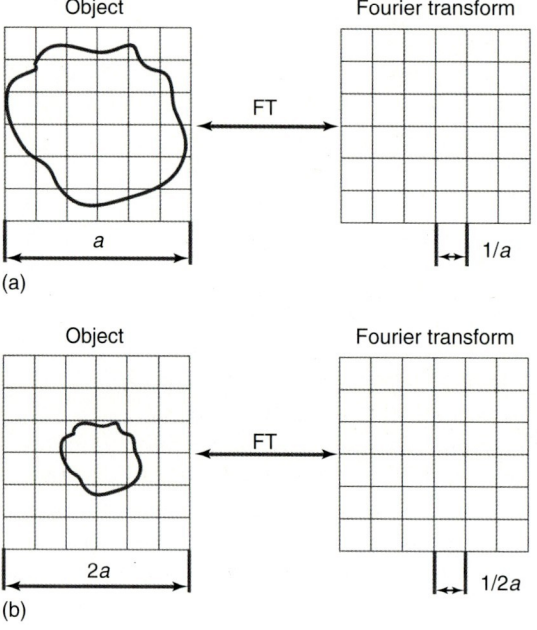

Figure 13.8 Sampling relation between an object and its discrete Fourier transform. (a) An object with a physical dimension of a is sampled in the Fourier space at an interval of $1/a$. (b) Sampling the Fourier space at an interval of $1/2a$ implies that the total field of view has a dimension of $2a$, resulting in a zero-padded object.

view should be twice the object size. Having a field of view larger than the actual dimension of the object is called *oversampling the object*.

The following argument advanced by Miao et al. [34] suggests that oversampling diffraction experiment gives the extra information that can be used to solve the phase problem. The inverse problem can be phrased as the following equation:

$$|F(k)| = \left| \sum_{x=0}^{N-1} f(x) \exp(2\pi i k \cdot x/N) \right| \qquad (13.9)$$

which is a set of equations, and the inverse problem is to solve these equations for $f(x)$ at each pixel. In the simplest case of a 1D real-valued object, the total number of equations in Eq. (13.9) is $N/2$ (*the factor of 2 comes from the symmetry of FT*), and the total number of unknown variables is N (all the pixels in the real space). Such a set of equations is mathematically underdetermined. However, if one oversamples the object by padding zero pixels around it and enlarging the field of view by a factor of 2, the number of unknown variables is then reduced to $N/2$ since *the other half of* the pixels are known to be zero. Therefore, by oversampling, the inverse problem can become mathematically overdetermined. The above argument also links the degree of oversampling with the overdetermination. Therefore, we can

13.4 Coherent Diffractive Imaging of Finite Objects

define an oversampling ratio σ:

$$\sigma = \frac{N_{\text{total}}}{N_{\text{object}}} \tag{13.10}$$

where N_{total} and N_{object} denote the number of pixels in the total field of view and in the object, respectively. The larger the σ, the more overdetermined is the phase problem.

13.4.3
Sampling Experimental Diffraction Patterns and the Field of View

Experimentally, the smallest measured frequency in the Fourier space is determined by the detection geometry. The first detection parameter is the camera length L, which is the equivalent distance between the object and the detector plane when diffraction patterns are recorded using TEM. The other parameters are the detector pixel size and the number of pixels. Figure 13.9 shows how to determine the sampling frequency in a diffraction experiment for a given detection geometry. In Figure 13.9, the diffraction pattern of an object with a lattice spacing of d is recorded at a distance of L away from the object. From the Bragg Law, we have

$$\lambda = 2d \sin \theta \simeq 2d\theta \tag{13.11}$$

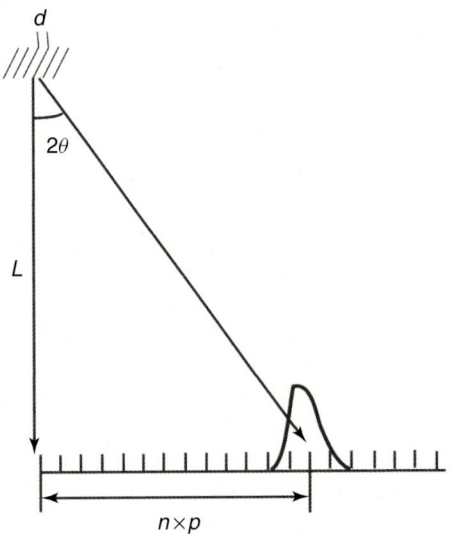

Figure 13.9 Camera length and sampling frequency in the detector plane. An object with a lattice spacing of d gives a Bragg reflection on the detector plane, which is a discrete array having a pixel size of p. The *camera length* is defined as the distance between the object and the detector plane.

where λ is the electron wavelength and 2θ is the angle between the diffracted and the forward beams. From the diffraction geometry, we also have

$$2\theta \simeq (n \cdot p)/L \tag{13.12}$$

where n is the number of pixels between the forward beam and the diffracted beam measured on the detector plane, and p is the physical pixel of the detector.

The combination of Eqs. (13.11) and (13.12) gives

$$\frac{1}{d} = \frac{np}{L\lambda} \tag{13.13}$$

This means that n pixels in the detector plane record a spatial frequency of $1/d$ and each pixel records a spatial frequency of $1/(nd)$, or

$$\text{Spatial frequency of each pixel} = \frac{p}{L\lambda} \tag{13.14}$$

The total field of view is the reciprocal of the spatial frequency of the pixel:

$$\text{Field of view} = \frac{L\lambda}{p} \tag{13.15}$$

For a 200 kV electron with a wavelength of 0.0251 Å, the camera length of 80 cm, and the pixel size is 50 μm, the total field of view according to Eq. (13.15) is 40 nm. The oversampling ratio can be calculated by dividing the total field of view by the size of the object for a perfectly coherent beam. A nanoparticle 10 nm in size, for example, has an oversampling ratio in one dimension of four under the above experimental conditions.

The minimum oversampling ratio required to solve the phase problem according to Miao et al. [34] is >2 for a 1D object, $>2^{1/2}$ in each dimension for a 2D square object, and $>2^{1/3}$ in each dimension for a 3D cubic object.

13.4.4
Requirements on Beam Coherence

Experimentally, the maximum oversampling is determined by the coherence length; the diffracting object must be fully contained in the coherence volume of the illuminating beam, which is defined by the coherence lengths in the lateral and temporal directions (Figure 13.10). The coherence length perpendicular to the electron beam, or the lateral coherence length, is related to the beam divergence angle from a finite source [35], according to the Van-Cittert-Zernike theorem:

$$X_c = \lambda/\theta_c \tag{13.16}$$

where X_c denotes the lateral coherence length and θ_c the beam divergent angle. The coherence length along the beam direction is measured by the temporal coherence length, which is determined by the monochromaticity of the beam [35]:

$$L_c = 2\lambda E/\delta E \tag{13.17}$$

where L_c denotes the temporal coherence length and δE the energy spread of the electrons. The temporal coherence length must be larger than the maximum

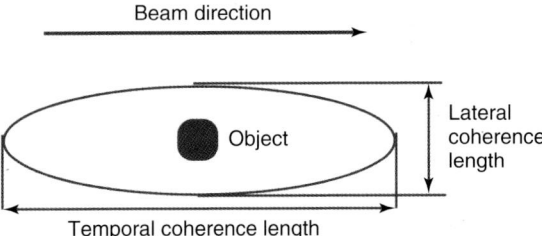

Figure 13.10 Schematic illustration of the coherence volume for electron diffraction. In order to form an oversampled diffraction pattern, the object must be contained within the coherence volume (solid oval). The dimensions of the coherence volume are characterized by its lateral and temporal coherence lengths.

path difference along the beam direction between any pair of interfering rays. The energy spread in a typical field emission microscope with a 200 kV accelerating voltage is about 1 eV. From Eq. (13.17), the temporal coherence length of such a beam is 1000 nm.

Partial coherence leads to a reduction of the oversampling ratio. In the ideal case, this has no effect as long as the oversampling ratio meets the minimum requirement. In practice, the combination of a reduced oversampling ratio and the noise in recorded experimental diffraction patterns leads to loss of information in reconstructed object function. This effect has been demonstrated by Huang et al. [36] in a simulation study.

13.4.5
Phase Retrieval Algorithms

Given that the phase problem is overdetermined by oversampling, the question becomes how to retrieve the phase from the recorded diffraction patterns and solve the inverse problem numerically. Several ITAs [25, 37–40] have been proposed to retrieve the missing phases from diffraction data. Common to all of these algorithms is the iteration between two domains, typically the real space and the reciprocal space, and the iterant is forced to satisfy what is known in each domain, called constraints in each domain. This iterative algorithm was first proposed by Gerchberg and Saxton [39]. The iteration continues until some error metric reaches a certain level.

Figure 13.11 illustrates a generic iterative phase retrieval algorithm. An estimate of the object function $g(\mathbf{x})$ is Fourier transformed into the Fourier domain, which yields $G(\mathbf{k})$:

$$G(\mathbf{k}) = FT\{g(\mathbf{x})\} \tag{13.18}$$

Generally, $G(\mathbf{k})$ are complex. The amplitudes of $G(\mathbf{k})$ are then made to satisfy the amplitude constraint by replacing the amplitudes with the experimentally

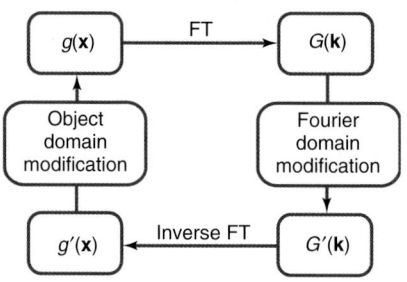

Figure 13.11 Flow chart of a generic iterative phase retrieval algorithm.

measured ones, $|F(\mathbf{k})|$, which leads to $G'(\mathbf{k})$:

$$G'(\mathbf{k}) = G(\mathbf{k}) \cdot \frac{|F(\mathbf{k})|}{|G(\mathbf{k})|} \qquad (13.19)$$

$G'(\mathbf{k})$ is then inversed Fourier transformed back to the object domain, giving $g'(\mathbf{x})$:

$$g'(\mathbf{x}) = \mathrm{FT}^{-1}\{G'(\mathbf{k})\} \qquad (13.20)$$

A set of constraints in the object domain is applied to modify $g'(\mathbf{x})$ into $g_{\text{new}}(\mathbf{x})$, which will then be fed into the next iteration cycle. The specific constraints, or modifications, change depending on the particular algorithm; below is a summary of the most popular algorithms:

1) **Gerchberg–Saxton (GS algorithm)**
   In GS, what is aimed to reconstruct is a complex wave field, and the algorithm assumes that the amplitudes are known in both the object and the Fourier domains [39]. Therefore, the last step in the loop is

$$g_{\text{new}} = g'(\mathbf{x}) \cdot \frac{|A(\mathbf{x})|}{|g'(\mathbf{x})|} \qquad (13.21)$$

   where $|A(\mathbf{x})|$ is the amplitude measured in the object domain.

2) **Error reduction (ER) algorithm**
   Fienup modified the GS algorithm to extend its application to situations where only the amplitudes in the reciprocal space are measured, such as in X-ray diffraction [25]. Instead of using the amplitude constraint in the object domain as the GS algorithm requires, he suggested using a support constraint:

$$g_{\text{new}}(\mathbf{x}) = g'(\mathbf{x}) \ \mathbf{x} \in S \quad \text{and} \quad g_{\text{new}}(\mathbf{x}) = 0 \ \mathbf{x} \notin S \qquad (13.22)$$

   where S denotes the support of the object, which is the region where the object has nonzero density. Equation (13.22) essentially applies the information gained by oversampling, which is the zero-valued region surrounding the support. For a real-valued object under kinematical diffraction, for example, two more constraints can be applied on top of the support constraint. They are the real constraint (Eq. (13.23)) and the positivity constraint (Eq. (13.24)):

$$g_{\text{new}}(\mathbf{x}) = \mathrm{Re}\{g'(\mathbf{x})\} \qquad (13.23)$$

$$g_{\text{new}}(\mathbf{x}) = 0 \text{ if } g'(\mathbf{x}) < 0 \qquad (13.24)$$

For kinematical X-ray diffraction without the absorption and refraction effects, the electron density is everywhere positive. Therefore, the conditions of Eq. (13.23) and (13.24) generally hold. For kinematical electron diffraction, the atomic potential of ionic materials can have both signs simultaneously. Fieunp [38] showed that the ER algorithm can work with a complex-valued object by removing the positivity constraint. The object function becomes complex as a result of dynamical scattering, or in the case of kinematical diffraction the projection of three-dimensional diffraction, or absorption, or refraction.

3) **Hybrid-input-output (HIO) algorithm**

   The ER algorithm suffers from slow convergence and a tendency to stagnate at local minima in the solution space [25]. To solve this problem, Fienup introduced a feedback mechanism into the ER as illustrated in Figure 13.12 The input and output of the Fourier domain modification are $g(\mathbf{x})$ and $g'(\mathbf{x})$, respectively. This operation produces the function $g'(\mathbf{x})$, which satisfies the amplitude constraint in the Fourier domain. The algorithm ultimately seeks a solution that satisfies the constraints in both the Fourier and the object (support) domains. To take account of this, Fienup treated the $g(\mathbf{x})$ as a driving function, rather than a solution, to drive the $g'(\mathbf{x})$ toward a solution that satisfies the object domain constraints. In general, however, this procedure is nonlinear. However, for a small change in $g(\mathbf{x})$, say $\Delta g(\mathbf{x})$, it produces approximately a linear response (Figure 13.12), $g'(\mathbf{x}) + \alpha \cdot \Delta g(\mathbf{x})$. To satisfy the support constraint, the property of the small change should have

$$\alpha \cdot \Delta g(\mathbf{x}) = 0 \quad \mathbf{x} \in S \quad \text{and} \quad \alpha \cdot \Delta g(\mathbf{x}) = -g'(\mathbf{x}) \quad \mathbf{x} \notin S \quad (13.25)$$

That is, the operation drives the output toward zero outside the support. Therefore, the desired input for the next iteration should take the following form:

$$g_{\text{new}}(\mathbf{x}) = 0 \quad \mathbf{x} \in S \quad \text{and} \quad g_{\text{new}}(\mathbf{x}) = g(\mathbf{x}) - \beta g'(\mathbf{x}) \quad \mathbf{x} \notin S \quad (13.26)$$

where β is α^{-1}. Since the new input mixes the previous input and output of the operation in a linear combination, this algorithm was named the HIO algorithm.

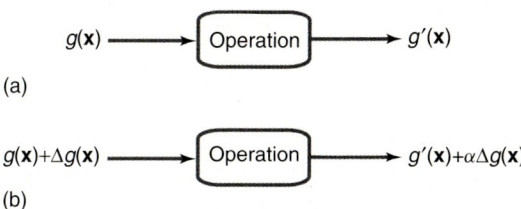

Figure 13.12 A schematic illustration of the HIO algorithm. (a) The input and output of the grouped operation. (b) The input with a small change and the corresponding linear response in the output.

4) **Charge flipping (CF) algorithm** [40]:
   Oszlanyi and Suto proposed a very simple phase retrieval scheme for X-ray crystallography, based on the fact that electron density is everywhere positive:

   $$g_{new}(\mathbf{x}) = -g'(\mathbf{x}) \quad \text{if } g'(\mathbf{x}) < \delta \tag{13.27}$$

   where δ is a positive threshold for flipping. Note that in Eq. (13.27), no support constraint is needed. Therefore, the CF algorithm can also be applied without oversampling. Wu and Spence [41] have extended the CF algorithm to phase a complex-valued object using a support constraint: for pixels outside the support, the algorithm flips the signs of their real parts.

5) **Difference map (DM) algorithm:**
   Mathematically, applying the Fourier amplitude constraint and the support constraint can be described as projections onto two subsets; each subset is a solution that satisfies one of the above two constraints [37]. If a solution exists, the two subsets must intersect, and the solution is found when the estimate falls into the intersection. On the basis of this concept, Elser [37] proposed an algorithm, called the *difference map*, for phase retrieval. Some new notations are needed to introduce this algorithm. Briefly, a solution should fall in the intersection between two subsets **A** and **B**, and π_1 and π_2 are the two projection operators to project the estimate $g_n(\mathbf{x})$ to the two subsets. In the DM algorithm, the new estimate is obtained by

   $$g_{n+1}(\mathbf{x}) = [1 + \beta(\pi_1(1 + \gamma_2\pi_2 - \gamma_2) - \pi_2(1 + \gamma_1\pi_1 - \gamma_1))]g_n(\mathbf{x}) \tag{13.28}$$

   where β, γ_1, and γ_2 are three parameters adjusted for best performance. For the phase retrieval problem under discussion, the two projection operators, respectively, map the estimate to a subset where the Fourier amplitudes match with the observed ones, and the other subset where the support constraint is satisfied (equivalent to Eq. (13.22)). It is found by Elser [37] that the HIO algorithm is a special case of DM. For example, if one sets $\gamma_1 = -1$ and $\gamma_2 = 1/\beta$, Eq. (13.26) becomes:

   $$g_{n+1}(\mathbf{x}) = [1 + (1 + \beta)\pi_1\pi_2 - \pi_1 - \beta\pi_2)]g_n(\mathbf{x}) \tag{13.29}$$

   where π_1 is the support constraint projector and π_2 is the Fourier amplitude projector.

13.4.6
Use of Image Information

In a TEM, an image of the diffraction object can be recorded directly up to the microscope resolution. At the minimum, the electron image provides accurate support information. At the maximum, the electron images can be used to obtain both the amplitude and phase of the exit wave function. In electron imaging, both the amplitude and the phase are affected by the microscope's CTF. In general, the phase is more reliably recorded up to the frequency where the CTF changes sign.

In electron diffraction, the recorded diffraction patterns are also far from ideal and contain only limited information. For example, weak intensities between diffraction spots are often lost because of the detector noise. The central peak is often missing or saturated in experimentally recorded diffraction patterns because of its strong intensities. Even if the central peak is recorded, its intensity is mixed with other small angle scatterings, such as inelastic scattering from apertures, which makes small angle scattering intensities less reliable for diffractive imaging. The effect of noise is twofold: it limits the amount information that can be recorded about the shape factors of the particles in the diffraction pattern and it also limits the amount of information about the weak interference originated from local defects.

The information obtained in electron images complements these recorded in diffraction patterns. The shape information of a nanometer-sized object, which is partially lost because of noise, can be compensated for by having an accurately determined boundary, or support, from the direct image. For certain types of defects, the change in structure introduces contrast variation within the objects. A direct image provides this contrast albeit at a lower resolution. Therefore, the loss of weak diffuse scattering can be compensated for in the real space by using a direct image in two aspects: (i) image provides an accurate determination of the object boundary as support and (ii) at a reduced resolution, the contrast of the reconstructed object should agree with that of the direct image. In addition, by knowing which parts of the diffraction pattern are noise and which are data using the image information, one can further reduce the amount of diffraction noise that does not play any meaningful role in the object function reconstruction.

The approach is to take limited information in an electron image and diffraction pattern and reconstruct the object function at a resolution of, or near, the diffraction limit based on the additional information obtained by oversampling. The authors of this chapter have developed techniques to achieve this. The information used for resolution improvement mainly comes from the diffraction intensity. The electron image recorded at resolutions available from the instrument is used: (i) to provide an initial set of phases for low-frequency diffraction intensities and (ii) to estimate the object boundary for real space constraint or support. To use image and diffraction information effectively, the electron image and the diffraction pattern are recorded from the same area of the sample. The recorded electron diffraction pattern and image are then aligned and scaled to match each other using image-processing techniques and used for image reconstruction using iterative phase retrieval techniques. Details about these procedures can be found in Ref. [42] and below.

13.4.7
Simulation Study of Phase Retrieval Algorithms

The nonlinear process used in phase retrieval algorithms makes it difficult to theoretically compare the effectiveness of these algorithms. Simulations are often carried for this purpose. In electron CDI, the following issues can be tested by simulations: (i) the effectiveness of the various iterative phase retrieval algorithms;

(ii) optimization of the phase retrieval parameters and the effective combinations of algorithms suited for electron diffraction patterns; and (iii) the effects of various experimental factors on the success of phase retrieval, including dynamical scattering, electron beam convergence, the missing central beam, and the experimental noise. Artificial objects have often been used to develop and test phase retrieval algorithms [34, 38, 41, 43, 44]. When it comes to studying specific objects, it is helpful to use objects that are close to the experiment and test how the iterative phasing and the quality of the reconstructed image are affected by experimental factors. Huang et al. [36] reported the use of Au nanoclusters as testing objects to address the issues outlined above. The numerical experiments carried out by Huang et al. included the following steps: (i) the construction of Au nanoclusters of different sizes for two structural models: truncated octahedron (TO) and icosahedron (Ih); the size range for TO is from 147 to 5341 atoms (about 1–6 nm in diameter) and for Ih is from 147 to 4000 atoms (about 1–4 nm in diameter). The two structures are chosen to represent crystalline and noncrystalline structures. (ii) Simulation of electron diffraction patterns using both kinematical and dynamical approximations based on the constructed structural models. (iii) The reconstruction of electron images by retrieving the missing phases in the simulated diffraction patterns using different iterative algorithms. The models of the truncated octahedra were built using the bulk gold bond distance (2.885 Å for the nearest Au–Au bond length) without structural relaxation, while the Ih were generated in accordance with the deformed Ih model proposed by Mackay [45]. The kinematical diffraction patterns were calculated by summing the scattered waves from all Au atoms in the cluster and the tabulated atomic scattering factor of Au [46]. The dynamical diffraction patterns were simulated using the multislice method [47]. For the dynamical diffraction simulation, the cluster model was rotated to a desired orientation, placed in a cell of σ (oversampling ratio) times of the particle size in each dimension and then divided into 1.2 Å thick slices. The effect of finite spatial and temporal coherence was simulated using a convergent beam with a specific energy spread. The simulation test results have been published [36]. On the basis of the simulations, Huang et al. suggested that the missing central beam can be replaced by using a low-resolution starting image obtained in the TEM and to provide the low-frequency phase information. The following emphasize the effect of dynamical scattering and the issue of limited coherence and noise.

Multiple scattering, or dynamical scattering, of electrons causes the diffraction intensities to deviate away from the Fourier relationship between the object potential and the diffraction pattern. The dynamical scattering effects are significant even for small objects such as Au nanoclusters. The deviation of the diffraction intensities from the kinematical limit can be quantitatively measured in simulation by comparing the dynamical and kinematical diffraction patterns calculated from the same object on a pixel-by-pixel basis using the R factor:

$$R = \frac{\sum |I_d - cI_k|}{\sum |cI_k|} \tag{13.30}$$

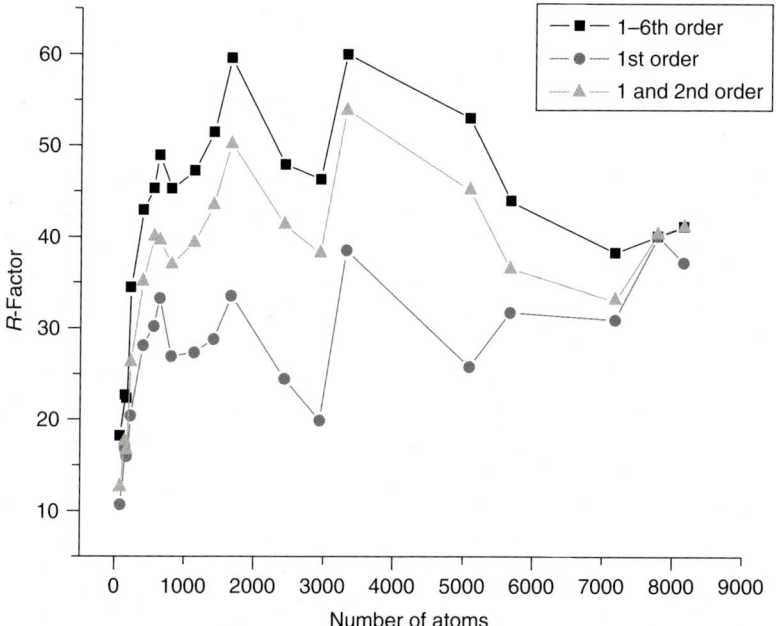

Figure 13.13 Intensity difference between dynamical and kinematical electron diffraction patterns calculated from Au-truncated octahedra of different sizes.

where I_k and I_d stand for the kinematical and dynamical diffraction intensities, respectively, and c is a scaling constant to match the two intensities (achieved by minimizing the R). Figure 13.13 plots the calculated R factor for the truncated octahedral Au nanoclusters as a function of the number of atoms in each cluster. When summing up the difference in intensity, a different set of reflections can be included in the comparison and plotted separately to examine the dynamic effects for different reflections. For example, comparisons were made for reflections of only the first order, or both the first and the second orders, or from the first to the sixth orders. The three calculated R factors all show a rapid increase at around 300 atoms in the nanocluster ($\sim$2.5 nm in diameter). The R factor starts to oscillate as the size increases beyond 1000 atoms ($\sim$3.5 nm), which we attribute to the Pendellösung effect. Note that even for an Au cluster as small as 2 nm in diameter ($\sim$250 atoms), the R factor calculated is already about 20%, indicating a nonnegligible dynamical character in the diffraction intensity of Au nanoclusters.

The effect of multiple scattering on image reconstruction can be understood in the following way. For a weak phase object, the exit wave function can be expressed according to the phase grating approximation as

$$\varphi_{\text{exit}} = \exp(-i\sigma V_p) \tag{13.31}$$

where σ is $\frac{me\lambda}{2\pi\hbar^2}$ and V_p is the projected potential of the nanocluster along the incident beam direction. If one represents the complex exit wave function in an Argand plot, the phase grating approximation should give a spiral pattern. In Figure 13.14, the Argand plots for the calculated exit wave function of three Au-truncated Ih of 405, 2300, and 8000 atoms are shown. There are two features worth noticing: first, most of the pixels have very small imaginary parts ($<5e^{-4}$ in absolute values); second, there is a change in the spiral pattern from small to larger particles. The Au cluster of 405 atoms shows perfect spiral patterns, while the cluster of 8000 atoms gives fuzzy spirals. This indicates that for the larger Au cluster, the phase grating approximation no longer holds and the exit wave function does not have a simple relationship with the projected potential. For a weakly scattering object, the weak phase approximation (or kinematical approximation) can be applied and the exit wave function can be rewritten as

$$\varphi_{exit} \approx 1 - i\sigma V_p \qquad (13.32)$$

which is made up of a constant background plus an imaginary part that is proportional to the projected potential. The constant background translates into the central peak in the diffraction pattern. With the constant background removed, the exit wave function is proportional to the potential and thus a real object can be assumed. For an object potential stronger than the weak phase object, higher order terms have to be considered in the expansion, which then will include both the real and imaginary parts. Therefore, a complex object is associated with dynamical diffraction. The same is true if the absorption potential is included. Reconstructing a complex object requires a much tighter support since the number of unknowns doubles, which has been pointed out by both Fienup [38] and Spence et al. [44]. The tight support required for complex object reconstruction causes difficulties sometimes in experimental iterative phase retrieval especially for X-ray diffraction, where the support is difficult to estimate directly. For electrons, it will be shown later that one can obtain a tight support by processing the as-recorded images.

Figure 13.15 compares the reconstructed images (b,c) with the projected potential (a) and their profiles (measured along the line marked in (a), and profiles are shown left to the reconstructed images). The reconstructed images shown in Figure 13.15b,c are both inverted from the same dynamical electron diffraction pattern, which is simulated from a Au TO of 405 atoms along its [110] zone axis, with and without applying the real object constraint, respectively. Both reconstructions are shown to recover the external shape and atom positions of the nanocluster. The reconstruction using the real object constraint, however, fails to reproduce the intensity distribution of the projected potential, which is clearly seen by comparing the intensity profiles. The discrepancy is due to the approximation of the real object constraint in a case where the dynamical effect is significant enough to require a complex object reconstruction. Figure 13.15c shows the modulus of the reconstructed complex image with the real object constraint removed, and the corresponding intensity profile on its left. In addition to the external shape and atom positions, the intensity profile is now closer to that of the projected potential. Specifically, the intensity maximum is located close to the center of the nanocluster. This indicates that complex object reconstruction is necessary to obtain the correct image contrast (i.e., proportional to the projected potential).

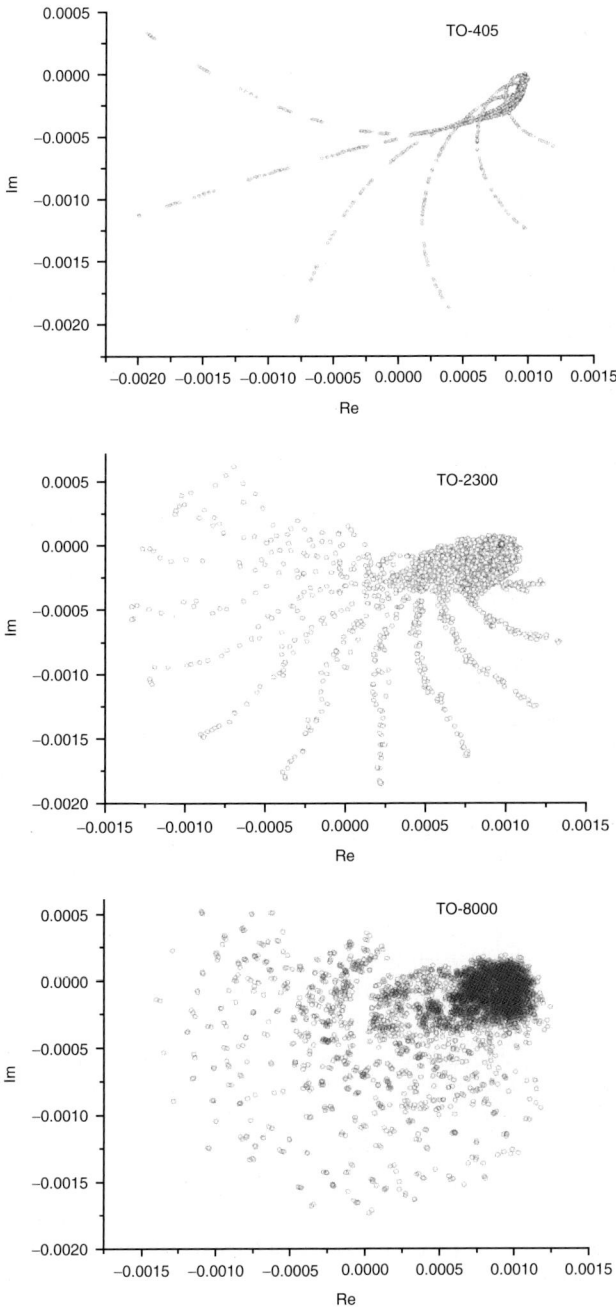

Figure 13.14 Argand plots for the calculated exit wave functions from three Au-truncated octahedral models: 405, 2300, and 8000 atoms.

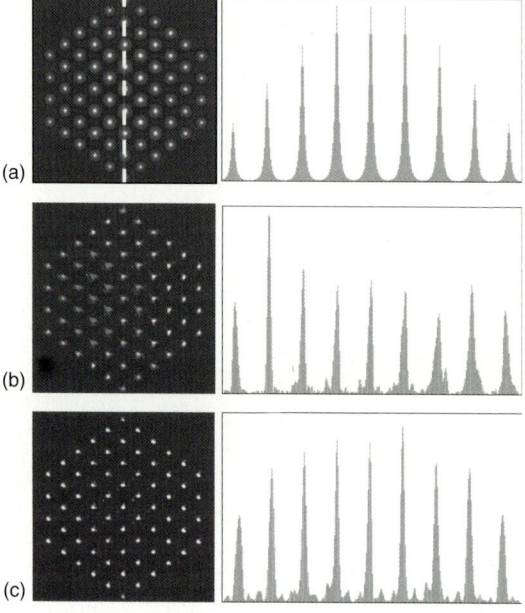

Figure 13.15 The reconstructed images from a simulated dynamical diffraction pattern of a 405-atom Au TO with (b) and without (c) the real object constraint. For comparison, the calculated projected potential is shown in (a). Their intensity profiles (along the dashed line in (a)) are shown on the right.

The success of complex object reconstruction requires that the object support is estimated with high accuracy. Real object constraint seems sufficient, however, as far as the cluster shape or the atom positions are concerned.

The amount of dynamical scattering as measured by the R factor of Eq. (13.30) grows with the cluster size up to 1000 atoms and then fluctuates. There is a point where image reconstruction fails to recover the atomic positions from the diffraction intensities using the real object constraint. The simulation study by Huang et al. [36] found that the dynamical diffraction patterns of the Au TOs of more than 4000 atoms (∼5 nm in diameter) are difficult to reconstruct and the images obtained have little resemblance to the projected potential. The upper limit is about 1000 atoms (∼3 nm in diameter) for Ih clusters.

As discussed in Section 13.4.4, phase retrieval requires a minimum ratio between the electron's lateral coherence width and the object size. The minimum oversampling ratio is 2 in each dimension for a two-dimensional object. This can be achieved by having a coherence width twice the object width (W). These arguments are based on the minimum oversampling requirement. In a general inversion situation, a higher oversampling ratio is required because of noises in the recorded diffraction patterns. By increasing the oversampling ratio, the inversion problem becomes mathematically overdetermined, thus yielding a more robust solution.

Huang et al. simulated the limited coherence length using probes with a semi-angle of 0.0, 0.25, and 0.5 mrad, respectively. The semi-angle of 0.0 mrad corresponds to a perfectly parallel beam. Their results show that compared to the parallel beam, the intensity away from the center of the cluster is dampened under a converged beam. The degree of damping also increases with the semi-angle. The damping is due to the broadening of the diffraction peaks in the reciprocal space, which is translated into an envelope function in the object space. The central part of the nanocluster is still recovered for a semi-angle of 0.5 mrad; however, the atoms located on the surface become less clear as the semi-angle increases because of the damping. The diffraction pattern of the crystalline TO structure can tolerate a 0.5 mrad beam convergence for reconstruction. Inversion from the noncrystalline Ih structure is found to be more difficult even for a probe with a 0.25 mrad convergence angle. In this case, the semi-angle of 0.125 mrad is the threshold under which a reasonable image can be reconstructed to reveal both the shape and the nonperiodic atom positions. The failure of reconstructing the nonperiodic atomic positions is because the diffraction intensities that carry the information about the shape and interference between different internal domains are broadened by the beam convergence. Unlike crystalline structures, which have highly peaked and repeated intensity distribution in the reciprocal space, the diffraction intensities are more continuous for the nonperiodic structures. This makes the reconstruction more sensitive to the beam convergence.

Experimentally, noise is introduced into as-recorded diffraction patterns from two sources: (i) the probability of a scattered electron falls into a pixel and (ii) the noise during the readout process. Huang et al. tested reconstruction of object function from the kinematical and dynamical diffraction patterns with different levels of noise. The reconstructed images show that as the amount of added noise increases, the blurriness of the atoms increases as a result of the loss of high-frequency diffraction information. In both kinematical and dynamical diffraction patterns, when the noise level is beyond ~60% (measured by the R factor) the patterns are difficult to invert. Diffraction intensity up to reflections of second order can survive this noise level. The increase in the level generally makes the recovery of the particle shape less reliable, since the shape information (additional diffraction features around the strong reflections) is weak and they can be buried by noise. It should be emphasized that the noise levels tested by Huang et al. [36] are significantly higher than experimental noise levels.

13.5
Phasing Experimental Diffraction Pattern

The experiments described here are to demonstrate that coherent electron diffraction patterns can be phased to obtain the object function of CdS quantum dots. The techniques, however, are general and not limited to nanoparticles such as quantum dots. Especially, they can be applied readily to coherent electron diffraction from a selected area of continuous samples using sharp apertures. Since a starting image significantly improves the results of diffraction pattern phasing as discussed before, the following procedures for phasing diffraction patterns assume that image information is available. The major steps of the phasing procedures are as follows

1) Processing of diffraction patterns and finding support in reciprocal space;
2) Processing of the direct image (start image) and finding the object support in real space;
3) Combining information from both the direct image and the diffraction pattern to reconstruct the object function at diffraction-limited resolution.

13.5.1
Processing of Experimental Diffraction Patterns and Images

Two diffraction patterns are typically recorded; one with the object (object pattern) and one without for recording diffraction background intensity (background pattern). The background intensities come from the tails of the central beam and electron diffraction of the supporting substrate, which can be subtracted from the object pattern. The background pattern is recorded slightly away from the object. For example, in the case of a nanoparticle sitting on graphene, the background pattern is recorded avoiding the nanoparticle but close enough to include a similar area of graphene. Since the two patterns are recorded under the same experimental conditions (illumination, exposure time, and similar support), they are assumed to contain the same background intensity from the supporting substrate. An automated program is used to align the two diffraction patterns with respect to their central beams. This is only necessary when imaging plates are used for diffraction pattern recording. The readout of imaging plates can introduce small image shifts and rotations between different plates, which must be corrected for the background subtraction purpose. The center alignment is done by finding the saturated pixels near the central beam in the diffraction patterns and locating the centers of that saturated region by calculating the center of weight. The small rotation between different imaging plates is taken into account by cross-correlating the two diffraction patterns using a common feature. After the pattern alignments via centering and rotation, the background pattern is subtracted. Figure 13.16 shows a diffraction pattern from a CdS quantum dot before and after subtraction.

Stray intensity near the center coming from the scattering of the aperture in the microscope is further removed using a noise-reduction procedure. In this procedure, a region of interest (ROI) is first selected, whose mean and standard deviation in intensity are then calculated. The script will set any pixel value within the ROI to be the mean value if the difference between the original value and the mean is larger than the standard deviation by a specified value.

The as-recorded image must be scaled to match its size with the field of view determined from Eq. (13.15). For example, a diffraction pattern recorded with an 80 cm camera length and a 50 μm pixel size will give a total field of view of 40 nm in each dimension. That means the as-recorded image needs to be resampled so it can be placed inside a 40×40 nm^2 area sampled.

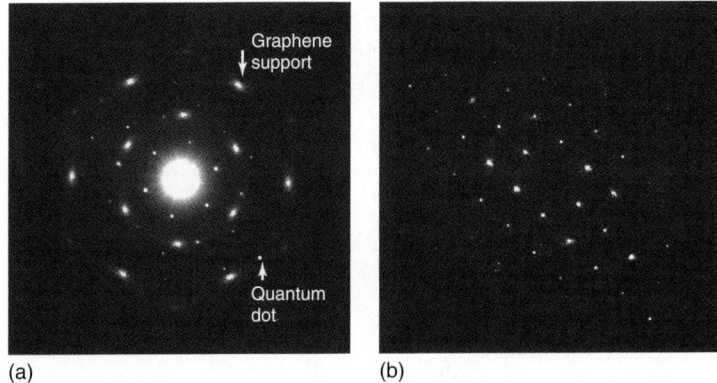

Figure 13.16 Background subtraction. (a) As-recorded electron diffraction pattern from a CdS quantum dot supported on graphene, (b) diffraction pattern after background subtraction.

The contrast of as-recorded images is transformed using the following procedures. First, the background surrounding the nanocluster of interest is subtracted by the mean value of the background. The contrast of the background-subtracted image is then inverted; under the Scherzer focus condition and the weak phase object approximation the atoms appear white and vacuum appears black in the inverted image. In addition, the CTF can be removed from the image using deconvolution procedures described in the literature [48–50]. The overall intensity of the image is also scaled such that its power spectrum matches the recorded diffraction intensity.

The boundary of the nanoparticles can also be estimated from the image by the digital image-processing method, implemented in a program called *maskfinder* developed by Zuo at the University of Illinois. The boundary finding method consists of four steps: (i) obtain a threshold intensity that differentiates the background from the peak in the intensity histogram, (ii) threshold the image and generate a binary image (0 and 1 for below or above the threshold), (iii) erode the binary image to remove isolated pixels, and (iv) dilate the image to obtain a continuous support. The erosion algorithm looks for the minimum among neighboring pixels within a given radius and then sets the intensity of neighboring pixels to its original intensity minus a given threshold or the minimum, whichever is higher. The effect of this is reduced intensity for the boundary pixels, which can then be removed by thresholding. In dilation, the process is reversed; intensity is added to neighboring pixels. The radius is an adjustable parameter.

In step 1, the threshold intensity is determined by finding a "fixed point" in the intensity histogram [51]. First, we start from an initial threshold intensity T, which can be taken as the midpoint of the histogram. Second, the image is segmented into two sets of pixels by the threshold, the pixels with intensity larger and smaller than T. Next, the average intensity values μ_1 and μ_2 of the two groups are calculated and a new threshold intensity is found by $T = \frac{1}{2}(\mu_1 + \mu_1)$. Finally, the second and

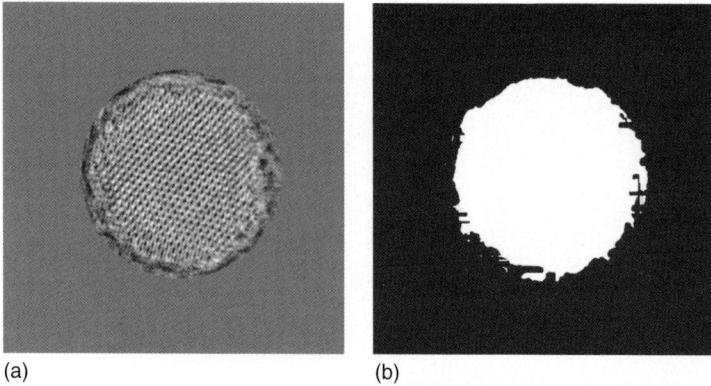

Figure 13.17 Estimation of the real space support. (a) Contrast inverted HREM image from a CdS quantum dot of ~9 nm in diameter and (b) the support obtained from the image using procedures described in text.

third steps are repeated until the difference in T found in successive iterations becomes smaller than a predefined value. This method normally takes about 10 iterations to converge.

The image is then converted into a binary image using the newly found threshold intensity T. The erosion process removes the isolated pixels in the image. We found that eroding by the width of three to five pixels is sufficient in most of simulations. The last step is to dilate the image by the same pixel width. The obtained mask sometimes contains isolated holes. Additional convolution with a disc is sometimes used to fill the holes. Figure 13.17 shows the support found by the *maskfinder* from an image. Similarly, a support for the diffraction pattern, defined as the region where meaningful data exists, can also be obtained using the same procedure.

13.5.2
Phasing CdS Quantum Dots

Cadmium selenide quantum dots were prepared by solution-based chemical methods by Dr. K.W. Kwon of Prof. Moonsub Shim's group at the University of Illinois. A solution containing 300 mg of trioctylphosphine oxide, 315 mg 1,2-hexadecanediol, and 10 ml of octyl ether was vacuum degassed at 100 °C for 30 min. Sulfur powder (15 mg) was added at 100 °C under N_2 and stirred for 5 min. After cooling to 80 °C, cadmium acetylacetonate (150 mg) was added and stirred for 10 min. The reaction mixture was heated to 280 °C and annealed for 30 min. The final CdS nanocrystals were precipitated with ethanol, centrifuged to remove excess capping molecules, and redissolved in chloroform. As synthesized, CdS nanocrystals were found by X-ray powder diffraction to be a mixture of wurzite and zinc blend structures.

The CdS quantum dots were supported on ultrathin graphene sheets or carbon nanotube bundles. Lacey carbon films attached to a copper TEM grid were used to support the graphene and the nanotubes. The TEM grid was first immersed in

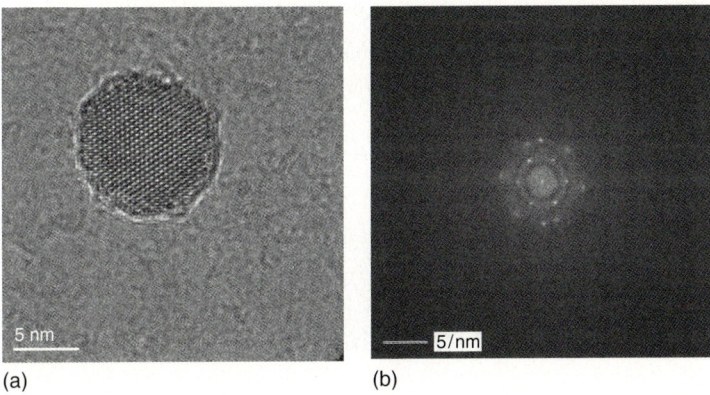

Figure 13.18 As-recorded HREM image (a) of a CdS quantum dot supported on graphene and its Fourier spectrum (b).

chloroform for 10 s to dissolve the formvar backing layer and then heated in Argon ambient to remove the residue formvar. Formvar was found to cause contamination problems under the nanoarea electron beam. Commercial double-walled carbon nanotubes were then dispersed onto the grid. Most of the nanotubes form bundles with each other. Graphene sheets were also dispersed onto the grid. The diluted solution of CdS quantum dots were finally dispersed onto the grid. The quantum dot density was controlled to be very low level, about one quantum dot in every 900–1600 nm².

Figure 13.18 shows a single-crystalline CdS quantum dot of 9 nm in diameter. The diffraction pattern from this quantum dot is shown in Figure 13.16. By indexing the diffraction pattern, the quantum dot was identified to have the wurzite structure, and the zone axis along which the pattern was recorded was near its c axis, or [0001]. The structure model suggests that along this orientation, the image should display a honeycomb-like structure, while it now appears to be the close-pack structure instead. This is because the resolution in the direct image is not enough to resolve the pair of atoms separated by 2.5 Å.

To phase the diffraction pattern, we found that suppressing the noise in between the Bragg peaks is very useful in improving the quality of the reconstructed image. To do that, we separate diffraction pattern into three different regions of G above or B below the background noise level and M where diffraction information is not available. We apply the diffraction intensity constraint in G. In region B, we place an upper limit of three times the background noise on the intensity. In region M, the intensity is allowed to float. In the reciprocal space, we replace the amplitudes of the Fourier transform according to

$$F(u,v) = \begin{cases} F(u,v) \cdot \sqrt{I^D(u,v)} / |F(u,v)| & \text{if } (u,v) \in G \\ F(u,v) \cdot \alpha \cdot \min\left(\sqrt{I^D(u,v)}, \sqrt{3\sigma}\right) / |F(u,v)| & \text{else if } (u,v) \in B \\ F(u,v) & \text{if } (u,v) \in M \end{cases}$$

(13.33)

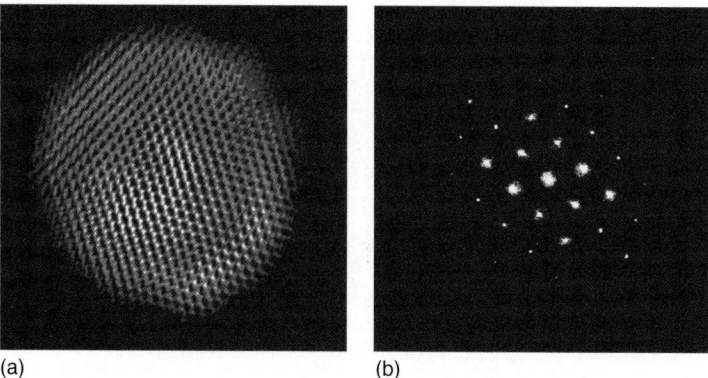

Figure 13.19 (a) Reconstructed object function and its power spectrum (b). The diffraction pattern and image used for reconstruction are shown in Figures 13.13 and 13.15.

where G, B, and M mark different regions in the reciprocal space mask, as described before. The I^D is the experimental diffraction intensity, α is a fractional number that was used to reduce the effect of background intensity on reconstructed image, and σ is the standard deviation of the background intensity. The maximum of 3σ is imposed on background intensity to remove artifacts from background subtraction. The α has a value between 0 and 1. It is used to reduce the amplitudes outside the diffraction support, where the background noise prevails. It is found that when α is between 0.1 and 0.4, satisfactory reconstructions can be achieved. When α is larger than 0.4 a noisy and blurred reconstruction resulted, while γ less than 0.1 generated artificial unphysical structures in the reconstruction.

The object function reconstructed from the diffraction pattern (shown in Figure 13.16 from the quantum dot imaged in Figure 13.18) is shown in Figure 13.9. The reconstruction process is as follows:

1) An initial estimate of the object function is obtained from the start image. A small background noise about one-fifth of the maximum in the starting image is generated with a random seed number and added to the estimated object function.
2) An iterative phase retrieval is performed starting with the estimated object function and using the HIO algorithm; real object constraint is applied during this step,
3) Step 2 is followed by iterations using the ER algorithm with the real object constraint.
4) Step 3 is followed by iterations using the HIO algorithm without the real object constraint to reconstruct the complex exit wave function.
5) Steps 1–4 are repeated using a different random seed number.
6) The object functions obtained from the above steps are averaged.

The experimental diffraction pattern always contains a certain amount of noise. The noise in the diffraction pattern is expected to transmit to the reconstructed object function. The background noise can be considered during the reconstruction process by defining a tolerance factor ε [26]. The function of $g'(\mathbf{x})$, whose Fourier transform satisfies the diffraction amplitude constraint (Eq. (13.20)), is considered to meet the real-space support constraint if it is below the tolerance factor. In HIO, a new estimate of the object function is obtained by driving the current estimate of the object function toward the support constraint if $g'(\mathbf{x})$ is above the tolerance factor (Eq. (13.26)). In all iterations, we use a factor of $\varepsilon = 1e - 4$.

Compared to the as-recorded HREM image, the reconstructed image clearly shows the honeycomb structure, thanks to the 0.72 Å information transfer in the recorded diffraction pattern. The resolution improvement can also be seen in the comparison between the power spectra of the as-recorded image (Figure 13.18b) and the reconstructed image (Figure 13.19b).

A highly debated issue about diffractive imaging is the uniqueness of the solution and whether global minimum is reached in ITA. The above reconstruction steps use noise perturbation to test the reproducibility of the reconstructed objection function. A sequence of random noises is generated starting from an integer seed. These noises are added to the object function estimated from the starting image. The effect of adding noise perturbs the starting phases for the spatial frequencies that are recorded in the image. For high spatial frequencies beyond the image information limit, the random noises introduce new starting phases. This differs from the practice used in X-ray CDI. Since an image is often not available, the starting image in X-ray CDI is usually synthesized from the observed Fourier amplitudes and randomly generated phases.

A major difficulty of object reconstruction without image information is that the results of ITA are quite often trapped in local minima, or iterations starting with different sets of random noises produce different reconstructions. Often, these constructions are similar as measured by merit metrics designed to monitor the progress of ITA, such as examining the difference between the calculated intensities and the experimental ones. Thus, none is accurate enough to be selected as the "right" reconstruction [52]. To overcome this problem, the guided HIO method is developed based on the idea that a set of images close to the sought-after solution can be used efficiently to guide ITA toward the solution. In the proposal by Chen *et al.*, the images used to guide the search are selected from a set of images reconstructed using different random started phases based on the level of fit to experimental diffraction intensities. The guided HIO method has been used successfully for the reconstruction of crystalline particles and cells without starting image information [53–55]. The role of the starting image is thus similar in guiding the search of ITA closer to the final solution. However, unlike the guided HIO where the initial selection of starting images can be subjective, there is no ambiguity in the starting image obtained experimentally with its phase and amplitude information, albeit limited in resolution.

Figure 13.20 compares four reconstructed object functions obtained using different sets of random noises and the average obtained over 10 different reconstructions. The overall contrast is similar among different object functions of Figure 13.20, the fine details can fluctuate in some regions of the quantum dot

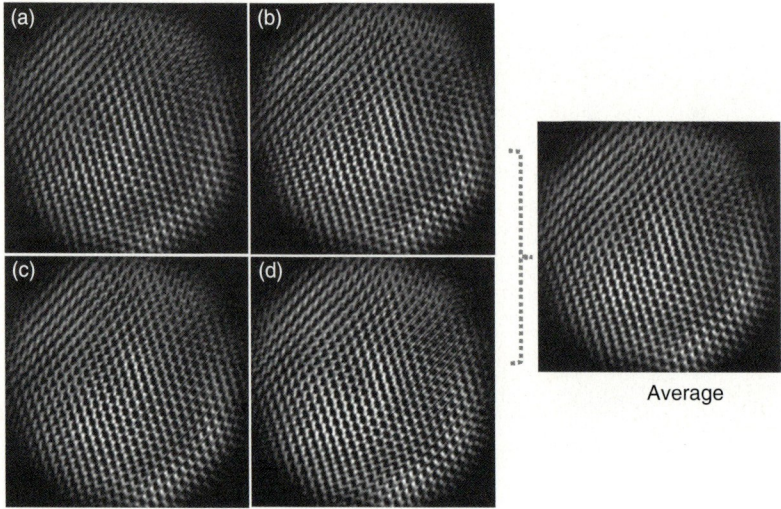

Figure 13.20 Reconstructed object functions using different random noise seeds (a–d) and their average. The average was over 10 reconstructed object functions including these shown here (image a–d). The images shown are the real part of the objection functions.

corresponding to higher resolutions. These fluctuations do not change significantly the fitting merit between the experimental and fitted diffraction intensities. Averaging over different runs of reconstruction using random noise for perturbations helps reduce the uncertainty caused by these fluctuations and can be also used to estimate the uncertainty of the reconstructed object function.

An error metric is needed to monitor the progress of iterative phase retrieval. There are several metrics that have been proposed. Fienup [25] used an object domain error metric by summing residual intensities outside the support as

$$E = \sqrt{\frac{\sum_{x \notin S} |g'_n(x)|^2}{\sum_{x} |g'_n(x)|^2}} \quad (13.34)$$

This error metric strongly depends on the estimate of the support. Another error metric, which is calculated in the Fourier domain, compares the amplitude difference between the Fourier transform amplitude of the estimated object function and the known amplitudes. This metric is popularly known as the *R factor*:

$$R = \left[\frac{\sum ||F(k)| - |G(k)||^2}{\sum |F(k)|^2} \right]^{1/2} \quad (13.35)$$

An example of the changes of the R factor during an iterative phase retrieval is shown in Figure 13.21. The R factor during the step 1 of the reconstruction with real object constraint decreases with successive iterations and eventually flattens out.

13.5 Phasing Experimental Diffraction Pattern

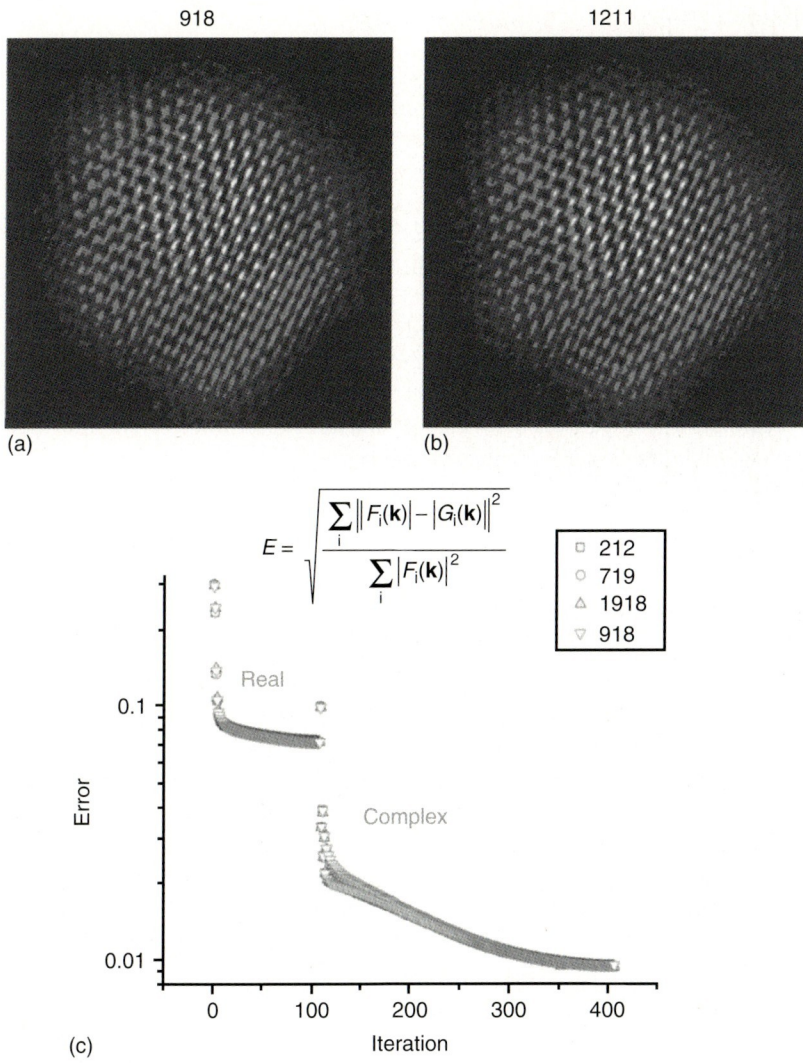

Figure 13.21 Noise perturbation. (a,b) Two reconstructed images (modulus is shown) started from different sets of random noise. The seed numbers generating the noise are shown above the images. The amplitude of the added noise is 1E-4. (b) Error evolutions from different starting points. Legends show the seed numbers generating each set of starting noise.

The R factor further decreases as the real object constraint is relaxed and a complex object is reconstructed. Figure 13.21 includes different reconstruction runs with different starting noises. The R factor differences between these different runs are negligible.

13.6
Conclusions

This chapter describes the methods for combining real and reciprocal space information for electron CDI to achieve diffraction-limited resolution. The image is used to estimate the starting phase, which, together with diffraction amplitudes, provides the starting point for image reconstruction. The image is also used to calculate the object support. The ITAs based on constraints in real and reciprocal spaces are used to reconstruct the object function using the combination of diffraction and imaging information. A diffraction mask to account for missing information and diffraction background noise in recorded electron diffraction patterns is used to selectively apply the diffraction information constraint. The method was demonstrated using quantum dots as examples. The methods described are also useful, in general, for quantitative comparison between real and reciprocal information and their combination for materials structure research.

Acknowledgments

The work reported here was made possible with the support by the U.S. Department of Energy Grant DEFG02–01ER45923. Microscopy performed on the JEOL 2010F was carried out at the Center for Microanalysis of Materials at the Frederick Seitz Materials Research Laboratory. We thank Jun Yamasaki for providing Figure 13.3 and Ke Ran for Figure 13.4.

References

1. Zuo, J.M. and Tao, J. (2011) Scanning electron nanodiffraction and diffraction imaging, in *Scanning Transmission Electron Microscopy* (eds S. Pennycook and P. Nellist), Springer, New York, 393–427.
2. Spence, J.C.H. (2003) *High-Resolution Electron Microscopy*, Oxford University Press, Oxford.
3. Spence, J.C.H. and Zuo, J.M. (1992) *Electron Microdiffraction*, Plenum, New York.
4. Rodenburg, J.M. (2008) Ptychography and related diffractive imaging methods. *Adv. Imaging Electron Phys.*, **150**, 87–184.
5. Hoppe, W. (1982) Trace structure-analysis, ptychography, phase tomography. *Ultramicroscopy*, **10** (3), 187–198.
6. Spence, J.C.H., Howells, M., Marks, L.D., and Miao, J. (2001) Lensless imaging: A workshop on "new approaches to the phase problem for non-periodic objects". *Ultramicroscopy*, **90** (1), 1–6.
7. Wen, J.G. et al. (2010) The formation and utility of sub-angstrom to nanometer-sized electron probes in the aberration-corrected transmission electron microscope at the university of illinois. *Microsc. Microanal.*, **16** (2), 183–193.
8. Zuo, J.M. et al. (2004) Coherent nano-area electron diffraction. *Microsc. Res. Tech.*, **64** (5–6), 347–355.
9. Dwyer, C., Kirkland, A.I., Hartel, P., Mueller, H., and Haider, M. (2007) Electron nanodiffraction using sharply focused parallel probes. *Appl. Phys. Lett.*, **90** (15), 151104.
10. Morishita S., Yamasaki J., Nakamura K., Kato T., and Tanaka N.

(2008) Diffractive imaging of the dumbbell structure in silicon by spherical-aberration-corrected electron diffraction. *Appl. Phys. Lett.* **93** (18), 183103.

11. Cowley, J.M. (1995) *Diffraction Physics*, North-Holland Personal Library, Amsterdam.

12. Lichte, H. (1986) Electron holography approaching atomic resolution. *Ultramicroscopy*, **20** (3), 293–304.

13. Orchowski, A., Rau, W.D., and Lichte, H. (1995) Electron holography surmounts resolution limit of electron-microscopy. *Phys. Rev. Letts.*, **74** (3), 399–402.

14. Podorov, S.G., Pavlov, K.M., and Paganin, D.M. (2007) A non-iterative reconstruction method for direct and unambiguous coherent diffractive imaging. *Opt. Express*, **15** (16), 9954–9962.

15. Karle, I.L. and Karle, J. (1964) The crystal and molecular structure of the alkaloid jamine from ormosia jamaicensis. *Acta Crystallogr.*, **17**, 1356.

16. Karle, J. and Karle, I.L. (1966) The symbolic addition procedure for phase determination for centrosymmetric and noncentrosymmetric crystals. *Acta Crystallogr.*, **21**, 849.

17. Haider, M. (1998) A spherical-aberration-corrected 200 kV transmission electron microscope. *Ultramicroscopy*, **75** (1), 53–60.

18. Haider, M., Muller, H., and Uhlemann, S. (2008) Present and future hexapole aberration correctors for high-resolution electron microscopy, *Advances in Imaging and Electron Physics*, vol. **153**, Elsevier Academic Press Inc., San Diego, pp. 43–+.

19. Krivanek, O.L., Dellby, N., and Lupini, A.R. (1999) Towards sub-angstrom electron beams. *Ultramicroscopy*, **78** (1–4), 1–11.

20. Zhang, Z. and Kaiser, U. (2009) Structural imaging of beta-Si3N4 by spherical aberration-corrected high-resolution transmission electron microscopy. *Ultramicroscopy*, **109**, 1114.

21. Gabor, D. (1948) A new microscopic principle. *Nature*, **161**, 777.

22. Spence, J.C.H. (1997) STEM and shadow-imaging of biomolecules at 6 eV beam energy. *Micron*, **28** (2), 101–116.

23. Spence, J.C.H., Vecchione, T., and Weierstall, U. (2010) A coherent photofield electron source for fast diffractive and point-projection imaging. *Philos. Mag.*, **90** (35–36), 4691–4702.

24. Sayre, D. (1952) Some implications of a theorem due to Shannon. *Acta Crystallogr.*, **5**, 843.

25. Fienup, J.R. (1982) Phase retrieval algorithms – a comparison. *Appl. Opt.*, **21** (15), 2758–2769.

26. Millane, R.P. and Stroud, W.J. (1997) Reconstructing symmetric images from their undersampled Fourier intensities. *J. Opt. Soc. Am. A: Opt. Image Sci. Vision*, **14** (3), 568–579.

27. Sayre, D., Chapman, H.N., and Miao, J. (1998) On the extendibility of X-ray crystallography to noncrystals. *Acta Crystallogr., Sect. A*, **54**, 232–239.

28. Miao, J.W., Charalambous, P., Kirz, J., and Sayre, D. (1999) Extending the methodology of X-ray crystallography to allow imaging of micrometre-sized non-crystalline specimens. *Nature*, **400** (6742), 342–344.

29. Robinson, I.K., Vartanyants, I., Williams, G.J., Pfeifer, M.A., and Pitney, J.A. (2001) Reconstruction of shapes of gold nanocrystals using coherent X-ray diffraction. *Phys. Rev. Lett.*, **87**, 195505.

30. Elser, V. (2003) Phase retrieval by iterated projections. *J. Opt. Soc. Am. A: Opt. Image Sci. Vision*, **20**, 40.

31. Marchesini, S. (2007) A unified evaluation of iterative projection algorithms for phase retrieval. *Rev. Sci. Instrum.*, **78** (1), 011301.

32. Nyquist, H. (1928) Certain topics in telegraph transmission theory. *Trans. AIEE*, **47**, 617.

33. Shannon, C.E. (1949) Communication in the presence of noise. *Inst. Radio Eng.*, **37**, 10.

34. Miao, J., Sayre, D., and Chapman, H.N. (1998) Phase retrieval from the magnitude of the Fourier transforms of non-periodic objects. *J. Opt. Soc. Am. A*, **15** (6), 1662–1669.

35. Spence, J.C.H., Weierstall, U., and Howells, M. (2004) Coherence and

sampling requirements for diffraction imaging. *Ultramicroscopy*, **101**, 149–152.

36. Huang, W.J., Jiang, B., Sun, R.S., and Zuo, J.M. (2007) Towards sub-a atomic resolution electron diffraction imaging of metallic nanoclusters: a simulation study of experimental parameters and reconstruction algorithms. *Ultramicroscopy*, **107** (12), 1159–1170.

37. Elser, V. (2003) Phase retrieval by iterated projections. *J. Opt. Soc. Am. A*, **20**, 40.

38. Fienup, J.R. (1987) Reconstruction of a complex-valued object from the modulus of its Fourier transform using a support constraint. *J. Opt. Soc. Am. A*, **6**, 118.

39. Gerchberg, R.W. and Saxton, W.O. (1972) A practical algorithm for the determination of phase from image and diffraction plane pictures. *Optik*, **35**, 237.

40. Oszlanyi, G. and Suto, A. (2004) Ab initio structure solution by charge flipping. *Acta Crystallogr., Sect. A*, **60**, 134–141.

41. Wu, J.S. and Spence, J.C.H. (2005) Reconstruction of complex single-particle images using charge-flipping algorithm. *Acta Crystallogr., Sect. A*, **61**, 194–200.

42. Zuo, J.M., Zhang, J., Huang, W.J., Ran, K., and Jiang, B. (2011) Combining real and reciprocal space information for aberration free coherent electron diffractive imaging. *Ultramicroscopy*, **111**, 817–823.

43. McBride, W., O'Leary, N.L., and Allen, L.J. (2004) Retrieval of a complex-valued object from its diffraction pattern. *Phys. Rev. Lett.*, **93**, 233902.

44. Spence, J.C.H., Weierstall, U., and Howells, M. (2002) Phase recovery and lensless imaging by iterative methods in optical, X-ray and electron diffraction. *Philos. Trans. R. Soc. London, Ser. A*, **360** (1794), 875–895.

45. Mackay, A.L. (1962) A dense non-crystallographic packing of equal spheres. *Acta Crystallogr.*, **15**, 916.

46. Peng, L.M., Ren, G., Dudarev, S.L., and Whelan, M.J. (1996) Robust Parameterization of elastic and absorptive electron atomic scattering factors. *Acta Crystallogr., Sect. A*, **52**, 257.

47. Cowley, J.M. and Moodie, A.F. (1957) The scattering of electrons by atoms and crystals. I. A new theoretical approach. *Acta Crystallogr.*, **10**, 609.

48. Thust, A., Coene, W.M.J., de Beeck, M.O., and Van Dyck, D. (1996) Focal-series reconstruction in HRTEM: simulation studies on non-periodic objects. *Ultramicroscopy*, **64** (1–4), 211–230.

49. Tang, D. and Li, F.H. (1988) A method of image-restoration for pseudo-weak-phase objects. *Ultramicroscopy*, **25** (1), 61–67.

50. Li, F.H., Wang, D., He, W.Z., and Jiang, H. (2000) Amplitude correction in image deconvolution for determining crystal defects at atomic level. *J. Electron Microsc.*, **49** (1), 17–24.

51. Gonzalez, R.C., Woods, R.E., and Eddins, S.L. (2002) *Digital Image Processing*, Prentice Hall.

52. Chen C.C., Miao J., Wang C.W., and Lee T.K. (2007) Application of optimization technique to noncrystalline X-ray diffraction microscopy: guided hybrid input-output method. *Phys. Rev. B*, **76** (6), 064113.

53. Dronyak R. et al. (2009) Electron diffractive imaging of nano-objects using a guided method with a dynamic support. *Appl. Phys. Lett.*, **95** (11) 111908.

54. Jiang, H.D. et al. (2010) Quantitative 3D imaging of whole, unstained cells by using X-ray diffraction microscopy. *Proc. Natl. Acad. Sci.*, **107** (25), 11234–11239.

55. Gulden J. et al. (2010) Coherent x-ray imaging of defects in colloidal crystals. *Phys. Rev. B*, **81** (22), 224105.

14
Sample Preparation Techniques for Transmission Electron Microscopy

Vasfi Burak Özdöl, Vesna Srot, and Peter A. van Aken

14.1
Introduction

Successful transmission electron microscopy (TEM) depends on many parameters, one of them being the preparation of high-quality specimens. With the tremendous advances in spatial and spectral resolution of state-of-the-art microscopes, the quality of the samples examined has become a major limiting factor for quantitative analysis at the nanometer scale. Most of the preparation techniques currently in use have been exploited for many years in the microscopy laboratories around the world. There is numerous literature available that describes the existing techniques in detail, such as the earlier book by Hirsch *et al.* [1], the books by Goodhew [2, 3], the textbook by Williams and Carter [4], a series of book compilation by the Materials Research Society during the 1990s [5–8], the Handbook of Microscopy [9] and, very recently, a comprehensive handbook set by Ayache *et al.* [10, 11]. Also, an interactive database has become available for sample preparation methods covering applications from materials science to biological and earth sciences [12].

Each material and each field of electron microscopy application may require its own specific preparation technique. In the following context, it will not be possible to cover all approaches systematically and in detail. Therefore, a short overview of the major preparation techniques will be followed by several examples of state-of-the-art applications, mainly for nanometer-scale characterization. For a more comprehensive and technical description of each particular method, the reader should follow the dedicated literature referred to throughout the specific chapters. Manufacturers producing sample preparation tools may provide useful and complete information for dedicated applications of the methods described here [13–16].

14.2
Indirect Preparation Methods

The indirect methods, often referred to as *replica techniques*, have been mainly used in earlier times to extract information about the surface topography or fracture surface of materials where the resolution of the nanometer-sized features was the limiting factor for scanning electron microscopy (SEM) investigation [17]. Replica techniques can be divided into four main classes as (i) *direct replicas*, (ii) *indirect replicas*, (iii) *extraction replicas*, and (iv) *freeze fracture replicas*.

Direct replicas rely on the shadowing of the material surfaces by deposition of thin metal film (usually Pt or W) at an oblique angle, which is then fixed by C deposition. The resulting shadowing metal–carbon surface film is then isolated by dissolution of the material investigated. The technique is mostly used for metals and semiconductors, but is also applicable for electron-beam-sensitive materials such as polymers. For characteristic film thicknesses such as 5 nm for Pt and 30 nm for C, the surface topography can be revealed as good as with 1 nm height and 3 nm lateral resolution [18].

Indirect replicas are prepared mainly in the absence of an appropriate solvent for the material investigated. Therefore, first, the topographical print of the surface is obtained by molding using liquid polymer. The direct replica procedure is then applied to the plastic mold, resulting in a negative impression of the surface morphology. The resolution is poorer compared to direct replicas, but surface features in the order of 10 nm are resolvable.

Extraction replicas are used mainly to investigate submicron-sized inclusions or precipitates in the metal matrix. It involves stepwise etching of the matrix material to expose the inclusions on the surface. A 10–30 nm thick carbon film is then deposited under vacuum as a supporting material. The matrix material is then further dissolved, revealing the inclusions trapped within the carbon film. The main limiting factor for this method is the availability of the solvent that not only dissolves the matrix selectively but also avoids the formation of new products such as oxide particles [19].

Freeze fracture replicas are prepared to investigate the ultrastructure of the rapidly frozen biological samples [20]. Four main steps of the method include (i) rapid freezing of the specimen, (ii) fracturing it at low temperatures (170 K or lower), (iii) making the direct replica of the newly exposed frozen surface by vacuum deposition of carbon or platinum, and (iv) cleaning the replica using bleach or acids to remove the biological material. The critical feature of the freeze fracture technique is the tendency of the fracture plane to follow a plane through the central core of frozen membranes, splitting them into half-membrane leaflets. The resulting views of the membranes give three-dimensional perspectives of cellular organizations, revealing the distribution of internal membrane proteins. Figure 14.1 displays an example of a freeze fracture overview of lipid droplets (LDs) in a lipid-laden macrophage and their association with endoplasmic reticulum (ER) membranes [21].

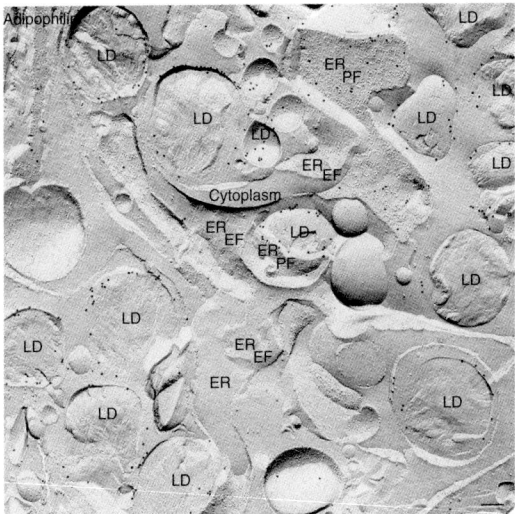

Figure 14.1 Freeze fracture replica image of lipid droplets (LDs) in a lipid-laden macrophage and their association with endoplasmic reticulum (ER) membranes. (Image reproduced from Ref. [21].)

14.3
Direct Preparation Methods

14.3.1
Preliminary Preparation Techniques

Preliminary preparation steps are necessary in most of the cases before thinning the specimen to electron transparency. The techniques suitable for individual materials systems vary depending on its properties, for example, whether it is ductile or brittle, soft or hard, porous or bulk, or conductive or nonconductive (Figure 14.2).

Sawing is necessary to cut thin slices from a bulk sample. Wheel or wire saws with abrasive blades or wires are used to obtain slices as thin as 100 μm. While cutting speed varies from millimeter per hour to millimeter per minute depending on the material, water cooling and lubricants help to prevent temperature rise. Alternatively, electrochemical saws for conducting materials or acid saws without abrasives are used to avoid mechanical damage on the sample.

Ultrasonic cutters are used to cut different shapes of small dimensions (few millimeters) from hard material specimens. A hollow cutting tool vibrates laterally at high frequency and cuts into the specimen with the aid of abrasive-containing lubricant or water.

Punching can be used for ductile metallic materials, where small discs (generally with a disc diameter of 3 mm) can be punched out from thin metal foils (100 μm).

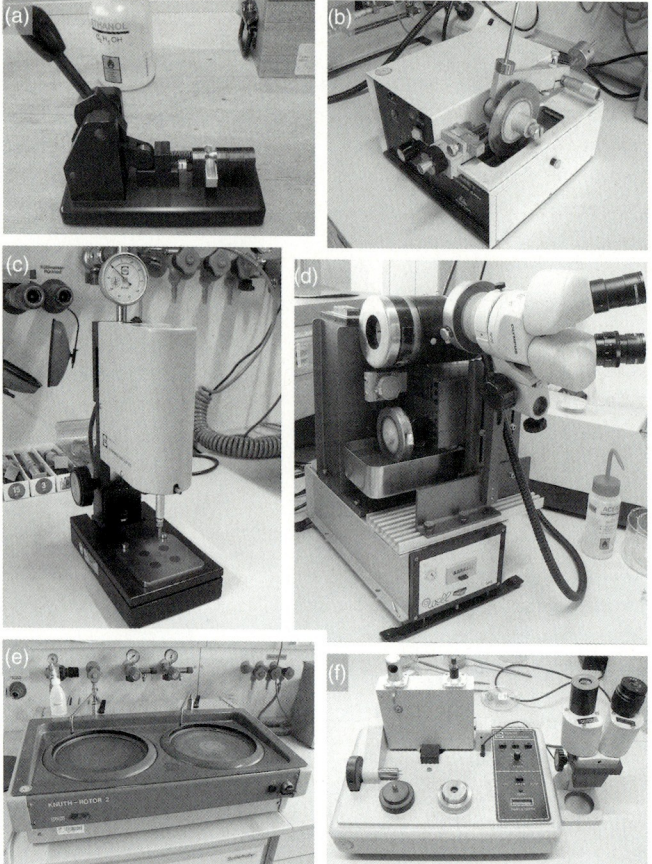

Figure 14.2 Several examples of preliminary preparation tools used in TEM specimen preparation: (a) punching tool, (b) wheel saw, (c) ultrasonic disc cutter, (d) wire saw, (e) rotary grinder/polisher, and (f) dimple grinder.

Mechanical polishing is necessary for thinning the specimen either in plan-view or cross-section geometry and to obtain smooth scratch-free and flat surfaces by removing the mechanical damage due to cutting or sawing. A rotating disc (20–30 cm in diameter) with variable speed, which holds the polishing platforms or cloths with abrasive powders spread on it, is used for polishing the specimen under water cooling. Several steps are performed starting from coarse grinding to final surface cleaning by using progressively finer sized abrasives (e.g., silicon carbide, diamond, alumina, or boron carbide, with grain sizes from 100 μm down to 1 μm). As an alternative to ultrasonic cutters or punching tools, discs can be prepared from crystals glued to 3 mm thick rods.

Dimpling is a widely used method to thin the center of a precut flat disc with a diameter of 3 mm before final thinning of the specimen by means of electrochemical polishing or ion milling. Using this procedure, the specimen thickness in the center of the disc is reduced generally from 100 to 5–10 µm. The specimen is centered and rotated on a variable-speed rotating axle, while a rotating grinding wheel made of bronze or steel grinds a dimple in the center. Progressively finer abrasives (3–0.1 µm in grain size), generally a paste containing a lubricant, are used to obtain a scratch-free final polish on both sides, while the depth of the dimple is controlled and regulated precisely by a micrometer sensor and counterweight load.

For fine particles or fibers, which are hardly possible to handle or are too soft, and for very brittle samples, *embedding* is necessary before other thinning methods. The sample is placed into a mold and immersed in a liquid resin, which is then hardened by polymerization. For porous materials, infiltration embedding can be performed, where the resin also fills into the gaps in the material. Embedding is specially needed for sample preparation using ultramicrotomy methods.

Chemical fixation is required mainly for hydrated biological samples before dehydration. Fixatives, either aldehydes or strong oxidants, are used to transform the protein gel into a cross-linked network so that the structure is preserved durin subsequent dehydration.

14.3.2
Cleavage Techniques

Cleavage is a relatively simple, but very quick and inexpensive method to prepare TEM samples without artifacts arising from further thinning procedures [22]. The technique requires materials to cleave or fracture; therefore, its application is mainly limited to hard materials such as semiconductors, glasses, silicon carbide, or sapphire. The procedure involves initiation of mechanical stress-induced microcracking (generally using a diamond tip) of a back-thinned substrate (100 – 150 µm) along weakly bonded atomic planes. For instance, for epitaxially grown compound semiconductors such as GaAs with [001] surface orientation, a 90° wedge is formed by initiating cleavage on the (110) and ($1\bar{1}0$) planes to obtain an electron-transparent edge [23]. The wedge is then mounted on a support grid using conductive epoxy as shown in Figure 14.3a.

A more delicate variation of cleaving, known as the small-angle cleaving technique (SACT), has been developed to achieve larger electron-transparent regions. In this case, the sample is first scribed on the prethinned side at a small angle to a standard cleavage plane (15–30°), and than cleaved along the scribe line. For a Si (001) wafer, for instance, the back of the wafer is scribed and cleaved along a {120} plane. The cleavage can preferably be performed under water, as the liquid will control the stress field during fracture, improving the quality of the cleaved edge [25]. Sufficient back-thinning is essential to minimize the artifacts such as cleavage steps or tears. A second cleave is made from the front of the wafer. The scribe is made along a {110} plane and cleaved along this standard cleavage plane, resulting in a small-angle wedge. For a more detailed description and various applications, the reader is referred to the following Refs. [22, 26, 27].

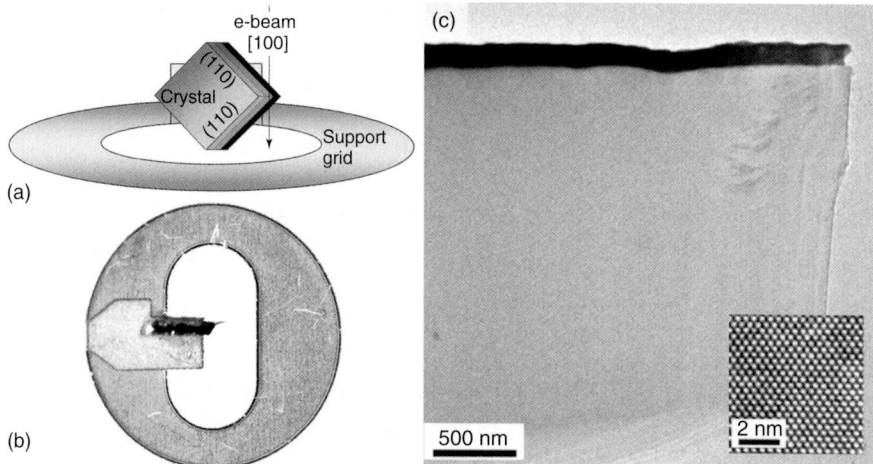

Figure 14.3 (a) Orientation of a 90° wedge crystal on a special support grid. (b) Visible light microscopy image of a Yb/Si film prepared using SACT. (c) Si wedge with a silicide film. The inset is an HRTEM image of the substrate. (Images reproduced from Ref. [24].)

While the short preparation time (less than an hour) enables to examine the films or coatings very soon after their growth, the technique can be used as a quick prethinning method before focused-ion milling as well [28]. It is particularly useful for samples that contain amorphous material, which may not be prepared using ion-assisted methods because of the severe artifacts induced by ion implantation. Figure 14.3b displays the view of a Yb/Si specimen prepared by SACT [24]. At the tip of the wedge, the specimen is thin enough even for high-resolution transmission electron microscopy (HRTEM), as shown in Figure 14.3c.

14.3.3
Chemical and Electrolytic Methods

Preparation of electron-transparent specimens by means of chemical and electrolytic methods has been developed mostly in the earlier years of electron microscopy. These methods have been applied to a broad range of materials [29–31], but mainly to metallic and semiconductor materials to avoid strain hardening and preparation artifacts due to high temperatures or ion irradiation [32]. The main principle of the methods relies on the polishing of the thin slice of material by chemical dissolution. In the case of electrolytic methods, the thinning occurs as an anodic electrochemical dissolution of a conductive material. For both chemical and electrolytic thinning, there are two basic techniques. In the *window technique*, the mechanically prethinned specimen is immersed into an appropriate solvent to obtain a thin foil with relatively flat surfaces. The area of interest is defined by isolating the rest of the specimen from the solvent. In the more widely

14.3 Direct Preparation Methods

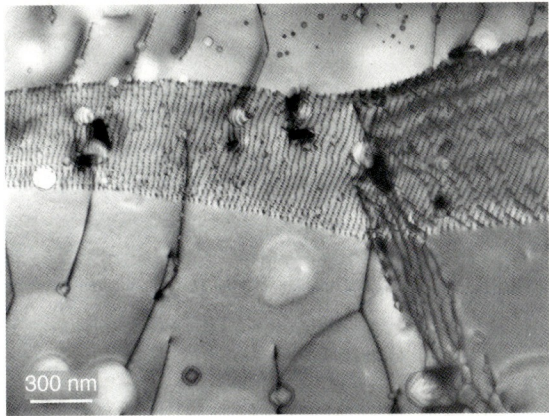

Figure 14.4 Bright-field TEM image of a dislocation network in a polycrystalline Cu specimen prepared by electrolytic etching. Potential difference of 35 V at 100 mA was applied to etch the specimen in HNO_3/Methanol solution (1 : 2 volume ratio) at $-50\,°C$. (Specimen courtesy of M. Kelsch, unpublished results.)

used *twin-jet polishing technique*, the electrolyte is introduced as a thin jet through a nozzle toward the surfaces of the precut specimen to localize the dissolution effect.

For either method, the availability of the appropriate solvents is the key factor. A broad range of recipes for etching solutions can be found in the literature [32, 33]. The solution temperature is kept low to control the etching rate. In the case of electrolytic polishing, the applied voltage and current as well as the electrode materials influence the thinning rate. In many cases, it is difficult to find the appropriate electrolytic solution, potential condition, current, and adequate temperature for complex or multiphase materials. However, if the correct conditions are fulfilled, this technique has been successfully applied to multiphase materials such as metal alloys, for example, Cu–Cr or Mo–C [34] and semiconductor multilayer materials, for example, heterostructure systems of GaAs, AlAs, InP, and HgTe [35, 36].

Since the thinning process does not introduce mechanical stress on the specimen, etching methods are particularly essential for the investigation of defects in deformed materials. The bright-field TEM image shown in Figure 14.4 reveals the dislocation network in a deformed polycrystalline Cu alloy prepared by electrolytic polishing. The original defect density is maintained by avoiding any preparation-related strain hardening.

14.3.4
Ion Beam Milling

One of the most widely spread specimen preparation techniques for TEM is ion beam milling [9, 37]. The basic setup for ion milling requires an ion gun attached to a vacuum chamber (with a pressure in the order of 10^{-5} mbar). It involves bombardment of a mechanically prethinned specimen with energetic ions (typically Ar^+), which are accelerated and formed into a focused ion beam (FIB) (Figures 14.5 and 14.6 display the basic setup and examples for ion milling equipment, respectively). Mechanical thinning (down to $10 - 30$ μm) is required before ion milling, which involves dimple grinding and polishing of the sample. The specimen surfaces should be smoothly polished, which is favorable for the final high quality of the sample.

In addition to the improvements in the specimen stage in terms of design and materials used [38], advances in modern electronics have allowed very precise manipulation of the ion beam in several ways such as modulating or focusing it tightly (e.g., $350 - 800$ μm full width at half-maximum (FWHM at 5 keV for broad beam guns) or controlling the incidence angle (typically from $10°$ to $1°$) by retarding fields [39]. Different setups, such as single- or double-sector thinning,

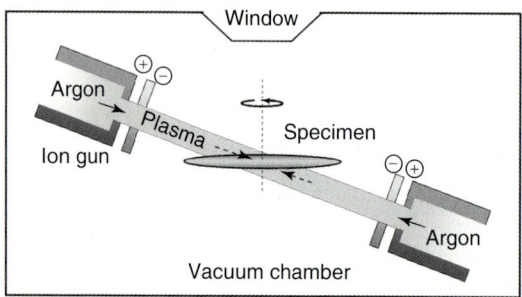

Figure 14.5 Schematic diagram of an ion milling machine. Ar gas is introduced into the ionization chamber where potential difference creates Ar ions accelerated toward the rotating specimen.

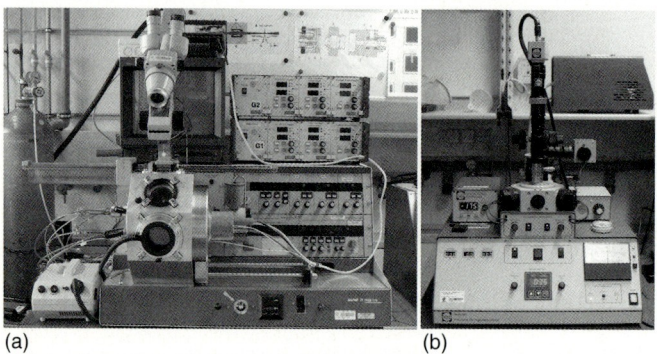

Figure 14.6 Low-energy ion milling equipment showing the evolution of device development: (a) a modified ion milling machine equipped with low-energy ion guns and cooling stage. (b) A commercially available ion milling device.

accurate alignment and positioning of ion beams on the specimen surface and rotation of the specimen (generally at a few revolutions per minute) have made ion milling a precise and effective method, and enabled not only plan-view but especially cross-section sample preparation.

In the earlier efforts at improving the final surface quality of the specimens, geometrical models for ion beam erosion revealed that the main characteristic features of the milling process (sputtering rate, topographical modifications, irradiation damage, etc.) depend on the incidence angle of the ion beam with respect to the specimen surface [40]. During preparation of cross-section TEM samples of multilayer structures, for instance, differential thinning effects can be minimized by milling at low angles (down to 1°), so that the layers with slow thinning rate protect the fast milling layers from the ion beam. The preferential thinning at the interfaces can be further reduced by rocking the specimen, so that the ion beam does not travel only along the interfaces [41]. Figure 14.7 shows defocused dark-field TEM images of a Si/SiGe heterostructure to demonstrate the influence of the ion

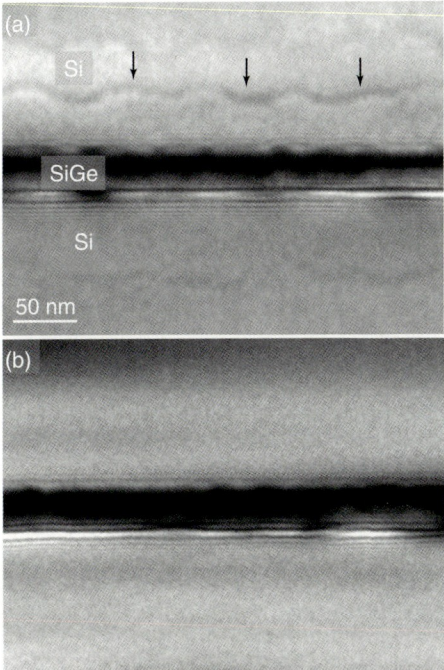

Figure 14.7 Largely defocused dark-field TEM images of a Si/SiGe heterostructure to demonstrate the topographical artifacts due to preferential thinning. (a) The specimen is thinned at both surfaces, top and bottom, using Ar ions and an accelerating voltage of 4 kV at an incidence angle of 8°, while oscillating at a sector field of ±20°. Preferential sputtering across the interface and redeposition of sputtered material modify the surface topography, as marked with black arrows. (b) The artifacts due to preferential thinning are minimized by reducing the incidence beam angle to 1°, while rotating the specimen continuously. The defocus in (a) and in (b) is identical.

beam incidence angle and of the rotation of the specimen on the topographical surface quality. Using an incidence angle of 8° at 4 kV accelerating voltage of the Ar ions and a specimen oscillation limited to ±20°, preferential thinning coupled with redeposition of sputtered material [42] results in topographical modification as marked with black arrows in Figure 14.7a. Reducing the incidence angle down to 2° and allowing complete rotation of the specimen, these artifacts can be minimized as shown in Figure 14.7b.

Another detrimental effect of ion beam sputtering is the formation of amorphized surface layers, which has been one of the major limiting factors for quantitative analysis of HRTEM images [42]. Many artifacts, such as radiation-induced defect

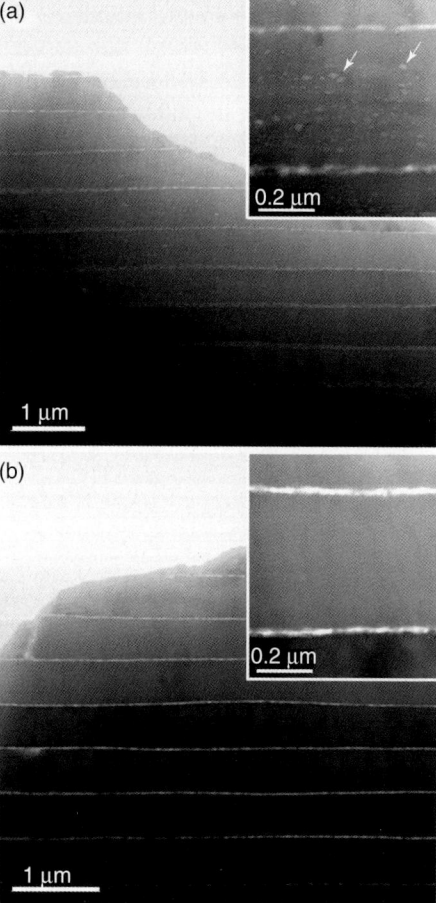

Figure 14.8 Bright-field STEM images of nacre consisting of lamellar material, where polygonal aragonite platelets are separated by thin organic layers: (a) conventional Ar ion milling (at 3 keV and with an incidence angle of 8° in a single-sector mode) introduces voidlike artifacts in aragonite, as marked with white arrows. (b) LN$_2$-cooling minimizes the void formation, revealing the uniform structure of the platelets.

agglomeration or segregation, are further triggered by beam heating of the TEM specimen during the thinning process [43]. Both experimental and theoretical studies have shown that under normal ion milling conditions, temperatures of up to several hundreds of degree celsius can be generated at the specimen surface [44]. The improvements in specimen stage design, including the specimen holders assisted with liquid nitrogen (LN_2) cooling, helped to avoid excessive heating of the specimen. The effect of beam-induced heating on sensitive materials, such as biological crystalline materials, is demonstrated in Figure 14.8. Figure 14.8a shows a bright-field scanning transmission electron microscopic (STEM) image of nacre consisting of aragonite platelets separated by thin organic layers. Excessive voids (as marked with arrows) are observed for conventional ion milling conditions at 3 keV without cooling. TEM specimens prepared using similar parameters but assisted with LN_2-cooling reveals relatively uniform aragonite platelets without the formation of large voidlike artifacts (Figure 14.8b) [45].

However, the most remarkable and pronounced improvement in the surface quality is achieved with the advent of stable ion sources operating at ultralow energies (in the energy regime from 100 to 500 eV) producing beam sizes as small as a few microns [46]. Figure 14.9a,b shows HRTEM images of Si/SiGe interfaces obtained from two specimens prepared under different conditions. The first specimen (Figure 14.9a) was thinned with 4 keV ions at an incident angle of 8° at both sides. In addition to the reduced contrast due to the additive noise from amorphized surface layers, the image clearly shows large contrast variations, which are more pronounced in the vicinity of the interface and prohibit both reliable qualitative and quantitative analysis. The second specimen is prepared by lowering the ion beam energy consecutively (from 4 to 0.2 keV), while thinning the specimen from both sides in a LN_2 cooled stage. By reducing the amorphous damage-layer thickness, high-contrast HRTEM images can be obtained as shown in Figure 14.9b. In addition, the pronounced contrast variations are eliminated revealing the defect-free structure of the interface.

Reactive ion techniques are also available to reduce the artifacts arising from Ar ion milling. Reactive ion beam etching (RIBE), where the inert gas is replaced by reactive gases (mainly halogen-containing gases), has shown to reduce the ion-milling-induced defects, for example, in II–VI compound semiconductors such as CdTe and ZnS by using iodine ions for thinning [47]. Chemically assisted ion beam etching (CAIBE), where the reactive gas is kept in contact with the specimen while it is being milled with Ar ions, can be applied to III–V group semiconductors such as InP, where the formation of In islands by Ar milling is avoided [48].

14.3.5
Tripod Polishing

The tripod polishing method was originally developed to prepare site-specific Si semiconductor device TEM specimens with large transparent areas [49]. The application of the method has been extended to a broader range of hard materials

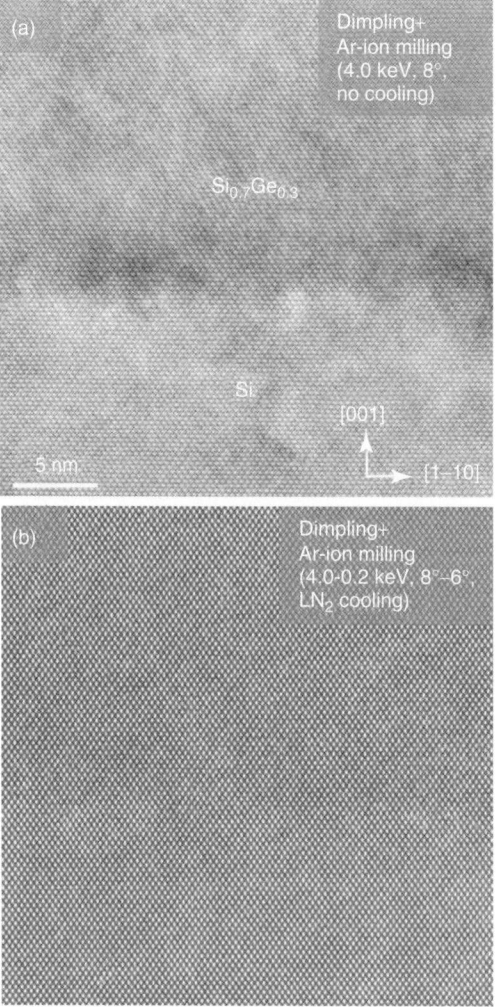

Figure 14.9 HRTEM images of a Si/SiGe interface obtained from two different specimens under similar imaging conditions using a JEOL 4000EX TEM operating at 400 kV. (a) The specimen is thinned with 4 keV ions at an angle of 8° from both sides simultaneously without cooling. (b) The sample is simultaneously thinned from both sides using a liquid nitrogen cooled stage with energies consecutively lowered from 4 keV down to 0.2 keV, while the incident beam angle was slightly decreased from 8° to 6° in the final low-energy ion milling step.

such as other types of semiconductors and ceramics or composite materials [50]. The basic setup relies on mechanical wet polishing of a precut specimen on a rotating wheel using a handheld tripod sample holder (Figure 14.10a). A sequence of progressively finer diamond lapping films (with a granularity ranging from 30 to 0.1 μm) is used as abrasive material while reducing the rotation speed

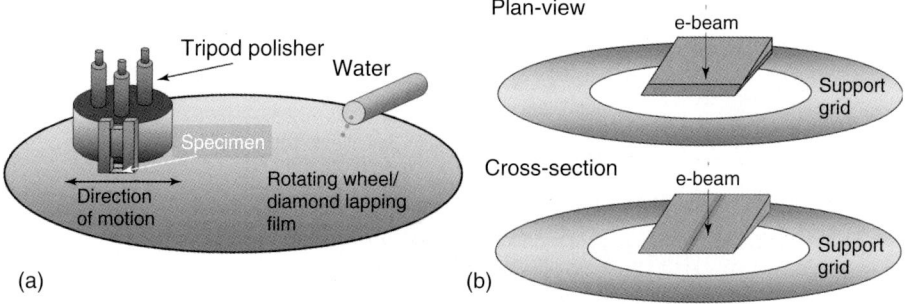

Figure 14.10 (a) Schematic diagram of tripod polishing. (b) Diagram of tripod wedge specimens prepared for plan-view and cross-section investigations.

(typically from 100 to 10 rpm) accordingly. The specimen surface is then briefly polished with a cloth wheel using colloidal silica slurries (with a grain size of 0.05 – 0.02 µm) to attain a scratch-free final polished surface. By introducing an angle between the first polished surface and the opposite surface during final polishing, it is possible to obtain an electron-transparent wedge. In addition to preparation of TEM specimens in either plan-view or cross-section format (Figure 14.10b), the method is widely used for prethinning before FIB milling or for backside polishing of integrated circuits [51].

Although mechanical thinning by tripod polishing is a delicate operation, the introduction of automated and mechanically more stable tripod polishing systems (Figure 14.11) has allowed controlling key parameters such as specimen thickness, polishing speed, applied load, and wedge angle with high precision and high reproducibility [52]. Consistent sample rotation, oscillation, and load provide uniform material removal and eliminate the artifacts that are associated with manual polishing. The desired wedge angle remains intact throughout the thinning process by letting only the sample contact with the abrasive.

In Figure 14.12, a cross-section bright-field image of an array of pMOS transistors extracted from an integrated circuit demonstrates the high precision in TEM sample preparation achieved for semiconductor devices with repeating structures. Much larger electron-transparent areas with uniform thickness are achievable compared to conventional preparation methods. In addition, by optimizing the wedge angle depending on the material system investigated, bending of the thinner regions of the specimen and, thus, artifacts such as bending contours are avoided.

14.3.6
Ultramicrotomy

Ultramicrotomy has been routinely used for the preparation of soft biological materials using glass knives since the 1950s [53]. The development of instrumentation

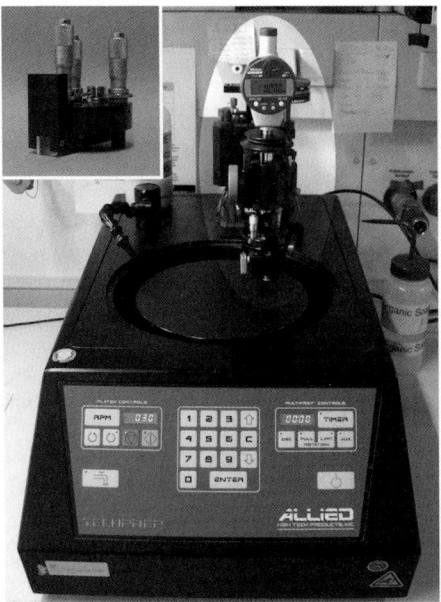

Figure 14.11 Automated tripod polishing machine for precise control of specimen thickness, wedge angle, and applied load. A simpler handheld version is shown in the upper inset.

and introduction of defect-free diamond knives in the early 1970s enabled the use of ultramicrotomy also for the preparation of different hard materials such as metals, ceramics, semiconductors, and composites [54, 55].

Ultramicrotomy is a technique used to prepare thin slices with a thickness <100 nm. The preparation process is purely mechanical, relying on plastic deformation and/or microcrack initiation that propagates and extends into the sample. Bulk materials that are strong enough to resist the cutting force and remain rigid can be prepared without any additional support. When the sample is soft, porous, small, or too brittle, embedding in resin or in epoxy is necessary. A typical basic setup of an ultramicrotome device is demonstrated in Figure 14.13. The procedure involves trimming of the sample block on one side to a pyramidal shape in order to produce a very small surface area (Figure 14.14a). Harder materials require smaller areas. The sample is mounted on an ultramicrotome holder to section thin slices onto a liquid, in most cases distilled water. The sections should be then collected onto the supporting grids (Figure 14.14b).

The technique can be used at room temperature or at low temperature (cryo-ultramicrotomy). Samples prepared by ultramicrotomy do not undergo any irradiation damage, chemical mixing, or preferential thinning. In addition, many thin sections can be prepared in a relatively short time. However, mechanical damage of the specimens, such as compression in soft samples or formation of cracks in brittle samples, can still be a limitation.

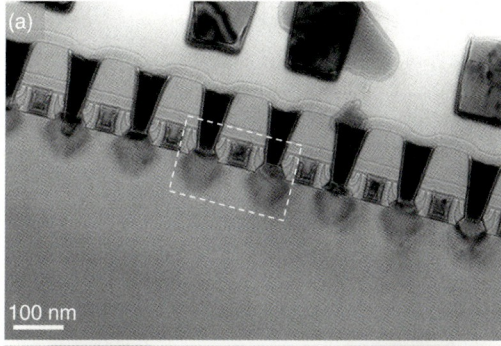
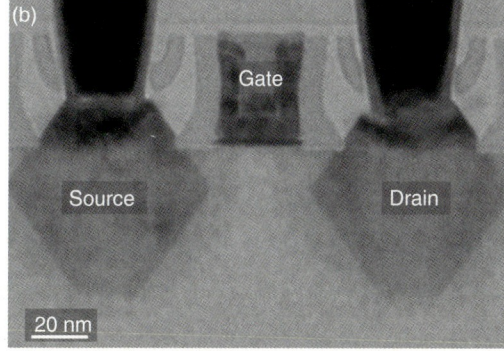

Figure 14.12 (a,b) Bright-field TEM image of a pMOS transistor array (the enlargement displays the main components of the single transistor) obtained from a cross-section specimen prepared by tripod polishing. Relatively large electron transparent areas with uniform thickness are achievable compared to conventional sample preparation, for example, using dimple grinding and ion milling.

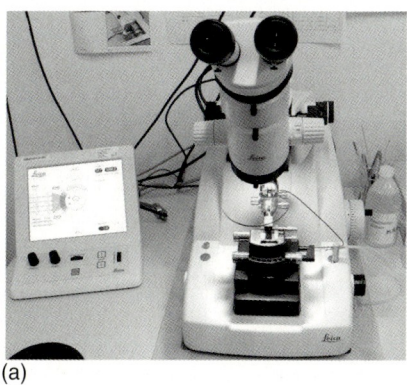
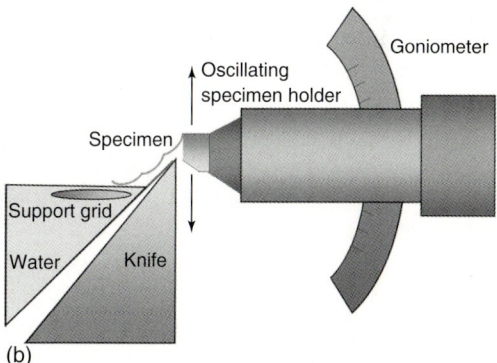

Figure 14.13 (a) Ultramicrotome for sectioning thin lamellas from biological and industrial materials. (b) Schematic diagram of ultramicrotomy: The specimen, generally embedded in resin or epoxy, is moved precisely across the knife edge with the help of an oscillating specimen holder. Thin slices of material are then collected from a liquid medium onto the specimen grid.

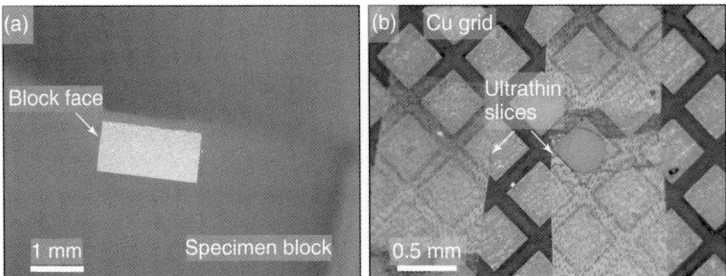

Figure 14.14 (a) Trimmed specimen block with a pyramidal shape. (b) Ultrathin slices of material on a mesh Cu grid. (Sample courtesy of B. Bussmann, unpublished results.)

14.3.7
Focused Ion Beam Milling

Although the early development of FIB workstations has been driven by the microelectronics industry because of their unique site specificity, performance for chip modification and circuit failure analysis, the FIB is being realized as a powerful characterization and sample preparation tool for almost any material system ranging from hard materials [56, 57] such as metals, semiconductors, or ceramics, to soft matter [58, 59] such as polymers or biological materials.

The principle of sample preparation in an FIB instrument, which generally consists of an ion gun attached to an SEM column (referred as *dual-beam* or *cross-beam*), relies on sputtering off the material by scanning the ion beam over the region of interest (Figure 14.15a). Originally, Ga was used as a liquid metal ion source (LMIS) because of its low melting point – which is near room temperature – and its ease of focusing it into a very fine probe. Over the years, the resolution of FIB milling has improved dramatically from 50–100 to 5 nm or better, whereas higher milling rates are achieved with the increased beam currents (higher than 50 nA) [60]. More recently, inductively coupled plasma (ICP) sources are incorporated into FIB systems [61]. The use of heavier ions such as Xe at higher beam currents enabled much higher sputter rates, allowing preparation of larger area cross sections [62].

Three fundamental FIB techniques to prepare TEM lamellae are (i) the H-bar or trench method [63], (ii) the *in situ* lift-out method [64], and (iii) the *ex situ* lift-out method [65]. Before cutting, a metal or a carbon layer should be deposited on the area of interest to protect the surface during the specimen milling process. In the *H-bar* or *trench technique*, a thin slice of material should be mechanically polished to a thickness of ∼50 μm, fastened to a special support grid (Figure 14.15b), and mounted vertically in the FIB.

The FIB is used to cut two trenches from each side and a thin electron-transparent slice or lamella remains connected to the bulk material. The *in situ* lift-out method relies on extraction of a thicker lamella (up to ∼5 μm) from a site-specific region, which is then transferred to a TEM half grid using a micromanipulator. The

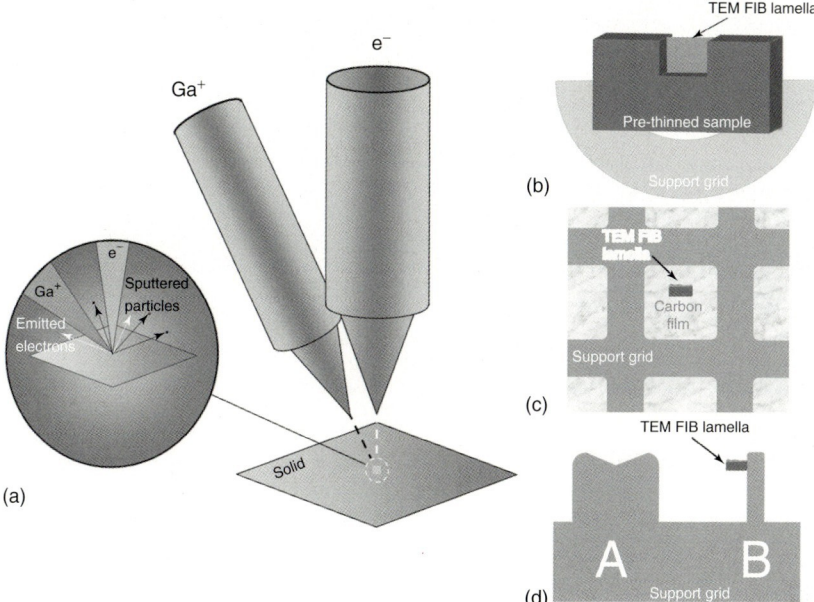

Figure 14.15 (a) Principle of a two-beam (ion–electron) FIB. Schematic diagrams of TEM samples prepared by different FIB thinning techniques: (b) H-bar or trench method, (c) *ex situ* lift-out, and (d) *in situ* lift-out.

probe of the micromanipulator is attached to the sample by the FIB metal deposition and then the sample that was lifted out is attached to a TEM grid (Figure 14.15d). Consecutively, the TEM sample is further thinned to electron transparency. TEM samples prepared in such a way can be additionally thinned in the FIB. For the *ex situ lift-out method*, a lamella from a site-specific region is milled to electron transparency. Both sides and the base of the sample are cut by the ion beam. The sample is then transferred from the FIB onto a carbon-coated TEM grid (Figure 14.15c) using an electrostatic micromanipulator under an optical microscope. Samples prepared in this way are not very suitable for further thinning. Lift-out methods are significantly faster compared to the H-bar technique, since no mechanical prethinning is required.

Despite the fast and precise microsampling achieved over the years, the application of FIB sample preparation for quantitative analysis at the nanometer scale has been limited because of severe sample damage resulting from heavy ion implantation during the sputtering process. Various procedures as well as different ion sources have been applied to reduce or repair these damages [66, 67]. For semiconductor materials, amorphized surface layers (typically 30 nm thick sidewall damage for Si samples prepared at 30 keV) are formed proportional to the accelerating energies used for FIB milling. Improved FIB columns operating at energies as low as 1–2 keV have been available, reducing the modified layer thickness to a

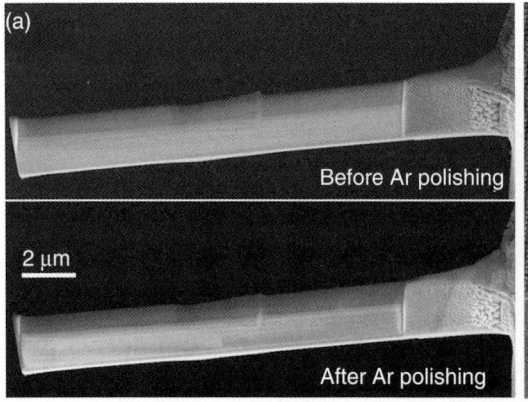
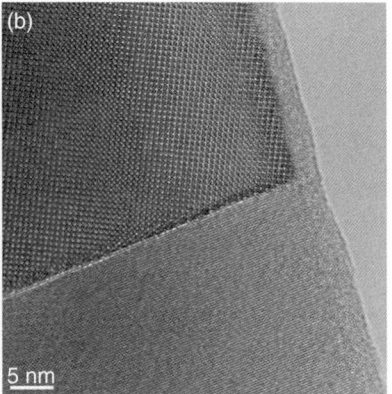

Figure 14.16 *In situ* low-energy Ar ion milling: (a) Lift-out lamella of a YAG bicrystal before and after 1 keV Ar polishing. Monitoring the polishing procedure by SEM allows precise control of the minimum thickness as well as the surface damage of the lamella. (b) HRTEM image of the bicrystal grain boundary after final polishing. (Sample courtesy of K. Hartmann, TEM images courtesy of S. Irsen, unpublished results.)

few nanometers [68]. In addition, new column designs such as triple beam systems are commercialized, combining a focused low-energy noble gas ion beam column (Ar^+, Xe^+) with a FIB and an SEM column. Small-diameter low-energy scanning Ar ion beams (500–1000 V, 10 nA, 100 µm) coincident with SEM imaging provides precise control of the lamella thickness during the final polishing process [69]. Figure 14.16a displays the lift-out lamella of a YAG-bicrystal material before and after the *in situ* Ar ion polishing at 1 keV. Reduced surface damage and roughness improves the HRTEM image quality, allowing the structural characterization of the bicrystal grain boundary (Figure 14.16b).

Although the FIB is more frequently used in the field of semiconductor materials, with the development of the *in situ* lift-out technique it has become a powerful tool to prepare thin sections from mechanically unstable material systems such as polymers, biological samples, or materials with porous structures. Figure 14.17 demonstrates the use of the FIB for the preparation of samples consisting of dense SiC fibers in a relatively porous and polycrystalline SiC matrix (Figure 14.17a). Different thinning properties of the fibers and the polycrystalline matrix lead to mechanically unstable samples, as has already been shown on the same samples that were prepared by classical ion milling or wedge-shape polishing methods [70]. Thickness differences across the fiber–matrix interface hindered detailed studies across that interface. The TEM sample preparation of this material with the FIB was a challenging enterprise. First, a platinum strap was deposited directly onto the region of interest in order to protect the sample and to additionally improve the mechanical support. FIB milling was then carried out using the gallium ion source operated at 30 kV (Figure 14.17b). The TEM lamella was lifted out using an *in situ* micromanipulator (Figure 14.17c) and fastened to a Cu grid. Pores in the matrix

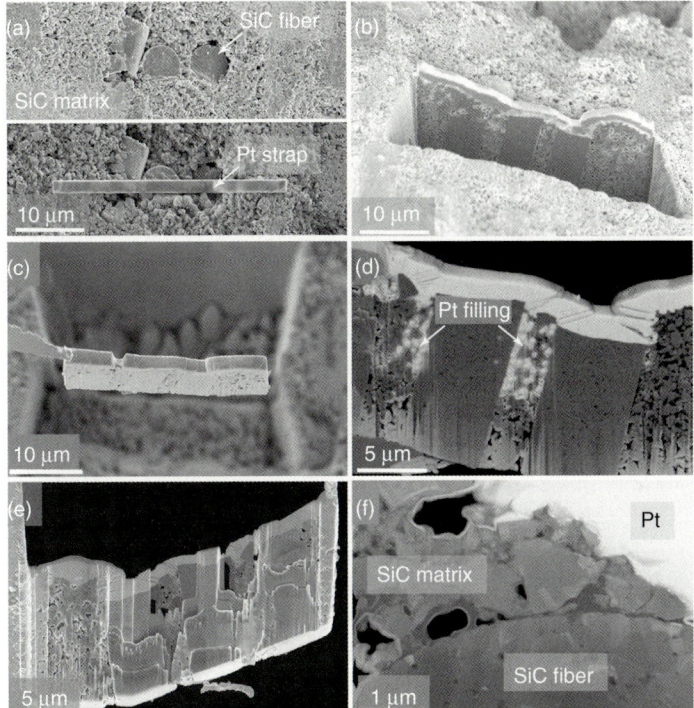

Figure 14.17 *In situ* lift-out technique to prepare cross section from a porous composite material: (a) Secondary electron image of SiC fibers in a porous SiC matrix and the Pt layer deposited onto the area of interest. (b) Sectioning of a thin lamella by sputtering of the material by focused Ga ions at 30 keV. (c) Extracting the lamella using a micromanipulator to mount on a support grid. (d) Filling the pores in the matrix with Pt before final thinning. (e) TEM lamella after the final thinning at 5 keV. (f) Annular dark-field STEM image of the cross section. (Specimen courtesy of T. Toplisek, unpublished results.)

were filled with Pt (Figure 14.17d) and only selected parts were thinned using Ga ions to a thickness of around 50–70 nm (Figure 14.17e). In this way, the prepared samples are robust enough to be handled and additionally they offer thin areas for a detailed TEM investigation (Figure 14.17f).

14.3.8
Combination of Different Preparation Methods

Over the past decades, sample preparation techniques have evolved remarkably in terms of precision, reliability, and reproducibility parallel to the requirements for state-of-the-art characterization techniques. However, depending on the physical, chemical, or structural nature of the material to be characterized, a combination of different preparation methods are generally applied to overcome the challenges as well as to reduce the preparation-related artifacts.

Tripod polishing is a useful technique to prepare wedge specimens from hard materials with relatively uniform and large electron-transparent areas. The control of mechanical stability at the final stage of thinning might be difficult for brittle materials. In addition, residual polishing artifacts rising from chemical slurries or fine particles may alter the surface quality. Alternatively, combining tripod thinning with successive final low-energy ion milling (typically 2.0–0.5 keV), it is even possible to obtain high-quality specimens for HRTEM, as shown in Figure 14.18. The figure displays an HRTEM image of a pMOS transistor, where uniform image contrast across the channel region allows quantitative analysis such as nanometer-scale strain mapping, for instance.

Modification in the composition of irradiation-sensitive alloys and compounds as a consequence of ion bombardment has been known for many years. Tripod polishing can be used as an alternative to dimple grinding to achieve thinner specimens before ion milling. Figure 14.19 displays an HRTEM image of the active region of a light-emitting diode consisting of ion-beam-sensitive InGaN quantum wells, where the specimen was prepared by tripod polishing [71]. Since the time necessary for final ion beam thinning is remarkably reduced (typically on the order of 5–10 min at 2.0 keV for Si), compared to conventional techniques (typically several hours at 4 keV for a dimpled Si specimen), sample preparation of such materials can be realized without alteration of the original structure, allowing reliable quantitative analysis.

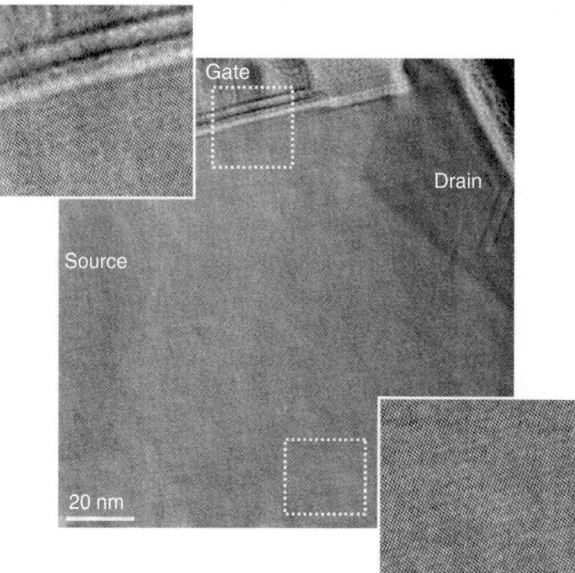

Figure 14.18 HRTEM image of a pMOS transistor channel region obtained from a specimen prepared by automated tripod polishing followed by low-energy Ar ion milling at 0.5 keV. A homogeneous specimen thickness and flat exit surfaces are necessary to obtain HRTEM images with uniform contrast across the field of view (see the enlargements from dashed squares).

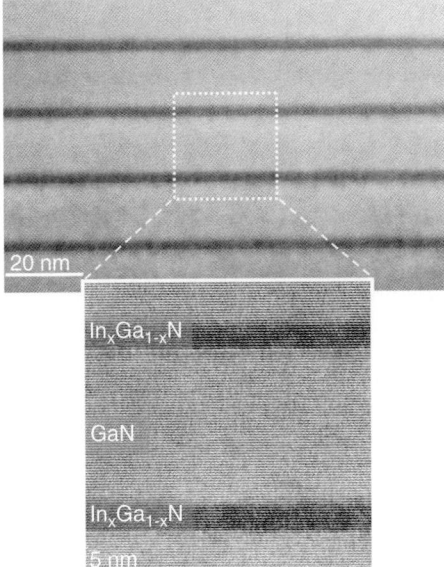

Figure 14.19 (0002) HRTEM image of the active region in a high-efficiency light-emitting diode. Two 2.5 nm thick InGaN quantum wells with sharp bottom and diffused upper interfaces are shown in the enlargement. Reduced ion milling times for the final ion thinning successively to tripod polishing minimizes the alteration of composition and morphology and reveals the original structure of the ion-beam-sensitive InGaN multiquantum wells for quantitative analysis.

The effects of different TEM sample preparation techniques on sensitive biological mineralized materials have been investigated through the preparation of human tooth enamel [72]. Crushed enamel deposited on a holey carbon grid and samples prepared by ultramicrotomy, FIB, as well as by tripod polishing with subsequent ion milling with cooling are shown in Figure 14.20. In samples prepared by ultramicrotomy, enamel material appears shattered (Figure 14.20b). Present investigations have shown that smaller voids are formed in enamel crystals in samples prepared by FIB (Figure 14.20c); in samples prepared by other techniques, such features were not observed. Therefore, it is reasonable to assume that the voids in enamel are artifacts produced by the FIB preparation. According to the present studies, a combination of tripod polishing and ion milling with LN_2 cooling (Figure 14.20d) is the most suitable technique for the preparation of human teeth enamel material.

14.4 Summary

Being an essential part of TEM, specimen preparation techniques have been evolved and refined over the years parallel to the advances and the requirements

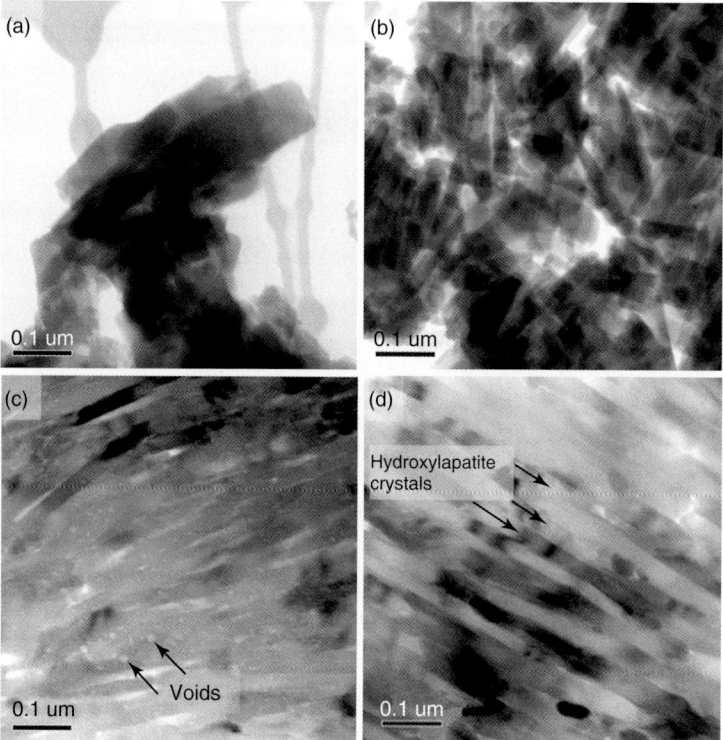

Figure 14.20 Bright-field STEM images of human tooth enamel prepared by different techniques: (a) crushed enamel on carbon holey support film, (b) shattering of enamel occurs during sectioning by ultramicrotomy, (c) artifacts such as void formation are observed during FIB milling, and (d) tripod polishing with subsequent ion milling with LN_2 cooling preserves the structure of hydroxylapatite crystals.

in state-of-the-art research. Depending on the nature of the material system and the specific problem to be investigated, different methods are generally employed to identify the preparation-related modification of the original structure. Moreover, it is often necessary to combine two or more methods to determine the intrinsic microstructure and chemical properties of the material without adding extrinsic defects or altering its chemistry. One has to keep in mind that the quality of the acquired experimental data is at least directly proportional to the quality of the specimen.

Acknowledgments

The authors are thankful to Ute Salzberger, Marion Kelsch, Birgit Bussmann, Bernhard Fenk, and Fritz Phillipp for their valuable contributions. The financial support from the European Union under the Framework 6 program under the contract for an Integrated Infrastructure Initiative, Reference 026019 ESTEEM, is gratefully acknowledged.

References

1. Hirsch, P.B., Nicholson, R.B., Howie, A., Pashley, D.W., and Whelan, M.J. (1965) *Electron Microscopy of Thin Crystals*, Butterworths, London.
2. Goodhew, P.J. (1985) Thin foil preparation for transmission electron microscopy, in *Practical Methods in Electron Microscopy*, vol. 11 (ed. A. Glauert), Elsevier, Amsterdam.
3. Goodhew, P.J. (1984) *Specimen Preparation for Transmission Electron Microscopy of Materials*, Bios Scientific Publishers, Oxford.
4. Wiliams, D.B. and Carter, C.B. (2009) *Transmission Electron Microscopy: A Text Book for Materials Science*, Springer, New York.
5. Brawman, J.C., Anderson, R.M., and McDonald, M.L. (eds) (1988) *Specimen Preparation for Transmission Electron Microscopy of Materials*, vol. 115, Materials Research Society, Pittsburgh.
6. Anderson, R.M. (ed.) (1990) *Specimen Preparation for Transmission Electron Microscopy of Materials II*, vol. 199, Materials Research Society, Pittsburgh.
7. Anderson, R.M., Tracy, B., and Bravman, J. (eds) (1992) *Specimen Preparation for Transmission Electron Microscopy of Materials III*, vol. 254, Materials Research Society, Pittsburgh.
8. Anderson, R.M. and Walck, S.D. (eds) (1997) *Specimen Preparation for Transmission Electron Microscopy of Materials IV*, vol. 480, Materials Research Society, Pittsburgh.
9. Barna, A., Radnoczi, G., and Pecz, B. (1997) in *Handbook of Microscopy: Applications in Materials Science, Solid-state Physics and Chemistry* (eds S. Amelinckx, D. van Dyck, J., van Landuyt, and G., van Tendeloo), Wiley-VCH Verlag GmbH, Weinheim, pp. 751–801.
10. Ayache, J., Beaunier, L., Boumendil, J., Ehret, G., and Laub, D. (2010) *Sample Preparation Handbook for Transmission Electron Microscopy: Techniques*, Springer, Berlin.
11. Ayache, J., Beaunier, L., Boumendil, J., Ehret, G., and Laub, D. (2010) *Sample Preparation Handbook for Transmission Electron Microscopy: Methods*, Springer, Berlin.
12. http://temsamprep.in2p3.fr/ (accessed 14 October 2010).
13. http://www.gatan.com/products/specimen_prep/ (accessed 15 October 2010) Gatan, Inc. Corporate Headquarters 5794 W. Las Positas Blvd. Pleasanton, CA 94588.
14. http://www.southbaytech.com/ (accessed 15 October 2010)South Bay Technology, Inc. West Coast Corporate Office 1120 Via Callejon San Clemente, CA 92673.
15. http://www.fishione.com/ (accessed 15 October 2010) E.A. Fischione Instruments, Inc. 9003 Corporate Circle Export, PA 15632.
16. http://www.leica-microsystems.com/products/electron-microscope-sample-preparation/ (accessed 15 October 2010) Leica Mikrosysteme Vertrieb GmbH Ernst-Leitz-Straße 17-37 Wetzlar, D-35578 Germany.
17. Willison, J.H.M. and Rowe, A.J. (1980) in *Practical Methods in Electron Microscopy*, vol. 8 (ed. A.M. Glauert), Elsevier, Amsterdam, pp. 171–301.
18. Adachi, K.K., Hojou, K., Katoh, M., and Kanaya, K. (1976) High resolution shadowing for electron microscopy by sputter deposition. *Ultramicroscopy*, **2**, 17–29.
19. Carpenter, G.J.C., Yelim, J.N.G., and Phaneuf, M.W. (1994) Extraction of second phases from magnesium and aluminum alloys for analytical electron microscopy. *Microsc. Res. Tech.*, **28**, 422–426.
20. Severs, N.J. (2007) Freeze-fracture electron microscopy. *Nat. Protoc.*, **2**, 547–576.
21. Robenek, H. and Severs, N.J. (2007) Recent advances in freeze-fracture electron microscopy: the replica immunolabeling

technique. *Biol. Proced. Online*, **10** (1), 9–19.

22. Ayache, J., Beaunier, L., Boumendil, J., Ehret, G., and Laub, D. (2010) *Sample Preparation Handbook for Transmission Electron Microscopy: Techniques*, Springer, Berlin, pp. 160–177.

23. Hetheringron, C.J.D. (1988) in *Specimen Preparation for Transmission Electron Microscopy*, vol. 115 (eds J.C. Bravman, R.M. Anderson, and M.L McDonald), Materials Research Society, Pittsburgh, pp. 143–148.

24. Nowak, J.D., Song, S.H., Campell, S.A., and Carter, C.B. (2008) in *Microscopy of Semiconducting Materials 2007*, vol. 120(5) (eds A.G. Cullis and P.A. Midgley), Springer, Bristol, pp. 333–336.

25. McCaffrey, J.P. (1991) Small-angle cleavage of semiconductors for transmission electron microscopy. *Ultramicroscopy*, **38**, 149–157.

26. Scott, D.W. and McCaffrey, J.P. (1997) in *Specimen Preparation for Transmission Electron Microscopy IV*, vol. 480 (eds R.M. Anderson and S.D. Walck), Materials Research Society, Pittsburgh, pp. 3–42.

27. http://www.southbaytech.com/appnotes/60pre-thinning for fib tem sample preparation using the small angle cleavage technique.pdf/ (accessed 15 October 2010)South Bay Technology, Inc. West Coast Corporate Office 1120 Via Callejon San Clemente, CA 92673.

28. Walck, S.D. and McCaffrey, J.P. (1997) The small angle cleavage technique applied to coatings and thin films, *Thin Solid Films*, **308–309**, 399–405.

29. Wheeler, R. (1991) Electropolishing of polycrystalline and single-crystal $YBa_2Cu_3O_{7-\delta}$ for TEM studies. *Ultramicroscopy*, **35**, 59–64.

30. Wang, N. and Fung, K.K. (1995) Preparation of TEM plan-view and cross-sectional specimens of ZnSe/GaAs epilayers by chemical thinning and argon ion milling. *Ultramicroscopy*, **60**, 427–435.

31. Aebersold, J.F., Stadelmann, P.A., and Matlosz, M. (1996) A rotating disk electropolishing technique for TEM sample preparation. *Ultramicroscopy*, **62**, 157–169.

32. Goodhew, P.J. (1985) in *Thin Foil Preparation for Electron Microscopy, (Practical Methods in Electron Microscopy)*, vol. 11 (ed. A. Glauert), Elsevier, Amsterdam, pp. 51–102.

33. Hirsch, P.B., Nicholson, R.B., Howie, A., Pashley, D.W., and Whelan, M.J. (1965) *Electron Microscopy of Thin Crystals*, Butterworths, London, p. 453, Appendix 1, and references therein.

34. Witcomb, M.J. and Dahmen, U. (1995) Method for jet polishing two-phase materials. *Microsc. Res. Tech.*, **32**, 70–74.

35. Howard, D.J., Paine, D.C., and Sacks, R.N. (1991) Large-area plan-view sample preparation for GaAs-based systems grown by molecular beam epitaxy. *J. Electron Microsc. Tech.*, **18**, 117–120.

36. Kim, Y., Ourmazd, A., Bode, M., and Feldman, R.D. (1989) Nonlinear diffusion in multilayered semiconductor systems. *Phys. Rev. Lett.*, **63** (6), 636–639.

37. Ayache, J., Beaunier, L., Boumendil, J., Ehret, G., and Laub, D. (2010) *Sample Preparation Handbook for Transmission Electron Microscopy: Techniques*, Springer, Berlin, pp. 125–135.

38. Strecker, A., Mayer, J., Baretzky, B., Eigenthaler, U., Gemming, T., Schweinfest, R., and Rühle, M. (1999) Optimization of TEM specimen preparation by double-sided ion beam thinning under low angles. *J. Electron Microsc.*, **48** (3), 235–244.

39. Barna, A., Gosztola, L., and Reisinger, G., (1989) Hungarian Patent, No. 205814.

40. Barna, A. (1992) in *Specimen Preparation for Transmission Electron Microscopy III*, vol. 254 (ed. R.M. Anderson), Materials Research Society, Pittsburgh, pp. 3–22.

41. Tagg, M.A., Smith, R., and Walls, J.M. (1986) Sample rocking and rotation in ion beam etching. *J. Mater. Sci.*, **21**, 123–130.

42. Müller, K.P. and Pelka, J. (1987) Redeposition in ion milling. *Microelectron. Eng.*, **7**, 91–101.

43. Kim, M.J. and Carpenter, R.W. (1987) TEM specimen heating during ion beam

thinning: microstructural instability. *Ultramicroscopy*, **21**, 327–334.
44. Viguier, B. and Mortensen, A. (2001) Heating of TEM specimens during ion milling. *Ultramicroscopy*, **87**, 123–133.
45. (a) Srot, V., Wegst, U.G.K., van Aken, P.A., Koch, C.T., and Salzberger, U. (2009) in *Proceedings Microscopy Conference 2009*, vol. 3 (eds W. Grogger, F. Hofer, and P.Pölt) Verlag der Technischen Universität Graz, pp. 287–288; (b) Srot, V., Wegst, U.G.K., van Aken, P.A., Koch, C.T., and Salzberger, U. (2010) in ELNES investigations of interfaces in abalone shell, *Microsc. Microanal.*, **16** (Suppl 2), 1218–1219.
46. Barna, A., Pecz, B., and Menyhard, M. (1998) Amorphisation and surface morphology development at low energy ion milling. *Ultramicroscopy*, **70**, 161–171.
47. Cullis, A.G., Chew, N.G., and Hutchinson, J.L. (1985) Formation and elimination of surface ion milling defects in cadmium telluride, zinc sulphide and zinc selenide. *Ultramicroscopy*, **17**, 203–212.
48. Alani, R., Jones, J., and Swann, P. (1990) in *Specimen Preparation for Transmission Electron Microscopy II*, vol. 199 (ed. R.M. Anderson), Materials Research Society, Pittsburgh, pp. 85–101.
49. Klepeis, S.J., Benedict, J.P., and Anderson, R.M. (1987) in *Specimen Preparation for Transmission Electron Microscopy*, vol. 115 (eds J.C. Bravman, R.M. Anderson, and M.L McDonald), Materials Research Society, Pittsburgh, pp. 179–184.
50. Ayache, J., Beaunier, L., Boumendil, J., Ehret, G., and Laub, D. (2010) *Sample Preparation Handbook for Transmission Electron Microscopy: Techniques*, Springer, Berlin, pp. 177–201.
51. Benedict, J.P., Anderson, R., Kelepeis, S.J., and Chaker, M. (1990) in *Specimen Preparation for Transmission Electron Microscopy II*, vol. 199 (ed. R.M. Anderson), Materials Research Society, Pittsburgh, pp. 189–204.
52. Okuno, H., Takeguchi, M., Mitsuishi, K., Xing, J.G., and Furuya, K. (2008) Sample preparation of GaN-based materials on a sapphire substrate for STEM analysis. *J. Electron Microsc.*, **57** (1), 1–5.
53. Claesson, S. and Svensson, A.A. (1956) A new ultramicrotome for electron microscopy. *Exp. Cell Res.*, **11**, 105–114.
54. Malis, T.F. and Steele, D. (1990) in *Specimen Preparation for Transmission Electron Microscopy II*, vol. 199 (ed. R.M. Anderson), Materials Research Society Pittsburgh, pp. 3–42.
55. Glauert, A.M. and Lewis, P.R. (1998) *Biological Specimen Preparation for Transmission Electron Microscopy*, Portland Press, London.
56. Gasser, P., Klotz, U.E., Khalid, F.A., and Beffort, O. (2004) site-specific specimen preparation by focused ion beam milling for transmission electron microscopy of metal matrix composites. *Microsc. Microanal.*, **10**, 311–316.
57. Stevie, F.A., Shane, T.C., Kahora, P.M., Hull, R., Bahnck, D., Kannan, V.C., and David, E. (1995) Applications of focused ion beams in microelectronics production, design and development. *Surf. Interface Anal.*, **23**, 61–68.
58. Loos, J., van Duren, J.K.J., Morrissey, F., and Janssen, R.A.J. (2002) The use of the focused ion beam technique to prepare cross-sectional transmission electron microscopy specimen of polymer solar cells deposited on glass. *Polymer*, **43**, 7493–7496.
59. Marko, M., Hsieh, C., Schalek, R., Frank, J., and Mannella, C. (2007) Focused ion beam thinning of frozen-hydrated biological specimens for cryo-electron microscopy, nature. *Methods*, **4** (3), 215–217.
60. Giannuzzi, L.A. (2010) Latest developments in FIB technology and applications. *Microsc. Microanal.*, **10**, 166–167.
61. Smith, N.S., Skoczylas, W.P., Kellogg, S.M., Kinion, D.E., Tesch, P.P., Sutherland, O., Aanesland, A., and Boswell, R.W. (2006) High brightness inductively coupled plasma source for high current focused ion beam applications. *J. Vac. Sci. Technol. B*, **24** (6), 2902–2906.
62. Kellogg, S.M., Schampers, R., Zhang, S.Y., Graupera, A.A., Miller, T., Laur, W.D., and Dirriwachter, A.B. (2010) High throughput sample preparation

and analysis using an inductively coupled plasma (ICP) focused ion beam source. *Microsc. Microanal.*, **16** (2), 222–223.

63. Young, R.J., Kirk, E.C.G., Williams, D.A., and Ahmed, H. (1990) in *Specimen Preparation for Transmission Electron Microscopy II*, vol. 199 (ed. R.M. Anderson), Materials Research Society, Pittsburgh, pp. 205–216.

64. Yaguchi, T., Urao, R., Kamino, T., Ohnishi, T., Hashimoto, T., Umemura, K., and Tomimatsu, S. (2001) A FIB micro-sampling technique and a site-specific TEM specimen preparation method for precision materials characterization. *Mater. Res. Soc. Proc.*, **636**, D9.35/1–D9.35/6.

65. Overwijk, M.H.F., van den Heuvel, F.C., and Bulle-Lieuwma, C.W.T. (1993) Novel scheme for the preparation of transmission electron microscopy specimens with a focused ion beam. *J. Vac. Sci. Technol., A B*, **11** (6), 2021–2024.

66. Mayer, J., Giannuzzi, L.A., Kamino, T., and Michael, J. (2007) TEM sample preparation and FIB-induced damage. *MRS Bull.*, **32**, 400–407.

67. Kato, N.I. (2004) Reducing focused ion beam damage to transmission electron microscopy samples. *Jpn. Soc. Microsc.*, **53** (5), 451–458.

68. Giannuzzi, L.A., Geurts, R., and Ringnalda, J. (2005) 2 keV Ga^+ FIB milling for reducing amorphous damage in silicon. *Microsc. Microanal.*, **11** (2), 828–829.

69. Stegmann, H., Ritz, Y., Utess, D., Engelmann, H.J., and Zschech, E. (2009) In-situ low energy argon ion milling of nanoelectronic structures using a triple beam system. *Microsc. Microanal.*, **15** (2), 170–171.

70. Gec, M., Toplisek, T., Srot, V., Drazic, G., Kobe, S., van Aken, P.A., and Ceh, M. (2008) in *Proceedings European Microscopy Congress 2008*, vol. 1 (eds M. Luysberg, K. Tillmann, and T. Weirich), Springer, Berlin, pp. 817–818.

71. Ozdol, V.B., Koch, C.T., and van Aken, P.A. (2010) A nondamaging electron microscopy approach to map In distribution in InGaN light-emitting diodes. *J. Appl. Phys.*, **108**, 056103–056105.

72. Srot, V., Bussmann, B., Salzberger, U., Koch, C., Cizmek, G., and van Aken, P.A. (2010) in *Proceedings International Microscopy Congress 2010* (eds G. Solorzano and W. de Souza), Sociadade Brasileira de Microscopia e Microanalise, pp. M15–M17. Srot, V., Bussmann, B., Salzberger, U., Koch, C., Cizmek, G., and van Aken, P.A. (2011) in Characterization of dentine, dentinal tubules and dentine-enamel junction in human teeth by advanced analytical TEM. *Microsc. Microanal.*, **12** (Suppl 2), 286–287.

15
Scanning Probe Microscopy – History, Background, and State of the Art
Ralf Heiderhoff and Ludwig Josef Balk

15.1
Introduction

In 1595, the eyeglasses polisher Hans Janssen developed the first optical microscope in Middelburg, Holland, extending optical inspections into the microworld. While the interest in the development of optics was commonly focused on telescopes during this period, Galileo Galilei developed an optical microscope in 1610 by modifying the distance of the optical tubes. The commercialization by Antonij van Leeuwenhoek took an additional 65 years. He fabricated over 200 optical microscopes in Delft from 1675 to 1723. It was not before 1872 that Carl Zeiss put Ernst Abbe in charge of calculating the requirements for the mass production of microscope lenses.

From the historical point of view, microscopy was mostly used just for morphological inspections. Today, the demands are continuously increasing. Especially in the characterization of semiconducting materials and devices, methods that allow comprehensive microscopic failure analyses and reliability investigations are mandatory. In engineering, more general methods, combining quantitative topographical analysis and the characterization of electrical/electronic, thermal/acoustic, and optical/optoelectronic properties, must be carried out. Nonradiating quantities such as material properties, doping profiles, potentials, static electric dipoles, temperature, strain, stress, forces, and radiating sources as well as waves such as Hertzian dipoles, light sources, thermal waves, and acoustic waves must be detected on the micro- and the nanoscale.

To get access to this variety of quantities, scanning microscopy was invented, where a probe/detector (tip, light beam, electron beam, X-ray) is moved from one point to another of the specimen under test and the reflected energy is detected. This signal is converted and illustrated by a monitor, xy-writer, and so on. Image processing is the advantage of scanning microscopes in comparison to classic microscopes, allowing the reflected energy to be of a different form from that of the source.

To execute these characterizations at highest spatial resolution, the investigations have to be performed under near-field conditions. This becomes obvious when

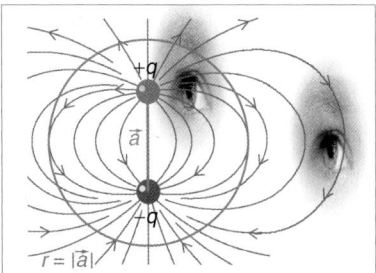

Figure 15.1 Near- and far-field investigations of a static electric dipole.

considering a static electric dipole, illustrated in Figure 15.1, with a dipole moment $\vec{p} = q \cdot \vec{a}$ where $\vec{a}$ is the distance of the charges. In the far field ($r \gg |\vec{a}|$), the electric field distribution is given by

$$\vec{E}(\vec{r}) = \frac{1}{4\pi\varepsilon r^3}\left(\frac{3\vec{p}\vec{r}}{r^2}\vec{r} - \vec{p}\right) \tag{15.1}$$

with the position vector $\vec{r}$ and the permittivity ε. The electric field distribution depends only on the dipole moment determined by the superposition of the electric field strength of each charge. Details of the source such as the electrostatic charge distribution or its values are only detectable in the near field ($r \ll |\vec{a}|$) because

$$\vec{E}(\vec{r}) = \pm \frac{q}{4\pi\varepsilon r^3}\vec{r} \tag{15.2}$$

The significance of this superposition becomes more evident when considering radiating sources emitting, for instance, light in the far field. Therefore, using the Huygens principle and Fraunhofer diffraction, the diffraction at a distance D of two small apertures spaced Δx and illuminated with a planar wave with a wavelength λ are estimated (Figure 15.2). As already introduced by Abbe in 1873 [1], the spatial resolution Δx of an optical microscope in the far field is limited to approximately half of the wavelength.

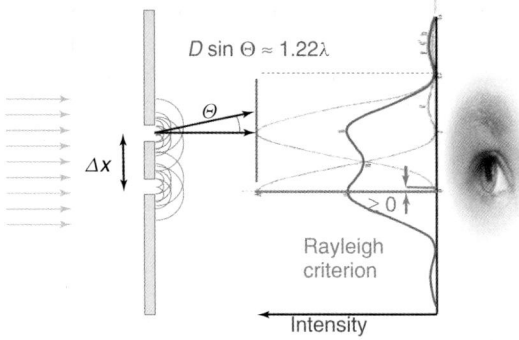

Figure 15.2 Illustration of the Rayleigh criterion within the Abbe limitation due to Huygens principle and Fraunhofer diffraction.

15.1 Introduction

Using the Rayleigh criterion [2], the difference between the first minimum of the diffraction pattern, generated by the first aperture, and the maximum of the diffraction pattern, generated by the second aperture, must be ≥ 0. The resolution

$$\Delta x \geq 0.61 \frac{\lambda}{n \sin \Theta} \qquad (15.3)$$

depends on the numerical aperture $n \sin \Theta$, where n is the diffraction index of the medium and Θ is the opening angle of the last lens. Many strategies for improving resolution have focused on an increase in the numerical aperture using solid immersion lenses [3], nonlinear patterned excitation microscopy [4], as well as the use of deep UV radiation with a wavelength of 193 nm [5], or even by combining confocal microscopy with synchrotron radiation [6].

In order to describe the near-field and the far-field properties of radiating sources emitting electromagnetic waves, the Hertzian dipole, introduced in 1887 by Hertz [7], is discussed in the following. The electric and magnetic field distributions can be found, for example, in Jackson's "Classic Electrodynamics" [8]:

$$\vec{E} = k^2 (\vec{n} \times \vec{p}) \times \vec{n} \frac{e^{jkr}}{r} + [3\vec{n}(\vec{n} \cdot \vec{p}) - \vec{p}] \left(\frac{1}{r^3} - \frac{jk}{r^2} \right) e^{jkr}$$

$$\vec{H} = k^2 (\vec{n} \times \vec{p}) \frac{e^{jkr}}{r} \left(1 - \frac{1}{jkr} \right) \qquad (15.4)$$

and are given by using the wave vector k. In the far field, at distances larger than 2λ, the factors with high exponents are negligible. $\vec{E}$ and $\vec{H}$ are in phase as well as rectangular to each other, and a propagating wave with light velocity c can be detected. Of course, we have to take the difference of an isotropic light source into consideration because no wave can be detected in the direction of the Hertzian dipole moment. In the near field, at distances smaller than $\lambda/2$, $\vec{E}$ and $\vec{H}$ have a phase difference of $\pi/2$. Instead of a propagating wave, evanescent fields are detectable: the electric field looks similar to that of a static dipole under far-field conditions and the magnetic field is analogous to Ampère's circuital law (Figure 15.3).

Concluding that a Hertzian dipole emits solely an electromagnetic wave at distances larger than $\lambda/2$ and that there is a limitation on the spatial resolution, the question has to be answered, in addition, whether there is also a limitation on the size of a radiating source.

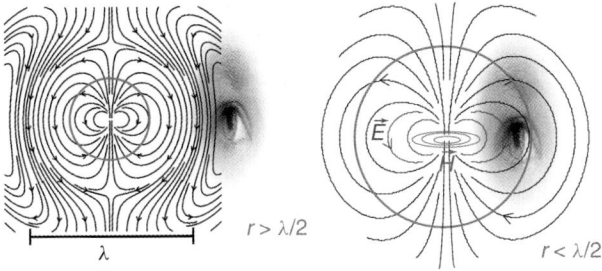

Figure 15.3 Far- and near-field investigations of a Hertzian dipole.

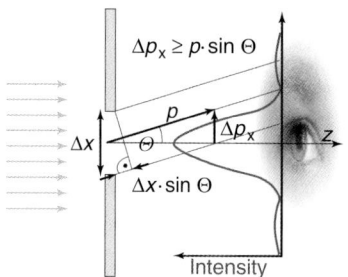

Figure 15.4 Diffraction at an infinitesimal small aperture in far field.

To be more general, the diffraction at an aperture size Δx of either waves or particles is investigated by introducing the de Broglie wavelength $p = \hbar \cdot k = \frac{h}{\lambda}$. As can be obtained from Figure 15.4, the uncertainty of the momentum in the direction of the screen is given by $\Delta p_x \geq p \sin \Theta$ and a diffraction pattern only exists for $\Delta x \cdot \sin \Theta \geq \frac{\lambda}{2}$. The principle of uncertainty, introduced in 1927 by Heisenberg [9], is delivered by the correlation of the inequalities above:

$$\Delta x \cdot \Delta p_x \underset{\sim}{>} \frac{h}{2} \tag{15.5}$$

Introducing the de Broglie wavelength and taking the diffraction index of the medium into account, this equation concludes that a propagating wave is only detectable in the far field if the minimum size of a radiating source is limited to

$$\Delta x \geq \frac{\lambda}{2n \cdot \sin \Theta} \tag{15.6}$$

Equation (15.6) looks quite similar to the Abbe limit. Considering an aperture much smaller than half the wavelength, the uncertainty of the momentum becomes $\Delta p_x \gg \frac{2h}{\lambda}$. As a consequence, the wave vector $k_z^2 \ll 0$ leads to $(k_z^{real} + jk_z^{imag})^2 \ll 0$. This inequality is only fulfilled for evanescent waves, where $k_z^{real} \ll k_z^{imag}$. Consequently, a higher brightness appears in near-field investigations. While in the far field, the diffracted light is spherical and, therefore, similar to a pointlike source, at this near field the distribution strongly depends on the source characteristic. The observer is able to detect the properties of the incident wave emitted from lines, pointlike, or even from planar light sources as illustrated in Figure 15.5. The exact pattern can be calculated by the strict theory of diffraction of an aperture, which Meixner and Andrejewski developed in 1950 [10].

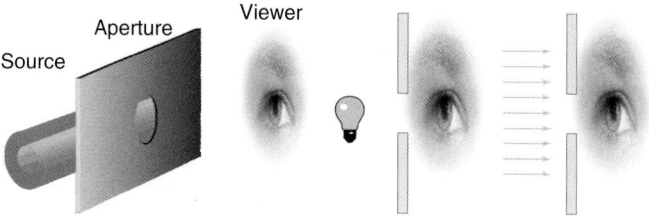

Figure 15.5 Detecting the source characteristic under near-field conditions.

15.2 Detecting Evanescent Waves by Near-Field Microscopy: Scanning Tunneling Microscopy

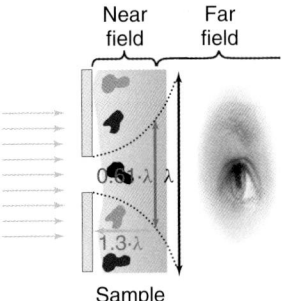

Figure 15.6 Overcoming the diffraction limit in the near field.

Synge already proposed in 1928 [11–13] that under this near-field condition, the diffraction limit of optical microscopes can be overcome by scanning an opaque screen with a small aperture in front of a sample (Figure 15.6). This becomes evident after careful consideration of the Fresnel diffraction and reciprocal principles. The technical realization took more than 40 years. It was not before 1972 that Ash and Nicholls [14] built a superresolution aperture microscope with resolutions far below the diffraction limit at microwave frequencies of 10 GHz, although the idea of a flying-spot microscope was already introduced by Young and Roberts in 1951 [15–17].

15.2
Detecting Evanescent Waves by Near-Field Microscopy: Scanning Tunneling Microscopy

Using fundamentals of light microscopy as summarized by Frank Mücklich in Chapter 1 of the first *Handbook of Microscopy*, it could be demonstrated in the paragraph above that microscopy with resolutions better than the diffraction limit are possible by detecting an evanescent wave or an evanescent field. In this context, a discussion of evanescence is incomplete without mentioning total internal reflections at a boundary where the incident angle of a wave ϑ_{in} is larger than the critical angle $\vartheta_{critical}$ (Figure 15.7).

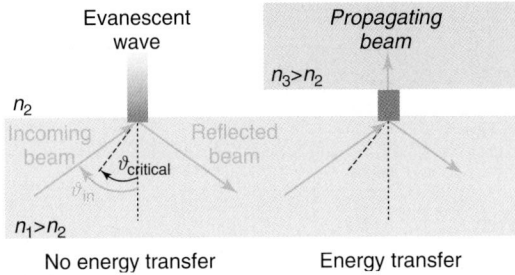

Figure 15.7 Photon tunneling at interfaces.

The critical angle $\vartheta_{critical}$ is given by the arcsine of the refractive indexes quotient n_2/n_1 of the different optically dense media using Snell's law. At this total internal reflection, the refracted wave becomes parallel to the interface and within the complex refracted wave vector $k_{real} < k_{imag}$ leads to an evanescent wave with no energy transfer. At the so-called frustrated total internal reflection, a medium with a refractive index $n_3 > n_2$ is brought in close proximity to this interface, and thus a propagating wave can be detected. This procedure is also called *photon tunneling*.

If a small potential difference $U_{Tunneling}$ is applied between two metals separated by a thin insulating film with a thickness d, a current will flow because of the quantum mechanical tunnel effect of electrons. For their experimental discoveries regarding tunneling phenomena in semiconductors and superconductors, respectively, Leo Esaki, Ivar Giaever, and Brian David Josephson finally got the Nobel Prize in 1973. Giaver and Mergerle already experimentally demonstrated in 1961 that the electron density of states can be inferred [18]. This behavior becomes obvious when it concerns an electron tunnels from a metal through an isolator into a semiconductor, as illustrated in Figure 15.8.

If the Fermi energy within the metal is higher than the energy of the valence band W_V and less than the energy of the lower edge of the conduction band W_C, the tunneling of electrons is prevented. For energies $W_{Fermi} > W_C$, the charge carrier density and Fermi function of the investigated semiconducting sample must be taken into account. The tunneling current $I_{Tunneling}$ depends on the local density of free states LDOS due to the Pauli principle. Considering fermions instead of photons, illustratively explained by Giaever in 1974 [19], the $I_{Tunneling}$ is finally given by

$$I_{Tunneling} = c_1 \cdot LDOS \cdot U_{Tunneling} \cdot e^{-c_2 \cdot \sqrt{\Phi \cdot d}} \tag{15.7}$$

where Φ is the work function.

This tunneling current differs significantly from a field emission current because of the overlap of the wave function of the surface electrons. While in a field emission microscope (FEM) high resolutions are detectable only because the electrons gain high energy, the energy of these tunneling electrons is low. Their wavelength will be larger than 1.2 nm applying a voltage of less than 1 V between a metal and a

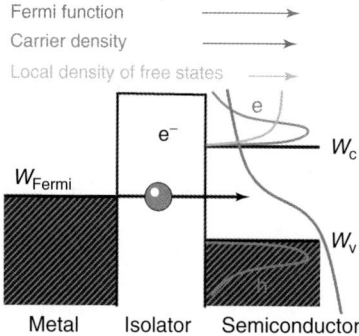

Figure 15.8 Electron tunneling from a metal into a semiconductor.

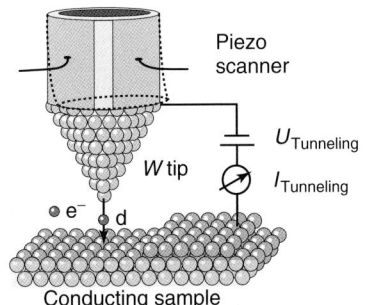

Figure 15.9 Principle of a scanning tunnelling microscope.

device under test (DUT). Resolutions better than the diffraction limit are possible if the distance of a probe is closer than approximately one-third of the wavelength, assuming the theory of detecting evanescent waves by near-field microscopy (Figure 15.6). These distances could already be adjusted with a subatomic resolution within tunneling experiments by Thompson and Hanrahan in 1976 [20]. A lateral atomic resolution was first demonstrated by Binning et al. in 1982 [21] using a system with a principle similar to a three-dimensional relocation profilometer stage [22]. Topographies in real space and work-function profiles on an atomic scale on Au(110), Si(111), and GaAs(111) surfaces were obtained by using this scanning tunneling microscopy (STM) [23, 24]. The design, schematically illustrated in Figure 15.9, led to the Nobel Prize for G. Binnig and H. Rohrer in 1985. At a constant height or at a constant current, a metal tip, for instance, tungsten, is scanned over the sample surface using piezoelectric actuators.

A feedback loop, already introduced by Young et al. in 1972 [24], is used to perform the constant-current mode STM. The surface microtopography has been investigated with a topografiner [25]. In Chapter 1 of the second *Handbook of Microscopy* and in Ref. [24], a more precise description of the STM is given.

Beside the inspection of surfaces with highest spatial resolution, this STM has the advantage of positioning single atoms [26]. Hereby, it is possible to characterize quantum mechanical effects that are obvious in Figure 15.10. The detected signal depends on the wavefunctions of the surface electrons and is not only a replica of the surface atoms.

15.3
Interaction of Tip–Sample Electrons Detected by Scanning Near-Field Optical Microscopy and Atomic Force Microscopy

Scanning near-field optical microscopes (SNOMs), with resolutions better than the diffraction limit, were developed simultaneously by three groups in 1983:

- Light was efficiently transmitted through subwavelength-diameter apertures by Lewis et al. [27].

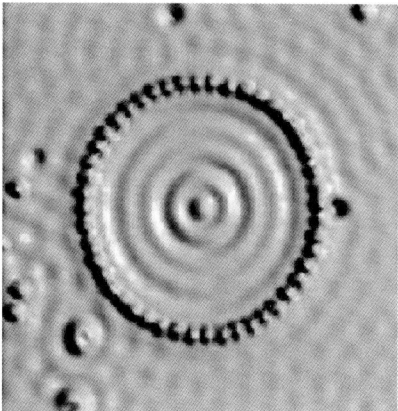

Figure 15.10 Quantum–mechanical interference patterns caused by scattering of a two-dimensional electron gas. (Image originally created by IBM Corporation.)

- For microscopy and pattern generation with scanned evanescent waves, apertures were fabricated either by a thin film on a transparent plate or optical fiber as introduced by Massey [28].
- Pohl *et al.* demonstrated subwavelength-resolution optical image recording by moving an extremely narrow aperture along a test object [29].

In the beginning, the distance between the aperture and the sample was regulated by STM. Using piezoelectric actuators for nanopositioning, SNOM imaging in transmission and reflection modes is carried out as shown in Figure 15.11.

Nowadays, further interactions of tip electrons and sample electrons are used for apertureless SNOMs. Bringing a metal tip in ultimate proximity with a sample, as illustrated in Figure 15.12, light is scattered at the tip end. Evanescent electromagnetic waves are transformed into propagating waves because of the overlap of the wavefunctions of the surface electrons. This new form of scanning optical

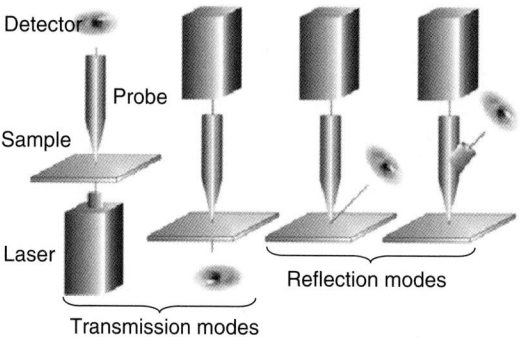

Figure 15.11 Transmission and reflection modes in scanning near-field optical microscopy.

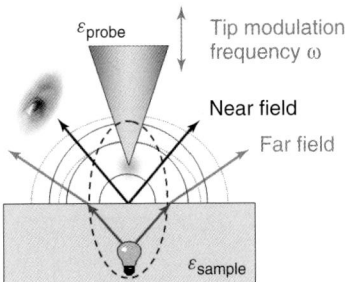

Figure 15.12 Light scattering from a metal tip end.

microscopy, introduced by Reddick *et al.* in 1989, allows optical imaging at 10 Å resolution by this apertureless near-field optical microscopy [30, 31] demonstrated by Zenhausern *et al.* in 1995 [32].

Up to now, propagating and evanescent waves as well as electric interactions of free electrons from conducting or semiconducting samples and probes have been investigated. But nonconducting samples can also be investigated considering the overlap of shell electron wavefunctions, the Pauli principle, and the van der Waals interaction. This becomes obvious from the force–distance curve, as illustrated in Figure 15.13.

A repulsive force occurs if electron wavefunctions with the same properties overlap because of the Pauli principle. An exchange of electrons will lead to attractive forces because of the van der Waals interaction. Using an atomic force microscope (AFM), first introduced by Binnig *et al.* in 1986 [33], surface atoms of nonconducting samples are detectable, even under an arbitrary environment. To achieve this, the bending of a tip mounted on a cantilever has to be measured. While in the first instrument an STM was used for detecting this bending, today the laser deflection technique is more common for distance regulation. A schematic setup of a force microscope is illustrated in Figure 15.14.

Similar to the STM, there are two operation modes. Constant force measurement can be performed using a feedback loop to keep the bending of the cantilever constant. In constant height mode, the deflection of the cantilever is measured, while the tip cantilever system is moved without regulation at a constant average distance over the sample surface. Using the theory of profilometric examinations [34], the topography can be obtained with atomic resolution.

Surface roughness and sliding friction, as already described by Bikerman in 1944 [35], can be obtained by simultaneously measuring the torsion of the cantilever

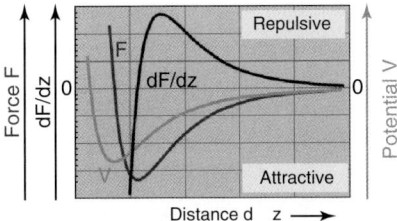

Figure 15.13 Force–distance curve.

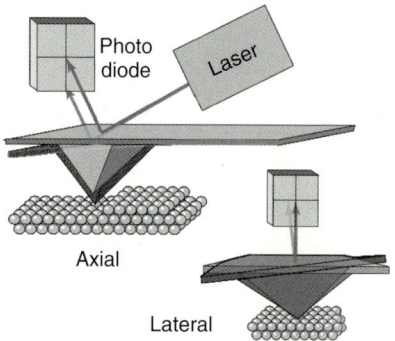

Figure 15.14 Axial and lateral force microscopy.

due to lateral forces and axial bending forces [36]. As an example, by detecting these friction forces, chemical phase separation studies in organic materials can be performed [37]. Consequently, an AFM is also called a scanning force microscope (SFM). Owing to the dF/dz characteristic of the force–distance curve, filigree surfaces can be investigated in the noncontact or intermediate mode. The theoretical aspects and different experimental setups of these techniques are considered in more detail in Chapter 2 of the second *Handbook of Microscopy*. Some application areas are shown in Figure 15.15.

A variety of probes is necessary because of the huge number of applications. Materials such as Si, Si_3Ni_4, diamond, metal, and so on are used. The designs such as single beam, V-shape or for noncontact applications piezoresistive and tuning forks, are described by their dynamic property – the Eigen frequencies ω_0:

$$\omega_0 = \sqrt{\frac{k}{m}} \tag{15.8}$$

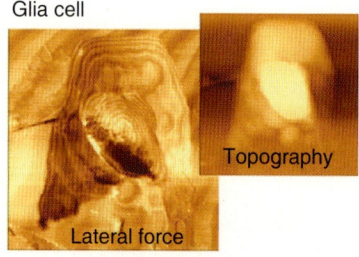

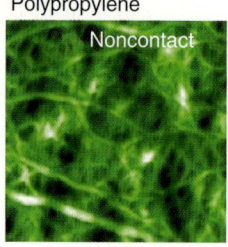

Figure 15.15 Application areas of lateral force and noncontact microscopy [38].

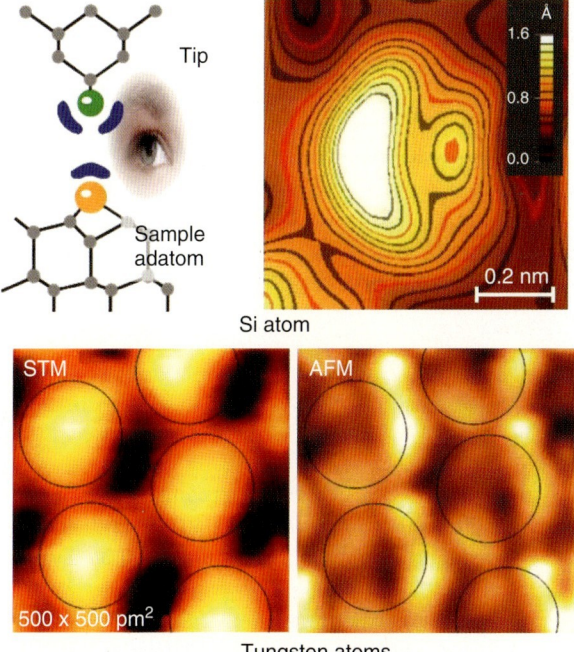

Figure 15.16 Highest resolution detectable by AFM using a tuning fork for distance regulation [39, 40].

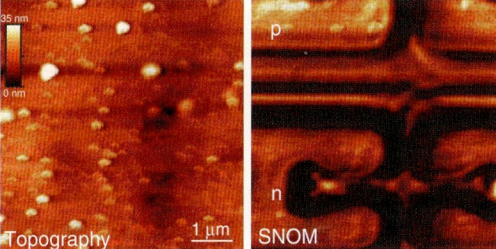

Figure 15.17 Reflection mode SNOM on IC [41].

where k is the spring constant and m is the mass of the system. This characteristic determines significantly the resolution beside the scan velocity and contact force. Using a quartz tuning fork for distance regulation, the high sensitivity of this technique is shown by the two crescents in the image of a single Si atom caused by the two sp^3 orbitals, and a high resolution of 77 pm is visualized for single tungsten atoms, as demonstrated by Giessibl [39, 40] (Figure 15.16).

Using SFM has the advantages of the detection of the authentic topography and simultaneous determination of a second interaction product. For instance, combining the above-mentioned SNOM with an SFM, optical inspections can be performed even on passivated integrated circuits (ICs), as illustrated in Figure 15.17. Buried

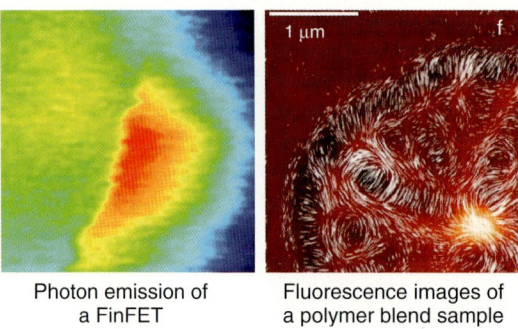

|Photon emission of a FinFET | Fluorescence images of a polymer blend sample|

Figure 15.18 Photon emission microscopy (spatial resolution <30 nm) [43] and fluorescence imaging [44] with resolutions far below the diffraction limit; FinFET, fin-shaped field effect transistor.

features of the DUT are visible with SNOM, while the remains of a polishing paste are detectable in the topographic image. A resolution of 60 nm, far below the diffraction limit of light, is achieved optically on the remaining 50 nm passivation layer.

Mostly, a quartz tuning fork is used for distance regulation of a tapered glass-fiber probe as introduced by Karrai and Grober [42]. Hereby, optical application of known failure analysis and reliability investigation techniques can be performed with high predefined resolutions, which are independent of the detected wavelength as illustrated in Figure 15.18.

Photon emission microscopy is a nondestructive technique that allows the exact localization of defects within devices. Fluorescence imaging is an established technique for the characterization of biological samples or polymers.

15.4
Methods for the Detection of Electric/Electronic Sample Properties

To detect the variety of electric/electronic quantities within a DUT, the probe must fulfill the near-field condition. In this connection, reference should again be made to the fact that just a miniaturization of the probe (to a pointlike source (a)) is insufficient because it is essential that the physical property of the probes changes in close approximation, as discussed above. This becomes obvious from Figure 15.19.

Near-field conditions can only be achieved by (b) suitable source geometries, for example, finite line source, by (c) aligned point sources, for instance multipoles, or by (d) use of external boundary conditions, such as refraction and diffraction or by their combinations.

As discussed in Ref. [45], a large field enhancement exists directly below a tip in close proximity to the sample surface if the probe and the sample have different electric potentials. Using a conductive tip in contact, a current flow is measurable

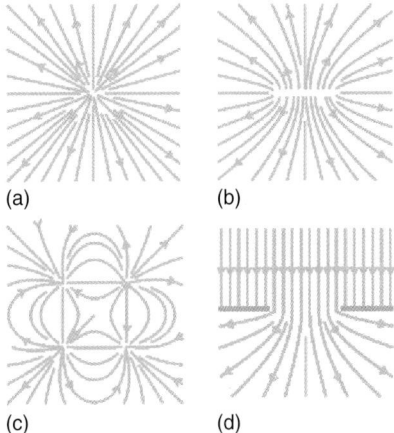

Figure 15.19 (a–d) Requirement for near-field conditions differ from a pointlike source.

depending on the resistance of the probe/surface contact. To keep the contact area constant, these measurements will be carried out in constant force mode and are called scanning spreading resistance microscopy (SSRM) [46]. Fulfilling the near-field condition by the described physical properties, doping profiles are detectable with highest resolution as illustrated here for the highly doped layer of a heterojunction bipolar transistor (HBT) in Figure 15.20. The lateral achievable resolution is in the nanometer region and depends on the effective contact area.

Furthermore, this technique is quite often used to determine the oxide thickness of gate contacts within modern devices. Hereby, a voltage is applied between the semiconducting material and the conducting tip and the tunneling current through the oxide layer, which changes exponentially with the distance according to Eq. (15.7), as detected.

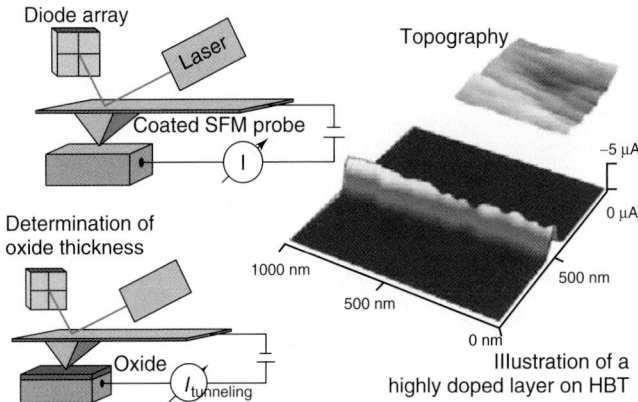

Figure 15.20 Scanning spreading resistance microscopy (SSRM) [47].

By the use of a metal-coated probe, a metaloxide-semiconductor (MOS) system is realized for an oxidized semiconductor. From this well-known MOS structure, it is possible to measure locally the doping profile by scanning capacitance microscopy (SCM) quantitatively. A schematic setup is illustrated in Figure 15.21. To increase the signal-to-noise ratio and to distinguish between n- and p-doped regions, the dC/dV characteristic is measured by the use of lock-in amplification. Differently doped regions and the depletion regions are detectable, as can be seen from Figure 15.22.

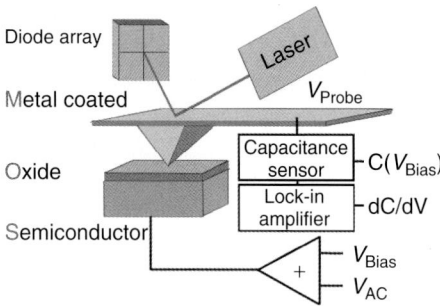

Figure 15.21 Setup for scanning capacitance microscopy (SCM).

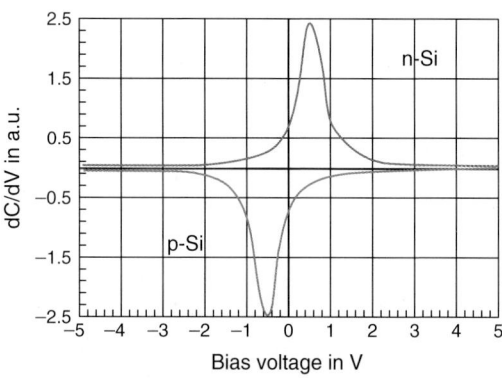

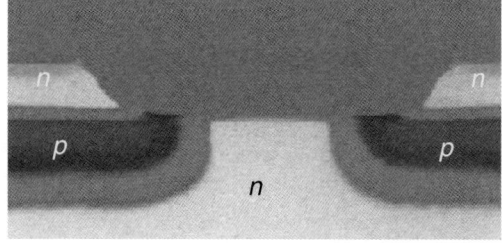

Figure 15.22 dC/dV image of depleted and doped regions. (Image originally created by Digital Instruments (now Veeco Instruments Inc.))

15.4 Methods for the Detection of Electric/Electronic Sample Properties

While high-resolution capacitance, measurement, and potentiometry by force microscopy was demonstrated first by Martin et al. in 1988 [48], Matey and Blanc had introduced SCM for detecting variations in surface topography already in 1984 [49]. Deposition and imaging of localized charge on insulator surfaces using a force microscope in noncontact were carried out by Stern et al. in 1988 [50].

Performing investigations in noncontact, electric forces $F_{el.}$ given by

$$F_{el.} = \frac{1}{2} V^2 \frac{\partial C}{\partial z} \tag{15.9}$$

must be taken into account in addition to the so-far mentioned repulsive and attractive forces. Analogous inspections can also be carried out in the case of a magnetic interaction of a magnetic tip and a locally magnetized sample. The principle of electric force microscopy (EFM) and magnetic force microscopy (MFM) is illustrated in Figure 15.23.

In this context, it must be mentioned that MFM, using a force microscope to measure the magnetic force between a magnetized tip and the scanned surface was developed by Martin and Wickramasinghe in 1987 [51]. Lateral domain resolutions of the order of 0.01 µm can be achieved with this technique, as demonstrated by Saenz et al. [52]. But it took 10 years longer to perform quantitative MFM on perpendicularly magnetized samples by Hug et al. [53].

Nowadays, EFM is well established to characterize modern devices and materials with highest resolution. As one example, electrostatic force microscopy of a polymer blend film for solar cells is given in Figure 15.24.

Beside localized charges, the contact potential difference, V_{CPD}, must be taken into account if two materials are arranged with a small spacing. This V_{CPD} depends on a variety of parameters such as the work function, adsorption layers, oxide layers, dopant concentration in semiconductors, or temperature changes in the sample. The Kelvin probe force microscopy (KPFM), as introduced by Nonnenmacher et al. in 1991 [55], is today a common method to measure this contact potential with the highest lateral resolution. Performing these investigations in noncontact, a DC voltage, V_{dc}, superimposed with an AC voltage, V_{ac}, is applied to the tip. Finally,

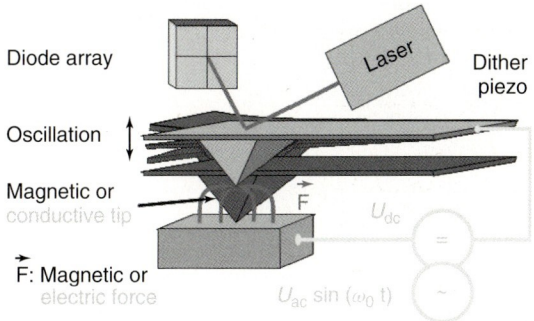

Figure 15.23 Principle of electric force microscopy and magnetic force microscopy.

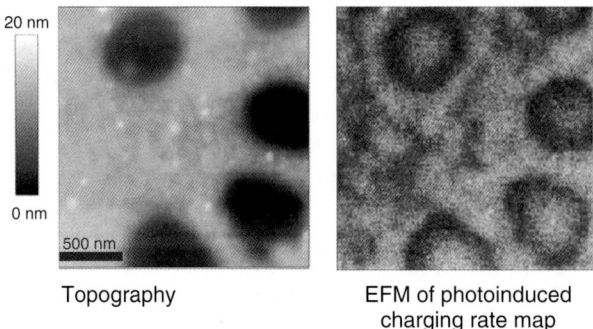

Figure 15.24 Charging rate image of a polymer blend film [54].

the resulting electric force, $F_{el.}$,

$$F_{el.} = \frac{1}{2} \cdot \frac{\partial C}{\partial z} \left[(V_{dc} - V_{cpd})^2 + \frac{1}{2} V_{ac}^2 + 2(V_{dc} - V_{cpd}) V_{ac} \sin(\omega_0 t) - \frac{1}{2} V_{ac}^2 \cos(2\omega_0 t) \right] \quad (15.10)$$

causes a first and second harmonic at the oscillating cantilever tip system. The changes in the local capacitance can be measured by the second harmonic. V_{CPD} is determined quantitatively by regulating the first harmonic to zero with a V_{dc} feedback loop. Local variations within the band structure can be detected as illustrated in Figure 15.25.

Applying an AC signal with a frequency f_P to the probe and a slightly higher frequency, f_{DUT}, to the DUT, signals at $f_{DUT} - f_P$, $2f_P$, $f_{DUT} + f_P$, and at $2f_{DUT}$ are mixed within the frequency domain. Here, EFM investigations can be performed at high frequencies if $f_{DUT} - f_P$ is below or at the eigenfrequency of the cantilever. Device inspections can be performed even on dynamically biased modern devices in the GHz region as it becomes obvious from Figure 15.26.

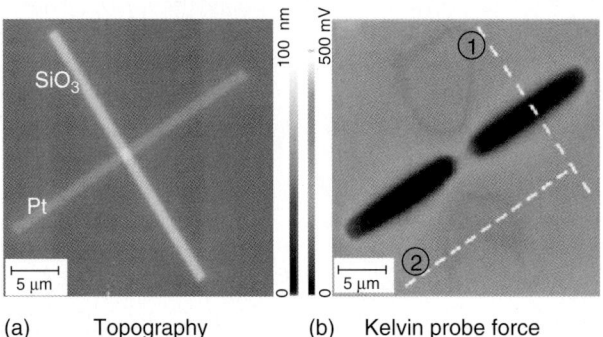

Figure 15.25 SiO$_3$ deposited on top of Pt. [56].
a) topography image b) Kelvin probe force microscopy measurement.

15.4 Methods for the Detection of Electric/Electronic Sample Properties | 515

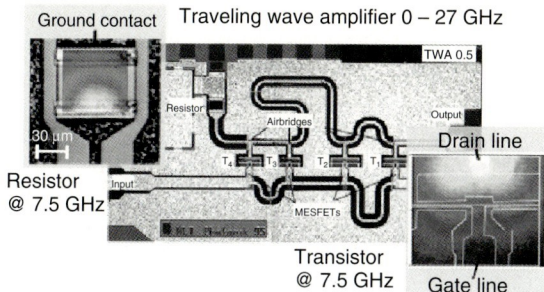

Figure 15.26 Electric force microscopy on a traveling wave amplifier working at high frequencies [57].

Diffusion lengths, barrier heights, and electric field distributions within devices can be detected by optical-beam-induced current (OBIC) examinations. The DUT is irradiated with laser light guided through an SNOM probe, as illustrated in Figure 15.27a [58]. Electron–hole pairs are created by inelastic scattering within the DUT, if the energy of the photon is higher than the band gap of the semiconductor.

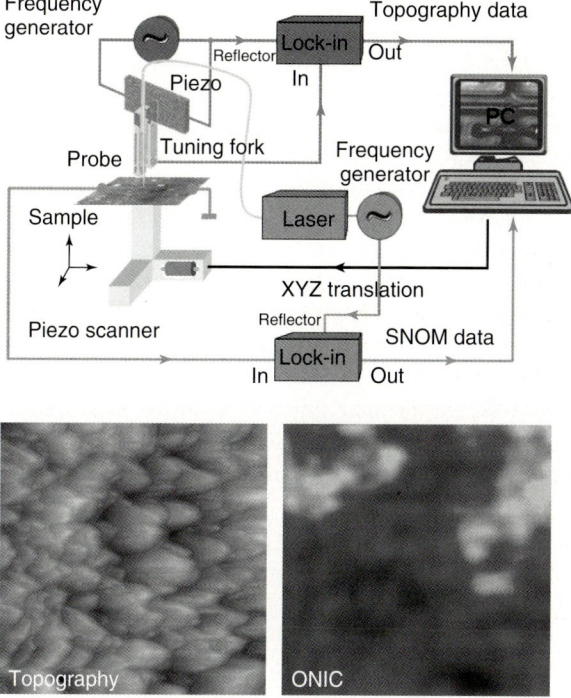

Figure 15.27 Near-field OBIC reliability investigation (a) setup and (b) local efficiencies on a CuInSe$_2$ (CIS) absorber layer for solar cells; ONIC, optical near-field-induced current.

With the built-in electric field, for example, of a depletion region, the excess carriers are separated and can be measured as a current at the DUT's electric contacts. Performing these investigations dynamically, even the quantitative determination of the local electric field strength is possible [59]. Therefore, reliability investigations, for instance, of the local efficiencies on an absorber layer for solar cells, as illustrated in Figure 15.27b, can be carried with highest spatial resolution.

15.5
Methods for the Detection of Electromechanical and Thermoelastic Quantities

Reliability and failure analysis should not strictly be limited to electric/electronic properties; more comprehensive investigations are mandatory to get access to all material properties. Beside optical/optoelectronic and thermal/mechanical quantities, electromechanical and thermoelastic quantities play a more and more important role in modern devises such as in microelectronic mechanical systems (MEMS).

Acoustic investigations can be carried out by using either STM or SFM. The determination of displacements in ultrasonic waves on surfaces of solids was performed by Heil *et al.* in 1988 for the first time [60]. Ultrasonic surface waves can be detected using the nonlinear characteristic of the gap of a tunneling microscope as described in Ref. [61]. Irradiating surfaces by acoustic waves using a scanning near-field acoustic microscope (SNAM) was introduced by Günther *et al.* in 1989 [62]. Using force modulations to image surface elasticities with the AFM was followed by Maivald *et al.* [63] two years later. While all the setups look quite different, the near-field conditions can be explained similarly concerning in addition the reciprocal principle, which will be discussed in the following.

Electromechanical interactions can be analyzed with highest lateral resolution when considering a piezoelectric (PZT) interface formed by a dielectric surface with permittivity ε_{rl} on top of the piezoelectric (Figure 15.28). A near-field piezoelectric interaction occurs directly below the biased tip illustrated by introducing image charges. To distinguish between topography and acoustic data, an AC signal is applied to the probe. The resulting periodic interaction can be measured as illustrated in Figure 15.29 either by the laser deflection method or by detecting the generated acoustic waves with a piezoelectric transducer on the back of the sample.

Ferroelectric materials have especially attracted attention in recent years because of the combination of these materials with IC technology in information and media engineering. Figure 15.30 shows the topography and the acoustic microscopy of ferroelectric domains in a transparent PLZT ceramic. The main features in the topography are surface scratches, and no information associated with the domain structure is revealed. Using the piezoresponse SFM mode, fingerprint patterns related to domains with antiparallel polarization are clearly exposed. Using a quartz tuning fork for distance regulation in SFMs, as demonstrated in Figure 15.16, the contact force is modulated continuously generating acoustic

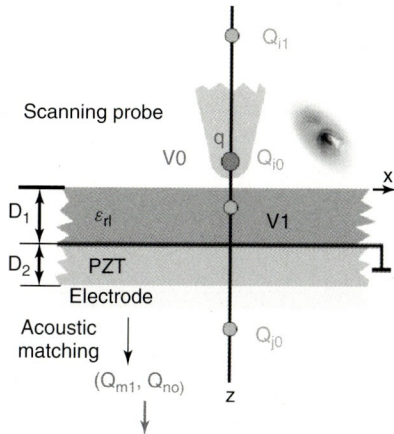

Figure 15.28 Modeling scanning near-field acoustic microscopy [64].

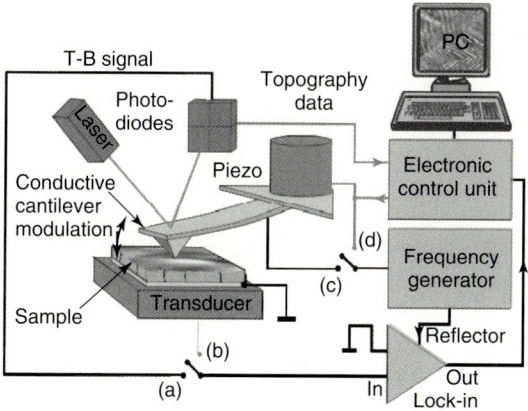

Figure 15.29 Setup of scanning near-field acoustic microscope.

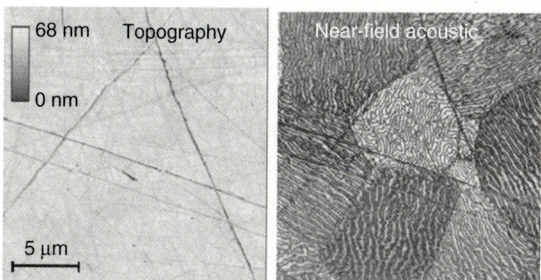

Figure 15.30 Piezoresponse microscopy of ferroelectric domains in transparent PLZT ceramics [65].

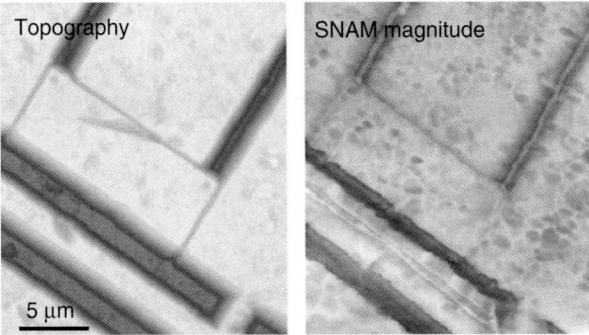

Figure 15.31 Scanning near-field acoustic microscopy on an electromigration test structure [66].

waves. Hereby, even on nonpiezoelectric materials, investigations can be performed with highest sensitivity detecting these waves with the transducer. Failure analyses and reliability investigations on electromigration can be performed as illustrated in Figure 15.31.

If a voltage is applied to an electrically conducting sample, the temperature rises because of Joule heating, for instance. Within actively biased devices, thermal expansion occurs resulting in three-dimensional surface displacement vectors. The so-called scanning Joule expansion microscopy (SJEM), introduced by Varesi and Majumdara in 1998 [67], can simultaneously image surface topography and the material expansion using conventional cantilevers. SJEM is a special form of scanning thermal expansion microscopy, which is a promising technique to analyze thermoelastic properties by detecting dynamic vertical expansions [68, 69]. Three-dimensional analyses can be performed by detecting vertical and lateral signals simultaneously, as was demonstrated recently [70] (Figure 15.32). By correlation of the three-dimensional thermal expansion vectors $\bar{u}$ and the device structure, thermally induced strains ε are detectable [71]. Compressive and tensile strains can be distinguished from line analyses (indicated in Figure 15.32) performed on a constriction of an interconnect structure, as illustrated in Figure 15.33. With a spatial resolution of 30 nm and a sensitivity of $10 \text{ fm}/\sqrt{Hz}$ defects can be localized by the detection of the lateral expansion, which can be obtained from Figure 15.34 [72].

The high sensitivity becomes obvious by comparison with the diameter of atom nucleus, which is 1 fm. Therefore, it is a powerful technique for advanced failure analysis on degradation processes.

ε is determined by the stress σ of the DUT and is a complex function of displacement and temperature distribution. Using measured surface data such as Dirichlet and Neumann, boundary conditions give access to the three-dimensional stress distribution by finite element calculations, for instance. The temperature distribution can be measured quantitatively by the so-called scanning near-field thermal microscopy (SThM) using a resistive probe. The thermal probe used is

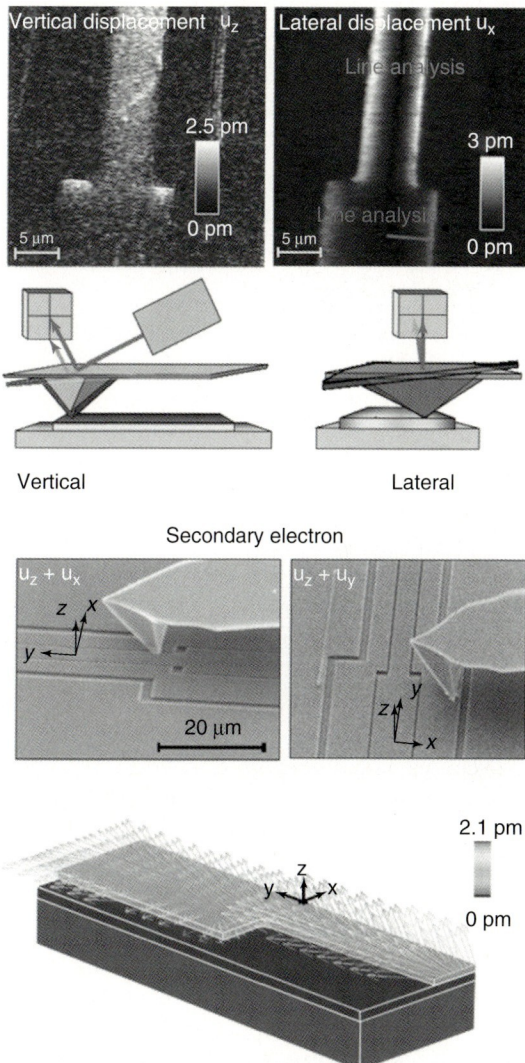

Figure 15.32 Scanning Joule expansion microscopy of the three-dimensional surface displacement vectors obtained at a constriction of an interconnect structure.

a Wollaston wire consisting of a thin Pt/Rh core surrounded by a thick silver sheath, as introduced by Dinwiddie *et al.* in 1994 [73]. The local temperature can be obtained by monitoring the probe resistance changes induced by variations of the sample surface temperature. These resistance variations shift the voltage balance of a Wheatstone bridge, resulting in a change in the output voltage of the circuit, as illustrated in Figure 15.35.

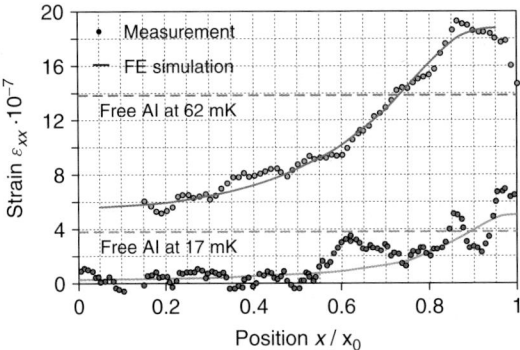

Figure 15.33 Thermally induced compressive and tensile strains measured at a constriction of an interconnect test structure [70, 71].

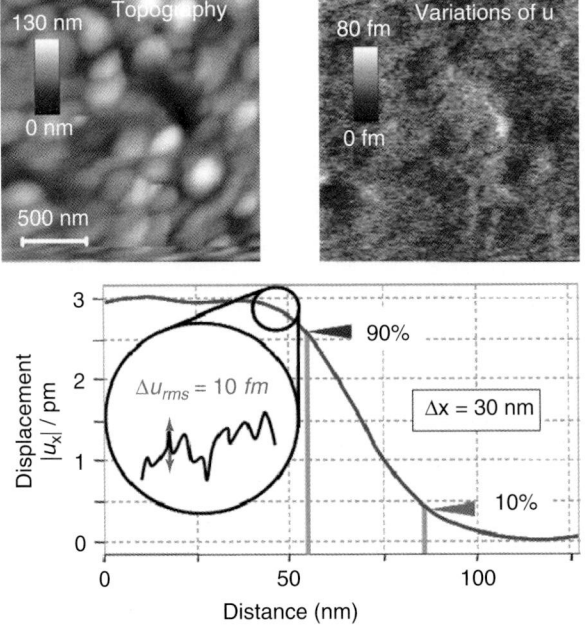

Figure 15.34 Localization of defects with highest resolution by SJEM [72].

Also attaching a thermocouple at the end of an SFM cantilever allows thermal measurements, as proposed by Majumdar et al. in 1995 [75]. Using microscopic thermocouples placed at the very end of an STM tip, high-resolution temperature measurements can be carried out on conducting samples even in noncontact [76].

These additional temperature measurements finally allow the calculation of the stress distribution using the introduced concept [72]. In reliability investigation,

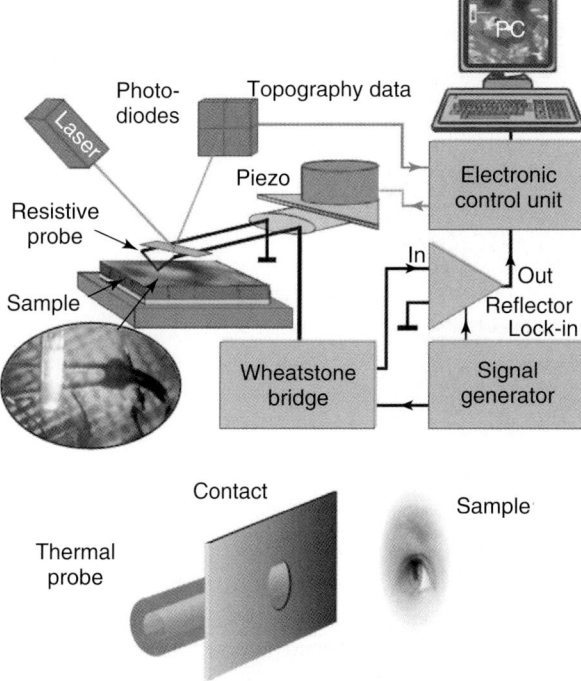

Figure 15.35 Performing thermal analysis with highest resolution with scanning near-field thermal microscopy [73, 74].

the stress tensor is transformed into the von Mises stress in order to take all stress components into account. Figure 15.36 shows the von Mises stress distribution of the constriction, using experimental data as boundary conditions, compared with conventional FEM simulations. While FEM simulations assume perfect layer bonding, the calculation of the stress using measured boundary values leads to peak regions. Finally, stress analysis without decapsulation of multilayers can be carried out by these complementary scanning thermal microscopy techniques.

In order to achieve high performances and reliabilities of devices, knowledge of all thermal properties, especially the mechanism of heat dissipation inside these devices, is indispensable. For the localization of defects in materials for future applications, the thermal conductivity λ_{sample} must be determined because static temperature distributions and thermal wave characteristics are influenced by this material property. To perform thermal analysis with highest resolution, the probe is considered as a line detector [74] and a line heat source, and the contact is considered as an aperture as illustrated in Figure 15.35. Quantitative thermal conductivity analysis, even on actively operated electronic devices or on thin film structures, can be carried out under near-field conditions [77] applying the 3 ω-method invented by Cahill and Pohl in 1987 [78, 79]. Both samples with high thermal conductivity and

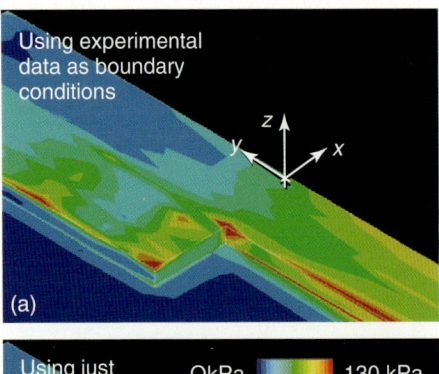
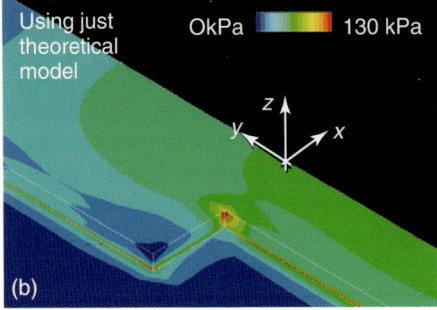

Figure 15.36 Von Mises stress distribution of the constriction, using a) experimental data as boundary conditions, compared with b) conventional FEM simulations.

low thermal conducting areas can be characterized as demonstrated in Figure 15.37. For this purpose, the output voltage of the Wheatstone bridge is detected at the third harmonic of the temperature modulation frequency, applied to the probe. Furthermore, as an idea, measurement of the thermal conductivity was used to read stored data in the so-called Millipede Project, as illustrated in Figure 15.38. An array of thermal probes can simultaneously write information by a melting process and detect the information by detecting the higher heat flow at these points.

Especially in nanolithography, the probe is used as an actuator as well as a sensor. The various forms of these techniques, such as shaving, grafting, and dip-pen lithography, are classified according to the different interactions between the AFM probe and the substrate during the nanolithography fabrication process in the review article by Rosa and Liang [81].

15.6
Advanced SFM/SEM Microscopy

SPM is already widely used in failure analysis and reliability investigations of modern devices, because in addition to the topography a lot of further interaction products are detectable with highest resolution. Beside the great number of

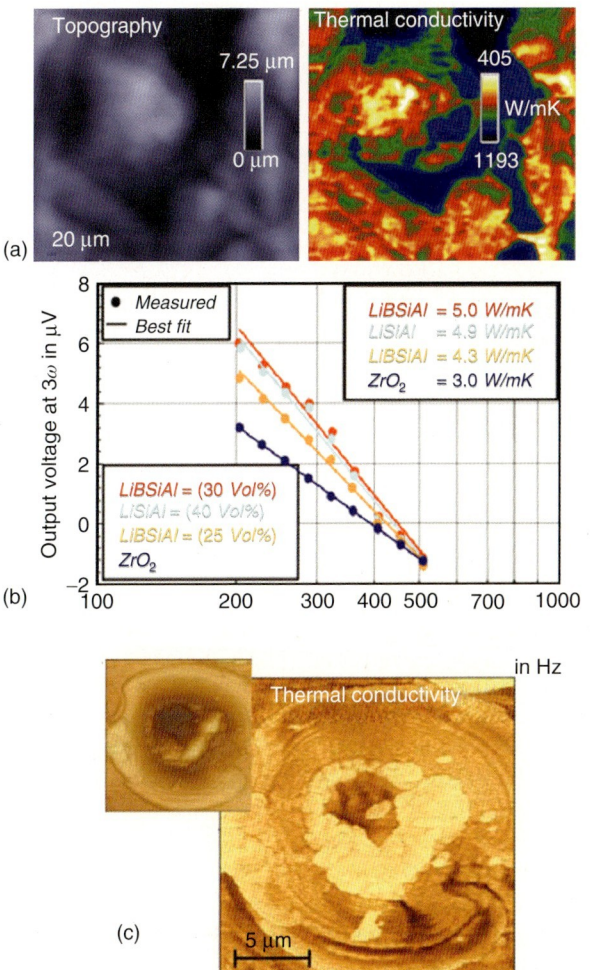

Figure 15.37 Quantitative thermal conductivity analysis on (a) high-conducting microwave-chemical vapor deposition (MW-CVD) diamond film [77], (b) low conducting Li–Si–Al glass ceramics [74], and (c) at a sensillum of a Melanophila [80]. (In collaboration with M. Muller from the Institute for Zoology, University of Bonn, Poppelsdorfer Schloss, D-53115 Bonn, Germany.)

well-known complementary analyses, the combination with a scanning electron microscope (SEM) in one hybrid system expands the field of applications.

Several SFM/SEM-based hybrid systems have been developed and are already commercialized. An MFM was combined with an SEM by Kikukawa *et al.* in 1993 [82]. A method for monitoring AFM cantilever deflection utilizing the focused electron beam of the SEM was introduced by Ermakov and Garfunkel in 1994 [83]. AFM using piezoresistive cantilevers and in combination with an SEM were performed by Stahl *et al.* in 1994 for the first time [84]. Heiderhoff *et al.* used an

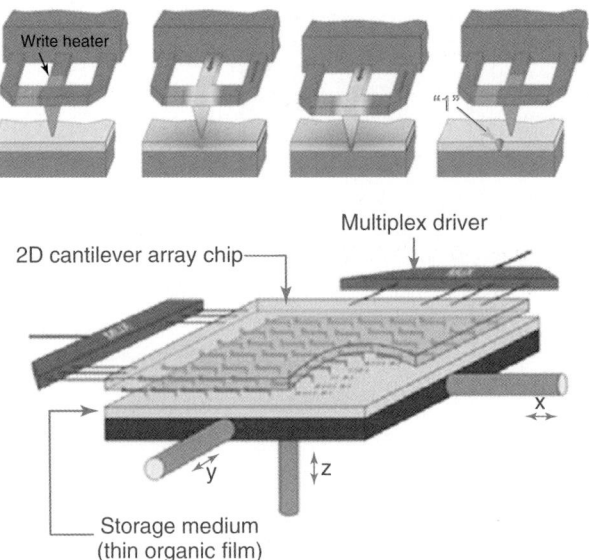

Figure 15.38 Millipede – a highly parallel, very dense AFM data storage system.

SFM equipped with a conducting tip to achieve electronic investigation in 1996 [85]. An SFM combined with an SEM for multidimensional data analysis was introduced by Troyon et al. in 1997 [86]. Dimensional metrology at the nanometer level was carried out by Postek et al. in the same year [87]. The difference between these systems is mostly given in the distance control and the implantation of the scanning probe microscope (SPM) inside the SEM analysis chamber.

Joachimsthaler et al. introduced a system that fits into most available SEM analysis chambers without modification (Figure 15.39) [88]. To get access to the wide field of SFM techniques, all special types of cantilevers can be used because a deflection detection method is chosen for SPM distance control. The SPM is tilted against the electron beam in order to see the tip in SEM. SEM working

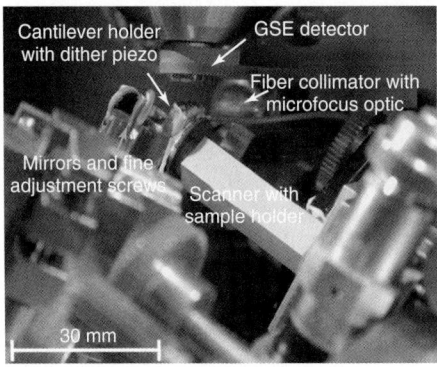

Figure 15.39 SPM/ESEM hybrid system.

15.6 Advanced SFM/SEM Microscopy

distances of below 10 mm are possible, which is necessary for high resolutions and investigations in an environmental scanning electron microscope (ESEM). This combination of SPM and an ESEM is even more interesting, as a huge number of sample properties can be analyzed by both techniques for insulating and vacuum-incompatible samples at variable pressures as well as at different gaseous environments.

A simple switching from a macroscopic field of view for orientation to a nanoscopic scan area for evaluation is not possible for a usual SPM, limiting its applicability. In an SPM/SEM hybrid system, the high resolution of the SPM and the wide field of view of the SEM allow a fast localization of the point of interest. Probing small structures, for example, becomes more and more difficult because of the continuous miniaturization progress. In an SPM/SEM hybrid system, the tip can be used as a free movable connector by morphologic and topographic imaging and successive zooming as illustrated in Figure 15.40.

Even the smallest structures can be probed quickly without physical damage, because in contrast to conventional probe stations the applied force can be controlled with nano-Newton precision. In hybrid systems, both probes can be used simultaneously either as sensor, which give access to a vast variety of material properties, or as an actuator, which can deliberately modify sample properties. This is the essential advantage of these hybrid systems, besides the great number of well-known complementary analyses that can be performed with each microscope at the same sample area (Figure 15.41).

Beside the accurate analysis of surface topography, the implementation of additional probe–specimen interaction mechanisms enhances a hybrid system to a nanocharacterization tool. An incident electron beam, for instance, causes a punctual heat source whose property varies with the well-known SEM parameters such as acceleration voltage and emission current. To measure the temperature distribution of this induced power-dependent heat source, an SThM is integrated into the analysis chamber of an SEM [89], as illustrated in Figure 15.42.

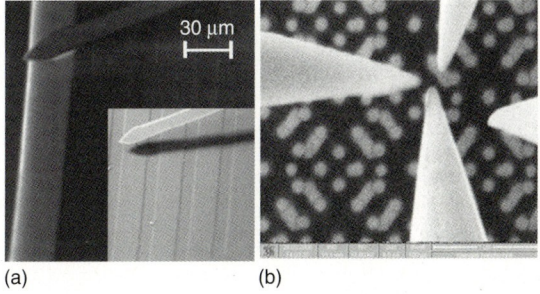

(a) (b)

Figure 15.40 Probing of microstructures; a) using conducting SFM tip and electron beam b) 4 probes conductivity measurement (Image originally created by Kleindiek Nanotechnik GmbH.)

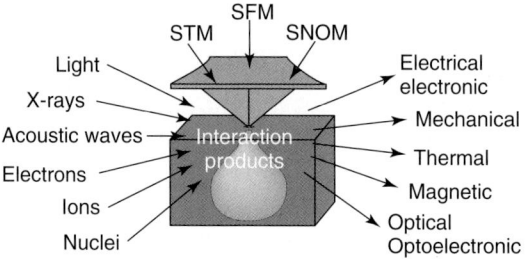

Figure 15.41 Illustration of interaction mechanisms within hybrid systems.

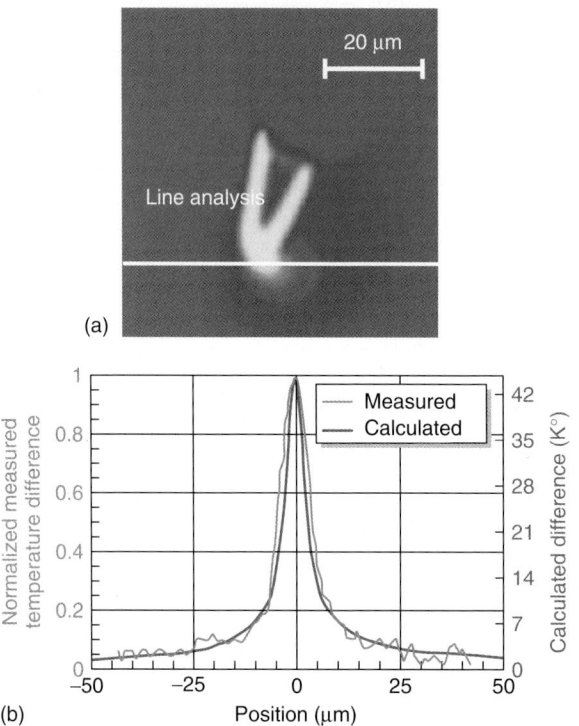

Figure 15.42 Detection of electron-beam-induced temperature distributions with an SThM tip; a) beam mode b) corresponding line analysis (measured and calculated) [89].

Hereby it is possible to get access to heat transport mechanisms even in highly integrated circuits. The thermal characterization of these ICs becomes more and more difficult because of the complex microscopic structures and uncertainties during manufacturing processes. The basic idea of this hybrid system is the simulation of variable punctual thermal hot spots at any interesting location by means of electron beam irradiation and simultaneous detection of induced temperature distribution by an SThM probe at a defined distance. The reliability and lifetime of semiconductor devices are determined by the efficiency of heat removal from hot spot regions. Therefore, the electron-beam-induced temperature oscillation is used for directional thermal conductivity analysis through interfaces and grain boundaries with high spatial resolution [90]. Thermal analysis can be performed on thin films, layered structures such as multiquantum wells (MQWs), passivations, and anisotropic materials, which play a major role in the functionality of modern electronic devices. To perform these investigations with highest resolution, near-field conditions, as described above, must be introduced.

This local periodic heating also generates acoustic waves. To implement acoustic near-field conditions, a line characteristic can be realized without using a beam blanker, as illustrated in Figure 15.43a,b [91] by fast beam scanning. Without the need for a sample mounting on a transducer, the resulting vertical and lateral displacements can be detected by the laser deflection method using an SFM (Figure 15.43c). Hereby, transport losses at interfaces within the bulk material are negligible.

Besides the thermal–mechanical interactions, the probes used can modify electronic properties within the sample, too. The electron beam is scattered inside a solid, and electron–hole pairs are generated because of inelastic collisions. The high number of generated free carriers changes the local electrical properties of the material and induces voltages even within an ESEM. The potential distribution derives from the change in Fermi energies due to diffusion processes inside the sample. These electron-beam-induced potentials are measured by an SEM/SPM hybrid system using a chopped electron beam and EFM in noncontact mode [92]. Potentials can be detected with sensitivities of microvolts on n-doped Si ($10 - 30\ \Omega$ cm) as illustrated in Figure 15.44. The electron beam is scanned and the position of the tip cantilever system is kept constant relative to the sample during this experiment.

Several electronic material parameters, such as the local measurement of diffusion lengths, can be obtained by this technique without contacting the sample. Carrier lifetimes can be analyzed within the frequency domain by the heterodyne mixing technique, applying an AC signal with a frequency f_P to the probe and a slightly higher frequency f_{DUT} to the beam blanker, as described above (Figure 15.26).

For many applications in SEM, the lateral resolution is limited by the energy dissipation volume and the diffusion length of the minority charge carriers. Considering a metal SPM tip in contact with a semiconducting sample, both probes interact electronically with the specimen. Using such a moveable Schottky contact high-resolution, electron-beam-induced current (EBIC) analysis can be performed. Excess charge carriers generated by the electron beam are separated by the tip-induced barriers. The induced current is determined by measuring the voltage drop across a serial resistor, A/D converted after lock-in amplification and transferred to the computer where the data is calculated and displayed (Figure 15.45a).

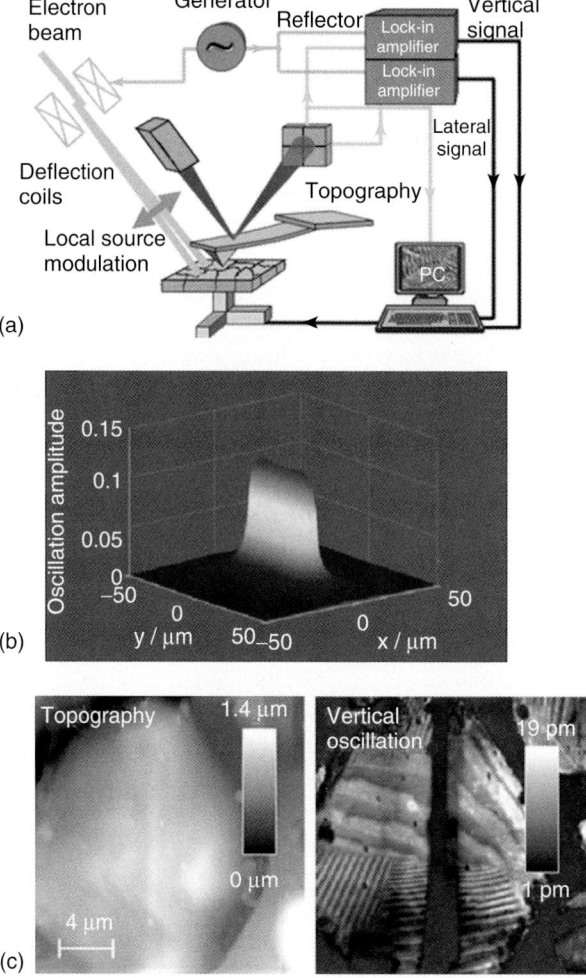

Figure 15.43 (a) Setup for near-field acoustic microscopy, (b) line source characteristic by local source modulation, and (c) detection of ferroelectric domains on BaTiO$_3$ [91].

This experimental method gives the opportunity to analyze structures either by means of the electron beam, while keeping the position of the induced barrier constant, or by scanning the tip/contact, while maintaining a homogeneous electron irradiation. For characterization of polycrystalline chemical vapor deposition (CVD) diamond films, high-resolution EBIC analyses by Au-tip-induced barriers as well as by nano Schottky contact displacement were carried out by Heiderhoff *et al.* already in 1996 [85] (Figure 15.45b). Besides a spatial resolution <5 nm, areas showing p- or n-characteristics were detected and confirmed by spreading resistance (Figure 15.46) and capacitance (not illustrated) measurements. Measuring the EBIC

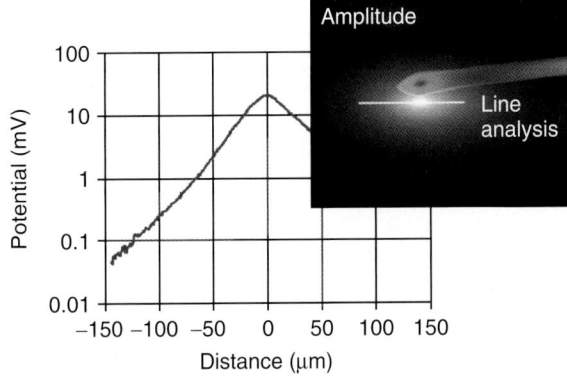

Figure 15.44 Analysis of electron-beam-induced potentials [92].

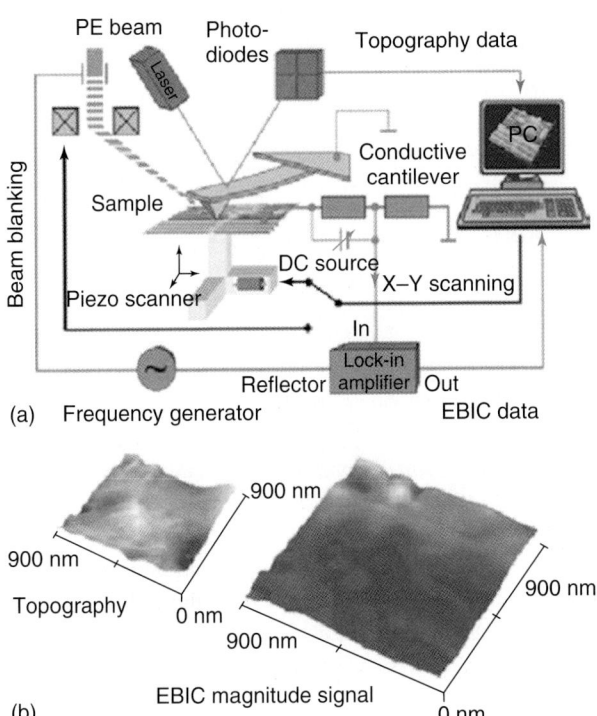

Figure 15.45 Nanoscopic EBIC analysis: (a) setup and (b) tip-induced nano-EBIC analysis on a CVD diamond surface [85].

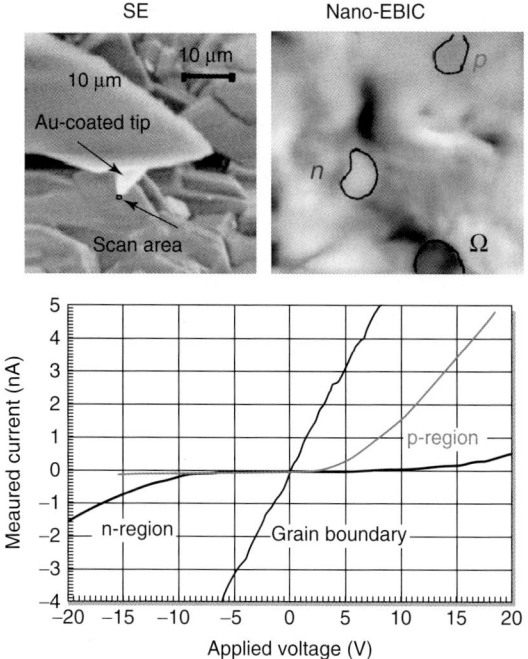

Figure 15.46 Nano-I/V-curve tracing by local spreading resistance microscopy [85].

versus the applied voltage, a local barrier height of 1.3 eV is determined for the p-region by electron beam scanning, as detected in Figure 15.47.

The application of nano-EBIC to the characterization of GaAs and InP homojunctions was demonstrated by Smaali and Troyon in 2008 [93]. The minority charge carrier diffusion length of InP and GaAs is measured and compared for different electron probe currents, and it is shown that the measurements are not perturbed by photon recycling, that is, the self-absorption of photons that gives rise to an extra generation of electron–hole pairs.

For the local detection of defects and their binding energy by complementary cathodoluminescence (CL) investigations, a SNOM/SEM hybrid system is mounted [94–96]. CL is well established as a nondestructive method for analyzing (opto)electronically active impurities and defects in semiconducting materials and devices at lowest concentrations. For high-resolution near-field analysis the vicinity of the SNOM probe area is irradiated homogeneously with primary electrons and the resulting luminescence is picked up directly above the recombination center by the glass-fiber probe. Introducing optical band pass filters into the optical path, spectrally resolved measurements can be performed, while the low average number of photon (<1000 per point) can be detected with significant improvement by using lock-in techniques or energy dispersive spectroscopy [97]. Spectral near-field detection CL analyses on diamond showed an interesting perspective for the recognition

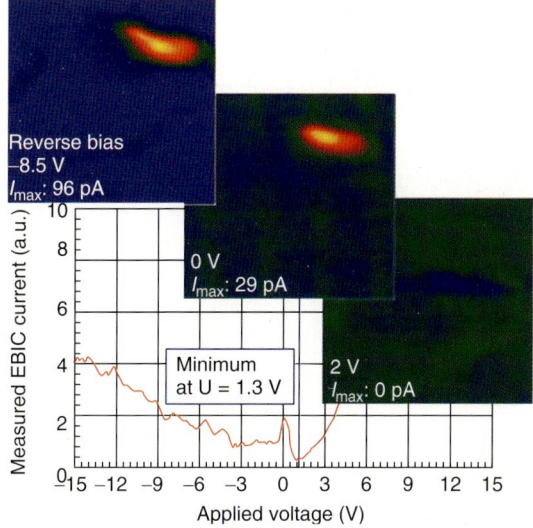

Figure 15.47 Determination of local barrier heights by electron beam scanning [85].

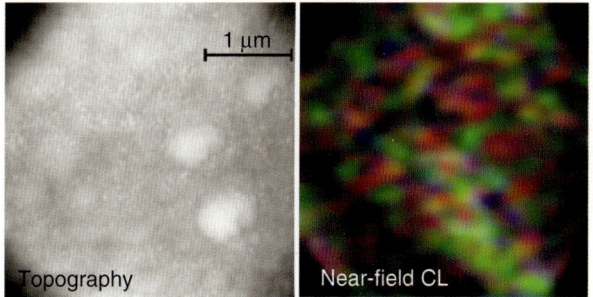

Figure 15.48 Spectrally resolved near-field cathodoluminescence analysis of defects on a diamond facette.

of even the smallest defects and their subsequent characterization, as can be seen from Figure 15.48 [98].

Optoelectronical properties of semiconducting materials have been localized for different parts of the visible spectrum at a lateral resolution of about 50 nm. No sample preparation is necessary, so the risk of changing relevant specimen properties during preparation is greatly reduced. From near-field CL measurements on n-doped gallium nitride films, the minority charge carrier diffusion length in different sample regions is estimated by the beam scanning mode [99] (Figure 15.49). A local variation of diffusion length can occur and be detected within one and the same DUT by 1 order of magnitude.

Thermal, optoelectronic, and electronic properties can be correlated with highest lateral resolution from complementary investigations. Figure 15.50 illustrates

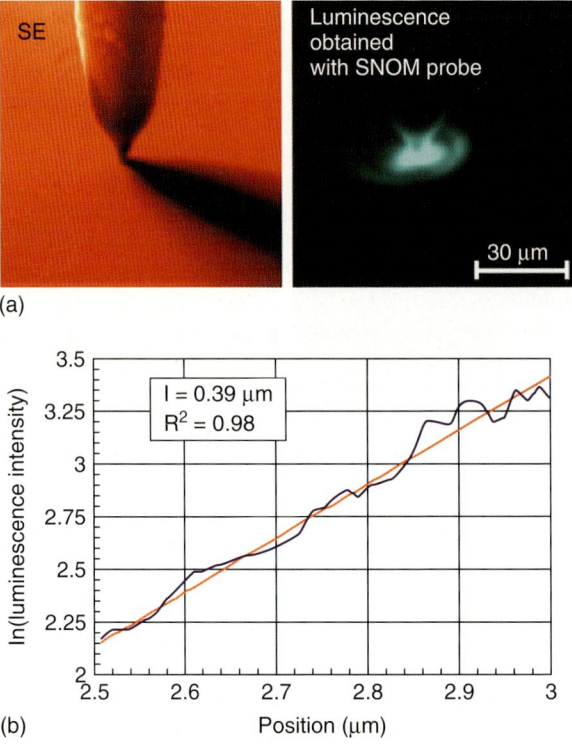

Figure 15.49 Determination of diffusion length on a GaN sample by near-field cathodoluminescence analysis [99]; a) luminescence obtained with SNOM probe by beam scanning mode b) corresponding line analysis of luminescence intensity.

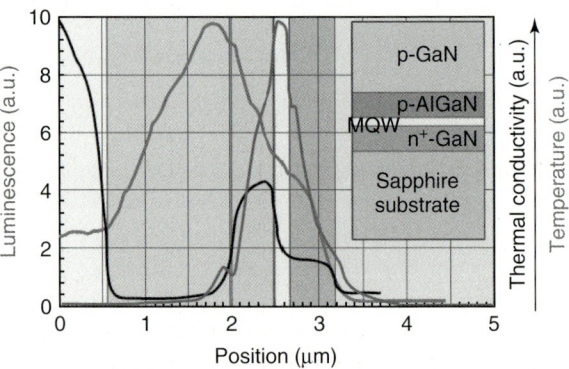

Figure 15.50 Correlation of near-field CL, temperature, and thermal conductivity analyses on a GaN light-emitting device (LED) [100].

near-field CL analyses and scanning thermal microscopy results obtained on a blue GaN light-emitting device [100].

The maximum of the luminescence is as expected at the MQW structure. The maximum temperature is found in the p-doped GaN layer near to AlGaN with its higher thermal conductivity. Consequently, a high potential difference is present at this interface because the current flow through the device is constant. This internal electrical field also leads to a decrease in the CL signal. Apart from the illustrated methods in the characterization of microelectronics, there are many other application fields of ESEM/SPM hybrid systems in microbiology/medicine, materials science/physics, and chemical/electrical engineering, because SPM as well as ESEM allow characterization of low conducting materials as well as living biomaterials. Further, quite often, no sample preparation is necessary.

Acknowledgments

The authors acknowledge all former colleagues of the Department of Electronics at the Faculty of Electrical, Information, and Media Engineering of the University of Wuppertal.

References

1. Abbe, E. (1873) Beiäge zur theorie des mikroskops und der mikroskopischen wahrnehemung. *Arch. Mikrosk. Anat.*, **9** (1), 413–468.
2. Rayleigh, F.R.S. (1879) Investigations in optics, with special reference to the spectroscope. *Philos. Mag. Ser.*, **5**, 261–274.
3. Mansfield, S.M. and Kino, G.S. (1990) Solid immersion microscope. *Appl. Phys. Lett.*, **57** (24), 2615–2616.
4. Heintzmann, R., Jovin, T.M., and Cremer, C. (2002) Saturated patterned excitation microscopy – a concept for optical resolution improvement. *J. Opt. Soc. Am. A*, **19** (8), 1599–1609.
5. Ehret, G., Pilarski, F., Bergmann, D., Bodermann, B., and Buhr, E. (2009) A new high-aperture 193 nm microscope for the traceable dimensional characterization of micro- and nanostructures. *Meas. Sci. Technol.*, **20** (8), DOI:10.1088/0957-0233/20/8/084010.
6. van der Oord, C.J.R., Gerritsen, H.C., Levine, Y.K., Myring, W.J., Jones, G.R., and Munro, I.H. (1992) Synchrotron radiation as a light source in confocal microscopy. *Rev. Sci. Instrum.*, **63** (l), 632–633.
7. Hertz, H. (1887) Ueber sehr schnelleelektrische schwingungen. *Ann. Phys.*, **267** (7), 421–448.
8. Jackson, J.D. (1998) *Classic Electrodynamics*, John Wiley & Sons, Inc., New York, Chichester, Weinheim, Brisbane, Singapore, Toronto, ISBN: 0-471-30932-X.
9. Heisenberg, W. (1927) Über den anschaulichen inhalt der quantentheoretischen kinematik und mechanik. *Z. Phys.*, **43** (3), 172–198.
10. Meixner, J. and Andrejewski, W. (1950) Strenge theorie der beugung ebener elektromagnetischer wellen an der vollkommen leitenden kreisscheibe und an der kreisförmigen öffnung im vollkommen leitenden ebenen schirm. *Ann. Phys.*, **6–7**, 157–168.
11. Synge, E.H. (1928) A suggested method for extending microscopic resolution into the ultramicroscopic region. *Philos. Mag. Ser. 7*, **6** (35), 356–362.

12. Synge, E.H. (1931) A microscopic method. *Philos. Mag. Ser. 7*, **11** (68), 65–80.
13. Synge, E.H. (1932) An application of piezo-electricity to microscopy. *Philos. Mag. Ser. 7*, **13** (83), 297–300.
14. Ash, E.A. and Nicholls, G. (1972) Superresolution aperture scanning microscope. *Nature*, **237**, 510–512.
15. Young, J.Z. and Roberts, F. (1951) A flying-spot microscope. *Nature*, **167**, 231.
16. Unknown (1951) The flying-spot microscope. *Lancet*, **257** (6657), 729.
17. Roberts, F. and Young, J.Z. (1952) High-Speed counting with the flying spot microscope. *Nature*, **169**, 963.
18. Giaver, I. and Mergerle, K. (1961) Study of superconductors by electron tunneling. *Phys. Rev.*, **122** (4), 1101–1111.
19. Giaver, I. (1974) Electron tunnelling and superconductivity. *Rev. Mod. Phys.*, **46** (2), 245–251.
20. Thompson, W.A. and Hanrahan, S.F. (1976) Thermal drive apparatus for direct vacuum tunneling experiments. *Rev. Sci. Instrum.*, **47** (10), 1303–1304.
21. Binning, G., Rohrer, H., Gerber, Ch., and Weibel, E. (1982) Surface studies by scanning tunneling microscopy. *Phys. Rev. Lett.*, **49** (1), 57–61.
22. Edmonds, M.J., Jones, A.M., O'Callaghan, P.W., and Probert, S.D. (1977) A three-dimensional relocation profilometer stage. *Wear*, **43**, 329–340.
23. Binnig, G. and Rohrer, H. (1983) Scanning tunneling microscopy. *Surf. Sci.*, **126**, 236–244.
24. Binnig, G. and Rohrer, H. (1987) Scanning tunneling microscopy-- from birth to adolescence. *Rev. Mod. Phys.*, **59** (3), 615–629.
25. Young, R., Ward, J., and Scire, F. (1972) The topografiner: an instrument for measuring surface microtopography. *Rev. Sci. Instrum.*, **43** (7), 999–1011.
26. Eigler, D.M. and Schweizer, E.K. (1990) Positioning single atoms with a scanning tunnelling microscope. *Nature*, **344**, 524–526.
27. Lewis, A., Isaacson, M., Harootunian, A., and Muray, A. (1984) Development of a 500 å spatial resolution light microscope: I. Light is efficiently transmitted through λ/16 diameter apertures. *Ultramicroscopy*, **13**, 227–232.
28. Massey, G.A. (1984) Microscopy and pattern generation with scanned evanescent waves. *Appl. Opt.*, **23** (5), 658–660.
29. Pohl, D.W., Denk, W., and Lanz, M. (1984) Optical stethoscopy: image recording with resolution λ/20. *Appl. Phys. Lett.*, **44** (7), 651–653.
30. Reddick, R.C., Warmack, R.J., and Ferrell, T.L. (1989) New form of scanning optical microscopy. *Phys. Rev. B*, **39** (1), 767–772.
31. Zenhausern, F., O'Boyle, M.P., and Wickramasinghe, H.K. (1994) Apertureless near-field optical microscope. *Appl. Phys. Lett.*, **65** (13), 1623–1625.
32. Zenhausern, F., Martin, Y., and Wickramasinghe, H.K. (1995) Scanning interferometric apertureless microscopy: optical imaging at 10 angstrom resolution. *Science*, **269** (5227), 1083–1085.
33. Binnig, G., Quate, C.F., and Gerber, Ch. (1986) Atomic force microscopy. *Phys. Rev. Lett.*, **56** (9), 930–934.
34. Williamson, J.B.P. (1967) Microtopography of surfaces. *Proc. Inst. Mech. Eng.*, **182** (3K), 21–30.
35. Bikerman, J.J. (1944) Surface roughness and sliding friction. *Rev. Mod. Phys.*, **16** (1), 53–68.
36. Cohen, S.R., Neubauer, G., and McClelland, G.M. (1990) Nanomechanics of a Au-Ir contact using a bidirectional atomic force microscope. *J. Vac. Sci. Technol. A*, **8** (4), 3449–3454.
37. Overney, R.M., Meyer, E., Frommer, J., Brodbeck, D., Lüthi, R., Howald, L., Güntherodt, H.-J., Fujihira, M., Takano, H., and Gotoh, Y. (1992) Friction measurements on phase-separated thin films with a modified atomic force microscope. *Nature*, **359**, 133–135.
38. Image originally created by Surface Science Western: Nie, H.-Y., Walzak, M.J., and McIntyre, N.S. (2000)

Drawratio-dependent morphology of biaxially oriented polypropylene films as determined by atomic force microscopy. *Polymer*, **41** (6), 2213–2218.

39. Giessibl, F.J., Bielefeldt, H., Hembacher, S., and Mannhart, J. (2001) Imaging of atomic orbitals with the atomic force microscope – experiments and simulations. *Ann. Phys.*, **10** (11–12), 887–910.

40. Hembacher, S., Giessibl, F.J., and Mannhart, J. (2004) Force microscopy with light-atom probes. *Science*, **305** (5682), 380–383.

41. Cramer, R.M., Balk, L.J., Boylan, R., Chin, R., Kaemmer, S., Reineke, F., and Utlaut, M. (1996) The use of near-field scanning optical microscopy (NSOM) for failure analysis of ultra large scale integrated circuits. Proceedings of the 22nd International Symposium for Testing and Failure Analysis, USA, pp. 19–24.

42. Karrai, K. and Grober, R.D. (1995) Piezoelectric tip-sample distance control for near field optical microscopes. *Appl. Phys. Lett.*, **66** (14), 1842–1844.

43. Isakov, D., Tio, A.A.B., Geinzer, T., Phang, J.C.H., Zhang, Y., and Balk, L.J. (2008) Near-field detection of photon emission from silicon with 30 nm spatial resolution. *Microelectron. Reliab.*, **48** (8–9), 1285–1288.

44. Goldner, L.S., Goldie, S.N., Fasolka, M.J., Renaldo, F., Hwang, J., and Douglas, J.F. (2004) Near-field polarimetric characterization of polymer crystallites. *Appl. Phys. Lett.*, **85** (8), 1338–1340.

45. Balk, L.J., Heiderhoff, R., Phang, J.C.H., and Thomas, Ch. (2007) Characterization of electronic materials and devices by scanning near-field microscopy. *Appl. Phys. A*, **87**, 443–449.

46. Shafai, C., Thomson, D.J., Simard-Normandin, M., Mattiussi, G., and Scanlon, P.J. (1994) Delineation of semiconductor doping by scanning resistance microscopy. *Appl. Phys. Lett.*, **64** (3), 342–344.

47. Maywald, M., Stephan, R.E., and Balk, L.J. (1994) Evaluation of an epitaxial layer structure by lateral force and contact current measurements in a scanning force microscope. *Micro. Eng.*, **24**, 99–106.

48. Martin, Y., Abraham, D.W., and Wickramasinghe, H.K. (1988) High resolution capacitance measurement and potentiometry by force microscopy. *Appl. Phys. Lett.*, **52** (13), 1103–1105.

49. Matey, J.R. and Blanc, J. (1985) Scanning capacitance microscopy. *J. Appl. Phys.*, **57** (5), 1437–1444.

50. Stern, J.E., Terris, B.D., Mamin, H.J., and Rugar, D. (1988) Deposition and imaging of localized charge on insulator surfaces using a force microscope. *Appl. Phys. Lett.*, **53** (26), 2717–2719.

51. Martin, Y. and Wickramasinghe, H.K. (1987) Magnetic imaging by 'force microscopy' with 1000 Å resolution. *Appl. Phys. Lett.*, **50** (20), 1455–1457.

52. Saenz, J.J., Garcia, N., Grütter, P., Meyer, E., Heinzelmann, H., Wiesendanger, R., Rosenthaler, L., Hidber, H.R., and Güntherodt, H.-J. (1987) Observation of magnetic forces by the atomic force microscope. *J. Appl. Phys.*, **62** (10), 4293–4295.

53. Hug, H.J., Stiefel, B., van Schendel, P.J.A., Moser, A., Hofer, R., Martin, S., Güntherodt, H.-J., Porthun, S., Abelmann, L., Lodder, J.C., Bochi, G., and O'Handley, R.C. (1998) Quantitative magnetic force microscopy on perpendicularly magnetized samples. *J. Appl. Phys.*, **83** (11), 5609–5620.

54. Coffey, D.C. and Ginger, D.S. (2006) Time resolved electrostatic force microscopy of polymer solar cells. *Nat. Mater.*, **5**, 735–740.

55. Nonnenmacher, M., O'Boyle, M.P., and Wickramasinghe, H.K. (1991) Kelvin probe force microscopy. *Appl. Phys. Lett.*, **58** (25), 2921–2923.

56. Berger, R., Butt, H.-J., Retschke, M.B., and Weber, S.A.L. (2009) Electrical modes in scanning probe microscopy. *Macromol. Rapid Commun.*, **30**, 1167–1178.

57. Mertin, W. (2002) Contactless probing of high-frequency electrical signals with scanning probe microscopy. Proceedings Microwave Symposium Digest, 2002 IEEE MTT-S International, Vol. 3, pp. 1493–1496.

58. Cramer, R.M., Heiderhoff, R., Selbeck, J., and Balk, L.J. (1997) Advanced scanning near-field optical microscopy of semiconducting materials and devices. *Inst. Phys. Conf. Ser.*, **157**, 685–688.
59. Geinzer, T., Heiderhoff, R., Phang, J.C.H., and Balk, L.J. (2010) Determination of the local electric field strength near electric breakdown. Proceedings of IPFA 2010, IEEE Catalog, Number CFP10777-PRT, pp. 285–290.
60. Heil, J., Wesner, J., and Grill, W. (1988) Determination of displacements in ultrasonic waves by scanning tunneling microscopy. *J. Appl. Phys.*, **64** (4), 1939–1944.
61. Rohrbeck, W., Chilla, E., Fröhlich, H.-J., and Riedel, J. (1991) Detection of surface acoustic waves by scanning tunneling microscopy. *Appl. Phys. A*, **52**, 344–347.
62. Günther, P., Fischer, U.Ch., and Dransfeld, K. (1989) Scanning near-field acoustic microscopy. *Appl. Phys. B*, **48**, 89–92.
63. Maivald, P., Butt, H.J., Gould, S.A.C., Prater, C.B., Drake, B., Guriey, J.A., Elings, V.B., and Hansma, P.K. (1991) Using force modulation to image surface elasticities with the atomic force microscope. *Nanotechnology*, **2**, 103–106.
64. Liu, X.X., Heiderhoff, R., Abicht, H.P., and Balk, L.J. (2002) Scanning near-field acoustic study of ferroelectric BaTiO3 ceramics. *J. Phys. D: Appl. Phys.*, **35**, 74–87.
65. Yin, Q.R., Li, G.R., Zeng, H.R., Liu, X.X., Heiderhoff, R., and Balk, L.J. (2003) Ferroelectric domain structures in (Pb,La)(Zr,Ti)O-3 ceramics observed by scanning force microscopy in acoustic mode. *Appl. Phys. A*, **78**, 699–702.
66. Cramer, R.M., Biletzki, V., Lepidis, P., and Balk, L.J. (1999) Subsurface analyses of defects in integrated devices by scanning probe acoustic microscopy. *Microelectron. Reliab.*, **39**, 947–950.
67. Varesi, J. and Majumdar, A. (1998) Scanning joule expansion microscopy at nanometer scales. *Appl. Phys. Lett.*, **72** (1), 37–39.
68. Dietzel, D., Meckenstock, R., Chotikaprakhan, S., Bolte, J., Pelzl, J., Aubry, R., Jacquet, J.C., and Cassette, S. (2004) Thermal expansion imaging and finite element simulation of hot lines in high power AlGaN HEMT devices. *Superlattice. Microst.*, **35**, 477–484.
69. Grauby, S., Lopez, L.-D.P., Salhi, A., Puyoo, E., Rampnoux, J.-M., Claeys, W., and Dilhaire, S. (2009) Joule expansion imaging techniques on microelectronic devices. *Microelectron. J.*, **40** (9), 1367–1372.
70. Tiedemann1, A.-K., Fakhri, M., Heiderhoff, R., Phang, J.C.H., and Balk, L.J. (2009) Advanced dynamic failure analysis on interconnects by vectorized scanning joule expansion microscopy. Proceedings of 16th IPFA, IEEE Catalog, Number: CFP09777-CDR, pp. 515–519.
71. Tiedemann1, A.-K., Kurz, K., Fakhri, M., Heiderhoff, R., Phang, J.C.H., and Balk, L.J. (2009) Finite element analyses assisted scanning joule expansion microscopy on interconnects for failure analysis and reliability investigations. *Microelectron. Reliab.*, **49**, 1165–1168.
72. Fakhri, M., Geinzer, A.-K., Heiderhoff, R., and Balk, L.J. (2010) Nanoscale thermally induced strain and stress analysis by complementary scanning thermal microscopy techniques. ESREF.
73. Dinwiddie, R.B., Pylki, R., and West, P.E. (1994) Thermal conductivity contrast imaging with a scanning thermal microscope. *Therm. Cond.*, **22**, 668–677.
74. Altes, A., Heiderhoff, R., and Balk, L.J. (2004) Quantitative dynamic near field microscopy of thermal conductivity. *J. Phys. D: Appl. Phys.*, **37** (6), 952–963.
75. Majumdar, A., Lai, J., Chandrachood, M., Nakabeppu, O., Wu, Y., and Shi, Z. (1995) Thermal imaging by atomic force microscopy using thermocouple cantilever probes. *Rev. Sci. Instrum.*, **66** (6), 3584–3592.

76. Stopka, M., Oesterschulze, E., and Kassing, R. (1994) Photothermal scanning near-field microscopy. *Mat. Sci. Eng.*, **B24**, 226–228.
77. Fiege, G.B.M., Altes, A., Heiderhoff, R., and Balk, L.J. (1999) Quantitative thermal conductivity measurements with nanometre resolution. *J. Phys. D: Appl. Phys.*, **32**, L13–L17.
78. Cahill, D.G. and Pohl, R.O. (1987) Thermal conductivity of amorphous solids above the plateau. *Phys. Rev. B*, **35**, 4067–4073.
79. Cahill, D.G., Fischer, H.E., Klitsner, T., Swartz, E.T., and Pohl, R.O. (1989) Thermal conductivity of thin films: measurements and understanding. *J. Vac. Sci. Technol. A*, **7**, 1259–1266.
80. Müller, M., Olek, M., Giersig, M., and Schmitz, H. (2008) Micromechanical properties of consecutive layers in specialized insect cuticle: the gula of Pachnoda marginata (Coleoptera, Scarabaeidae) and the infrared sensilla of Melanophila acuminata (Coleoptera, Buprestidae). *J. Exp. Biol.*, **211**, 2576–2583.
81. Rosa, L.G. and Liang, J. (2009) Atomic force microscope nanolithography: dip-pen, nanoshaving, nanografting, tapping mode, electrochemical and thermal nanolithography. *J. Phys.: Condens. Matter*, **21** (48), 483001.
82. Kikukawa, A., Hosaka, S., Honda, Y., and Koyanagi, H. (1993) Magnetic force microscope combined with a scanning electron microscope. *J. Vac. Sci. Technol. A*, **11** (6), 3092–3098.
83. Ermakov, A.V. and Garfunkel, E.L. (1994) A novel AFM/STM/SEM system. *Rev. Sci. Instrum.*, **65** (9), 2853–2854.
84. Stahl, U., Yuan, C.W., de Lozanne, A.L., and Tortonese, M. (1994) Atomic force microscope using piezoresistive cantilevers and combined with a scanning electron microscope. *Appl. Phys. Lett.*, **65** (22), 2878–2880.
85. Heiderhoff, R., Cramer, R.M., and Balk, L.J. (1996) High resolution electron beam induced current measurements in an SEM-SPM hybrid system by tip induced barriers. *Inst. Phys. Conf. Ser.*, **149**, 189–194.
86. Troyon, M., Lei, H.N., Wang, Z., and Shang, G. (1997) A scanning force microscope combined with a scanning electron microscope for multidimensional data analysis. *Microsc. Microanal. Microstruct.*, **8**, 393–402.
87. Postek, M.T., Ho, H.J., and Weese, H.L. (1997) Dimensional metrology at the nanometer level: combined SEM and PPM. *Proc. SPIE*, **3050**, 250–263.
88. Joachimsthaler, I., Heiderhoff, R., and Balk, L.J. (2003) A universal scanning probe- microscope-based hybrid system. *J. Meas. Sci. Technol.*, **14**, 87–96.
89. Altes, A., Joachimsthaler, I., Zimmermann, G., Heiderhoff, R., and Balk, L.J. (2002) SEM / SThM-Hybrid-system: a new tool for advanced thermal analysis of electronic devices. Proceedings of the 9th IPFA, IEEE Catalog, No. 02TH8614, pp. 196–200.
90. Tiedemann, A.-K., Heiderhoff, R., Balk, L.J., and Phang, J.C.H. (2009) Electron beam induced temperature oscillation for qualitative thermal conductivity analysis by an SThM / SEM-hybrid system. Proceedings 47th IRPS, IEEE Catalog, No. CFP09RPS-CDR, pp. 327–332.
91. Thomas, Ch., Heiderhoff, R., and Balk, L.J. (2007) Nanoscale acoustic near-field imaging in an SEM/SPM hybrid with sub-picometer sensitivity. *J. Scan. Probe Microsc.*, **2**, 15–18.
92. Thomas, Ch., Joachimsthaler, I., Heiderhoff, R., and Balk, L.J. (2004) Electron beam- induced potentials in semiconductors: calculation and measurement with an SEM/SPM hybrid system. *J. Phys. D: Appl. Phys.*, **37**, 2785–2794.
93. Smaali, K. and Troyon, M. (2008) Application of nano-EBIC to the characterization of GaAs and InP homojunctions. *Nanotechnology*, **19**, 155706.
94. Cramer, R.M., Heiderhoff, R., and Balk, L.J. (1998) Scanning near-field cathodoluminescence investigations. *Scanning*, **20** (6), 433–435.
95. Cramer, R.M., Ebinghaus, V., Heiderhoff, R., and Balk, L.J. (1998)

Nearfield detection cathodoluminescence investigations. *J. Phys. D: Appl. Phys.*, **31**, 1918–1922.
96. Troyon, M., Pastré, D., Jouart, J.P., and Beaudoin, J.L. (1998) Scanning near-field cathodoluminescence microscopy. *Ultramicroscopy*, **75**, 15–21.
97. Geinzer, T., Heiderhoff, R., Phang, J.C.H., and Balk, L.J. (2010) Determination of the local electric field strength by energy dispersive photo emission microscopy. Proceedings IRPS 2010, ISBN 978-4244-5431-0, 271-276.
98. Cramer, R.M., Sergeev, O.V., Heiderhoff, R., and Balk, L.J. (1999) Spectrally resolved cathodoluminescence analyses in the optical near-field. *J. Microsc. (Oxford)*, **194** (Part 2–3), 412–414.
99. Nogales, E., Joachimsthaler, I., Heiderhoff, R., Piqueras, J., and Balk, L.J. (2002) Near-field cathodoluminescence studies on n-doped gallium nitride films. *J. Appl Phys.*, **92** (2), 976–978.
100. Heiderhoff, R., Palaniappan, M., Phang, J.C.H., and Balk, L.J. (2000) Correlation of scanning thermal microscopy and near-field cathodoluminescence analyses on a blue GaN light emitting device. *Microelectron. Reliab.*, **40**, 1383–1388.

16
Scanning Probe Microscopy – Forces and Currents in the Nanoscale World

Brian J. Rodriguez, Roger Proksch, Peter Maksymovych, and Sergei V. Kalinin

16.1
Introduction

Achieving the full potential of nanoscience and nanotechnology requires the capability to image, manipulate, and control matter and energy on the nanometer, molecular, and ultimately, atomic levels. Scanning probe microscopy (SPM) techniques provide unparalleled access to the nanoscale world through structural, functional, and chemical imaging and manipulation on nanometer and atomic scales [1–4]. Beyond imaging surface topography, SPMs have found an extremely broad range of applications for probing electronic, transport, optical [5–7], magnetic, mechanical, and electromechanical properties – often at the level of several tens of nanometers and below. A number of excellent reviews [8–10] and books summarize applications of SPM for polymers [11], biological systems [12–14], ferroelectrics [15, 16], and other materials and devices [17]. Significant interest for SPM stems from the fact that these techniques can be readily implemented in liquid environments (including physiological buffers and electrolytes) [18].

To quote Freeman Dyson, one of the visionary physicists of the twentieth century, *"Major events in the history of science are called scientific revolutions. There are two kinds of scientific revolutions, those driven by new concepts and those driven by new tools"* [19]. SPM is a paradigmatic example of a scientific revolution driven by new tools, which has resulted in a vast array of new measurements and scientific insights. One notable example of a scientific area enabled by SPM is macromolecular unfolding spectroscopy [20–22], which has subsequently given rise to a broad discipline of force-driven single-molecule reactions [23–27] and to novel areas of statistical physics [23]. Another prominent example is scanning tunneling microscopy (STM)-based current spectroscopy [28] of band gaps and charge ordering [29, 30], superconductive gaps [31–34], and vibrational levels [35, 36], which have provided new insights into the physical properties of strongly correlated oxides and molecular systems. Similarly, piezoresponse force microscopy (PFM) [37, 38] and electrochemical strain microscopy (ESM) [39, 40] have opened pathways for probing bias-induced phase transitions and electrochemical reactions at the nanometer scale, ultimately to the level of a single structural defect.

Handbook of Nanoscopy, First Edition. Edited by Gustaaf Van Tendeloo, Dirk Van Dyck, and Stephen J. Pennycook.
© 2012 Wiley-VCH Verlag GmbH & Co. KGaA. Published 2012 by Wiley-VCH Verlag GmbH & Co. KGaA.

This impressive progress has occurred in the three decades since the pioneering work by Binnig and Rohrer [41] demonstrating the STM (Figure 16.1a) and 25 years since Binnig, Quate, and Gerber [42] introduced the original concept of atomic force microscopy (AFM) (Figure 16.1b). SPM tools now range from simple ambient machines for topographic imaging, to increasingly advanced platforms for electrical characterization and biological imaging in physiological environments, and massive low-temperature, high-magnetic field machines being developed for atomic, vibrational, and spin imaging. Here, we (i) summarize the operating principles of the primary SPM modes, explaining the current state of the SPM field, and the role it is playing in nanoscience, (ii) describe the current state-of-the-art

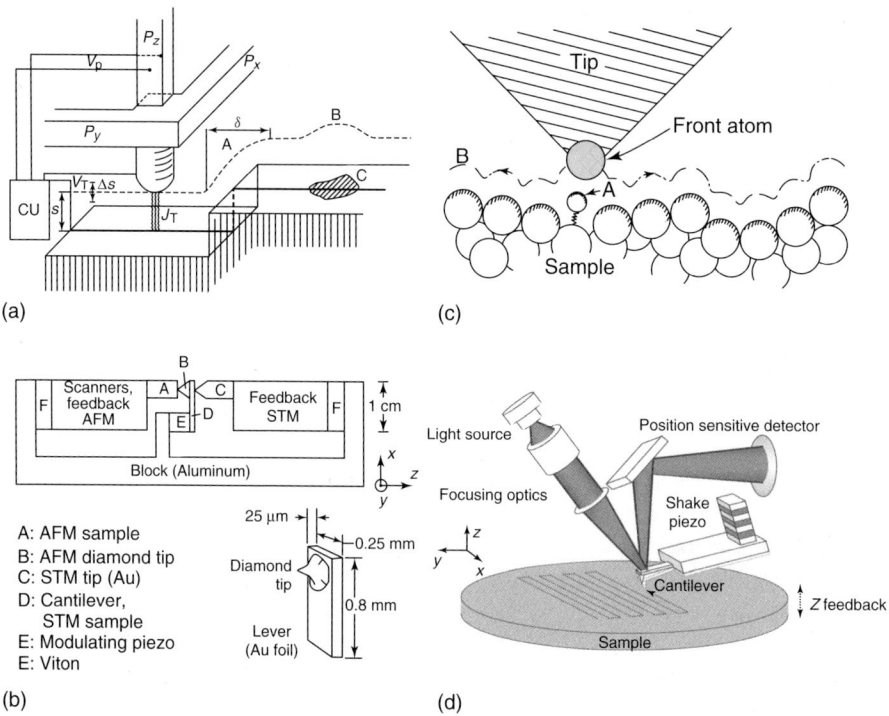

Figure 16.1 Principles of SPM. (a) Experimental setup of the original STM, wherein piezo drives (P_x and P_y) scan a metal tip over a surface and the control unit (CU) applies a bias to the piezo drive P_z in order to maintain a constant tunneling current (J_T). (b) Experimental setup of the original AFM, which utilized an STM-based detection scheme. (c) General description of the principle of operation for STM and AFM: the tip follows the sample surface while trying to maintain constant current (STM) or force (AFM). Note that STM requires conducting samples, while any sample can be investigated by AFM. (d) Example of a modern AFM experimental setup showing the main elements of an AFM. (Panels (a) and (b,c) reproduced with permission from the American Physical Society from [43] and [42], copyright 1982 and 1986, respectively. Panel (d) is reproduced courtesy of Asylum Research).

SPM methods, and (iii) discuss promising future areas of instrument design and scientific inquiry.

16.2
Scanning Probe Microscopy – the Science of Localized Probes

The quintessence of an SPM approach is the combination of a local probe bearing a specific aspect of the functionality to be studied (mechanical, magnetic [44], charge [45], chemical [46], etc.), combined with a precise actuator system, which is able to position the probe at a selected location with respect to the surface of interest, and a detection system that measures forces, currents, or optical signals indicative of the probe-surface interactions. This experimental framework links local materials structure, properties, and functionalities, probed through nanoscale interactions in the probe-surface junction, to the macroscopic world [41, 42]. The progress in the capability of SPM to probe relevant phenomena is directly linked to the performance and functionality of each of these elements, as well as to the data acquisition and processing electronics that acquire, condition, visualize, and analyze the detected signal.

16.2.1
Local Probe Methods before SPM

The history of the local probe approach can be traced to the dawn of the industrial age. The industrial revolution brought along with it the development of various instruments to determine the surface finish quality of machined components. Of these instruments, the surface profilometer was a direct ancestor of the AFM, and an extensive description of this history is given by Thomas [47]. Stylus profilometers have a sharp probe, which is brought into contact with a surface and then dragged across it. The vertical motion of the probe is measured and recorded, providing a record of the surface topography. Similar to AFMs developed later, a feedback loop was developed for a profilometer to maintain the tip–sample interaction at a constant value [48]. This feedback allowed forces to be controlled and was developed to enable profiling of soft samples. A profilometer that used an optical lever to measure vertical probe deflections, in a striking preview of similar techniques used in AFM [49, 50], was developed by Schmaltz in 1929 (Figure 16.2) [51].

In anticipation of the oscillating cantilever AFM techniques (e.g., tapping mode AFM), in 1950 Becker et al. [52] and later Lee and Harrison [53] solved problems caused by unstable probe bending and "whipping" as it was scanned over rough surfaces by oscillating (using piezoactivation) the probe perpendicular to the surface at the resonant frequency of the probe. Because the probe only contacted the surface for a small fraction of the time, it tended to minimize lateral forces and better avoid becoming stuck to the sample surface.

In addition to these examples of surface topographic mapping techniques, there were precursors to nontopographic, functional property imaging modes that occurred well before the invention of the STM. One example comes from

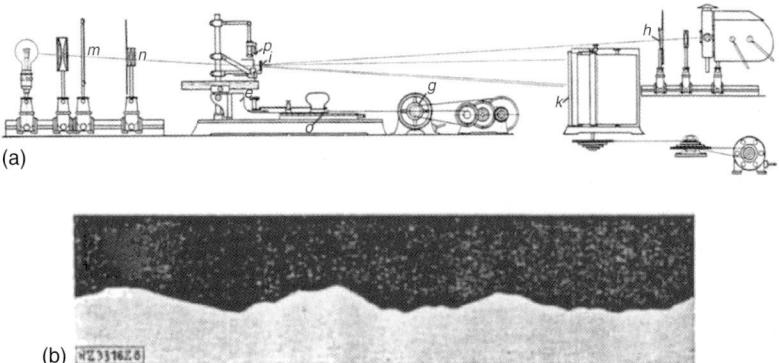

Figure 16.2 Optical lever profilometer. (a) A schematic diagram of the optical detection scheme used to measure the motion of a stylus probe over a surface. This instrument recorded surface topography using photographic film, an example of which is shown in (b). (Reproduced from [51], copyright, 1929, VDI-Z.)

the field of magnetism. Kaczér [54, 55] developed a scanning probe system for observing domain structures in ferromagnetic materials just a few short years after the first experimental verification of the existence of ferromagnetic domains. Kaczér's microscope measured the magnetic state of a small permalloy probe using induction (Figure 16.3). As the probe was rastered over a sample surface, the magnetically soft magnetic state of the permalloy probe changed in response to the localized field from the sample. The resulting images correlated well with colloid images of the same sample.

The recognized precursor of modern SPMs, the "Topografiner" was developed by Young between 1966 [56] and 1971 [57]. According to Young, the name originated from the Greek word $\tau o \pi o \gamma \rho \alpha \phi \varepsilon \iota \nu$ – "to describe a place" [56]. The Topografiner was a type of noncontact (NC) profilometer developed with the stated goal *"to characterize the so-called 'single-crystal surfaces' on an atomic level"* [56]. Instead of sensing mechanical forces, the tip–probe distance was measured using the electron field emission current. In retrospect, this instrument had some striking similarities to the STM, including the use of piezoelectric actuators to control the tip–sample position, the use of an electronic feedback circuit controlling the z-piezo to keep the tip–sample distance constant, and the ability to produce high-resolution (albeit not atomic resolution) 3D images. Note also, that similar to the STM, the Topografiner required conducting samples.

In the remainder of this chapter, we focus on the progress in the field of SPM since the invention of STM and AFM, highlighting fundamental operating principles, imaging modes, and breakthroughs in design and technique development, which have provided insight into the nanoworld. A timeline highlighting key events in the history of SPM is shown in Figure 16.4 and the specific references are discussed throughout the text.

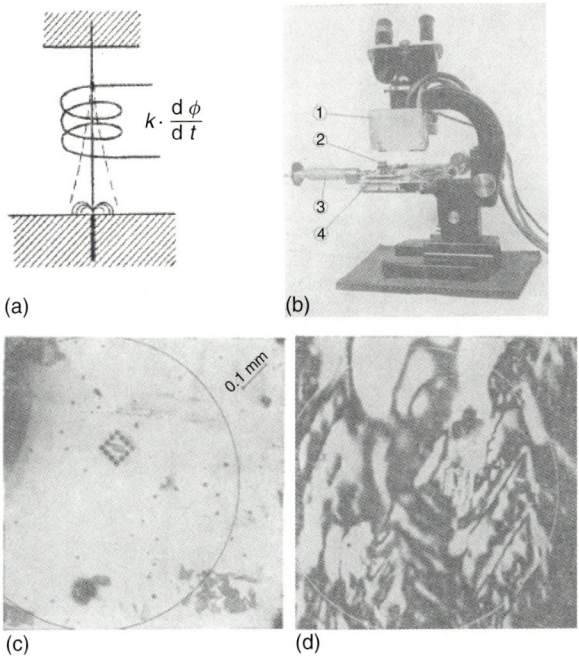

Figure 16.3 Vibrating permalloy probe. (a) Schematic showing the oscillating magnetic probe that induced a current in a detection coil. (b) A photograph of the complete instrument from Kaczér and Gemperle's 1956 paper. (c) An optical micrograph of the topography of the sample surface and (d) magnetic fields from the associated domain structures as mapped with the scanning probe. Note the square topographic feature visible in both the optical (c) and magnetic (d) images. (Reproduced from [55], copyright 1956, Springer Netherlands.)

16.2.2
Positioning, Probes, and Detectors

16.2.2.1 Positioning

The key element of SPM is a positioning system that maintains the probe at a predefined location with respect to the surface and enables scanning. The requirements for positioning systems can be very stringent – for example, enabling stable tip–sample contact to measure tunneling current in STM requires a relative stability of the probe and surface at about the picometer level (i.e., $\sim 10^{-12}$ m, well below an atomic radius), while high-resolution spectroscopic imaging requires that the lateral drift is $<\sim 1$ nm per hour, depending on the application. While sometimes presenting a serious challenge to engineering, the positioning systems are by now well understood and developed, in some cases allowing closed-loop, atomic-resolution imaging. High-stability platforms have been developed in the context of low-temperature STM and AFM [58–61]. In modern commercial systems, lateral drift rates are routinely approaching 10 nm/C. Operation in a temperature-controlled environment then translates to a lateral drift

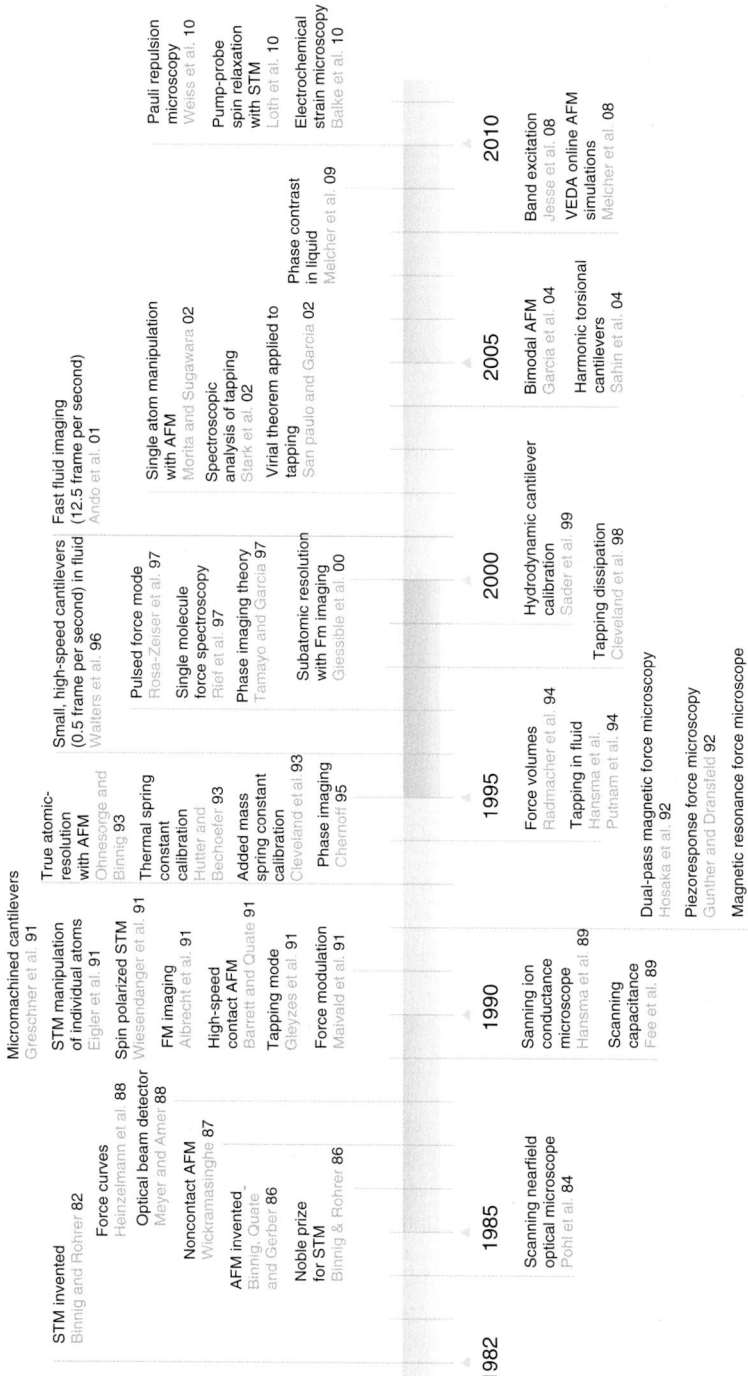

Figure 16.4 Timeline of the progress in SPM methods over the past three decades, discussed in this chapter.

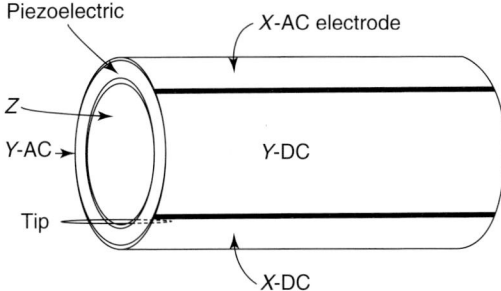

Figure 16.5 Piezo-tube scanner for SPM. (Reproduced with permission from [62], copyright 1986, American Institute of Physics.)

<1 nm h^{-1}. This enables extremely high spatial resolution and stable measurements, including spectroscopic measurements at single points. The detection sensitivity is commonly <100 fm Hz$^{-1/2}$. In conventional 1 kHz imaging bandwidths, this translates to a z-noise level of <30 pm. Many commercial systems are based on a piezoelectric tube scanner first demonstrated by Binnig and Smith [62], which allows for scanning in x and y directions (sample plane), and for tip–sample position control in the z-direction, perpendicular to sample surface (Figure 16.5). Note that in a tube scanner, the x-y motion is inherently coupled to a z-displacement. Some AFMs decouple motion in three directions using three separate piezo stacks.

16.2.2.2 Detectors

The detector system is the second key aspect of SPM. In STM, the heart of the detector system is the current amplifier that measures the tunneling current in the tip–surface junction, which is generated in response to a bias applied between the probe and the surface. This measured current is then used as a feedback signal to maintain the tip–surface distance and enable topographic imaging (Figure 16.1a), or as a basis for spectroscopic and modulation measurements such as current–voltage spectroscopies. Similar amplifiers are used in current-based SPM methods including scanning ion conductance microscopy [63], scanning spreading resistance microscopy (SSRM) [64], conductive atomic force microscopy (cAFM), electrochemical SPMs, and SPM-based impedance measurements and their variants. In these techniques, the current is detected in parallel with the force signal that is used for topographic and functional imaging of the surface.

A wide variety of techniques have been developed to measure the force acting on the cantilever tip (see e.g., Sarid [65]). The original AFM used the electron tunneling STM as a detector for the cantilever deflection (Figure 16.1b). The highly sensitive, exponential decay of the tunneling current allowed detection of extremely small deflections, on the order of 10^{-3} nm (1 pm). However, this same exquisite sensitivity proved to be a detriment as well, as the drift between the STM detection tip and the cantilever, which was separated by a large mechanical loop, caused variations in the sensitivity and resulted in deflection artifacts. In addition, the STM detection method proved to be very sensitive to contaminants on the back surface of the cantilever. As a consequence, this detection scheme is rarely, if ever, used in modern AFMs.

Currently, cantilever detection is almost universally based on the optical lever method illustrated in Figure 16.1d [49]. The *optical lever* is also a common method used in ultraprecision instruments [66], and had been used in stylus profilometers for over 80 years [51]. In this method, a light source (typically a diode laser), is used to illuminate the backside of a cantilever beam (with the probe tip attached to the opposite side of the beam). The laser beam is reflected into the center of a position-sensitive photodetector. When the probe interacts with a surface, the tip, and thus the cantilever beam are deflected. The deflection of the beam can be monitored by measuring the change in position of the reflected beam via the photodetector. The unique aspect of this approach is that it allows both the normal and torsional components of the force acting on the tip to be measured, enabling a broad variety of SPM modes.

The disadvantage of the standard cantilever configuration is that it does not allow the longitudinal (acting along the cantilever axis) and normal modes, which both result in flexural cantilever oscillations, to be separated. Recently, a number of sensor configurations using membrane-type detectors have been suggested to circumvent this problem [67–69]. Similarly, cantilever sensors for optical detection are constantly being developed, with notable examples including small single-crystal Si cantilevers [70], paddle beam preamplifying cantilevers [71], paddle beam cantilevers to tailor the harmonic resonance responses [72], microfabricated tribolevers for friction measurements [68], small cantilevers [73] for force spectroscopy and high-speed imaging, and dual-frequency (two cantilevers in series) cantilevers [74]. Although the optical beam detector [75] has not been displaced as the dominant form of cantilever detection, there have been many attempts at producing cantilevers that are self-sensing – that is, that have integrated detection schemes, most notably piezolevers, which do not require laser beam alignment. Development of force-based SPMs operating in high magnetic fields or extreme environments will likely lead to future progress in this area.

Current amplifiers in STM and cAFM and optical position sensor/cantilever/electronics force detectors in scanning probe microscopes have intrinsic limitations on bandwidth and sensitivity imposed by fundamental physical limits (e.g., thermomechanical noise [76] of a cantilever with a defined spring constant [77], laser shot noise, Johnson noise, etc.) [65]. However, these fundamental limits are typically not achieved, and the noise is dominated by the electronic or environmental noise.

The noise level of an SPM system is the fundamental factor that determines both the detection limit and the spatial resolution of the image (or more precisely, the information limit of the technique) [78–82]. While the relationships between the noise level and spatial resolution are seldom analyzed in the context of SPM because of the fact that imaging is often nonlinear, a detailed formalized resolution theory is available in the context of electron and optical microscopy [78, 81, 83, 84].

Correspondingly, the strategies for improving SPM performance invariably include low acoustic and electronic noise environments, ranging from acoustic hoods and active vibration damping systems, to specially designed low-noise buildings. Minimization of the electronic noise of SPM electronics and laser noise present another avenue for technique development. As an example of such developments, the recent introduction of low-noise detector systems by the Yamada group [85, 86] has significantly increased the resolution in NC-AFM.

16.2.2.3 Probes

The probe is at the heart of every SPM experiment. Probe functionality determines both the measured functionality (e.g., structure, conductivity, polarization, chemical properties, magnetization, charge, temperature, etc.), and the spatial resolution at which the properties can be measured. The interactions between the probe and the surface provide information on the structure and functionality of the materials under investigation. Scanning electron microscopy (SEM) images of an STM probe are shown in Figure 16.6.

One of the traditional goals in SPM has been to increase the rate of imaging, ideally to a "video" rate of at least a few frames per second. Since SPMs are mechanical devices, there are significant limitations to increasing the scan rate, including (i) the response time or the bandwidth of the cantilever probe system, including the detector [89], (ii) the response time or the bandwidth of the mechanical and electronic components of the feedback loop [90, 91], (iii) the inherently serial nature of point-to-point scanning of a tip over a surface, and (iv) the speed of SPM control, acquisition, and display electronics.

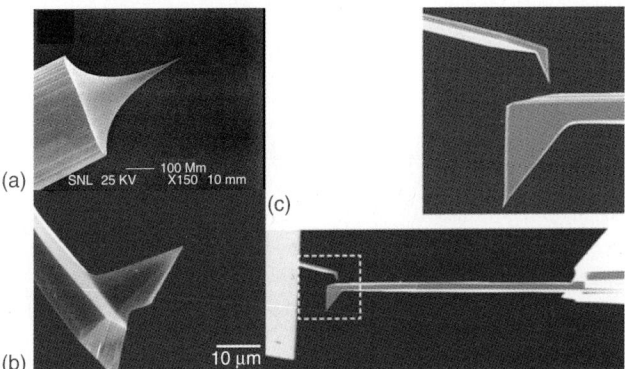

Figure 16.6 SEM images of STM and AFM tips. (a) SEM image of an etched tungsten STM tip. (b) SEM image of an AFM tip. (c) SEM images of large and small cantilevers. The small cantilever has the same nominal spring constant as the large cantilever (~10 N/m) but has a resonance frequency of 4.6 MHz, ×15 higher than the larger cantilever. Because the Q-factors for these two levers are the same, the smaller cantilever responds and can image ×15 faster than the large lever. ((a) Reproduced with permission from [87], copyright 2008, Institute of Physics. (b) Reproduced with permission from [88], copyright 2010, Elsevier. (c) Courtesy of Asylum Research.)

Image acquisition rates can also be increased by using multiple cantilevers to simultaneously image different regions of the same sample. This parallel processing approach requires a composite image to be constructed. In addition, there are challenges associated with registration of the cantilever tip position. Finally, since the mechanical properties of the cantilevers limit the time resolution, the use of multiple levers does not improve the intrinsic time resolution of the measurements [92].

Barrett and Quate [93] demonstrated high imaging rates of over three frames per s in 1991 using contact mode AFM. This instrument had a respectable scan range of 18 μm × 18 μm and demonstrated real-time operator control of the scanning parameters and data processing. One limitation of this pioneering instrument is that the piezo tube had a fundamental resonance of ∼2 kHz, which limited the response time of the vertical feedback loop. To avoid this issue, it was operated in a "constant height mode," meaning that the deflection of the cantilever was allowed to vary along each scan line such that the forces between the tip and the sample were not constant. One can conjecture that these variable forces are one reason this microscope did not attain wider use; while the varying forces did not affect the integrated circuits imaged in the paper, other softer samples are more susceptible to damage if the force is not carefully controlled. Carrying the concept of Barrett and Quate to a remarkable extreme, Humphris et al. [94–96] have similarly avoided the use of an active z-feedback loop and in the process developed an extremely high-speed system, reporting acquisition rates as high as 1300 frames per s.

Improvement of time resolution can be accomplished for cyclic processes using a stroboscopic approach to AFM data acquisition [97]. In this technique, sequential AFM images are acquired while the phase with respect to some periodic process is varied. When the process is truly periodic, the details of the dynamics, limited only by the control of the phase resolution, can be reconstructed. This requirement, however, is quite a significant limitation since many of the dynamic processes of interest at the nanoscale are dissipative and not necessarily stable over many cycles.

By increasing the speed of all of the system components, it is possible to make a high-speed AFM that maintains the benefits of gentle tip–sample forces from tapping mode. One of the most challenging aspects has been the fabrication of small cantilevers with high resonant frequencies and moderate spring constants, as shown in Figure 16.3c [98]. Along with improvements in the response time of the electronic and mechanical feedback systems (Schitter et al. [99, 100], Fantner et al. [101], and Rost et al. [102]) and an increase in scanning frequency (see, for example: Viani et al. [73, 103], Sakamoto et al. [104], and Ando et al. [105–108]), high-speed AFM imaging has been achieved.

As compared to the remarkable range of technologies emerging in the context of force-based SPM probing, progress in STM methods has been significantly more linear. In STM methods, the paradigmatic probe is an atomically sharp needle fabricated from tungsten (Figure 16.3a) or a Pt–Ir alloy wire. Atomic-resolution imaging requires that the apex of the probe has a well-defined atomic configuration with a single atom at the end of the tip. Fortuitously, such terminations can form during the scanning process, enabling routine atomic-resolution STM imaging of semiconductors and metals. Some insight into the atomic structure of the STM

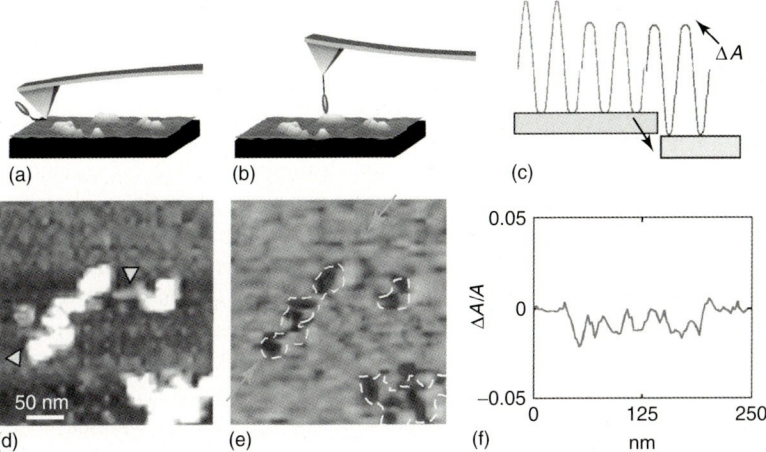

Figure 16.7 Recognition imaging. (a) Schematic showing an antibody tethered to a tip. When there is (b) a recognition event, that is, when the antibody binds to the an antigen, there is (c) a reduction in the tip oscillation amplitude, which can be mapped simultaneously with the (d) topography imaging as (e) a change in the peak signal. (f) A line profile of the peak signal between the arrows in (e). (Reproduced with permission from [132], copyright 2004, National Academy of Sciences, USA.)

probe has been obtained from combined STM-scanning transmission or transition electron microscopy ((S)TEM) experiments, providing an atomically resolved view of the tip structure [109–111]. In cases where the surface is unstable and transfer of atoms between the tip and the surface is possible, high-resolution imaging becomes significantly more challenging. Some insight into the atomic structure of STM probes can be achieved from *in situ* high-resolution electron microscopy experiments [112].

Modification of STM probes is highly nontrivial, given the need for atomic-level control of the apex structure. A well-known example is the use of spin-polarized probes fabricated from magnetic or antiferromagnetic metals, which enable spin-polarized STM [113–118]. A second example is scanning tunneling hydrogen microscopy, in which a single deuterium molecule confined in the STM tip–sample junction is used to obtain ultrahigh-resolution images [119–121].

In comparison, force-based SPM methods offer a wider range of opportunities for the development of the cantilever probe system. Paradigmatic examples include the electrically biased conductive probes that enable Kelvin probe force microscopy (KPFM) [45] and electrostatic force microscopy (EFM) [122–124], and magnetic probes that enable magnetic force microscopy (MFM) [125, 126]. Note that electrostatic interactions can be measured on an arbitrarily small length scale if a conductive metallic sphere with a known and controlled potential can be placed at an arbitrary distance from the surface, and the force on it can be measured. Practically, the probe represents the apex of the conductive cone of a fabricated AFM tip. Hence, the local component of the electrostatic force is measured jointly with the signal from the nonlocal conical part. The optimal probe size, R, can be determined as the one for which the electrostatic force is larger than the response from the conical part for the experimentally accessible range of probe–surface separations (on the order of R) [127].

Chemical functionalization of the probe allows high-veracity imaging of chemical interactions [128–130] by affecting short-range hydrogen bonding and specific interactions. By combining chemical functionalization for single molecule detection based on specific antibody–antigen recognition binding (i.e., a tip-tethered antibody) with high-resolution topography imaging and advanced signal detection, the topography and recognition (TREC) imaging mode [131–133] has been developed, as shown in Figure 16.7. In a way, this method can be considered an SPM analog of fluorescent tagging in optical microscopy [134]. In TREC, the deflection signal of the cantilever during normal scanning is processed to detect rapid changes in the signal. When there is a recognition event, the oscillation amplitude of the cantilever changes (Figure 16.7c). The tip–sample separation is adjusted to restore the amplitude, but there is a change in the value of the peak signal (a DC offset), which can be used to create a map of antibody–antigen recognition events, as shown in Figure 16.7e.

Finally, a number of approaches (Figure 16.8) are now emerging to employ probes carrying more complex local functionalities, including field-effect transistors [135, 136] for probing local electric fields, magnetoresistive and Hall sensors, microfluidic delivery probes [137], and shielded/insulated probes for electrochemical studies [138–141] and microwave imaging. An important class of functional SPM imaging is thermal SPM based on probes with integrated local heaters, as described in Section 16.7.1.

16.2.3
Data Processing and Acquisition

16.2.3.1 Classical Detection Strategies in SPM

In static SPM methods, such as contact mode SPM or force–distance measurements, a static signal that changes at rates (~ms) comparable to the pixel acquisition/feedback times is measured. In these measurements, the full time evolution of the signal is collected at appropriate sampling rates. In comparison, the operation of dynamic SPM methods is based on the interaction, mediated by short- and long-range tip–surface interactions, between an oscillating cantilever and the surface [145, 146]. The direct acquisition of the full signal (e.g., for a 100 kHz resonant frequency, ~1–10 MHz processing rates are required to detect sufficiently high-order nonlinearities) is in most cases impractical or impossible. Hence, the data processing electronics in SPM merit specific discussion, since they link the ~0.1–10 MHz time scale of the cantilever oscillations to the ~1 kHz time scale of feedback loop operation or single pixel acquisition. This stage selects relevant time-averaged aspects of the cantilever response (e.g., amplitude and phase for lock-in detection, resonant frequency and amplitude for phase-locked loops (PLL)) [147], and plays the dual role of making the volume of the acquired information

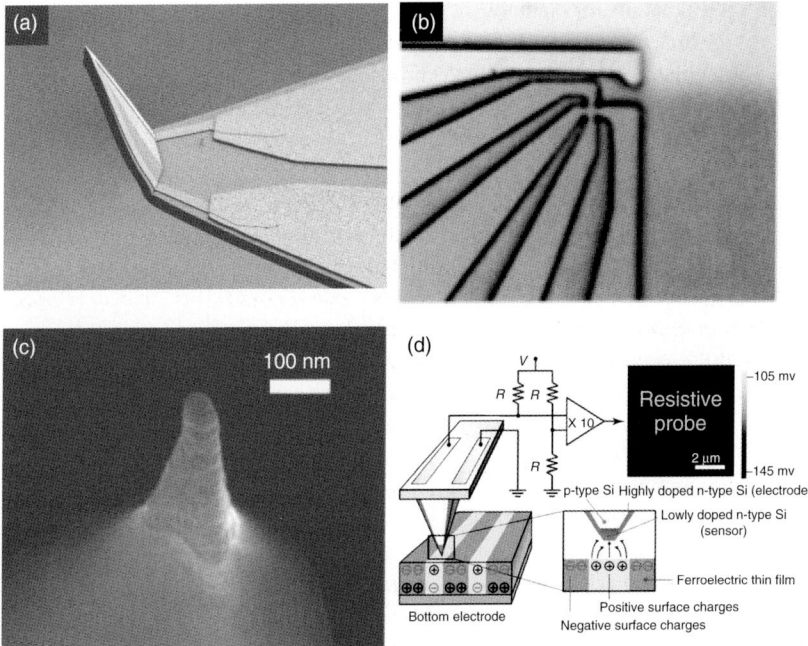

Figure 16.8 Novel multifunctional SPM probes. Examples of (a) thermal [142], (b) Hall [143], (c) insulated [144], and (d) field-effect [135] probes. (Images reproduced with permission. (a) Courtesy of Anasys Instruments. (b) Copyright, 2002, IEEE. (c) Copyright, 2010, Institute of Physics. (d) Copyright, 2011, American Chemical Society.)

manageable (e.g., 2 data points per pixel, rather than $10^4 - 10^5$), while limiting the information obtained from the experiment.

The mainstay of dynamic SPM methods are lock-in amplifiers (LIAs) and PLL-based detection methods. An LIA is essentially a narrow band filter that can extract a signal near a particular frequency from a multifrequency waveform, and convert the amplitude and phase (compared to a reference signal) of the desired frequency to DC values. LIAs are universally used in constant frequency SPM methods ranging from standard tapping mode imaging to atomic force acoustic microscopy (AFAM), KPFM, and PFM. A PLL compares the excitation and response signals and uses the measured phase lag as in a feedback loop to maintain a constant 90° phase at cantilever resonance by adjusting the excitation frequency. PLLs allow changes in resonance frequency due to tip–surface interactions to be tracked, and are routinely used in NC-AFM, MFM, EFM, and some versions of KPFM. The strong attention to PLL-based frequency-tracking methods results from the fact that they allow precise control of tip–surface interaction forces, enabling atomic-resolution imaging in ultrahigh vacuum (UHV) and liquid using frequency-modulation atomic force microscopy (FM-AFM) [85, 86].

16.2.3.2 Multifrequency Methods

The use of single frequency LIA and PLL methods significantly limits data that can be acquired by SPM. This limitation can be understood as follows: in the absence of nonlinearities, the tip–surface interactions can be represented by a simple mass-spring model (simple harmonic oscillator (SHO)), characterized by three independent parameters – resonant frequency, amplitude at resonance, and the quality factor (Q-factor, or peak width). Whereas single frequency measurements give only two time-averaged parameters, three are needed to uniquely determine conservative and dissipative interactions even assuming prior knowledge of the SHO behavior.

This limitation can be circumvented if (i) the force driving the system is constant and (ii) the system follows the SHO dynamics (i.e., tip–surface interactions are linear), thus providing an additional constraint on the dynamic behavior of the system. In this case, the response amplitude is inversely proportional to the peak width. In techniques with acoustic excitation [65], the effective force is determined by the driving voltage applied to the piezoactuator, and hence this condition is (semiquantitatively) satisfied. This allows dissipation measurements using the Garcia-Cleveland approach [148, 149] and also enables PLL frequency-tracking techniques, which rely on the defined relationship between the phase of the excitation force and the response signal.

However, in methods based on the electrical excitation of the tip such as KPFM [45] and PFM [37, 38], the relationship between the phase of the excitation force and the driving voltage strongly depends on the local material properties of the sample. In these cases, the amplitude and phase of the local response are a convolution of the response of the material to an external field, and the cantilever response to the material-dependent local force, which cannot be unambiguously separated. Even for techniques with mechanical excitation, the transfer function of the cantilever or oscillator couples to the signal, resulting in systematic errors in PLL detection. In other words, the voltage sent to the oscillator is *constant* as imposed by microscope electronics. However, the driving force is determined by voltage and oscillator and cantilever transfer functions, which have nonzero frequency dispersion, and depend on tip–surface interactions. Hence, the effective force is frequency-dependent and specific for a particular cantilever and microscope, necessitating a lengthy calibration process [150]. While seemingly minor, this limitation has long precluded reproducible measurements of dissipation in techniques such as MFM. The situation becomes even more complex when tip–surface interactions are nonlinear and the SHO model is inapplicable.

These considerations have stimulated the development of alternative signal detection schemes based on measurements at two or several frequencies or within well-defined frequency bands, and include the following:

1) **Thermal (White Noise) Excitation.** A number of approaches have been developed for SPM imaging using thermally excited cantilevers [151]. In this case, the cantilever is exited by thermal fluctuations that can be modeled as an excitation force with uniform amplitude and a random phase in the frequency domain.

Changes in the cantilever response as a function of tip–surface separation allow the response to be used as a feedback signal. The obvious limitation of this method is that the excitation force amplitude is not controlled, and the phase information of the response is lost.

2) **Ring Down**. This approach is based on measuring the cantilever response following a step or delta function excitation. During the relaxation of the oscillator to the excitation, the resonance frequency and Q-factor can be measured. This approach for measuring the cantilever transfer function was suggested by Proksch and Dahlberg [152]. Stark *et al.* [153] developed a related technique using cantilever instabilities during the cantilever approach and retract in order to excite oscillations in the cantilever motion, which are then analyzed in the Fourier domain to obtain the system response (including the dissipation).

3) **Dual-AC Resonance Tracking (DART)**. An alternative to single-frequency detection has been developed based on dual-AC measurements. This approach overcomes some fundamental limitations of LIA and PLL detection – including sensitivity to excitation phase uncertainties and an insufficient number of detected parameters – by performing excitation and detection simultaneously at two frequencies [154]. DART allows one to implement an amplitude-based feedback loop for maintaining resonance and determining the dissipation from peak width or response phases directly [155]. Extending this approach, a number of multifrequency techniques have recently been developed, which hold promise for separating long- and short-range forces and measuring nonlinearities in the tip–sample interaction forces and properties of the sample [156–163].

4) **Fast Lock-In Sweep**. Very recently, an approach for fast sequential frequency scanning has been developed by Kos and Hurley [164]. This approach allows the full amplitude–frequency and phase–frequency response curves to be determined at each point of the image, from which the resonance frequency can be determined, obviating the need for a PLL-based feedback. Equivalent results can be obtained using fast SPM imaging at different frequencies, as pioneered by Huey *et al.* [165].

5) **Band Excitation (BE)**. The BE approach provides an alternative to standard LIA and PLL methods and frequency sweeps by exciting and detecting response at all frequencies *simultaneously* [166]. BE introduces a synthesized digital signal that spans a continuous band of frequencies, and monitors the response within the same (or larger) frequency band. The cantilever response is detected using high-speed data acquisition methods and is then Fourier transformed. The resulting amplitude–frequency and phase–frequency curves are collected at each point and stored in 3D data arrays. This data is analyzed to extract relevant parameters of the cantilever behavior. For example, in the SHO approximation, the resonance frequencies, response amplitude, and Q-factors are deconvoluted and stored as images, and in the case of adaptive control can be used as a microscope feedback signal. The BE method allows one to obtain the transfer function of the system directly. BE-MFM [166–168], band excitation piezoresponse (force) spectroscopy (BEPS) [169], BE-AFAM measurements

with a heated probe [170, 171], and band excitation scanning Joule thermal expansion microscopy (BE-SJThEM) based on periodic tip heating [171] have all been demonstrated.

The underpinnings of these methods can be conveniently illustrated through the intensity and phase distributions of the excitation signal in Fourier space, as depicted in Figure 16.9.

Note that the excitation/detection methods discussed above (single frequency, DART, lock-in sweeps, BE) or more complex frequency mixing [172] and intermodulation modes [173], refer to the mode of excitation and can be applied to a broad variety of SPM imaging and spectroscopic modes. For example, KPFM or MFM measurements can be performed using LIA (constant frequency) and PLL (frequency-tracking) methods, as well as using BE or DART excitation. Similarly, these excitations can be used in spectroscopies and spectroscopic imaging modes, as illustrated in Figure 16.10 for the BEPS mode. Often, spectroscopic imaging modes give rise to complex multidimensional (3, 4, and even 5D) data sets, necessitating the development of new approaches to visualize and interpret this data. Some of these challenges are not dissimilar from areas such as hyperspectral imaging in optical microscopy and electron energy loss spectroscopy in STEM, suggesting future synergies between these areas.

16.2.3.3 Real-Space Methods

An alternative to the detection in the Fourier domain is the analysis of the real-space trajectory of the probe. Similarly to Fourier methods, this approach relies on the rapid acquisition of spectroscopic data at each spatial point of the image, and subsequent deconvolution of the resulting 3D data set to yield 2D maps of relevant parameters. The prototype of this approach is the force–volume imaging mode in contact AFM, in which force–distance curves are acquired at each pixel and are analyzed to derive local elasticity, adhesion, and other properties [174–177]. The development of fast data acquisition electronics has allowed the development of a pulsed force mode, in which a specially engineered signal allows semiquantitative probing of adhesion and elasticity [178]. Further development of this concept was achieved in a recent work by Sahin [179–181], which utilizes the decoupling between the vertical and lateral responses of an asymmetric cantilever to collect information on tip–surface forces. The deconvolution algorithm is designed to directly reconstruct the force–distance curve at each tip oscillation at standard imaging rates [182].

16.3
Scanning Tunneling Microscopy and Related Techniques

STM is the oldest SPM technique, stemming from the pioneering experiment of Binnig and Rohrer in 1981 [41], who demonstrated topographical imaging of the Au(110)–2 × 1 reconstructed surface. Since then, STM has been and will likely remain the primary scanning probe technique in surface physics and chemistry,

16.3 Scanning Tunneling Microscopy and Related Techniques

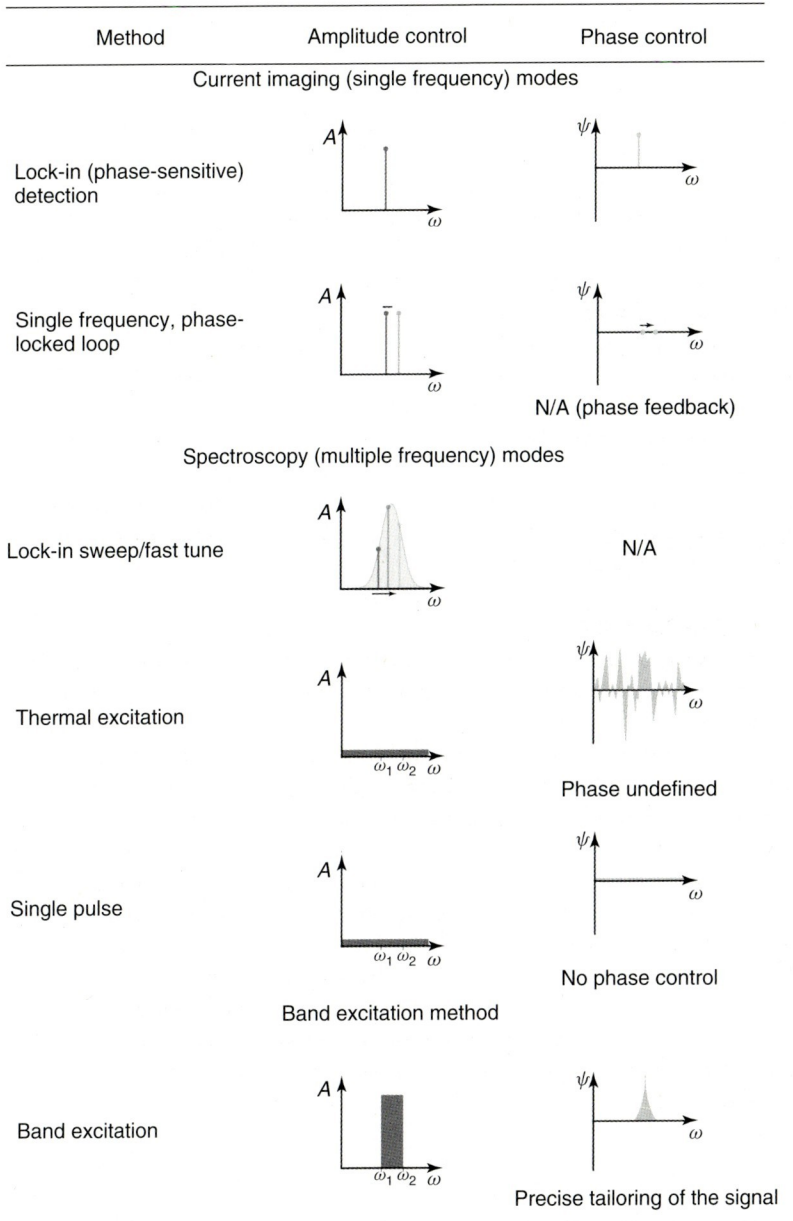

Figure 16.9 A comparison of the existing methods for detection and excitation in SPM through representations in the Fourier domain. Of practical interest is the system response (amplitude and phase) in a frequency region close to the resonance. The table illustrates that conventional SPM excitation/detection methods that operate either at one single frequency (LIA, PLL) or excite all frequencies within the bandwidth of the system (thermal excitation or single pulse).

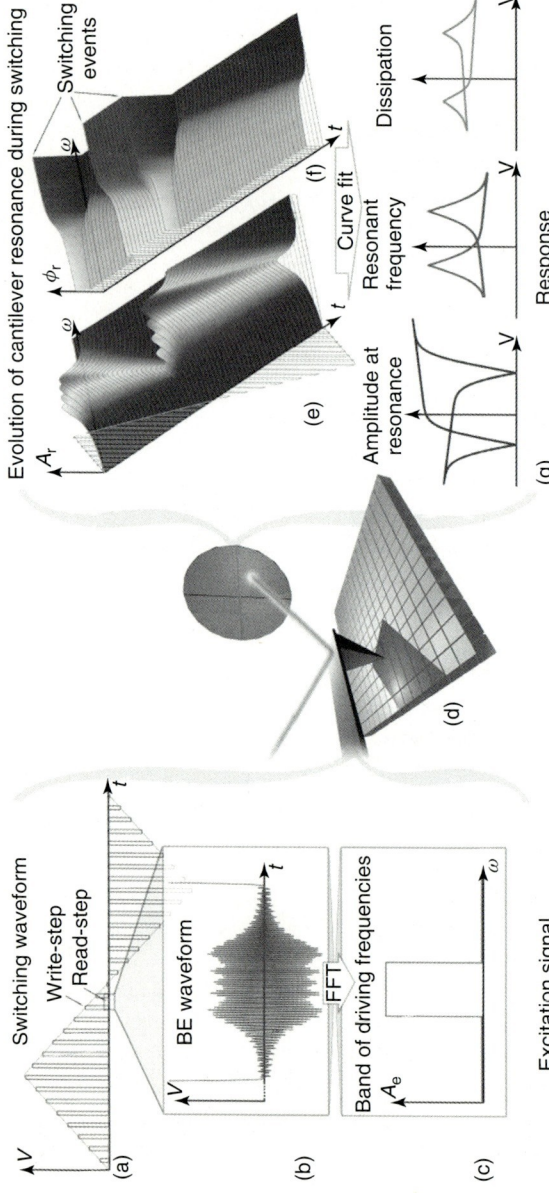

Figure 16.10 Data acquisition and processing in band excitation piezoresponse force spectroscopy (BEPS). (a) Ferroelectric switching is induced by a pulse train of increasing DC bias while the changes in the piezoresponse, contact stiffness, and dissipation are measured by exciting the cantilever with a narrow frequency band around its contact resonance (b,c). The ferroelectric hysteresis is measured across a grid of points (d) resulting in a spatially resolved 4D data set, where each point represents the cantilever's resonant response along the local hysteresis loop (e,f). The resonant response (amplitude and phase) is then fit by an SHO model to yield the resonance amplitude, resonance frequency, and Q-factor (g) as a function of the DC bias. FFT, fast Fourier transform. (Reproduced with permission from [169], copyright 2008, American Institute of Physics.)

whose diverse applications range from atomic imaging, to quasiparticle electron spectroscopy, electronic interferometry, chemistry and catalysis, DNA sequencing, and atomic manipulation. Over the years, a number of books and comprehensive reviews have been written on various aspects of STM [183–185]. The scope of this section is to provide a brief overview of the key capabilities of STM and to highlight several topical areas that offer potential for significant development in the coming years.

In general, STM samples a 3D data volume of current as a function of applied voltage and geometric distance between the tip and the surface. The overwhelming power of the technique stems from the tunneling mechanism of the tip–surface junction conductance, which allows one to: (i) confine the spatial extent of the tunneling current to subatomic dimensions and therefore resolve atoms in real space; (ii) tune the energy of tunneling electrons, thus enabling the measurement of the energy-resolved density of electronic states and the creation of hot-electron and hot-hole excitations; (iii) tune the tunneling current strength by changing the tip–surface distance and at the same time measuring tunneling barrier height (which is related to local work function); and (iv) probe the surface in the regime of very weak perturbation. In essence, STM combines the power of tunneling spectroscopy measurements, developed in fixed tunnel junctions at least 60 years before the invention of STM in the first experiments by Esaki and Giaever (see Ref. [186] for an overview), with the capability to tune the tunnel junction size perpendicular to the surface and its position along the surface.

The most widely employed experimental measurement modes in STM are topographic and spectroscopic imaging. In topographic imaging modes, the tip–surface distance is adjusted to maintain constant current during a raster scan, or, alternatively, the current is imaged at constant tip–surface distance, yielding a 2D data set representing electronic properties of the surface. In spectroscopic modes, the shape, gaps, and features of the current–voltage $(I - V)$ curves are analyzed by acquiring the first (dI/dV) and/or second (d^2I/dV^2) derivatives as a function of position on the surface and tunneling parameters. While it is impossible to adequately represent the advances achieved by STM in a short summary, some of the seminal results include real-space atomically resolved imaging of the Si(7 × 7) reconstruction (Figure 16.11a) [187], which served as the basis for the Nobel Prize in Physics awarded to Binnig and Rohrer in 1986 for the invention of the STM; manipulation of a single atom with atomic precision [188]; a vibrational spectrum and image of a single molecule, obtained in the d^2I/dV^2 spectroscopic regime, which detects subtle changes in tunneling conductance induced by inelastic scattering of a tunneling electron off a vibrational mode of the adsorbate and enables chemical imaging and chemical reaction (Figure 16.11b) studies at the atomic scale; a "quantum corral" image of an electron standing wave (a dramatic visualization of the wave-nature of quantum electronic states) confined in a circular pattern of single Co atoms arranged on a surface using STM manipulation (Figure 16.11c) [189]; inhomogeneity in the superconducting gap in a high-T_c superconductor [190], emphasizing the strong spatial variability of these materials on the nanometer scale [191]; magnetic hysteresis and spectroscopy of single atoms (Figure 16.11d)

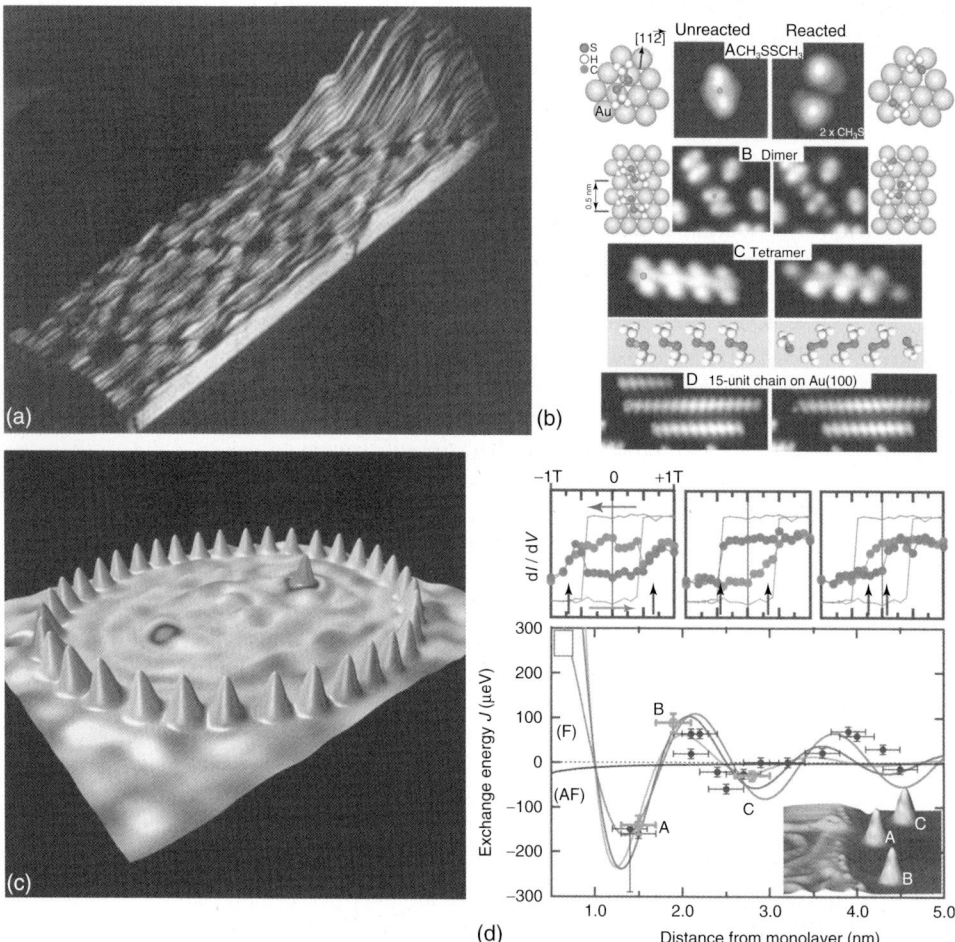

Figure 16.11 STM imaging. (a) First image of the Si(111)-(7 × 7) surface reconstruction. (b) STM images before and after electron-induced dissociation of a single molecule. (c) Quantum mirage effect mediated by a quantum corral structure constructed from 36 Co atoms on a Cu(111) surface. A single Co atom placed in one of the foci of the elliptical corral (dark gray circle on the left) can be spectroscopically identified in an opposite focus even though the atom is physically not there. This is a most vivid manifestation of the wave nature of electrons. (d) Magnetic exchange interaction between individual atoms (measured using single-atom magnetization curves, top panels) and the corresponding distance dependence of the exchange energy measured using spin-polarized tunneling microscopy (bottom panels). ((a) Reproduced with permission from 187, copyright, 1983, American Physical Society. (b,d) Reproduced with permission from [193] and from [192], copyright 2008, American Association for the Advancement of Science. Image in (c) Reproduced from [194] and originally created by IBM Corporation.)

[118, 192], among many others. In addition to these spectacular results illustrating cutting edge achievements of STM, equally important is the fact that in the 30 years since the invention of STM, it has become the workhorse of surface science labs, routinely providing high-resolution images of semiconductors, metals, low-dimensional materials, molecules (Figure 16.11b), and oxides.

Given the mature state of STM development, we delineate below several areas of surface physics where initial experiments have already shown promise and further advancements will be needed in the years to come.

16.3.1
The Role of Tip Effects in Tunneling Conductance and Inelastic Phenomena

Despite decades of STM research, the source of all signals – the metal tip – is still incompletely described. A sizeble range of empirical procedures have been developed, many of them successful and reproducible, which give a "good" imaging tip. For example, field emission, electron bombardment, and argon sputtering of the tip can often make it geometrically sharper, free from adsorbates or an oxide layer, and make it appropriate for atomically resolved STM imaging. When in tunneling contact, slight dipping into a metallic surface, short-term pulsing, and scanning at high bias also can make the imaging "cleaner," more stable, and allow higher spatial resolution to be achieved (Figure 16.12). However, as the range of questions being addressed, and the experiments themselves, go beyond topographic scanning, the issue of tip effects is elevated to a qualitatively new level. STM tips yielding perfectly good topography can exhibit vastly different electronic states in the scanning tunneling spectroscopy (STS) mode and can affect the shape and linewidth of the individual states. Some of these features originate from the

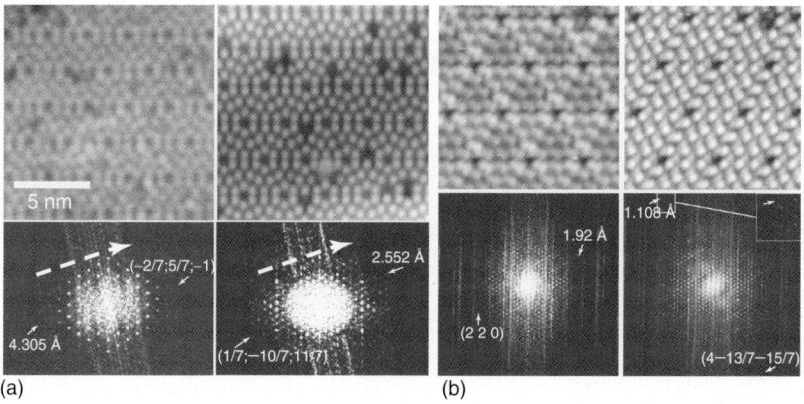

Figure 16.12 Effect of tip state and bias on STM resolution. (a) Resolution in the STM imaging of a Si(7 × 7) surface with a blunt and sharp tip. (b) Normal and anomalous contrast of a Si(7 × 7) surface for the same bias. Note the higher resolution and anisotropic diffractogram, indicative of contribution of rotationally nonsymmetric (e.g., $P_x - P_y$) tip states in the tunneling process. (Images and data analysis courtesy of J. Guo (UT Knoxville) and A. Borisevich (ORNL), respectively.)

highly localized electronic states appearing on the tip apex itself, but a more subtle origin is the momentum resolution of the tunneling electrons. Even though it is traditional to assume that the tip, as a point source, samples the entire k-space of the surface, it is now known from studies of impurity-induced quantum wells [195], that certain tips can sample the bulk-derived states with higher density than the surface states [196]. The problem becomes even more complicated in the inelastic regime, where in addition to tip-derived vibrations in the spectrum [197], there is a strong likelihood that the propensity for electron–phonon coupling is influenced subtly by the electronic structure of the tip apex. Finally, the role of electron interference between tunneling channels in both STS and inelastic electron tunneling spectroscopy (IETS) is necessarily important, as it has an influence on topographic imaging in STM [198].

A plausible route to resolving these tip effects is to more actively explore adsorbate-induced tip modification. Picking up single atoms and molecules has become a well-defined procedure [199]. An attractive alternative "soft" modification of the tunneling junction is through weakly bonded adsorbates. Recently, the group of Temirov and Tautz demonstrated that topographic imaging was dramatically enhanced by adsorbing a layer of hydrogen on the surface (Figure 16.13), which according to the authors, gives rise to Pauli repulsion between a hydrogen molecule trapped in the tunneling junction and the surface atoms (see Ref. [120] and references therein). A third possibility is cross-correlation between STS, IETS, and force–distance measurements. The latter channel has recently become accessible when mounting STM tips on quartz-resonator-based force sensors [200]. Combined with spatially resolved analysis of spectroscopic variations, multichannel spectroscopy can provide a qualitative benchmark for the tip performance, and an equally useful input into theoretical calculations, which can take into account the full complexity and the electronic and vibrational structures of the tip apex. Pioneering examples of studies along these lines can already be found in the literature, but a more dedicated experimental effort, combined with improved theoretical methods, will advance our understanding of tip effects, and ultimately take advantage of the ability to tune and amplify desired characteristics of the tunnel junction.

16.3.2
Electron Tunneling in Correlated Electron Materials

Superconductivity, colossal magnetoresistance, and more broadly, electron–electron and electron–phonon interactions lie at the heart of condensed matter physics, and hundreds of STM measurements have been directed toward understanding the phase transitions on the surfaces of the respective materials. Arguably the effort has primarily been directed toward superconductivity [201] and many fascinating aspects, such as heterogeneous melting of the superconducting order parameter [32], pseudogap phases [202], and tunneling into vortex cores [203] have been discovered and explored over the years. Yet, compared to the STM effort on semiconducting surfaces in the 1990s, less attention has

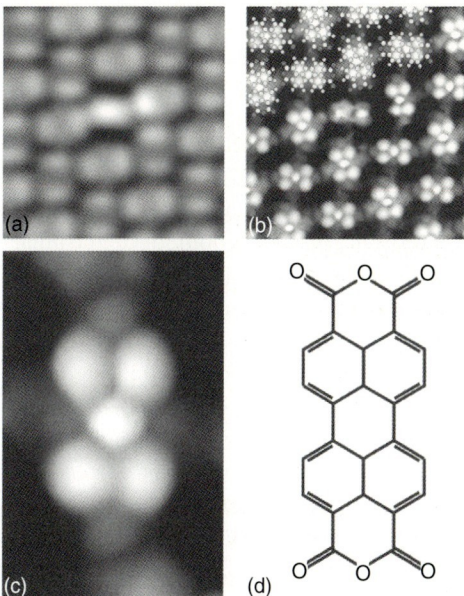

Figure 16.13 Imaging of perylene-3,4,9,10-tetracarboxylic-3,4,9,10-dianhydride (PTCDA)/Ag(111) in the STM "geometric" mode, assisted by a hydrogen molecule between the tip and the surface. (a) shows the STM image without the hydrogen molecule, while (b,c) are large- and small-scale images with the hydrogen molecule, correspondingly. (d) is the structural formula of the PTCDA molecule. (Reproduced with permission from [119], copyright 2008, Institute of Physics.)

been drawn to the systematic analysis of the STS and STM images, such as distance-dependent spectroscopy, tip–surface interactions, including tip-induced band bending, the normalization of STS, and general tip effects. Fundamentally, tunneling into a correlated electron state perturbs the order parameter and can be associated with nontrivial collective electronic excitations, making the process significantly more complicated than tunneling into metals and degenerate superconductors, which are both amenable to mean-field theory treatments. While superconductivity is relatively immune to these subtleties, simply because the superconducting gap is a rather well-defined and characteristic feature in the STS spectrum, other phenomena, including the pseudogap state and metal-insulator transitions in colossal magnetoresistance materials, are less straightforward to measure and analyze. Without going into further detail, STM studies in colossal magnetoresistance materials have so far not been fully consistent, with reports of both the presence [204] and absence [205, 206] of phase separation, the existence of pseudogap states [205], and even the lack of a phase transition [207]. It is probably fair to say that none of the works so far has revealed a clear correlation between the macroscopic conductivity (metal-insulator transition) and the local signatures of colossal magnetoresistance, such as zero-bias conductance, the size of the apparent gap, and so on. Similar uncertainties can be found for other metal-insulator systems, necessitating the push toward more insight into the tunneling properties of correlated electron materials.

16.3.3
Tunneling into Low Conducting Surfaces

Traditionally, the electrostatic potential is assumed to drop almost entirely across a tip–surface gap in STM because the resistance of the vacuum junction is much higher than the in-series resistance of the bulk of the film and the macroscopic leads in the system setup. This is not generally true, however, and if the resistance of the surface or bulk becomes comparable, tunneling measurements will reflect this and can be used to investigate the volume of material beyond the tunnel junction. Recent work [208] has highlighted such a possibility in the context of imaging deep levels formed due to Si adatoms adsorbed onto a clean Si surface. The nonresonant deep level in the Si band gap originating from a Si adatom is relatively long lived. The tunneling rate from this state into the bulk was found to be comparable to tunneling rate into the state from the STM tip. In this case, the dependence of the tunneling current on the tip–surface distance deviates from the anticipated exponential, and one can then infer the actual escape rate. Feenstra et al. have recently demonstrated the ability to measure the mobility of charge carriers in a thin film of pentacene, a prototypical organic semiconductor (R.A. Feenstra, private communication). Given the importance of low conducting surfaces in a variety of energy-related fields (solar energy harvesting, energy storage, etc.), as well as the ubiquity of insulating materials in bad metals in condensed matter physics, further exploration of this approach seems worthwhile.

16.3.4
Time-Resolved Studies

Bringing together the ultimate time and space resolutions in the STM experiment is a long coveted goal. An excellent review by Graftstrom [209] recounts numerous attempts at this milestone using light-irradiated and gated junctions. Undoubtedly, there will be many approaches appearing in the coming years that aim to achieve this goal. A particularly attractive approach to attaining nanosecond time resolution, in that it can be implemented without introducing extra modifications to the STM setup (in the form of lasers, gates, microwave circuitry, etc.), has recently been achieved by the IBM Almaden group [210] using tip bias as both the pump and probe signal. The authors have succeeded in flipping the spin of a single atom adsorbed on a surface using spin-polarized tunneling current, and then monitoring the relaxation of the spin-flipped state with a time resolution of <50 ns. They could then probe the dependency of the relaxation rate on the immediate neighboring atom. In principle, this scheme can be extended to any combination of pump-probe voltage signals, and can prospectively address dynamics on the single atom level in many materials.

Undoubtedly, the sheer range of the problems that can be addressed by STM, combined with the increased availability and capability of commercial setups for doing both STM and force tunneling microscopies, and, on the other hand, the novel developments in exotic materials where tunneling measurements are particularly useful, heralded by graphene and topological insulators, indicate that STM will retain its crown as the ultimate scanning probe measurement in surface physics experiments for decades to come.

16.4
Force-Based SPM Measurements

In force-based SPM measurements, material structure and functionality are determined from the short- and long-range forces acting on an SPM probe. The two primary parameters that can be dynamically controlled in an SPM experiment are the tip–surface separation and the probe bias. The signals detected are the probe current and the force components acting on the probe. Hence, the SPM imaging mechanism can be represented as a force-distance-bias surface $\mathbf{F} = \mathbf{F}_c(h, V_{tip}, \mu)$, where h is the tip–surface separation (for NC methods) or indentation depth (for contact), V_{tip} is probe bias, and μ are parameters describing the chemical functionality of the probe (Figure 16.14) [211]. Note that the cantilever detection system adopted in most commercial SPMs implies that the full force vector, $\mathbf{F}_c$, cannot be measured directly. Rather, flexural cantilever displacements measure a superposition of normal and longitudinal (along the cantilever axis) force components, while torsion is sensitive to lateral force components.

One of the primary challenges in SPM is the decoupling between multiple interactions simultaneously present in the tip–surface junction that all contribute to the force, $\mathbf{F}_c$. One approach for decoupling different interactions is based on measuring the force at several separations from the surface. As an example, in typical double-pass methods the first scan line is used to determine the position of the surface through the detection of strong short-range van der Waals (vdW) and elastic forces, while weaker, but long-range magnetic and electrostatic interactions are measured when the tip is retracted 10–1000 nm from the surface.

An alternative is offered by modulated methods, in which a specific functionality of the probe is modulated, and the oscillatory response is detected. This allows both the subset of tip–surface interactions (e.g., oscillating electrostatic potential does not affect vdW forces) and linear response dynamics through measurement of both the amplitude and the phase of the response to be probed. This approach further allows higher signal-to-noise ratios by moving away from the $1/f$ noise corner (~1–10 kHz) and resonance enhancement of the signal. Practically, only some aspects of probe functionality can be modulated at the rates required for imaging (>1 kHz). These include position, h (e.g., acoustic driving), or force, $\mathbf{F}_c$ (e.g., magnetic driving), and electrical bias, V_{tip}. Chemical functionality, hydrophobicity, and so on, and other chemical functionalities, μ, do not offer obvious universal strategies for modulation, even though optical- and bias-induced transformations

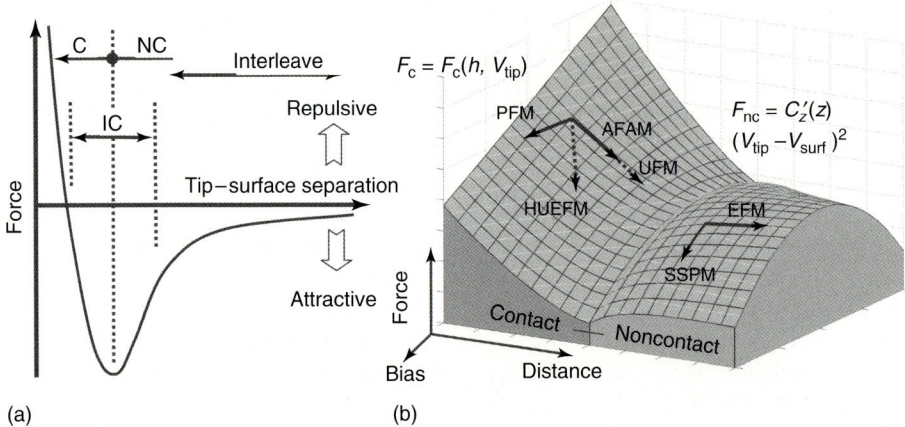

Figure 16.14 Force-distance-bias surface. (a) Force-based SPM can be conveniently described using force-distance curve, showing the regimes in which contact (C), noncontact (NC), intermittent contact (IC), and interleave imaging are performed. Also shown are domains of repulsive and attractive tip-surface interactions. (b) Voltage modulation SPMs can be described using a force-distance-bias surface. In the small signal limit, the signal in techniques such as PFM, AFAM, EFM, scanning surface potential microscopy (SSPFM (i.e., KPFM)), ultrasonic force microscopy (UFM), and heterodyne ultrasonic and electrostatic force microscopy (HUEFM) is directly related to the derivative in bias or distance direction. (Reproduced from Ref. [212], copyright 2006, IEEE.)

provide some possibilities [213]. Below, we briefly discuss SPM techniques based on whether the separation between different interactions is achieved through force-based measurements or voltage-modulated measurements.

16.4.1
Contact, Intermittent, and Noncontact AFM

The most popular application of SPM is topographic imaging of surface structures, that is, plotting constant-force contours corresponding to elastic or vdW forces. These are used for direct imaging of surfaces, and as an initial step for force spectroscopies and functional (magnetic, electrostatic, thermal, etc.) SPM imaging. The primary methods for topographic imaging include contact, NC, and tapping mode AFM, as detailed below.

The contact mode is historically the first AFM mode, and is based on the detection of strong repulsive tip–surface forces. In this mode, the cantilevered tip approaches the surface until repulsive mechanical contact is established, as can be detected from the deflection of the cantilever. The feedback loop is used to maintain the constant static force (i.e., cantilever deflection) by adjusting the verticle position of the cantilever base while scanning. The control signal of the feedback loop then provides surface topography. The significant advantages of contact mode AFM are the simplicity and the sensitivity to lateral (frictional forces) of the technique, while disadvantages include potentially significant tip and surface damage and sometimes low resolution. The latter is limited by the tip–surface contact area determined in turn by tip–surface interaction forces. In air, the tip–surface forces are dominated by strong capillary interactions that limit contact radii to several nanometers and typically preclude atomic-resolution imaging. Lateral resolution in contact mode AFM is on the order of several nanometers, albeit multiple reports of lattice resolution and some reports on individual defect imaging [214] are available. This limitation is absent in liquids and UHV, and multiple examples of atomic (UHV) and molecular (biological liquid imaging) are available [215–218].

As mentioned above, a significant advantage of this mode is that lateral forces acting on the cantilever can also be detected [219], giving rise to lateral or friction force microscopy [75, 220, 221] that enabled the emergence of the field of nanotribology [222, 223].

One of the most popular AFM imaging modes, "tapping mode" (also known as "AC" or "*tapping mode*") was first described by Finlan and McKay [224], apparently independently discovered by Gleyzes *et al.* [225], and later broadly popularized and commercialized by others [226, 227]. In this mode, the cantilever is dynamically excited at a frequency close to the free resonance of the beam. The amplitude and phase of cantilever oscillations are detected using LIA methods. On approaching the surface, the tip–surface interactions reduce the amplitude of the oscillation, and the amplitude can thus be used as a feedback signal for topographic imaging. Significant advantages of tapping mode SPM include strongly reduced tip and surface damage, which allows imaging of much softer samples compared to contact mode. Second, the possibility for detection of the phase of the cantilever response, referred to as *"phase imaging,"* was a source of much excitement in the late 1990s. In particular, the first phase images (of a wood pulp sample) were presented at a meeting of Microscopy and Microanalysis (Figure 16.15a,b) [228]. Since then, phase imaging has become the mainstay in a number of AFM application areas, most notably in polymers (Figure 16.15c), where the phase channel is often capable of resolving fine structural details [229].

The phase response has been interpreted in terms of the mechanical [230–235] and chemical [236] properties of the sample surface. Notable progress has been made in quantifying energy dissipation between the tip and sample [237–240], which can be linked to specific material properties. Even with these advances, obtaining quantitative material or chemical properties remains problematic, partially due to the aforementioned uncertainty in system dynamics inevitable in single-frequency methods and partially due to a poorly controlled tip–surface contact area. Only recently the limitations of single-frequency phase images have been realized concurrently with the emergence of multiple frequency modes, as discussed in Section 16.30.

Tapping mode imaging in liquid environments [241, 242] has led to the imaging of a wide variety of high-resolution, soft polymeric, and biological samples (Figure 16.16). Fluid AFM, both in contact and in tapping mode, has become a standard imaging tool for label-free, high-resolution imaging of living cells. Some examples of AFM cell studies include glial cells [243], platelets [244], lung cells

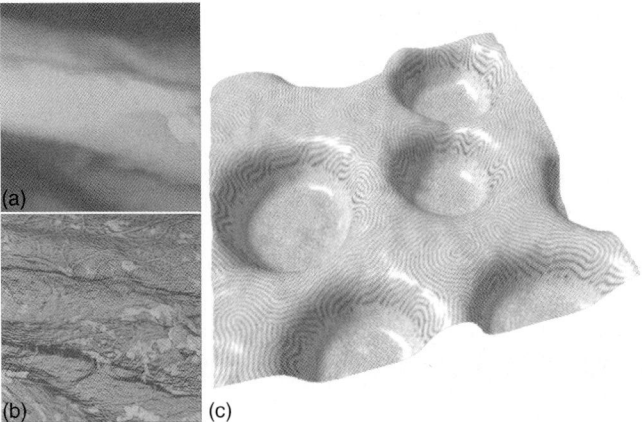

Figure 16.15 Phase imaging. (a) The topography of Chernoff's wood pulp sample and (b) the phase channel showing bright contrast associated with lignin inclusions. (c) Phase channel overlaid on the topography of a styrene-ethylene/butylene-styrene block copolymer sample spincoated onto a Si wafer substrate.[1] The different domain structures and their correlation with sample thickness are clearly visible in the composite image. Scan size is 2 µm, vertical scale is 20 nm and phase color scale ranges 10°. ((a,b) Courtesy of Advanced Surface Microscopy. (c) Courtesy of Asylum Research.)

[245], Madin-Darby canine kidney and R5 cells [246], myocytes [247], fibroblasts [248], leukemia (RBL) cells [249], and liver endothelial cells [250].

Some examples of AFM images taken of fixed and living cells in physiological conditions are shown in Figure 16.17. Fixed MRC-5 fibroblast cells labeled with DAPI (nucleus) and Alexa Fluor 488 (actin filaments) imaged with a fluorescent optical microscope are shown in Figure 16.17a. Simultaneously imaged AFM deflection data (Figure 16.17b) is shown overlaid onto the fluorescence image in Figure 16.17c, allowing mechanical and optical contrast to be compared. Figure 16.17d shows merged topography and fluorescent images of a rat cardiac myocyte, which has been cultured on a microcontact-printed square island of extracellular matrix (ECM) protein. The cell was fixed and stained for actin (green), sarcomeric alpha-actinin (red), and DNA (blue). After fixation, the cell was scanned by AFM. The merged topography and fluorescent images reveal that the actin/alpha-actinin myofibrils and the nucleus are responsible for specific AFM topographic features. Figure 16.17e shows a 90 µm fluid AC mode AFM image of a live rat mesenchymal stem cell showing both linear and geodesic arrangements of the F-actin cytoskeleton [252, 253].

Finally, in NC-AFM (also called *FM-AFM*), the tip is excited dynamically and the resonance frequency, rather than the amplitude, is measured [254]. This approach allows much more precise measurements of tip–surface interaction, and is broadly used for atomic-resolution imaging in UHV [255] and atomic and submolecular resolution in liquid environments [256–260]. Recently, constant current STM and

---

1) Sample courtesy of R. Segalman and A. Hexemer, Kramer Group, UCSB.

Figure 16.16 High-resolution imaging in liquid. (a) DNA origami triangles, self-assembled in a single step from over 200 DNA strands. Each is a single 5 MDa molecular complex, incorporating 15 000 nucleotides, ~120 nm per edge, 1 μm scan. (b) Images of viruslike particles taken in a physiological buffer solution, 200 nm scan [251]. (c,d) Closed-loop images of the cytoplasmic face of bacteriorhodopsin in buffer, 90 nm (c) and 50 nm (d) scans. (c) Shows defects in the lattice from a missing protein.[2] (e,f) AC mode topography images of the cleavage plane of calcite showing repeatable point defects, scan size 20 nm. Vertical grayscale is 3 Å and scan size 20 nm. ((a) Sample courtesy of Paul W.K. Rothemund, California Institute of Technology. All images courtesy of Asylum Research.)

AFM imaging has been used to visualize organic molecules [261] and to determine the structure of unknown compounds [262].

16.4.2
Force–Distance Spectroscopy and Spectroscopic Imaging

In a mode analogous to indentation measurements [263, 264], measuring the deflection of the flexible cantilever as a function of the tip–sample separation allows single-point spectroscopy and a spectroscopic mode referred to as *"force curves"* or *"force spectroscopy."* Often, two-dimensional arrays of force curves are

---

2) Images courtesy of G.M. King Lab, University of Missouri-Columbia and Asylum Research.

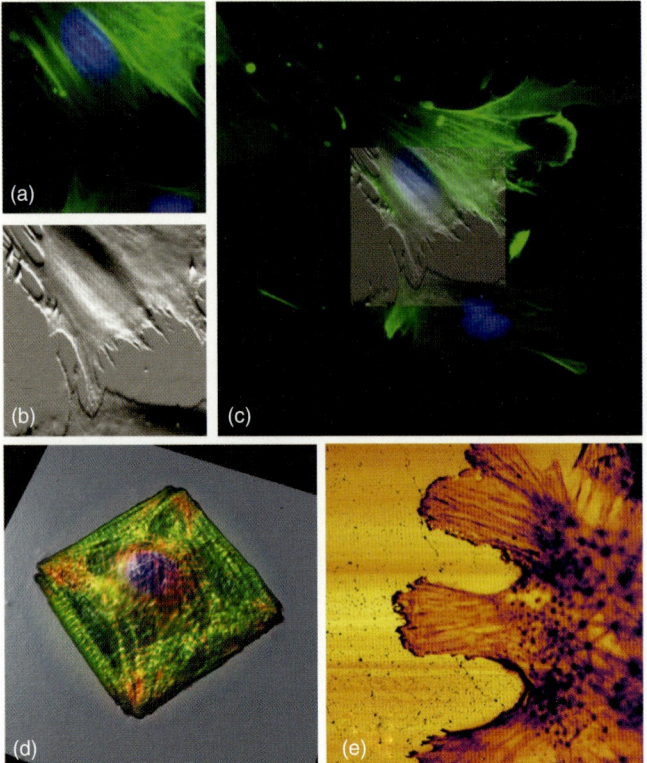

Figure 16.17 AFM imaging of cells in liquid. (a–c) Fixed MRC-5 fibroblast cells. (a) Fluorescent image of cells labeled with 4′,6-diamidino-2-phenylindole (nucleus) and Alexa Fluor 488 (actin filaments) and viewed with wide fluorescence using a Nikon TE2000U inverted optical microscope (×40). (b) AFM deflection image, 60 μm scan. (c) AFM deflection data ((b) with 50% transparency) overlaid onto merged (a) fluorescence optical image allowing mechanical and optical contrast to be compared. (d) AFM image overlaid with the epifluorescence image of a patterned cardiac myocyte. (e) Dynamic AFM image of a live rat mesenchymal stem cell showing both linear and geodesic arrangements of the F-actin cytoskeleton, 90 μm scan. ((a–c) Courtesy of Asylum Research. (d) Image courtesy of K. Parker and N. Geisse, Harvard University. (e) Courtesy of Suzi Jarvis and colleagues [253], Conway Institute for Biomolecular and Biomedical Research, University College Dublin, and copyright, 2007, Taylor & Francis.)

made, in a technique referred to as *force volume imaging*. In contrast to earlier techniques such as indentation, the experimentally achievable force sensitivity levels for an AFM is on the order of 1–10 pN, sufficient to probe a single hydrogen bond [20]. The projections made in the pioneering work by Binnig and Rohrer suggest that forces down to 10^{-18} N can be measured, and steady progress in this direction has been achieved in the context of, for example, magnetic resonance force microscopy [265].

The first single-point force measurement was reported by Heinzelmann et al. [266] in 1988, two years after the invention of the AFM (Figure 16.18a). This

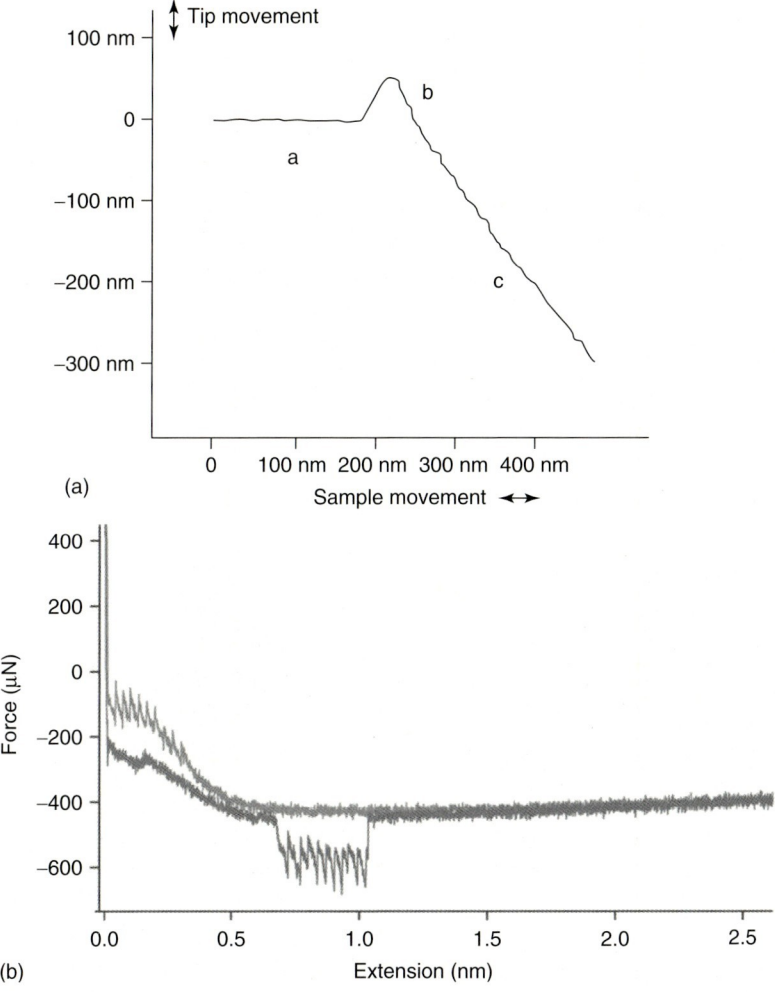

Figure 16.18 Evolution of force–distance measurements. (a) First reported force–distance measurement. There is no interaction force at a, and attractive force at b and a repulsive force at c. (Reproduced with permission from [266], copyright 1988, American Institute of Physics.) (b) Characteristic sawtooth pattern responses in both the retraction (black) and the approach (gray) force traces measured from the amyloid-based adhesive of *Prasiola linearis*. (Reproduced with permission from [267], copyright 2007, John Wiley & Sons, Inc.)

quickly grew into a popular measurement mode, allowing characterization of the full spectrum of tip–sample forces, ranging from the long-range, typically attractive interactions to the short-ranged repulsive interactions.

One of the most exciting examples of this approach is molecular unfolding spectroscopy (Figure 16.18b), in which force–distance curves obtained at different rates contain information on the thermodynamics and kinetics of force-induced reactions

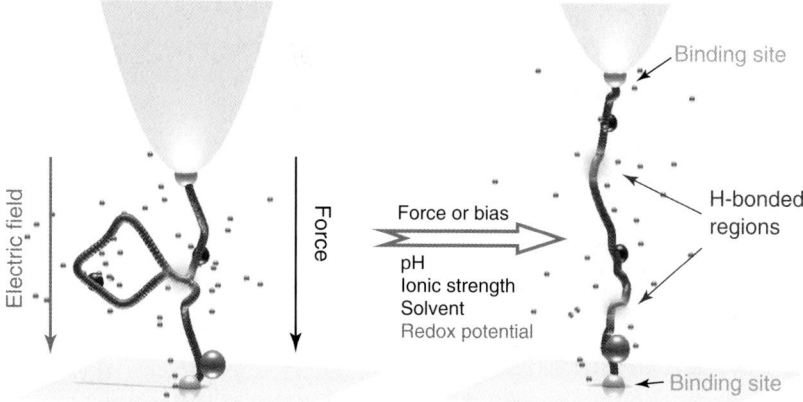

Figure 16.19 Artistic rendition of a molecular transformational change in the electric field at the tip–surface junction. The redox potential of the molecule is expected to be dependent on the force acting on the molecule. (Image courtesy of S. Jesse, ORNL.)

on a single molecule level [20–22]. Some of the early examples of single-molecule force experiments came in the mid-1990s [268–270]. However, the seminal work by Rief and Gaub [20] established the relationship between force–distance curves and the thermodynamics of single-molecule transformations, laying the foundation for a broad field of single-molecule chemistry. Since then, multiple studies of structure–function relationships in single macromolecules has been reported, including protein structure determination by mechanical triangulation [271], probing the unfolding pathways of individual bacteriorhodopsin molecules [272], probing the folding energy landscape of a single protein molecule [273], and many others. Novel modifications of force pulling techniques based on improved detection are constantly emerging, including force-clamp methods [274], dynamic stiffness measurements during force–distance curve acquisition [275], and unfolding with electrochemical control (Figure 16.19) [276, 277].

Alternatively, force–distance curves can be used to determine local adhesion, indentation modulus, or long-range electrostatic interactions. Chemical functionalization of the probe allows controlling the nature of tip–surface interaction and probing hydrogen bonding and acid–base interactions. The review by Butt et al. [8] provides an in-depth and detailed account of force-based studies of materials.

The utility of the AFM for measuring forces in fluids has allowed colloidal forces to be measured with extremely high resolution [278, 279]. A recent example includes investigations measuring the forces between droplets in emulsions. In both foams and emulsions, droplets and bubbles are dynamic in nature as they move through the suspending liquid and collide with each other. During collisions, the droplets or bubbles may bounce off each other causing the emulsion or foam to remain stable or coalesce. In addition, the outcome of these collisions is related to the types of molecules present at the interface between the two phases. AFM

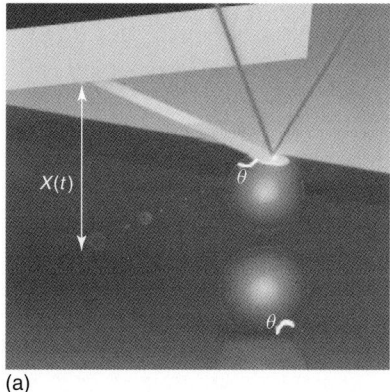
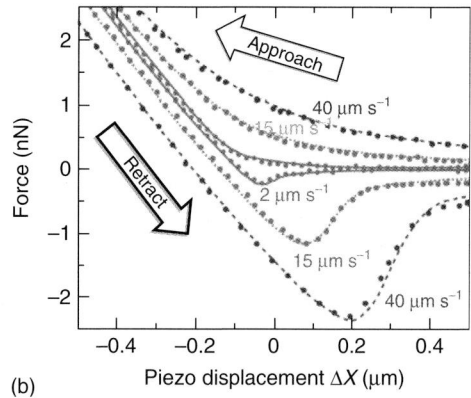

Figure 16.20 Measuring the forces between droplets in emulsions. (a) A schematic of an AFM fluid cell with two oil droplets in water. The oil droplets are immobilized on a hydrophobic surface and a custom-manufactured tipless AFM cantilever with a circular gold structure patterned at the end of the cantilever. This gold pattern can be chemically modified to be hydrophobic, providing a well-defined contact area for the drop immobilized on the cantilever. The radii of the droplets are typically 30–50 μm. (b) Typical force curves showing the repulsive interaction between two tetradecane oil droplets in a 10 mM solution of surfactant as a function of changing approach and retract speeds. The increase of repulsion on approach and the smoothly varying minima on retract are the result the hydrodynamic drainage forces in the thin film between the drops. ((a) Courtesy of R. Dagastine. (b) Adapted with permission from [283], copyright 2011, American Chemical Society.)

has been used to measure repulsive and attractive forces in controlled collisions between two droplets or two bubbles with sizes between 30 and 100 μm redundant (Figure 16.20) [280–283]. The intervening thin films range in thickness from a micrometer to a nanometer. Experimental measurements have been integrated with the development of an accurate quantitative model to analyze these results. The approach provides a "front seat" view of how drops or bubbles collide in aqueous solution and the physical mechanisms behind the coupling of the fluid flow between the droplets, the nanoscale deformation of the droplets or bubbles, and the interaction forces depending on the types of molecules coating the oil droplet interfaces [284, 285]. The agreement between the experimental data and theory has resulted in an improved understanding of droplet and bubble interactions as well as in identifying counterintuitive coalescence behavior from coupling between deformation and hydrodynamic drainage.

Force–distance spectroscopy can be readily generalized for spectroscopic imaging modes. In these, the force–distance curves are measured at each point on a spatial grid yielding a 3D data set [286, 287], which can be subsequently analyzed to provide 2D maps of relevant materials parameters (Figure 16.21). Recent work on atomic-level force-field mapping demonstrates tremendous potential for understanding the atomic origin of dissipation and friction, as well as for single-atom chemical recognition [288].

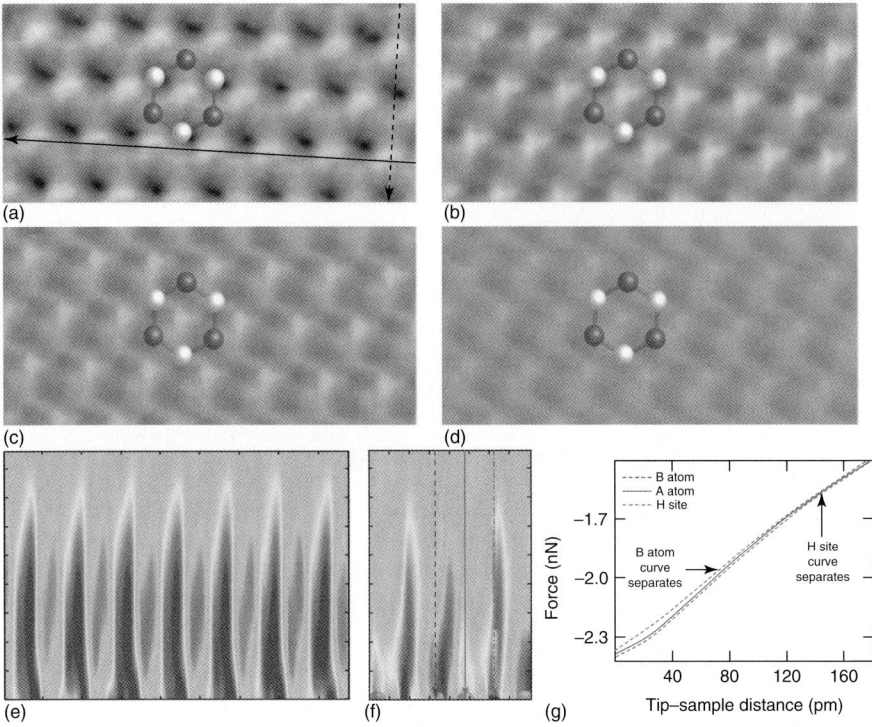

Figure 16.21 3D spectroscopic imaging. (a–d) 1750 × 810 pm force maps determined 12, 52, 97, and 132 pm above the surface, respectively. (e,f) Orthogonal vertical force maps along the direction indicated by the arrows in (a). (g) Force curves obtained at different atom locations indicated in (f). (Reproduced with permission from [287], copyright 2010, WILEY-VCH Verlag GmbH & Co.)

16.4.3
Decoupling Interactions

While comprehensive understanding of surface functionality can be achieved from 3D spectral force–distance imaging, the practical considerations of image acquisition time and microscope drift generally limit this approach. Alternatively, decoupling topography from the measured force component can be achieved either through (i) height variation, (ii) use of force-modulation approaches, or (iii) detection through complementary mechanical degrees of freedom of the force sensor.

16.4.3.1 Dual-Pass Methods

An important example of height variation as a method of decoupling short- and long-range forces is the well-known double-pass or interleave mode ("lift" or "nap"), first pioneered by Hosaka *et al.* [289] (Figure 16.22).

In these modes, the first AFM point, line, or full two-dimensional image scan is used to determine the position of the surface (measure the topography), that

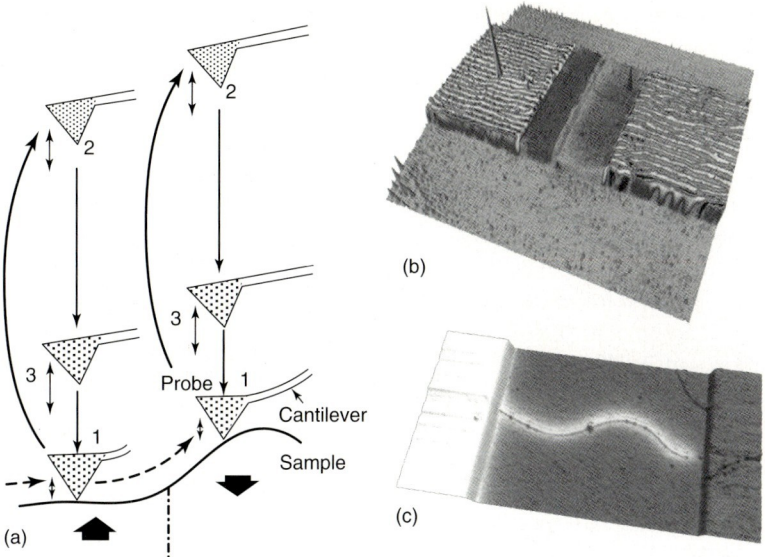

Figure 16.22 Interleave modes. (a) Schematic of Hosaka's interleave method for separating the long-ranged and short-ranged forces. Briefly, the surface is referenced in position 1 and then the probe is raised to position 2 where it measures the long-ranged magnetic interactions. Optionally, the interaction can be measured at a different height 3 before it is moved on to the next position. (b) Shows the magnetic fields above an Iomega Zip™ 1 GB drive write head. The MFM phase signal was overlaid on top of the topography, 20 μm scan. (c) EFM of a carbon nanotube attached to an electrode.[3] In the image, color represents the EFM phase channel overlaid on the topography, scan size 5 μm × 2.5 μm. ((a) Reproduced with permission from [289], copyright, 1992, Japan Society of Applied Physics. (b,c) Courtesy of Asylum Research.)

is, the condition at which the measured signal, $R(h, V_0) = R_0$, where R_0 is a set-point value. The feedback signal, R, can be a static deflection for contact mode AFM, oscillation amplitude for an amplitude-based detection signal, or frequency shift for frequency-tracking methods. The second scan is performed to determine interactions at a constant distance or bias condition to measure $R = R(h + \delta, V_1)$. As a typical example, MFM and EFM utilize the fact that these forces are relatively long ranged and weak compared to the vdW interactions. Hence, once the position of the surface has been determined, force measurement at significant ($\delta = 10$–500 nm) separations yields magnetic (if the probe is magnetized) or electrostatic (if the probe is biased) force components.

16.4.3.2 Modulation Approaches

An alternative approach to force detection is based on AC modulation approaches, for example, AFAM [290], force modulation [291, 292], and similar methods. In these, the condition of $R(h, V_0) = R_0$ is used to determine the position of the surface,

---

3) CNT sample courtesy Minot Lab, Oregon State University.

and modulation of the probe height or bias is used to obtain additional information on the distance- or bias-derivative of the force-distance-bias surface. For example, in the small signal approximation, the AFAM signal is related to $(\partial h/\partial F)_{V=\text{const}}$. In addition to providing additional information (i.e., not only the force curve but also the derivative of the force), the modulation methods allow the sensitivity to be increased by moving away from the $1/f$ noise corner, and the effective cantilever stiffness to be controlled through the inertial in dynamic measurements. This decoupling can be performed dynamically, for example, response phase and amplitude in AFM phase imaging (amplitude yields information on topography, and phase yields information on elasticity and adhesion). The emergence of dual-modulation modes [154] has allowed the sensitivity of modulation methods to be increased by lifting the restrictions imposed by weak dissipation, which confines the phase changes to a relatively small region of the phase space of the system. These modes have led to the development of compositional mapping and subsurface tomography imaging modes [172, 293–295].

16.4.3.3 Decoupling through Different Mechanical Degrees of Freedom

Finally, different interactions can be decoupled using flexural and torsional degrees of freedom of the probe, with normal force mapping the topography and friction forces sensitive to adhesion and chemical interactions. In lateral force microscopy [220], the topography is detected from the normal force signal, while the friction force detected from the lateral signal provides information on the short-range tip–surface interactions sensitive to local chemical composition, molecular orientation, and so on [296]. The recent development of harmonic detection [179, 297, 298] and torsional resonance [299] modes allows the reconstruction of tip–surface interactions based on decoupling between torsional and flexural oscillations.

16.5
Voltage Modulation SPMs

The efficient approach for decoupling tip–surface interactions is based on the use of static and dynamic biasing of a conductive tip, giving rise to a broad spectrum of contact and NC voltage modulation modes. Electrically modulated SPM modes in ambient and UHV environments are well established by now and include KPFM (Figure 16.23b) with amplitude (measures force) and frequency (measures force gradient) feedback, and PFM/ESM for detecting bias-induced deformations in a variety of materials.

16.5.1
Kelvin Probe Force Microscopy and Scanning Impedance Microscopy

The static electrostatic SPMs are based on the detection of long-range tip–surface electrostatic forces. In KPFM, the electric potential on the conductive SPM probe

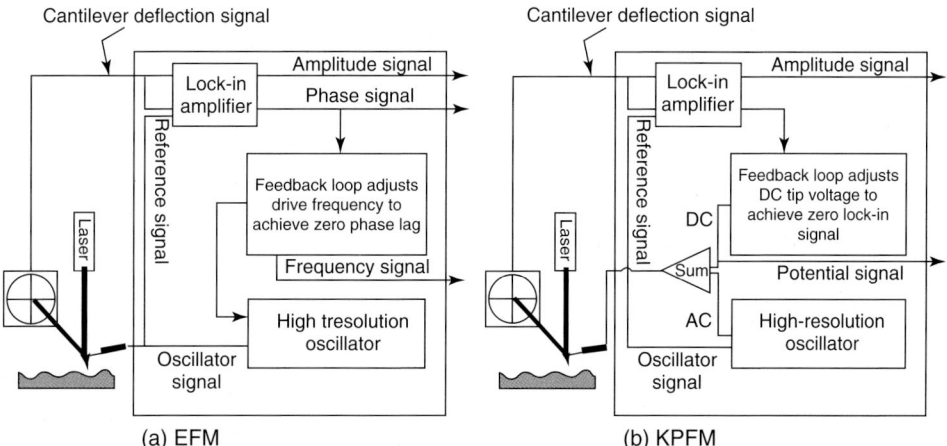

Figure 16.23 Schematics of EFM and KPFM. (a) In EFM, the electrostatic force is measured, while (b) in KPFM, the force is measured and simultaneously minimized using a feedback loop that applies an appropriate DC tip voltage. The EFM amplitude signal is then effectively the error signal for the feedback loop. (Reproduced with permission from [305], copyright 2008, Institute of Physics.)

is modulated as $V_{tip} = V_{dc} + V_{ac}\cos(\omega t)$, where V_{dc} is the static potential offset, V_{ac} is the driving voltage, and the driving frequency, ω, is typically chosen close to the free cantilever resonance. The tip bias results in the capacitive tip–surface force, $F_{el} = C'_z(V_{tip} - V_s - \Delta CPD)^2$, where C'_z is the (unknown) tip–surface capacitance gradient, V_s is the electrostatic surface potential, and ΔCPD is the contact potential difference between the tip and surface. Depending on the experimental configuration, voltage modulation can be applied either during the interleave scan (i.e., when the tip retraces the predetermined surface topography while maintaining constant separation), or during the acquisition of topographic information (at a different frequency).

In KPFM, an LIA is used to select the first harmonic component of the force, $F_{el}(1\omega) = C'_z V_{ac}(V_{dc} - V_s - \Delta CPD)$, and the feedback loop is engaged to keep it zero by adjusting the static offset of tip potential, V_{dc}. The condition $F_{el}(1\omega) = 0$ is satisfied for $V_{dc} = V_s + \Delta CPD$, that is, when the microscope-controlled compensation potential is equal to the (unknown) local surface potential. On a (nominally) grounded surface, KPFM allows direct detection of the materials-specific contact potential containing the contributions from surface dipole layers, nonequilibrium surface charges, etc. The image formation mechanism in KPFM for conductive and semiconductive materials is studied extensively [300, 301]. The typical energy and spatial resolution for constant frequency KPFM is ~1–5 mV and ~30 nm. Frequency-modulated KPFM utilizes the feedback to nullify $\Delta\omega_r$, allowing direct detection of surface potential. Notably, recent reports indicate that frequency-modulated KPFM [302–304] can achieve atomic resolution (it should be noted that the nature of measured contrast is actively discussed), suggesting tremendous potential for this technique for probing materials structures.

A complementary approach for electrostatic force imaging is EFM (Figure 16.23a) [122–124], in which the electrostatic force gradient acting on the tip is determined from the resonant frequency shift of the DC-biased vibrating cantilever, $\Delta \omega_r \sim C_z''(V_{dc} - V_s - \Delta CPD)^2$. Unlike KPFM, the absolute strength of the signal in EFM is difficult to quantify and the method is more sensitive to surface topography through the tip–surface force gradient, C''.

To date, KPFM and EFM have been broadly applied to the characterization of dipole moments and surface work functions in systems such as self-assembled monolayers, III–V semiconductors, well-defined crystalline edges, and ferroelectric domains [306]. In semiconductors and devices, KPFM and EFM have been used for dopant profiling [307], studies of surface defects [308], and for mapping potentials in operating devices [309, 310]. Other studies include multiple observations of charge relaxation on dielectric [311] and ferroelectric surfaces [312], and temperature-induced phase transitions in ferroelectric materials [313, 314]. The notable recent work by the Prinz group [315] demonstrates the use of the variable temperature-controlled environment KPFM to study charge injection and oxygen vacancy diffusion in oxide ion conductors such gadolinia-doped ceria. They were able to extract surface diffusion coefficients for oxide ion vacancies with an activation energy of 0.56 eV. EFM and KPFM have been used extensively to study organic molecules [316–321] and biomolecular systems [322–327] including lipid bilayers [328–330]. As with polarization switching in ferroelectrics [331] and photoinduced phenomena in semiconductors [332–334], KPFM has been used to track dynamic changes of potential in biosystems after the formation of biomolecular complexes [335, 336] and due to changes in photovoltage [337–341].

Overall, while EFM and KPFM have been applied successfully for studies of solar materials [332, 342], and semiconductor and electroceramic devices [306], further progress in areas such as high-resolution imaging, probing macromolecular transformations, or cellular and subcellular electrophysiology necessitates implementation of electrically modulated SPM in liquid environments. Early studies of electrostatics in solution dealt with measuring electrostatic force by varying the electrolyte ion concentration or pH [343–345]. The key task here is the capability to control DC and AC electric potential on small length scales. The conductivity of the solution, stray currents, and electrochemical reactions present significant obstacles. Experimentally, it was shown that the use of sufficiently high AC frequencies accessed through direct imaging or frequency mixing down conversion allows probing AC behavior (local AC field is localized) [346–348]. At the same time, DC fields are not localized in most solvents [349]. The development of insulated and shielded probes [138–140, 350] offers a pathway to future progress and will open broad access to force-coupled electrochemical and electrophysiologic phenomena in liquids.

16.5.2
Transport Imaging of Active Device Structures

KPFM and EFM can be readily extended for probing potential distributions in active devices. In these measurements, electric bias is applied laterally across the device through macroscopic or microscopic electrodes, and the SPM tip acts as a moving voltage sensor, as shown in Figure 16.24. These measurements are thus equivalent to four-probe transport measurements, when the SPM tip acts as a moving force-based voltage electrode. This approach was pioneered by Vatel and Tanimoto [351], and subsequently adopted by several groups for applications ranging from semiconductor structures to electroceramic interfaces (Figure 16.25) [352], carbon nanotubes, and quantum systems [353, 354]. The advantage of this approach is that the effective tip–surface impedance is extremely large, since there is no electron transfer between the tip and the surface. Correspondingly, KPFM is effectively the infinite-impedance probe, and the measurement process does not affect the potential distribution within the device. In EFM, the large tip–surface potential difference can induce changes of local electronic structure beneath the probe, giving rise to scanning gate microscopy (SGM). However, the drawbacks of this approach include limited dynamic range (1 mV to 10 V, as compared to $\sim$nV sensitivity of modern voltmeters), general lack of knowledge on local currents (since local current can be determined from macroscopic properties only for 1D systems), the presence of conductive layers on oxide surfaces, and the contribution of surface charges.

An alternative approach for studies of lateral current transport in active devices is based on scanning impedance microscopy (SIM) [356, 357]. In this method, the

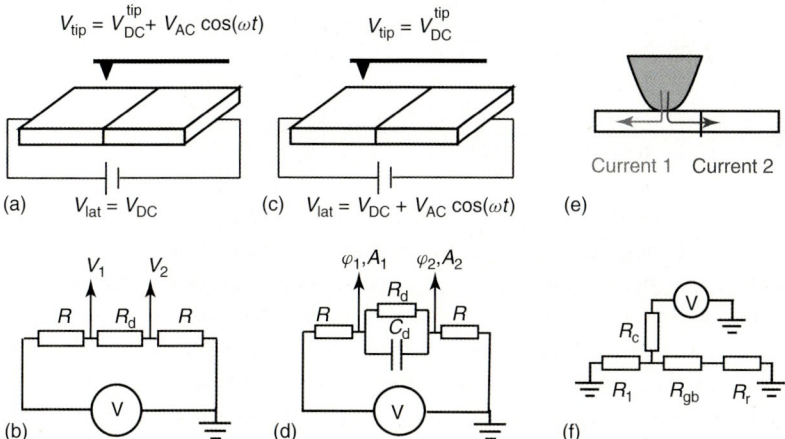

Figure 16.24 Comparison of KPFM, SIM, and cAFM. Schematic diagram (a,c,e) and corresponding equivalent circuit (b,d,f) of KPFM (a,b), SIM (c,d), and cAFM (e,f). (Reprinted with permission from [355], copyright 2004, American Physical Society.)

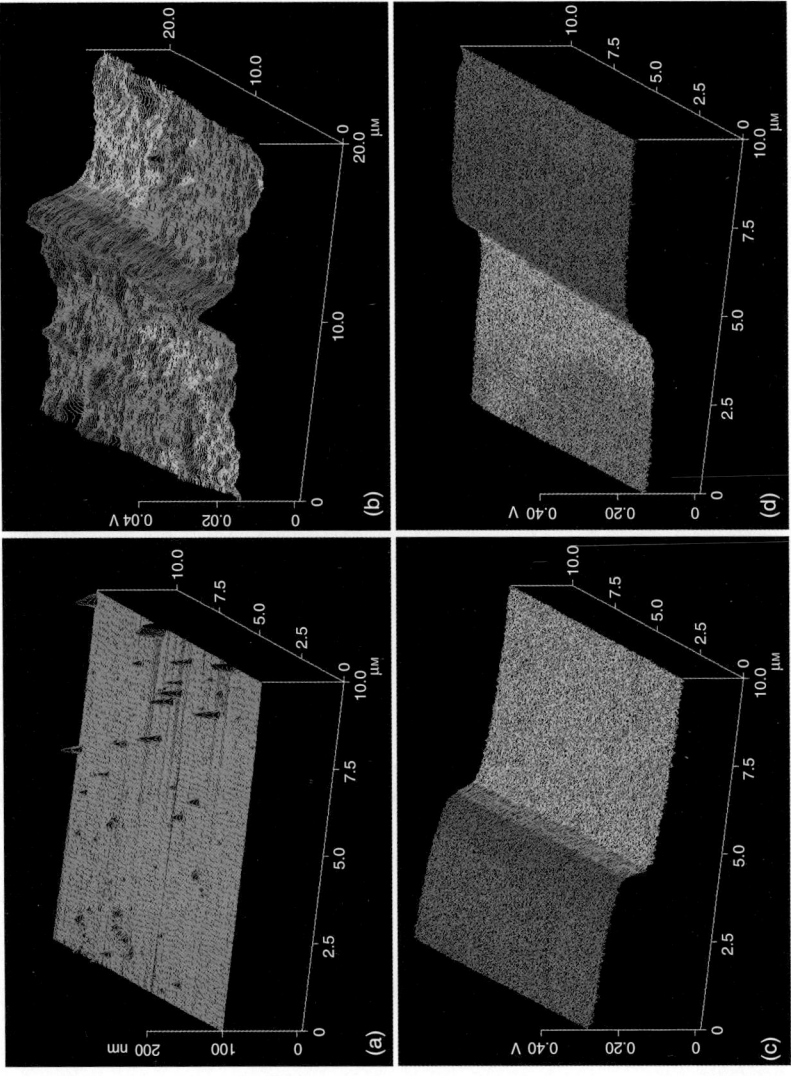

Figure 16.25 KPFM of an electroceramic interface. Surface topography (a), surface potential of the grounded surface (b) and surface potential for forward (c) and reverse (d) bias. The scale for (b) is 40 mV and for (c), (d) is 400 mV. (Reprinted with permission from [355], copyright 2004, American Physical Society.)

AC bias is applied laterally across the device, while the SPM tip is DC biased. The detection of the amplitude and phase of the tip deflection allows measurement of not only the amplitude, but also of the phase of the voltage oscillations in the sample. In this manner, both resistive and capacitive elements of equivalent circuits of a material can be determined. Hence, SIM is essentially equivalent to the dynamic version of impedance spectroscopy. SIM has been applied for direct measurement of the capacitances of grain boundaries in semiconducting strontium titanate [355] and calcium–copper titanate [358]. Superimposing the DC potential gradient and simultaneous measurement of SIM phase shift and potential drop allows the full capacitance–voltage curve to be reconstructed, as demonstrated for a model Schottky diode [357].

Finally, in cases when conductance in a material is confined to a narrow 1D channel, the moving SPM tip can act as a moving gate electrode, giving rise to SGM. This technique has been extensively used for characterization of carbon nanotubes [359], graphene [360], nanowires [361], quantum point contacts [362], semiconductor heterostructures [363], and quantum Hall edge states [364], to name a few.

16.5.3
Piezoresponse Force Microscopy and Electrochemical Strain Microscopy

Voltage modulation of the SPM tip in the contact regime results in both long-range electrostatic interactions and field penetration into the material. The latter can induce intrinsic electromechanical response of a material, that is, a bias-induced surface deformation. While electromechanical coupling coefficients ($\sim$2–100 pm/V) are typically small, they are well within the detection limit of SPM. Furthermore, the high stiffness of the tip–contact junction (100–1000 N/m) in contact mode typically leads to the prevalence of electromechanical response compared to long-range electrostatic forces.

Electromechanical SPM imaging and spectroscopy on ferroelectric and piezoelectric materials is typically referred to as *piezoresponse force microscopy* (PFM) and piezoresponse force spectroscopy (PFS) (Figure 16.26). In PFM, an electrically conductive cantilevered tip traces surface topography using standard deflection-based feedback (contact mode). During the scanning, a sinusoidally varying electrical bias is applied to the tip, and the electromechanical response of the surface is detected as the first harmonic component of the bias-induced tip deflection. The response amplitude and phase provide a measure of the local electromechanical activity of the surface. Notably, PFM is the first method that allowed reliable and high-veracity electromechanical measurements on the nanoscale. Traditionally, these measurements were challenging even for piezoelectric crystals, since typical piezoelectric displacements induced by moderate (1–100 V) biases are typically in the low nanometer range. The properties of thin films have become accessible only in the past two decades with the advent of single- and double-beam interferometry techniques [365–369].

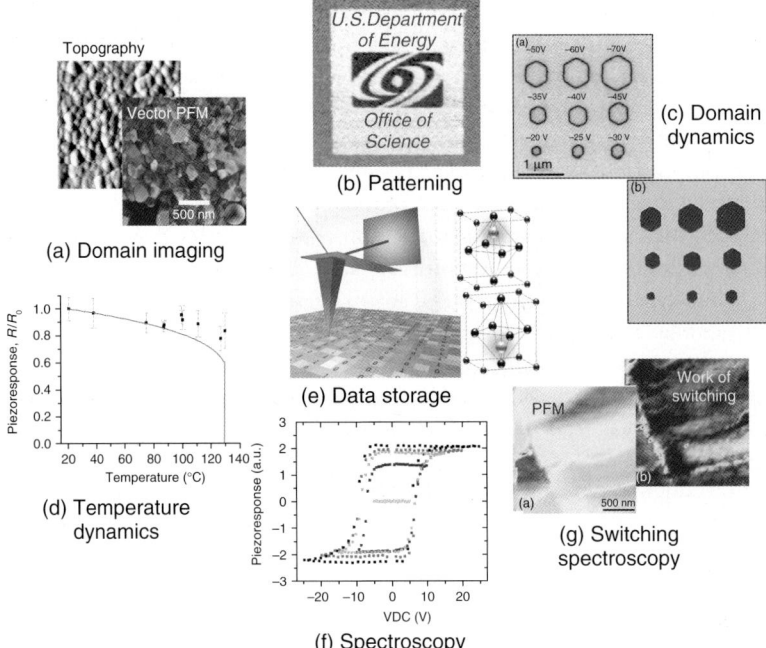

Figure 16.26 Piezoresponse force microscopy. Applications of PFM for (a) domain imaging, (b) domain patterning, (c,d) dynamic phenomena, (e) data storage, (f) spectroscopy, and (g) spectroscopic imaging. (Adapted from [211], copyright 2007, Annual Reviews.)

The operation of PFM is complementary to that of conventional SPMs. As an example, in AFM methods the contact stiffness of the tip–surface junction scales proportionally with the contact radius, resulting in the topographic cross-talk inherent to many force-based SPM imaging modes. In contrast, the electromechanical response in PFM depends only weakly on the tip–contact area, thus minimizing direct topographic cross-talk.

PFM has emerged as the primary tool for studying polarization dynamics in ferroelectric materials, bringing about advances such as sub-10 nm resolution imaging of domain structures, domain patterning for nanostructure fabrication, and local spectroscopic studies of bias-induced phase transitions. The development of high-sensitivity PFM has allowed probing of piezoelectric phenomena in weakly piezoelectric compounds such as III–V nitrides and biopolymers [370] (∼1–5 pm/V), providing spectacular sub-10 nm resolution images of material structures based on the difference in the local piezoelectric response.

Strong coupling between polarization and electromechanical response allows use of the latter as a functional basis for nanoscale probing of polarization distribution and switching processes in ferroelectrics. This approach can be directly extended to probe local polarization dynamics. The probe concentrates an electric field to a nanoscale volume of material (diameter ~10 nm), and induces local domain formation. Simultaneously, the probe detects the onset of nucleation and the size of a forming domain via detection of the electromechanical response of the material to a small AC bias [371]. In this manner, local hysteresis loops can be measured, which contain information on the domain nucleation and growth processes.

In switching spectroscopy piezoresponse force microscopy (SS-PFM), the hysteresis loops are acquired at each point of the image, and analyzed to yield 2D maps of coercive and nucleation biases, imprint, work of switching, and switchable polarization. Maps of switching behavior can be correlated with surface structure and morphology. In the past two years, *spectroscopic* 3D [371–373] and 4D [169] PFM modes have been implemented on ambient SPM platforms to study switching processes in ferroelectrics. In these, the measured local electromechanical response as a function of DC tip bias provides information on the size of the domain formed below the tip [374]. The variation of nucleation biases along the surface has been used to map the random field and random bond components of disorder potential (Figure 16.27) [375]. Furthermore, the effect of a *single* localized defect on the thermodynamics of local polarization switching can be determined (Figure 16.27) [376]. A recent work has demonstrated that the ferroelastic domain walls strongly affect the polarization switching in one direction, and virtually do not affect it in the other. Similarly, grain boundaries result in anomalous switching behavior with asymmetric nucleation contrast. These studies illustrate the potential of this methodology to study deterministic mesoscopic switching mechanisms on well-defined defect structures.

Recently, voltage modulation contact mode SPM has been extended for mapping ionic dynamics in solids [40, 377–379]. In ESM, a biased SPM tip concentrates an electric field in a nanometer-scale volume of the material, inducing interfacial electrochemical processes at the tip–surface junction and ionic currents through the solid. The intrinsic link between the concentration of ionic species and/or the oxidation states of the host cation and the molar volume of the material results in electrochemical strain and surface displacement.

The ESM imaging mode is based on detecting the strain response of a material to an applied electric field through a blocking or electrochemically active SPM tip. This imaging mode can be further extended to a broad set of spectroscopic imaging modes. An example of BE-ESM decoupling of electrochemical reactivity and transport in electrochemical systems is shown in Figure 16.28. At each point, a 2D amplitude-phase spectrogram representing the frequency response of the cantilever in the proximity of the resonant peak as a function of DC bias offset (5D data set) is acquired. This data can be analyzed to yield a series of first-order reversal curves at each location. Mapping the hysteresis loop opening as a function of bias allows reaction and diffusion to be separated, providing striking images of electrochemical reactivity with sub-10 nm resolution.

To date, ESM has been demonstrated for a variety of lithium-ion materials, including layered transition metal oxide cathodes, Si anodes, and electrolytes such as LISICON; oxygen electrolytes, including yttria-stabilized zirconia and samarium-doped ceria; mixed electronic-ionic conductors for fuel cell cathodes;

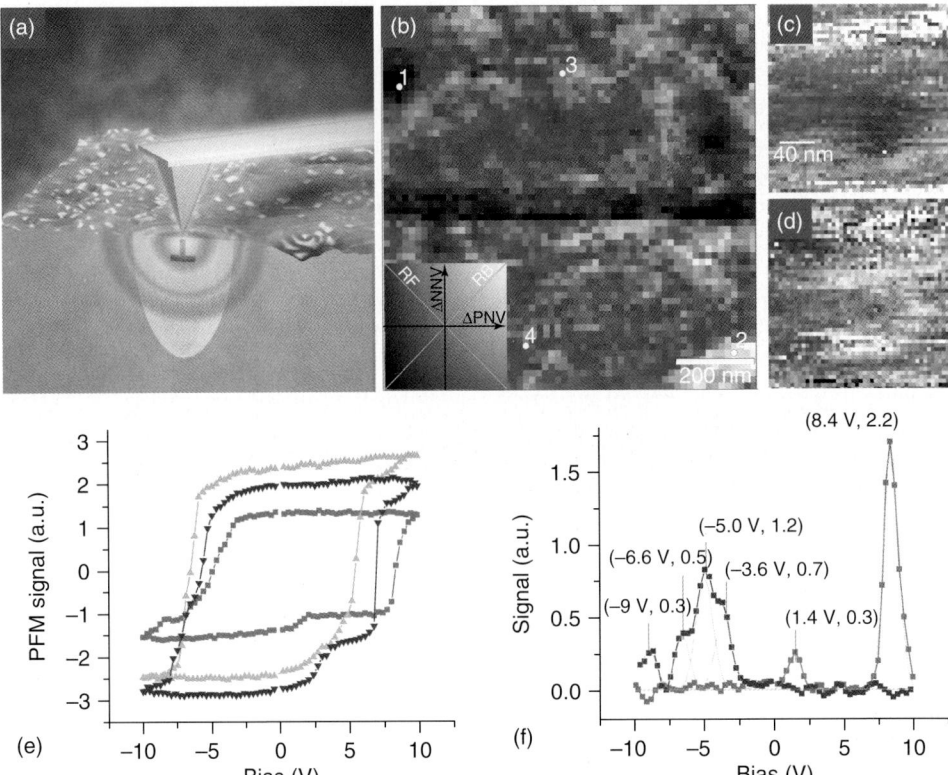

Figure 16.27 PFM of defects. (a) The confinement of an electric field by an AFM probe allows bias-induced phase transitions within a defect-free volume or at a given separation from defects to be probed. (b) The map illustrates the random field (RF) and random bond (RB) disorder potential in an epitaxial PZT film based on the positive and negative nucleation voltages (PNV and NNV). (c,d) A single defect in multiferroic BiFeO$_3$ determined from the nucleation bias and fine structure features on a (e,f) hysteresis loop. ((a) Courtesy of Stephen Jesse (ORNL). (b) Reproduced with permission from [375], copyright, 2008, Nature Publishing Group. (c–f) Reproduced with permission from [376], copyright, 2008, American Physical Society.)

and some proton conductors. The ability to probe electrochemical processes and ionic transport in solids is invaluable for a broad range of applications for energy generation and storage (e.g., batteries and fuel cells). In addition, because electrochemical strains are ubiquitous in virtually all solid-state ionics, ESM is applicable to all battery and fuel cell materials in energy technologies and electroresistive/memristive materials in information technologies.

The remarkable results obtained by PFM and ESM for ferroelectric, multiferroic, and recently solid-state ionics and biomolecular systems suggest the potential for significant discoveries in the future. Electromechanical coupling is ubiquitous in biological systems, underpinning processes from hearing to motion to cardiac activity, as well as in soft condensed matter systems such as polyelectrolytes,

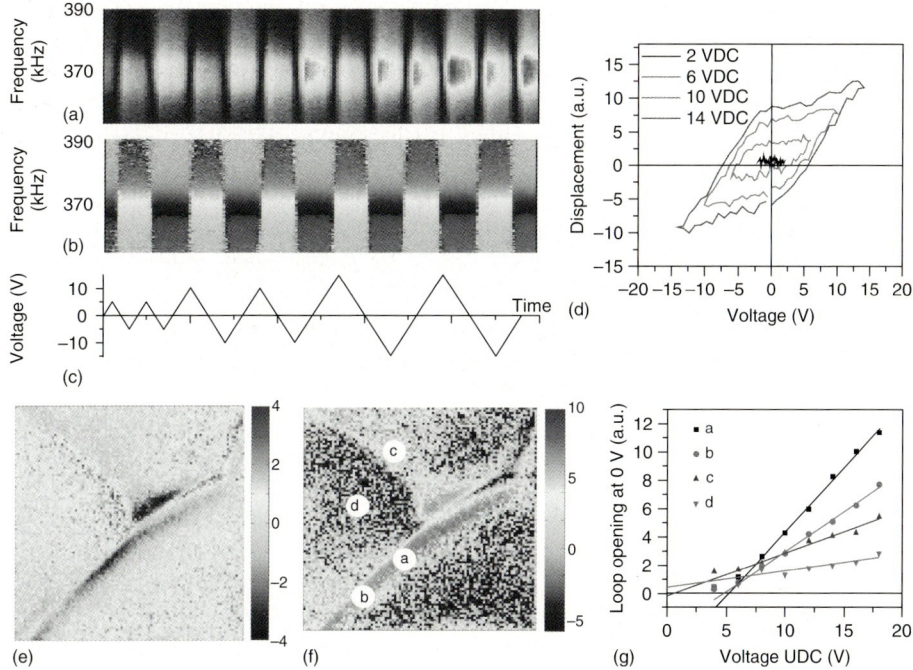

Figure 16.28 Separation of reaction and transport phenomena by electrochemical strain microscopy in amorphous Si. (a) Amplitude and (b) phase spectrogram for (c) triangular bias sweep with increasing amplitude (first-order reversal curve). (d) ESM hysteresis loop opening for six subsequent bias sweeps with increasing envelope voltage (signal is averaged over 20 points on a grain boundary). (e) Slope a and (f) Critical bias (i.e., x-axis intercept from linear fit of loop opening) as function of maximum DC voltage V_f. (g) Example curves with fit from four different points in different regions from (g). The anomalous response at the grain boundaries in Si is clearly seen. (Reproduced with permission from [380], copyright 2010, American Chemical Society.)

redox-active molecules, ferroelectric polymers, and so on. In the future, PFM/ESM will open the pathway for probing and controlling electromechanical conversion on a single molecule level and harnessing this conversion for device applications.

16.6 Current Measurements in SPM

16.6.1 DC Current

The biased SPM tip can be used as a moving current electrode, providing information on local conductivity. In the most direct application, current induced by the potential difference between the tip and the surface is measured. These sets of techniques have been developed in multiple contexts, primarily for applications for semiconductor characterization and basic physical studies. Depending on the type of current–voltage converter (linear vs logarithmic) and community, these techniques are referred to as cAFM, tunneling AFM, and SSRM. Despite the overall simplicity of the setup, it provides a unique window to material surfaces where a significant heterogeneity in electronic conductance exists, including pinholes, dislocations, and more exotic phase-separated electronic states on solid-state surfaces, multicomponent materials such as polymer blends, junctions, organic semiconductors, and many others. Fundamentally, the key defining signature of the cAFM approach is a significant penetration of the electric field into the subsurface volume.

The applications of electrical SPM techniques and in particular cAFM and SSRM of semiconductors has been summarized by Oliver [305]. cAFM can be used as a sensitive probe to map functional changes in a material induced by the application of bias or pressure, as has been demonstrated recently via ferroelectric polarization-dependent tunneling current [381–383]. cAFM has also been used for lateral transport measurements, and nanopotentiometry has been implemented in STM [384].

Recently, cAFM was extended in a novel research direction, where the electric field drives the dynamics of the material under the tip, and the electronic conductance is used to monitor the dynamics. The relevant processes involve bias-induced phase transitions, for example, ferroelectric switching [382, 383, 385] and motion of domain walls [386], migration of oxygen vacancies or other ionic species [387], and the cross-hybrids of the two, that is, migration of ionic species inducing a local phase transition. The resulting current–voltage characteristics are highly nonlinear, and are often collectively referred to as *resistive switching*, owing to the presence of several conductance states at a given bias depending on the history of the dynamic changes in the probed volume (Figure 16.29). Resistive switching characteristics are attractive for simple memory applications, and more broadly for memristive logic, but it is likewise important that the $I - V$ curves themselves, which exhibit nonlinearities, hysteresis, and often strong time dependence, are a rich source of information about the governing dynamics [388].

Overall, the DC-current-based SPM methods offer a powerful tool for exploring the electronic transport properties of semiconductors, metals, ionic materials, and devices. The primary limitation of these methods is the extreme sensitivity to surface state and minute amounts of contamination (which may be both an advantage and a disadvantage) and the high propensity for inducing surface electrochemical damage (reactions, vacancy injection) due to extremely large current densities. Furthermore, in an ambient environment, the presence of the water droplets in the tip–surface junction can be a significant factor affecting measurements, necessitating transition to controlled environments, and vacuum systems.

16.6.2
AC Current

A number of authors [389, 390] have demonstrated the use of AFM as a probe for local impedance spectroscopy. However, a simple comparison of the tip–surface

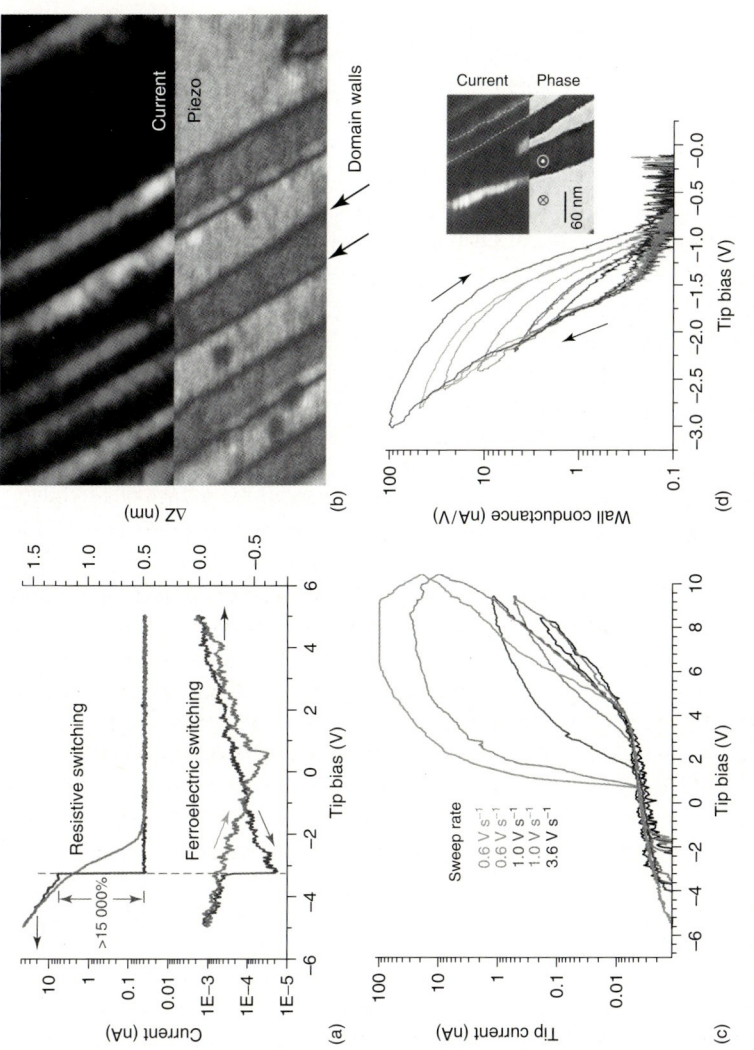

Figure 16.29 Electroresistance measurements on ferroelectrics. (a) Giant electroresistance in 30 nm thick ferroelectric Pb(Zr$_{0.2}$Ti$_{0.8}$)O$_3$ film showing a simultaneous change in local strain and conductivity measured as a function of tip bias (−5 to 5 V and back correspond to dark gray and light gray arrows, respectively). (b) cAFM and PFM images of conducting 109° domain walls in a 100 nm thick La-doped BiFeO$_3$ film measured in UHV. (c) Local current–voltage spectroscopic measurements of the bias-induced metal–insulator transition in Ca-doped BiFeO$_3$. (d) A memristive series of hysteretic I–V curves obtained sequentially on the surface of an annealed BiFeO$_3$ film. Inset shows simultaneously acquired cAFM and PFM at 109° domain walls. (Reproduced with permission from [388], copyright 2011, American Chemical Society.)

junction and cantilever surface impedances illustrates that direct measurements are possible only for well-defined mesoscopic objects (i.e., single-crystalline conductive grain with insulating grain boundaries), but not local volume of material below the tip. Correspondingly, while reported for a small number of systems, these measurements are still in their infancy. For battery-related materials, scanning impedance measurements were performed on electrolytes. Layson et al. investigated conducting polymer electrolytes and have been able to differentiate regions of high and low ion conductivity in the films consistent with macroscale measurements [391]. Similar measurements by O'Hayre et al. [390] demonstrated SIM on a Nafion electrolyte. Recent work by the Haile group [392] has demonstrated direct electrochemical impedance probing of a solid-state electrolyte enabled by reactions at the tip–electrolyte junction.

16.6.3
Scanning Capacitance Microscopy

In scanning capacitance microscopy (SCM), an AC bias is applied to the tip, which generates an AC electric field that causes carriers to be attracted and repelled by the tip. SCM incorporates a capacitance sensor to measure the capacitance between the tip and the sample through an insulating layer [393]. The thin insulating layer is formed either on the semiconductor surface or on the surface of a highly conducting AFM tip so that the tip–sample system becomes a metal-insulator-semiconductor (MIS) structure when the tip is in contact with the sample. The capacitance of the MIS structure, C, arising because of a space change in the depletion layer in the semiconductor at the interface with the insulator, is a function of the voltage, V, applied between the tip and semiconductor bulk. Alternatively, a conducting tip and a clean, unpassivated semiconductor can form a Schottky barrier capacitor. During imaging, the capacitance of the structure is varied by an AC voltage of a frequency ~10 kHz applied to the tip, and the tip–sample capacitance variations are detected by a capacitance sensor with use of a lock-in technique. The images are maps of the differential capacitance dC/dV. The local dopant concentrations and charge carrier sign can be extracted from these maps after calibration of the measurement circuits using numerical procedures, which makes SCM a quantitative as well as qualitative technique. In a standard configuration, the capacitance sensor is a lumped LC resonant circuit where the capacitance of the tip–sample system is a part of the capacitance in the resonant circuit. The resonant frequency of the LC circuit is close to 1 GHz, and it shifts in response to the changes of the tip–sample capacitance. The tip–sample capacitance variations are detected by the changes in the amplitude of the oscillations in the LC circuit driven at a constant frequency near the resonance. Systems with distributed resonators have also been demonstrated [394].

SCM was first demonstrated using an electrode attached to a moveable insulating stylus [395] and later by combining STM with a capacitance sensor [396]. Modern SCM is still based on the AFM-based technique developed by Barrett and Quate [397] and Abraham et al. [398], which employs contact mode constant-force feedback.

The primary purpose of SCM is dopant profiling and failure analysis for semiconductor integrated circuits, and SCM has been widely used in semiconductors and dielectrics [399] and devices [400, 401] for high-resolution two-dimensional imaging of dopant distributions and for determining the effect of dislocations on carrier mobility [402, 403]. Spectroscopic dC/dV techniques have also been developed [404].

16.7
Emergent SPM Methods

In recent years, a number of SPM modes have emerged that introduce novel excitation (e.g., thermal, microwave) and detection modes (e.g., chemical mass spectrometry). In many cases, these developments were envisioned at the dawn of the SPM era, but only the progress in large-scale fabrication of SPM platforms and probes has allowed practical implementation and eventually (e.g., for thermal probes) large-scale adoption. Below, we discuss several emergent SPM techniques that are poised to gain broad acceptance in the near future.

16.7.1
Thermal SPM Imaging

A broad spectrum of novel SPM capabilities is enabled by introducing thermal tips [405, 406] to develop a method for characterizing the thermomechanical behavior of materials on the nanoscale. The origins of this field can be traced to the introduction of a Wollaston wire (Pr/Rh thermocouple) with a sharpened junction. These probes could heat a sample up to several hundred degrees Celsius and measure the temperature of the probe–surface junction. Wollaston probes suffered from poor reproducibility of probe characteristics (probe stiffness, probe angle, reflectivity, etc.) because of the manual manufacturing of each probe. Recently, a process has been developed for manufacturing heatable Si AFM probes [407]. This process has provided the scientific community with probes that have mechanical properties and specifications (reflectivity, probe angle, etc.) similar to the those for conventional probes used for standard AFM imaging. This development has enabled a broad spectrum of novel SPM capabilities [405, 406], such as temperature-assisted modification of materials [408], and the characterization of thermomechanical behavior at the nanoscale [170, 171, 409, 410].

SPM methods that utilize heatable AFM probes can be divided into classes by the type of signal used either for feedback or for properties measurements: static – transition temperature mapping (TTM),[4] single-frequency scanning thermal expansion microscopy (SThEM) [411], multiple-frequency Ztherm,[5] and band excitation nanothermal analysis (BE-NanoTA). TTM uses the probe deflection as

---

4) Anasys Instruments.
5) Ztherm Modulated Thermal Analysis Data Sheet, Asylum Research

a function of the tip temperature to measure localized thermal expansion. When these probes are heated, the region under the tip usually expands, causing the cantilever to deflect upward. When the material under the tip warms enough to soften, this process can begin to reverse as the tip penetrates the softer, but still expanding, material. The transition temperature is then determined as the temperature maximum on the probe deflection versus temperature curve. This approach was further developed to combine acoustically broadband excitation using an external oscillator (similar to AFAM). In this case, the evolution of the dynamic properties of the tip–surface junction as a function of tip temperature are determined [170], providing information on mechanical and loss moduli. In addition to increased signal-to-noise ratio and significantly higher resolution, this approach allows the glass transitions in polymers to be probed.

An alternative approach for local thermomechanical behavior is given by SThEM [412], Ztherm, and BE-NanoTA [171, 409], which use periodic heating of the AFM probe with AC current. In these techniques, the tip temperature is ramped following a triangular wave. Superimposed on this slowly modulated (~10–100 s) waveform is a small-amplitude sinusoidal or BE waveform that results in periodic changes in the tip temperature. The latter results in the thermomechanical expansion and contraction of the surface, which are detected as tip oscillations. Such a temperature waveform allows elastic properties of the material under the tip at different temperatures to be probed. Measuring the temperature dependence of the elastic properties of materials provides information about the temperature of phase transitions (glass transition, melting) in the materials under study. This approach obviates the need for a piezoelectric actuator and allows for significantly higher sensitivity and spatial resolution. Recently, this analysis was extended to sub-zeptoliter volumes, as demonstrated by Nikiforov *et al.* [410] and Proksch,[2] and performed with 50 nm point-to-point spatial resolution (Figure 16.30) [409].

16.7.2
Microwave Microscopy

The terms "*near-field microwave microscopy*," "*evanescent microwave microscopy*," and "*near-field microwave impedance microscopy*" refer to an ensemble of SPM techniques in which electromagnetic waves of frequencies in the microwave frequency range, that is, from about 0.1 to 100 GHz, are delivered to a sample and brought into interaction with it through an aperture, constriction, or a field-concentrating feature such as a sharp tip with a size much smaller than the wavelength of the radiation both in air and in the sample (at a frequency f = 1 GHz, the wavelength of electromagnetic radiation in vacuum is $\lambda = 30$ cm). An aperture can be made in a wall of a resonant cavity or a waveguide, or it can be formed by the open end of a tapered transmission line. The spatial resolution is determined by the size of the aperture (rather than by the radiation wavelength), which allows one to overcome the diffraction limit of resolution in classical geometrical optics. An aperture or a field-concentrating feature should be scanned over the sample to obtain an image. In the case of aperture probes, the probe–sample interaction takes place through

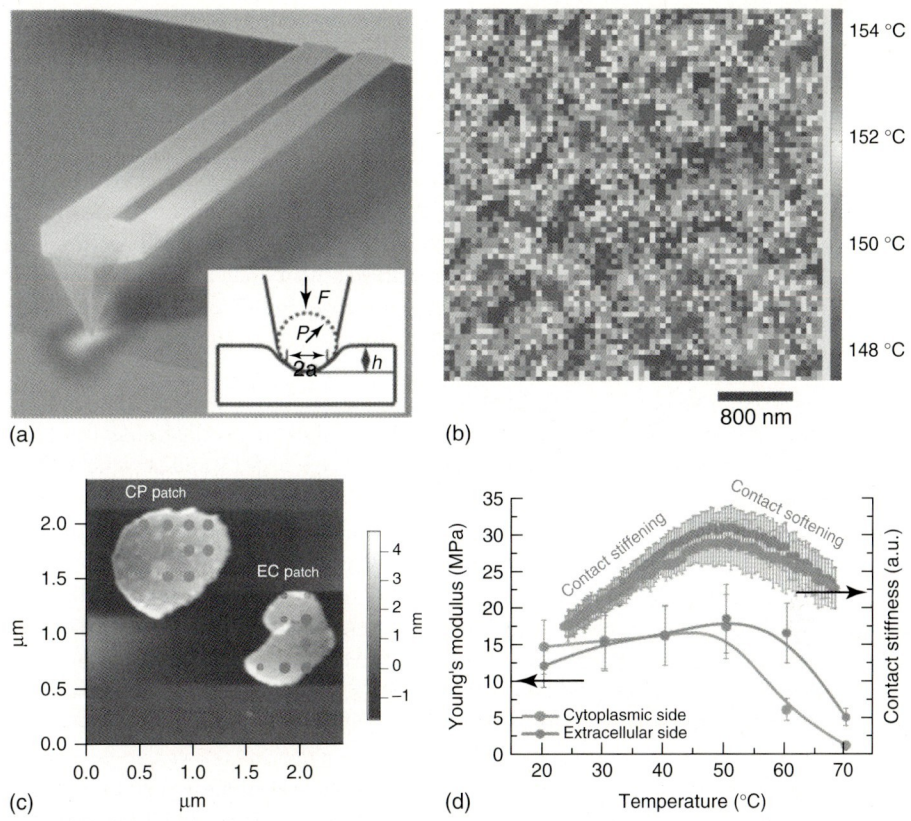

Figure 16.30 SPM with a heatable probe. (a) Schematics of a heatable AFM probe in contact with a sample. (b) Glass transition temperature map of a phase-separated poly(styrene-co-acrylonitrile)/poly(methyl methacrylate) polymer blend. (c) Localized heating curves made on the cytoplasmic (CP) and extracellular (EC) sides of bacteriorhodopsin membranes on mica and (d) localized (contact stiffness, right axis) and conventional (Young's modulus, left axis) stiffness measurements showing a differentiation in the thermal properties. ((a,b) Adapted from [409], copyright 2010, American Chemical Society. (c,d) Reproduced with permission from [410], copyright 2011, Institute of Physics.)

fields, which can be mathematically decomposed into a series of evanescent waves exponentially decaying away from the aperture at the length scale of the aperture size, $\sim R_0$. To overcome the sensitivity limitations of aperture-based schemes, which necessarily operate below cutoff frequencies, Fee *et al.* [413] proposed that a sharp tip terminating the central conductor of a coaxial transmission line or a signal contactor of a microstrip line can be used to increase the spatial resolution. While evanescent waves form around sharp tips as well, it is more convenient to look at such fields in a different way. Electrodynamically, the fringing (near) electric and magnetic fields and electric currents at distances $r < D \ll \lambda$, where D

is the characteristic size of the probe–tip structure, can be treated as quasi-static, determined by the charge and electric current distributions in the proximate probe conductors at a given moment in time, neglecting field retardation and derivatives of the fields over time. The fringing fields store reactive energy, and therefore near-field probes behave like capacitors or inductors depending on the predominant component of the electromagnetic field determined by the probe configuration. The value of the probe capacitance or inductance is a function of the electric properties of a sample in the near vicinity of the probe, with higher sensitivity obtained from regions with strongest fields. Therefore, measurement of the probe impedance using an appropriate model for the probe–sample system opens up a path to quantitative characterization of local electrodynamic properties [414–418].

During imaging, the impedance of the probe–sample system can be monitored by measuring the amplitude and phase of waves reflected directly from it through a transmission line as in the original proposal by Fee et al. [413] and in some modern systems (with use of impedance-matching techniques [419, 420]). To increase the system sensitivity and maximize the signal-to-noise ratio, a probe can be made as a load or part of a resonator. Systems with coaxial resonators [414], transmission line resonators of different types [416, 421, 422], and dielectric resonators [423, 424] have been demonstrated. These systems take advantage of the strong frequency dependence of the resonator response in reflection or transmission, and either work at a fixed frequency close to the frequency of the largest slope of the response curve or follow and monitor the changes of the resonant frequency and Q-factor of the resonator.

Since sharp tip probes currently provide the highest resolution and sensitivity, they are used in most of the modern scanning microwave microscopy (SMM) systems (Figure 16.31a). Probes of this type behave like electric probes with capacitive properties, where the reactive energy is stored overwhelmingly in the electric field. Magnetic probes can be realized by utilizing a short circuited loop probe [421, 425]. Electric probes are sensitive to the sample dielectric function. Dielectric constant measurements are most advanced among other applications of SMM for quantitative characterization of materials [426]. Microscopes with electric probes can be used for dopant profiling in semiconductors through the measurement of local conductivity [427] or in the manner typical for conventional SCM – through measurements of differential capacitance dC/dV_{bias}, since capacitance of the probe–sample system is essentially measured during imaging with electric probes. Simultaneous measurements of dielectric constant and conductivity (sheet resistance) can be accomplished by measuring the probe–sample response in a broad frequency range to properly take into account sample-dependent coupling between the probe and sample [428].

Reliable imaging with high spatial resolution requires bringing sharp probe tips in contact with, or in very close proximity to the sample surface. Therefore, the tip–sample distance control is a key component of any SMM system. STM-based distance control systems, shear force-based systems, and cantilever-based probes incorporated in an AFM have been used. As was recently demonstrated, AFM-based

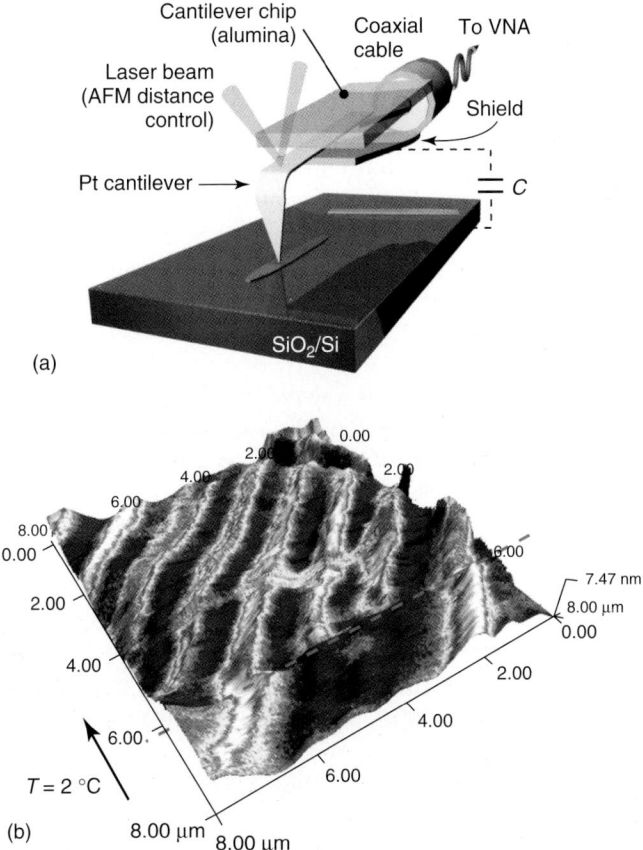

Figure 16.31 Scanning microwave microscopy. (a) Schematic of the SMM probe tip in contact with a sample. The Pt cantilever tip is connected to the central conductor of a coaxial cable. VNA stands for vector network analyzer. C is the capacitance between the conducting substrate and the probe shield. (b) SMM image of mesoscopically conductive ferroelastic domain walls in VO_2. (Reproduced with permission from [429], copyright 2010, American Chemical Society.)

SMM provides great flexibility and high spatial resolution, remaining, however, primarily as a qualitative imaging characterization technique [429–434]. Comprehensive reviews of near-field microwave microscopy have been provided by Rosner and van der Weide [435], Anlage et al. [436], and Paulson and van der Weide [437].

Note that microscopes based on lumped element resonant circuits operating at microwave frequencies can be included in the class of near-field microscopes as well. Thus, conventional SCM and scanning nonlinear dielectric microscopy (developed by Cho et al. [438]), can be considered highly specialized versions of SMM, where the measurement techniques are shaped by specific properties of interest inherent to the sample under investigation. An example SMM image of mesoscopically conductive ferroelastic domain walls in VO_2 is shown in Figure 16.31b.

16.7.3
Mass Spectrometric Imaging in SPM

One of the key limitations of classical SPM techniques is the lack of chemical information. Mass spectrometric techniques are among the most powerful chemical analysis techniques; however, they often rely on low pressures or liquid media and are poorly compatible with SPM. Recently, significant advances have been achieved in atmospheric-pressure mass spectrometry [439–443], which provides an approach to combine mass spectrometry with SPM to yield a family of new methods for studying systems such as cell membranes, energy-related materials, material junctions, and so on, which require both nanoscale spatial resolution as well as high chemical specificity.

To date, several groups have successfully coupled mass spectrometry with near-field tip-enhanced laser desorption [444–448]. These techniques rely on using tip-enhanced laser ablation to create submicron-sized (∼200–50 nm) ablation craters [446, 447] in a surface and subsequently sampling the ablated material into a mass spectrometer for ionization and detection. The ionization of the ablated material has been accomplished in these cases using electron ionization [444], inductively coupled plasma [445, 446], or direct laser desorption ionization (LDI) [447]. These experiments demonstrated the potential of acquiring mass spectrometric data from nanometer-sized craters in vacuum and at atmospheric pressure as well as obtaining structural surface information through the use of SPM. However, owing to the difficult nature of tip-enhanced laser desorption experiments and the ability to effectively extract and detect material from submicron craters produced through the tip-enhanced laser desorption at atmospheric pressure, more recent attempts have been made to combine AFM and mass spectrometry for multimodal chemical and topographic imaging with a 1 μm spatial resolution of a surface using direct LDI on an AFM platform [449].

In addition to the laser-based techniques, another approach to high-resolution surface sampling at atmospheric pressure, which combines the chemical specificity of mass spectrometry with the physical characterization allowed by AFM has been demonstrated by the use of local thermal desorption from Wollaston wire AFM tips [450]. Reading *et al.* [451–453] have used Wollaston wire AFM tips with a tip radius of 5 μm to carry out local thermal desorption on a surface and then to analyze the desorbed material from polymers and plant tissue, as well as pharmaceuticals, using gas-chromatography mass spectrometry. Using this setup, they were able to desorb material from craters that were conical in shape and approximately 6 μm wide and 1.7 mm deep. These works demonstrate the potential for a nanometer spatially resolved thermal desorption-based mass spectrometry surface analysis technique.

These methods illustrate the promise of combined SPM/mass spectrometry imaging and the future will undoubtedly see the emergence of commercially available tools for simultaneous mass-based chemical imaging and topographical imaging at atmospheric pressure.

16.8
Manipulation of Matter by SPM

The unique aspect of SPM techniques compared to optical and electron microscopies is the strong interaction between the probe and the surface. This not only limits SPM imaging to surface and near-surface regions but also enables manipulation of matter down to atomic and molecular levels. This can provide insights into the fundamental mechanisms of molecular and mesoscopic interactions, development of unique atomic and mesoscopic devices, as well as large-scale microfabrication and data storage.

Historically, one of the factors that attracted very broad attention to SPM methods in the early 1990s was the demonstration of atomic manipulation by STM achieved by Eigler and Schweizer at IBM Almaden [188]. By lowering the tip onto a nickel surface dosed with xenon atoms, picking up a xenon atom, dragging it to a chosen location, and releasing it, they were able to position individual xenon atoms on a nickel surface at low temperatures to spell out IBM, and later a variety of other atomic arrangements including the Japanese kanji characters for atom (Figure 16.32a). This demonstration has been heralded as a remarkable achievement in the advancement of nanotechnology and nanoscience [454], and beyond the obvious impact of demonstrating the smallest artificially fabricated structure, this opened the pathway for the fabrication of new atomic and device structures. In a series of seminal papers, Eigler and colleagues [189] went on to demonstrate the construction and confinement of an electron in a

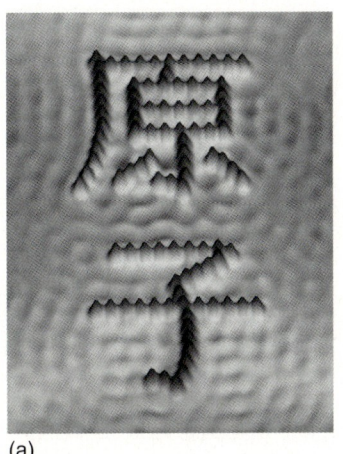
(a)

(b)

Figure 16.32 Manipulation of matter by current and force. (a) Japanese kanji characters for "atom" written with iron atoms on copper (image originally created by IBM Corporation). (Reproduced from [463].) (b) "Scratch" lithography on a polycarbonate surface 5 μm. (Courtesy of Asylum Research.)

quantum corral, demonstrated STM imaging of standing waves [455], atomic switches [456], and molecular cascades for atom-based computing [457]. Additional structures have been demonstrated by Eigler *et al.* [458], Manoharan *et al.* [459], and Moon *et al.* [460], and the manipulation of individual atoms has been extended to near-contact AFM [461, 462] and contact mode AFM via tip lithography (Figure 16.32b).

In NC-AFM measurements, the forces acting on the atom during manipulation can be measured directly, allowing the potential energy landscape of the surface to be reconstructed [464]. An equally broad range of manipulation is possible using AFM on the mesoscopic scale, with applications including scratching, reordering, dissecting, wearing, and other vertical- and lateral-force-induced modifications of the surface [465].

Some of the notable applications of this approach include haptic control, allowing direct translation of the mechanical motion of the human hand to the nanoscale, and applications for data storage using multiple probe arrays [466]. Further manipulation functionality can be achieved using AFM tweezers [467] and dip-pen lithography methods [468, 469].

Biasing SPM probes offers additional opportunities for surface modification and lithography through electrochemical reactions and polarization switching. The use of an SPM probe often allows confinement of the reaction zone to the 3–30 nm scale of the tip–surface junction. The early developments in this area include electrochemical writing in oxides [470–472]. Applications of this technology have often been spectacular, including recent resistive writing of nanocircuits, manipulation of 2D electron gas in oxide heterostructures, and many others, providing in certain cases direct competition to e-beam writing methods [473–478]. Similarly, local studies of ferroelectric switching have attracted much effort because of applications in ferroelectric data storage and lithography [479–481]. Recent results from the Cho group [482] demonstrate that the process can be confined to 2 nm domains (i.e., an area <25 unit cells), following their earlier demonstration of 8 nm domain arrays [483].

16.9
Perspectives

In formulating the possible pathways in which the SPM field will develop, it is useful to distinguish (i) the short-term improvements, the need for which is obvious and which are within current engineering capabilities (albeit the financial aspect of pursuing those is not immediately clear), (ii) the medium-range goals, which represent significant experimental and engineering challenges, but which are (a) either necessitated by practical needs or (b) enabled by engineering developments, and (iii) serendipitous directions that are either unpredictable, or are believed to be impossible in the context of existing technologies. A recent analysis of the direction of the field is given by Morita [484].

16.9.1
Platforms, Probes, and Detectors

The short-term goals for the field of SPM include increasingly stable SPM platforms, reduced noise detectors, and functional SPM probes, including, for example, shielded and insulated probes, and probes with multiple-electrode geometries. Novel probe concepts and prototypes appear regularly in academic and national laboratory environments; however, scaling to mass production is always a general problem. Similarly, detection systems beyond the cantilever platform (longitudinal and normal force components are coupled) which allow 3D vector force measurements have been reported [67, 485]. Note that the importance of mass production should not be underestimated – for example, the explosive growth of AFM applications since the 1990s is related to the introduction of etched Si probes that can be reproducibly fabricated to specification in large volumes (Figure 16.33).

The development of fast, flexible controllers and significant progress in computer technologies opens the pathway for imaging and spectroscopy of objects with highly

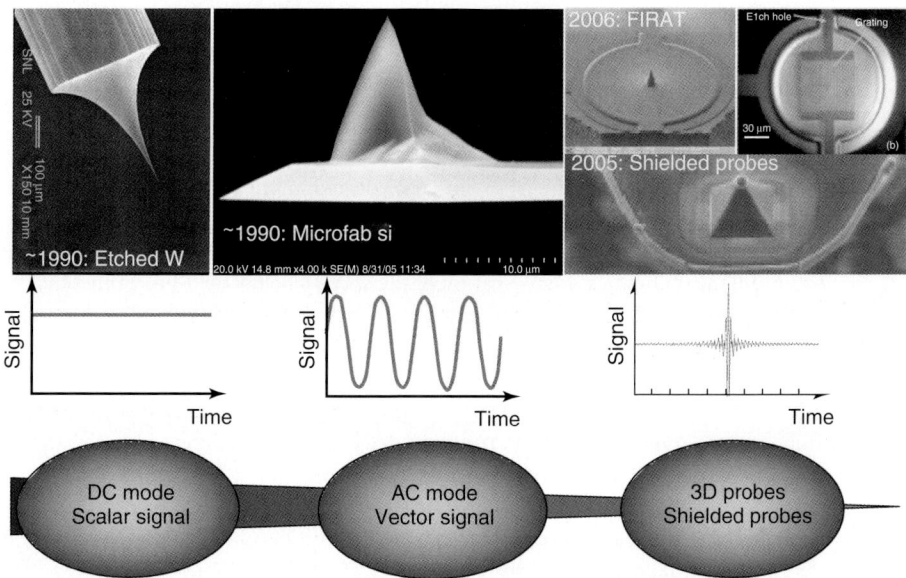

Figure 16.33 Roadmap for the development of SPM probes and imaging modes. Shown is the probe evolution from simple etched STM tip (current probe) to AFM cantilever (force and current) to 3D probes with a force-sensing integrated readout and active tip (FIRAT) and shielded probes. In parallel, data acquisition methods evolved from static detection (STM and contact AFM) to constant frequency (Tapping mode AFM, AFAM, etc.) and frequency tracking (NC-AFM) to more complex waveforms. STM tip. (Reproduced with permission from [87], copyright 2008, Institute of Physics. FIRAT probe reproduced with permission from [67], copyright 2006, American Institute of Physics. Shielded probe reproduced with permission from [140], copyright 2005, Institute of Physics.)

nonuniform information density. Examples include objects with high aspect ratios such as carbon nanotubes, nanowires, sparse nanoparticles, domain walls, and grain boundaries. In these cases, an insignificant number of pixels are used to represent the object, limiting the information on the system of interest. For example, the determination of fractal (and especially multifractal properties) of grain boundaries and domain walls requires mapping the scaling over multiple orders of magnitude, clearly impossible in $\sim 10^2$–10^3 pixel images. Some of the approaches to alleviate this may include scanning over nonrectangular [486], or intelligent grids, potentially adapted to have a higher concentration of sampling points across objects of interest, or where a measured signal or functionality changes rapidly.

Finally, the development of multiple-probe SPM systems has truly spectacular potential for probing nanoscale phenomena. Independently controlled SPM probes are currently being developed for STM applications, including classical four-probe conductivity measurements of surfaces and nanostructures, SGMs, or piezoresistive and piezogalvanic measurements in nanowires and nanotubes. The adoption of force-based multiple-probe SPMs will provide multiple opportunities for studying long-range interactions, including many body effects, local potentials and fields, and so on.

16.9.2
Dynamic Detection

The obvious medium-range goals for SPM development necessitated by applications are high-frequency probing beyond the ~ 10–100 kHz limit imposed by the amplifier bandwidth-gain product for STM and ~ 1–10 MHz for AFM. The second requirement is chemical identification of the species below the SPM probe, that is, chemically specific imaging. The possible pathways for these include combining SPM with microwave and optical measurements, in which the SPM tip acts as a local passive antenna (e.g., amplifies signal in tip-enhanced Raman scattering (TERS) mode, or breaks the evanescent wave condition in total internal reflection fluorescence microscopy), or is used for detection. In the latter case, the use of amplitude-modulated or pumped probe measurements allows linking the time scales of optics or microwaves to the time scale of the SPM detection system. These methods can bring into reality the capability to probe ultrafast phenomena and chemical identification below the tip.

Complementary to the aforementioned are developments enabled by technological advances. A remarkable example is the development of fast data analysis algorithms made available by the introduction of commercial fast data acquisition cards. In SPM, the typical frequency range of operation for the cantilever is ~ 0.1–10 MHz (and is expected to grow once ultrasmall levers for video-rate imaging are introduced), while pixel time and feedback times are on the order of ~ 1 kHz. The typical data conversion process from the real-time trace of the cantilever oscillations to the measured signal includes either LIA processing to extract amplitude and phase or PLL processing to extract resonance frequency and amplitude. In this process, information about singular or aperiodic events is essentially lost. Similarly, the capability to respond to and control events in the tip–surface junction is limited – that is, the response occurs on the time scale of a single oscillation, and the feedback becomes active only after ∼1000 oscillations. The development of fast electronics now provides control of a single oscillation of the cantilever, as well as the ability to use much more complex signal processing schemes. While the full potential of such methods is impossible to predict, some of the examples will likely include adaptive control of molecular transformations, probing inaccessible regions of the free energy surface, controlling conformation pathways, and so on, analogous to of recent studies of spin control by Mamin and Rugar [487].

Last but not least, serendipitous discoveries are impossible to predict, and often bridge an impasse between necessity and invention. Some of the many examples include the capability to modulate chemical functionality of the probe in time (note that light-induced processes are slow, since cross sections are small unless surface enhanced Raman spectroscopy is involved, and electrical modulation also affects electrostatic forces), or using active feedback to lower the "temperature" of a probe which is weakly thermally coupled to the environment.

16.9.3
Multidimensional Data Analysis and Interpretation

Recent years have seen tremendous growth in spectroscopic imaging techniques in SPM with examples including continuous imaging tunneling spectroscopy (CITS), dI/dV (density of states), dI/dz (work function), and d^2I/dV^2 (vibrational) imaging in STM [28], and force–volume [488], BE [166], fast frequency sweeps [164], and SS-PFM [371] in force-based SPMs. In these spectroscopic imaging techniques, the signal is measured as a function of position over a closely spaced grid of points (x, y), and an external parameter, η, yielding a 3D data array, $R(x, y, \eta)$, as summarized in Table 16.1. Beyond these 3D modes, a number of approaches for 4D and 5D measurements have been demonstrated, including first-order reversal curve measurements [169].

The broad application of spectroscopic imaging studies requires the development of data analysis methods to link experimental observations and physical models,

Table 16.1 Spectroscopic imaging methods in scanning probe microscopy.

Method	Response, R	Parameter, η	Dense
Force volume	Deflection	Tip–surface distance	Y
Nanoindentation	Force	Indentation	N
CITS	Current	Voltage	Y
BE	Amplitude	Frequency	Y
SS-PFM	Piezoresponse	Voltage	Y

and hence extract relevant physical behaviors. Traditionally, 3D data arrays have been interpreted as a collection of 1D spectra analyzed independently using an appropriate analytical model. Model parameters are then plotted individually to yield 2D spatially resolved maps. As an example, in force–volume imaging, each force–distance curve is analyzed to extract parameters describing long-range interactions, indentation moduli, and adhesion [489, 490]. While this analysis enables the extraction of relevant information from a 3D data set, it requires a priori knowledge of a physical model describing the local interactions. The use of improper models can lead to unintended intermixing of parameters and result in spurious contrast in the 2D maps. Second, in many cases, SPM measurements can be taken with pixel spacing well below the signal generation volume (*dense regime*). In such cases, spectra from adjacent locations are not independent and the information of local behavior is contained not only in the spectrum acquired from a selected location, but from adjacent locations as well. The model parameter extraction from 1D spectra ignores the lateral correlations between adjacent locations. Finally, the noise level in spectroscopic data per 3D voxel (i.e., volumetric pixel) is in most cases higher than that per 2D pixel because of the relatively smaller acquisition times, and may result in amplified noise in the deconvoluted data.

An alternative is provided by multidimensional statistical projection and clustering methods. A basic example of the former is principal component analysis (PCA) [491]. This analytical approach is now broadly applied in electron microscopy [492–494]. The alternative to projection methods are clustering techniques, for example, based on self-organized feature maps. Correlative models based on artificial neural networks [495, 496] may offer a solution.

16.9.4
Combined Methods

One of the primary limitations of SPM methods is the limited structural and chemical information they provide. While forces and currents can often be measured on the level of a single hydrogen bond or atom and bias- and strain-induced transformation can be measured on the level of a single defect, identification of atomic and molecular species or defect type presents a significant challenge. Correspondingly, a tremendous potential for future studies is offered by combining local electrical and mechanical excitation by the SPM probe with high-resolution structural and electronic structure imaging with high-resolution electron microscopy and focused X-ray microscopy. While these techniques impose a number of restrictions on sample geometry (e.g., the need for a thin transmission electron microscopy (TEM) sample) or still have not achieved sufficient sensitivity, the applications are likely to grow in the near future.

Similarly, a broad range of advances can be expected for the synergistic implementation of various SPM and other analytic modalities. Recent examples include the combination of PFM and ESM, which yield new insights into nanoscale electrochemical processes with microwave and cAFM methods, (offering powerful capabilities for probing local metal-insulator transitions) and the combination of heated probes with high-resolution contact resonance and multidimensional data acquisition schemes that allow the thermodynamic processes of individual molecules to be probed. It seems likely that this synergistic trend will continue and in fact, increase. Overall, SPM has been and will continue to be a field in which both detailed, quantitative studies and serendipitous discoveries await, illustrating Dyson's premise.

Acknowledgments

This research was conducted at the Center for Nanophase Materials Sciences, which is sponsored at Oak Ridge National Laboratory by the Office of Basic Energy Sciences, U.S. Department of Energy. The authors gratefully acknowledge Olga Ovchinnikova (UT) for contributing the section on mass spectral methods, Alexander Tselev (ORNL) for the discussion on microwave imaging, and Maxim Nikiforov (Argonne) for the discussion on thermal analysis. BJR would also like to acknowledge support from Science Foundation Ireland (grant no. 10/RFP/MTR2855) and UCD Research.

References

1. Meyer, E., Hug, H.J., and Bennewitz, R. (2004) *Scanning Probe Microscopy: The Lab on a Tip*, Springer Berlin.
2. Bonnell, D.A. (ed.) (2000) *Scanning Probe Microscopy and Spectroscopy: Theory, Techniques, and Applications*, John Wiley & Sons, Inc., New York.
3. Kalinin, S.V. and Gruverman, A. (eds) (2007) *Scanning Probe Microscopy: Electrical and Electromechanical Phenomena on the Nanoscale*, vol. I, II, Springer Science + Business Media, LLC, New York.
4. Gross, L. (2011) *Nat. Chem.*, **3**, 278.
5. Pohl, D.W., Denk, W., and Lanz, M. (1984) *Appl. Phys. Lett.*, **44**, 651.
6. Lewis, A., Isaacson, M., Harootunian, A., and Muray, M. (1984) *Ultramicroscopy*, **13**, 227.
7. Novotny, L. and Hecht, B. (2006) *Principles of Nano-Optics*, Cambridge University Press, Cambridge.
8. Butt, H.J., Cappella, B., and Kappl, M. (2005) *Surf. Sci. Rep.*, **59**, 1.
9. Hofer, W.A., Foster, A.S., and Shluger, A.L. (2003) *Rev. Mod. Phys.*, **75**, 1287.
10. Giessibl, F.J. (2003) *Rev. Mod. Phys.*, **75**, 949.
11. Tsukruk, V. (ed.) (2001) *Advances in Scanning Probe Microscopy*, Macromolecular Symposia, Vol. 167, John Wiley & Sons, Inc., New York.
12. Rademacher, M., Tillman, R.W., Fritz, M., and Gaub, H.E. (1992) *Science*, **257**, 1900–1905.
13. Hansma, H.G. and Hoh, J.H. (1994) *Annu. Rev. Biophys. Biomol. Struct.*, **23**, 115–139.
14. Müller, D.J. and Dufrêne, Y.F. (2008) *Nat. Nanotechnol.*, **3**, 261–269.
15. Alexe, M. and Gruverman, A. (eds) (2004) *Ferroelectrics at Nanoscale: Scanning Probe Microscopy Approach*, Springer-Verlag, Berlin.
16. Eng, L.M., Grafstrom, S., Loppacher, Ch., Schlaphof, F., Trogisch, S., Roelofs, A., and Waser, R. (2001) *Adv. Solid State Phys.*, **41**, 287.
17. Foster, A. and Hofer, W. (2006) *Scanning Probe Microscopy: Atomic Scale Engineering by Forces and Currents*, Springer, Berlin.

18. Drake, B., Prater, C.B., Weisenhorn, A.L., Gould, S.A.C., Albrecht, T.R., Quate, C.F., Cannell, D.S., Hansma, H.G., and Hansma, P.K. (1989) *Science*, **243**, 1586–1589.
19. Dyson, F. (2005) The Future of Evolution, http://www.metanexus.net/magazine/essay/future-evolution/. (accessed 6 January 2011).
20. Rief, M., Oesterhelt, F., Heymann, B., and Gaub, H.E. (1997) *Science*, **275**, 1295.
21. Rief, M., Gautel, M., Oesterhelt, F., Fernandez, J.M., and Gaub, H.E. (1997) *Science*, **276**, 1109.
22. Noy, A. (ed.) (2010) *Handbook of Molecular Force Spectroscopy*, Springer.
23. Jarzynski, C. (1997) *Phys. Rev. Lett.*, **78**, 2690.
24. Crooks, G.E. (1999) *Phys. Rev. E*, **60**, 2721.
25. Tinoco, I. (2004) *Annu. Rev. Biophys Biomol. Struct.*, **33**, 363.
26. Bustamante, C., Chemla, Y.R., Forde, N.R., and Izhaky, D. (2004) *Annu. Rev. Biochem.*, **73**, 705.
27. Ritort, F. (2006) *J. Phys. C*, **18**, R531.
28. Stroscio, J.A. and Kaiser, W.J. (eds) (1993) *Scanning Tunneling Microscopy*, Academic Press, Boston.
29. Renner, Ch., Aeppli, G., Kim, B.G., Soh, Y.A., and Cheong, S.-W. (2002) *Nature*, **416**, 518.
30. Matzdorf, R., Fang, Z., Ismail, Zhang, J., Kimura, T., Tokura, Y., Terakura, K., and Plummer, E.W. (2000) *Science*, **289**, 746.
31. McElroy, K., Simmonds, R.W., Hoffman, J.E., Lee, D.H., Orenstein, J., Eisaki, H., Uchida, S., and Davis, J.C. (2003) *Nature*, **422**, 592.
32. Gomes, K.K., Pasupathy, A.N., Pushp, A., Ono, S., Ando, Y., and Yazdani, A. (2007) *Nature*, **447**, 569.
33. McElroy, K., Lee, J., Slezak, J.A., Lee, D.H., Eisaki, H., Uchida, S., and Davis, J.C. (2005) *Science*, **309**, 1048.
34. Vershinin, M., Misra, S., Ono, S., Abe, Y., Ando, Y., and Yazdani, A. (2004) *Science*, **303**, 1995.
35. Gawronski, H., Mehlhorn, M., and Morgenstern, K. (2008) *Science*, **319**, 930.
36. Wu, S.W., Ogawa, N., and Ho, W. (2006) *Science*, **312**, 1362.
37. Gruverman, A. and Kholkin, A. (2006) *Rep. Prog. Phys.*, **69**, 2443.
38. Kalinin, S.V., Morozovska, A.N., Chen, L.Q., and Rodriguez, B.J. (2010) *Rep. Prog. Phys.*, **73**, 056502.
39. Balke, N., Jesse, S., Morozovska, A.N., Eliseev, E., Chung, D.W., Kim, Y., Adamczyk, L., Garca, R.E., Dudney, N., and Kalinin, S.V. (2010) *Nat. Nanotechnol.*, **5**, 749.
40. Balke, N., Jesse, S., Kim, Y., Adamczyk, L., Tselev, A., Ivanov, I.N., Dudney, N.J., and Kalinin, S.V. (2010) *Nano Lett.*, **10**, 3420.
41. Binnig, G. and Rohrer, H. (1982) *Helv. Phys. Acta*, **55**, 726–735.
42. Binnig, G., Quate, C.F., and Gerber, Ch. (1986) *Phys. Rev. Lett.*, **56**, 930.
43. Binning, G., Rohrer, H., Gerber, Ch., and Weibel, E. (1982) *Phys. Rev. Lett.*, **49**, 57.
44. Martin, Y. and Wickramasinghe, H.K. (1987) *Appl. Phys. Lett.*, **50**, 1455.
45. Nonnemmacher, M., O'Boyle, M.P., and Wickramasinghe, H.K. (1991) *Appl. Phys. Lett.*, **58**, 2921.
46. Noy, A., Frisbie, C.D., Rozsnyai, L.F., Wrighton, M.S., and Lieber, C.M. (1995) *J. Am. Chem. Soc.*, **117**, 7943.
47. Thomas, T.R. (1999) *Rough Surfaces*, 2nd edn, Imperial College Press, London.
48. Schnell, A. and Oepen, H. US Patent 4, 359, 892, (Priority Date December 1979, Issued November 1982).
49. Meyer, G. and Amer, N.M. (1988) *Appl. Phys. Lett.*, **53**, 1045–1047.
50. Amer, N.M. and Meyer, G. (1988) *Bull. Am. Phys. Soc.*, **33**, 319.
51. Schmaltz, G. (1929) *Z. Ver. Dtsch. Ing.*, **73**, 144.
52. Becker, H., Bender, O., Bergmann, L., Rost, K.H., Zobel, A. Apparatus for measuring surface irregularities. US Patent 2, 728, 222 (Priority Date Oct. 1950, Issued Dec. 1955).
53. Lee, D. and Harrison, R.G. (1979) UK Published Patent Application GB 2009409 A.
54. Kaczér, J. (1955) *Czech. J. Phys.*, **5**, 239–244.

55. Kaczér, J. and Gemperle, R. (1956) *Czech. J. Phys.*, **6**, 173–184.
56. Young, R.D. (1966) *Rev. Sci. Instrum.*, **37**, 275.
57. Young, R., Ward, J., and Scire, F. (1972) *Rev. Sci. Instrum.*, **43**, 999.
58. Eigler, D.M., Weiss, P.S., Schweizer, E.K., and Lang, N.D. (1991) *Phys. Rev. Lett.*, **66**, 1189–1192.
59. Meyer, G. (1996) *Rev. Sci. Instrum.*, **67**, 2960.
60. Pan, S.H., Hudson, E.W., and Davis, J.C. (1999) *Rev. Sci. Instrum.*, **70**, 1459.
61. Song, Y.J., Otte, A.F., Shvarts, V., Zhao, Z., Kuk, Y., Blankenship, S.R., Band, A., Hess, F.M., and Stroscio, J.A. (2010) *Rev. Sci. Instrum.*, **81**, 121101.
62. Binnig, G. and Smith, D.P.E. (1986) *Rev. Sci. Instrum.*, **57**, 1688.
63. Hansma, P.K., Drake, B., Marti, O., Gould, S.A.C., and Prater, C.B. (1989) *Science*, **243**, 641–643.
64. De Wolf, P., Snauwaert, J., Hellemans, L., Clarysse, T., Vandervorst, W., D'Olieslaeger, M., and Quaeyhaegens, D. (1995) *J. Vac. Sci. Technol. A*, **13**, 1699.
65. Sarid, D. (1991) *Scanning Force Microscopy*, Oxford University Press, New York.
66. Smith, S.T. and Chetwynd, D.G. (1992) *Ultraprecision Mechanism Design*, Gordon & Breach, Amsterdam.
67. Onaran, A.G., Balantekin, M., Lee, W., Hughes, W.L., Buchine, B.A., Guldiken, R.O., Parlak, Z., Quate, C.F., and Degertekin, F.L. (2006) *Rev. Sci. Instrum.*, **77**, 023501.
68. Zijlstra, T., Heimberg, J.A., van der Drift, E., Glastra von Loon, D., Dienwiebel, M., de Groot, L.E.M., and Frenken, J.W.M. (2000) *Sens. Actuators*, **84**, 18–24.
69. Dienwiebel, M., de Kuyper, E., Cram, L., Frenken, J.W.M., Heimberg, J.A., Spaanderman, D.-J., Glastra van Loon, D., Zijlstra, T., and van der Drift, E. (2005) *Rev. Sci. Instrum.*, **76**, 043704.
70. Nakagawa, K., Hashiguchi, G., and Kawakatsu, H. (2009) *Rev. Sci. Inst.*, **80**, 095104.
71. Zeyen, B., Virwani, K., Pittenger, B., and Turner, K.L. (2009) *Appl. Phys. Lett.*, **94**, 103507.
72. Felts, J.R. and King, W.P. (2009) *J. Micromech. Microeng.*, **19**, 115008.
73. Viani, M.B., Schäffer, T.E., Chand, A., Rief, M., Gaub, H.E., and Hansma, P.K. (1999) *J. Appl. Phys.*, **86**, 2258–2262.
74. Solares, S.D. and Chawla, G. (2008) *Meas. Sci. Technol.*, **19**, 055502.
75. Mate, C.M., McClelland, G.M., Erlandsson, R., and Chiang, S. (1987) *Phys. Rev. Lett.*, **59**, 1942.
76. Butt, H.-J. and Jaschke, M. (1995) *Nanotechnology*, **6**, 1–7.
77. Neumeister, J.M. and Ducker, W.A. (1994) *Rev. Sci. Instrum.*, **65**, 2527–2531.
78. Rayleigh, L. (1896) *Phil. Mag.*, **S 5**, 167.
79. Fertig, J. and Rose, H. (1979) *Optik*, **54**, 165.
80. O'Keefe, M.A. (1992) *Ultramicroscopy*, **47**, 282.
81. Sherzer, O. (1949) *J. Appl. Phys.*, **20**, 20.
82. Kalinin, S.V., Jesse, S., Rodriguez, B.J., Shin, J., Baddorf, A.P., Lee, H.N., Borisevich, A., and Pennycook, S.J. (2006) *Nanotechnology*, **17**, 3400.
83. Nellist, P.D., Chisholm, M.F., Dellby, N., Krivanek, O.L., Murfitt, M.F., Szilagyi, Z.S., Lupini, A.R., Borisevich, A., Sides, W.H., and Pennycook, S.J. (2004) *Science*, **305**, 1741.
84. Peng, Y., Oxley, M.P., Lupini, A.R., Chisholm, M.F., and Pennycook, S.J. (2008) *Microsc. Microanal.*, **14**, 36–47.
85. Fukuma, T., Kimura, M., Kobayashi, K., Matsushige, K., and Yamada, H. (2005) *Rev. Sci. Instrum.*, **76**, 053704.
86. Fukuma, T., Kobayashi, K., Matsushige, K., and Yamada, H. (2005) *Appl. Phys. Lett.*, **87**, 034101.
87. Ekvall, I., Wahlström, E., Claesson, D., Olin, H., and Olsson, E. (1999) *Meas. Sci. Technol.*, **10**, 11–18.
88. Sitterberg, J., Özcetin, A., Ehrhardt, C., and Bakowsky, U. (2010) *Eur. J. Pharm. Biopharm.*, **74**, 2–13.
89. Sulchek, T., Yaralioglu, G.G., Quate, C.F., and Minne, S.C. (2002) *Rev. Sci. Instrum.*, **73**, 2928–2936.
90. Schitter, G., Åström, K.J., DeMartini, B.E., Thurner, P.J., Turner, K.L., and

Hansma, P.K. (2007) *IEEE Trans. Control Syst. Technol.*, **15**, 906–915.
91. Ando, T. (2008) Control techniques in high-speed atomic force microscopy, Proceedings of the American Control Conference, Seattle, WA, June 2008, pp. 3194–3200.
92. Minne, S.C., Yaralioglu, G., Manalis, S.R., Adams, J.D., Zesch, J., Atalar, A., and Quate, C.F. (1998) *Appl. Phys. Lett.*, **72**, 2340–2342.
93. Barrett, R.C. and Quate, C.F. (1991) *J. Vac. Sci. Technol. B*, **9**, 302–306.
94. Humphris, A.D.L., Hobbs, J.K., and Miles, J.M. (2003) *Appl. Phys. Lett.*, **83**, 6–8.
95. Humphris, A.D.L., Miles, M.J., and Hobbs, J.K. (2005) *Appl. Phys. Lett.*, **86**, 034106.
96. Picco, L.M., Bozec, L., Ulcinas, A., Engledew, D.J., Antognozzi, M., Horton, M.A., and Miles, M.J. (2007) *Nanotechnology*, **18**, 044030.
97. Anwar, M. and Rousso, I. (2005) *Appl. Phys. Lett.*, **86**, 014101.
98. Walters, D.A., Cleveland, J.P., Thomson, N.H., Hansma, P.K., Wendman, M.A., Gurley, G., and Elings, V. (1996) *Rev. Sci. Instrum.*, **67**, 2580–3590.
99. Schitter, G., Allgower, F., and Stemmer, A. (2004) *Nanotechnology*, **15**, 108–114.
100. Schitter, G. and Rost, M.J. (2008) *Mater. Today*, **11**, 40–48.
101. Fantner, G.E., Hegarty, P., Kindt, J.H., Schitter, G., Cidade, G.A.G., and Hansma, P.K. (2005) *Rev. Sci. Instrum.*, **76**, 026118.
102. Rost, M.J., Crama, L., Schakel, P., van Tol, E., van Velzen-Williams, G.B.E.M., Overgauw, C.F., ter Horst, H., Dekker, H., Okhuijsen, B., Seynen, M., Vijftigschild, A., Han, P., Katan, A.J., Schoots, K., Schumm, R., van Loo, W., Oosterkamp, T.H., and Frenken, J.W.M. (2005) *Rev. Sci. Instrum.*, **76**, 053710.
103. Viani, M.B., Pietrasanta, L.I., Thompson, J.B., Chand, A., Gebeshuber, I.C., Kindt, J.H., Richter, M., Hansma, H.G., and Hansma, P.K. (2000) *Nat. Struct. Biol.*, **7**, 644–647.
104. Sakamoto, T., Amitani, I., Yokota, E., and Ando, T. (2000) *Biochem. Biophys. Res. Commun.*, **272**, 586–590.
105. Ando, T., Kodera, N., Takai, E., Maruyama, D., Saito, K., and Toda, A. (2001) *Proc. Natl. Acad. Sci. U.S.A.*, **98**, 12468–12472.
106. Ando, T., Uchihashi, T., Kodera, N., Miyagi, A., Nakakita, R., Yamashita, H., and Sakashita, M. (2006) *Jpn. J. Appl. Phys.*, **45**, 1897–1903.
107. Ando, T., Uchihashi, T., and Fukuma, T. (2008) *Prog. Surf. Sci.*, **83**, 337–437.
108. Ando, T., Uchihashi, T., Kodera, N., Yamamoto, D., Miyagi, A., Taniguchi, M., and Yamashita, H. (2008) *Pflügers Arch.: Eur. J. Physiol.*, **456**, 211–225.
109. http://www.nanofactory.com (accessed 6 January 2011).
110. Ohnishi, H. et al., Patent Number 6242737 (Filed Aug. 19, 1998, Issued June 5, 2001).
111. Erts, D., Lõhmus, A., Lõhmus, R., and Olin, H. (2001) *Appl. Phys. A*, **72**, S71–S74.
112. Kuwahara, S., Sugai, T., and Shinohara, H. (2009) *Nanotechnology*, **20**, 225702.
113. Wiesendanger, R., Güntherodt, H.-J., Güntherodt, G., Gambino, R.J., and Ruf, R. (1990) *Phys. Rev. Lett.*, **65**, 247–250.
114. Kubetzka, A., Bode, M., Pietzsch, O., and Wiesendanger, R. (2002) *Phys. Rev. Lett.*, **88**, 057201.
115. Bucher, J.-P. (2010) *Nat. Nanotechnol.*, **5**, 315–316.
116. Serrate, D., Ferriani, P., Yoshida, Y., Hla, S.-W., Menzel, M., von Bergmann, K., Heinze, S., Kubetzka, A., and Wiesendanger, R. (2010) *Nat. Nanotechnol.*, **5**, 350–353.
117. Heinze, S., Bode, M., Kubetzka, A., Pietzsch, O., Nie, X., Blügel, S., and Wiesendanger, R. (2000) *Science*, **288**, 1805–1808.
118. Pietzsch, O., Kubetzka, A., Bode, M., and Wiesendanger, R. (2001) *Science*, **292**, 2053–2056.
119. Temirov, R., Soubatch, S., Neucheva, O., Lassise, A., and Tautz, F. (2008) *New J. Phys.*, **10**, 053012.

120. Weiss, C., Wagner, C., Kleimann, C., Rohlfing, M., Tautz, F.S., and Temirov, R. (2010) *Phys. Rev. Lett.*, **105**, 086103.
121. Weiss, C., Wagner, C., Temirov, R., and Taut, F.S. (2010) *J. Am. Chem. Soc.*, **132**, 11864–11865.
122. Martin, Y., Abraham, D.W., and Wickramasinghe, H.K. (1988) *Appl. Phys. Lett.*, **52**, 1103.
123. Stern, J.E., Terris, B.D., Mamin, H.J., and Rugar, D. (1988) *Appl. Phys. Lett.*, **53**, 2717.
124. Terris, B.D., Stern, B.D., Rugar, D., and Mamin, H.J. (1989) *Phys. Rev. Lett.*, **63**, 2669.
125. Martin, Y. and Wickramasinghe, H.K. (1987) *J. Appl. Phys.*, **61**, 4723.
126. Hartmann, U., Göddenhenrich, T., and Heiden, C. (1991) *J. Magn. Magn. Mater.*, **101**, 263–270.
127. Jacobs, H.O., Leuchtmann, P., Homan, O.J., and Stemmer, A. (1998) *J. Appl. Phys.*, **84**, 1168.
128. Frisbie, C.D., Rozsnyai, L.F., Noy, A., Wrighton, M.S., and Lieber, C.M. (1994) *Science*, **265**, 2071–2074.
129. Noy, A., Vezenov, D.V., and Lieber, C.M. (1997) *Annu. Rev. Mat. Sci.*, **27**, 381–421.
130. Vezenov, D.V., Noy, A., Rozsnyai, L.F., and Lieber, C.M. (1997) *J. Am. Chem. Soc.*, **119**, 2006–2015.
131. Stroh, C.M., Ebner, A., Geretschläger, M., Freudenthaler, G., Kienberger, F., Kamruzzahan, A.S.M., Smith-Gill, S.J., Gruber, H.J., and Hinterdorfer, P. (2004) *Biophys. J.*, **87**, 1981–1990.
132. Stroh, C., Wang, H., Bash, R., Ashcroft, B., Nelson, J., Gruber, H.J., Lohr, D., Lindsay, S.M., and Hinterdorfer, P. (2004) *Proc. Natl. Acad. Sci. U.S.A.*, **101**, 12503–12507.
133. Ebner, A., Kienberger, F., Kada, G., Stroh, C.M., Geretschläger, M., Kamruzzahan, A.S.M., Wildling, L., Johnson, W.T., Ashcroft, B., Nelson, J., Lindsay, S.M., Gruber, H.J., and Hinterdorfer, P. (2005) *Chem. Phys. Chem.*, **6**, 897–900.
134. E.g., Nie, S., Chiu, D.T., and Zare, R.N. (1994) *Science*, **266**, 1018–1021.
135. Ko, H., Ryu, K., Park, H., Park, C., Jeon, D., Kim, Y.K., Jung, J., Min, D.-K., Kim, Y., Lee, H.N., Park, Y., Shin, H., and Hong, S. (2011) *Nano Lett.*, **11**, 1428–1433.
136. Park, H., Jung, J., Min, D.K., Kim, S., Hong, S., and Shin, H. (2004) *Appl. Phys. Lett.*, **84**, 1734.
137. Meister, A., Gabi, M., Behr, P., Studer, P., Vörös, J., Niedermann, P., Bitterli, J., Polesel-Maris, J., Liley, M., Heinzelmann, H., and Zambelli, T. (2009) *Nano Lett.*, **9**, 2501–2507.
138. Smith, T. and Stephenson, K. (2007) Electrochemical SPM: Fundamentals and Applications, in *Scanning Probe Microscopy: Electrical and Electromechanical Phenomena on the Nanoscale* (eds S.V. Kalinin and A. Gruverman), Springer Science + Business Media, LLC, New York, pp. 280–314.
139. Rosner, B.T. and van der Weide, D.W. (2000) *Rev. Sci. Instrum.*, **73**, 2505.
140. Frederix, P.L.T.M., Gullo, M.R., Akiyama, T., Tonin, A., de Rooij, N.F., Staufer, U., and Engel, A. (2005) *Nanotechnology*, **16**, 997.
141. Rodriguez, B.J., Jesse, S., Seal, K., Baddorf, A.P., Kalinin, S.V., and Rack, P. (2007) *Appl. Phys. Lett.*, **91**, 093130.
142. http://www.anasysinstruments.com/nano-TA2.pdf (accessed 6 January 2011).
143. Oral, A., Kaval, M., Dede, M., Masuda, H., Okamoto, A., Shibasaki, I., and Sandhu, A. (2002) *IEEE Trans. Magn.*, **38**, 2438.
144. Noh, J.H., Nikiforov, M., Kalinin, S.V., Vertegel, A.A., and Rack, P.D. (2010) *Nanotechnology*, **21**, 365302.
145. García, R. and Pérez, R. (2002) *Surf. Sci. Rep.*, **47**, 197–301.
146. García, R. (2010) *Amplitude Modulation Atomic Force Microscopy*, Wiley-VCH Verlag GmbH & Co. KGaA, Weinheim.
147. Gershenfeld, N. (2002) *Physics of Information Technology*, Cambridge University Press, Cambridge.
148. Cleveland, J.P., Anczykowski, B., Schmid, A.E., and Elings, V.B. (1998) *Appl. Phys. Lett.*, **72**, 2613.
149. Tamayo, J. and Garcia, R. (1998) *Appl. Phys. Lett.*, **73**, 2926.
150. Proksch, R. and Kalinin, S.V. (2010) *Nanotechnology*, **21**, 455705.

151. Gannepalli, A., Sebastian, A., Cleveland, J., and Salapaka, M. (2005) *Appl. Phys. Lett.*, **87**, 111901.
152. Proksch, R. and Dahlberg, E.D. (1993) *Rev. Sci. Instrum.*, **64**, 912.
153. Stark, M., Guckenberger, R., Stemmer, A., and Stark, R.W. (2005) *J. Appl. Phys.*, **98**, 114904.
154. Proksch, R. (2006) *Appl. Phys. Lett.*, **89**, 113121.
155. Rodriguez, B.J., Callahan, C., Kalinin, S.V., and Proksch, R. (2007) *Nanotechnology*, **18**, 475504.
156. Baumann, M. and Stark, R.W. (2010) *Ultramicroscopy*, **110**, 578–581.
157. Chawla, G. and Solares, S.D. (2009) *Meas. Sci. Technol.*, **20**, 015501.
158. Stark, M., Stark, R.W., Heckl, W.M., and Guckenberger, R. (2002) *Proc. Natl. Acad. Sci. U.S.A.*, **99**, 8473–8478.
159. Rodriguez, T.R. and Garcia, R. (2004) *Appl. Phys. Lett.*, **84**, 449.
160. Li, J.W., Cleveland, J.P., and Proksch, R. (2009) *Appl. Phys. Lett.*, **94**, 163118.
161. Lozano, J.R. and Garcia, R. (2008) *Phys. Rev. Lett.*, **100**, 076102.
162. Stark, R.W. (2009) *Appl. Phys. Lett.*, **94**, 063109.
163. Platz, D., Tholén, E.A., Pesen, D., and Haviland, D.B. (2008) *Appl. Phys. Lett.*, **92**, 153106.
164. Kos, A.B. and Hurley, D.C. (2008) *Meas. Sci. Technol.*, **19**, 015504.
165. Nath, R., Chu, Y.H., Polomoff, N.A., Ramesh, R., and Huey, B.D. (2008) *Appl. Phys. Lett.*, **93**, 072905.
166. Jesse, S., Kalinin, S.V., Proksch, R., Baddorf, A.P., and Rodriguez, B.J. (2007) *Nanotechnology*, **18**, 435503.
167. Jesse, S. and Kalinin, S.V. (2009) *Nanotechnology*, **20**, 085714.
168. Jesse, S., Mirman, B., and Kalinin, S.V. (2006) *Appl. Phys. Lett.*, **89**, 022906.
169. Jesse, S., Maksymovych, P., and Kalinin, S.V. (2008) *Appl. Phys. Lett.*, **93**, 112903.
170. Jesse, S., Nikiforov, M.P., Germinario, L.T., and Kalinin, S.V. (2008) *Appl. Phys. Lett.*, **93**, 073104.
171. Nikiforov, M.P., Jesse, S., Morozovska, A.N., Eliseev, E.A., Germinario, L.T., and Kalinin, S.V. (2009) *Nanotechnology*, **20**, 395709.
172. Tetard, L., Passian, A., and Thundat, T. (2010) *Nat. Nanotechnol.*, **5**, 105.
173. Hutter, C., Platz, D., Tholen, E.A., Hansson, T.H., and Haviland, D.B. (2010) *Phys. Rev. Lett.*, **104**, 050801.
174. Radmacher, M., Cleveland, J.R., Fritz, M., Hansma, H.G., and Hansma, P.K. (1994) *Biophys. J.*, **66**, 2154.
175. Baselt, D.R. and Baldeschwieler, J.D. (1994) *J. Appl. Phys.*, **76**, 33.
176. Koleske, D.D., Lee, G.U., Gans, B.I., Lee, K.P., DiLella, D.P., Wahl, K.J., Barger, W.R., Whitman, L.J., and Colton, R.J. (1995) *Rev. Sci. Instrum.*, **66**, 4566.
177. Cappella, B., Baschieri, R., Frediani, C., Miccoli, P., and Ascoli, C. (1997) *Nanotechnology*, **8**, 82.
178. Krotil, H.U., Stifter, T., and Marti, O. (2000) *Rev. Sci. Instrum.*, **71**, 2765.
179. Sahin, O., Magonov, S., Su, C., Quate, C.F., and Solgaard, O. (2007) *Nat. Nanotechnol.*, **2**, 507–514.
180. Dong, M.D., Husale, S., and Sahin, O. (2009) *Nat. Nanotechnol.*, **4**, 514–517.
181. Sahin, O. (2008) *Phys. Rev. B*, **77**, 115405.
182. Dong, M.D. and Sahin, O. (2011) *Nat. Commun.*, **2**, 247.
183. Chen, C.J. (2007) *Introduction to Scanning Tunneling Microscopy*, Oxford University Press, New York.
184. Neddermeyer, H. (1993) *Scanning Tunneling Microscopy*, Springer, New York.
185. Wiesendanger, R. and H.J., Guntherodt (eds) (1993) *Scanning Tunneling Microscopy I–III*, Springer, New York.
186. Wolf, E.L. (1984) *Principles of Electron Tunneling Spectroscopy*, Oxford University Press, New York.
187. Binnig, G., Rohrer, H., Gerber, Ch., and Weibel, E. (1983) *Phys. Rev. Lett.*, **50**, 120.
188. Eigler, D.M. and Schweizer, E.K. (1990) *Nature*, **344**, 524.
189. Crommie, M.F., Lutz, C.P., and Eigler, D.M. (1993) *Science*, **262**, 218–220.
190. Pan, S.H., O'Neal, J.P., Badzey, R.L., Chamon, C., Ding, H., Engelbrecht, J.R., Wang, Z., Eisaki, H., Uchida, S., Gupta, A.K., Ng, K.-W., Hudson, E.W., Lang, K.M., and Davis, J.C. (2001) *Nature*, **413**, 282–285.

191. Stipe, B.C., Rezaei, M.A., and Ho, W. (1998) *Science*, **280**, 1732–1735.
192. Meier, F., Zhou, L., Wiebe, J., and Wiesendanger, R. (2008) *Science*, **320**, 82–86.
193. Maksymovych, P., Sorescu, D.C., Jordan, K.D., and Yates, J.T. Jr. (2008) *Science*, **322**, 1664–1667.
194. http://www.almaden.ibm.com/almaden/media/image_mirage.html (accessed 6 January 2011).
195. Schmid, M., Hebenstreit, W., Varga, P., and Crampin, S. (1996) *Phys. Rev. Lett.*, **76**, 2298–2301.
196. Sprodowski, C. and Morgenstern, K. (2010) *Phys. Rev. B*, **82**, 165444.
197. Vitali, L., Borisovam, S.D., Rusina, G.G., Chulkov, E.V., and Kern, K. (2010) *Phys. Rev. B*, **81**, 153409.
198. Bocquet, M.L., Cerda, J., and Sautet, P. (1999) *Phys. Rev. B*, **59**, 15437.
199. Hahn, J.R. and Ho, W. (2001) *Phys. Rev. Lett.*, **87**, 196102.
200. Giessibl, F.J. (2003) *Rev. Mod. Phys.*, **75**, 949–983.
201. Fischer, ., Kugler, M., Maggio-Aprile, I., Berthod, C., and Renner, C. (2007) *Rev. Mod. Phys.*, **79**, 353.
202. Timusk, T. and Statt, B. (1999) *Rep. Prog. Phys.*, **62**, 61.
203. Hoffman, J.E., Hudson, E.W., Lang, K.M., Madhavan, V., Eisaki, H., Uchida, S., and Davis, J.C. (2002) *Science*, **295**, 466–469.
204. Fäth, M., Freisem, S., Menovsky, A.A., Tomioka, Y., Aarts, J., and Mydosh, J.A. (1999) *Science*, **285**, 1540.
205. Mitra, J., Paranjape, M., Raychaudhuri, A.K., Mathur, N.D., and Blamire, M.G. (2005) *Phys. Rev. B*, **71**, 094426.
206. Seiro, S., Fasano, Y., Maggio-Aprile, I., Koller, E., Kuffer, O., and Fischer, . (2008) *Phys. Rev. B*, **7**, 020407(R).
207. Rnnow, H.M., Renner, Ch., Aeppli, G., Kimura, T., and Tokura, Y. (2006) *Nature*, **440**, 1025–1028.
208. Berthe, M., Stiufiuc, R., Grandidier, B., Deresmes, D., Delerue, C., and Stiévenar, D. (2008) *Science*, **319**, 436.
209. Grafstrom, S. (2002) *J. Appl. Phys.*, **91**, 1717.
210. Loth, S., Etzkorn, M., Lutz, C.P., Eigler, D.M., and Heinrich, A.J. (2010) *Science*, **329**, 1628.
211. Kalinin, S.V., Rodriguez, B.J., Jesse, S., Mirman, B., Karapetian, E., Eliseev, E.A., and Morozovska, A.N. (2007) *Annu. Rev. Mat. Sci.*, **37**, 189.
212. Kalinin, S.V., Rar, A., and Jesse, S. (2006) *IEEE Trans. Ultrason. Ferroelectr. Freq. Control*, **53**, 2226.
213. Aburaya, Y., Nomura, H., Kageshima, M., Naitoh, Y., Li, Y.J., and Sugawara, Y. (2011) *J. Appl. Phys.*, **109**, 064308.
214. Thundat, T., Sales, B.C., Chakoumakos, B.C., Boatner, L.A., Allison, D.P., and Warmack, R.J. (1993) *Surf. Sci. Lett.*, **293**, L863.
215. Yamada, H., Kobayashi, K., Fukuma, T., Hirata, Y., Kajita, T., and Matsushuge, K. (2009) *Appl. Phys. Expr.*, **2**, 095007.
216. Fukuma, T., Ueda, Y., Yoshioka, S., and Asakawa, H. (2010) *Phys. Rev. Lett.*, **104**, 016101.
217. Frenken, J.W.M. and Oosterkamp, T.H. (2010) *Nature*, **464**, 38–39.
218. Gan, Y. (2009) *Surf. Sci. Rep.*, **64**, 99–121.
219. Meyerand, G. and Amer, N.M. (1990) *Appl. Phys. Lett.*, **57**, 2089–2091.
220. Meyer, E., Howald, L., Overney, R., Brodbeck, D., Lüthi, R., Haefke, H., Frommer, J., and Güntherodt, H.-J. (1992) *Ultramicroscopy*, **42–44**, 274–280.
221. Overney, R.M., Takano, H., Fujihira, M., Paulus, W., and Ringsdorf, H. (1994) *Phys. Rev. Lett.*, **72**, 3546.
222. Krim, J. (1996) *Sci. Am.*, **275**, 74.
223. Carpick, R.W. and Salmeron, M. (1997) *Chem. Rev.*, **97**, 1163–1194.
224. Finlan, M.F. and McKay, I.A. (1989) US Patent 5, 047, 633, UK filing May 8, 1989.
225. Gleyzes, P., Kuo, P.K., and Boccara, A.C. (1991) *Appl. Phys. Lett.*, **58**, 2989–2991.
226. Elings, V.B. and Gurley, J. (1992) US Patent 5, 412, 980, filed Aug. 7, 1992.
227. Zhong, Q., Inniss, D., Kjoller, K., and Elings, V.B. (1993) *Surf. Sci. Lett.*, **290**, L688.
228. Chernoff, D.A. (1995) in *Proceedings of Microscopy and Microanalysis* (eds G.W. Bailey et al.), Jones & Begell Publishing, New York, pp. 888–889.

229. Achalla, P., McCormick, J., Hodge, T. et al. (2005) *J. Poly. Sci. B*, **44**, 492–503; (b) Gheno, S.M., Passador, F.R., and Pessan, L.A. (2010) *J. Appl. Poly. Sci.*, **117**, 3211–3210 (c) Hobbs, J.K., Farrance, O.E., and Kailas, L. (2009) *Polymer*, **50**, 4281 (d) Park, J.H., Sun, Y., Goldman, Y. et al. (2009) *Macromolecules*, **42**, 1017–1023 (e) Djurisic, A.B., Wang, H., Chan, W.K. et al. (2006) *J. Scanning Probe Microsc.*, **1**, 21–31 (f) Wang, D., Fujinami, S., Nakajima, K. et al. (2010) *Polymer*, **51**, 2455–2459 (g) Qu, M., Deng, F., Kalkhoran, S.M. et al. (2011) *Soft Matter*, **7**, 1066–1077.

230. Pethica, J.B. and Oliver, W.C. (1987) *Phys. Scr.*, **T19A**, 61–66.

231. Garcia, R., Tamayo, J., and San Paulo, A. (1999) *Surf. Interface Anal.*, **27**, 312–316.

232. Zhao, Y., Cheng, Q., Qian, M., and Cantrell, J.H. (2010) *J. Appl. Phys.*, **108**, 094311.

233. Xu, W., Wood-Adams, P.M., and Robertson, C.G. (2006) *Polymer*, **47**, 4798.

234. Dubourg, F., Aime, J.P., Maursaudon, S., Boisgard, R., and Leclère, P. (2001) *Eur. Phys. J.*, **E6**, 49–55.

235. Braithwaite, G.J.C. and Luckham, P.F. (1999) *J. Colloid Interface Sci.*, **218**, 917.

236. Noy, A., Sanders, C.H., Vezenov, D.V., Wong, S.S., and Lieber, C.M. (1998) *Langmuir*, **14**, 1508–1511.

237. Garcia, R. and Martinez, N.F. (2006) *Nanotechnology*, **17**, S167–S172.

238. Cleveland, J.P., Anczykowski, B., Schmid, A.E., and Elings, V. (1998) *Appl. Phys. Lett.*, **72**, 2613–2615.

239. Gomez, C.J. and Garcia, R. (2010) *Ultramicroscopy*, **110**, 626–633.

240. San Paulo, A. and Garcia, R. (2002) *Phys. Rev. B*, **66**, 041401–041404.

241. Hansma, P.K., Cleveland, J.P., Rademacher, M., Walters, D.A., Hillner, P.E., Bezanilla, M., Fritz, M., Vie, D., Hansma, H.G., Prater, C.B., Massie, J., Fukunaga, L., Gurley, J., and Elings, V. (1994) *Appl. Phys. Lett.*, **64**, 1738–1740.

242. Putman, C.A.J., van der Werf, K.O., de Grooth, B.G., van Hulst, N.F., and Greve, J. (1994) *Appl. Phys. Lett.*, **64**, 2454–2456.

243. Henderson, E., Haydon, P.G., and Sakaguchi, D.S. (1992) *Science*, **257**, 1944–1946.

244. Radmacher, M., Tillmann, R.W., Fritz, M., and Gaub, H.E. (1992) *Science*, **257**, 1900–1905.

245. Kasas, S., Gotzos, V., and Celio, M.R. (1993) *Biophys. J.*, **64**, 539–544.

246. Hoh, J.H. and Schoenenberger, C.A. (1994) *J. Cell Sci.*, **107**, 1105–1114.

247. Shroff, S.G., Saner, D.R., and Lal, R. (1995) *Am. J. Physiol. Cell Physiol.*, **269**, 286–292.

248. Lal, R., Drake, B., Blumberg, D., Saner, D.R., Hansma, P.K., and Feinstein, S.C. (1995) *Am. J. Physiol. Cell Physiol.*, **269**, 275–285.

249. Braunstein, D. and Spudich, A. (1994) *Biophys. J.*, **66**, 1717–1725.

250. Braet, F.C., Rotsch, C., Wisse, E., and Radmacher, M. (1998) *Appl. Phys. A.*, **66**, 575–578.

251. Ohtake, N., Niikura, K., Suzuki, T., Nagakawa, K., Mikuni, S., Matsuo, Y., Kinjo, M., Sawa, H., and Ijiro, K. (2010) *ChemBioChem*, **11**, 959–962.

252. Lazarides, E. (1976) *J. Cell Biol.*, **68**, 202–219.

253. Maguire, P., Kilpatrick, K.I., Kelly, G., Prendergast, P.I., Campbell, V.A., O'Connell, B.C., and Jarvis, S.P. (2007) *HFSP J.*, **1**, 181–191.

254. Albrecht, T.R., Grutter, P., Horne, D., and Ruger, D. (1991) *J. Appl. Phys.*, **69**, 668–673.

255. Fukui, K., Onishi, H., and Iwasawa, Y. (1997) *Phys. Rev. Lett.*, **79**, 4202–4205.

256. Fukuma, T., Higgins, M.J., and Jarvis, S.P. (2007) *Phys. Rev. Lett.*, **98**, 106101.

257. Fukuma, T., Higgins, M.J., and Jarvis, S.P. (2007) *Biophys. J.*, **92**, 3603–3609.

258. Loh, S.H. and Jarvis, S.P. (2010) *Langmuir*, **26**, 9176–9178.

259. Ferber, U.M., Kaggwa, G.B., and Jarvis, S.P. (2011) *Eur. Biophys. J.*, **40**, 329–338.

260. Sheikh, K.H., Giordani, C., Kilpatrick, J.I., and Jarvis, S.P. (2011) *Langmuir*, **27**, 3749–3753.

261. Gross, L., Mohn, F., Moll, N., Liljeroth, P., and Meyer, G. (2009) *Science*, **325**, 1110–1114.

262. Gross, L., Mohn, F., Moll, N., Meyer, G., Ebel, R., Abdel-Mageedm, W.M., and Jaspars, M. (2010) *Nat. Chem.*, **2**, 821.
263. Pharr, G.M. and Oliver, W.C. (1992) *MRS Bull.*, **17**, 28.
264. Oliver, W.C. and Pharr, G.M. (1992) *J. Mat. Res.*, **7**, 1564.
265. Rugar, D., Budakian, R., Mamin, H.J., and Chui, B.W. (2004) *Nature*, **430**, 329.
266. Heinzelmann, H., Meyer, E., Grutter, P., Hidber, H.R., Rosenthaler, L., and Guntherodt, H.J. (1988) *J. Vac. Sci. Technol. A*, **6**, 275–278.
267. Jarvis, S.P. and Mostaert, A.S. (2007) *GIT Imaging Microsc.*, **51**, 415–417.
268. Florin, E.-L., Moy, V.T., and Gaub, H.E. (1994) *Science*, **264**, 415.
269. Lee, G.U., Chris, L.A., and Colton, R.J. (1994) *Science*, **266**, 771.
270. Hinterdorfer, P., Baumgartner, W., Gruber, H.J., Schilcher, K., and Schindler, H. (1996) *Proc. Natl. Acad. Sci. U.S.A.*, **93**, 3477–3481.
271. Dietz, H. and Rief, M. (2006) *Proc. Natl. Acad. Sci. U.S.A.*, **103**, 1244–1247.
272. Oesterhelt, F., Oesterhelt, D., Pfeiffer, M., Engel, A., Gaub, H.E., and Muller, D.J. (2000) *Science*, **288**, 143–146.
273. Gebhardt, J.C.M., Bornschlogl, T., and Rief, M. (2010) *Proc. Natl. Acad. Sci. U.S.A.*, **107**, 2013–2018.
274. Oberhauser, A.F., Hansma, P.K., Carrion-Vazquez, M., and Fernandez, J.M. (2001) *Proc. Natl. Acad. Sci. U.S.A.*, **98**, 468–472.
275. Marszalek, P.E., Oberhauser, A.F., Pang, Y.P., and Fernandez, J.M. (1998) *Nature*, **396**, 661.
276. Erdmann, M., David, R., Fornof, A.R., and Gaub, H.E. (2010) *Nat. Chem.*, **2**, 745.
277. Erdmann, M., David, R., Fornof, A., and Gaub, H.E. (2010) *Nat. Nanotechnol.*, **5**, 154.
278. Ducker, W.A., Senden, T.J., and Pashley, R.M. (1991) *Nature*, **353**, 239–241.
279. Butt, H.-J. (1994) *J. Colloid Interface Sci.*, **166**, 109.
280. Gunning, A.P., Mackie, A.R., Wilde, P.J., and Morris, V.J. (2004) *Langmuir*, **20**, 116.
281. Gromer, A., Penfold, R., Gunning, A.P., Kirby, A.R., and Morris, V.J. (2010) *Soft Matter*, **6**, 3957.
282. Dagastine, R.R., Manica, R., Carnie, S.L., Chan, D.Y.C., Stevens, G.W., and Grieser, F. (2006) *Science*, **313**, 210–213.
283. Lockie, H.J., Manica, R., Stevens, G.W., Grieser, F., Chan, D.Y.C., and Dagastine, R.R. (2011) *Langmuir*, **27**, 2676–2685.
284. Vakarelski, I.U., Manica, R., Tang, X.S., O'Shea, S.J., Stevens, G.W., Grieser, F., Dagastine, R.R., and Chan, D.Y.C. (2010) *Proc. Natl. Acad. Sci. U.S.A.*, **107**, 11177–11182.
285. Manor, O., Vakarelski, I.U., Tang, X., O'Shea, S.J., Stevens, G.W., Grieser, F., Dagastine, R.R., and Chan, D.Y.C. (2008) *Phys. Rev. Lett.*, **101**, 024501.
286. Albers, B.J., Schwendemann, T.C., Baykara, M.Z., Pilet, N., Liebmann, M., Altman, E.I., and Schwarz, U.D. (2009) *Nat. Nanotechnol.*, **4**, 307.
287. Baykara, M.Z., Schwendemann, T.C., Altman, E.I., and Schwarz, U.D. (2010) *Adv. Mater.*, **22**, 2838–2853.
288. Sugimoto, Y., Pou, P., Abe, M., Jelinek, P., Pérez, R., Morita, S., and Custance, Ó. (2007) *Nature*, **446**, 64–67.
289. Hosaka, S., Kikukawa, A., Honda, Y., Koyanagi, H., and Tanaka, S. (1992) *Jpn. J. Appl. Phys.*, **31**, L904–L907.
290. Rabe, U. and Arnold, W. (1994) *Appl. Phys. Lett.*, **64**, 1493.
291. Maivald, P., Butt, H.-J., Gould, S.A.C., Prater, C.B., Drake, B., Gurley, G., Elings, V.B., and Hansma, P.K. (1991) *Nanotechnology*, **2**, 103–106.
292. Florin, E.L., Radmacher, M., Fleck, B., and Gaub, H.E. (1994) *Rev. Sci. Instrum.*, **65**, 639.
293. Dietz, C., Zerson, M., Riesch, C., Gigler, A.M., Stark, R.W., Rehse, N., and Magerle, R. (2008) *Appl. Phys. Lett.*, **92**, 143107.
294. García, R., Magerle, R., and Perez, R. (2007) *Nat. Mater.*, **6**, 405–411.
295. García, R. (2010) *Nat. Nanotechnol.*, **5**, 101–102.

296. Burns, A.R., Houston, J.E., Carpick, R.W., and Michalske, T.A. (1999) *Langmuir*, **15**, 2922.
297. Sahin, O. (2007) *Rev. Sci. Instrum.*, **78**, 103707.
298. Sahin, O. and Erina, N. (2008) *Nanotechnology*, **19**, 445717.
299. Huang, L. and Su, C. (2004) *Ultramicroscopy*, **100**, 277–285.
300. Baumgart, C., Helm, M., and Schmidt, H. (2009) *Phys. Rev. B*, **80**, 085305.
301. Glatzel, T., Lux-Steiner, M.C., Strassburg, E., Boag, A., and Rosenwaks, Y. (2007) in *Scanning Probe Microscopy: Electrical and Electromechanical Phenomena on the Nanoscale*, vol. 1 (eds S.V. Kalinin and A. Gruverman), Springer Science and Business Media, New York, pp. 113–131.
302. Kitamura, S. and Iwatsuki, M. (1998) *Appl. Phys. Lett.*, **72**, 3154.
303. Loppacher, C., Zerweck, U., and Eng, L.M. (2004) *Nanotechnology*, **15**, S9–S13.
304. Loppacher, C., Zerweck, U., Teich, S., Beyreuther, E., Otto, T., Grafström, S., and Eng, L.M. (2005) *Nanotechnology*, **16**, S1–S6.
305. Oliver, R.A. (2008) *Rep. Prog. Phys.*, **71**, 076501.
306. Sadewasser, S. and Glatzel, T. (eds) (2011) *Kelvin Probe Force Microscopy*, Springer Science + Business Media, New York.
307. Henning, A.K., Hochwitz, T., Slinkman, J., Never, J., Hoffman, S., Kaszuba, P., and Daghlian, C. (1995) *J. Appl. Phys.*, **77**, 1888.
308. Xu, Q. and Hsu, J.W.P. (1999) *J. Appl. Phys.*, **85**, 2465.
309. Shikler, R., Meoded, T., Fried, N., and Rosenwaks, Y. (1999) *Appl. Phys. Lett.*, **74**, 2972.
310. Vatel, O. and Tanimoto, M. (1995) *J. Appl. Phys.*, **77**, 2358.
311. Cunningham, S., Larkin, I.A., and Davis, J.H. (1998) *Appl. Phys. Lett.*, **73**, 123.
312. Shvebelman, M.M., Agronin, A.G., Urenski, R.P., Rosenwaks, Y., and Rosenman, G.I. (2002) *Nano Lett.*, **2**, 455.
313. Kalinin, S.V. and Bonnell, D.A. (2001) *Appl. Phys. Lett.*, **78**, 1116–1118.
314. Kalinin, S.V. and Bonnell, D.A. (2000) *J. Appl. Phys.*, **87**, 3950–3957.
315. Lee, W., Lee, M., Kim, Y.-B., and Prinz, F.B. (2009) *Nanotechnology*, **20**, 445706.
316. Palermo, V., Palma, M., and Samorì, P. (2006) *Adv. Mater.*, **18**, 145.
317. Sugimura, H., Hayashi, K., Saito, N., Nakagiri, N., and Takai, O. (2002) *Appl. Surf. Sci.*, **188**, 403.
318. Lü, J., Delamarche, E., Eng, L., Bennewitz, R., Meyer, E., and Güntherodt, H.J. (1999) *Langmuir*, **15**, 8184.
319. Fujihira, M. and Kawate, H. (1994) *J. Vac. Sci. Technol. B*, **12**, 1604.
320. Chi, L.F., Jacobi, S., and Fuchs, H. (1996) *Thin Solid Films*, **284**, 403.
321. Fujihira, M. (1999) *Annu. Rev. Mater. Sci.*, **29**, 353.
322. Lee, I. and Greenbaum, E. (2007) in *Scanning Probe Microscopy: Electrical and Electromechanical Phenomena on the Nanoscale*, vol. 2 (eds S.V. Kalinin and A. Gruverman), Springer Science and Business Media, New York, pp. 601–614.
323. Gil, A., de Pablo, P.J., Colchero, J., Gómez-Herrero, J., and Baró, A.M. (2002) *Nanotechnology*, **13**, 309.
324. Kwak, K.J., Yoda, S., and Fujihira, M. (2003) *Appl. Surf. Sci.*, **210**, 73.
325. Leung, C., Kinns, H., Hoogenboom, B.W., Howorka, S., and Mesquida, P. (2009) *Nano Lett.*, **9**, 2769.
326. Fumagalli, L., Ferrari, G., Sampietro, M., and Gomila, G. (2009) *Nano Lett.*, **9**, 1604–1608.
327. Gramse, G., Casuso, I., Toset, J., Fumagalli, L., and Gomila, G. (2009) *Nanotechnology*, **20**, 395702.
328. Leonenko, Z., Rodenstein, M., Döhner, J., Eng, L.M., and Amrein, M. (2006) *Langmuir*, **22**, 10135.
329. Leonenko, Z., Gill, S., Baoukina, S., Monticelli, L., Doehner, J., Gunasekara, L., Felderer, F., Rodenstein, M., Eng, L.M., and Amrein, M. (2007) *Biophys. J.*, **93**, 674.
330. Hane, F., Moores, B., Amrein, M., and Leonenko, Z. (2009) *Ultramicroscopy*, **109**, 968–973.
331. Chen, X.Q., Yamada, H., Horiuchi, T., Matsushige, K., Watanabe, S., Kawai,

M., and Weiss, P.S. (1999) *J. Vac. Sci. Technol. B*, **17**, 1930–1934.
332. Coffey, D.C. and Ginger, D.S. (2006) *Nat. Mater.*, **5**, 735–740.
333. Kronik, L. and Shapira, Y. (2001) *Surf. Interface Anal.*, **31**, 954–965.
334. Chavez-Pirson, A., Vatel, O., Tanimoto, M., Ando, H., Iwamura, H., and Kanbe, H. (1995) *Appl. Phys. Lett.*, **67**, 3069.
335. Sinensky, A.K. and Belcher, A.M. (2007) *Nat. Nanotechnol.*, **2**, 653.
336. Gao, P. and Cai, Y. (2009) *Anal. Bioanal. Chem.*, **394**, 207.
337. Lee, I., Lee, J.W., Stubna, A., and Greenbaum, E. (2000) *J. Phys. Chem. B*, **104**, 2439.
338. Frolov, L., Rosenwaks, Y., Carmeli, C., and Carmeli, I. (2005) *Adv. Mater.*, **17**, 2434–2437.
339. Kuritz, T., Lee, I., Owens, E.T., Humayun, M., and Greenbaum, E. (2005) *IEEE Trans. Nanobiosci.*, **4**, 196.
340. Knapp, H.F., Mesquida, P., and Stemmer, A. (2002) *Surf. Interface Anal.*, **33**, 108.
341. Lee, I., Greenbaum, E., Budy, S., Hillebrecht, J.R., Birge, R.R., and Stuart, J.A. (2006) *J. Phys. Chem. B*, **110**, 10982–10990.
342. Glatzel, T., Fuertes Marron, D., Schedel-Niedrig, T., Sadewasser, S., and Lux-Steiner, M.C. (2002) *Appl. Phys. Lett.*, **81**, 2017–2019.
343. Butt, H.-J. (1991) *Biophys. J.*, **60**, 777–785.
344. Butt, H.-J. (1991) *Biophys. J*, **60**, 1438–1444.
345. Butt, H.-J. (1992) *Nanotechnology*, **3**, 60–68.
346. Lynch, B.P., Hilton, A.M., and Simpson, G.J. (2006) *Biophys. J.*, **91**, 2678.
347. Rodriguez, B.J., Jesse, S., Baddorf, A.P., and Kalinin, S.V. (2006) *Phys. Rev. Lett.*, **96**, 237602.
348. Rodriguez, B.J., Jesse, S., Seal, K., Baddorf, A.P., and Kalinin, S.V. (2008) *J. Appl. Phys.*, **103**, 014306.
349. Rodriguez, B.J., Jesse, S., Baddorf, A.P., Kim, S.H., and Kalinin, S.V. (2007) *Phys. Rev. Lett.*, **98**, 247603.
350. Rodriguez, B.J., Jesse, S., Seal, K., Baddorf, A.P., Kalinin, S.V., and Rack, P.D. (2007) *Appl. Phys. Lett.*, **91**, 093130.
351. Vatel, O. and Tanimoto, M. (1995) *J. Appl. Phys.*, **77**, 2050.
352. Huey, B.D. and Bonnell, D.A. (2000) *Appl. Phys. Lett.*, **76**, 1012.
353. Woodside, M.T. and McEuen, P.L. (2002) *Science*, **196**, 1098.
354. Topinka, M.A., LeRoy, B.J., Westervelt, R.M., Shaw, S.E.J., Fleischmann, R., Heller, E.J., Maranowski, K.D., and Gossard, A.C. (2001) *Nature*, **410**, 183.
355. Kalinin, S.V. and Bonnell, D.A. (2004) *Phys. Rev. B*, **70**, 235304.
356. Kalinin, S.V. and Bonnell, D.A. (2001) *Appl. Phys. Lett.*, **78**, 1306.
357. Kalinin, S.V. and Bonnell, D.A. (2002) *J. Appl. Phys.*, **91**, 832.
358. Kalinin, S.V., Shin, J., Veith, G.M., Baddorf, A.P., Lobanov, M.V., Runge, H., and Greenblatt, M. (2005) *Appl. Phys. Lett.*, **86**, 102902.
359. Freitag, M., Johnson, A.T., Kalinin, S.V., and Bonnell, D.A. (2002) *Phys. Rev. Lett.*, **89**, 216801.
360. Connolly, M.R., Chiou, K.L., Smith, C.G., Anderson, D., Jones, G.A.C., Lombardo, A., Fasoli, A., and Ferrari, A.C. (2010) *Appl. Phys. Lett.*, **96**, 113501.
361. Zhou, X., Dayeh, S.A., Wang, D., and Yu, E.T. (2007) *Appl. Phys. Lett.*, **90**, 233118.
362. Aoki, N., Cunha, C.R., Akis, R., Ferry, D.K., and Ochiai, Y. (2005) *Phys. Rev.*, **B 72**, 155327.
363. Hsu, J.W.P., Weimann, N.G., Manfra, M.J., West, K.W., Lang, D.V., Schrey, F.F., Mitrofanov, O., and Molnar, R.J. (2003) *Appl. Phys. Lett.*, **83**, 4559.
364. Hackens, B., Martins, F., Faniel, S., Dutu, C.A., Sellier, H., Huant, S., Pala, M., Desplanque, L., Wallart, X., and Bayot, V. (2010) *Nat. Commun.*, **1**, 39.
365. Kholkin, A.L., Wuthrich, C., Taylor, D.V., and Setter, N. (1996) *Rev. Sci. Instrum.*, **67**, 1935.
366. Maeder, T., Muralt, P., Sagalowicz, L., Reaney, I., Kohli, M., Kholkin, A., and Setter, N. (1996) *Appl. Phys. Lett.*, **68**, 776.
367. Kholkin, A.L., Colla, E.L., Tagantsev, A.K., Taylor, D.V., and Setter, N. (1996) *Appl. Phys. Lett.*, **68**, 2577.

368. Muralt, P., Kholkin, A., Kohli, M., and Maeder, T. (1996) *Sens. Actuators A*, **53**, 398.
369. Kholkin, A.L., Brooks, K.G., and Setter, N. (1997) *Appl. Phys. Lett.*, **71**, 2044.
370. Kalinin, S.V., Rodriguez, B.J., Jesse, S., Thundat, T., and Gruverman, A. (2005) *Appl. Phys. Lett.*, **87**, 053901.
371. Jesse, S., Lee, H.N., and Kalinin, S.V. (2006) *Rev. Sci. Instrum.*, **77**, 073702.
372. Jesse, S., Baddorf, A.P., and Kalinin, S.V. (2006) *Appl. Phys. Lett.*, **88**, 062908.
373. Rodriguez, B.J., Jesse, S., Baddorf, A.P., Zhao, T., Chu, Y.H., Ramesh, R., and Kalinin, S.V. (2007) *Nanotechnology*, **18**, 405701.
374. Kalinin, S.V., Rodriguez, B.J., Jesse, S., Chu, Y.H., Zhao, T., Ramesh, R., Eliseev, E.A., and Morozovska, A.N. (2007) *Proc. Natl. Acad. Sci. U.S.A.*, **104**, 20204.
375. Jesse, S., Rodriguez, B.J., Baddorf, A.P., Vrejoiu, I., Hesse, D., Alexe, M., Eliseev, E.A., Morozovska, A.N., and Kalinin, S.V. (2008) *Nat. Mater.*, **7**, 209.
376. Kalinin, S.V., Jesse, S., Rodriguez, B.J., Chu, Y.H., Ramesh, R., Eliseev, E.A., and Morozovska, A.N. (2008) *Phys. Rev. Lett.*, **100**, 155703.
377. Morozovska, A., Eliseev, E., and Kalinin, S.V. (2010) *Appl. Phys. Lett.*, **96**, 222906.
378. Morozovska, A., Eliseev, E., Balke, N., and Kalinin, S.V. (2010) *J. Appl. Phys.*, **108**, 053712.
379. Balke, N., Jesse, S., Morozovska, A.N., Eliseev, E., Chung, D.W., Kim, Y., Adamczyk, L., Garcia, R.E., Dudney, N., and Kalinin, S.V. (2010) *Nat. Nanotechnol.*, **5**, 749.
380. Balke, N., Jesse, S., Kim, Y., Adamczyk, L., Ivanov, I.N., Dudney, N.J., and Kalinin, S.V. (2010) *ACS Nano*, **4**, 7349–7357.
381. Gajek, M., Bibes, M., Fusil, S., Bouzehouane, K., Fontcuberta, J., Barthélémy, A., and Fert, A. (2007) *Nat. Mater.*, **6**, 296–302.
382. Maksymovych, P., Jesse, S., Yu, P., Ramesh, R., Baddorf, A.P., and Kalinin, S.V. (2009) *Science*, **324**, 1421.
383. Gruverman, A., Wu, D., Lu, H., Wang, Y., Jang, H.W., Folkman, C.M., Zhuravlev, M.Y., Felker, D., Rzchowski, M., Eom, C.B., and Tsymbal, E.Y. (2009) *Nano Lett.*, **9**, 3539.
384. Baddorf, A.P. (2007) Scanning tunneling potentiometry: the power of STM applied to electrical transport, in *Scanning Probe Microscopy: Electrical and Electromechanical Phenomena on the Nanoscale* (eds S.V. Kalinin and A. Gruverman), Springer Science and Business Media, New York, pp. 11–30.
385. Garcia, V., Fusil, S., Bouzehouane, K., Enouz-Vedrenne, S., Mathur, N.D., Barthelemy, A., and Bibes, M. (2009) *Nature*, **460**, 81.
386. Seidel, J., Martin, L.W., He, Q., Zhang, Q., Chu, Y.-H., Rother, A., Hawkridge, M.E., Maksymovych, P., Yu, P., Gajek, M., Balke, N., Kalinin, S.V., Gemming, S., Wang, F., Catalan, G., Scott, J.F., Spaldin, N.A., Orenstein, J., and Ramesh, R. (2009) *Nat. Mater.*, **8**, 229.
387. Yang, C.H., Seidel, J., Kim, S.Y., Rossen, P., Yu, P., Gajek, M., Chu, Y.H., Martin, L.W., Holcomb, M.B., He, Q., Maksymovych, P., Balke, N., Kalinin, S.V., Baddorf, A.P., Basu, S.R., Scullin, M.L., and Ramesh, R. (2009) *Nat. Mater.*, **8**, 485.
388. Maksymovych, P., Seidel, J., Chu, Y.H., Wu, P.P., Baddorf, A., Chen, L., Kalinin, S.V., and Ramesh, R. (2011) *Nano Lett.*, **11**, 1906–1912.
389. Shao, R., Kalinin, S.V., and Bonnell, D.A. (2003) *Appl. Phys. Lett.*, **82**, 1869–1871.
390. O'Hayre, R., Lee, M., and Prinz, F.B. (2004) *J. Appl. Phys.*, **95**, 8382–8392.
391. Layson, A., Gadad, S., and Teeters, D. (2003) *Electrochim. Acta*, **48**, 2207–2213.
392. Louie, M.W., Hightower, A., and Haile, S.M. (2010) *ACS Nano*, **4**, 2811.
393. Williams, C.C. (1999) *Annu. Rev. Mater. Sci.*, **29**, 471.
394. Tran, T., Oliver, D.R., Thomson, D.J., and Bridges, G.E. (2001) *Rev. Sci. Instrum.*, **72**, 2618.
395. Matey, J.R. and Blanc, J. (1985) *J. Appl. Phys.*, **57**, 1437.

396. Williams, C.C., Hough, W.P., and Rishton, S.A. (1989) *Appl. Phys. Lett.*, **55**, 203.
397. Barrett, R.C. and Quate, C.F. (1991) *J. Appl. Phys.*, **70**, 2725.
398. Abraham, D.W., Williams, C., Slinkman, J., and Wickramasinghe, H.K. (1991) *J. Vac. Sci. Technol. B*, **9**, 703.
399. Kopanski, J.J. (2007) in *Scanning Probe Microscopy: Electrical and Electromechanical Phenomena on the Nanoscale* (eds S.V. Kalinin and A. Gruverman), Springer Science and Business Media, New York, pp. 88–112.
400. Nakakura, C.Y., Tangyunyong, P., and Anderson, M.L. (2007) in *Scanning Probe Microscopy: Electrical and Electromechanical Phenomena on the Nanoscale* (eds S.V. Kalinin and A. Gruverman), Springer Science and Business Media, New York, pp. 634–662.
401. Nakakura, C.Y., Hetherington, D.L., Shaneyfelt, M.R., Shea, P.J., and Erickson, A.N. (1999) *Appl. Phys. Lett.*, **75**, 2319.
402. Hansen, P.J., Strausser, Y.E., Erickson, A.N., Tarsa, E.J., Kozodoy, P., Brazel, E.G., Ibbetson, J.P., Mishra, U., Narayanamurti, V., DenBaars, S.P., and Speck, J.S. (1998) *Appl. Phys. Lett.*, **72**, 2247.
403. Schaadt, D.M., Miller, E.J., Yu, E.T., and Redwing, J.M. (2001) *Appl. Phys. Lett.*, **78**, 88.
404. Edwards, H., McGlothlin, R., San Martin, R., Gribelyuk, E.U.M., Mahaffy, R., Shih, C.K., List, R.S., and Ukraintsev, V.A. (1998) *Appl. Phys. Lett.*, **72**, 698–700.
405. Nelson, B.A. and King, W.P. (2007) *Rev. Sci. Instrum.*, **78**, 023702.
406. Hammiche, A., Bozec, L., Conroy, M., Pollock, H.M., Mills, G., Weaver, J.M.R., Price, D.M., Reading, M., Hourston, D.J., and Song, M. (2000) *J. Vac. Sci. Technol. B*, **18**, 1322.
407. Lee, J., Beechem, T., Wright, T.L., Nelson, B.A., Graham, S., and King, W.P. (2006) *J. Microelectromech. Syst.*, **15**, 1644.
408. Wei, Z.Q., Wang, D.B., Kim, S., Kim, S.Y., Hu, Y.K., Yakes, M.K., Laracuente, A.R., Dai, Z.T., Marder, S.R., Berger, C., King, W.P., de Heer, W.A., Sheehan, P.E., and Riedo, E. (2010) *Science*, **328**, 1373.
409. Nikiforov, M.P., Gam, S., Jesse, S., Composto, R.J., and Kalinin, S.V. (2010) *Macromolecules*, **43**, 6724.
410. Nikiforov, M.P., Hohlbauch, S., King, W.P., Voitchovsky, K., Contera, S.A., Jesse, S., Kalinin, S.V., and Proksch, R. (2011) *Nanotechnology*, **22**, 055709.
411. Hammiche, A., Price, D.M., Dupas, E., Mills, G., Kulik, A., Reading, M., Weaver, J.M.R., and Pollock, H.M. (2000) *J. Microsc. (Oxford)*, **199**, 180.
412. Lee, J. and King, W.P. (2007) *Rev. Sci. Instrum.*, **78**, 3.
413. Fee, M., Chu, S., and Hänsch, T.W. (1989) *Opt. Commun.*, **69**, 219.
414. Gao, C. and Xiang, X.D. (1998) *Rev. Sci. Instrum.*, **69**, 3846.
415. Steinhauer, D.E., Vlahacos, C.P., Wellstood, F.C., Anlage, S.M., Canedy, C., Ramesh, R., Stanishevsky, A., and Melngailis, J. (2000) *Rev. Sci. Instrum.*, **71**, 2751.
416. Steinhauer, D.E., Vlahacos, C.P., Dutta, S.K., Feenstra, B.J., Wellstood, F.C., and Anlage, S.M. (1998) *Appl. Phys. Lett.*, **72**, 861.
417. Talanov, V.V., Scherz, A., Moreland, R.L., and Schwartz, A.R. (2006) *Appl. Phys. Lett.*, **88**, 192906.
418. Talanov, V.V., Barga, C.D., Wickey, L., Kalichava, I., Gonzales, E., Shaner, E.A., Gin, A.V., and Kalugin, N.G. (2010) *ACS Nano*, **4**, 3831.
419. Lai, K., Kundhikanjana, W., Kelly, M., and Shen, Z.X. (2008) *Rev. Sci. Instrum.*, **79**, 063703.
420. Huber, H.P., Moertelmaier, M., Wallis, T.M., Chiang, C.J., Hochleitner, M., Imtiaz, A., Oh, Y.J., Schilcher, K., Dieudonne, M., Smoliner, J., Hinterdorfer, P., Rosner, S.J., Tanbakuchi, H., Kabos, P., and Kienberger, F. (2010) *Rev. Sci. Instrum.*, **81**, 113701.
421. Tabib-Azar, M., Su, D.P., Pohar, A., LeClair, S.R., and Ponchak, G. (1999) *Rev. Sci. Instrum.*, **70**, 1725.
422. Talanov, V.V., Scherz, A., Moreland, R.L., and Schwartz, A.R. (2006) *Appl. Phys. Lett.*, **88**, 134106.

423. Abu-Teir, M., Golosovsky, M., Davidov, D., Frenkel, A., and Goldberger, H. (2001) *Rev. Sci. Instrum.*, **72**, 2073.
424. Kim, J., Lee, K., Friedman, B., and Cha, D. (2003) *Appl. Phys. Lett.*, **83**, 1032.
425. Lee, S.-C., Vlahacos, C.P., Feenstra, B.J., Schwartz, A., Steinhauer, D.E., Wellstood, F.C., and Anlage, S.M. (2000) *Appl. Phys. Lett.*, **77**, 4404.
426. Chen, G., Hu, B., Takeuchi, I., Chang, K.-S., Xiang, X.-D., and Wang, G. (2005) *Meas. Sci. Technol.*, **16**, 248.
427. Wang, Z., Kelly, M.A., Shen, Z.-X., Shao, L., Chu, W.-K., and Edwards, H. (2005) *Appl. Phys. Lett.*, **86**, 153118.
428. Tselev, A., Anlage, S.M., Ma, Z., and Melngailis, J. (2007) *Rev. Sci. Instrum.*, **78**, 044701.
429. Tselev, A., Meunier, V., Strelcov, E., Shelton, W.A., Luk'yanchuk, I.A., Jones, K., Proksch, R., Kolmakov, A., and Kalinin, S.V. (2010) *ACS Nano*, **4**, 4412.
430. Kundhikanjana, W., Lai, K., Wang, H., Dai, H., Kelly, M.A., and Shen, Z.-X. (2009) *Nano Lett.*, **9**, 3762.
431. Lai, K., Peng, H., Kundhikanjana, W., Schoen, D.T., Xie, C., Meister, S., Cui, Y., Kelly, M.A., and Shen, Z.-X. (2009) *Nano Lett.*, **9**, 1265.
432. Lai, K., Nakamura, M., Kundhikanjana, W., Kawasaki, M., Tokura, Y., Kelly, M.A., and Shen, Z.-X. (2010) *Science*, **329**, 190.
433. Wu, S. and Yu, J.-J. (2010) *Appl. Phys. Lett.*, **97**, 202902.
434. Plassard, C., Bourillot, E., Rossignol, J., Lacroute, Y., Lepleux, E., Pacheco, L., and Lesniewska, E. (2011) *Phys. Rev. B*, **83**, 121409.
435. Rosner, B.T. and van der Weide, D.W. (2002) *Rev. Sci. Instrum.*, **73**, 2505.
436. Anlage, S.M., Talanov, V.V., and Schwartz, A.R. (2007) in *Scanning Probe Microscopy: Electrical and Electromechanical Phenomena on the Nanoscale* (eds S.V. Kalinin and A. Gruverman), Springer Science and Business Media, New York, pp. 215.
437. Paulson, C.A. and van der Weide, D.W. (2007) in *Scanning Probe Microscopy: Electrical and Electromechanical Phenomena on the Nanoscale* (eds S.V. Kalinin and A. Gruverman), Springer Science and Business Media, New York, p. 315.
438. Cho, Y., Kazuta, S., and Matsuura, K. (1999) *Appl. Phys. Lett.*, **75**, 2833.
439. Van Berkel, G.J., Pasilis, S.P., and Ovchinnikova, O. (2008) *J. Mass Spectrom.*, **43**, 1161–1180.
440. Harris, G.A., Nyadong, L., and Fernandez, F. (2008) *Analyst*, **133**, 1297–1301.
441. Venter, A., Nefliu, M., and Cooks, R.G. (2008) *Trends Anal. Chem.*, **27**, 284–290.
442. Ifa, D.R., Wu, C., Ouyang, Z., and Cooks, R.G. (2010) *Analyst*, **135**, 669–681.
443. Huan-Wen, C., Bin, H., Xie, Z., and Chin, Z. (2010) *J. Anal. Chem.*, **38**, 1069–1088.
444. Stockle, R., Decker, V., Lippert, T., Wokaun, A., and Zenobi, R. (2001) *Anal. Chem.*, **73**, 1399–1402.
445. Becker, J.S., Gordunoff, A., Zoriy, M., Izmer, A., and Kayser, M. (2006) *J. Anal. At. Spectrom.*, **21**, 19–25.
446. Zoriy, M.V., Kayser, M., and Becker, J.S. (2008) *Int. J. Mass Spectrom.*, **273**, 151–155.
447. Meyer, K.A., Ovhinnikova, O., Ng, K., and Goeringder, D.E. (2008) *Rev. Sci. Instrum.*, **79**, 123710.
448. Schmitz, T., Gamez, G., Setz, P., Zhu, L., and Zenobi, R. (2008) *Anal. Chem.*, **80**, 6537–6544.
449. Bradshaw, J.A., Ovchinnikova, O.S., Meyer, K.A., and Goeringer, D.E. (2009) *Rapid Commun. Mass Spectrom.*, **23**, 3781–3786.
450. Reading, M., Price, D.M., Grandy, D.B., Smith, R.M., Bozec, L., Conroy, M., Hammiche, A., and Pollock, H.M. (2001) *Macromol. Symp.*, **167**, 45–62.
451. Price, D.M., Reading, M., Hammiche, A., and Pollock, H.M. (1999) *Int. J. Pharmacol.*, **192**, 85–96.
452. Price, D.M., Reading, M., Lever, R.J., Hammiche, A., and Hubert, A.P. (2001) *Thermochim. Acta*, **367–368**, 195–202.
453. Craig, D., Kett, V., Andrews, C., and Royall, P. (2002) *J. Pharm. Sci.*, **91**, 1201–1213.

454. Toumey, C. (2010) *Nat. Nanotechnol.*, **5**, 239–241.
455. Crommie, M.F., Lutz, C.P., and Eigler, D.M. (1993) *Nature*, **363**, 524–527.
456. Eigler, D.M., Lutz, C.P., and Rudge, W.E. (1991) *Nature*, **352**, 600.
457. Heinrich, A.J., Lutz, C.P., Gupta, J.A., and Eigler, D.M. (2002) *Science*, **298**, 1381.
458. Eigler, D.M., Lutz, C.P., Crommie, M.F., Manoharan, H.C., Heinrich, A.J., and Gupta, J.A. (2004) *Philos. Trans. R. Soc. Lond. A*, **362**, 1135–1147.
459. Manoharan, H.C., Lutz, C.P., and Eigler, D.M. (2000) *Nature*, **403**, 512–515.
460. Moon, C.R., Lutz, C.P., and Manoharan, H.C. (2008) *Nat. Phys.*, **4**, 454–458.
461. Sugimoto, Y., Abe, M., Hirayama, S., Oyabu, N., Custance, Ó., and Morita, S. (2005) *Nat. Mater.*, **4**, 156.
462. Custance, Ó., Perez, R., and Morita, S. (2009) *Nat. Nanotechnol.*, **4**, 803–810.
463. IBM, Lutz and Eigler http://www.almaden.ibm.com/vis/stm/atomo.html. (accessed 6 January 2011).
464. Schirmeisen, A., Weiner, D., and Fuchs, H. (2006) *Phys. Rev. Lett.*, **97**, 136101.
465. Quate, C.F. (1997) *Surf. Sci.*, **386**, 259–264.
466. Pantazi, A., Lantz, M.A., Cherubini, G., Pozidis, H., and Eleftheriou, E. (2004) *Nanotechnology*, **15**, S612.
467. Brown, K.A. and Westervelt, R.M. (2009) *Nanotechnology*, **20**, 385302.
468. Piner, R.D., Zhu, J., Xu, F., Hong, S., and Mirkin, C.A. (1999) *Science*, **283**, 661–663.
469. Lee, K.-B., Park, S.-J., Mirkin, C.A., Smith, J.C., and Mrksich, M. (2002) *Science*, **295**, 1702–1705.
470. Martinez, R.V. and Garcia, R. (2005) *Nano Lett.*, **5**, 1161.
471. Martinez, R.V., Losilla, N.S., Martinez, J., Tello, M., and Garcia, R. (2007) *Nanotechnology*, **18**, 084021.
472. Lee, M., O'Hayre, R., Prinz, F.B., and Gur, T.M. (2004) *Appl. Phys. Lett.*, **85**, 3552.
473. Cen, C., Thiel, S., Mannhart, J., and Levy, J. (2009) *Science*, **323**, 1026.
474. Garcia, R., Martinez, R.V., and Martinez, J. (2006) *Chem. Soc. Rev.*, **35**, 29.
475. Kuramochi, H., Ando, K., Tokizaki, T., Yasutake, A., Perez-Murano, F., Dagata, J.A., and Yokoyama, H. (2004) *Surf. Sci.*, **566**, 343.
476. Chien, F.S.S., Hsieh, W.F., Gwo, S., Vladar, A.E., and Dagata, J.A. (2002) *J. Appl. Phys.*, **91**, 10044.
477. Matsumoto, K., Gotoh, Y., Maeda, T., Dagata, J.A., and Harris, J.S. (2000) *Appl. Phys. Lett.*, **76**, 239.
478. Tseng, A.A., Jou, S., Notargiacomo, A., and Chen, T.P. (2008) *J. Nanosci. Nanotechnol.*, **8**, 2167.
479. Kalinin, S.V., Bonnell, D.A., Alvarez, T., Lei, X., Hu, Z., Ferris, J.H., Zhang, Q., and Dunn, S. (2002) *Nano Lett.*, **2**, 5892002.
480. Kalinin, S.V., Bonnell, D.A., Alvarez, T., Lei, X., Hu, Z., and Ferris, J.H. (2003) *Adv. Mater.*, **16**, 7952004.
481. Hanson, J.N., Rodriguez, B.J., Nemanich, R.J., and Gruverman, A. (2006) *Nanotechnology*, **17**, 49462006.
482. Tanaka, K., Kurihashi, Y., Uda, T., Daimon, Y., Odagawa, N., Hirose, R., Hiranaga, Y., and Cho, Y. (2008) *Jpn. J. Appl. Phys.*, **47**, 33112008.
483. Cho, Y., Hashimoto, S., Odagawa, N., Tanaka, K., and Hiranaga, Y. (2006) *Nanotechnology*, **17**, S1372006.
484. Morita, S. (ed.) (2007) *Roadmap of Scanning Probe Microscopy*, Springer, New York.
485. Dienwiebel, M., de Kuyper, E., Crama, L., Frenken, J.W.M., Heimberg, J.A., Spaanderman, D.J., Glastra van Loon, D., Zijlstra, T., and van der Drift, E. (2005) *Rev. Sci. Instrum.*, **76**, 043704.
486. Ovchinnikov, O.S., Jesse, S., and Kalinin, S.V. (2009) *Nanotechnology*, **20**, 255701.
487. Mamin, H.J., Budakian, R., Chui, B.W., and Rugar, D. (2005) *Phys. Rev.*, **B 72**, 024413.
488. Bonnell, D.A. (ed.) (2008) *Scanning Probe Microscopy and Spectroscopy: Theory, Techniques, and Applications*, Wiley-VCH Verlag GmbH, New York.
489. Lin, D.C. and Horkay, F. (2008) *Soft Matter*, **4**, 669.

490. Sokolov, I.Y. (1994) *Surf. Sci.*, **311**, 287.
491. http://www.snl.salk.edu/~shlens/pca.pdf (accessed 6 January 2011).
492. Bonnet, N. (2004) *Micron*, **35**, 635.
493. Bonnet, N. (2002) *J. Micros.*, **190**, 2.
494. Bosman, M., Watanabe, M., Alexander, D.T.L., and Keast, V.J. (2006) *Ultramicroscopy*, **106**, 1024.
495. Haykin, S. (1998) *Neural Networks: A Comprehensive Foundation*, 2nd edn, Prentice Hall.
496. Hagan, M.T., Demuth, H.B., and Beale, M.H. (2002) *Neural Network Design*, PWS Publishing, Boston.

17
Scanning Beam Methods
David Joy

17.1
Scanning Microscopy

17.1.1
Introduction

The scanning electron microscope (SEM) is the most widely used of all electron beam imaging instruments. It owes its popularity to the versatility of its many modes of imaging, the excellent spatial resolution of its images, the ease with which the micrographs that are generated can be interpreted, the modest demands that are made on specimen preparation, and its "user-friendliness." At one end of its operating range the SEM provides images that can readily be compared to those of conventional optical microscopes, while at the other end, it offers performance that is complementary to that of atomic force microscopes (AFMs) or high-energy transmission electron microscopes (TEMs).

There is also growing interest in scanning microscopy using ion, rather than electron, beams. A scanning ion microscope with a beam of He^+ ions helium ion microscope (HIM) can currently match the performance of a conventional SEM while offering the ultimate promise of significantly enhanced resolution and novel contrast and operating modes. It is also now common practice to combine electron and ions columns into one machine to form a "dual-beam" instrument providing both imaging and material preparation capabilities.

The SEM had its origins in the work of von Ardenne [1, 2] who added scanning coils to a TEM. A photographic plate beneath the electron-transparent sample was mechanically scanned in synchronism with the beam to capture the image. The first recognizably modern SEM, including such features as a cathode-ray tube display, a secondary electron (SE) detector, and nanometer-scale resolution, was described by Zworykin *et al.* [3]. In 1948, Oatley [4] and his group commenced their developments of what eventually became the first commercial SEM, the Cambridge Scientific Instruments Mark 1 "Stereoscan." There are now manufacturers of such instruments in Europe, Japan, and the USA, and it is estimated that more

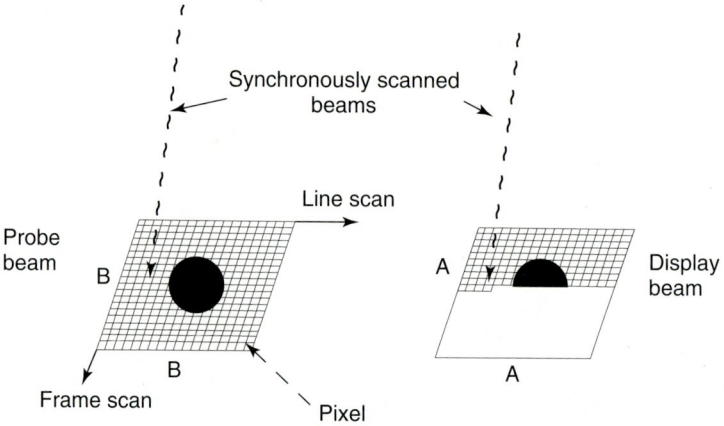

Figure 17.1 Schematic principle of scanning microscope operation.

than 70 000 SEMs are in use worldwide ranging in size from tabletop "personal" devices to multimillion dollar tools for the semiconductor industry.

The SEM is a mapping rather than an imaging device (Figure 17.1) and so belongs to the same class of instruments as the AFM, confocal optical microscopes, and FAX machines. The sample is probed by a beam of ions or electrons scanned in a square or rectangular raster pattern across the surface. Radiations, or other responses, stimulated by the incident beam are detected, amplified, and used to modulate the brightness of a display device such as a flat-panel TV scanned in synchronism with the beam. The scanning action breaks down the object being observed into discrete "pixels" (picture elements) a typical image containing of the order of 1–2 million pixels. This procedure sets both an absolute lower limit to the size of detail that can be observed – the dimension of the pixel – and a softer upper limit since variations in signal that occur over many hundreds of pixels provide just a general change in display brightness rather than discrete information. If the area rastered on the display is $A \times A$ in size and the corresponding area on the sample is $B \times B$, then the image magnification is $M = A/B$. The magnification is therefore purely geometric and may be changed by varying the magnitude of either the raster on the specimen or on the screen. (It should be noted that the values of the lengths A and B involved in the scanning procedure are neither known nor constant, so the "magnification" of an image is not a useful parameter; the image should therefore be characterized by the absolute width of the horizontal field of view (HFOV)). This arrangement makes it possible to vary the "magnification" over a very wide range, rapidly and without any need to refocus or adjust the instrument, and even to display two side-by-side images at different "magnifications." Finally, multiple detectors can be used to collect several signals simultaneously, which can then be displayed individually, or combined in perfect register with each other. It is these capabilities that make the scanning so versatile and productive in both research and industrial environments.

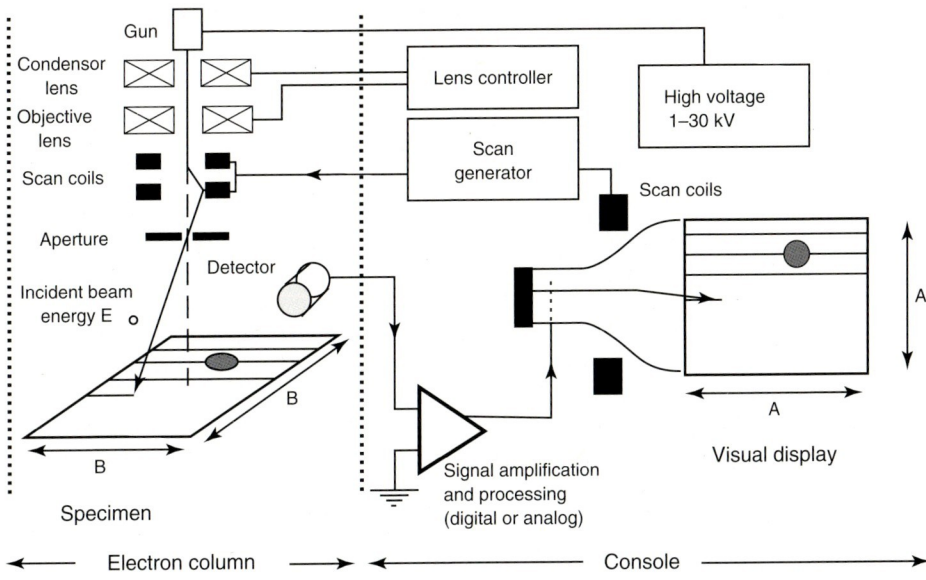

Figure 17.2 Basic components of scanning microscope.

17.1.2
Instrumentation

Figure 17.2 shows the basic components of a scanning microscope. These can be divided into two main categories: the optical system required to form the beam and usually described as the "column"; and the beam scanning and image processing and display systems, which comprise the "console." The charged particle beam is generated by a "gun" in which electrons are produced by thermionic emission from a tungsten "hairpin" filament or a LaB_6 cathode, by field emission from a sharp tungsten tip or from a Schottky field-enhanced thermionic emitter. The electrons are then accelerated to an energy typically in the range 500 eV to 30 keV and focused on to the specimen by one or more electromagnetic lenses. Ion beams can be produced in several ways of which the most effective choice for high-resolution imaging is field-enhanced ionization at the apex of a sharp tungsten tip immersed in cooled helium gas [5]. For either choice of particle, the beam is then focused on to the sample by one or more lenses which for electrons can be either electrostatic or electromagnetic in type but which for ions are always electrostatic.

The column is generally held at a fairly high vacuum ($10^{-4} - 10^{-6}$ Pa) to minimize scattering during transit of the beam to the specimen. Increasingly, however, the sample chamber may now be operated in an "environmental" or "variable pressure" mode in which the region around the specimen, or the entire volume of the specimen chamber, is held at a pressure in the range 1–1000 Pa. This environmental scanning electron microscope (ESEM) or variable pressure scanning electron microscope (VPSEM) option is of great value for the observation of samples that charge under the beam or are fragile or sensitive to their surroundings; over 50% of all current SEMs now offer some capability of this type.

In lower-cost SEMs, the final, probe-forming, lens is often of the pinhole type with the sample placed outside of the magnetic field of the lens because this gives good physical access to the specimen. However, this configuration places the specimen at a working distance of 10–20 mm from the lens, which must therefore be of long focal length and correspondingly high aberration coefficients. In higher-performance instruments, it is more common to use a "snorkel" lens [6] in which the magnetic field extends beyond the lens to envelop the sample or an immersion lens [7] in which the sample is mounted inside the lens gap. The highest performance contemporary SEMs often employ a combination of both electromagnetic and electrostatic elements [8] to optimize performance across a wide range of beam-landing energies. The beam scan coils are usually within, or close to, the probe-forming objective lens. A double scan arrangement is used in which the first set of scan coils deflects the beam through some angle θ, while the second set deflects the beam through an angle of 2θ in the opposite sense. This permits a small defining aperture to be placed at the point through which all scanned beams pass without restricting the field of view. This is of particular importance for VPSEM and ESEM operation where the final aperture must serve to separate regions of high (1–1000 Pa) and low (10^{-4}–10^{-6} Pa) pressure.

The lenses of the HIM are electrostatic because the typical velocity **v** of ions as compared to electrons of the same energy is so low that the Lorentz force $q\mathbf{v} \times \mathbf{B}$ experienced from a magnetic field **B** and the charge on the ion q is too small to permit significant focusing to be performed for realistic fields. The lenses are typically of the simple "Einzel" type, which consists of a circular electrode, held at some potential relative to ground and containing a single aperture. Because the convergence α of the ion beam can be made very small (<1 mrad) without causing diffraction limiting, the usual effects of spherical and chromatic aberration are negligible, and the depth of field is high.

Both electron and ion beams generate SEs from samples, and without exception the basic operational mode in scanning microscopy is the secondary imaging. In addition, current SEMs also typically offer detectors for backscattered electrons (BSEs), for fluorescent X-rays to permit chemical microanalysis, for cathode luminescence, and even for specialized modes of operation such as scanning transmission electron microscopy. Scanning ion microscopes offer both SE and ion detectors and can also support detectors for cathode-luminescent light. Further details of these and other detectors are given below.

Signals from any of the detectors are amplified, processed, and presented to the display screens on the console. The details of the electronics differ surprisingly little from those of the Oatley microscope [5] as they basically provide simple analog amplification and adjustments of gain and offset. Fully digital SEMs and HIMs would offer many advantages but the extraordinarily wide dynamic range (10^5 : 1 or more) of signals encountered in even routine operations make this difficult to implement. Thus at some point before its display on the viewing screen, the signal is digitized for presentation, so it can be enhanced, displayed on a monitor, and archived. Beam scan rates can vary from 100 s for a single high-quality scan to "TV rate" scans that give 15–20 images per second for searching the specimen or recording dynamic phenomena. Despite the fact that it is the quality of the scan pattern that ultimately defines the reliability and credibility of the image that is displayed and stored, the linearity and geometrical accuracy of contemporary scanning microscope rasters is often poor with nonlinearity and orthoganality errors of 5–10% being common.

17.1.3
Performance

It is usual to define the performance of a microscope as being the spatial resolution that it can demonstrate, but because scanning microscopes typically operate over a very wide range of image field and perform many different tasks, this is rarely the most important number. In reality, the most significant parameter is the available beam current, and then the resolution under various operating conditions, and the depth of field available.

As first shown by Rose [9], meaningful information can only be extracted from an image if the signal-to-noise-ratio (SNR) exceeds a threshold that is typically taken to be 5 : 1 The magnitude of the SNR depends directly on the square root of the number of electrons (or other signal specie) collected while the beam is positioned at a particular pixel in the image field, so increasing the SNR requires generating a beam containing a larger current, increasing the collection time τ spent at each pixel, providing more efficient detectors, or any combination of these options. The current in the incident beam at each pixel in turn depends on the brightness of the source β given in units of (amps cm^{-2} sr^{-1}). Brightness is conserved in the beam column, so the brightness measured as the beam strikes the sample surface is equal to the value that would be measured at the gun. If (Figure 17.3) the beam diameter at focus (ignoring the effects of lens aberrations) is d_0 and the beam convergence angle is α, then the brightness at the source, and thus at the specimen

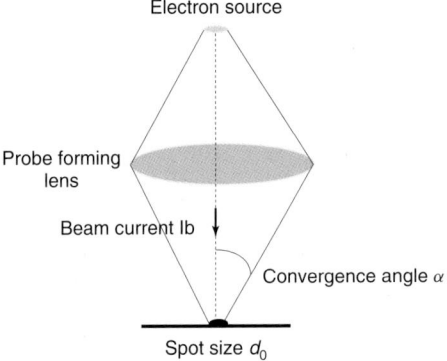

Figure 17.3 Beam parameters for brightness equation.

surface, is readily seen to be

$$\beta = \frac{4I_b}{\pi^2 d_0^2 \alpha^2} \text{ amps cm}^{-2} \text{ str}^{-1} \tag{17.1}$$

The magnitude of β is a property of the type of source used, and, in all cases, scales directly with the energy E_0 of the beam. The performance of the SEM depends crucially on the choice of the electron source and possibly the beam energy.

A tungsten "hairpin" thermionic electron emitter has a low brightness ($\sim 10^5$ amps cm^{-2} sr^{-1} at 20 keV) and an effective source size of about 50 μm in diameter. With this scenario, the best performance will be obtained by constantly adjusting the demagnification of the focused beam so as to match the probe size to the pixel size of the image. As the field of view of the image is reduced, the probe size must also be reduced and the current in the beam will decrease as d_0^2. This reduction in beam current cannot be compensated for by increasing the convergence angle α because this will increase the effective probe size as spherical and chromatic aberrations in the lenses (proportional to α^3 and α respectively) become significant. Under high-resolution conditions where the field of view is a few micrometers or less, it is no longer possible to reduce the probe size so as to match the pixel size because the incident beam current will become too small to be acceptable. At low beam energies, this limit will be reached sooner because of the lower source brightness. Resolution in thermionic SEMs is therefore typically of the order of 10 nm or so at 30 keV, and 25 or 30 nm at 5 keV, as a result of the elevated probe size, signal-to-noise restrictions, and electron–solid interaction effects.

In an SEM with a cold field emitter gun (CFEG), the situation is very different. The source brightness at 20 keV is $\sim 10^8$ amps cm^{-2} sr^{-1} or more and the effective source size is of the order of 25 nm. As a consequence, it is neither possible nor necessary to increase the available current at low imaging magnifications by increasing the probe size because adequate current is almost always available. (It should be noted that while this generally provides high-quality images, the wide difference in scale between the probe size and the pixel size can give rise to Moire interference patterns [10]. This artifact is readily distinguished because small changes in the area scanned or the orientation of the sample to the beam will result in very large changes in the appearance of the image). In practice, both the probe size and convergence angle are usually held fixed at predetermined compromise values. When higher currents are required, for example, for wide-field imaging or X-ray microanalysis, the source brightness can be raised by increasing the total emission current from the CFEG without any penalty in probe size and with no need to adjust the beam convergence. Although the effect of aberrations in the optical system does become more significant at lower beam energies this is partially offset by the reduction in the interaction volume and the improvement in SE yield that occurs, yielding essentially constant imaging performance from below 1 to 30 keV or higher. Figure 17.4 shows an example of high image quality now available from a CFEG SEM even at low energies.

The behavior of the HIM He$^+$ ion beam microscope is similar to those of a field emission gun SEM because the brightness of the "trimer" gun [5] is comparable.

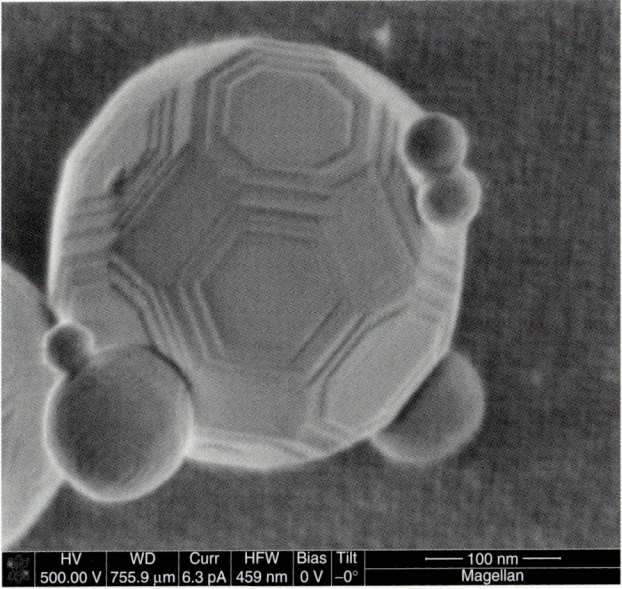

Figure 17.4 Secondary electron image of alumina powder grains, recorded at 500 eV beam-landing energy in a Magellan (FEI Inc.) SEM. Horizontal field of view 459 nm. (Image courtesy of Dr. J. Michael, Sandia National Laboratory.)

Thus within the typical 30–40 keV operating energy for a HIM, the resolution of the ion-generated secondary electron (iSE) signal extends down to the subnanometer range. There are also important differences, however. The wavelength λ of a helium ion beam of energy E_0 is smaller than that of an electron beam of the same energy by a factor of 86. As a result, the beam is not diffraction limited even if its convergence angle α is reduced to a value below 1 mrad. Although the brightness of the He^+ ion source is comparable with that of a field emitter electron source, this choice of small convergence angle results in a beam current of the order of a few picoamperes. However, as discussed later, because ions produce SEs much more efficiently than do electrons, such a current is still sufficient to produce an acceptable SNR and the depth of field that results from this choice is significantly better than that of a corresponding e-beam image. Finally, the beam range of He^+ ions in a solid is about less than that of electrons of the same energy in the same material by a factor of 30; so there is no reason to operate a helium-scanning microscope at low energy, and 40 keV and higher is the normal choice. Figure 17.5 shows a He^+ ion-induced SE image of the same alumina spheres similar to those imaged in Figure 17.4 with a low-energy electron beam. The quality of image detail, the signal to noise even for a current of just 1 pA, and the depth of field clearly demonstrate the benefits of this new imaging approach.

Ultimately, the performance of the SEM is limited by the large wavelength associated with electrons with energies of 30 keV or less, and by the relatively large

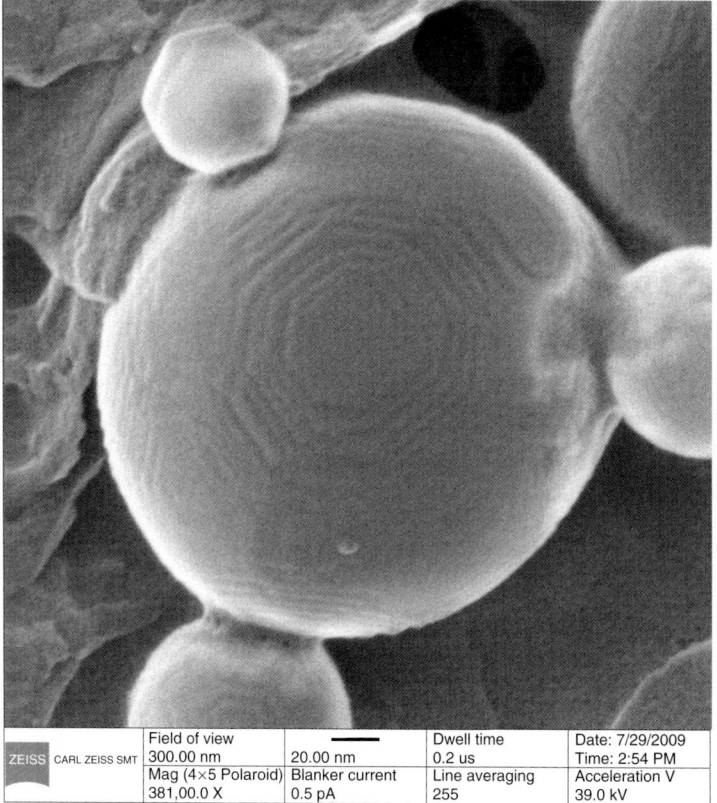

	Field of view	———	Dwell time	Date: 7/29/2009
ZEISS CARL ZEISS SMT	300.00 nm	20.00 nm	0.2 us	Time: 2:54 PM
	Mag (4×5 Polaroid) 381,00.0 X	Blanker current 0.5 pA	Line averaging 255	Acceleration V 39.0 kV

Figure 17.5 Helium ion–induced secondary electron image of alumina powder grains, recorded at 39 keV in an Orion (Zeiss SMT) helium ion scanning microscope. Horizontal field of view 300 nm.

aberrations of even the best probe-forming lens. These fundamental aberrations of lenses can now be reduced to vanishingly small values by "aberration correctors" [11]. This in turn allows the convergence angle α of the beam to be increased up to 30 or 40 mrad from the few milliradians usually found; in addition, the probe diameter is much reduced, and the current in the beam (which varies as α^2) is greatly increased. The inescapable drawback is that the depth of field, the vertical distance over which the image is in focus and which depends on the spot size and inversely on the convergence angle, becomes so small that much of the surface detail is always out of focus. This development, although of great interest, is thus likely to be of limited benefit in most circumstances. An alternative approach by Mullerova and Frank [12] relies on an abrupt deceleration of the beam as the result of a local "retarding field." The effect of this is to effectively replace the aberration coefficients of the probe lens with much smaller factors determined only by the parameters of the retarding field. Although simple, this procedure works well on a variety of specimens and requires no setting up or adjustment. Retarding fields have therefore been widely adopted for low-voltage, high-performance instruments such as those used for biology, and in the semiconductor industry for critical dimension (CD) metrology.

Even under the best circumstances, demonstrating and verifying the performance of a scanning microscope is a difficult task. The usual approach of quantifying performance in terms of a claimed spatial resolution derived from measurements of edge sharpness, or the minimum discernable gaps between particles, is unreliable because use is made of only a tiny fraction of the available pixels in the image. A better approach is to determine the contrast transfer function (CTF) of the microscope, which measures how different spatial frequency components in the imaged are transferred through the microscope and displayed on the screen [13]. This is done by computing the two-dimensional Fourier transform of the complete image of interest, radially averaging this over 2π, and then plotting the normalized Fourier intensity as a function of feature size (1/spatial frequency). Figure 17.6a shows the CTF derived from the image in Figure 17.4. For large-scale features (10 nm or greater), the CTF is close to 100% confirming that those spatial frequencies are transferred perfectly; but as the object size falls toward 1 nm the CTF profile falls and eventually at about 0.6 nm the CTF dissolves into random noise. Thus for this example, under the operating conditions employed, the resolution limit, as determined by the beam size, electron–solid interactions, and the SNR, is 0.6 nm. From the slope of the CTF at this point, it can also be deduced that improving the SNR by a factor of 2 (for example, by slowing the scan) would potentially improve the resolution to better than 0.5 nm. The smooth profile of the CTF as it falls toward the noise shows that the system is optimally adjusted. Problems such as defocus, astigmatism, drift, or less than optimum aperture choices, result in the profile displaying steps and local maxima and minima. For comparison, Figure 17.6b shows the corresponding CTF from the He^+ ion beam iSE image in Figure 17.5 of a similar specimen. The CTF is generally quite similar to the SEM version although the slope is a little different and the noise is also higher.

17.1.4
Modes of Operation

17.1.4.1 Secondary Electron Imaging

SEs (eSEs) are those emitted by the specimen as the result of irradiation by an electron beam and which emerge with energies below 50 eV. Because of their low energy, SEs can travel only relatively small distances (3–10 nm) in the specimen and so are associated with a shallow "escape" region just below the surface of the specimen. There are two circumstances in which an SE can be generated and subsequently escape from a specimen: first, when an incident electron passes downward through the escape depth region, and secondly when a BSE exits from the specimen and again passes through the escape region. Secondaries produced in the first type of event are labeled *SE1* and, because they are generated at or close

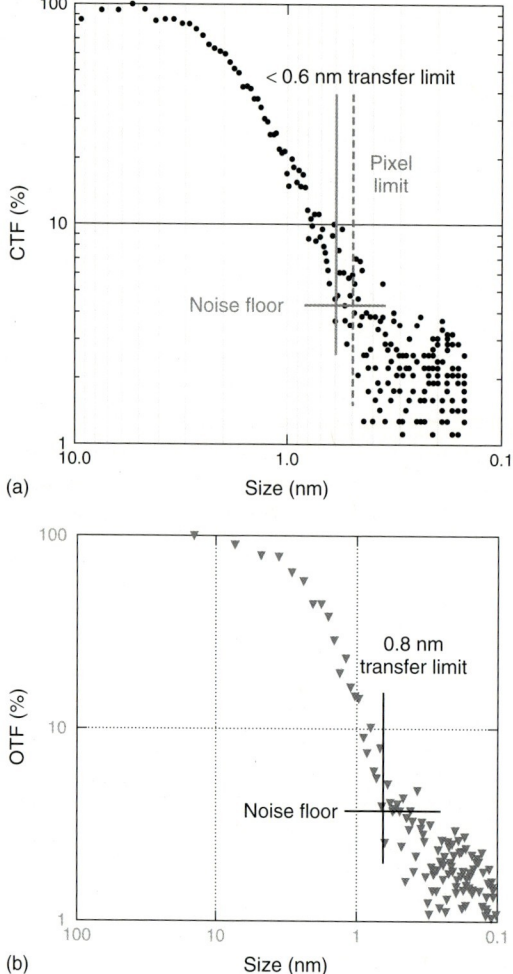

Figure 17.6 (a) Contrast transfer function derived from the image in Figure 17.4. The noise-limited resolution is 0.6 nm. (b) The corresponding contrast transfer plots for the helium ion beam image in Figure 17.5. The noise-limited resolution is 0.8 nm.

to the point where the incident beam enters, it is these which carry high-resolution, surface-specific information. The other secondaries called *SE2* come from the region within which the BSEs are generated. Since this can be micrometers in diameter, it can be seen the SE2 only carry low-resolution subsurface information. Because the yield δ of SE increases as the electron energy falls, the eSE2 signal is stronger than the eSE1 component by a factor of 2–3. It is usually taken that eSE1 and eSE2 signals cannot be separated because they have identical properties but recent evidence [14] suggests that the eSE1 energy spectrum lies predominately in the energy range below 10 eV while the SE2 spectrum contributes a long "tail" that extends from 10 up to 50 eV. Energy-filtering SE detectors could therefore enhance eSE1 visibility. When the SEM is operated under conditions such that the field of view is comparable with the diameter of backscattered interaction volume, then the eSE2 signal will be mostly independent of beam position and so form only a constant background to the eSE1 component.

SEs (iSE) are also generated by the helium ion beam. As for electrons, both iSE1 and iSE2 components can be generated but there are important differences between the two cases. The backscatter yield of He$^+$ ions is generally smaller than for electrons and falls rapidly as the energy increases, so the iSE2 contribution is less significant [15] and because the range of ions is an order of magnitude smaller than that of electrons of the same energy, both the iSE1 and iSE2 components will carry high-resolution, more surface-specific information.

The yield δ of SEs is directly proportional to the stopping power (in electron volts per nanometer) of the generating particle in the materials of interest [16]. Typically, the stopping power for electrons is low (10 eV nm^{-1}), reaches its maximum magnitude at an energy E_0 around 300 eV and then falls away as $1/E_0$; for He$^+$ ions, the stopping power is typically 50 eV nm^{-1} and reaches its maximum at an energy E_0 of about 700 keV [15]. Thus for electron beams the eSEs typically reach a maximum yield δ of about 1–1.5 eSE per incident electron at some energy below 1 keV and then falls steadily with increasing energy above this value. For the helium beam, the yield δ in the low kiloelectron volts region 10–50 keV is greater than unity and rises with increasing energy reaching a maximum of four to six depending on the material. SE imaging is therefore optimized at low energies in the SEM but at high energies in the HIM.

SE imaging is the most important mode of operation for the SEM and HIM and it has been estimated that 95% of all SEM/HIM images are recorded using this signal. This dominance is the result of several factors:

1) SEs are easy to collect
2) SEs carry information about the surface topography of the specimen. In addition, information about surface chemistry, and magnetic and electric fields may also be available on suitable samples
3) SE images are readily understood without the need for specialist knowledge
4) The eSE and iSE modes are capable of a spatial resolution below 1 nm on bulk samples, and Zhu et al. [17] have demonstrated single-atom resolution under optimized conditions and at high beam energy (200 keV).

The practical key to the success of the SE imaging has been the "ET" detector originally described by Everhart and Thornley [18] and illustrated in Figure 17.7. SEs are accelerated on to a scintillator material, such as a doped YAG crystal, by a bias voltage of about 10 keV. The light generated travels down the light pipe to a photomultiplier tube (PMT) which produces a greatly magnified electron signal as its output. To prevent the bias potential deflecting the incident beam, the scintillator is often surrounded by a Faraday cage biased to about +250 V. Because of the logarithmic gain characteristic of the PMT, the ET detector has a wide dynamic

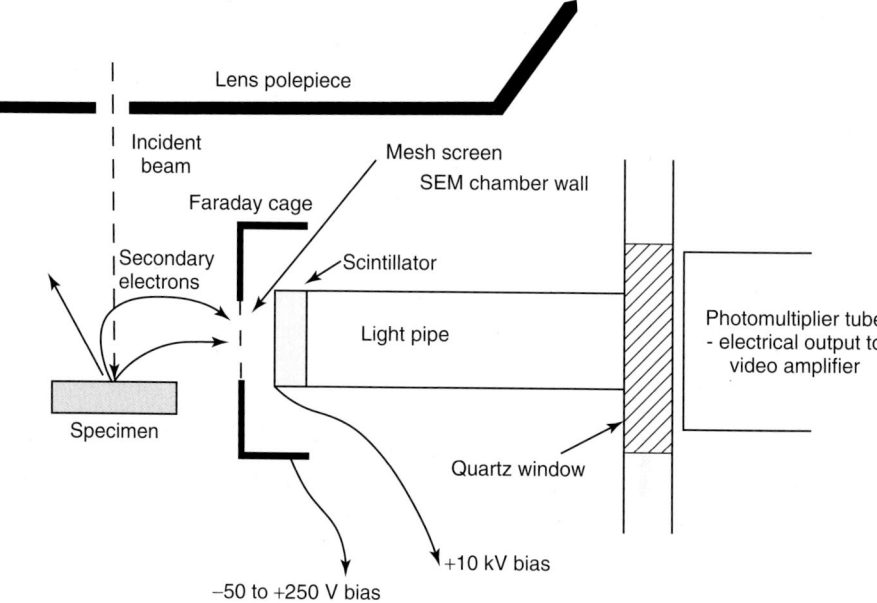

Figure 17.7 Schematic diagram of the Everhart–Thornley secondary electron detector.

range, a rapid response to changes in signal intensity, and collects 30–50% of the available SE signal. Increasingly, additional SE detectors are also provided. In particular, positioning a detector above the probe-forming lens and out of line of sight of the sample has been shown to greatly reduce the quantity of spurious scattered SEs collected by the detector [19]. These "SE3" electrons are the result of SE production at the walls of specimen chamber, the lens, X-rays detector, and so on, and can represent as much as 50% of the total SE flux, so minimizing this component significantly enhances image contrast and quality.

The general characteristics of electron beam–generated SE signals are illustrated by the example in Figure 17.8a (and see also Figures 17.4 and 17.5), which shows gold particles deposited on to a carbon substrate. The image has a marked three-dimensional appearance caused by the lighting and shadows that decorate each object. The ability of an observer to visualize topographic detail occurs because the yield of SE varies with the angle between the incident beam and the local surface normal [20]. Areas at a high angle to the beam are bright (large signal) compared to those faces which are normal to the beam, and faces looking toward a detector are in general brighter than those facing away. This behavior is analogous to the scattering of visible light as described by Lambert's cosine law, a situation which makes it easy to interpret SE images even without knowing anything of the mechanisms involved. Conventionally, the ET detector is positioned along the top edge of the field of view so that the "lighting" appears to come from above. Moving the detector, or changing to another detector in a different position (e.g., a through-the-lens detector), effectively changes the illumination. Thus a detector

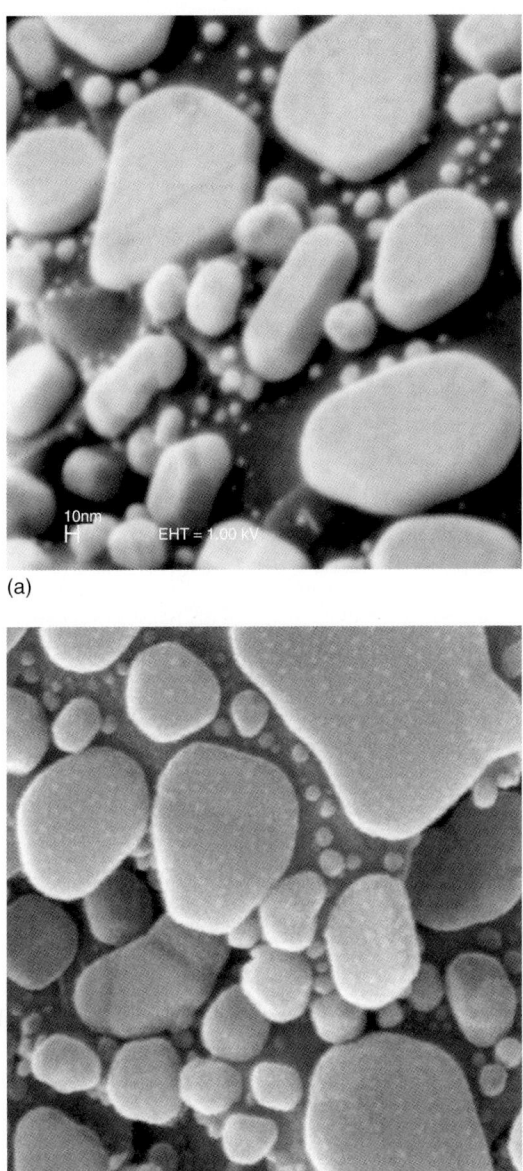

Figure 17.8 SE images of gold on carbon recorded (a) at 1 keV with an electron beam and (b) at 39 keV with a helium ion beam. Horizontal field of view 200 nm.

placed to one side of the sample and level with it will show strong shadowing effects, while the same detector positioned directly above the same specimen will give a "high noon" shadow-free view. The fact that this model for SE image interpretation applies over a wide range of scales and is reliable down to the nanometer scale has been a factor contributing to the popularity of the SEM because the ease of image interpretation compares well to the significantly greater complexity of, for example, the situation in an AFM.

The corresponding behavior of the He^+ iSE signal is similar to that of the electron beam case as is shown in Figure 17.18b. The appearance, and consequently the interpretation, of the image is similar in both cases although the enhanced visibility of surface detail and better edge definition is notable in the iSE case as is the improved depth of field.

SE images can also visualize other properties of the sample such as the presence, magnitude, and polarity of surface potentials (voltage contrast) or the presence of leakage magnetic fields (magnetic contrast). Such observations are generally qualitative rather than quantitative because the experimental conditions are not well defined nor are they reproducible; however, they can be of practical utility.

17.1.4.2 Backscattered Electrons and Ions

BSEs are defined as being those emitted from the sample with energies between 50 eV and the incident energy E_0. Unlike eSEs, which are viewed as electrons emitted from the sample as the result of the interaction of the incident beam, BSEs are viewed as being incident electrons that have been scattered through a total angle approaching 180° and so have once again left the sample. This is over simplistic as it is certain that there are eSEs with energies greater than 50 eV and BSEs with energies less than 50 eV but the distinction drawn is useful for discussion purposes. The yield η of BSEs per incident electron generally rises with the atomic number Z of the sample and for normal beam incidence, η is approximately equal to $Z/100$. Above $Z \sim 40$, the slope decreases and the yield ultimately saturates at a value of $\eta = 0.5$ for the heaviest elements. The BSE yield varies little as a function of beam energies in the range 5–50 keV but begins to fall at higher energies. For beam energies below 5 keV, the situation is more complex because η for heavy elements begins to fall, while η for light elements rises, with falling energy. As a result at 1 keV, gold and copper have the same BSE yield. The mean energy of a BSE is of the order of half the incident beam energy, but the most probable energy is higher for high Z materials and slightly lower for low Z.

For most materials at most energies, the BSE yield is higher than the eSE yield but backscattered imaging has received much less attention than eSE imaging. This is because of the practical problem of efficiently collecting the BSE signal. As the energy is high, it becomes difficult to deflect the electrons toward a detector, so the device must itself be made physically large enough to intercept and collect a significant fraction of the available signal. Typically, the BSE detector is a large-area Schottky diode, often subdivided into segments to provide a choice of collection geometries, but microchannel plates and Everhart–Thornley detectors with no bias voltage applied are also used.

Because the BSE yield varies with Z, the most common use of this mode is atomic number, or Z-contrast, imaging. This provides a quick and convenient method of examining the distribution of chemistry within a material and qualitatively distinguishing regions of high and low atomic number. The lateral, and the depth, spatial resolution of the BSE image are both some fraction of the electron range in the material of interest, a useful approximation for which is

$$\text{Range (nm)} = \frac{0.78 E_0^{1.67}}{\rho} \tag{17.2}$$

where E_0 is the beam energy (in keV) and ρ is the density of the material (in grams per cubic centimeter), so usefully good resolution is available. Monte Carlo simulations show that the typical BSE has reached a maximum depth of about 0.3 of the beam range before turning back toward the surface, so the visible contrast represents an average over depth. The vertical resolution of the BSE signal can however be enhanced by recording a sequence of images of the same while stepping upward in beam energy, and then subtracting sequential images from each other to form a simple tomogram. Z contrast is also widely used in biology because nanoscale particles of gold, or other heavy metals, can be attached to active molecular groups that preferentially bind to specified locations in a cell. A SEM with sufficiently good resolution and signal to noise can then distinguish several different labels by their size and signal level.

Backscattering is also of importance in the HIM. The yield of "Rutherford backscattered ions" (RBI) generated from the incident helium ions varies with the target atomic number in a manner similar to that for electrons, although, as shown in Figure 17.9, in this case, there are distinct minima in the yield curve that

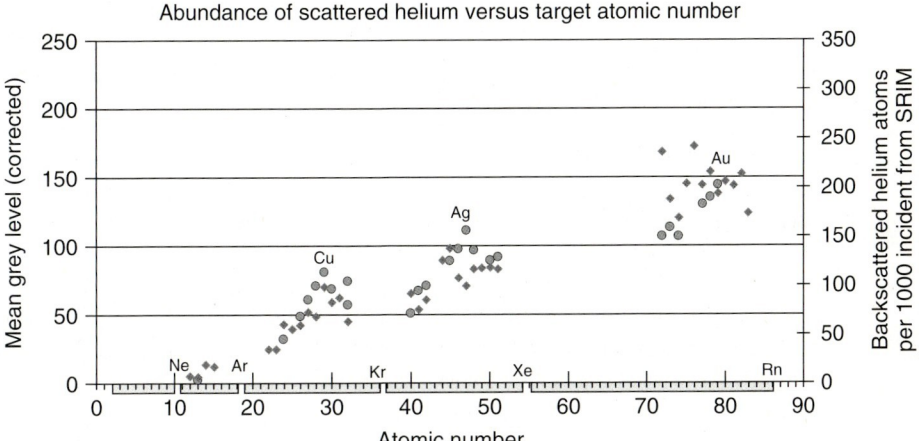

Figure 17.9 Experimental relative yield of Rutherford backscattered He$^+$ ions as a function of the scattering atom, and the corresponding sputter yield computed from Stopping and Range of Ions in Matter (SRIM), for an incident energy of 40 keV.

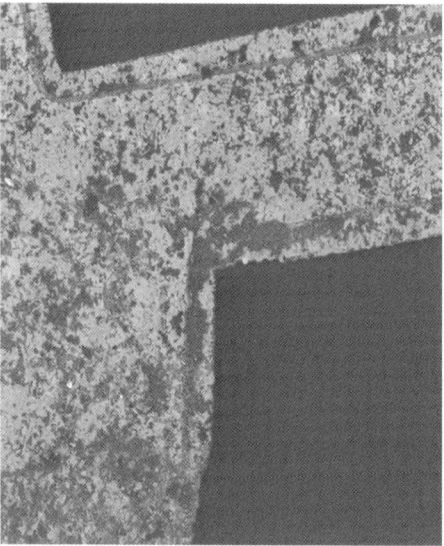

Figure 17.10 RBI image of a copper grid contaminated with carbon recorded with a 30 keV He$^+$ ion beam. Horizontal field of view 500 μm.

occur each time a shell of electrons in the atom is filled, an event which occurs for $Z = 2, 8, 18, 34$, and so on. As a result, although ion backscatter images are rich in chemical information as seen in Figure 17.10, it is not possible to perform chemical analysis of an unknown material because multiple materials produce the same ion backscatter yield. The ion backscatter yields for all elements fall with increasing beam energy, typically falling by a factor of 5 between 10 and 50 keV, but are generally lower than the BSE yield from the same material. The beam range for He$^+$ ions can be estimated as

$$\text{Range (nm)} = \frac{0.78 E_0^{0.7}}{\rho} \tag{17.3}$$

so the range of He$^+$ ions is smaller than that of electrons at the same energy by a factor of about "E_0." In addition, the He$^+$ beam, as noted above, has a low backscatter coefficient and less lateral scatter, so the form of the interaction volume is less diffuse than that for electrons (Figure 17.11). Consequently, the RBI image offers both higher spatial resolution and more surface sensitivity.

The backscattered signal also carries information about the crystalline nature of the sample. The origin of this contrast is illustrated schematically in Figure 17.12. If the incident electron beams enters a crystal at a random angle of incidence then the electron will approach an atom sufficiently closely so that backscattering will occur in the usual way. But if the incident beam is aligned along a direction of symmetry of the lattice, then the incident electron flies in between and parallel to the atom planes and as a result penetrates more deeply into the crystal and the backscattering yield is reduced. This simple particle model is not physically realistic but a detailed

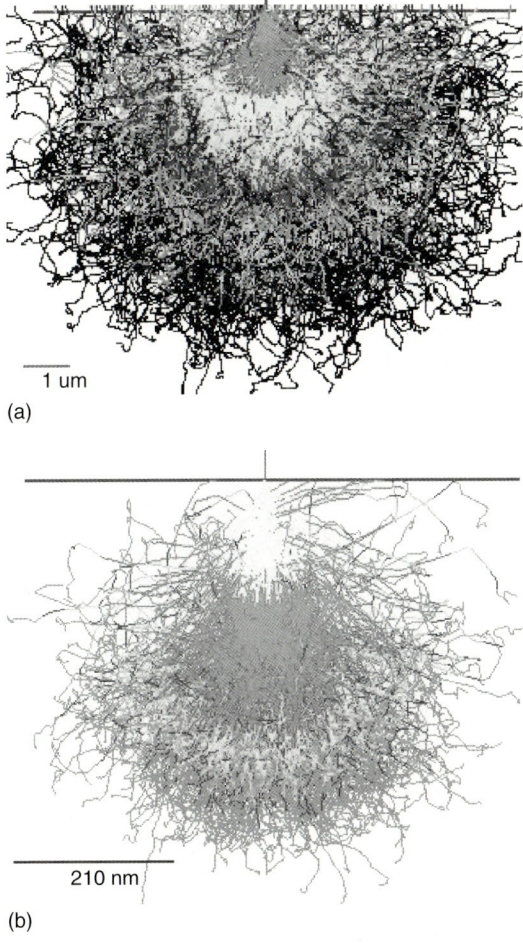

(a)

1 um

(b)

210 nm

Figure 17.11 Monte Carlo simulations of electron (a) and He+ (b) interaction volumes in silicon at 30 keV. Note the major difference in the size, scale, and shape of the volumes.

analysis using dynamical diffraction theory [21] confirms the existence of such a phenomenon. If the angle of incidence between the electron beam and the crystal is varied, then the backscattering yield is modulated at angles corresponding to the direction of symmetry of the crystals. An example of an "electron-channeling pattern" produced from a silicon wafer in this manner is shown in Figure 17.13. Here the variation in the angle of incidence has been achieved by scanning the large crystal at low magnification so that between the extremes of the field of view, the incidence angle varies by about $\pm 8°$. The micrograph therefore contains both spatial and angular information. These two can readily be separated because lateral motion of the crystal will vary topographic contrast but leave the channeling pattern

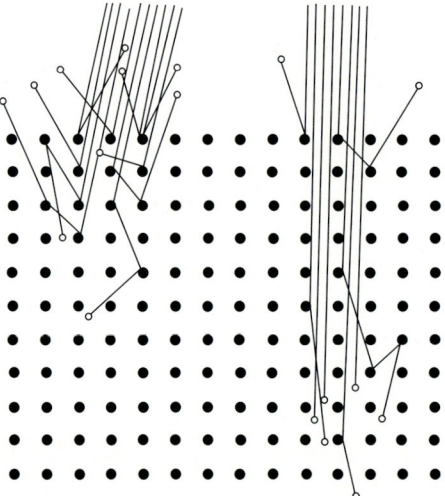

Figure 17.12 The origin of channeling for electrons traveling (left) in a random direction and (right) along a symmetry direction of the lattice.

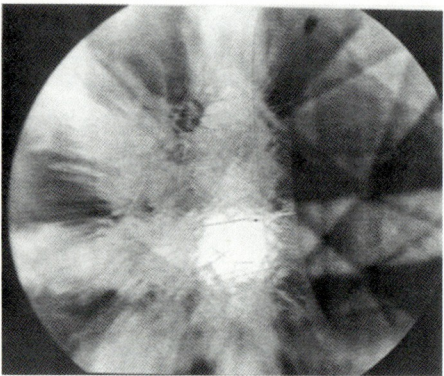

Figure 17.13 Electron channeling pattern recorded across a boundary between CaF$_2$ and silicon. Recorded at 20 keV with a backscattered electron detector.

untouched. The prominent bands in the pattern each have an angular width of $2\theta_{Bragg}$ where θ_{Bragg} is the solution of the Bragg equation

$$\lambda = 2d_{(hkl)} \sin \theta \tag{17.4}$$

where λ is the electron wavelength and $d_{(hkl)}$ is the lattice spacing of the planes whose Miller indices are (hkl). At 30 keV beam energy, typical Bragg angles are 1° or 2°.

By modifying the scan coil arrangement, the beam can be made to rock about a point, generating a channeling pattern from a region 1 μm or less in diameter.

The contrast level of channeling patterns is low, typically less than 5%, because channeling only extends for a short distance (30–100 nm) beneath the surface. Beyond that depth, the beam has been scattered too much and channeling ceases. As a result, the weak channeling contrast information is superimposed on the usual backscattered intensity and the visibility is poor and high beam currents and efficient detectors are necessary.

When the image field of view becomes sufficiently small, the scanned beam moves essentially normal to the specimen surface and no channeling pattern is generated. But if the crystal lattice of the specimen is polycrystalline, or imperfect because of the presence of dislocations or stacking faults, then as the beam scans, the relative orientation of the beam to the lattice will vary, producing contrast changes. The visibility of this variant of channeling contrast can be greatly enhanced by tilting the sample to 70° or so and placing the backscatter detector just on the horizon of the specimen. In that position, the BSEs reaching the sample have lost little energy before being deflected out of the sample in a single high-angle scattering event. As shown in Figure 17.14, this can be employed to image individual defects within the crystal [22] because if the sample is oriented with respect to the beam so as avoid strong reflections, then the lattice distortion around the dislocations may rotate the lattice into a strong channeling condition. The sample and beam geometry used here is identical to that for the collection of electron backscatter diffraction (EBSD) discussed below. Furthermore, when both channeling imaging and EBSD are used together, the dislocations, stacking faults, and other crystal defects can be characterized in detail.

Crystallographic contrast is also readily evident in He^+ ion beam images and, in particular, is a feature of the iSE image mode. The HIM does not generate visible channeling patterns because the short wavelength of the ions (Eq. (17.4)) results in Bragg angles that are below 1 mrad (0.05°) in size. But HIM images of

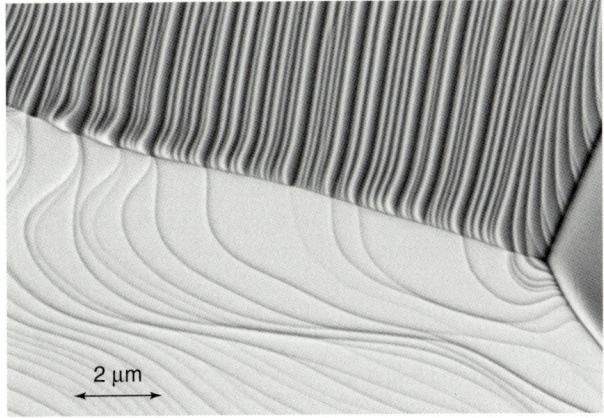

Figure 17.14 Electron channeling image of dislocation arrays in deformed copper recorded at 20 keV in backscatter mode. Horizontal field of view 15 μm. (Image courtesy Dr. Heiner Jachs, ZeissSMT.)

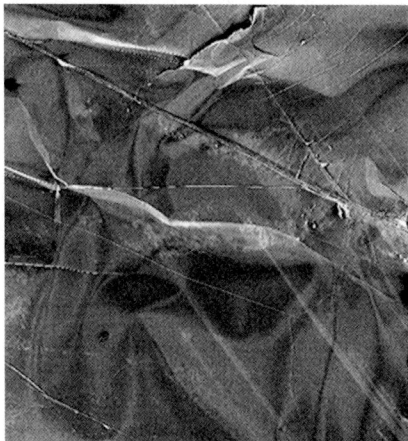

Figure 17.15 Bend contours and line defects (dislocations) in a bulk sample of MoS2 recorded in an ion-induced secondary electron image generated by a 39 keV helium ion beam. Field of view 25 μm.

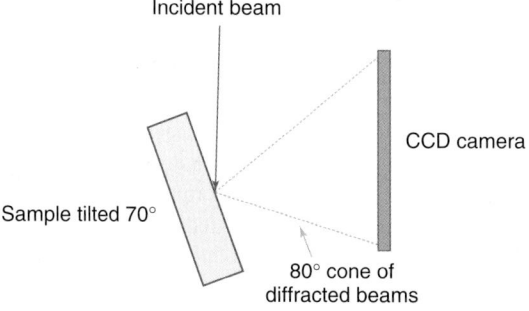

Figure 17.16 Schematic of geometry used to generate electron backscatter diffraction patterns.

polycrystalline specimens show strong (30%) crystallographic contrast details such as the bend contours and dislocations imaged on the surface of a bulk crystal of MoS2 (Figure 17.15) by a 40 keV He$^+$ beam. The fact that the sample can be bulk rather than thin, the excellent spatial resolution of the image, the very limited beam range in the specimen, and the strong contrast generated make this technique a valuable tool for the crystallographic study of surfaces at the nanoscale.

17.1.4.3 Electron Backscatter Diffraction Patterns

EBSD patterns [23] are generated using the geometry shown in Figure 17.16. The electron beam is focused on to the surface of a crystalline material at an incidence of 70° or more. In this condition backscattering is strong and occurs close to the sample surface before any significant energy loss has occurred. As the BSEs travel back through the crystal lattice, they are diffracted producing on a CCD camera, or

Figure 17.17 EBSD pattern recorded at 30 keV from a CdTe sample using a CCD camera. The angular width of the field of view is 80°. The central pole is (111), the upper is (001), and the lower is (110). (Image courtesy Dr. Joe Michael.)

a viewing screen, placed as shown, a complex geometric pattern of parallel lines often referred to as *Kikuchi lines* (Figure 17.17). The intersection of two or more pairs of lines is a zone axis and so related to a specific crystallographic direction in the crystal. The line pairs are associated with specific lattice planes in the crystal, and their spacing is proportional to the Bragg angle (Eq. (17.4)) and hence inversely proportional to the crystal spacing. An analysis of the pattern therefore permits the orientation of the crystalline region beneath the beam to be determined absolutely and with high precision. This information comes from an area of the specimen defined by the beam spot size horizontally but extended as a result of the specimen tilt in the vertical direction. Typically, minimum area a few hundred nanometers in size are achievable.

Using specialized computer programs the crystal orientation represented by an EBSD pattern can be determined in a few milliseconds, so it is possible to scan the incident beam across a material at almost normal imaging rates to produce an orientation map of the material. If an energy-dispersive X-ray analyzer is also installed on the SEM and is operating simultaneously with the EBSD system, then a map identifying the chemical phases present, by comparing their chemistry with a library of known materials, and their crystallography is generated. This technique is of great value and wide application and represents an increasingly important application of the SEM.

17.1.4.4 Electron Beam–Induced Conductivity (EBIC)

In addition to the major modes of imaging provided by the SEM, there are a variety of other techniques that are of more specialized interest and application, for example, in semiconductor materials containing junctions [24]. When energetic electrons strike a semiconductor material such as silicon, they promote electrons from the valence band across the bandgap leaving behind a "hole" in the valence band. The energy required to generate an electron–hole pair is typically between 2 and 4 eV, so a single incident electron with an energy of, for example, 10 keV generates about $10\,000/3 \sim 3300$ electron–hole pairs. In the absence of any external field, the electron and holes will quickly recombine. However, a semiconductor can sustain a potential difference across itself, so if a voltage is applied then the field will cause holes to move toward the negative potential and electrons to move toward the positive potential. The motion of these charge carriers constitutes a current flow and so the incident electron beam has produced conductivity in the semiconductor. If the electron beam is turned off, the current flow will cease. This phenomena is referred to as *electron beam–induced conductivity* (EBIC).

If the semiconductor contains a junction region connected as shown in the sketch of Figure 17.18a, then the junction is short-circuited through the external conductor and with no incident beam of electrons, no current will flow although there is a potential difference between the p and n sides of the junction of 0.5–1.0 V. The field associated with this potential difference extends for a distance which depends on the resistivity of the material but which is typically a few micrometers on either side of the physical position of the junction. This so-called "depletion zone" contains no mobile charge carriers. If the electron beam is allowed to strike the semiconductor far from the junction then there is no electric field and the electron–hole pairs generated will recombine and disappear. But if the electron beam is placed within the depletion zone, the electrons and holes will separate and generate a current flow. If the distance between the beam point and the junction is X then the fraction $f(X)$ of the carriers that diffuse to the junction is

$$f(X) = \exp\left(\frac{-X}{L}\right) \tag{17.5}$$

where L is the minority carrier diffusion length. The charge collected EBIC signal therefore falls away exponentially on either side of the junction, as shown in Figure 17.18b, at a rate dependent on the value of L, which is typically a few micrometers. This behavior was first observed experimentally by Everhart, Wells, and Oatley [25] and is now widely employed as a diagnostic for semiconductor devices and as a tool to measure minority carrier diffusion lengths. Alternatively, the requirement for a p–n junction can be avoided by forming a Schottky barrier on the surface (Figure 17.18c). This creates a depleted region beneath the surface so that during irradiation by the electron beam, the electron and holes are once again separated and produce a measurable current flow. Figure 17.18d shows an application of this technique to a GaAs wafer. The circular feature is the Schottky barrier, the finger structure is the top electrical contact to the Schottky barrier, and the contrast that is visible arises from differences in oxygen distribution within the wafer, which leads to local variations in the depth of the depleted region.

17.1.4.5 Image Artifacts

Charging, Charge Control The electron beam in an SEM injects a current I_b into the specimen. The emitted SEs are a current flow of δI_b, and the BSE signal is

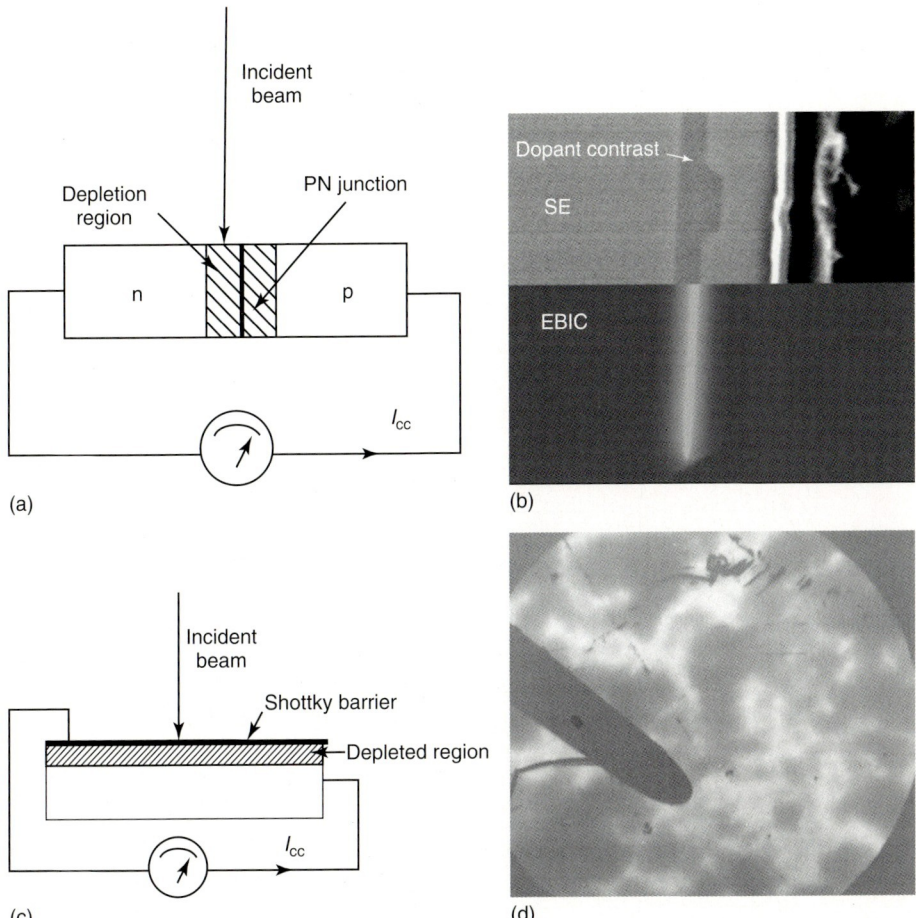

Figure 17.18 (a) Arrangement for EBIC imaging of a p–n junction, (b) EBIC image of junction in silicon, (c) arrangement for Schottky barrier imaging of semiconductor material, and (d) Schottky EBIC image of GaAs wafer showing variations in oxygen content.

ηI_b, and, in general, a specimen current I_{sc} can flow to or from ground. If these contributions match, that is, if

$$I_b = \delta I_b + \eta I_b + I_{sc} \tag{17.6}$$

then the specimen will remain electrically neutral and stable imaging will be possible under all conditions. At high beam energies (>5 keV) the secondary yield δ and the backscatter yield η are both significantly less than unity, so the excess current will flow to ground as specimen current. At low beam energies (<2 keV), the secondary yield δ may be higher than unity, so current must flow from ground to the sample to achieve charge balance.

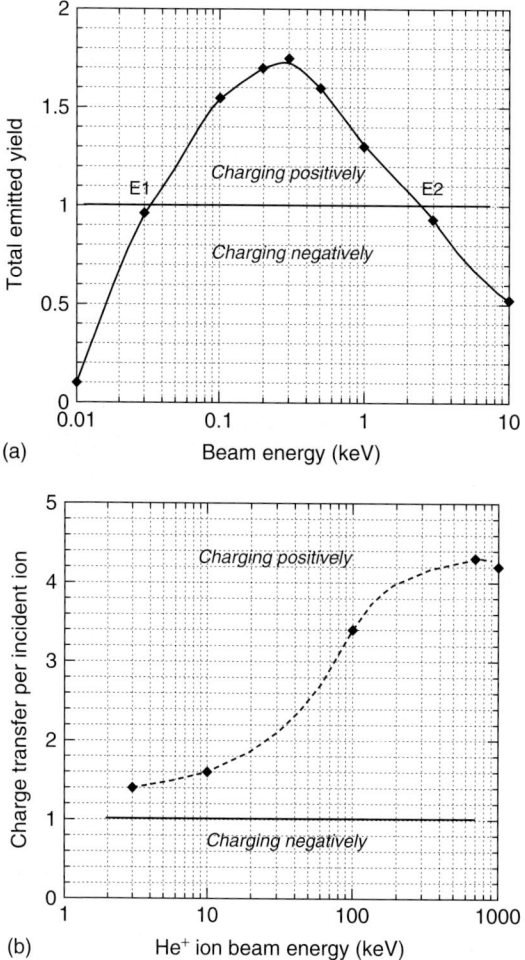

Figure 17.19 (a) Total yield ($\delta + \eta$) plot for quartz and charging behavior for electrons as a function of energy and (b) variation of net charge transfer to quartz as a function of energy for a beam of helium ions.

If the specimen is not electrically conductive, then the situation is more complex as shown in Figure 17.19a. The sum of the secondary and BSE signals δI_b and ηI_b is low at low energies, then rises with increasing energy before reaching a maximum at an energy typically around 1 keV before falling again. The energies E1 and E2 at which ($\delta + \eta$) is equal to unity are now charge balance points at which, in effect, each incident electron is reemitted as either a secondary or a BSE. Between E1 and E2 the specimen will charge positively, below E1 and above E2, it will charge positive, but a sample maintained at E1 or E2 would not charge although it is not conductive. Typically E1 energies are in the range 50–250 eV which is too low for

Table 17.1 E2 energies in kiloelectron volts for typical materials.

Resist	0.55	Kapton	0.4
Resist on Si	1.10	Polysulfone	1.1
PMMA	1.6	Nylon	1.2
Pyrex glass	1.9	Polystyrene	1.3
Cr on glass	2.0	Polyethylene	1.5
GaAs	2.6	PVC	1.65
Sapphire	2.9	PTFE	1.8
Quartz	3.0	Teflon	1.8

successful operation on many SEMs, but $E2$ is usually between 500 eV and 3 keV so the use of an appropriate choice of beam in this range in principle allows stable imaging of nonconductive specimens [26]. Table 17.1 lists $E1$ and $E2$ energies for a variety of materials.

In the case of the He^+ ion microscope, the charging situation is very different. The incident beam of positive ions generates a yield δ of SEs that is usually greater than unity and rises with increasing beam energy, and there is a small backscattered ion component that falls with increasing beam energy, so in an insulating specimen, each incident He ion implants positive charge into the sample and the SEs that are generated carry away negative charge. Consequently, at all ion energies of interest, the charging induced in the specimen by the He^+ ion beam is always positive in sign and comparatively high in magnitude (Figure 17.19b). The most successful method of compensating for ion-induced charging is to periodically irradiate the scanned area of the sample with a flood beam of very low-energy electrons [5], adjusting the intensity of this beam so as to exactly balance the ion-induced positive charging.

Low-voltage operation has now become the most popular mode of SEM usage as a result of the adoption of the idea that nonconducting samples can be imaged in a stable manner by operating at the $E2$ energy. This trend has been further encouraged by the improvements, discussed earlier in this chapter, which have greatly enhanced SEM performance at low energies. In practice, however, although low-voltage operation can stabilize the imaging of some insulating specimens, in many other cases, the method does not work well. The value of the $E2$ energy for a given material varies with the angle of incidence of the beam to the surface because this changes both the secondary and the backscattered yields and so on a specimen exhibiting a rough surface topography, the value of $E2$ will change from point to point and the image will again be unstable.

An alternative to this problem, which permits a nonconducting specimen to be observed in a stable manner at any energy of choice, is to fill the specimen chamber with low-pressure gas, typically, 30–50 Pa of air. Alternatively, the sample chamber itself may be held at high vacuum with gas being injected from a hypodermic needle aimed toward and close to the sample surface and locally providing a gas pressure of a 10–30 Pa. SEMs operated in this mode are referred to as *variable pressure scanning electron microscopes* as distinct from ESEMs in which the specimen

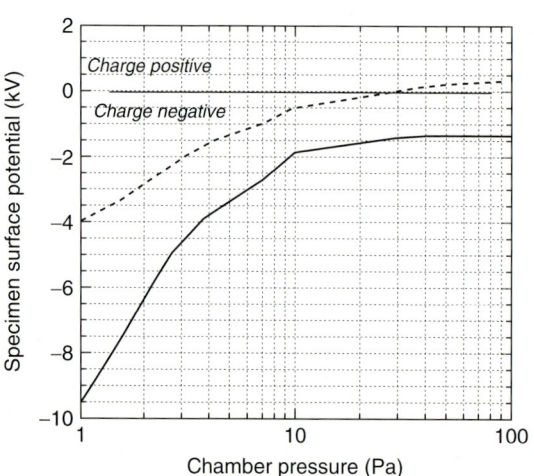

Figure 17.20 Experimental measurement of the surface potential of a glass slide under electron bombardment at 10 keV (dotted line) and 30 keV (solid line) as a function of gas pressure in the specimen chamber.

chamber pressure can exceed 5000 Pa [20]. In either case, when the specimen is irradiated by the incident beam, the SEs that are generated ionize the gas molecules closest to the specimen surface producing both positive and negative species (Figure 17.20). Positive ion species drift toward the negatively charged specimen gradually neutralizing its charge while the negatively charged ions drift toward the gun. This mechanism is self-correcting, works over a wide range of beam energies and gas pressures, and can hold the specimen surface potential at or close to zero potential (Figure 17.20). The presence of gas in the VPSEM specimen chamber causes electrical breakdowns, which prevent the use of the standard ET-detector so alternative strategies – such as solid-state BSE detectors or devices that rely on photoluminescence – are employed. The presence of the gaseous atmosphere also leads to some scattering of the incident electron beam. This reduces the contrast of the image but has little effect on the imaging resolution that can be obtained.

Damage Electrons beams are the most intense source of ionizing radiation; consequently, SEM observations of samples such as polymers, some ionic materials, and most biological materials, results in mass loss, shrinkage, and, in some cases, the complete destruction of the specimen. It is often suggested that low-energy SEM observation minimizes or avoids such damage. However, since even an SE with just a few electron volts of energy can disrupt bonding in a polymer, this belief is incorrect. A more complete analysis as illustrated in Figure 17.21 shows that at very low energies (<100 eV) beam damage is severe but, because the beam

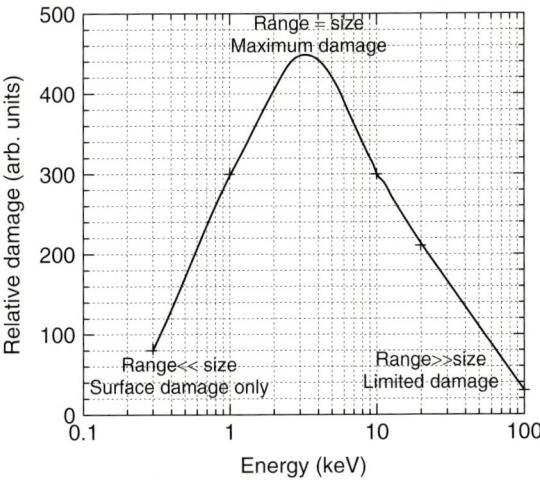

Figure 17.21 Schematic plot of variation of sample damage by electrons as a function of beam energy. (After Ref. [27]).

range is short, is confined to a region within a few nanometers of the surface. As the beam energy is increased, the damage zone is extended in depth and ultimately the damage reaches a maximum intensity when the beam range is comparable with the thickness of the specimen. Further increases in beam energy result in a reduction in the damage because an increasing fraction of the electrons deposit their energy somewhere outside the sample. Thus, while it is not possible to eliminate beam induced damage, its severity and manner in which it affects the specimen can be managed by choosing the most appropriate beam energy.

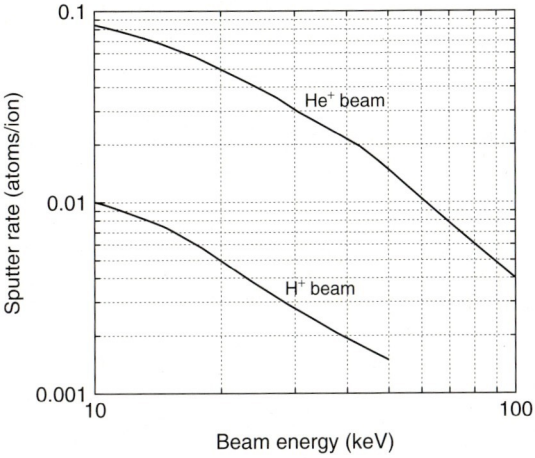

Figure 17.22 Predicted sputter damage rate (atoms per ion) from H^+ and He^+ ion beams as a function of energy.

For ions, the nature and the severity of damage depends on the mass of the ion. For light ions such as H^+ and He^+, the rate (atoms per incident ion) at which material is removed from the surface by the beam is modest at 10 keV and falls further as the beam energy is increased as shown in Figure 17.22. Since in the case of He^+, the iSE yield rises quickly with increasing beam energy, the operation at an elevated energy, say 100 keV, would decrease the damage rate by a factor of 10× and increase the signal by a factor of 5, so for a constant signal-to-noise ratio, the damage could be reduced by a large factor indicating the damage is not significantly more serious for He^+ than for electron beams under comparable signal-to-noise conditions.

17.2
Conclusions

Scanning microscopy is a uniquely versatile and powerful tool for the characterization and visualization of materials. For beams of both electrons and light ions such as He^+, current scanning instruments offer spatial resolution at the subnanometer level, operation over a wide range of energies, information from both the surface and the interior of samples, a wide variety of imaging and analytical modes, and the ability to handle real-world samples rather than just nanoparticles and ultrathin foils. When used separately, or when combined into a dual-beam instrument, the future of both the SEM and the HIM look promising.

References

1. von Ardenne, M. (1938) Das Elektronen-Rastermikroscop. *Z. Phys.*, **109**, 553.
2. von Ardenne, M. (1938) Das Elektronen-Rastermikroscop. *Z. Tech. Phys.*, **19**, 407.
3. Zworykin, V.K., Hillier, J., and Snyder, R.L. (1942) A scanning electron microscope. *ASTM Bull.*, **117**, 15.
4. Oatley, C.W. (1982) The early history of the scanning microscope. *J. Appl. Phys.*, **53**, R1–R13.
5. Ward, B.M., Notte, J., and Economou, N.P. (2006) Helium ion microscopy. *J. Vac. Sci. Technol.*, **B24**, 2871–2874.
6. Mulvey, T.E. and Newman, C.D. (1973) New electron-optical systems for SEM and STEM. *Inst. Phys. Conf. Ser.*, **18**, 16.
7. Nagatani, T., Saito, S., Sato, M., and Yamada, M. (1987) Development of an ultra high resolution scanning electron microscope by means of a field-emission source and in-lens systerm. *Scanning Microsc.*, **1**, 901.
8. Jachs, H. (2004) High performance SEM. *Imaging Microsc.*, **4**, 34–35.
9. Rose, A. (1948) in *Advances in Electronics* (ed. A. Marton), Academic Press, New York, pp. 131–184.
10. Read, D.T. and Dally, W.R. (1996) Theory of electron beam Moire. *J. Res. Natl. Inst. Stand. Technol.*, **101**, 47–52.
11. Joy, D.C. (2008) in *Biological Low-Voltage Scanning Electron Microscopy* (eds H. Schatten and J. Pawley), Springer Science, New York, pp. 107–128.
12. Mullerova, I. and Frank, L. (1992) Low voltage scanning electron microscopy. *Scanning*, **15**, 93–99.
13. Joy, D.C., Michael, J., and Griffin, B.J. (2010) Measuring the performance of the SEM. *Proc. SPIE*, **7638** (3J), 1–8.
14. Cazaux, J. (2003) On secondsry electron emission. *J. Microsc.*, **214**, 341–345.

15. Ramachandra, R., Griffin, B.J., and Joy, D.C. (2009) A model of secondary electron imaging in the Helium Ion scanning microscope. *Ultramicroscopy*, **109**, 748–757.
16. Bethe, H.A. (1941) Secondary electron emission. *Phys. Rev.*, **59**, 940–942.
17. Zhu, Y., Inada, H., Nakamura, K., and Wall, J. (2009) Imaging single atoms using secondary electrons with an aberration-corrected electron microscope. *Nat. Mater.*, **8**, 808–812.
18. Everhart, T.E. and Thornley, R.F.M. (1960) Wide-band detector for micro-microampere low-energy electron currents. *J. Sci. Instrum.*, **37**, 246–248.
19. Kimura, H. and Tamura, H. (1967) A new detector for the SEM, Proceedings of the 9th Annual Symposium on Electron, Ion and Laser Beams, pp. 198–204.
20. Goldstein, J., Newbury, D.E., Joy, D.C., Lyman, C.E., Echlin, P., Lifshin, E., Sawyer, L., and Michael, J. (eds) (2006) *Scanning Electron Microscopy and X-ray Microanalysis*, 3rd edn, 2nd reprint, Kluwer/Plenum, New York.
21. Joy, D.C., Newbury, D.E., and Davidson, D.E. (1982) Electron channeling patterns in the SEM. *J. Appl. Phys.*, **53**, R81–103.
22. Trager-Cowan, C., Sweeney, F., Edwards, P.R., Dynowski, F.L., Wilkinson, A.J., Winkleman, A., Day, A.P., Wang, T., Parbrook, P.J., Watson, I.M., and Joy, D.C. (2008) Electron channeling and ion channeling contrast imaging of dislocations in nitride thin films. *Microsc. Microanal.*, **14**, 1194–1195.
23. Schwartz, A.J., Kumar, M., and Adams, B.L. (eds) (2000) *Electron Backscatter Diffraction in Materials Science*, Kluwer/Academic Press, New York.
24. Holt, D.B. and Joy, D.C. (eds) (1989) *SEM Microcharacterization of Semiconductors*, Academic Press, London.
25. Everhart, T.E., Wells, O.C., and Oatley, C.W. (1959) Electron beam induced currents. *J. Electron. Control*, **7**, 9–16.
26. Joy, D.C. and Joy, C.S. (1996) Low voltage scanning electron microscopy. *Micron*, **27**, 247–263.
27. Egerton, R.F., Li, P., and Malac, M. (2004) Electron beam damage. *Micron*, **35**, 399–403.

18
Fundamentals of the Focused Ion Beam System
Nan Yao

18.1
Focused Ion Beam Principles

18.1.1
Introduction

As a relatively new tool, the focused ion beam (FIB) system offers a broad range of potential nanotechnology applications. Curiously, the FIB originated from research intended for developing space thrusters. While looking to develop thrusters that used charged metal droplets, Krohn first documented ion emission from a liquid metal source in 1961 [1]. His research on liquid metal ion sources (LMISs) for applications in outer space [2] quickly proved to be useful for many other areas, especially semiconductors and materials science. Commercialization of the FIB was not far behind in the 1980s, mainly oriented toward the rapidly growing semiconductor industry. However, it also found a niche in the field of materials science. Materials scientists perpetually seek imaging and analysis on ever-shrinking scales to obtain a more complete understanding of material structure, properties, performance, and other phenomena, as well as material modification via micro- and nanomachining.

The modern FIB system utilizes an LMIS at the top of its column to produce ions (Figure 18.1). Gallium (Ga) is usually used for its low melting point (30 °C), high mass, low volatility, and low vapor pressure, and because it is easily distinguished from other elements. The gallium source is heated to near evaporation, allowing to liquefy, and flow down the tungsten needle where it can remain in a liquid state for weeks without further heating, owing to its supercooling properties [3]. The extractor, an annular electrode centered just below the tip of the needle, is held at a voltage on the order of −6 kV relative to the source, drawing the liquid Ga into a Taylor cone. An electric field on the order of 10^{10} V m^{-1} between the tip and aperture induces field evaporation, causing ion emission and acceleration down the column. The source is typically operated at low emission currents of 1−3 µA to produce a stable beam [3]. Another electrode, the suppressor, works alongside the extractor at around +2 kV to maintain a constant beam current. The suppressor

Handbook of Nanoscopy, First Edition. Edited by Gustaaf Van Tendeloo, Dirk Van Dyck, and Stephen J. Pennycook.
© 2012 Wiley-VCH Verlag GmbH & Co. KGaA. Published 2012 by Wiley-VCH Verlag GmbH & Co. KGaA.

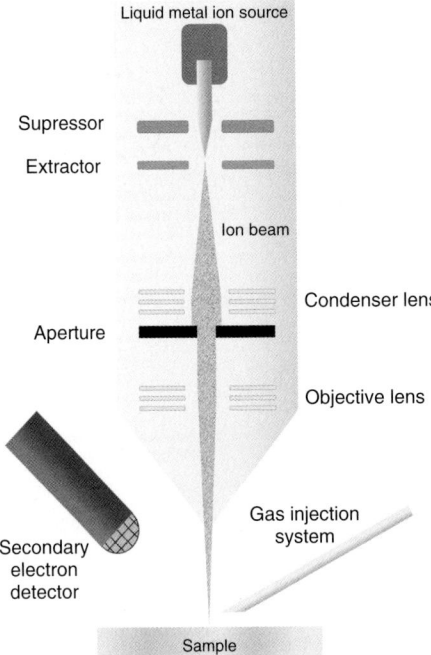

Figure 18.1 Schematic diagram of the FIB system.

and extractor act as the "fine" and "coarse" controls in regulating the ion extraction current.

18.1.2
The Ion Beam

18.1.2.1 Focusing the Ion Beam

The ion beam must be focused by a series of lenses to obtain the desired spot size. A common setup for an FIB system uses a condenser lens, aperture, and objective lens in series. Unlike the scanning electron microscope (SEM), which uses magnetic fields as lenses, the ion beam requires electrostatic lenses (for kiloelectron volt ion beams) because force of a magnetic field depends directly on the velocity of the charged particle, and the much more massive ions cannot be accelerated to the same speeds as electrons. The lenses consist of two or three electrodes with alternating polarities, which create electric fields that accelerate or decelerate the ion beam while constricting the beam diameter. As shown in Figure 18.2, when Ga^+ ions enter a three-electrode lens, they first accelerate under the influence of an increasingly negative field between the first and second electrode. The beam also constricts as the ions follow the field lines. Since the ions gain momentum after passing through the first field, the beam is pulled outward to a lesser degree by the negative electrode. The beam constricts again and the ions decelerate as they pass

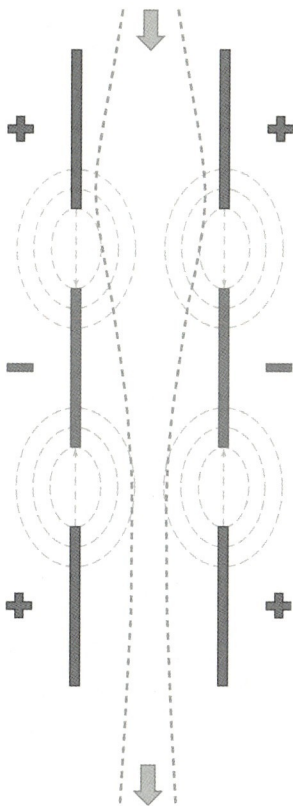

Figure 18.2 Diagram of the operating principle of an electrostatic lens. The dark dotted lines represent the boundaries of the ion beam, the dotted light lines represent the electric field lines.

through the increasingly positive field between the second and third electrodes. This process is repeated and yields the net effect of focusing the beam with no net change in beam energy. For higher-voltage megaelectron volt ion beams, magnetic quadrupole lenses are used because their field lines are perpendicular to the ion trajectories [4].

18.1.2.2 Beam Overlap

At this point, a short discussion of how the ion beam scans the sample surface is in order. First, the ion beam current acceleration voltage, focus, and stigmatism are set by the operator in order to define the beam diameter (D) (Figure 18.3). The operator then draws the pattern to be scanned by the beam spot. This pattern consists of digital pixels indicating the location of the beam spot. The ion beam will stay at each pixel for a predetermined period of time, known as the *dwell time*, after which the beam is blanked and moves with step size S to the next pixel. On

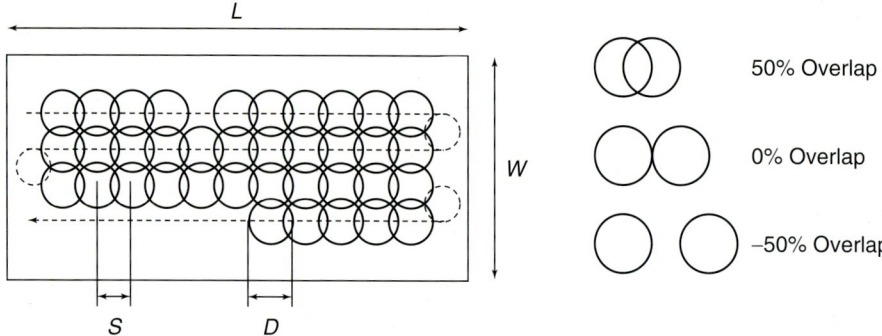

Figure 18.3 Relationship of step size S, beam diameter D, and overlap [5].

the basis of beam diameter D and step size S, we can define the beam overlap (OL) as follows:

$$\text{OL} = \frac{D - S}{D} \quad (18.1)$$

Beam overlap is one of the critical parameters to achieving successful milling or deposition in FIB. As can be seen from the above formula, negative overlap can occur if the step size is greater than the spot size. In general, positive overlap is used for milling, zero overlap is used for etching, and negative overlap is used for deposition.

18.1.3
Interactions of Ions with Matter

A number of interactions occur when the beam of energetic particles strikes the sample surface. The collisions of energetic ions on a solid creates a complicated many-body problem. Some of the particles are backscattered from the surface layers, while the others are slowed down within the solid. Unlike electrons, the relatively large ions have difficulty penetrating the sample surface because it is much harder for them to pass through individual atoms. Their size increases their probability of interactions with atoms, causing a rapid loss of energy as they slow down to a halt. Ions are many times more massive than electrons; a Ga^+ ion can carry 370 times more momentum than an electron. Because of this large mass difference, the collisions between ions and atomic nuclei (called *elastic collisions* or *nuclear collisions*) are distinct from the collisions between ions and electrons (called *inelastic collisions* or *electronic collisions*).

In a nuclear collision, kinetic energy and momentum should be conserved (hence the name *elastic*). When the incident ions strike the target atoms, the energy and momentum transferred to the atoms on the surface of the material cause a sufficient disturbance to remove them from their aligned positions. This sputtering effect, which does not occur when an electron beam is used, allows the FIB to precisely remove atoms from the surface of a material in a controlled manner. The

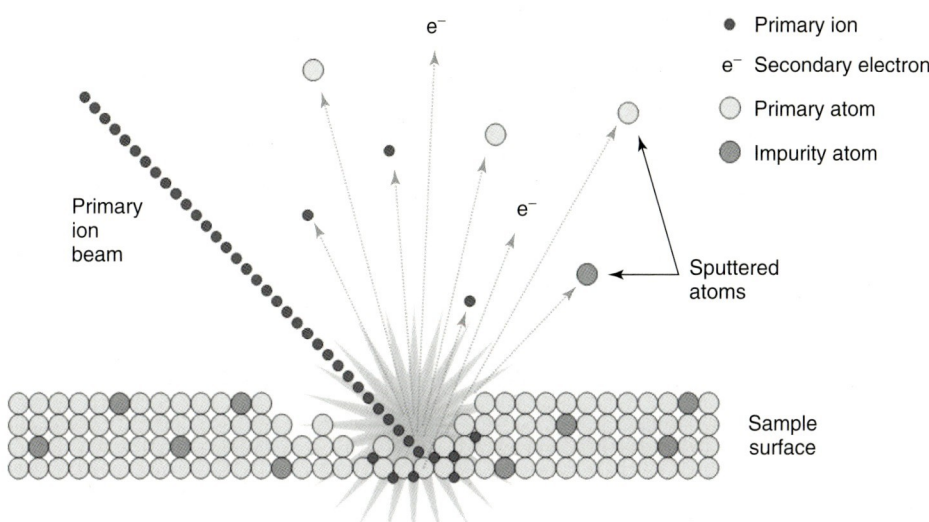

Figure 18.4 Schematic diagram of ion beam sputtering process on sample surface.

incident ions are also scattered, some becoming implanted in the material and others getting backscattered. Furthermore, collisions that result from ion beam bombardment induce many secondary processes such as recoil and sputtering of constituent atoms, electron excitation and emission, defect formation, and photon emission (Figure 18.4). These collisions lead to thermal and radiation-induced diffusion, which contributes to various phenomena involving the constituent elements, including phase transformations, crystallization, amorphization, track formation, and so on. Also, processes such as ion implantation and sputtering alters the surface morphology of the sample, resulting in the formation of craters, facets, pyramids, grooves, ridges, blisters, or a porous surface.

Aside from sputtering effects from nuclear collisions, the electronic collisions produce secondary electrons and X-rays. In the inelastic ion–electron collisions, the electrons are excited and ionized. Additionally, the incident ions transfer energy to the surface atoms and electrons, producing secondary electrons, ions, and so on. If the incident ion has insufficient energy to penetrate deep inside the atom, the inner-shell electrons screen the nuclear charge; this screening effect by electrons must be taken into account for these nuclear collisions. In the *following* subsections, some theoretical models describing these collision interactions are considered [6].

18.1.3.1 Interatomic Potentials

When two particles with charges $Z_1 e$ and $Z_2 e$ collide, an accurate and versatile potential that is often used is the Coulomb potential multiplied by the screening function Φ_{TF} deduced on the basis of the Thomas–Fermi atomic model:

$$V(r) = \frac{Z_1 Z_2 e^2}{r} \Phi_{TF}(x) \tag{18.2}$$

where r is the distance between two particles, x is given by $x = r/a$ and $e^2 = 1.44$ eV nm. The screening length a can be approximated by

$$a \approx \frac{0.8853 a_B}{\left(Z_1^{1/2} + Z_2^{1/2}\right)^{2/3}} \tag{18.3}$$

where a_B is the Bohr radius. The Thomas–Fermi screening function Φ_{TF} can be calculated from the equation

$$\frac{d^2 \Phi_{TF}(x)}{dx^2} = \frac{\Phi_{TF}^{3/2}(x)}{x^{1/2}} \tag{18.4}$$

On the basis of the above equations, several approximations have been developed, including a universal screening function proposed by Ziegler et al.[6]:

$$V(r) = \frac{Z_1 Z_2 e^2}{r[0.1818 \exp(-3.2x) + 0.5099 \exp(-0.9423x) + 0.2802 \exp(-0.4029x) + 0.02817 \exp(-0.2016x)]} \tag{18.5}$$

where $x = r/a_p$ with $a_p = \left(\frac{1}{(Z_1^{0.23} + Z_2^{0.23})}\right) \times 0.8854 a_B$

18.1.3.2 Binary Scattering and Recoil

The collision of atoms of masses M_1 and M_2 and charges Z_1 and Z_2 can be treated in the center-of-mass system as the classical motion of a particle with mass $\mu = \frac{(M_1 M_2)}{(M_1 + M_2)}$ in the potential of $V(r)$. The scattering angle χ is obtained as a function of the impact parameter p from the integral:

$$\chi = \pi - 2 \int_{r_{min}}^{\infty} \frac{p \, dr}{r \sqrt{\left(1 - \frac{V(r)}{E}\right) r^2 - p^2}} \tag{18.6}$$

If we consider the case of a hard sphere with radius r_h, and the Coulomb potentials are given by $V(r) = \frac{Z_1 Z_2 e^2}{r_h}$ and $V'(r) = \frac{-Z_1 Z_2 e^2}{r_h^2}$, then the scattering angles can be reduced to $\chi = \pi - 2 \arcsin\left(\frac{p}{r_h}\right)$, $\chi = \pi - 2 \arctan\left(\frac{2pE}{Z_1 Z_2 e^2}\right)$, $\chi = \pi \left[1 - \left(1 + \left(\frac{Z_1 Z_2 e^2}{p^2 E}\right)\right)^{-\frac{1}{2}}\right]$. Given the kinetic energy E_0 of the incident ion, the energies E_1 and E_2 of the scattered and recoiled particles, respectively, are

$$E_1 = \left(\frac{(M_1 - M_2)^2 + 4 M_1 M_2 \cos^2\left(\frac{\chi}{2}\right)}{(M_1 + M_2)^2}\right) E_0 \tag{18.7}$$

$$E_2 = \frac{4 M_1 M_2 E_0 \sin^2\left(\frac{\chi}{2}\right)}{(M_1 + M_2)^2} \tag{18.8}$$

18.1.3.3 Cross Section

The cross section of a particle is a way of describing the probability of collision with incident ions. Lighter materials tend to have lower cross section, thus allowing for deeper ion penetration. For a particle scattered to the polar angle χ, the cross section can be obtained from the following relation:

$$\sigma(\chi) = \frac{p(\chi)}{\sin \chi} \left| \frac{dp}{d\chi} \right| \tag{18.9}$$

Again, for the case of Coulomb potentials, $V(r) = \frac{Z_1 Z_2 e^2}{r}$ and $V'(r) = \frac{-Z_1 Z_2 e^2}{r^2}$, the results can be expressed as

$$\sigma(\chi) = \left(\frac{Z_1 Z_2 e^2}{4E} \right)^2 \sin^{-4}\left(\frac{\chi}{2} \right) \tag{18.10}$$

and

$$\sigma(\chi) = \frac{\left(\frac{\pi^2 Z_1 Z_2 e^2}{E} \right) (\pi - \chi)}{\chi^2 (2\pi - \chi)^2 \sin \chi} \tag{18.11}$$

Using the Thomas–Fermi potential for the low-energy range, Lindhard et al. approximated the differential scattering cross section as

$$d\sigma = \pi a^2 \frac{dt}{2t^{\frac{3}{2}}} f(\sqrt{t}) \tag{18.12}$$

where $t = \varepsilon^2 \sin^2\left(\frac{\chi}{2} \right)$. The reduced energy ε, the ratio between the screening radius a and the collision diameter b is given by

$$\varepsilon = \frac{a}{b} = \frac{M_2 E_0}{\frac{(M_1 + M_2) Z_1 Z_2 e^2}{a}} \tag{18.13}$$

$f(\sqrt{t})$ can be approximated as

$$f(\sqrt{t}) = \frac{\lambda t^{1/6}}{\left[1 + 2\left(\lambda t^{2/3} \right)^{2/3} \right]^{3/2}} \tag{18.14}$$

where $\lambda = 1.309$.

18.1.3.4 Energy Loss

As ions travel through matter, the electronic and nuclear collisions cause them to lose energy. These energy losses are known respectively as *electronic* or *inelastic* energy loss and *nuclear* or *elastic* energy loss. The most familiar formula for energy loss is the Bethe–Bloch equation, which is applicable in the high-velocity range where screening of nuclear charge and nuclear energy loss can be neglected:

$$\frac{dE}{Ndx} = \frac{4\pi Z_1^2 Z_2 e^4}{mv^2} \left\{ \ln \frac{2mv^2}{I} - \left[\ln\left(1 - \left(\frac{v}{c} \right)^2 \right) + \left(\frac{v}{c} \right)^2 + \frac{C}{Z_2} + \frac{\delta}{2} \right] \right\} \tag{18.15}$$

where dE/dx is energy loss per unit length, N is the number of atoms in unit volume, m is the mass of an electron, v is the ion velocity, I is the mean energy of excitation and ionization, c is the speed of light, C/Z_2 is a shell correction term (important in the relatively lower-velocity range where the ion cannot ionize or excite inner-shell electrons), and $\delta/2$ a density effect.

However, the Bethe–Bloch formula is not applicable in the kiloelectron volt energy range, which is typical for most FIB systems, because the velocity of ions is generally very low (especially heavy ions). In this velocity range, ions capture electrons from constituent atoms, making the screening of nuclear charge a significant factor. Furthermore, with decreasing ion velocity, nuclear collisions increasingly predominate over electronic collisions. Universal electronic and nuclear energy loss formulae have been derived by Lindhard, Scharff, and Schiott (LSS), based on the Thomas–Fermi atomic model in the low-velocity range. The electronic energy loss is given by

$$S_e = \frac{dE}{Ndx} = \eta \frac{8\pi Z_1 Z_2 e^2 a_B v}{N Z v_B} \tag{18.16}$$

where $v < v_B Z^{\frac{2}{3}}$, $\eta \approx Z_1^{\frac{1}{6}}$, and a_B and v_B are the Bohr radius and velocity, respectively. The nuclear energy loss is given by

$$\left(\frac{d\varepsilon}{d\rho}\right)_n = \frac{1}{\varepsilon} \int_0^s f(\sqrt{t}) d\sqrt{t} \tag{18.17}$$

where ε, t, and $f(\sqrt{t})$ are the same as those defined in the previous subsection on cross section. Through experimentally obtained values, however, it has been found that energy loss depends on the incident ion type and energy, as well as the type of target atom. Ziegler et al. compiled data from a vast array of parameters which cover all ion–atom combinations over the range of 1 keV/u and 2 GeV/u, and have developed a program named SRIM which calculates energy losses, ranges, range stragglings, damage distributions, sputtering yields, and so on.

Because these processes are so interrelated, no single phenomenon can be fully understood without considering several at once. Therefore, it is vital to have a quantitative understanding of the experimental observations as well as creativity in design so that new and more sophisticated combinations of these versatile processes can be applied in the field of materials science and nanotechnology. With it, we can aim at more advanced material modification, nanofabrication, surface analysis, and countless other applications in the future. Table 18.1 summarizes some comparisons between the beams and particle interactions of the ion beam and electron beam.

18.1.4
Detection of Electron and Ion Signals

The products of ion–atom collisions are collected, amplified, and analyzed as signals to form an image. Secondary electrons are the primary signals used by both SEM and FIB to generate high-resolution images, although backscattered electrons and secondary ions can also be used to form images. For secondary electron detection, the two main types of detectors are the multichannel plate and the electron multiplier. The multichannel plate is usually mounted directly above the sample, and therefore cannot provide much topographical information. The most common design used today for secondary electron detection is the Everhart–Thornly

Table 18.1 Quantitative comparison of FIB and SEM.

Particle	FIB	SEM	Ratio
Type	Ga$^+$ ion	Electron	–
Elementary charge	+1	−1	–
Particle size	0.2 nm	0.00001 nm	20 000
Mass	1.2×10^{-25} kg	9.1×10^{-31} kg	130 000
Velocity at 30 kV	2.8×10^5 m s^{-1}	1.0×10^8 m s^{-1}	0.0028
Velocity at 2 kV	7.3×10^4 m s^{-1}	2.6×10^7 m s^{-1}	0.0028
Velocity at 1 kV	5.2×10^4 m s^{-1}	1.8×10^7 m s^{-1}	0.0028
Momentum at 30 kV	3.4×10^{-20} kg m s^{-1}	9.1×10^{-23} kg m s^{-1}	370
Momentum at 2 kV	8.8×10^{-21} kg m s^{-1}	2.4×10^{-23} kg m s^{-1}	370
Momentum at 1 kV	6.2×10^{-21} kg m s^{-1}	1.6×10^{-23} kg m s^{-1}	370
Beam			
Size	Nanometer range	Nanometer range	–
Energy	Up to 30 kV	Up to 30 kV	–
Current	Picoampere to nanoampere range	Picoampere to microampere range	–
Penetration depth			
In polymer at 30 kV	60 nm	12 000 nm	0.005
In polymer at 2 kV	12 nm	100 nm	0.12
In iron at 30 kV	20 nm	1800 nm	0.011
In iron at 2 kV	4 nm	25 nm	0.16
In silicon at 30 kV	36 nm	7500 nm	0.005
In silicon at 10 kV	13 nm	1200 nm	0.011
Average signal per 100 particles at 20 kV			
Secondary electrons	100–200	50–75	1.3–4.0
Backscattered electron	0	30–50	0
Substrate atom	500	0	∞
Secondary ion	30	0	∞
X-ray	0	0.7	0

electron multiplier detector, also known as a *scintillator-photomultiplier tube (PMT)* detector.

The scintillator-PMT detector has three main parts: the collector grid and screen, the scintillator, and the PMT. The collector grid and screen is positioned to the side of the sample stage at a 45° angle to the beam. It attracts secondary electrons with a potential of hundreds of volts, and most of the electrons are then accelerated into the scintillator. Captured electrons cause the scintillator to emit visible-light photons due to its cathodoluminescent properties; for a typical scintillator voltage of 10 kV, on the order of 100 photons is generated per electron. A light pipe similar to a fiber optic cable extends from the scintillator and transmits the photons to the PMT by internal reflection. The PMT consists of a sealed glass tube under high vacuum. Incoming photons strike the photocathode and liberate valence electrons that are accelerated as photoelectrons toward a series (usually eight) of dynode electrodes, which are biased positively to the photocathode. Each dynode is also biased 100–200 V positively to the preceding one. The photoelectrons generate secondary electrons on striking the first dynode, which are amplified by a factor of 10^6 after striking the remaining dynodes. Combined with the 100 photoelectrons per secondary electron produced in the scintillator, the overall magnification of the scintillator-PMT detector can be as high as 10^8, depending on the dynode voltage. Raised areas of the sample produce more collectable secondary electrons, while depressed areas produce fewer, resulting in a contrast. A viewing monitor synced to the scan coils controls the beam so that as it scans across the sample surface, the image of the sample is reproduced on the screen. The scintillator-PMT can also be used to detect backscattered electrons by reversing the bias of first grid, thereby repelling lower-energy secondary electrons but not backscattered electrons.

The signals from the ejected ions can be amplified and displayed to show the detailed structure of the sample surface, including information unavailable with electron beams. Images obtained using the FIB and secondary ions can exceed 10 nm resolution, showing topographic data and materials contrast that complement the information obtained from an SEM image. Compared to secondary electron images, secondary ion images can be used to more readily distinguish materials contrast resulting from differences in specimen chemistry. While secondary electron images provide uniformly good depth of field, secondary ion images reflect more selective depths that are dependent on the sample material and structure. Because of the dependency of the ion–atom interaction on the crystal grain orientation, information about the grain size and crystal orientation can also be obtained using an FIB (channeling contrast).

Detection of secondary ions can be categorized as microprobe mode or microscope (direct) mode. The microprobe mode is analogous to the process used in SEM: the primary ion beam is rastered while the secondary ion signal is simultaneously detected. The main difference is that a strongly negative bias is applied to the particle detector grid to repel both secondary and backscattered electrons and attract the positive ions for amplification. In general, secondary ion microprobe images are of the "total detected positive ion" type, in which practically all positive ions are collected and amplified without distinguishing the different masses. Recently, mass-resolved ion imaging has been developed, present in FIB systems equipped with secondary ion mass spectrometers (SIMSs). This allows for elemental analysis of the sample to be combined with imaging. In the microscope mode, ion image formation is nonraster based and relies on electrostatic lenses in the secondary ion column. Several types of position-sensitive detectors can be used with this process, the most common of which is a microchannel plate connected to a phosphor screen, with a highly sensitive charge-coupled device (CCD) to capture the resulting image. Another option to capture ion images is to use a direct ion-imaging detector such as the resistive anode encoder (RAE). However, the resolution of microscope mode imaging is limited by the electric field strengths of the lenses to about 1 μm – much lower than the microprobe mode.

18.2 FIB Techniques

18.2.1 Milling

Milling is a natural consequence of the heavy ions used in the FIB, and it is the most basic and one of the most important capabilities of the FIB. This process adds a third dimension to microscopy, enabling the local modification of the material surface, preparation of cross sections, and shaping of sample materials into any desired configuration (Figure 18.5). By controlling aspects of the sputtering such as rate, location, and depth, we can take advantage of the FIB system's remarkable milling abilities.

For a given beam energy, the sputtering rate (the rate at which surface atoms are excavated) is generally proportional to the primary beam current for a given energy. Therefore, it is advisable to use a higher beam current initially to save time and then to use a lower current at the end of the operation to create a fine polish. On the other hand, a low-current beam must be used for imaging with FIB. Otherwise, the sample will be excessively eroded. The sputtering rate can be easily and accurately controlled by altering the beam current with apertures or using smaller spot sizes. The typical limiting feature size using direct milling is around 10 nm [8], although nanopores with diameters as small as 3 nm have been drilled through 20 nm thick membranes, as well as nanowires and sharpened nanotips with critical dimensions below 10 nm [9].

The physical structure of the sample is altered when incident ions, after losing their energy in atomic collisions, come to rest as implanted impurities. The dislocation of surrounding atoms will often convert a crystalline structure to an amorphous one. The range and straggle of incident ions depend not only on their initial energy and angle of incidence, but also on the type of incident ion. Ions that lose their energy within a short distance will produce less damage at greater

Figure 18.5 The S3 logo milled into an MgO substrate [7].

depths. From an analysis [10] of the dynamics of scattering, it can be shown that there is a maximum scattering angle, Θ_{MAX}, of the incident ion given by

$$\Theta_{MAX} = \cos^{-1}\sqrt{1 - \frac{M_{atom}^2}{M_{ion}^2}} \quad 0 \leq \Theta_{MAX} \leq \pi/2 \quad (18.18)$$

The sputtering process can also be made material-selective through a process known as *gas-assisted etching* (GAE). In this technique, a halogen gas is introduced to the work surface in the immediate vicinity of the desired milling site. Under ion bombardment, the material-specific absorbency rates enhance the formation of volatile reaction products that can be removed under vacuum. This allows oxides to be cleared without damage to conductors, and for conductors to be cut without damage to underlying insulators. Typical milling resolution of around 0.1 μm is achieved with GAE. The holes and cuts can be very accurately placed (usually within 20 nm) [11] and can reach buried layers provided that the depth-to-width aspect ratio is kept below 4:1. Using I_2 or XeF_2, GAE can increase the aspect ratio to 8:1 or better. The milling efficiency associated with this technique is typically a few micrometer cubed per nanocoulomb, but the actual rate depends on the mass of the target atom, its binding energy to the atom matrix, and its matrix orientation with respect to the incident direction of the ion beam.

One disadvantage of extensive FIB milling due to a limited of ion beam current is that the process can be extremely slow. Fortunately, as we have seen, the GAE technique offsets this problem by providing much higher efficiency for removing large volumes of material. Also, GAE is very specific to the chemistry of the materials involved, adding the capability to safely etch one material without significantly affecting the others. For example, the removal rate of Si, Al, and GaAs is increased by a factor of 20–30 with GAE, while oxides such as SiO_2 and Al_2O_3 are not affected [12]. In the presence of water vapor, carbon-based materials show a similarly enhanced removal rate; on the other hand, however, the water drastically reduces the etching rate of materials that oxidize, such as Al metallization lines and silicon. Aside from faster etching rates, another advantage of GAE over ordinary FIB milling is the absence of unwanted redeposition of etched material. The technique can be considered "neater" for this reason.

GAE, however, is not without its disadvantages. First of all, it produces more permanent contamination deeper below the sample surface [13]. Furthermore, while ordinary ion milling can be used to remove almost all kinds of materials, GAE is only compatible with certain substrates because a precursor gas must exist that forms volatile products with the substrate. Also, to have the desired dimensional control, a spontaneous reaction between the precursor gas and the substrate should be avoided or minimized. Ultimately, GAE sacrifices some of the pinpoint precision available with unassisted ion milling for higher speed. To fabricate nanoscale structures, the material removal rate should have sub-nanoscale resolutions, which is achieved by adjusting the material removal rate in a controllable manner. Since GAE has relatively high material removal rates, the smallest feature sizes it has been able to achieve have been on the order of 50 nm, such template lines etched on quartz with XeF_2 as the precursor gas [14]. For smaller structures, the irregularities induced from etching can be much larger than the 10 nm resolution required for true nanofabrication capability and ruin the intended shape or design with unacceptably low precision [15–17]. Thus, GAE is rarely used alone in nanofabrication; instead, a two-stage process of initial GAE followed by ion milling is frequently used when an operation requires bulk volumes of material removal as well as high precision.

The FIB system allows control of the beam in the lateral and depth directions. By holding the beam over an area for an extended period of time, more materials will be removed, and a deeper hole will be excavated. This technique can be used to expose buried lines for eventual cutting or attaching to other circuit elements, depending on the desired modification of the structure. However, unlimited depths cannot be milled with an ion beam through a material. The maximum depth can be increased by injecting a reactive gas into the milling site. Depending on such variables as sample composition, mill area, beam parameters, and whether an enhanced etch process is used, the aspect ratio of the ion beam's cutting depth can be dramatically altered. This makes it possible to reach greater depths through a narrow opening in the sample without altering the upper layers – not unlike minimally invasive surgery. Having such precision and control allows applications unique to the FIB system.

18.2.1.1 TEM Sample Preparation

Because FIB milling has become increasingly popular for preparation of ultrathin electron-transparent samples for TEM analysis, we will devote this subsection to address this technique in greater detail. The first step of TEM sample preparation is pre-thinning of the sample to a thickness of 50 – 100 µm either using FIB milling or a mechanical techniques such as polishing or slicing, depending on the material. Nonconductive materials should be coated before FIB milling with 20–30 nm of carbon or metal (usually Pt, Au, or Pd) to avoid charge buildup. The pre-thinned sample can then be cut into a plate with dimensions of around 1–2 mm by 2.5–3 mm and mounted directly onto the FIB sample holder. Smaller samples should be sectioned into small plates and glued (using conductive paste such as carbon or silver) to a semicircular metal disk as a sample support. If the sample consists of particles or fibers, it can be embedded in epoxy and then sectioned into foils and glued to the edge of a support.

For reasons that are discussed below, it is important to have a smooth sample surface before FIB milling in order to get a final specimen of uniform thickness. Polishing and slicing can be employed in some cases, but if the top surface is of interest, it is recommended to bypass mechanical polishing and deposit a 20–40 nm layer of carbon using a vacuum evaporator. This can be followed with FIB deposition of a 0.5–2 µm layer of metal as the final treatment to smooth out the sample surface. The process of thinning itself is generally executed in three steps, going from high beam current initially to low beam current for the final polish. For a 30 kV Ga ion beam, a typical initial beam current of 10 nA can be used for rough milling until the sample has been reduced to a thickness of 2–5 µm. The beam current is then decreased to around 2 nA for medium milling, and finally lowered to 30 pA for the final polish. Typical thicknesses of samples for TEM observation range from 50 nm to 0.2 μm, but ultimately the best thickness depends on the desired image from the sample. For high-resolution TEM imaging, a thickness of less than 0.1 μm is required; for EDX or EELs analysis, a thickness of 0.1–0.3 μm will suffice [18].

Using the FIB, TEM lamellae samples can be prepared to within 50 nm of a feature of interest. This technique has been claimed to possess the lowest preferential sputtering rate of all conventional TEM sampling methods. One drawback of using the FIB, however, is that the high-energy Ga ions used for milling leads to the formation of damaged and amorphous layers near the surfaces of the sample. The processes involved in the formation of these layers and methods for damage mitigation have been studied in detail by several authors, and it has become established that these layers are a result of direct damage from the ion beam and subsequent redeposition of sputtered material. When the sample surface has uneven topography or nonuniform chemical composition, FIB milling results in the formation of striations. This is known as the *curtain effect*, which leads to samples of nonuniform thickness. Bottom and side cleaning procedures have been proposed to avoid this problem.

A recent study performed by Montoya *et al.* provides detailed insight of the process of TEM sample preparation using the FIB [19]. This experiment used a sample of $LaAlO_3/SrTiO_3$ multilayers grown on a single-crystal $SrTiO_3$ substrate. Before FIB preparation, a piece of the multilayer sample was glued onto a TEM grid, and a 60 nm layer of fine-grained gold particles was sputtered onto the sample, followed by a second 500 nm layer of coarse-grained gold. These gold layers were intended to protect the LAO/STO multilayer of interest (Figure 18.6). The TEM grid could be mounted on the FIB system's rotatable grid holder, which, combined with tilting of the sample stage, allowed the ion beam to reach all necessary angles for the cleaning schedules. A 30 kV ion beam was used for all steps with the exception of the final cleaning, which used an energy of 5 kV. It was found that the curtain effect was strongest for the so-called "top-cleaned" specimens,

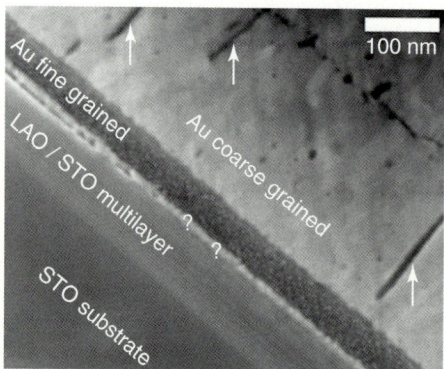

Figure 18.6 Cross-section of multilayer sample with gold covering [19].

in which the direction of the component of the ion beam in the plane of the lamella was perpendicular to the direction of the STO/LAO layers. The curtain effect was negligible for side- and bottom-cleaned specimens. "Cross sections of the cross section" (double cross-sectional lamellae) were prepared after the final cleaning step in order to study the amorphous layers still present at the faces of the prepared TEM sample (Figure 18.7). The two faces of the grid were again covered by sputtering gold to protect the lamella, and then FIB deposition (which is discussed in the next section) was used to encase the gold-covered lamella with platinum. It was also found that the sputtered gold layer successfully protected the lamella from damage from FIB-induced platinum deposition. This experiment also showed that the amorphous layers remaining on the faces of "top-angle cleaned" specimens had an average thickness of roughly 5 nm (Figure 18.8).

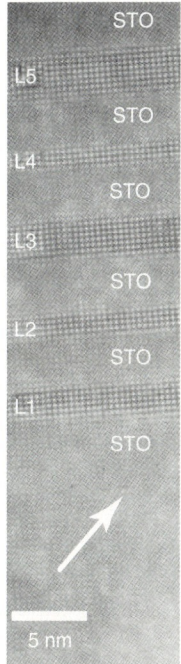

Figure 18.7 TEM image of multilayer. L1, L2, ..., LN refer to the LAO layers [19].

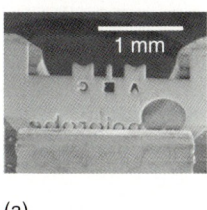
(a)

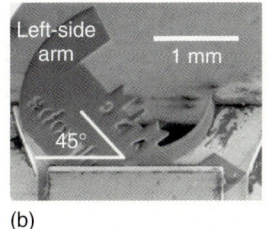
(b)

(c)

Figure 18.8 Different ways of mounting the sample holder to access different cleaning angles [19].

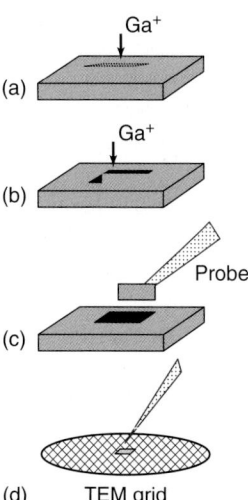

Figure 18.9 Diagram of lift-out technique [18].

One of the greatest advantages of using the FIB over other conventional TEM sample preparation methods is the superior positional accuracy in setting the area to be thinned. The FIB lift-out technique was developed for milling and lifting out a thinned specimen directly from a bulky sample (Figure 18.9). First, the sample is placed in the FIB system and the area surrounding the site of interest is trench milled, and the site is thinned to 100 nm or less. Then, the sample is removed from the FIB chamber and the thinned sample area is lifted out using a glass rod and deposited onto a carbon-coated TEM grid. This simple procedure can prepare samples with clean surfaces and uniform thickness such that even atomic-level high-resolution TEM images can be observed (Figure 18.10). The time required for the whole process is also greatly shortened compared to the conventional FIB technique since mechanical polishing and dicing is rendered unnecessary. A more advanced variation of the lift-out technique is the FIB microsampling

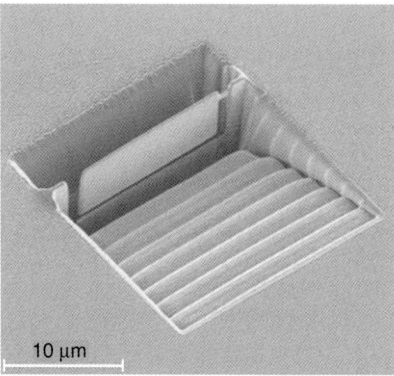

Figure 18.10 SEM image of trench-milled area before lift out [18].

technique, which employs an elaborate mechanical probe system that allows the entire procedure to be conducted under the vacuum of the FIB chamber.

18.2.2
Deposition

The FIB can be converted from a milling machine to a deposition system with one simple adjustment, a testament to its remarkable versatility. To enable the deposition of material instead of sputtering, a gas delivery system is activated in order to supply a chemical compound immediately above the sample, in the path of the ion beam. The gas, usually an organometallic compound, initially adsorbs onto the sample surface before being struck by the incident gallium ion beam (and by secondary emission products), which cause it to decompose. The volatile organic impurities are released and removed by the FIB's vacuum system. The remaining metal is heavy enough to remain deposited on the surface, creating a thin film that can then act as an electrical conductor.

There are two other techniques used for FIB deposition. Although they are less common than chemical deposition, they may offer advantages in certain situations. The first method is direct deposition, typically used to deposit thin films of gold or other materials. A broad-diameter beam with low current is used for the deposition; however, this process is very slow. The other method is nucleation with chemical vapor deposition (CVD), usually employed to deposit carbon, aluminum, and iron. This technique extends the conventional chemical deposition concept by using the same gas injection system to create nucleation on the substrate and then growing a thin film in the nucleation region (Figure 18.11). The main advantage of this method is that it produces a film of uniform thickness [20].

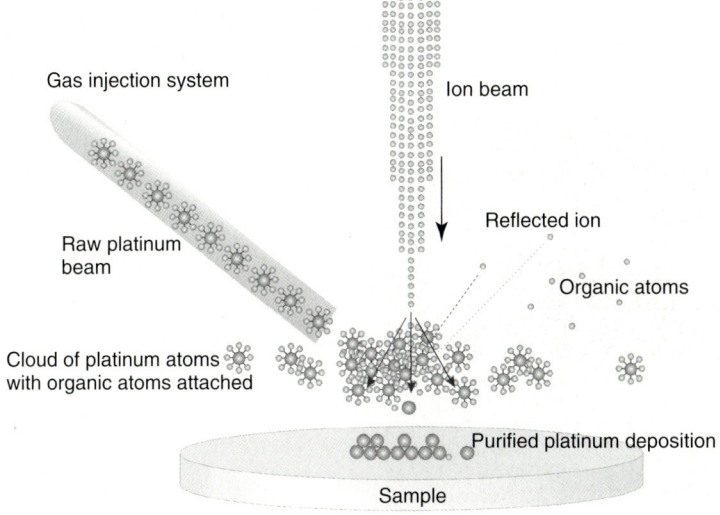

Figure 18.11 Diagram of the platinum deposition process [3].

When an organometallic gas such as $W(CO)_6$ is introduced in the area of the ion beam, the ion beam can be controlled to decompose the molecules that adsorb onto the surface and form a metal layer. This metal conduction layer can function as a probe pad or be used to make a new connection for nanocircuitry. One disadvantage of this technique is that it produces impure films with high resistivity. Long wiring runs requiring lower resistivity can be achieved by combining FIB local connections with long laser CVD-deposited gold conductors. In the CVD technique, a gold carrier gas is decomposed by a scanning laser beam, leaving a $5 - 10$ µm wide, low sheet resistance gold line on the work surface. In addition to being able to deposit a metallic conducting thin film, the FIB can also deposit a high-quality oxide insulator. Usually, this is accomplished by introducing a siloxane-type gas and oxygen into the path of the ion beam at the impact area, where the decomposition of the complex silicon-bearing molecule in the presence of oxygen leads to absorption into the surface and formation of a silicon dioxide insulator layer. Oxide deposition enables the re-insulation of cut integrated circuit wiring and FIB-deposited conductors, allowing for the complex multilevel wiring repair.

Although the FIB deposition system is very useful, it does suffer from some minor drawbacks. Cracking of the organometallic molecules by the ion beam is rarely complete, leaving residual organic impurities deposited in the thin film. The lingering gallium ions from the ion beam can also compromise the insulating ability of a deposited oxide layer. However, despite the fact that CVD deposits are often more pure, the FIB system does offer more precise, localized deposition, and its capability for controlling different thicknesses of depositions is unrivalled. Furthermore, research has shown that in certain cases, electron nanofabrication has been able to match the deposition results of FIB without the impurities that result from using ions [21].

Despite its shortcomings, the FIB is still unparalleled in its patterning capability. By precisely adjusting the deposition parameters of time, gas flux, and ion current, one has complete control over size, position, and height, and can fabricate remarkable three-dimensional structures (Figure 18.12). For example, Khizroev

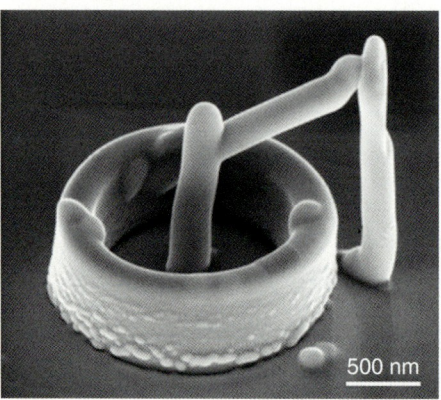

Figure 18.12 Microelectrochemical cell fabricated by ion beam platinum deposition [9].

successfully deposited non-magnetic nanoscale tungsten to aid in the fabrication of nanomagnetic probes [22]. In another set of experiments, ion-induced deposition yielded self-assembled nanofins (nanoscale hollow bulk structures) deposited perpendicular to the substrate [23]. Indeed, with both deposition and milling capabilities, the FIB can precisely construct almost any nanostructure, opening exciting new channels in today's nanotechnology research.

18.2.3
Implantation

Implantation is another important function of the FIB system, especially for fabricating integrated circuits and other semiconductor devices. The conventional fabrication method for semiconductors uses broad ion beam (BIB) systems and involves patterned films, or masks. The masks serve to protect certain areas of the wafer from ion bombardment, so that only the unmasked areas become implanted. The FIB's level of localized control and precision, makes it possible to forego the use of a mask and simply direct the beam at the areas where implantation is required. This provides entirely new abilities to control doping gradients and the depth of the implantation [24].

Ion implantation has been used for materials modification as well as semiconductor production. Although defects from ion implantation when using the FIB system are usually undesirable, they can occasionally offer advantages. For example, recent experiments have shown that implanted atoms can be induced to self-assemble into nanoclusters through thermal annealing or radiation. To understand these properties, a Monte Carlo model has been developed with the ability to simulate diffusion, precipitation, and interaction kinetics of implanted ions [25]. Ion implantation has also been combined with deep reactive ion etching to form complex silicon microstructures [26]. Researchers found that Si implanted with Ga^+ ions behaved as an etch mask, resulting in the selective etching of unimplanted regions.

Another application for ion implantation is the localized modification of material properties on the nanoscale. In an experiment to modulate the resonant frequencies of ZnO nanowires [27], the nanowires were exposed to Ga^+ ion bombardment on one side, producing asymmetrically implanted regions. The implanted nanowires exhibited ultralow harmonic resonances that deviated from classical elastic theory, believed to be the result of charge imbalances introduced by the implanted ions (Figure 18.13). Refinement of this resonance tuning technique could open the way to a plethora of nanodevice applications, such as sensors and energy-scavenging devices.

Like many other FIB techniques, in exchange for its accuracy and precision, FIB-induced ion implantation suffers from a slow processing rate. However, the trade-off between speed and precision is inherent in almost any technology, and there are many applications in which speed is less important than achieving high resolution. In these cases, the FIB becomes a valuable tool.

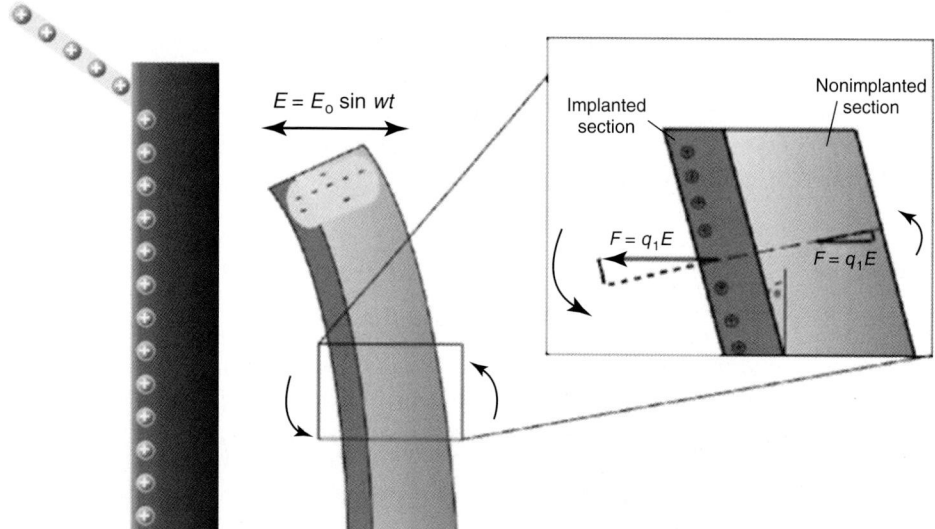

Figure 18.13 Diagram of a ZnO nanowire exposed to FIB on its left face. The accumulation of positive charge on the implanted side leads to charge imbalance, which in turns leads to a bending moment along the length of the nanowire and the appearance of FIB-induced superharmonic resonance [24].

18.2.4
Imaging

The FIB can achieve imaging resolution below 10 nm. For the best resolution and signal, the beam voltage and current are typically set to 30 kV and 40 pA, respectively. As previously discussed, however, a significant drawback to FIB imaging is the unavoidable sputtering damage to the sample caused by the ion beam. This greatly limits the range of imaging applications for a single-ion beam system. Fortunately, a new tool has been developed that combines the milling capability of the ion beam with nondestructive imaging capabilities of the electron beam. This machine, often referred to as the *dual-beam system* or *two-beam system*, places the ion beam and electron beam in fixed positions, with the ion beam tilted at an angle of 45–52° for ideal performance. The two beams share focal points at what is called the *coincidence point*, an optimized position for the majority of operations, including FIB sample direct writing. The stage can be controlled to tilt, allowing changes in the sample–beam orientation. Like most FIB systems, integrated software with a single user interface controls the dual-beam system.

By incorporating FIB and SEM technology together in a single machine, we can use the two symbiotically to achieve tasks beyond the limitations of either individual system (Figures 18.14 and 18.15). The two-beam system simplifies the

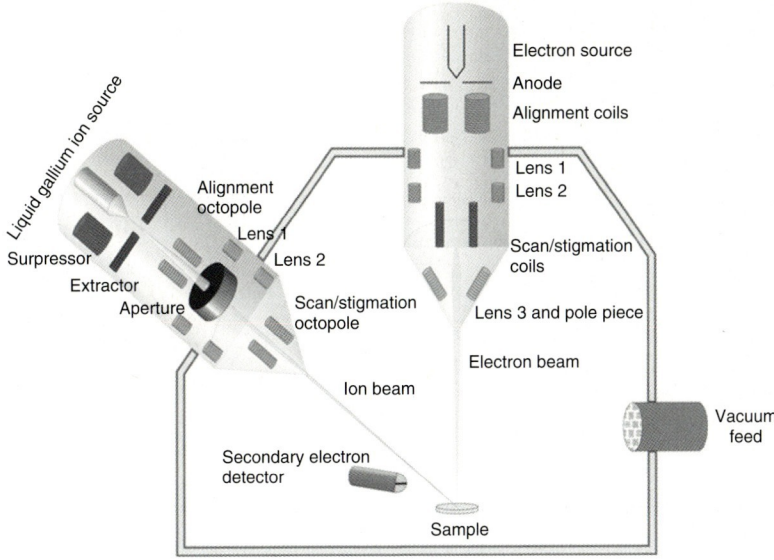

Figure 18.14 A schematic showing the configuration of a dual-beam FIB system [3].

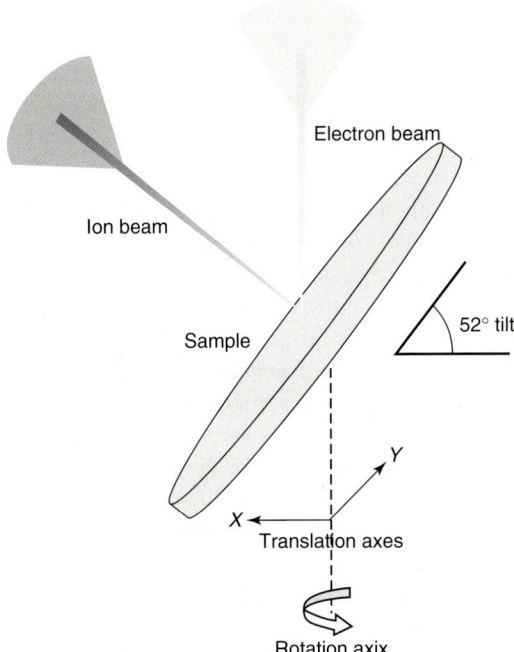

Figure 18.15 Schematic of an FIB dual-beam system, in which both e-beam and ion-beams are co-focused at the same point on the sample surface [3].

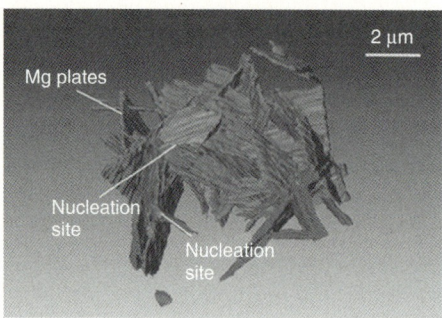

Figure 18.16 3-D reconstructed image of the $Mg_{77}Cu_{12}Y_6Zn_5$ alloy showing Mg-rich plates nucleating at various particles and forming an interwoven network structure of crystalline phase within the amorphous matrix [29].

reconstruction of the three-dimensional structure and chemistry of a sample by interpolating two-dimensional SEM/FIB images and SIMS chemical maps of layers that have been successively exposed through ion milling [28]. In one experiment, an Mg-based metallic glass composite was imaged using the dual-beam system, which milled cross sections of the sample with the FIB and captured an SEM image of the cross section every 0.1 μm to create a three-dimensional elemental distribution map (Figure 18.16) [29]. In another study, the microstructure of a Ni-YSZ solid-oxide fuel cell was also reconstructed in 3-D using an FIB–SEM system (Figure 18.17), allowing numerical simulations to be conducted for the anode overpotential [30].

The dual-beam system also makes possible precise monitoring of FIB operation through the SEM. The operator can use the slice-and-view technique to observe the progress of the ion beam cross section and stop the milling process at a precise point in order to obtain local information. Also, the state-of-the-art "CrossBeam" system now allows for the use of both the ion beam and the electron beam simultaneously without interference, eliminating the necessity to switch back and forth. The system allows the sample to be imaged in real time with the SEM while the FIB is milling, depositing, and so on, providing for higher levels of accuracy when creating cross sections [31]. The SEM's damage-free imaging is especially useful in the final phase of sample preparation for the TEM, since using an FIB alone causes unavoidable sputtering damage. It can also be very useful in the localization of integrated circuit failures, so that excess damage is not done; Zimmermann presented a completely *in situ* two-beam technique that combines SEM monitored ion milling, XeF_2 staining, and SEM imaging to characterize the failures [32].

Evidently, the two-beam FIB system is a remarkable multifunctional tool for sample imaging, modification, and analysis. It offers integrated high-resolution 3-D nondestructive imaging and can carry out precise, highly controlled, and automated ion milling, GAE, deposition, and implantation. Although damage to

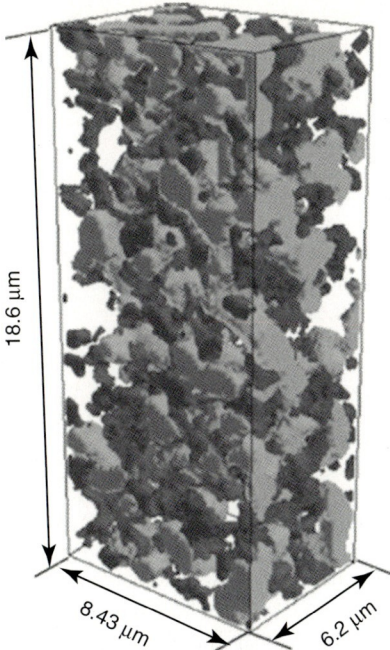

Figure 18.17 Reconstructed Ni-YSZ anode microstructure used for simulations. Light represents nickel and dark represents YSZ. The empty spaces are porous voids [30].

the sample is an ever-present problem when using ion beams, the two-beam system minimizes it by offering SEM imaging. Thus, the two-beam FIB stands to be one of the most important tools in the modern study of nanotechnology.

Acknowledgments

The author is grateful to Anton Li, Alexander Epstein, and Austin Akey who have partially contributed in preparation of the manuscript. The National Science Foundation-MRSEC program through the Princeton Center for Complex Materials (DMR-0819860) and the New Jersey Commission of Science and Technology are gratefully acknowledged.

References

1. Krohn, V.E. and Ringo, G.R. (1975) Ion source of high brightness using liquid metal. *Appl. Phys. Lett.*, **27**, 479.
2. Krohn, V.E. (1961) in *Progress in Astronautics and Rocketry*, Academic Press, New York, pp. 73–80.
3. Yao, N. (2007) in *Focused Ion Beam Systems Basics and Applications* (ed. N. Yao), Cambridge University Press, New York, pp. 1–30.
4. Breese, M. (2009) Focusing keV and MeV ion beams, in *Ion Beams in Nanoscience and Technology* (eds R.

Hellborg, H.J. Whitlow and Y. Zhang), Springer-Verlag, Heidelberg.
5. Kang, H.H., Chandler, C., and Weschler, M. (2007) in *Focused Ion Beam Systems Basics and Applications* (ed. N. Yao), Cambridge University Press, New York, pp. 67–86.
6. Imanishi, N. (2007) in *Focused Ion Beam Systems Basics and Applications* (ed. N. Yao), Cambridge University Press, New York, pp. 31–66.
7. Volkert, C.A. and Minor, A.M. (2007) Focused ion beam microscopy and micromachining. *MRS Bull.*, **32**, 389–399.
8. Gierak, J., Madouri, A., Biance, A.L., Bourhis, E., Patriarche, G., Ulysse, C., Lucot, D., Lafosse, X., Auvray, L., Bruchhaus, L., and Jedde, R. (2007) Sub-5 nm FIB direct patterning of nanodevice. *Microelectron. Eng.*, **84**, 779–783.
9. Frabboni, S. and Gazzadi, G.C. Focused ion Beam, National Center of CNR-INFM S3, http://www.s3.infm.it/fib.html.
10. Jamison, R. (2000) *Computational and Experimental Quantification of Focused Ion Beam Damage in Silicon During TEM Sample Preparation*, University of California, Berkeley.
11. Phaneuf, M.W. (1999) Applications of focused ion beam microscopy to materials science specimens. *Micron*, **30**, 277.
12. Van Doorselaer, K., Van den Reeck, M., Van den Bempt, L., Young, R., and Whitney, J. (1993) How to prepare golden devices using lesser materials, Proceedings of the 19th International Symposium for Testing and Failure Analysis.
13. Phaneuf, M.W. and Li, L. (2000) FIB techniques for analysis of metallurgical specimens. *Microsc. Microanal.*, **6S2**, 524.
14. Kettle, J., Hoyle, R.T., and Dimov, S. (2009) Fabrication of step-and-flash imprint lithography (S-FIL) templates using XeF_2 enhanced focused ion-beam etching. *Appl. Phys. A*, **96**, 819–825.
15. Taniguchi, J., Ohno, N., Takeda, S., Miyamoto, I., and Komuro, M. (1998) Focused-ion-beam-assisted etching of diamond in XeF_2. *J. Vac. Sci. Technol. B*, **16**, 2506.
16. Stanishevsky, A. (2001) Patterning of diamond and amorphous carbon films using focused ion beams. *Thin Solid Films*, **560**, 398–399.
17. Adams, D.P., Vasile, M.J., Mayer, T.M., and Hodges, V.C. (2003) Focused ion beam milling of diamond: Effects of H_2O on yield, surface morphology and microstructure. *J. Vac. Sci. Technol. B*, **21**, 2334.
18. Kamino, T. and Yaguchi, T. (2007) in *Focused Ion Beam Systems Basics and Applications* (ed. N. Yao), Cambridge University Press, New York, pp. 250–267.
19. Montoya, E., Bals, S., Rossell, M.D., Schryvers, D., and Van Tendeloo, G. (2007) Evaluation of top, angle, and side cleaned FIB samples for TEM analysis. *Microsc. Res. Tech.*, **70**, 1060–1071.
20. Gerlach, R. and Utlaut, M. (2001) Focused ion beam methods of nanofabrication: room at the bottom. *Proc. SPIE Int. Soc. Opt. Eng.*, **96**, 4510.
21. Mitsuishi, K., Shimojo, M., Han, M., and Furuya, K. (2003) Electron-beam-induced deposition using a subnanometer-sized probe of high-energy electrons. *Appl. Phys. Lett.*, **83**, 2064.
22. Khizroev, S., Bain, J.A., and Litvinov, D. (2002) Fabrication of nanomagnetic probes via focused ion beam etching and deposition. *Nanotechnology*, **13**, 619.
23. Allameh, S.M., Yao, N., and Soboyejo, W.O. (2004) Synthesis of self-assembled nanoscale structures by focused ion-beam induced deposition. *Scr. Mater.*, **50**, 915.
24. Russell, P.E., Stark, T.J., Griffis, D.P., and Gonzales, J.C. (2001) Chemically assisted focused ion beam micromachining: overview, recent developments and current needs. *Microsc. Microanal.*, **7S2**, 928.
25. Strobel, M., Heinig, K.H., and Möller, W. (1999) Can core/shell nanocrystals be formed by sequential ion implantation? Predictions from kinetic lattice Monte Carlo simulations. *Nucl. Instr. Meth. B*, **148**, 104.

26. Qian, H.X., Zhou, W., Miao, J., Lim, L.E.N., and Zeng, X.R. (2008) Fabrication of Si microstructures using focused ion beam implantation and reactive ion etching. *J. Micromech. Microeng.*, **18**, 035003.
27. Cohen-Tanugi, D., Akey, A., and Yao, N. (2010) Ultralow superharmonic resonance for functional nanowires. *Nano Lett.*, **10**, 852–859.
28. Hull, R., Dunn, D., and Kubis, A. (2001) Nanoscale tomographic imaging using focused ion beam sputtering, secondary electron imaging and secondary ion mass spectrometry. *Microsc. Microanal.*, **7S2**, 34.
29. Robin, L., Laws, K.J., Xu, W., Kurniawan, G., Privat, K., and Ferry, M. (2009) The three-dimensional structure of Mg-rich plates in As-cast Mg-based bulk metallic glass composites. *Metall. Mater. Trans. A*, **41**, 1691–1698.
30. Shikazono, N., Kanno, D., Matsuzaki, K., Teshima, H., Sumino, S., and Kasagi, N. (2010) Numerical assessment of SOFC anode polarization based on three-dimensional model microstructure reconstructed from FIB-SEM images. *J. Electrochem. Soc.*, **157**, B665–B672.
31. Gnauck, P., Zeile, U., Rau, W., and Schumann, M. (2003) Real time SEM imaging of FIB milling processes for extended accuracy in cross sectioning and TEM preparation. *Microsc. Microanal.*, **9S3**, 524.
32. Zimmermann, G. and Chapman, R. (1999) In-situ dual-beam (FIBSEM) techniques for probe pad deposition and dielectric integrity inspection on 0.2 mm technology DRAM single cells. Proceedings of the 25th International Symposium for Testing and Failure Analysis.

Further Reading

Anzalone, P.A., Mansfield, J.F., and Giannuzzi, L.A. (2004) Dual-beam milling and deposition of complex structures using bitmap files and digital patterning. *Microsc. Microanal.*, **10S2**, 1154.

Arshak, K., Mihov, M., Arshak, A., McDonagh, D., and Sutton, D. (2004) Novel dry-developed focused ion beam lithography scheme for nanostructure. *Microelectron. Eng.*, **73**, 144.

Bischoff, L., Teichert, J., Kitova, S., and Tsvetkova, T. (2002) Optical pattern formation in a-SiC: H films by Ga^+ ion implantation. *Vacuum*, **69**, 73.

Cairney, J.M., Harris, S.G., Munroe, P.R., and Doyle, E.D. (2004) Transmission electron microscopy of TiN and TiAlN thin films using specimens prepared by focused ion beam milling. *Surf. Coat. Technol.*, **183**, 239.

Casey, J.D., Phaneuf, M., Chandler, C., Megorden, M., Noll, K., Schuman, R., Gannon, T.J., Krechmer, A., Monforte, D., Antoniou, N., Bassom, N., Li, J., Carleson, P., and Huynh, C. (2002) Copper device editing: strategy for focused ion beam milling of copper. *J. Vac. Sci. Technol. B.*, **20**, 2682.

Chen, X., Wang, R., Yao, N., Evans, A.G., Hutchinson, J.W., and Bruce, R.W. (2003) Foreign object damage in a thermal barrier system: mechanisms and simulations. *Mater. Sci. Eng. A*, **352**, 221.

Dunn, D.N. and Hull, R. (2004) Three-dimensional volume reconstructions using focused ion beam serial sectioning. *Microsc. Today*, **12** (4), 52.

Flierl, C., White, I.H., Kuball, M., Heard, P.J., Allen, G.C., Marinelli, C., Rorison, J.M., Penty, R.V., Chen, Y., and Wang, S.Y. (1999) Focused ion beam etching of GaN. *MRS Internet J. Nitride Semicond. Res.*, **4S1** (G6), 75.

Fujii, T. (2007) Focused ion beam system as a multifunctional tool for nanotechnology, in *Focused Ion Beam System: Basics and Applications* (ed. N. Yao), Cambridge University Press, Cambridge. 355.

Giannuzzi, L.A., Drown, J.L., Brown, S.R., Irwin, R.B., and Stevie, F.A. (1997) Focused ion beam milling and micromanipulation lift-out for site specific cross-section TEM specimen preparation. *Mater. Res. Soc. Symp. Proc.*, **480**, 19.

Harris, G. and Zhou, P. (2001) The growth and characterization processes of gallium nitride (GaN) nanowires. *Natl. Nanofabricat. User Netw.*, **2001**, 34.

Haythornthwaite, R., Nxumalo, J., and Phaneuf, M.W. (2004) Use of the focused ion beam to locate failure sites within electrically erasable read only memory microcircuits. *J. Vac. Sci. Technol. A*, **22**, 902.

Huey, B.D. and Langford, R.M. (2003) Low-dose focused ion beam nanofabrication and characterization by atomic force microscopy. *Nanotechnology*, **14**, 409.

Inkson, B.J. and Möbus, G. (2001) 3D determination of grain shape in FeAl by focused ion beam (FIB) tomography. *Microsc. Microanal.*, **7S2**, 936.

Kang, K.J., Yao, N., He, M.Y., and Evans, A.G. (2003) A Method for in-situ measurement of the residual stress in thin films by using the focused ion beam. *Thin Solid Films*, **443**, 71.

Langford, R.M. (2007) Preparation for physico-chemical analysis, in *Focused Ion Beam System: Basics and Applications* (ed. N. Yao), Cambridge University Press, Cambridge. 215.

Li, H.W., Kang, D.J., Blamire, M.G., and Huck, W.T.S. (2003) Focused ion beam fabrication of silicon print masters. *Nanotechnology*, **14**, 220.

Lipp, S., Frey, L., Lehrer, C., Frank, B., Demm, E., Pauthner, S., and Ryssel, H. (1996) Tetramethoxysilane as a precursor for focused ion beam and electron beam assisted insulator (SiO_x) deposition. *J. Vac. Sci. Technol. B*, **14**, 3920.

Liu, Y., Longo, D.M., and Hull, R. (2003) Ultrarapid nanostructuring of poly(methylmethacrylate) films using Ga^+. *Appl. Phys. Lett.*, **82**, 346.

Lou, J., Shrotriya, P., Allameh, S., Yao, N., Buchheit, T., and Soboyejo, W.O. (2002) Plasticity length scale in LIGA nickel MEMS structure. *Mater. Res. Soc. Symp. Proc.*, **687**, 41.

Mavrocordatos, D., Steiner, M., and Boller, M. (2003) Analytical electron microscopy and focused ion beam: complementary tool for the imaging of copper sorption onto iron oxide aggregates. *J. Microsc. (Oxford)*, **210**, 45.

Orloff, J., Utlaut, M., and Swanson, L. (2003) *High Resolution Focused Ion Beams: FIB and Its Applications*, Kluwer Academic/Plenum Publishers, New York.

Principe, E.L. (2007) Advances in real-time SEM imaging during focused ion beam milling: from 3D reconstruction to end point detection, in *Focused Ion Beam System: Basics and Applications* (ed. N. Yao), Cambridge University Press, Cambridge. 146.

Rice, L. (2001) Semiconductor failure analysis using EBIC and XFIB. *Microsc. Microanal.*, **7S2**, 514.

Rossie, B.B., Shofner, T.L., Brown, S.R., Anderson, S.D., Jamison, M.M., and Stevie, F.A. (2001) A method for thinning FIB prepared TEM specimens after lift-out. *Microsc. Microanal.*, **7S2**, 940.

Rubanov, S. and Munroe, P.R. (2004) FIB-induced damage in silicon. *J. Microsc. (Oxford)*, **214**, 213.

Russell, P.E., Stark, T.J., Griffis, D.P., Phillips, J.R., and Jarausch, K.F. (1998) Chemically and geometrically enhanced focused ion beam micromachining. *J. Vac. Sci. Technol. B.*, **16**, 2494.

Sivel, V.G.M., Van den Brand, J., Wang, W.R., Mohdadi, H., Tichelaar, F.D., Alkemade, P.F.A., and Zandbergen, H.W. (2004) Application of the dual-beam FIB/SEM to metals research. *J. Mircrosc. (Oxford)*, **214**, 237.

Sutton, D., Parle, S.M., and Newcomb, S.B. (2001) Focused ion beam damage: its characterization and minimization. *Inst. Phys. Conf. Ser.*, **168**, 377.

Takanashi, K., Shibata, K., Sakamoto, T., Owari, M., and Nihei, Y. (2003) Analysis of non-woven fabric fibre using an ion and electron multibeam microanalyser. *Surf. Interface Anal.*, **35**, 437.

Volkert, C.A., Busch, S., Heiland, B., and Dehm, G. (2004) Transmission electron microscopy of fluorapatite-gelatine composite particles prepared using focused ion beam milling. *J. Microsc. (Oxford)*, **214**, 208.

Yaguchi, T., Konno, M., Kamino, T., Hashimoto, T., Onishi, T., and Umemura, K. (2003) FIB micro-pillar sampling technique for 3-D stem observation and its application. *Microsc. Microanal.*, **9S2**, 118.

Yao, N., Epstein, A., and Akey, A. (2006) Screw dislocations and spiral growth in abalone nacre. *J. Mater. Res.*, **21** (8), 1939–1946.

Yao, N., Evans, A.G., and Cooper, C.V. (2004) Wear mechanisms operating in diamond like carbon coatings in contact with machined steel surfaces. *Surf. Coat. Technol.*, **179**, 306.

Young, R.J. and Moore, M.V. (2005) Dual beam (FIB-SEM) systems, in *Introduction to Focused Ion Beams: Instrumentation, Theory, Techniques and Practice*, Chapter 12 (eds L.A. Gianuzzi and F.A. Stevie), Springer, New York.

Utlaut, M. (2007) Micro-machining and mask repair, in *Focused Ion Beam System: Basics and Applications* (ed. N. Yao), Cambridge University Press, Cambridge. 268.

Walker, J.F. (1997) Preparing TEM sections by FIB: stress relief to straighten warping membranes. *Inst. Phys. Conf. Ser.*, **157**, 469.

Wang, Z.G., Kato, N., Sasaki, K., Hirayama, T., and Saka, H. (2004) Electron holographic mapping of two-dimensional doping areas in cross-sectional device specimens prepared by the lift-out technique based on a focused ion beam. *J. Electron Microsc.*, **53**, 115.

Wirth, R. (2009) Focused Ion Beam (FIB) combined with SEM and TEM: advanced analytical tools for studies of chemical composition, microstructure and crystal structure in geomaterials on a nanometre scale. *Chem. Geol.*, **261**, 217–229.